Microwave Photonics

IEEE Press
445 Hoes Lane
Piscataway, NJ 08854

Microwave Photonics

Jianping Yao
University of Ottawa
Ottawa, ON
Canada

José Capmany
Universitat Politécnica de Valencia
Valencia
Spain

Ming Li
Institute of Semiconductors
Chinese Academy of Sciences
Beijing
China

IEEE Press

WILEY

Published by John Wiley & Sons, Inc., Hoboken, New Jersey.
Published simultaneously in Canada.

Library of Congress Cataloging-in-Publication Data Applied for:

Hardback ISBN: 9781394205288

Cover Design: Wiley
Cover Image: © Panther Media GmbH/Alamy Stock Photo

Set in 9.5/12.5pt STIXTwoText by Straive, Pondicherry, India

Contents

About the Authors

Jianping Yao is a Distinguished University Professor and University Research Chair in microwave photonics in the School of Electrical Engineering and Computer Science, University of Ottawa, Canada. He was editor-in-chief of *IEEE Photonics Technology Letters* (2017–2021), a topical editor of *Optics Letters* (2015–2017), an elected member of the Board of Governors of the IEEE Photonics Society (2018–2021), and a past chair of the International Steering Committee for Microwave Photonics (2019–2020). Dr. Yao has published 700+ papers including 400+ in refereed journals and 300+ in conference proceedings. He was an IEEE MTT-S Distinguished Microwave Lecturer 2013–2015 and was the recipient of the 2018 R.A. Fessenden Silver Medal from IEEE Canada. Dr. Yao is a Fellow of IEEE (2012), the Optica (formerly Optical Society of America) (2010), the Canadian Academy of Engineering (CAE) (2012), and the Royal Society of Canada (RSC) Academy of Science (2018).

Dr. José Capmany is a full professor in Photonics and leader of the Photonics Research Labs (www.prl.upv.es) at the institute of Telecommunications and Multimedia Applications (www.iteam.upv.es), Universitat Politècnica de Valencia, Spain. He holds BSc + MSc degrees and doctorates in electrical engineering and physics.

He has published over 650 papers in international refereed journals and conferences and has been a member of the Technical Program Committees of the European Conference on Optical Communications (ECOC), the Optical Fiber Conference (OFC). He is a fellow of the Optical Society of America (OSA) and the Institute of Electrical and Electronics Engineers (IEEE). He is also a founder and chief innovation officer of the spin-off companies VLC Photonics (acquired by Hitachi in 2020) dedicated to the design of photonic integrated circuits and iPronics (www.ipronics.com) dedicated to programmable photonics.

Professor Capmany is the 2012 King James I Prize Laureate on novel technologies and the National Research Award in Engineering 2020, the two highest scientific distinctions in Spain, for his outstanding contributions to the field of microwave photonics. He has also received the Engineering Achievement Award from the IEEE Photonics Society and the Innovation Prize from the Royal Society of Physics in Spain. He is a double ERC Advanced and a double Proof of Concept grantee and was a distinguished lecturer of the IEEE Photonics Society for the 2013–2014 term. Associate editor of *IEEE Photonics Technology Letters* (2010–2016) and the *IEEE Journal of Lightwave Technology* (2016–2018). He served as editor in chief of the *IEEE Journal of Selected Topics in Quantum Electronics* from 2018 to 2022.

Ming Li received the Ph.D. in electrical and electronics engineering from the University of Shizuoka, Hamamatsu, Japan, in 2009. In 2009, he was with the Microwave Photonics Research Laboratory, School of Electrical Engineering and Computer Science, University of Ottawa, Ottawa, ON, Canada, as a postdoctoral research fellow. In 2011, he was with the Ultrafast Optical Processing Group under the supervision of INRS-EMT, Montreal, QC, Canada, as a postdoctoral research fellow. In 2013, he was with the Institute of Semiconductors, Chinese Academy of Sciences, Beijing, China, as a full professor. He served as an associate editor of *IEEE Journal of Quantum Electronics* and an executive editor of *Journal of Semiconductors*. He also served as a chair or TPC member for multiple international conferences. He has authored more than 220 high-impact journal papers. His research interests include optoelectronic integrated circuits, microwave photonics and optical signal processing.

About the Companion Website

This book is accompanied by a companion website.

www.wiley.com/go/yao/microwavephotonics

This book includes learning objectives, and questions and/or problems.

1

Introduction to Microwave Photonics

In the realm of modern information and communications, the demand for high-speed, highly secure, and reliable systems is ever-increasing. As a result, researchers and engineers are constantly exploring novel solutions to transmit and manipulate data at higher speeds and higher frequencies. Microwave photonics has emerged as a promising interdisciplinary field that combines the advantages of photonics and microwave engineering to overcome the limitations of traditional electronic systems. Microwave photonics is a field that encompasses the interaction of microwave and optical waves and leverages the inherent advantages of both domains to achieve enhanced system performance in terms of bandwidth, speed, signal quality, and energy efficiency. In essence, microwave photonics aims to bridge the gap between optics and microwaves by utilizing optical techniques to generate, process, control, and distribute microwave signals. This book explores the fundamentals of microwave photonics including photonic generation, processing, control, and distribution of microwave signals or waveforms. Specifically, the book covers the following key concepts and techniques.

1.1 Photonic Generation of Microwave Signals

Photonic generation of microwave signals refers to the use of photonic techniques for the generation of microwave signals. There are two major approaches to generating a microwave signal, including heterodyne detection of two coherent optical waves at a photodetector in which the two coherent optical waves can be generated by either using a dual-wavelength laser source or through optical modulation, and the use of an optoelectronic oscillator (OEO). Photonic RF generation is particularly useful in applications where high-frequency signals need to be generated with excellent performance in terms of stability, phase noise, and frequency agility. Photonic microwave generation offers several advantages over traditional electronic microwave generation techniques. It can provide wide bandwidth, low phase noise, high-frequency resolution, and immunity to electromagnetic interference (EMI). It also allows for integration with other photonic components, enabling compact and efficient microwave systems.

1.2 Photonic Microwave Signal Processing

Microwave photonic filters are essential devices for photonic microwave signal processing. Microwave photonic filters are implemented by a combination of microwave and photonic technologies to achieve various filtering functions. The filters utilize the advantages of photonic systems,

Microwave Photonics, First Edition. Jianping Yao, José Capmany, and Ming Li.

Companion website: www.wiley.com/go/yao/microwavephotonics

such as wide bandwidth, low loss, and high-speed signal processing. In a microwave photonic filter, the microwave signal is converted into an optical signal, processed using photonic techniques, and then converted back to the microwave domain. This approach allows for the implementation of advanced filtering functions with improved performance compared to conventional electronic filters such as wider bandwidth, higher selectivity, lower loss, and better tunability. In addition, photonic implementation has an added advantage by processing high-frequency and wideband microwave signals without suffering from EMI. These advantages make them suitable for applications in radar and wireless communications systems.

1.3 Photonic Distribution of Microwave Signals

Photonic distribution of microwave signals is implemented by microwave photonic links. A microwave photonic link, also known as analog photonic link, is a technology that combines microwave and photonic technologies to enable the transmission of wideband and high-frequency microwave signals over an optical fiber. In a microwave photonic link, the microwave signal (usually digital data modulated on a microwave carrier) is converted into an optical signal using an electro-optic modulator (EOM) such as a Mach–Zehnder modulator (MZM) or a phase modulator (PM). The EOM imposes the microwave signal onto an optical carrier, which is transmitted over an optical fiber at an ultra-low loss. At the receiving end, the optical signal is converted back into a microwave signal at a photodetector (PD). The use of optical fibers in microwave communications offers several benefits. First, optical fibers have a much wider bandwidth compared to traditional copper cables, allowing for the transmission of wideband microwave signals. Second, optical fibers have much lower loss compared to traditional copper cables, allowing for the transmission over a long distance. In addition, optical fibers are immune to EMI, a feature that is absent in copper cables and is extremely important when being used in an electromagnetic complex environment. This makes microwave photonic links ideal for applications where high-speed, long distance, and high-quality microwave transmission is required, such as in wireless communications systems, radar systems, and satellite communication systems.

1.4 Photonic Generation of Ultra-wideband Signals

Photonic generation of Ultra-wideband (UWB) signals is a technology to use photonic techniques to generate UWB signals thanks to the ultra-wide bandwidth offered by modern photonics. The FCC has allocated a specific frequency range for UWB operations in the US from 3.1 to 10.6 GHz with a power spectral density (PSD) defined to prevent interference with other wireless services. The photonic generation of UWB signals offers several advantages including large bandwidths, high data rates, and low power consumption. In addition, the inherent characteristics of UWB, such as resistance to multipath fading and precise ranging capabilities, can be combined with the benefits of optical communication systems. Applications of photonic UWB signals include wireless communications, ranging and localization systems, radar imaging, and high-resolution sensing. By leveraging the capabilities of both photonics and UWB, these systems can achieve advanced performance in terms of high-speed data transfer, high positioning accuracy, and high imaging resolution.

1.5 Photonic Generation of Microwave Arbitrary Waveforms

Photonic Arbitrary Waveform Generation (AWG) refers to the generation of arbitrary microwave waveforms using photonic techniques. It involves the use of optical components and devices to create complex waveforms in the optical domain with specified amplitude, frequency, and phase characteristics, and converted to the electrical domain at a photodetector. Photonic AWG provides several advantages, including wide bandwidth, fast processing speeds, and precise control over waveform parameters. Photonic arbitrary waveform generation can find applications in various fields, including optical communications, to generate complex waveforms with advanced modulation formats, signal processing, and optical signal shaping in optical communication systems; radar, to generate versatile waveforms for radar imaging, target detection, and signal processing. It also finds applications in sensing systems, such as optical coherence tomography (OCT), where precise waveform control is crucial. Photonic arbitrary waveform generation offers unique capabilities for creating customized waveforms with wide bandwidth, high accuracy, and large flexibility.

1.6 Microwave Photonic Beamforming Networks for Phased Array Antennas

Microwave photonic beamforming based on photonic true time delays is a technique used in phased array antenna systems to steer the beam of an antenna array by applying precise time delays to the microwave signal at each individual antenna element. It is commonly used in applications such as radar systems, wireless communications, and satellite communications. A phased array antenna system consists of an array of antenna elements arranged in an array or a grid pattern. Each antenna element can transmit and receive signals independently, enabling beam steering and forming of the radiation pattern. By controlling the time delays of the signals across the antenna elements, the main beam of the antenna array can be steered thanks to the constructive interference of the time delayed microwave signals at the far field, resulting in a focused beam in the desired direction. True time delay beamforming offers several unique advantages over conventional phase-shift-based beamforming techniques in terms wider instantaneous bandwidth and free of beam squint. True time delay can be implemented using photonic delay lines such as dispersive fibers, fiber Bragg gratings, and photonic integrated resonators such as ring resonators and microdisk resonators. True time delay phased array beamforming is a powerful technique which can find applications where squint-free microwave beamforming with wide instantaneous bandwidth is required.

1.7 Photonic-Assisted Instantaneous Microwave Frequency Measurements

Photonic-assisted microwave frequency measurements refer to the use of photonic techniques to perform measurements and analysis in the microwave frequency range. It involves the conversion of microwave signals into optical signals and the subsequent processing and analysis of these optical signals in the optical domain using various photonic devices, such as optical delay lines, fiber Bragg gratings, and dispersive elements. The field of microwave photonic measurements has gained significant research interest due to the advantages offered by photonics including high-frequency measurements, wide bandwidth, high sensitivity, and large dynamic range, which contribute to the advancements in microwave signal and system characterization. The immunity to EMI of photonic systems is an added advantage of performing microwave measurements using photonics.

1.8 Microwave Photonic Sensors

Microwave photonic sensors are a class of sensors that utilize microwave photonic techniques to perform optical sensing in which the optical sensing information is translated to the microwave domain with much higher sensing sensitivity and precision. In a microwave photonic sensor, an optical sensing signal in which the sensing information is encoded as a frequency or wavelength change is converted to a microwave signal. By measuring the microwave frequency change, the sensing information is obtained. Since the frequency of a microwave signal is much lower than that of an optical signal, the sensing sensitivity and precision are significantly increased. In addition to the increased sensing sensitivity and precision, microwave photonic sensors have other advantages such as immunity to EMI and ability to achieve distributed and remote sensing. Microwave photonic sensors have found applications in various fields, including telecommunications, aerospace, defense, and civil engineering.

1.9 Photonic Analog-to-Digital Conversion

Photonic analog-to-digital conversion (ADC) is a technique that utilizes photonic techniques to convert analog signals into their digital representations. It leverages the advantages of photonics, such as wide bandwidth, high-speed processing, and immunity to EMI, to achieve high-performance ADCs with enhanced capabilities. The basic principle of photonic ADC involves the conversion of an analog signal into an optical signal, followed by the processing and digitization of the optical signal using photonic techniques. Some common approaches include (i) optical sampling ADC, to utilize ultrafast photonic sampling techniques to directly sample an analog signal at a very high sampling rate. The analog signal is first converted into an optical signal and then a train of optical pulses is used to sample the signal in the time domain. The optical samples are subsequently processed and digitized using electronic components; (ii) photonic time-stretch ADC, to time-stretch an analog signal in the time domain using a dispersive medium, and the time-stretched analog signal is then optically sampled and digitized using high-speed photodetectors and electronic ADCs. The time-stretching approach allows for slower electronic ADCs to be used, enabling higher resolution and improved performance.

1.10 Novel Optoelectronic Oscillators

Novel OEOs refer to recently reported OEO schemes that overcome the limitations of traditional OEOs. Novel OEOs have been widely investigated in the past few years, and a diversity of new insights and breakthroughs have been proposed and demonstrated. For instance, a novel microwave photonic iterative nonlinear gain model has been reported, which can be used to analyze different oscillation states of an OEO. A novel PT-symmetric OEO has been proposed and demonstrated by using PT symmetry for mode selection, where single-mode oscillation can be achieved without using any narrowband optical/electrical filters. Novel OEOs that are capable of producing complex microwave waveforms have been developed, including novel Fourier domain mode-locked OEO for chirped signal generation, optoelectronic parametric oscillator for stable multi-mode oscillation, and broadband random OEO for random signal generation. Moreover, novel integrated OEOs featuring compact size and low power consumption have also been demonstrated, which are key steps toward a new generation of compact microwave sources for demanding applications.

1.11 Quantum Microwave Photonics

Quantum microwave photonics is to use quantum technology to improve the performance of classical microwave photonic systems. By combining microwave photonics and quantum technology, quantum microwave photonics can achieve features or functions that are very complex or even not possible using traditional microwave photonic or quantum technology. The key to the current quantum microwave photonic systems is the quantum single-photon detection scheme, which replaces the signal detection scheme in a traditional microwave photonic link to the quantum single-photon detection scheme. By doing so, the bandwidth and sensitivity limitations of a traditional microwave photonic system can be solved due to the wide processing bandwidth and high sensitivity of a quantum single-photon detection scheme. Quantum microwave photonic systems for weak microwave signal detection, RF phase shifting and filtering, nonlocal frequency-to-time mapping, RF compressed sensing, as well as quantum key distribution have been proposed and demonstrated in recent years, which show clearly that quantum technology brings new opportunities to microwave photonics.

1.12 Integrated Microwave Photonics

Integrated microwave photonics (IMWP) is the study of the integration of photonic devices and microwave components on a single chip or substrate through monolithic or heterogeneous integration. IMWP systems typically consist of optical waveguides, which guide and manipulate light, a set of active and passive photonic devices, and microwave components. Active devices include lasers, modulators, and photodetectors, which generate, modulate, and detect optical signals, respectively. Passive devices include optical couplers, optical filters (resonators), and optical delay lines, which route and process the optical signals. Microwave components include amplifiers and electronic couplers. The most important materials used for photonic integration include InP, Silicon, Silicon Nitride, and Lithium Niobate. Among them, silicon is being heavily studied thanks to its unique feature of compatibility with the mature CMOS technology, to leverage existing infrastructure and manufacturing processes of the semiconductor industry, allowing for cost-effective and scalable production of photonic devices. However, silicon is a non-direct bandgap material, which cannot produce light amplification and lasing. To have a fully integrated system, a light source is always needed. A solution is to implement hybrid integration, by integrating InP-based light sources on a silicon substrate.

1.13 Applications of Microwave Photonics

Microwave photonics has numerous applications in telecommunications systems, radar and sensing systems, radio astronomy, and biomedical imaging. For telecommunications systems, it enables the transmission of wide-bandwidth microwave signals over optical fibers, significantly extending the communications distance for wireless systems. In addition, microwave photonics enables the integration of wireless and optical networks, facilitating seamless connectivity and improved network performance. In radar and sensing systems, microwave photonics enhances radar capabilities by enabling the processing and distribution of microwave signals with higher frequency and wider bandwidth. It also enables the integration of multiple radar systems, improving the efficiency and accuracy of surveillance and sensing applications. In radio astronomy, radio telescopes use microwave photonics to capture and analyze radio frequency signals from celestial objects. The high-frequency stability, accuracy, and wide bandwidth of microwave photonics enable precise measurements and imaging of distant stars, galaxies, and cosmic microwave background radiation, and finally, in biomedical imaging, microwave photonics has potential applications in biomedical imaging techniques, such as microwave tomography and microwave-induced thermoacoustic imaging. These techniques leverage the unique properties of microwave signals to create detailed images of biological tissues, aiding in medical diagnostics and treatment planning.

2

Optical Devices for Microwave Photonics

2.1 Introduction

The construction and assembly of microwave photonic systems require the use of diverse optoelectronic and photonic components. In the following chapters of this book, we will present different configurations enabling the main functionalities of microwave photonics (MWP) including, among others, signal generators, optical beamforming networks, signal processors, and radio over fiber links.

The objective of this chapter is to provide a summarized description of the main optical devices that are employed in MWP systems including the operation principles and most common configurations. The list of photonic components currently being employed in MWP systems is certainly very extensive and a detailed and exhaustive treatment is beyond the scope of this chapter. Here we have just concentrated on those we consider essential. This includes optical fibers and planar waveguides, light sources, detectors and modulators, and some passive optical components. Wherever possible, we have provided as well as some indicative performance figures, especially for commercial devices. Due to space restrictions, we just cover the essential properties of these components but there is, of course, a vast literature on photonic devices and we direct the reader to these references for further details and also for an introduction to other active and passive components that are not covered in the chapter.

We provide a short but useful reference section to this chapter that is composed exclusively of textbooks. In doing so, we have aimed to provide the reader with a short but comprehensive list of references where he or she will be able to get the relevant information models and advanced features not covered in the chapter at a minimum time cost and without needing to surf the very extensive journal literature that already exists in the field. For example, detailed treatments of fiber and planar waveguide propagation properties can be found in [1–4]. Optical sources are described in depth in [4–7], while modulators are studied in [1, 2]. FBGs are covered in the authoritative references [8, 9] while for array waveguide grating (AWG) devices, the reader is instructed to [10]. The main optical passive components are covered in many textbooks with [2, 3, 11, 12] being a good selection. Finally, for a comprehensive description of active components such as fiber and semiconductor amplifiers and in-line optical filters like ring cavities, the reader is pointed to [13–15].

Microwave Photonics, First Edition. Jianping Yao, José Capmany, and Ming Li.

Companion website: www.wiley.com/go/yao/microwavephotonics

2.2 Optical Fibers and Planar Waveguides

2.2.1 Structure and Geometry of Optical Fibers and Planar Waveguides

Although there is a wide variety of optical fiber types and geometries, in MWP, it is customary to use the so-called standard optical fiber which is a cylindrical structure formed by an inner core of dielectric material and radius a surrounded by an outer cladding also made of dielectric material and external radius b as shown in Fig. 2.1. Upon several conditions, the fiber behaves as a dielectric waveguide for lightwave signals. This means that input signals are coupled to internal propagation modes which travel along the fiber confined inside the core without being radiated outside. Depending on the number of modes that a fiber propagates, one speaks of singlemode (if it only propagates one) or multimode (if it propagates more than one). Typical values of a range from 3 to 5 μm for singlemode fibers and 5–25 μm for multimode fibers while a typical value for b is 125 μm.

Referring to Fig. 2.1, we define and introduce several conventions that will be employed throughout the chapter. First of all, we assume that signal propagation is in the direction of the x, y axis and therefore the transversal plane to the direction of propagation will be characterized by means of the x, y coordinates in a Cartesian system or by the r, θ coordinates in a cylindrical system. The cylindrical system is usually employed in the description of the optical fiber although, as we will later see, the Cartesian system is useful in the description of the propagation under the weakly guiding approximation, as well as in the description of the propagation through planar waveguides.

The dielectric materials composing the core and the cladding are usually compounds that use silica glass (SiO_2) as the basic constituent, combined with a series of dopants like germanium (Ge), phosphorus (P), and boron (B) and are characterized by their respective refractive indices. The cladding refractive index n_2 is constant along the radial coordinate r while the core refractive index can be constant or variable along the radial coordinate and, therefore, we shall employ the notation $n_1(r)$ to describe it.

Optical fibers can be classified into two groups attending to how $n_1(r)$ varies with r. If the refractive index is constant with the radial coordinate, then the optical fiber is known as *step index* or SI since the index is constant until reaching the interface between the core and the cladding at $r = a$, where the refractive index changes abruptly. In SI fibers, it is customary to express the core refractive index as n_1.

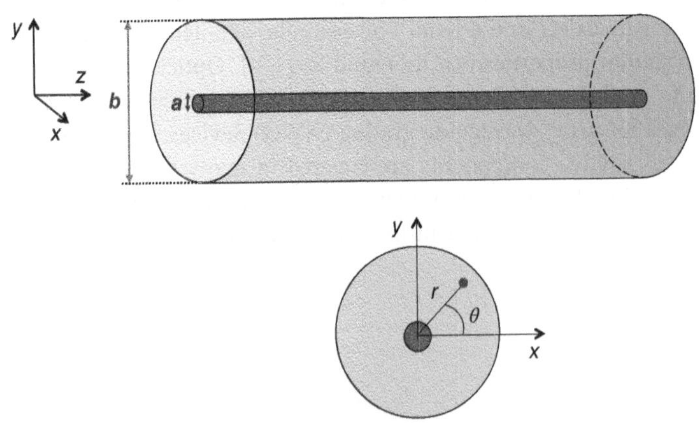

Figure 2.1 Geometry of an optical fiber.

The second type of fibers, known as graded index or GI, is characterized by a smooth and decreasing variation of the refractive index within the core, from a maximum value in the core axis $n_1(0)$ to a minimum value in the core-cladding $r = a$ interface that coincides with the refractive index in the cladding $n_2 = n_1(a)$. Figure 2.2 shows the transversal section and the index profile variation for both types of fibers. As can be observed in both cases, the value of the refractive index in the core is higher than that of the cladding. This is a necessary condition for the structure to behave as a cylindrical optical waveguide.

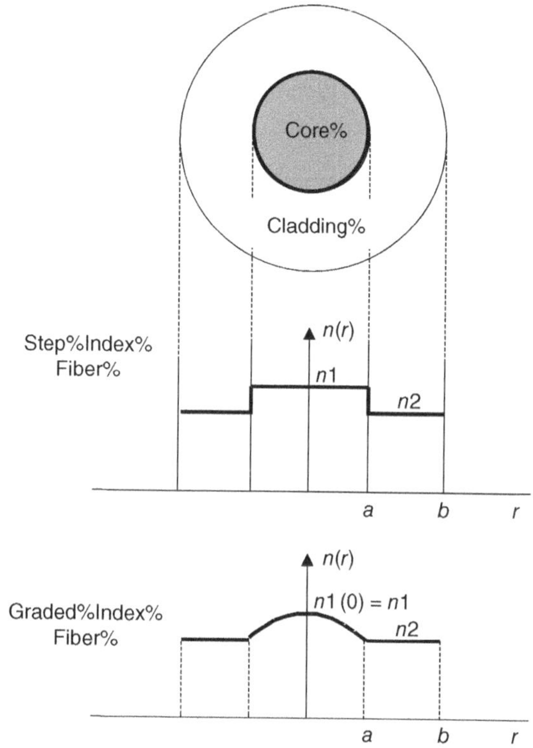

Figure 2.2 Transversal section and index profile variation for SI and GI fibers.

Figure 2.3 shows several examples of planar dielectric waveguides which are employed in practice to build devices, circuits, and subsystems employed in MWP applications.

As can be observed, these are formed by a central zone with a typical rectangular geometry (although other designs are also possible) and a higher refractive index than that of the rest of surrounding regions. The central zone acts in a similar way as the core in the optical fiber while the surrounding regions play a similar role to that of the cladding. A first difference however with the cylindrical waveguide is that the cladding region does not need to be composed of a uniform material. Furthermore, it can be implemented by different regions, each one composed of a different material and, hence, featuring a different refractive index. A second difference is that to analyze planar waveguide it will be more convenient to employ Cartesian coordinates.

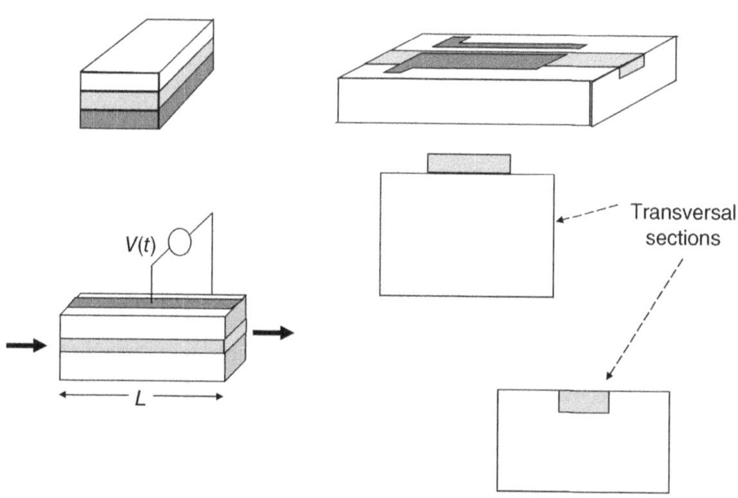

Figure 2.3 Several configurations of planar waveguides.

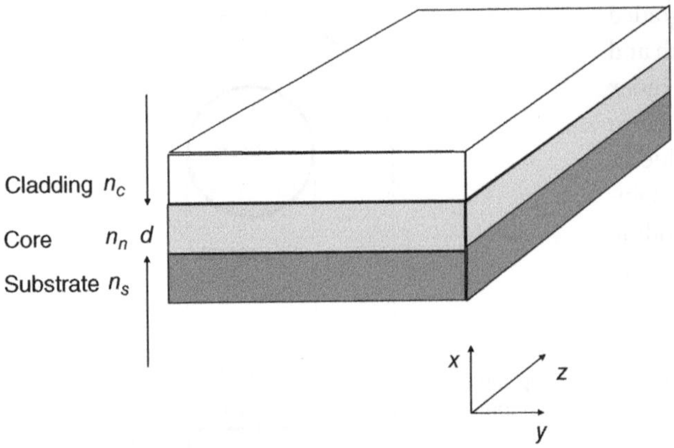

Figure 2.4 Layout of the asymmetric planar waveguide.

The electromagnetic analysis of planar dielectric waveguides is considerably complex and, in general, requires numerical treatment. Analytic solutions can only be obtained for ideal configurations such as the asymmetric infinite planar waveguide shown in Fig. 2.4.

As can be observed, this structure is assumed to be unlimited in the z (direction of propagation) and the y axis and is composed of three different regions. Referring to Fig. 2.4, the inferior region, which usually represents the basic material upon which the waveguide is grown, is called *substrate* and has a refractive index given by n_s. The central region, which represents the waveguide core, has a refractive index $n_n > n_s$. The core is limited to a width of d on the x axis. The layer on top of the core is the cladding which is composed of a material with refractive index $n_c < n_n$. Furthermore, in many cases, the cladding is just air.

2.2.2 Basic Electromagnetic Theory for Optical Fibers and Planar Waveguides

The starting point for the analysis of the propagation through optical fibers and waveguides are Maxwell equations which, for nonconducting and charge-free media such as dielectrics have the following form in the time domain:

$$\nabla \times \mathbf{E}(\overline{r}, t) = -\frac{\partial \mathbf{B}(\overline{r}, t)}{\partial t} \tag{2.1}$$

$$\nabla \times \mathbf{H}(\overline{r}, t) = \frac{\partial \mathbf{D}(\overline{r}, t)}{\partial t} \tag{2.2}$$

$$\nabla \cdot \mathbf{D}(\overline{r}, t) = 0 \tag{2.3}$$

$$\nabla \cdot \mathbf{B}(\overline{r}, t) = 0 \tag{2.4}$$

We use bold typing to describe the vectorial characteristics of the fields. Thus, $\mathbf{E}(\overline{r}, t)$ and $\mathbf{H}(\overline{r}, t)$ are the electric and magnetic field vectors, respectively, at a given time instant t and a given point in space identified by its position vector $\overline{r}$, while $\mathbf{D}(\overline{r}, t)$, $\mathbf{B}(\overline{r}, t)$ represent the electric displacement and magnetic induction vectors, respectively, which are related to the electric and magnetic field vectors by means of the *constitutive relations*.

$$\mathbf{D}(\overline{r}, t) = \varepsilon_0 \mathbf{E}(\overline{r}, t) + \mathbf{P}(\overline{r}, t) \tag{2.5}$$

$$\mathbf{B}(\overline{r}, t) = \mu_0 \mathbf{H}(\overline{r}, t) + \mathbf{M}(\overline{r}, t) \tag{2.6}$$

where ε_0 is the dielectric permittivity of vacuum, μ_0 the magnetic permeability of vacuum, and $\mathbf{P}(\bar{r},t)$ and $\mathbf{M}(\bar{r},t)$ represent the induced electric and magnetic polarization, respectively. Due to the fact materials usually employed in the fabrication of optical fibers and waveguides are nonmagnetic, we will assume that $\mathbf{M}(\bar{r},t) = 0$. On the other hand, $\mathbf{P}(\bar{r},t)$ represents the internal response of the fiber material to the action of an external electric field. In general, the relationship between $\mathbf{P}(\bar{r},t)$ and $\mathbf{E}(\bar{r},t)$ is nonlinear, as it depends on the intensity of the optical fields of the signals which excite the dielectric. This relationship is the origin of the so-called nonlinear effects in fibers and waveguides, such as Raman and Brillouin scattering, self-phase modulation, and four-wave mixing among others. In many cases, however, and especially when the injected signals are below a given power threshold, the relationship between $\mathbf{P}(\bar{r},t)$ and $\mathbf{E}(\bar{r},t)$ can be linearized by means of a convolution integral, where the impulse response of the dielectric medium is its *linear susceptibility* $h(\bar{r},t) = \bar{\chi}(\bar{r},t)$.

$$\mathbf{P}(\bar{r},t) = \varepsilon_0 \int_{-\infty}^{\infty} \bar{\chi}(\bar{r},t-\tau) \cdot \mathbf{E}(\bar{r},\tau)\, d\tau \tag{2.7}$$

In general, $\bar{\chi}(\bar{r},t)$ is a rank 2 tensor which reduces to a scalar if the medium, as it happens in optical fiber dielectrics, is isotropic.

Equations (2.1)–(2.7) provide part of the required framework for the electromagnetic analysis of the propagation in optical fibers and waveguides. Together with other relationships and boundary conditions, it will allow us to determine the exact solutions for the so-called propagation modes.

First of all, it is convenient to reduce the complexity of the problem. A first step is to consider the determination of only one of the vector fields, either **E** or **H**, since knowing one immediately gives the other by direct application of Maxwell equations. If, for example, we apply the rotational operator to both members of (2.1) and we take into account (2.2) and (2.5), we get:

$$\nabla \times (\nabla \times \mathbf{E}) = -\frac{1}{c^2}\frac{\partial^2 \mathbf{E}}{\partial t^2} - \mu_0 \frac{\partial^2 \mathbf{P}}{\partial t^2} \tag{2.8}$$

where $c = 1/\sqrt{\mu_0 \varepsilon_0}$ is the speed of light in vacuum.

Our next step is to express (2.8) in the frequency domain by taking its Fourier transform:

$$\nabla^2 \tilde{\mathbf{E}} + \varepsilon(\bar{r},\omega)\frac{\omega^2}{c^2}\tilde{\mathbf{E}} = 0 \tag{2.9}$$

Here $\tilde{\mathbf{E}}(\bar{r},\omega) = TF[\mathbf{E}(\bar{r},t)]$ is the Fourier transform of the electric field vector. We have made use of the following vectorial identity $\nabla \times \nabla \times \mathbf{A} = \nabla(\nabla \cdot \mathbf{A}) - \nabla^2 \mathbf{A}$ and we have taken into account (2.3), (2.5), and the Fourier transform of (2.7) to express the divergence of the electric field vector as $\nabla \cdot \tilde{\mathbf{E}} = \nabla \cdot (\tilde{\mathbf{D}}/\varepsilon) \approx (1/\varepsilon)\nabla \cdot \tilde{\mathbf{D}} = 0$. In (2.9), $\varepsilon(\bar{r},\omega)$ represents the dielectric constant of the medium:

$$\varepsilon(\bar{r},\omega) = 1 + \tilde{\chi}(\bar{r},\omega) \tag{2.10}$$

which is directly related to its refractive index $n(\omega)$ and its absorption coefficient $\alpha(\omega)$ through [6]:

$$\varepsilon(\bar{r},\omega) = \left(n + j\frac{\alpha c}{2\omega}\right)^2 \tag{2.11}$$

Although optical materials are lossy and the absorption mechanisms are well known and understood, we will assume from this point onward that, unless otherwise stated, the losses in the core and the cladding dielectrics are negligible. Thus, $\alpha = 0$ and $\varepsilon(\bar{r}, \omega) = n^2(\bar{r}, \omega)$ and, therefore, Eq. (2.9) can be written as:

$$\nabla^2 \tilde{\mathbf{E}} + n^2(\bar{r}, \omega) k_o^2 \tilde{\mathbf{E}} = 0 \tag{2.12}$$

where $k_o = \omega/c = 2\pi/\lambda$ represents the propagation constant of plane waves in vacuum. Equation (2.12) is known as the *Helmholtz equation* for the electric field and an identical one can be obtained following a similar process for the magnetic field.

$$\nabla^2 \tilde{\mathbf{H}} + n^2(\bar{r}, \omega) k_o^2 \tilde{\mathbf{H}} = 0 \tag{2.13}$$

It is important to note that Eqs. (2.12) and (2.13) are linear. This implies that, for simplicity, we can consider, in a first step, the propagation of monochromatic signals centered at a given frequency ω_o. The propagation of more complex nonmonochromatic signals can then be readily obtained by applying the superposition principle to all its frequency components. We will thus consider the following dependencies: $\mathbf{E}(\bar{r}, t) = \tilde{\mathbf{E}}(\bar{r}, \omega_o)e^{j\omega_o t}$; $\mathbf{H}(\bar{r}, t) = \tilde{\mathbf{H}}(\bar{r}, \omega_o)e^{j\omega_o t}$, and $n(\bar{r}, \omega) = n(\bar{r}, \omega_o)$.

2.2.3 Propagation in Optical Fibers

We will now briefly outline the process leading to the solution of (2.12) and (2.13) for the case of step index fibers (the case of GI fibers is more elaborated, but the interested reader can find a detailed treatment in the literature), where the refractive indices in the core and cladding can be considered independent of the spatial coordinates, i.e. $n_1(\bar{r}) = n_1$ and $n_2(\bar{r}) = n_2$, respectively. We will also assume (and justify later) that the cladding can be considered infinite, extending in the region $a \le r \le \infty$.

The Helmholtz equations to be solved for the electric and magnetic field vectors after the above simplifications are:

$$\begin{aligned} \nabla^2 \tilde{\mathbf{E}} + k^2 \tilde{\mathbf{E}} &= 0 \\ \nabla^2 \tilde{\mathbf{H}} + k^2 \tilde{\mathbf{H}} &= 0 \end{aligned} \tag{2.14}$$

where $k = nk_o$ and physically represents the propagation constant for plane waves in the dielectric. Note that $k = k_1 = n_1 k_o$ in the core and $k = k_2 = n_2 k_o$ in the cladding. So, there are actually two different Helmholtz equations to be solved in each of the two physically different regions of the fiber.

The first step is to choose the correct coordinate system which for optical fibers is the cylindrical system. This means that the Laplacian operator in (2.14) must be expressed in such a coordinate system:

$$\nabla^2 = \frac{\partial^2}{\partial r^2} + \frac{1}{r}\frac{\partial}{\partial r} + \frac{1}{r^2}\frac{\partial^2}{\partial \theta^2} + \frac{\partial^2}{\partial z^2} \tag{2.15}$$

and, therefore, the Helmholtz equations are:

$$\left[\frac{\partial^2}{\partial r^2} + \frac{1}{r}\frac{\partial}{\partial r} + \frac{1}{r^2}\frac{\partial^2}{\partial \theta^2} + \frac{\partial^2}{\partial z^2}\right] \cdot \begin{bmatrix} \tilde{\mathbf{E}}(r, \theta, z) \\ \tilde{\mathbf{H}}(r, \theta, z) \end{bmatrix} + k^2 \begin{bmatrix} \tilde{\mathbf{E}}(r, \theta, z) \\ \tilde{\mathbf{H}}(r, \theta, z) \end{bmatrix} = 0 \tag{2.16}$$

Equation (2.16) is really a system of six differential equations in the core ($k = k_1$) and six differential equations in the cladding ($k = k_2$) as each field vector has three components. For instance, the electric field vector can be expressed as $\tilde{\mathbf{E}}(r,\theta,z) = E_r(r,\theta,z)\hat{r} + E_\theta(r,\theta,z)\hat{\theta} + E_z(r,\theta,z)\hat{z}$ and a similar expression holds for the magnetic field. To solve (2.16), it is customary to start with the equations corresponding to the z components of the field. Once solved, the radial and azimuthal components can be readily obtained by direct application of Maxwell equations. Hence, the equations to solve are:

$$\left[\frac{\partial^2}{\partial r^2} + \frac{1}{r}\frac{\partial}{\partial r} + \frac{1}{r^2}\frac{\partial^2}{\partial \theta^2} + \frac{\partial^2}{\partial z^2}\right] \cdot \begin{bmatrix} \tilde{E}_z(r,\theta,z) \\ \tilde{H}_z(r,\theta,z) \end{bmatrix} + k^2 \begin{bmatrix} \tilde{E}_z(r,\theta,z) \\ \tilde{H}_z(r,\theta,z) \end{bmatrix} = 0 \tag{2.17}$$

To solve (2.17), we apply a technique known as separation of variables where the solution is expressed as:

$$\begin{bmatrix} \tilde{E}_z(r,\theta,z) \\ \tilde{H}_z(r,\theta,z) \end{bmatrix} = f(r)e^{jl\theta}e^{-j\beta z} \tag{2.18}$$

In (2.18), $l = 0, 1, 2, 3, \ldots$ is an integer, as the part of the solution corresponding to the azimuthal variation must replicate after a 2π variation in the angle coordinate θ. On the other hand, β represents the phase variation of the solution upon propagation in the z direction, which is known as the propagation constant. Substituting (2.18) in (2.17) leads to:

$$\frac{\partial^2 f(r)}{\partial r^2} + \frac{1}{r}\frac{\partial f(r)}{\partial r} + \left(k^2 - \beta^2 - \frac{l}{r^2}\right)f(r) = 0 \tag{2.19}$$

which is a Bessel equation.

The solutions of (2.19) are linear combinations of Bessel functions of order l and the particular functions that enter that combination depend on the value of the term $k^2 - \beta^2$. If $k^2 - \beta^2 > 0$, then the solution is:

$$\begin{aligned} f(r) &= C_1 J_l(hr) + C_2 Y_l(hr) \\ h &= \sqrt{k^2 - \beta^2} \end{aligned} \tag{2.20}$$

where $J_l(hr)$ and $Y_l(hr)$ are the Bessel functions of first and second kind and order l. If, however, $k^2 - \beta^2 < 0$, then:

$$\begin{aligned} f(r) &= C_3 I_l(qr) + C_4 K_l(qr) \\ q &= \sqrt{\beta^2 - k^2} \end{aligned} \tag{2.21}$$

where $I_l(hr)$ and $K_l(hr)$ are the modified Bessel functions of first and second kinds and order l.

For guided modes, the field must oscillate in the core and rapidly decrease in the cladding. According to the variation of Bessel functions with their argument [· · ·], this means that (2.20) must be the solution for the core region while (2.21) has to be the solution for the cladding region. Furthermore, the field must be bounded in the core and in the cladding. Since $Y_l(0) \to \infty$ and $I_l(\infty) \to \infty$, this means that $C_2 = C_3 = 0$. Thus, the complete solution for the z-component fields is:

$$\begin{aligned} E_z(\bar{r},t) &= AJ_l(hr)e^{jl\theta}e^{j(\omega_o t - \beta z)} \\ H_z(\bar{r},t) &= BJ_l(hr)e^{jl\theta}e^{j(\omega_o t - \beta z)} \\ h &= \sqrt{n_1^2 k_o^2 - \beta^2} \end{aligned} \tag{2.22}$$

in the core and:

$$\begin{aligned} E_z(\bar{r},t) &= CK_l(qr)e^{jl\theta}e^{j(\omega_o t-\beta z)} \\ H_z(\bar{r},t) &= DK_l(qr)e^{jl\theta}e^{j(\omega_o t-\beta z)} \\ q &= \sqrt{n_2^2k_o^2-\beta^2} \end{aligned} \tag{2.23}$$

in the cladding.

It should be noted that according to (2.22) and (2.23):

$$n_2k_o \leq \beta \leq n_1k_o \tag{2.24}$$

In other words, the propagation constants of guided modes lie in between those of the bulk materials composing the cladding and the core, respectively. Furthermore, also from (2.22) and (2.23), one gets:

$$(ha)^2+(qa)^2 = \left(\frac{2\pi a}{\lambda}\right)^2 \left(n_1^2-n_2^2\right) = V^2 \tag{2.25}$$

V, known as the *normalized frequency* parameter, depends on the core and cladding refractive indices, the core radius, and the signal wavelength, and plays an important role in the description of the propagation of guided modes.

To obtain the values of A, B, C, D, and β, one must first obtain $E_\theta(\bar{r},t)$ and $H_\theta(\bar{r},t)$ from $E_z(\bar{r},t)$ and $H_z(\bar{r},t)$ and then apply the boundary conditions of the tangential electromagnetic fields at the core-cladding interface which is given by the equation $r = a$. This leads to a homogeneous system of four equations and four unknowns. Forcing the coefficient determinant to be zero results in the so-called dispersion equation:

$$\begin{gathered} \left[\frac{J_l'(ha)}{hJ_l(ha)}+\frac{K_l'(qa)}{qK_l(qa)}\right]\cdot\left[\frac{n_1^2}{h}\frac{J_l'(ha)}{J_l(ha)}+\frac{n_2^2}{q}\frac{K_l'(qa)}{K_l(qa)}\right] = \left(\frac{l\beta}{k_o}\right)^2\left[\left(\frac{1}{qa}\right)^2+\left(\frac{1}{ha}\right)^2\right] \\ J_l'(x) = dJ_l(x)/dx \qquad K_l'(x) = dK_l(x)/dx \end{gathered} \tag{2.26}$$

Equation (2.26) is solved numerically in ha by fixing the value of l (which sets the azimuthal variation) and then expressing qa and β in terms of ha. Note that according to (2.25), $qa = \sqrt{V^2-(ha)^2}$. So, it is necessary to fix V as well as a value of the normalized frequency. Once the value of ha is determined, one gets the value of the propagation constant using the last expression in Eq. (2.22). For fixed l and V values, the dispersion equation usually renders more than one solution. Each of them has the same azimuthal dependence but a different radial variation and it is customary to use a second subindex m to label this variation. Following this solution process by changing the discrete values of $l = 0, 1, 2...$ and sweeping the value of V, one gets the *dispersion curves* for all the guided modes of an optical fiber that yield the value of the propagation constant β versus V which is shown in Fig. 2.5.

Note that each curve characterizes a given mode and that there are four types of guided modes according to the strength of the z components of their electric and magnetic fields. Hybrid magnetic-electric or HE_{lm} modes are those for which $E_z \neq 0$, $H_z \neq 0$, $|H_z| > |E_z|$, hybrid electric-magnetic or EH_{lm} modes are those for which $E_z \neq 0$, $H_z \neq 0$, $|E_z| > |H_z|$, transversal electric or TE_{0m} modes are those for which $E_z = 0$, $H_z \neq 0$ and, finally, transversal magnetic or TM_{0m} modes

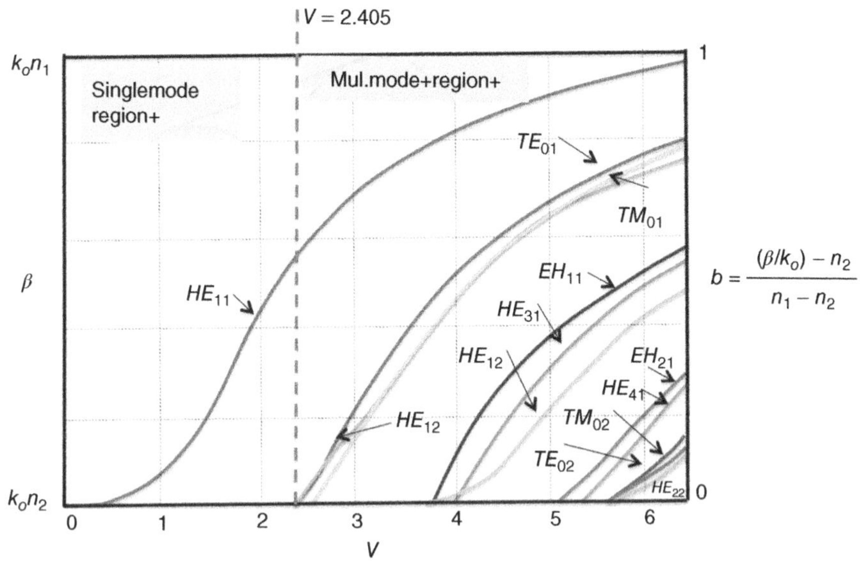

Figure 2.5 Propagation constant versus normalized frequency for an SI optical fiber.

are those for which $H_z = 0$, $E_z \neq 0$. The meaning of the subindices is directly connected to the azimuthal and radial variations of the field profile. l represents the number of zeros in the field profile when $0 \leq \theta \leq \pi$ while $m$gives the number of zeros in the field profile in the $0 \leq r \leq a$ range.

Note that for every node except for the HE_{11}, there is a cutoff value for Vbelow which $\beta_{lm} < k_o n_2$ and, therefore, the mode is no longer guided by the optical fiber. Indeed, theHE_{11} is known as the *fundamental mode* as it is always guided by the fiber. For instance, if $V < 2.405$, then only the fundamental mode is propagated, and the fiber is in singlemode operation regime. Otherwise, the fiber is multimode.

The exact solution for the modes even in the simplest case of optical fiber renders complex solutions for which it is very difficult to obtain a physical interpretation as, in general, the electric and magnetic vector fields have three nonzero components. Fortunately, in practice, the relative difference between the refractive indices of the core and the cladding $\Delta = \sqrt{n_1^2 - n_2^2}$ is very small $(0.1\% \leq \Delta \leq 1\%)$. So, one can apply the well-known weakly guiding approximation which considerably simplifies the solutions for the vector fields. The basic concept behind this approximation can be understood by referring to Fig. 2.5 where one can notice that there are several groups of modes with a very similar propagation constant. One example, for instance, is the group formed by the modes HE_{12}, TE_{01}, and TM_{01}. Under the weakly guiding approximation, these modes become degenerate. In other words, they have the same propagation constant and instead of considering their propagation as separate and independent modes, one can consider the propagation of their linear combinations. It results that these linear combinations are described by extremely simple vector fields, which feature only one nonzero component for the electric field and one nonzero component for the magnetic field. Furthermore, these nonzero components are mutually orthogonal and lay always in the transversal plane to the direction of propagation. Accordingly, these modes are similar to linearly polarized plane waves and thus are known as *linearly polarized* modes. Figure 2.6 shows the vector components of the electric field for an exact and linearly polarized mode.

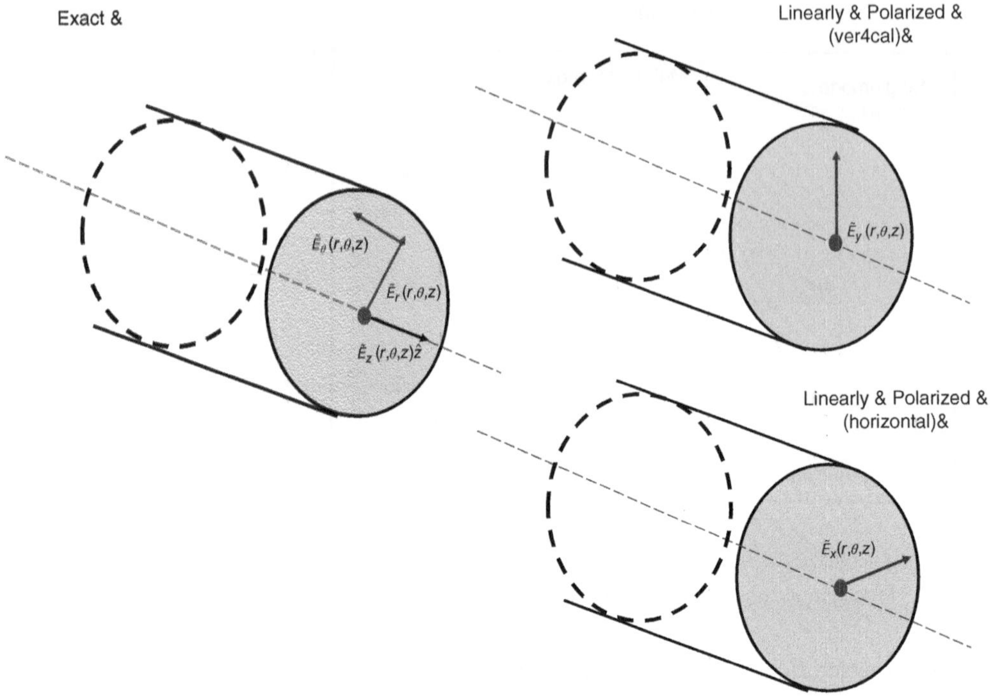

Figure 2.6 Vector field components for exact and linearly polarized modes in an SI optical fiber.

As it can be observed, there are two options for each linearly polarized mode; vertical or horizontal polarization depending on the direction to which the electric field vector points out (the magnetic field vector is orthogonal). In fact, the solutions of the simplified Maxwell equations for the electric field in this case are:

$$\mathbf{E}_{core}(\overline{r},t)=\begin{cases}E_x(\overline{r},t)=0\\E_y(\overline{r},t)=AJ_l(hr)\begin{Bmatrix}\cos l\theta\\\sin l\theta\end{Bmatrix}e^{j(\omega_o t-\beta z)}\\E_z(\overline{r},t)=0\end{cases}$$
$$\mathrm{E}_{cladding}(\overline{r},t)=\begin{cases}E_x(\overline{r},t)=0\\E_y(\overline{r},t)=BK_l(qr)\begin{Bmatrix}\cos l\theta\\\sin l\theta\end{Bmatrix}e^{j(\omega_o t-\beta z)}\\E_z(\overline{r},t)\approx 0\end{cases}\tag{2.27}$$

for the vertical polarized modes and:

$$\mathbf{E}_{core}(\overline{r},t)=\begin{cases}E_x(\overline{r},t)=AJ_l(hr)e^{jl\theta}e^{j(\omega_o t-\beta z)}\\E_y(\overline{r},t)=0\\E_z(\overline{r},t)=0\end{cases}$$
$$\mathrm{E}_{cladding}(\overline{r},t)=\begin{cases}E_x(\overline{r},t)=BK_l(qr)e^{jl\theta}e^{j(\omega_o t-\beta z)}\\E_y(\overline{r},t)=0\\E_z(\overline{r},t)\approx 0\end{cases}\tag{2.28}$$

for the horizontally polarized modes. Similar expressions follow for the magnetic field in each case but point in each case toward the orthogonal axis (x or y, respectively).

The determination of the propagation constant for the linearly polarized modes follows a similar procedure to that of the case of exact modes, but in this case, the dispersion equation is considerably simpler:

$$ha\frac{J_{l-1}(ha)}{J_l(ha)} = -qa\frac{K_{l-1}(qa)}{K_l(qa)} \tag{2.29}$$

Linearly polarized modes are designated by LP_{lm}, where the meaning of the subindices $l$and m is the same as in the case of the exact modes. LP_{lm} modes are degenerate and the order of degeneracy is 4 for $l \neq 0$ as there are two different linear polarizations and for each one, it is possible to have an azimuthal variation in cosine or sine functions. On the other hand, when $l = 0$, the order of degeneracy is 2.

The spatial variation of the intensity of LP_{lm} modes can be easily represented, and this allows one to gather a considerable understanding of their properties. For instance, Fig. 2.7 shows the typical intensity profile versus radial and azimuthal coordinates of the LP_{01} and LP_{02} modes in a $a = 5$ μ moptical fiber, while Fig. 2.8 shows a similar plot for LP_{11} modes.

The solution of (2.29) renders similar curves of the propagation constants for LP_{lm} modes than those obtained for exact modes represented in Fig. 2.5. For instance, in Fig. 2.9, we plot the *normalized propagation* constant defined as:

$$b_{lm} = \frac{(\beta_{lm}/k_o) - n_2}{n_1 - n_2} \tag{2.30}$$

as a function of the normalized frequency V for LP_{lm} modes.

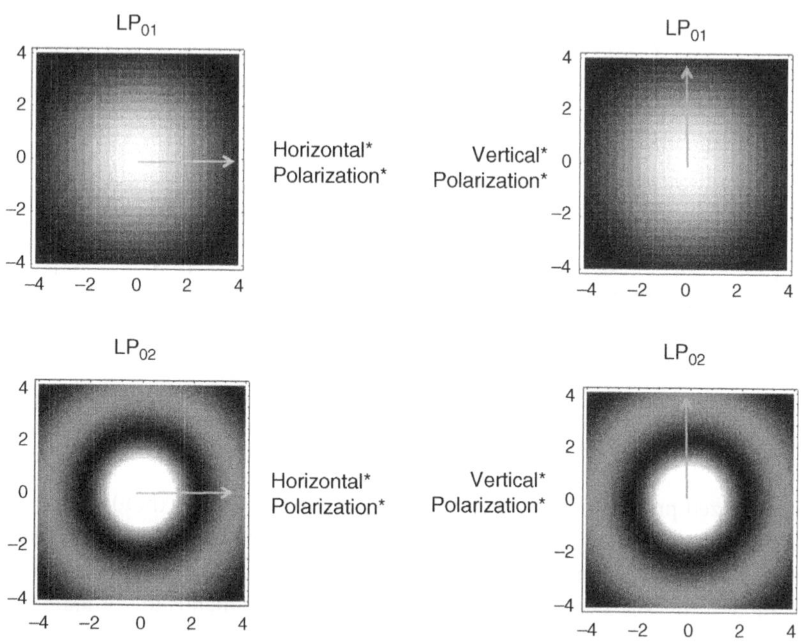

Figure 2.7 Typical intensity profiles for LP_{01} and LP_{02} modes in a SI optical fiber.

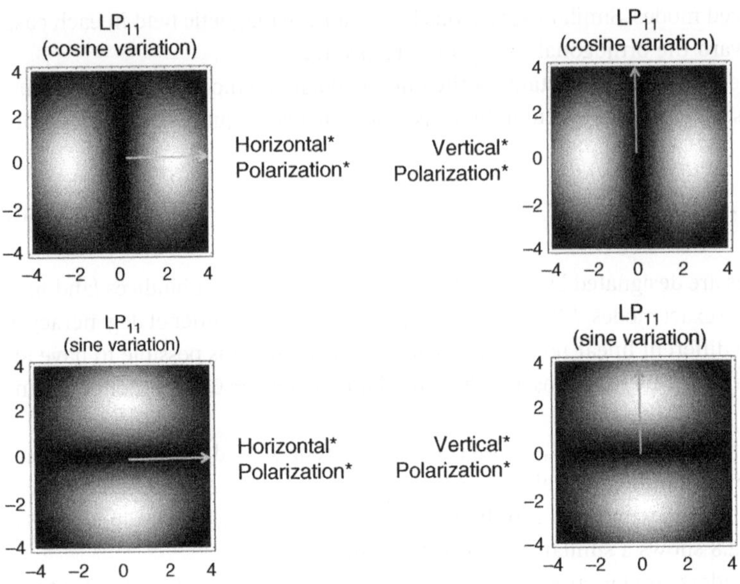

Figure 2.8 Typical intensity profiles for LP_{11} modes in an SI optical fiber.

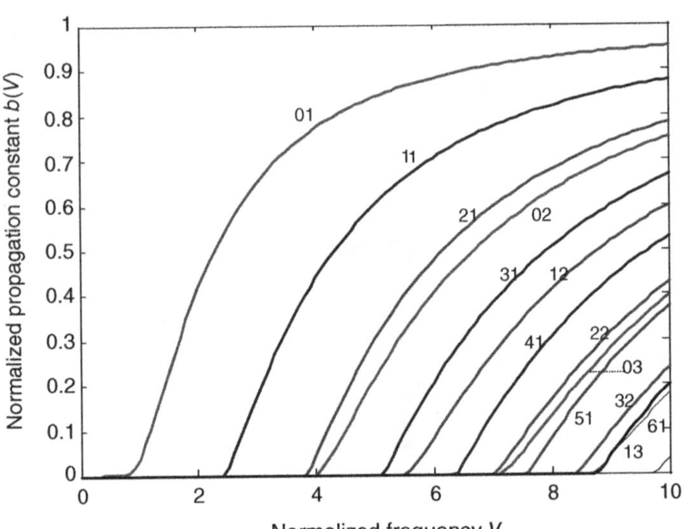

Figure 2.9 Normalized propagation constant versus normalized frequency for linearly polarized modes in an SI optical fiber.

The value of the normalized propagation constant for any LP_{lm} mode verifies that$0 \leq b_{lm} \leq 1$. If $b_{lm} \rightarrow 0$, the mode is close to cutoff and energy is poorly confined in the core. If $b_{lm} \rightarrow 0$, then well confined in the core and only a minimum fraction of its power is guided in the cladding. Note that for all LP_{lm} modes, there is a cutoff frequency except for LP_{01}, which is the fundamental linearly polarized mode. If $V < 2.405$, then only this mode propagates, and the fiber is singlemode.

Singlemode optical fibers are the preferred guided wave propagation medium in most microwave photonic and radio over fiber systems as they provide the maximum possible transmission bandwidth. In this case and according to (2.27) and (2.28), there are two possible fundamental modes LP^y_{01}, LP^x_{01}, and each one featuring a different linear polarization. The power-normalized (to unity) electric fields for each polarization are given by:

$$
\begin{aligned}
LP^y_{01} \rightarrow E_y = E_o \begin{cases} \dfrac{J_0(hr)}{J_0(ha)} e^{j\left(\omega_o t - \beta^y_{01} z\right)} \hat{y} & r \leq a \\ \dfrac{K_0(qr)}{K_0(qa)} e^{j\left(\omega_o t - \beta^y_{01} z\right)} \hat{y} & r > a \end{cases} \\
LP^x_{01} \rightarrow E_x = E_o \begin{cases} \dfrac{J_0(hr)}{J_0(ha)} e^{j\left(\omega_o t - \beta^x_{01} z\right)} \hat{x} & r \leq a \\ \dfrac{K_0(qr)}{K_0(qa)} e^{j\left(\omega_o t - \beta^x_{01} z\right)} \hat{x} & r > a \end{cases}
\end{aligned}
\tag{2.31}
$$

In principle, for ideal fibers, both fundamental modes are degenerate, i.e. $\beta^y_{01} = \beta^x_{01} = \beta_{01}$. Figure 2.10 represents the normalized propagation constant of the fundamental mode as a function

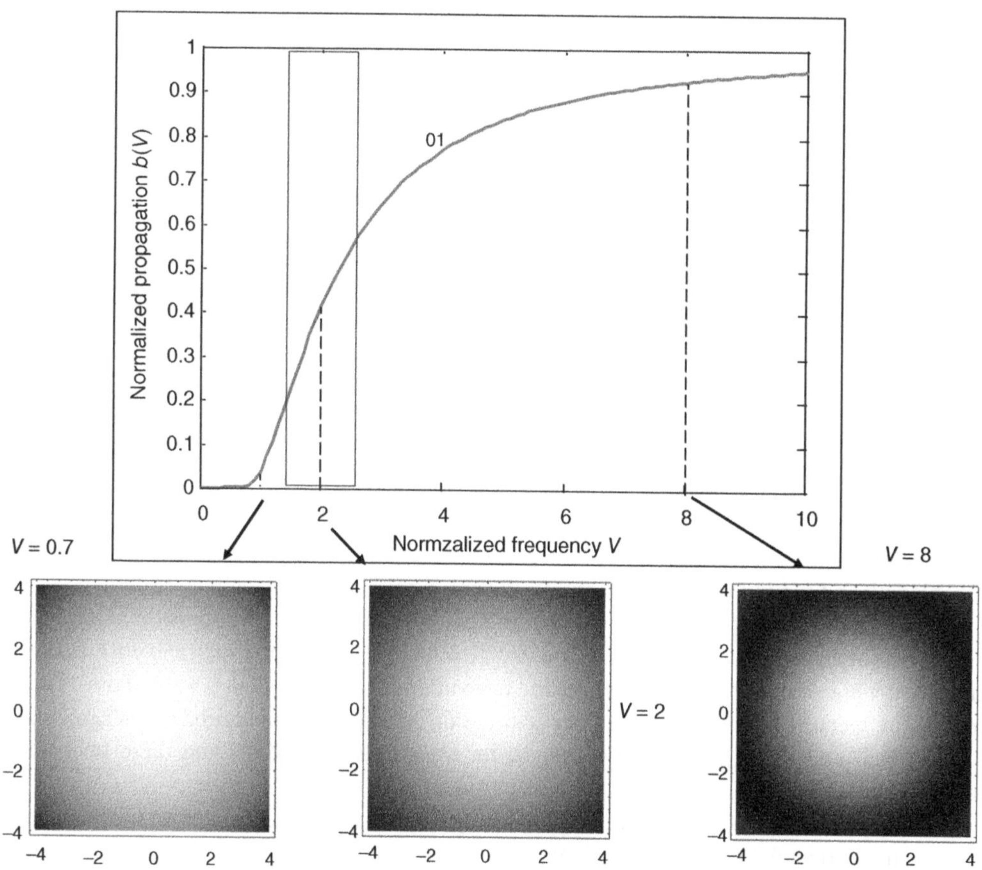

Figure 2.10 Normalized propagation constant of the fundamental mode of a SI fiber as a function of the normalized frequency. The spatial extension of the mode is shown in the bottom part.

of the normalized frequency. The insets represent the mode intensity profiles for different values of the normalized frequency with the purpose of illustrating that the mode energy is increasingly confined inside the core as the value of b_{01} approaches to unity.

The dependence of b_{01} with the normalized frequency is rather complex but, inside the region defined by $1.5 \leq V < 2.405$ (boxed in Fig. 2.10), the following approximation can be employed:

$$b_{01}(V) = \left(1.1428 - \frac{0.996}{V}\right)^2 \tag{2.32}$$

It is clear, however, that, in general, one will have a complex dependency of the propagation constant with frequency. This dependency is behind the well-known mechanism of *chromatic dispersion* in optical fibers which limits the transmission bandwidth. Chromatic dispersion yields a frequency selective group delay which results in pulse broadening in the transmission of digital signals through singlemode fibers. In the case of analog photonic applications, chromatic dispersion renders a nonlinear frequency-dependent phase delay that results as we will see in Chapter 5 in the fading of different radio and microwave frequencies conveyed by the optical carrier.

There are two mechanisms behind chromatic dispersion. The first one is *material dispersion* and results from the fact that even for bulk material, the linear susceptibility is a frequency-dependent function (see Eq. (2.10)). Thus, the material is dispersive *per se*, even if it is not forming a waveguide. The second mechanism is waveguide dispersion. Even if we assume that the bulk materials are not dispersive, there is a frequency dependence of the mode propagation constants resulting from the fact that they are forming a waveguide. This follows immediately, for instance, from Fig. 2.10. Note that the propagation constant was computed assuming that neither n_1 nor n_2 were frequency dependent.

It is customary to approximate the complex dependence of the propagation constant $\beta_{01}(\omega)$ (we drop the subscript $_{01}$ from now on) with the frequency ω by means of a Taylor series centered at the optical carrier frequency ω_o:

$$\begin{aligned}\beta(\omega) &= \beta(\omega_o) + \left.\frac{d\beta}{d\omega}\right|_{\omega_o}(\omega-\omega_o) + \frac{1}{2}\left.\frac{d^2\beta}{d\omega^2}\right|_{\omega_o}(\omega-\omega_o)^2 + \frac{1}{6}\left.\frac{d^3\beta}{d\omega^3}\right|_{\omega_o}(\omega-\omega_o)^3 + \cdots \\ &= \beta_o + \beta_1(\omega-\omega_o) + \frac{\beta_2}{2!}(\omega-\omega_o)^2 + \frac{\beta_3}{3!}(\omega-\omega_o)^3 + \cdots\end{aligned} \tag{2.33}$$

In the above expression, β_o (unit in rad/km) is a frequency-independent phase shift term per unit of fiber length and β_1 (unit in ps/km) represents the group delay per unit length. β_2 (unit in ps^2/km) and β_3 (unit in ps^3/km) represent, respectively, the first- and second-order dispersion parameters. In standard singlemode fibers, $\beta_2 \gg \beta_3$ in the first ($\lambda = 850$ nm) and third ($\lambda = 1550$ nm) telecommunication windows, while $\beta_2 \approx 0$ in the second telecommunication window ($\lambda = 1310$ nm). These windows are spectral regions where the fiber attenuation constant $\alpha(\omega)$ measured in dB/Km is particularly small and hence are widely employed for transmission in MWP and radio over fiber systems.

In summary, the effect of a dispersive singlemode optical fiber link of z km can be represented by the following optical field transfer function:

$$H(\omega, z) = 10^{-(\alpha(\omega_o)z)/10} e^{-j\left[\beta_o + \beta_1(\omega-\omega_o) + ((\beta_2)/(2!))(\omega-\omega_o)^2 + ((\beta_3)/(3!))(\omega-\omega_o)^3\right]z} \tag{2.34}$$

2.2.4 Propagation in Planar Dielectric Waveguides

The analysis of planar waveguides follows a similar approach to that of optical fibers. In this case, however, it is more convenient to employ a Cartesian coordinate system. For an infinite asymmetric waveguide as shown in Fig. 2.4, the equivalent to (2.16) is:

$$\left[\frac{\partial^2}{\partial x^2}+\frac{\partial^2}{\partial y^2}+\frac{\partial^2}{\partial z^2}\right]\cdot\begin{bmatrix}\tilde{\mathbf{E}}(x,y,z)\\ \tilde{\mathbf{H}}(x,y,z)\end{bmatrix}+k^2\begin{bmatrix}\tilde{\mathbf{E}}(x,y,z)\\ \tilde{\mathbf{H}}(x,y,z)\end{bmatrix}=0 \tag{2.35}$$

which is again a system of six partial derivative equations for each of the material regions that compose the waveguides. The approach to solving these equations is identical to that followed in the case of the cylindrical waveguide. In particular, we can employ the method of separation of variables and solve only for two of the six vectorial components (the z components for instance). We will employ the following terminology:

$$\begin{aligned}\tilde{E}(x,y,z)&=\tilde{E}(x,y)e^{-j\beta z}\\ \tilde{H}(x,y,z)&=\tilde{H}(x,y)e^{-j\beta z}\end{aligned} \tag{2.36}$$

to represent any of the Cartesian components of interest.

Fortunately, as shown in Fig. 2.4, the waveguide is infinite in the y direction. Hence, we can assume that $\partial/\partial y=0$. So, after substituting (2.36) in (2.35), we get the following set of equations which must be solved in each region:

$$\begin{aligned}&\frac{d^2}{dx^2}\begin{bmatrix}\tilde{E}(x,y)\\ \tilde{H}(x,y)\end{bmatrix}+\left(k_o^2n_s^2-\beta^2\right)\begin{bmatrix}\tilde{E}(x,y)\\ \tilde{H}(x,y)\end{bmatrix}=0 \quad -\infty<x<-d\\ &\frac{d^2}{dx^2}\begin{bmatrix}\tilde{E}(x,y)\\ \tilde{H}(x,y)\end{bmatrix}+\left(k_o^2n_n^2-\beta^2\right)\begin{bmatrix}\tilde{E}(x,y)\\ \tilde{H}(x,y)\end{bmatrix}=0 \quad -d<x<-0\\ &\frac{d^2}{dx^2}\begin{bmatrix}\tilde{E}(x,y)\\ \tilde{H}(x,y)\end{bmatrix}+\left(k_o^2n_c^2-\beta^2\right)\begin{bmatrix}\tilde{E}(x,y)\\ \tilde{H}(x,y)\end{bmatrix}=0 \quad 0<x<\infty\end{aligned} \tag{2.37}$$

To solve (2.37), we need to apply the corresponding boundary conditions for the electromagnetic fields which require the continuity of the tangential fields at the interface surfaces between the different media. Before we proceed, however, we will qualitatively investigate the physical nature of its solutions capitalizing on the knowledge acquired in the solution of the cylindrical waveguide. In particular, we will assume that $n_n > n_s > n_c$ and classify the solutions according to the value of the mode propagation constant β as compared to that of the plane waves in the different media.

Case 1: If $\beta > k_on_n$, then the solution of (2.37) is an exponential function in the three media (substrate, core, and cladding). Furthermore, it is exponentially increasing in the substrate and the cladding as we depart from the core. This solution is physically meaningless as it requires infinite energy.

Case 2: If $k_on_c, k_on_s < \beta < k_on_n$, then the solution is oscillatory in the core and exponentially decreasing (evanescent) in the substrate and cladding. The solutions correspond to the *guided modes*. As it happens with the case of the cylindrical waveguide, the allowed values for β are (as we will see) a discrete set. The number of guided modes will depend on the values of the refractive indices of the three regions and the core width d.

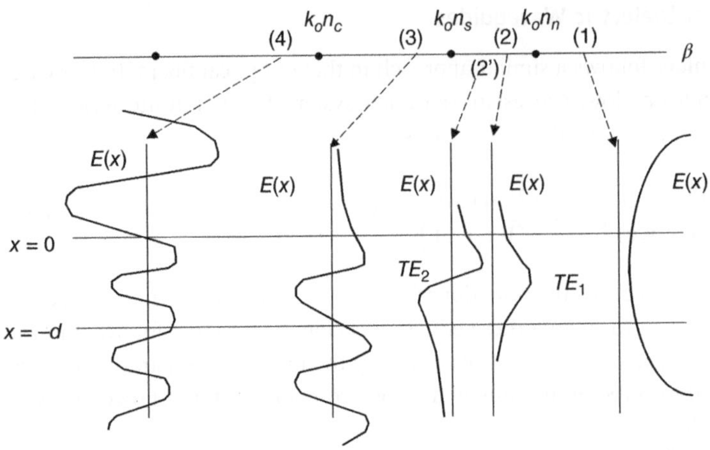

Figure 2.11 Typical mode profiles for the different solution cases of the Helmholtz equation in a planar waveguide.

Case 3: If $k_o n_c < \beta < k_o n_s$, then the solution is exponentially decreasing in the cladding and oscillatory in the core and the substrate. These are known as *substrate radiation modes* and the allowed values for β are a continuum.

Case 4: If $0 < \beta < k_o n_c$, then the solutions are oscillatory in the three regions. These solutions are called *waveguide radiation modes* and the allowed values for β are a continuum. Figure 2.11 summarizes the typical solutions for the four cases previously discussed.

We now consider the solution for case 2 (guided modes). Assuming the conditions for that case and $n_c < n_s$, we only have transversal electric (*TE;* (0, E_y, 0)) and magnetic (*TM;*(0, H_y, 0)) guided modes. In the case of TE modes, the *y* component for the electric field must satisfy:

$$\begin{aligned} &\frac{d^2E_y}{dx^2} - q^2E_y = 0 \quad 0 < x < \infty \\ &\frac{d^2E_y}{dx^2} + h^2E_y = 0 \quad -d < x < 0 \\ &\frac{d^2E_y}{dx^2} - p^2E_y = 0 \quad -\infty < x < -d \end{aligned} \tag{2.38}$$

where:

$$\begin{aligned} h &= \sqrt{k_o^2 n_n^2 - \beta^2} \\ q &= \sqrt{\beta^2 - k_o^2 n_c^2} \\ p &= \sqrt{\beta^2 - k_o^2 n_s^2} \end{aligned} \tag{2.39}$$

the solution of (2.38) after applying the boundary conditions is:

$$E_y = \begin{cases} Ce^{-qx} & 0 < x < \infty \\ C\left[\cos(hx) - \frac{q}{h}\sin(hx)\right] & -d < x < 0 \\ C\left[\cos(hd) + \frac{q}{h}\sin(hd)\right]e^{p(x+d)} & -\infty < x < -d \end{cases} \tag{2.40}$$

where C is a constant. From (2.40), we can obtain the other tangential component of the TE mode, which is H_z

$$H_z = -\frac{1}{j\omega\mu}\frac{\partial E_y}{\partial x} = \begin{cases} -\dfrac{jqC}{\omega\mu}Ce^{-qx} & 0 < x < \infty \\ -\dfrac{jC}{\omega\mu}[h\sin(hx) + q\cos(hx)] & -d < x < 0 \\ \dfrac{jCp}{\omega\mu}\left[\cos(hd) + \dfrac{q}{h}\sin(hd)\right]e^{p(x+d)} & -\infty < x < -d \end{cases} \tag{2.41}$$

Now, if we impose the boundary condition for the continuity of H_z in the surface $z = -d$, we get the following *dispersion equation*:

$$\tan(hd) = \frac{p+q}{h(1-((pq)/h^2))} \tag{2.42}$$

Equation (2.42) together with the relationships given by (2.39) gives the values of the propagation constants for the guided TE modes. In a similar way as with linearly polarized modes in cylindrical waveguides, the solutions to (2.39) and (2.42) yield a discrete set of propagation constants (i.e. guided transversal electric modes). Each mode is labeled using a discrete index m such that mode TE_m, $m = 1, 2, 3 \ldots$ is characterized by $(p_m, q_m, h_m, \beta_m^{TE}, C_m)$. The mode TE_m is known as the fundamental TE mode as it propagates for any operating frequency. The rest have cutoff frequencies below which they no longer propagate as guided modes. The value of C_m is in principle arbitrary, but it is customary to compute it by normalizing the power per unit length carried by the mode to 1 watt. Note as well that the modes are orthogonal because they verify the relationship:

$$\int_{-\infty}^{\infty} E_{y,m}E_{y,n}dx = \frac{2\omega\mu}{\beta_m}\delta_{m,n} \tag{2.43}$$

where $\delta_{m,\,n}$ is the Kronecker delta function.

A similar procedure can be followed for TM modes which result in:

$$H_y = \begin{cases} -\dfrac{hCn_s^2}{qn_n^2}e^{-qx} & 0 < x < \infty \\ C\left[-\dfrac{hn_s^2}{qn_n^2}\cos(hx) + \sin(hx)\right] & -d < x < 0 \\ -C\left[\dfrac{hn_s^2}{qn_n^2}\cos(hd) + \sin(hd)\right]e^{p(x+d)} & -\infty < x < -d \end{cases} \tag{2.44}$$

from which we can compute $E_z = (j\omega\varepsilon_o n^2)^{-1}\partial H_y/\partial x$ in each region and, applying the boundary conditions for the tangential fields to the interface surfaces, arrive at the *dispersion equation* for TM modes:

$$\tan(hd) = \frac{\overline{p}+\overline{q}}{h(1-((\overline{p}\overline{q})/h^2))} \tag{2.45}$$

where:

$$\overline{p} = p\frac{n_n^2}{n_c^2}$$
$$\overline{q} = q\frac{n_n^2}{n_c^2} \quad (2.46)$$

The solutions to (2.45)–(2.46) are also discrete sets, which are classified according to an integer index m. Thus, TM modes follow a similar nomenclature to that of TE_m and we employ the terminology TM_m to identify them. The fundamental mode in this case is TM_1.

In practice, planar waveguides are also limited in the y coordinate and an approximate analytical solution of (2.35) can be obtained by means of the so-called Marcatili's method. It is customary, however, to employ a numerical approach in most cases. Here, instead of transversal electric and magnetic modes, we have hybrid electric and hybrid magnetic guided modes, respectively, as in the optical fiber. Nevertheless, the fundamental physical concepts remain very close to those presented in the case of the infinite asymmetric waveguide. So, we will not elaborate further on this topic.

2.3 Light Sources, Detectors, and Modulators

Lasers are the preferred optical sources in most MWP applications as they bring unrivaled advantages in terms of output power, linearity, and low-intensity and phase noise. In particular, semiconductor laser diodes are most widely employed as they are compact, only require a few mA of drive for operation, can be directly or externally modulated, and also offer a potential route to monolithic integration together with other electronic and photonic devices on the same chip. Other alternatives that have been proposed for their use in microwave photonic systems include solid-state and fiber-based lasers. Optical detectors for MWP rely almost exclusively on semiconductor structures and more precisely on reverse-biased pin junctions. In this section, we will briefly review the fundamental operation principles of optical sources and detectors based on the interaction mechanisms between radiation and matter, the specialization of these principles to semiconductor materials and the basic structures, equations and parameters of optical sources, and detectors and modulators.

2.3.1 Fundamentals of the Interaction Between Radiation and Matter

To understand the operation of optical sources and detectors, it is instructive to review the mechanisms of interaction between radiation and matter which account for the different ways in which an electromagnetic field is absorbed or amplified when propagating through a given material.

We take as a starting point some very basic results of the quantum theory of materials by which the allowed energy states of electrons, atoms, and molecules[1] in a material can only take discrete values which are known as *energy levels*.

1 We will from now on in this subsection refer only to electrons as these particles are responsible for the interaction between radiation and matter in semiconductors. It must be understood however that what is written can also apply to atoms or molecules in other materials. For instance, in fiber lasers, atoms are responsible for the transitions leading to the generation and absorption of light.

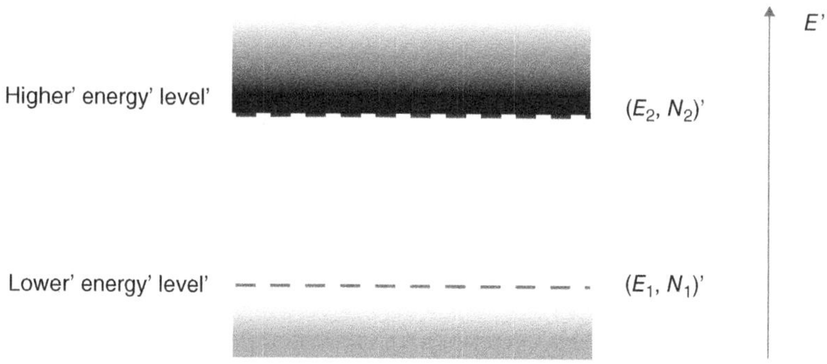

Figure 2.12 Diagram showing adjacent energy levels and populations in a material.

Let us assume two adjacent energy levels in a material given by E_1 and E_2, respectively, as shown in Fig. 2.12, such that N_1 and N_2 characterize their respective electron numbers or populations.

Under thermodynamic equilibrium, the population of a given energy level *N*(*E*) follows the Boltzmann distribution:

$$N(E) = e^{-((E)/(k_B T))} \tag{2.47}$$

where $k_b = 1.38 \times 10^{-23}$ J°/K is the Boltzmann constant and T is the temperature expressed in Kelvin degrees. From (2.47), it follows that under thermodynamic equilibrium, $N_1 > N_2$. We now describe referring to Fig. 2.12 the three main interaction mechanisms between radiation and matter due to the transfer of electrons from adjacent energy levels E_1 y E_2.

In the first place, as shown in Fig. 2.13, an electron in E_2 can experience a downward transition to E_1. Energy conservation requires the emission of a photon of energy $E_2 - E_1$ or, in other words, a quantum of electromagnetic radiation of frequency:

$$\nu = \frac{E_2 - E_1}{h} \tag{2.48}$$

where $h = 6.626 \times 10^{-23}$ J seg is the *Planck constant*. This process is known as spontaneous emission as it happens randomly and does not require the presence of any external electromagnetic field. Photons generated by spontaneous emission have the same frequency but their phases are

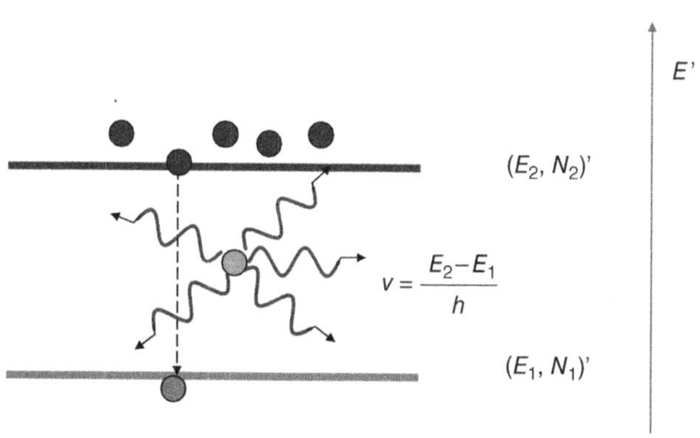

Figure 2.13 Schematic illustration of the mechanism of spontaneous emission.

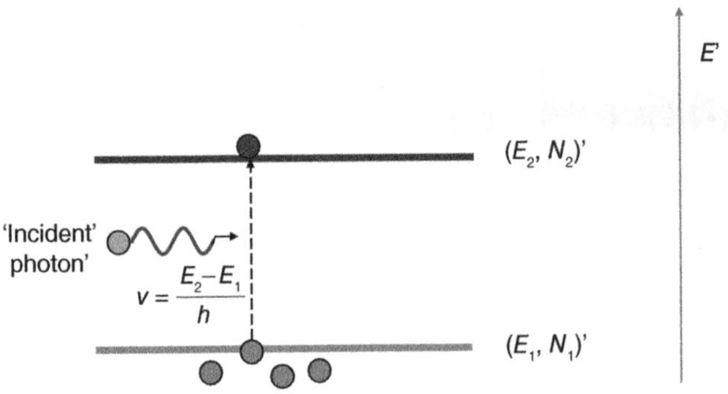

Figure 2.14 Schematic illustration of the mechanism of absorption.

uncorrelated. They are therefore incoherent. Their emission is performed with equal probability in all the directions of space and their state of polarization is arbitrary.

The rate of change of N_2 due to spontaneous emission is given by:

$$R_{spon} = \left.\frac{dN_2}{dt}\right|_{spon} = AN_2 \tag{2.49}$$

where A is the Einstein coefficient for spontaneous emission that will be derived later.

The second process known as *absorption*, shown in Fig. 2.14, is produced when the energy of an incident photon of frequency $\nu = (E_2 - E_1)/h$ coming from external electromagnetic radiation is employed to transfer an electron from level E_1 to level E_2. The incident photon is said to be absorbed and disappear, thus reducing the energy of the external electromagnetic radiation as it propagates through the material.

If $\rho(\nu)$ represents the energy spectral density of the electromagnetic radiation incident to the material, then the rate of change of N_1 due to absorption is given by:

$$R_{abs} = \frac{dN_1}{dt} = B_{12}N_1\rho(\nu) \tag{2.50}$$

where B_{12} is *the Einstein absorption coefficient.*

Finally, under certain conditions, a photon of frequency $\nu = (E_2 - E_1)/h$ coming from external electromagnetic radiation can stimulate the transition of an electron from level E_2 to level E_1, as illustrated in Fig. 2.15, generating a new photon with the same frequency, phase, state of polarization, and direction of propagation. This process is known as *stimulated emission* and results in the generation of coherent electromagnetic radiation as all the newly generated photons have a fixed phase relationship among them as opposed to the case of those generated by spontaneous emission.

The rate of change of N_2 due to stimulated emission is given by:

$$R_{stim} = \left.\frac{dN_2}{dt}\right|_{stim} = B_{21}N_2\rho(\nu) \tag{2.51}$$

where B_{21} is the Einstein-stimulated emission coefficient.

Under thermodynamic equilibrium conditions, the rate of transitions from E_1 to E_2 must be equal to the rate of transitions from E_2 and E_1, or equivalently:

$$\begin{aligned} R_{stim} + R_{spon} &= R_{abs} \\ AN_2 + B_{21}N_2\rho(\nu) &= B_{12}N_1\rho(\nu) \end{aligned} \tag{2.52}$$

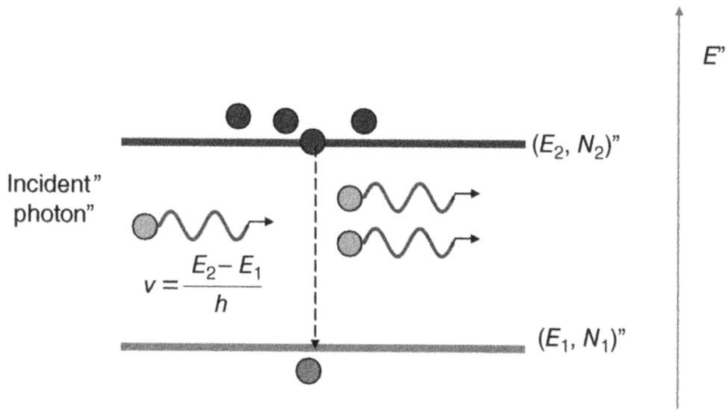

Figure 2.15 Schematic illustration of the mechanism of stimulated emission.

From (2.62) and taking into account the Boltzmann statistics given by (2.47), one obtains the following expression for the energy spectral density of the electromagnetic radiation:

$$\rho(\nu) = \frac{(A/B_{21})}{(B_{12}/B_{21})e^{((h\nu)/(k_BT))} - 1} \tag{2.53}$$

Equation (2.53) must hold for any electromagnetic radiation in general and, in particular, for the *blackbody radiation* whose energy spectral density is given by:

$$\rho(\nu) = \frac{8\pi h\nu^3/c^3}{e^{(h\nu/k_BT)} - 1} \tag{2.54}$$

Identifying (2.53) and (2.54), we get $B_{12} = B_{21} = B$ and $A = (8\pi h\nu^3/c^3)B$. From these values and taking into account (2.49) and (2.51), we obtain the following expression for the quotient between stimulated and spontaneous emission rates under thermodynamic equilibrium:

$$\frac{R_{spon}}{R_{stim}} = \frac{A}{B\rho(\nu)} = e^{((h\nu)/(k_BT))} - 1 \tag{2.55}$$

Under normal operating conditions, that is for $T = 298°\text{K}$ (25 °C), $k_BT = 25$ meV and for $\lambda = 1$ μm then $h\nu = hc/\lambda = 1.42$ eV, which implies that $h\nu \gg k_BT$ and, therefore $R_{spon} \gg R_{stim}$. So, the spontaneous emission rate is much higher than the stimulated emission rate under thermodynamic equilibrium. Hence, to revert this situation in order for stimulated emission to dominate, thermodynamic equilibrium must be broken, which requires that $N_2 > N_1$, a condition known as *population inversion*. Both spontaneous and stimulated emissions are *radiative* processes since they generate electromagnetic radiation or photons. In all materials, there are other processes called *nonradiative* that can take place with nonzero probability, leading to the transition of electrons from E_2 to E_1. The excess energy is conferred to *phonons* which are energy quanta corresponding to the mechanical vibration modes of the material structure (atoms, electrons, and molecules). In general, radiative and nonradiative processes compete in a given material and therefore to make efficient optical sources, it is desirable to reduce the probability of nonradiative processes in the selected material.

2.3.2 Semiconductor Materials for Optical Sources and Detectors

Since most of the optical sources and detectors employed in MWP are based on semiconductors, we will review some of the salient characteristics of these materials.

Figure 2.16 Energy-level diagram for the valence and conduction bands in a semiconductor.

Semiconductors have electrical properties that lie in between those of isolators and conductors. They feature a crystalline structure where atoms or atom groups are distributed following a periodic spatial pattern or lattice. The spatial repetition period is given by a lattice constant a_o with typical values of a few Angstroms. One of the most popular semiconductor materials is silicon (Si) that belong to group IV of the periodic system. The silicon atom has four electrons in its outer energy level which can be employed to form covalent bonds with other Si atoms forming a regular crystalline lattice. It is useful to use *energy band diagrams* to study the behavior of semiconductors. The two relevant energy levels known as the *valence and the conduction bands* respectively are shown in Fig. 2.16 and following the nomenclature given in the previous section, (E_1, N_1) will characterize the energy and population of the valence band while (E_2, N_2) will characterize the energy and population of the conduction band.

At very low temperatures (close to the absolute zero), the valence band is completely full of electrons, while the conduction band is completely empty. These bands are separated by an *energy gap* $E_g = E_2 - E_1$ inside which intermediate energy levels are not allowed. If the temperature is risen, then some electrons in the valence band get enough energy to make a transition to the conduction band. The origin of this energy is thermal and, in particular, for Si, has to be around 1 eV. As a consequence of such transitions, the population in the conduction band is no longer zero or, in other words, there will be a certain concentration (population per unit of volume) n of electrons in the conduction band called *free carriers*. Each transition also generates an electron vacancy or *hole* in the valence band, which is equivalent to a virtual positive charge. It is customary to use p to denote the hole concentration in the valence band.

Both free carriers and holes can move inside the material and, as a consequence, can increase its conductivity since, for example, the presence of an internal or external electric field can move an electron in the valence band to occupy the position of a hole or vacant, moving this hole in opposite direction as shown in Fig. 2.17. As it can be appreciated in the figure, the free carrier and hole flows in a semiconductor material are produced in opposite directions.

Semiconductors with no impurities are called intrinsic and the free electron and hole concentrations are given by:

$$n = N_c e^{-((E_2 - E_f)/(k_B T))} \quad p = N_p e^{-((E_f - E_1)/(k_B T))} \tag{2.56}$$

here $E_f \approx (E_1 + E_2)/2$ denotes the *Fermi Level* placed in the middle of the energy gap and $N_c = 2(2\pi m_e k_B T/h^2)^{3/2}$, $N_v = 2(2\pi m_h k_B T/h^2)^{3/2}$ where m_e and m_h, respectively, represent the effective masses of electrons in the conduction band and holes in the valence band. The product of both concentrations is known as the *intrinsic carrier concentration* n_i.

$$np = n_i^2 = N_c N_v e^{-((E_g)/(k_B T))} \tag{2.57}$$

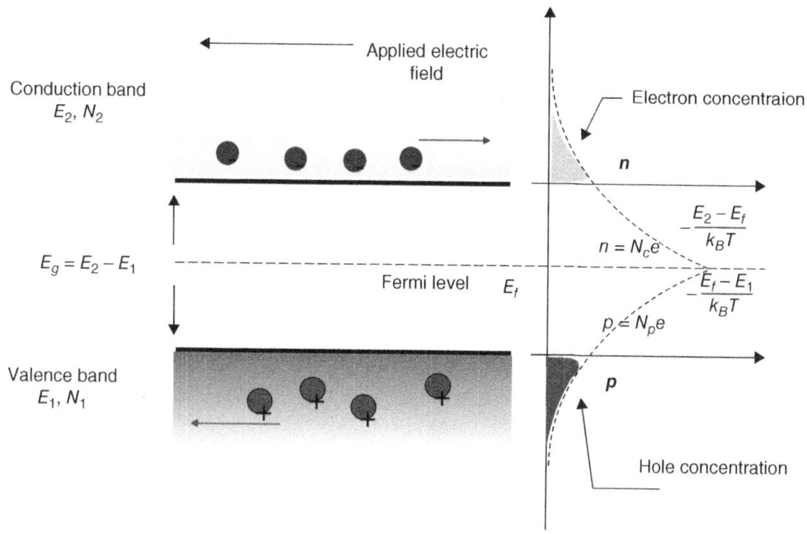

Figure 2.17 Free carrier and hole flow directions under an applied electric field in a semiconductor.

Intrinsic semiconductors have very limited conductivity and one way to substantially increase it is by doping the material with impurities to form an *extrinsic semiconductor*. Semiconductor doping can be done in two different ways as shown in Figs. 2.18 and 2.19.

If the objective is to increase the carrier concentration in the conduction band, then it is convenient to introduce elements of group V of the periodic system (P, As, Sb) in the Si atom lattice. These elements have five electrons in their outer energy level. Four of them can be employed to form covalent bonds with the four outer electrons of surrounding Si atoms while the fifth electron, which is

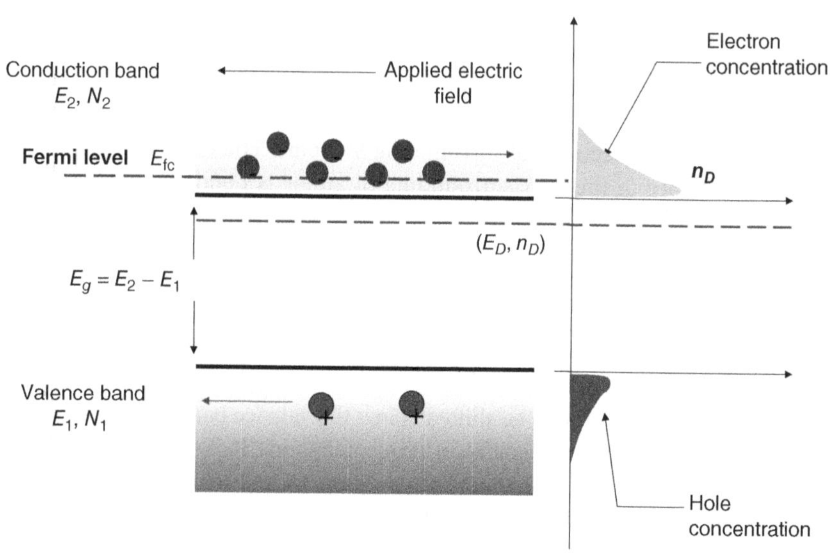

Figure 2.18 Energy band diagram for a n-doped semiconductor.

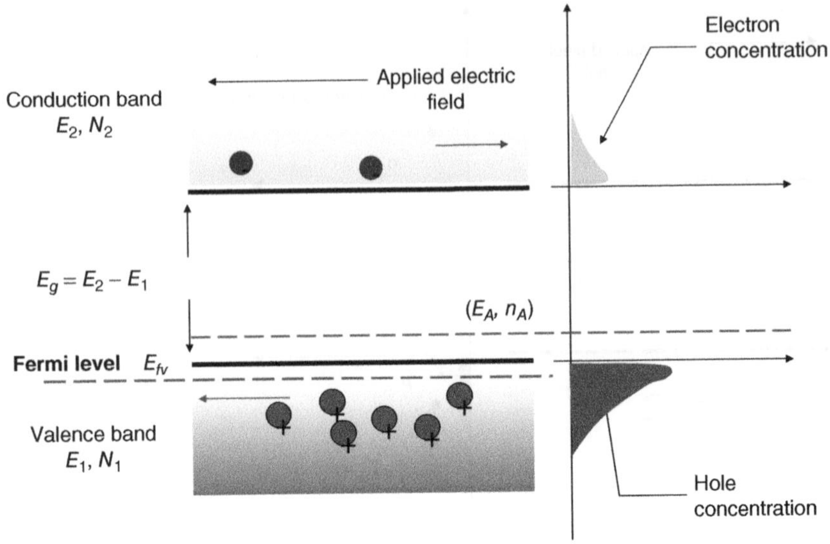

Figure 2.19 Energy band diagram for a *p*-doped semiconductor.

not employed for bonding, and remains weakly confined to the lattice, is available for conduction. This process is illustrated in Fig. 2.18 and results in the appearance of a new energy level populated by electrons inside the energy gap close to the conduction band. This is known as the *donor energy level*, and we will use (E_D, n_D) to represent its energy and concentration, respectively. Impurities are called donor atoms. This doping process leads to the so-called *n-type semiconductors* since the current in them will be due majorly to electron flow. Another consequence of the doping process is that the Fermi level is no longer placed in the middle of the energy gap but is displaced toward the conduction band. Furthermore, if the semiconductor is heavily doped, the Fermi level can be shifted above the conduction band.

In a similar way, it is also possible to dope the semiconductor to increase the hole concentration in the valence band. This leads to the so-called *p-doped semiconductors.* In this case, one introduces elements of group III of the periodic system (B, Ga) in the Si atom lattice. These elements have three electrons in their outer energy level which are employed to form covalent bonds with the four outer electrons of surrounding Si atoms. The fourth electron necessary to fully bond the impurity to the four surrounding Si atoms is obtained by creating a weakly confined hole that is available for conduction. This process leads to the appearance of an acceptor energy level (E_A, n_A) inside the energy gap and close to the valence band as shown in Fig. 2.19.

The conduction process is *p*-type semiconductors is achieved when electrons in the valence band are excited to the acceptor level. As a consequence of *p*-doping, the Fermi level is no longer in the middle of the energy gap, but shifted toward the valence band and as in the case of *n*-type semiconductors, if doping is strong enough, it can be shifted below the valence band.

To study the absorption and emission processes in semiconductors, it is useful to consider their energy band structure, in particular, the energy-momentum or *E–K diagrams.* Figure 2.20 shows a typical *E–K* diagram where both the valence and the conduction bands feature a parabolic shape. The maximum of the valence band is represented by E_v, while E_c is the minimum of the

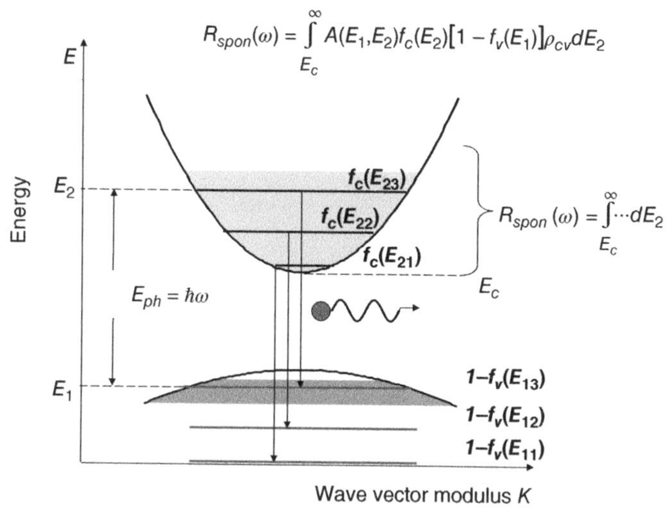

Figure 2.20 Typical parabolic *E–K* diagram for the valence and conduction bands in a semiconductor.

conduction band. Figure 2.20 also illustrates the spontaneous emission process which we are going to characterize.

In a semiconductor, spontaneous emission can only take place if the conduction band E_2 has one available electron and the valence band E_1 an available hole. The occupation probabilities in the conduction (electrons) and valence (holes) bands follow Fermi-Dirac rather than Boltzmann distributions:

$$f_c(E_2) = \frac{1}{1 + e^{(E_2 - E_{fc})/(k_B T)}}; \quad f_v(E_1) = \frac{1}{1 + e^{(E_1 - E_{fv})/(k_B T)}} \tag{2.58}$$

where E_{fc} and E_{fv} are the Fermi levels for free electrons and holes, respectively ($E_{fc} = E_{fv} = E_f$ under thermodynamic equilibrium). The spontaneous emission rate produced at a frequency ω is obtained by summing all the possible transitions between the conduction and valence energy band that verify $E_2 - E_1 = E_{ph} = \hbar\omega$, where $\omega = 2\pi\nu$, $\hbar = h/2\pi$, and E_{ph} represents the energy of the emitted photon. The result of this sum is:

$$R_{spon}(\omega) = \int_{E_c}^{\infty} A(E_1, E_2) f_c(E_2)[1 - f_v(E_1)]\rho_{cv} dE_2 \tag{2.59}$$

where $A(E_1, E_2)$ is the Einstein spontaneous emission coefficient, and ρ_{cv} represents the joint density of states, defined as the joint number of energy states separated by $\hbar\omega$ per unit of volume and unit of energy:

$$\rho_{cv} = \frac{(2m_r)^{3/2}}{2\pi^2\hbar^3}\sqrt{\hbar\omega - E_g} \tag{2.60}$$

$E_g = E_c - E_v$ is the energy gap and $m_r = m_e m_h/(m_e + m_h)$. From (2.60), one can observe that ρ_{cv} does not depend on E_2 and therefore can be taken out of the integral (2.59).

In a similar way, one can obtain the rates for the stimulated emission and absorption:

$$R_{stim}(\omega) = \int_{E_c}^{\infty} B(E_1, E_2) f_c(E_2)[1 - f_v(E_1)]\rho_{cv}\rho(\omega)dE_2$$
$$R_{abs}(\omega) = \int_{E_c}^{\infty} B(E_1, E_2) f_v(E_1)[1 - f_c(E_2)]\rho_{cv}\rho(\omega)dE_2 \quad (2.61)$$

where $\rho(\omega)$ is the energy spectral density of the incident radiation field and $B(E_1, E_2)$ represents the relevant Einstein coefficient. Comparing (2.61) and (2.59), we obtain the condition for population inversion in a semiconductor:

$$E_{fc} - E_{fv} > E_2 - E_1 > E_g \quad (2.62)$$

In other words, to obtain population inversion in a semiconductor, the separation between the Fermi levels for free electrons and holes must exceed the energy gap between the conduction and valence bands. Since at thermodynamic equilibrium $E_{fv} = E_{fc}$, population inversion can only be achieved by means of applying an external energy source. The most straightforward way to do this is by means of a forward-biased *pn* junction.

Figure 2.21 shows the energy band diagrams of a *pn* junction under thermodynamic equilibrium and under the appliance of a forward bias voltage. With applied forward bias voltage, the Fermi level must be continuous along the *pn* junction. This is actually achieved by majority carrier diffusion from both sides of the *pn* junction. Electrons coming from the *n*-zone fill the holes in the *p*-zone generating, at the same time, holes in the *n*-zone. A similar process happens with the holes coming from the *p*-zone. A potential barrier is created as a consequence of the accumulation of minority carriers at both sides of the junction (holes in the *n*-zone and electrons in the *p*-zone) preventing further diffusion flow once equilibrium is reached.

The final result is that the central zone of the *pn* junction is depleted from mobile carriers (depletion zone). By applying a forward bias voltage *V* to the *pn* junction, the equilibrium is destroyed, the potential barrier is reduced, and, as a result, majority carriers can overcome it

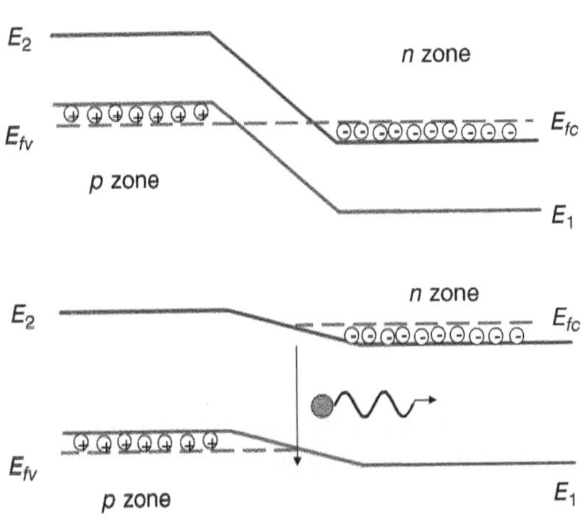

Figure 2.21 Energy band diagrams for a *pn* junction in thermodynamic equilibrium (top) and under a forward bias voltage (bottom).

and diffuse to the other side of the *pn* junction where they create an excess of minority carriers. This diffusion process generates a current given by:

$$I = I_s\left[e^{(eV)/(k_BT)} - 1\right] \tag{2.63}$$

where I_s is the *saturation current*.

The most relevant aspect of carrier diffusion in connection to light generation is that majority carriers injected from one part of the junction create an excess of minority carriers in the other part where they can *recombine* with majority carriers to generate light photons provided that these recombinations are radiative. The final concentration of minority carriers at each side of the junction is the result of an equilibrium between the carrier injection rate and the recombination rate. In consequence, it is close to the *pn* junction and at each side of it where electron and hole recombination can be produced and therefore the generation of light. Naturally, asymmetric doping is usually employed to favor light generation in one of the two zones. For example, if the *p*-zone is doped more heavily than the *n*-zone an np^+ junction is formed where hole diffusion dominates and therefore light is generated predominantly in the *n*-zone of the junction. Furthermore, as it can be observed in the bottom part of Fig. 2.21 when the equilibrium is broken, the Fermi level is unfolded into two levels, one for electrons E_{fc} in the conduction band and another E_{fv} for holes in the valence band. If the bias voltage applied to the junction is high enough, then E_{fc} will be inside the conduction band and E_{fv} inside the valence band and, therefore the condition for population inversion will be verified.

Light generation in semiconductors is thus the result of a radiative recombination process between an electron in the conduction band and a hole in the valence band. The overall result is the transition of the electron from the conduction to the valence band generation as a result of a photon. However, the electron-hole recombination mechanisms can be radiative or nonradiative. In this last case, as mentioned in the previous section, a phonon or quantum of mechanical oscillation is produced, the energy of which is employed in making the crystal lattice vibrate. The three most important nonradiative recombination mechanisms are *defect*, *surface*, and *Auger* recombinations. The first two are similar in nature and are based on the fact that any defect or discontinuity in the semiconductor material generates localized zones where the energy level concentration is so high that they behave, in all aspects, as sinks for the injected diffusion current. Of particular importance due to its practical implications is superficial recombination, which is produced at the interface between two different semiconductors. The change in the surface current density due to this mechanism is given by:

$$\Delta J = -qsn \tag{2.64}$$

Here n represents minority carrier excess concentration la concentration and s is the surface recombination speed with values ranging from $1 - 10^3$ m/s for a semiconductor–air interface and $0.01 - 0.1$ m/s for interfaces between semiconductors. Most of the optical sources made of semiconductors employ a sandwich of layers composed of semiconductor materials with slightly different material compositions, yet having a very similar lattice constant. This is due to the fact that a semiconductor separation interface where the two materials have lattice constants differing by an amount Δa behaves as a surface recombination center with recombination speed given by $s(\text{m/s}) = 2.10^5(\Delta a/a_o)$

Auger recombination is especially important for optical sources emitting in the second and third spectra windows. In this process, the energy released after an electron-hole recombination is transferred to a second electron or hole increasing its kinetic energy.

In a semiconductor material, radiative and nonradiative processes compete among themselves. In order to characterize the quality of a given material in terms of its capacity for light generation, it is customary to define a parameter called *internal quantum efficiency* η_i that is the quotient between the rate of radiative recombinations and the total recombination rate.

$$\eta_i = \frac{R_{rr}}{R_{rr} + R_{nr}} \tag{2.65}$$

where R_{rr} represents the radiative recombination rate and the nonradiative recombination rate. If n represents the injected carrier concentration, then the above rates can be expressed in terms of two time constants: τ_{rr} for radiative recombinations and τ_{nr} for nonradiative recombinations, in which case:

$$\eta_i = \frac{1}{1 + \tau_{rr}/\tau_{nr}} \tag{2.66}$$

From (2.66), it is clear that a suitable material for light generation must verify $\tau_{rr} \leq \tau_{nr}$. Semiconductors that verify this relationship are known as *direct-gap* and are characterized as shown in the left part of Fig. 2.22 by the fact that the minimum of their conduction band and the maximum of their valence band are produced for the same value of the wave vector k. The electron transition from the conduction to the valence band only requires an energy matching performed by the emission of a photon since the conservation of moment $p = \hbar k$ is guaranteed as k is conserved. On the other hand, for indirect-gap semiconductors, the minimum of their conduction band and the maximum of their valence band are not produced for the same value of the wave vector k. Hence, the transition of the electron from the conduction to the valence band requires first the matching of moments which is a nonradiative recombination. Hence, as radiative and nonradiative transitions compete, $\tau_{rr} \gg \tau_{nr}$ and $\eta_i \ll 1$ as it is not probable that a nonradiative recombination is followed by a radiative one.

Semiconductor materials traditionally employed in electronics such as Ge and Si are indirect gaps featuring intrinsic quantum efficiency values of $\eta_i \approx 10^{-5}$ and therefore are not suitable for the implementation of optical sources, at least within the spectral range employed in optical communications. To achieve efficiency values of $\eta_i \geq 0.5$ binary, ternary, and quaternary compounds based

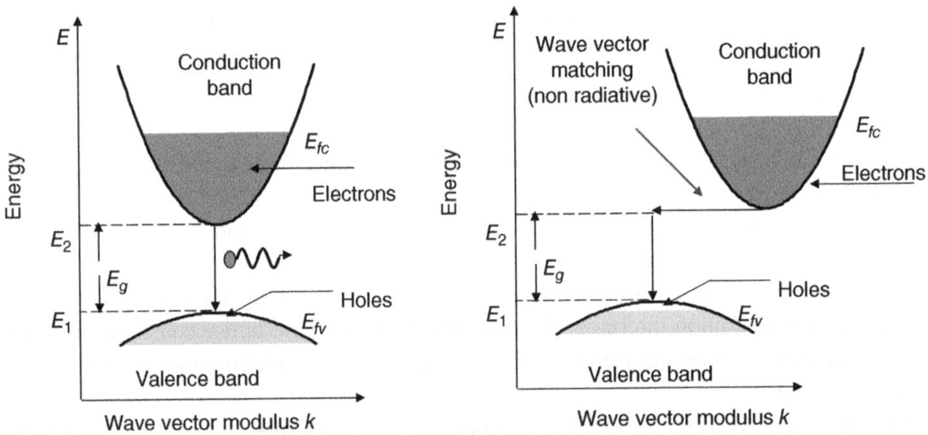

Figure 2.22 Energy versus wave vector variations for the conduction and valence band in a direct gap semiconductor (left) and in an indirect gap semiconductor (right).

on materials from groups III (Al, Ga, In) and V (As, P, B) of the periodic system are employed leading to the so-called *III–V compounds*.

As previously mentioned, a material is more suitable for optical sources the closer to unity is its value of η_i. In general, the contribution to the radiative recombination rate will be due to the combined effect of stimulated and spontaneous emissions. Therefore, $R_{rr} = R_{stim} + R_{spon}$. For LEDs, for instance, spontaneous emission dominates, and hence, $R_{spon} \gg R_{stim}$. However, for lasers, the dominating mechanism is stimulated emission $R_{spon} \ll R_{stim}$. Since R_{spon} is usually comparable to the nonradiative recombination rate R_{nr}, $\eta_i \approx 0.5$ for LED sources, while $\eta_i \to 1$ for lasers. It is convenient to define a *time constant for injected carriers* τ_c representing the average time that a free carrier stays in the conduction band before recombination in absence of stimulated emission:

$$R_{spon} + R_{nr} = \frac{n}{\tau_c} \tag{2.67}$$

In general, if $R_{spon} + R_{nr}$ change linearly with n and τ_c is a time constant. In practice, however, $R_{spon} + R_{nr} = A_{nr}n + Bn^2 + Cn^3$, where A_{nr} is the nonradiative recombination coefficient in defects and surfaces, B the recombination coefficient due to spontaneous emission, and C the Auger recombination coefficient. Using these definitions:

$$\frac{1}{\tau_c} = A_{nr} + Bn + Cn^2 \tag{2.68}$$

To fabricate semiconductor devices in general and optical sources and detectors in particular, the starting point is a crystalline substrate that provides the required mechanical consistency to support the rest of the device. Different layers are then grown or deposited by means of chemical reactions. To avoid surface recombinations and defects, it is necessary that the employed materials have a crystalline structure identical to that of the substrate and that adjacent layers have a lattice constant as close as possible. The fabrication process based on the deposition of successive semiconductor material layers is known as epitaxial growth and can be performed in liquid (LPE) or vapor (VPE) phases or through the use of molecular beams (MBE).

As mentioned above, III–V compounds are the most adequate option for the implementation of optical communication sources. For sources operating in the first spectral window (0.8 μm ≤ λ ≤ 0.9 μm), one of the most suitable materials is a ternary compound made from a partial substitution of Gallium atoms in Gallium Arsenide (AsGa) by Aluminum atoms. The resulting compound Al_xGa_{1-x} As, where x represents the fraction of Gallium atoms replaced by Aluminum atoms has the feature of providing a tunable emission wavelength by changing the value of x, as shown in Fig. 2.23. This is due to the fact that by changing the value of x, the energy gap between the conduction and valence band is changed, as it will be shown below.

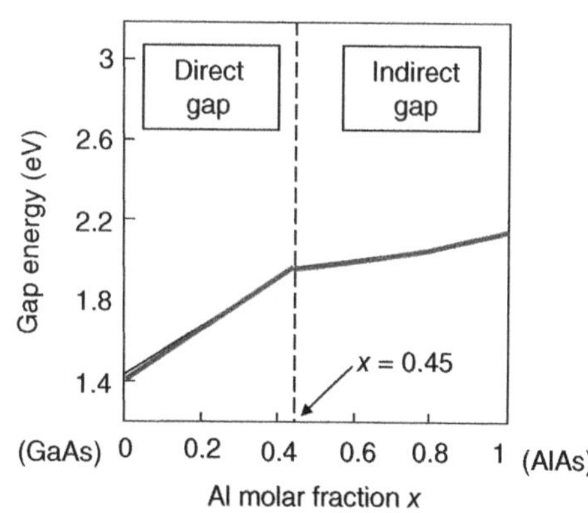

Figure 2.23 Gap energy in eV as a function of the Al molar fraction x in $Al_xGa_{1-x}As$.

From Fig. 2.23, it can be observed that this family of compounds can be obtained to implement a source emitting in any wavelength within the range of 0.65 ($x = 0.34$) and 0.88 μm ($x = 1$). Note that x cannot be less than 0.34, since in that case, the compound is no longer a direct-gap semiconductor.

For the second and third spectral windows, it is customary to employ compounds obtained by substitution of Indium and Phosphor atoms in Indium Phosphide (InP) by Gallium and arsenic atoms, respectively. The resulting compound is $In_{1-x}Ga_xAs_yP_{1-y}$, where x represents the fraction of Indium atoms substituted by Gallium atoms and y the fraction of Phosphor atoms substituted by arsenic atoms. The emission wavelength can be tailored within the 1 − 1.7 μm region by changing the values of x and y.

The most widely employed semiconductor structure for the fabrication of optical sources is not the standard *pn* junction or homojunction but a rather more complex configuration known as *heterostructure* or *heterojunction*. A heterostructure is a structural configuration that allows both electron and photon confinement within a reduced zone inside the device featuring dimensions compatible with those of optical fiber or integrated waveguide cores. In essence, a heterostructure is a sandwich structure where successive semiconductor material layers surround a central *pn* junction. Figure 2.24 shows, as an example, a heterostructure configuration for a $Al_xGa_{1-x}As$ source.

Assuming that the junction is of the type p^+n, then to confine the injected carriers in the active zone (n-zone of the junction), materials with higher energy gaps are employed in the surrounding layers. This can be achieved by changing the value of the Aluminum fraction x in the $Al_xGa_{1-x}As$ compound. Electrons and holes injected in the active zone cannot escape as this zone is surrounded by others with a higher energy gap. Therefore, they remain confined in this region where they are forced to recombine. A second advantage of heterostructures is that the layers that surround the active zone have a lower refractive index and therefore the active zone behaves as an optical waveguide for the photons generated by recombination.

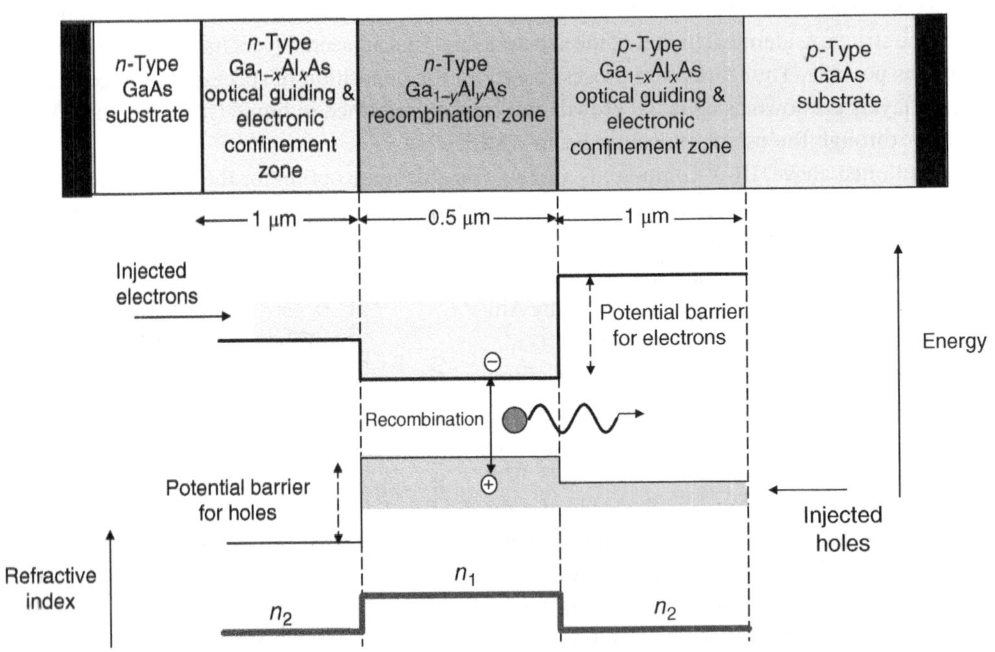

Figure 2.24 Heterostructure configuration for a $Al_xGa_{1-x}As$ source.

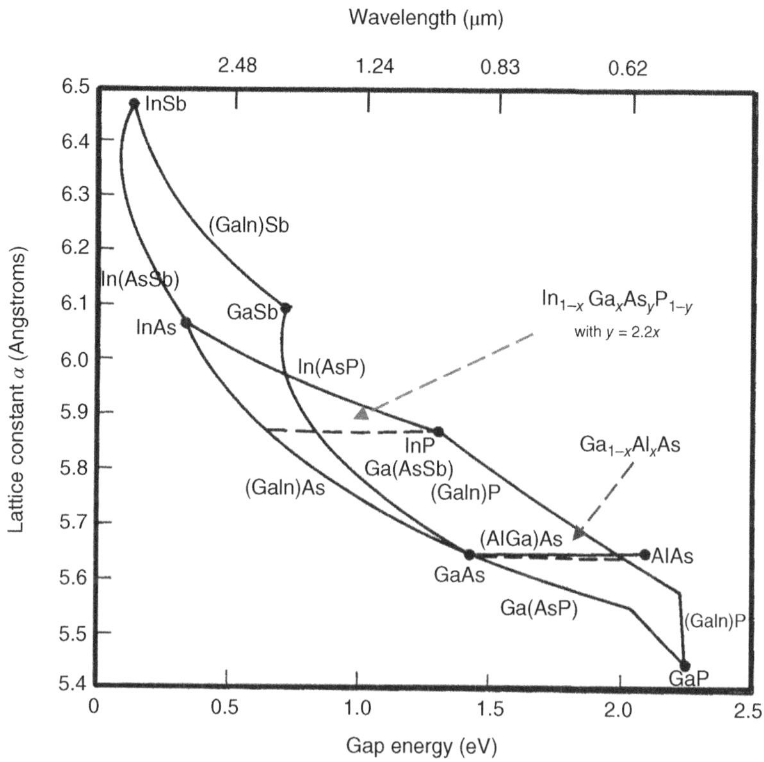

Figure 2.25 Lattice constant versus Gap energy for several III–V semiconductor alloys.

The main problem when implementing a heterostructure lies in the fact that the different layer materials to be employed need to have a lattice constant as similar as possible to that of the substrate material. Figure 2.25 shows, as an example, the value of the lattice constants for different compositions of $Al_xGa_{1-x}As$ and $In_{1-x}Ga_xAs_yP_{1-y}$. For the case of using $Al_xGa_{1-x}As$, the lattice constants of the heterostructure materials have to be almost identical to that of the GaAs substrate. Fortunately, as can be observed in Fig. 2.25, the line joining GaAs ($Al_xGa_{1-x}As$ with $x = 0$) and AlAs($Al_xGa_{1-x}As$ with $x = 1$) is horizontal, which means that any compound of the form $Al_xGa_{1-x}As$ independent of the value of x will have a lattice constant value almost equal to that of GaAs. For second and third window sources, the substrate material is InP and, as it can be noticed, the horizontal line characterizing $In_{1-x}Ga_xAs_yP_{1-y}$ compound with equal lattice constant to that of InP requires that $y = 2.2x$.

The value of the energy gap for $Al_xGa_{1-x}As$ compounds is a function of x:

$$E_g(\text{eV}) = 1.424 + 1.266x + 0.266x^2 \tag{2.69}$$

While for $In_{1-x}Ga_xAs_yP_{1-y}$, compound one has:

$$\begin{aligned} E_g(\text{eV}) &= 1.35 - 0.72y + 0.12y^2 \\ y &= 2.2x \end{aligned} \tag{2.70}$$

From (2.69) and (2.70), the value of the emitted wavelength from a $Al_xGa_{1-x}As$ or a $In_{1-x}Ga_xAs_yP_{1-y}$ based source can be obtained since:

$$\lambda(\mu m) = \frac{1.24}{E_g(eV)} \tag{2.71}$$

2.3.3 Optical Sources

As mentioned previously, in most microwave photonic applications, the preferred optical source is a laser and in particular a semiconductor laser diode. We will therefore concentrate on describing the basic operation principles of lasers and, in particular of semiconductor-based structures.

A laser is an optical oscillator and therefore requires a medium providing gain and a negative feedback mechanism. Optical gain requires the existence of stimulated emission which, in a semiconductor material, is the result of achieving a population inversion between the valence and the conduction bands. This population inversion is achieved as described in the previous section by injecting electrons into a forward-biased *pn* junction. When the density of injected carriers is above a certain value known as *transparency value*, population inversion is obtained and the medium exhibits *optical gain*. A monochromatic optical signal of frequency ω injected with power P_i in the active zone is amplified upon propagating through the medium according to:

$$\frac{dP}{dz} = gP \Rightarrow P(z) = P_i e^{g(w)z} \tag{2.72}$$

where $g(\omega)$ is the material gain expressed in units of inverse length (normally cm^{-1}) and obtained according to:

$$g(\omega) = \frac{(R_{stim}(\omega) - R_{abs}(\omega))}{v_g} \tag{2.73}$$

Here $v_g = c/n_g$ represents the group velocity in the active zone material and $R_{stim}(\omega)$, $R_{abs}(\omega)$ are the stimulated emission and absorption rates that have been defined in (2.61). In general, $g(\omega)$ is evaluated numerically and represented as a function of the incoming radiation energy expressed in eV (or equivalently in frequency units) for different values of the carrier density as shown in Fig. 2.26 for an InGaAsP laser operating in the second and third spectral windows.

As it can be observed, for carrier densities below a given *transparency value* n_o (close to 1×10^{18} cm^{-3} for the curves in Fig. 2.26), the optical gain is negative and therefore the medium predominantly absorbs radiation. If, however, the carrier density is above the transparency value, then there is a wavelength region within which stimulated emission dominates over absorption an optical gain is achieved. As it can also be observed from Fig. 2.26, the higher the injected carrier concentration the broader the spectral over which optical gain is obtained. The inspection of Fig. 2.26 also reveals that the maximum value for the material gain shifts toward shorter wavelengths (higher energy values) when the injected carrier density is increased. For a fixed wavelength (or energy) value, these curves show that from a given value of injected carrier density, the material gain changes linearly with *n* (in this case, for material gain coefficients above 100 cm^{-1}). This behavior is shown in more detail in Fig. 2.27.

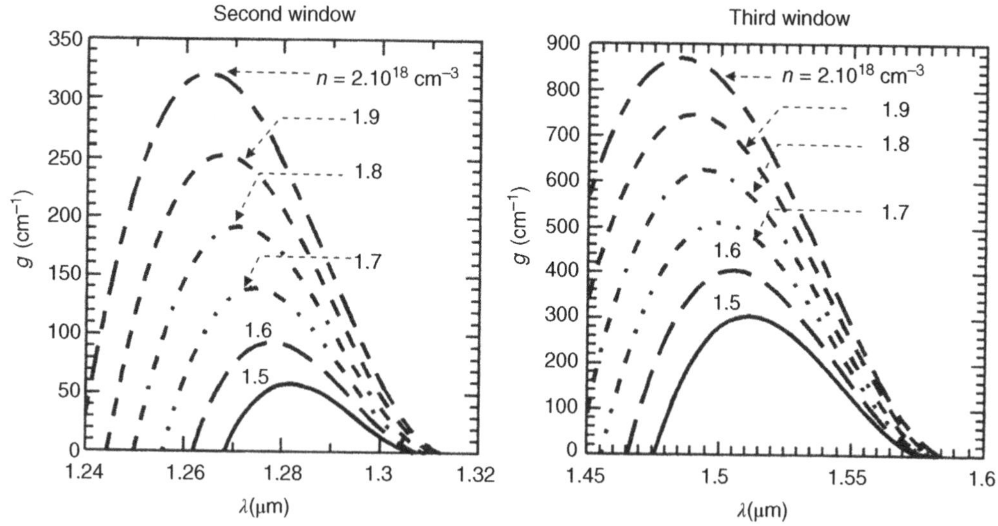

Figure 2.26 Material gain coefficient versus wavelength for different values of carrier density in an InGaAsP laser operating in the second and third spectral windows.

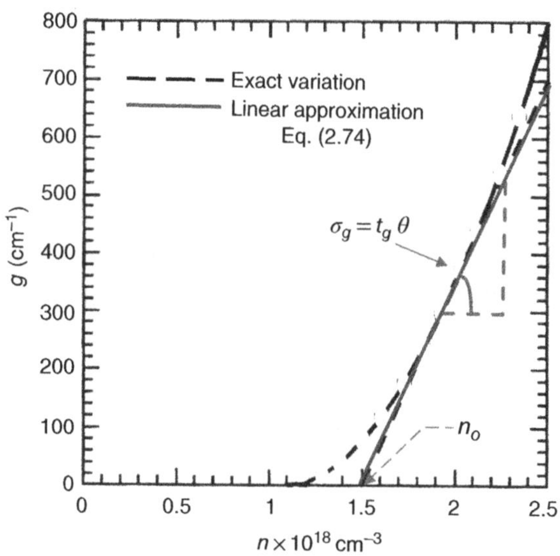

Figure 2.27 Material gain coefficient versus injected carrier density for a fixed operation wavelength.

In fact, it is possible to approximate this relationship by:

$$g = \sigma_g(n - n_o) \tag{2.74}$$

where σ_g is known as the gain cross-section which is expressed in units of area (normally cm^2). Typical values for the parameters in (2.74) are $n_o = 1 - 1.5 \times 10^{18}\ \text{cm}^3$, $\sigma_g = 200 - 300 \times 10^{-18}\ \text{cm}^2$.

It is useful to employ a complex refractive index to describe the field propagation through the material exhibiting optical gain:

$$\overline{n} = n' - jn'' \tag{2.75}$$

We, hence, represent the electric or magnetic field propagating through the material as:

$$\psi = \psi_o e^{-jk\tilde{n}z} = \psi_o e^{-kn''z} e^{-jkn'z} \tag{2.76}$$

Since the optical power is proportional to $|\psi|^2$, we obtain from (2.76) the following relationship:

$$\frac{dP}{dz} = -2kn''P \tag{2.77}$$

And by identifying (2.77) with (2.72):

$$n'' = -\frac{g}{2k} = -\frac{\sigma_g}{2k}(n - n_o) \tag{2.78}$$

It is very important to note here that the real n' and the imaginary n'' parts of the refractive medium are not independent but are linked by the Kramers–Kroenig relationships. This means that according to (2.78), a change in the injected carrier density will not only produce a variation in n'', but that it will also affect n'. The relationship between the two parts of the refractive index is usually expressed through the so-called *α parameter* defined by:

$$\alpha = \frac{\Delta n'}{\Delta n''} \tag{2.79}$$

the values of which are within $\alpha = -3... -7$. The α parameter is extremely important when describing the laser operation under direct current intensity modulation as it is behind the residual and unavoidable frequency modulation or *chirp*.

The second element to build the oscillator is the negative feedback structure. A very simple way to achieve this in a semiconductor active zone is by means of a Fabry–Perot cavity as shown in Fig. 2.28.

The Fabry–Perot cavity is composed of two mirrors with optical field reflection coefficients given by $r_1 = \sqrt{R_1}$ y $r_2 = \sqrt{R_2}$, respectively, and separated by a distance L. The material surrounded by the two mirrors provides the optical gain, featuring a complex-valued propagation constant given by $\gamma = (\alpha_c - g)/2 + j\beta$, where g is the optical gain and β the phase constant. A linearly polarized field (for example, in the x direction) incident to the cavity with amplitude E_{ix} suffers successive reflections in the mirrors that form the cavity. So, the output wave from the structure is the result of the multiple interference of waves that exit the cavity after each roundtrip. Hence, if t_1, t_2 represent the

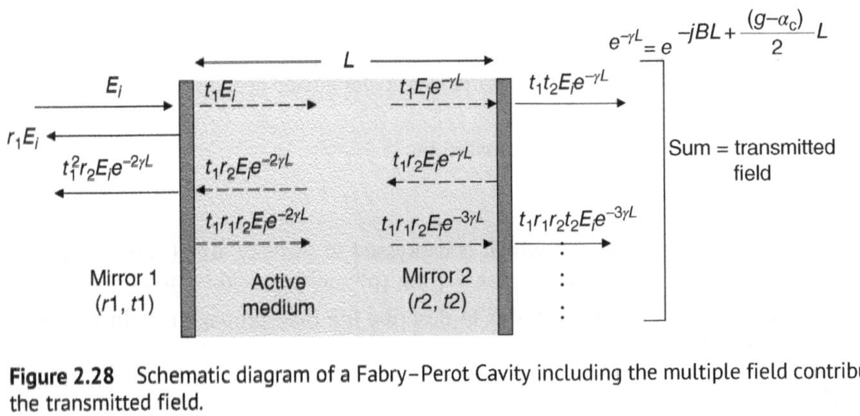

Figure 2.28 Schematic diagram of a Fabry–Perot Cavity including the multiple field contributions that form the transmitted field.

field transmission coefficients of the two respective mirrors, then the output field from the Fabry–Perot cavity is given by:

$$E_{ox} = t_1 t_2 E_i e^{-\gamma L} \sum_{i=0}^{\infty} \left[(r_1 r_2)^i e^{-2i\gamma L} \right] = E_{ix} \frac{t_1 t_2 e^{-\gamma L}}{1 - r_1 r_2 e^{-2\gamma L}} \tag{2.80}$$

The implementation of a Fabry–Perot cavity in a semiconductor heterostructure is shown in Fig. 2.29.

The cavity is formed by embedding the heterostructure in two mirrors which are formed by cleaving the outer surfaces of the active zone along the semiconductor crystal planes. Field reflection is due to the refractive index discontinuity between the semiconductor medium and the air. The intensity reflectivity of the semiconductor-air mirror can be computed using the Fresnel formula for normal incidence:

$$R_1 = R_2 = \left(\frac{\overline{n} - \overline{n}_{ext}}{\overline{n} + \overline{n}_{ext}} \right)^2 \tag{2.81}$$

where $\overline{n}_{ext}$ represents the refractive index of the external medium which in most cases will be air ($\overline{n}_{ext} \approx 1$). In general, the refractive index of most semiconductor materials is a round $\overline{n} \approx 3.5$, hence a typical value for the mirror reflectivity is.

As can be observed in Fig. 2.29, the light propagation inside the cavity is parallel to the *pn* junction plane and is physically localized in the active zone of the semiconductor. On the other hand, the population inversion that is required in the active zone is achieved by injecting electrons into the active zone in a perpendicular direction to that of the *pn* junction plane. Typical dimensions for the active zone are around 100–500 μm for the cavity length (z-coordinate), 5–15 μm for the width (x-coordinate) and 0.1–0.5 μm for the height (y-coordinate). Hence, the active zone can be considered an *integrated optical waveguide* subject to the propagation equations and design conditions described in Section 2.2.4. In particular, there are three types of modes in the cavity: (i) *Lateral modes*, which are stationary waves formed in the y-coordinate of the active zone and conform the spatial profile of propagating modes along such axis, (ii) *Transversal modes*, that are stationary waves formed in the x-coordinate of the active zone and determine the spatial profile of the propagating modes along such axis, and (iii) *Longitudinal modes*, that are stationary waves formed in the z-coordinate and determine the spectrum of frequencies emitted by the laser. The combination of (i) and (ii) modes results in the spatial propagation modes of the waveguide studied in Section 2.2.4. Usually, the lateral and transversal dimensions are designed so the laser operates as a spatial singlemode device (i.e. only the x-polarized TE_0 mode is allowed to propagate and exit the cavity). (iii) Modes determine the cavity oscillation frequencies.

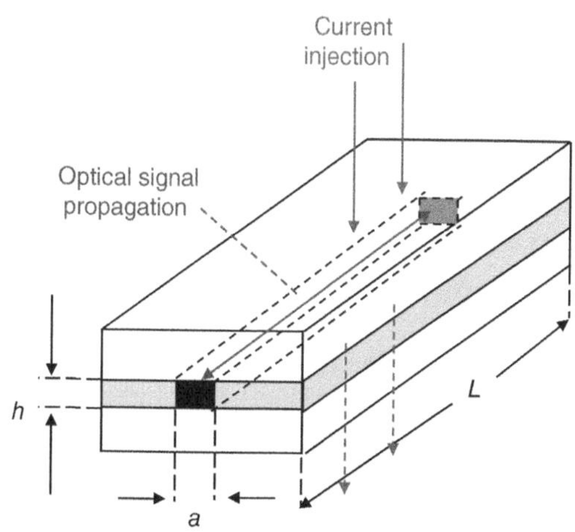

Figure 2.29 Fabry–Perot cavity implementation using a semiconductor heterostructure and the relevant height, width, and length parameters.

The Fabry–Perot laser oscillation conditions are obtained by forcing the existence of an output field under zero input field. According to (2.80), this requires that:

$$\sqrt{R_1R_2}e^{(-2j\beta L+(g-\alpha_c)L)}=1 \tag{2.82}$$

which is a complex-valued equation and thus unfolds into two equations, one for its real part and another for its imaginary part. The real part of (2.82) gives the so-called *threshold gain condition* required for laser operation:

$$\Gamma g_{th}=\alpha_c+\frac{1}{2L}\ln\left(\frac{1}{R_1R_2}\right)=\alpha_c+\alpha_{mirror} \tag{2.83}$$

In (2.83), a confinement factor Γ has been introduced as not all the spatial profile of the TE mode is subject to amplification (the evanescent field is outside the active zone).

The imaginary part of (2.82) yields the phase condition:

$$\beta L=q\pi,\quad q\in Z \tag{2.84}$$

From (2.84), the laser oscillation frequencies or longitudinal modes can be obtained. Since $\beta=2\pi\nu n'/c$, the oscillation frequencies are given by:

$$\nu_q=q\frac{c}{2n'L},\quad q\in Z \tag{2.85}$$

which is an infinite discrete set where two consecutive oscillation frequencies are separated by:

$$\delta\nu=\frac{c}{2L\overline{n}_g}=\frac{\mathrm{v}_g}{2L}=\frac{1}{\tau_L} \tag{2.86}$$

In (2.86), $\overline{n}_g=n'+\nu dn'/d\nu$ is the group index of the active zone material, $\mathrm{v}_g=c/\overline{n}_g$ the corresponding group velocity and τ_L the cavity roundtrip transit time. The values of $\overline{n}_g$ are usually in range between 3.5 and 5. So, for a cavity length L in the range of 300–400 μm, $\tau_L\approx 10$ p seg.

Not all the oscillation frequencies will verify the threshold gain condition. So, only a finite set of those in (2.85) will be finally emitted by the laser. This is illustrated in Fig. 2.30, where the oscillation frequencies of the Fabry–Perot cavity are displayed together with the material gain and the cavity loss spectra. Only those frequencies for which $g\geq g_{th}$ will be finally emitted.

Let us now recall that both the gain and the real part of the active zone refractive index n' depend on the injected carrier density. Furthermore, n', depends itself on the frequency. So, in general, $n'=n'(n,\nu)$. The value of the refractive index when the devices operate in the laser regime is given by $n'(n_{th},\nu_{th})$, where:

$$g(n_{th})=g_{th}\Rightarrow n_{th}=n_o+\frac{g_{th}}{\sigma_g} \tag{2.87}$$

Hence, (2.85) can be re-expressed to obtain the value of ν_{th} for a current injection given byn_{th}.

$$\nu_{th}=q\frac{c}{2Ln'(n_{th},\nu_{th})} \tag{2.88}$$

It is useful to know how much the emission frequency deviates from the value given in (2.88) when the injection current is changed around the value of n_{th}. In particular, this result is important

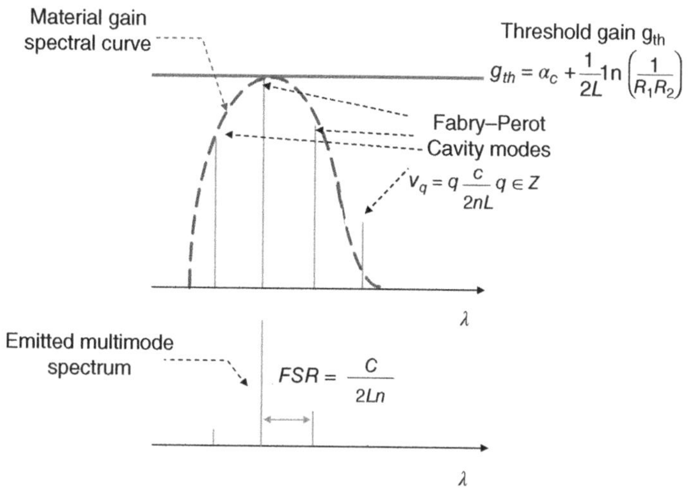

Figure 2.30 Oscillation spectrum of a Fabry–Perot cavity composed of longitudinal modes satisfying both the gain and oscillation frequency conditions.

when characterizing the direct modulation of the laser by changing its injection current. To compute this variation, we use a Taylor series approximation:

$$n' = n'(n_{th}, \nu_{th}) + \frac{\partial n'}{\partial n}(n - n_{th}) + \frac{\partial n'}{\partial \nu}(\nu - \nu_{th}) \tag{2.89}$$

where, taking into account (2.85), (2.88), and (2.89) [6]:

$$\nu - \nu_{th} = -\frac{\nu_{th}}{\overline{n}_g}\frac{\partial n'}{\partial n}(n - n_{th}) \tag{2.90}$$

which indicates that a modulation of the laser injection current always produces a frequency modulation or *chirp* of the frequency emitted by the laser.

Though Fabry–Perot lasers were widely employed in initial microwave photonic systems nowadays and with the exception of the systems based on laser combs and laser arrays that will be treated at the end of this section, the vast majority of configurations employ singlemode lasers where a single transverse (TE) and a single longitudinal mode is emitted. In these lasers, a relevant parameter is the *modal suppression ratio* (MSR) defined by $MSR = 10\log(P_{ms}/P_{ma})$, where P_{ms} represents the optical power of the selected mode for emission and P_{ma} the power of the mode adjacent to that selected. MSRs above 40 dB are required for singlemode lasers, and several alternatives are shown, which we now review.

The first structure, shown in Fig. 2.31, is the distributed feedback (DFB) laser. This laser is obtained by incorporating a Bragg diffraction grating into the active zone of the laser. Front and rear-end mirrors are usually suppressed by adding an antireflection coating to the semiconductor–air interfaces.

As is explained in Section 2.4 of this chapter, a Bragg grating is a periodic variation of the refractive index along one direction in a given zone of a material or structure. In the case of the DFB laser, the index variation is along the longitudinal coordinate in the active zone. Λ defines the period of the refractive index variation and $K = 2\pi/\Lambda$ its associated wave vector. The periodic perturbation of the refractive index leads to the generation of two modes propagating in the $+z$ and $-z$ directions

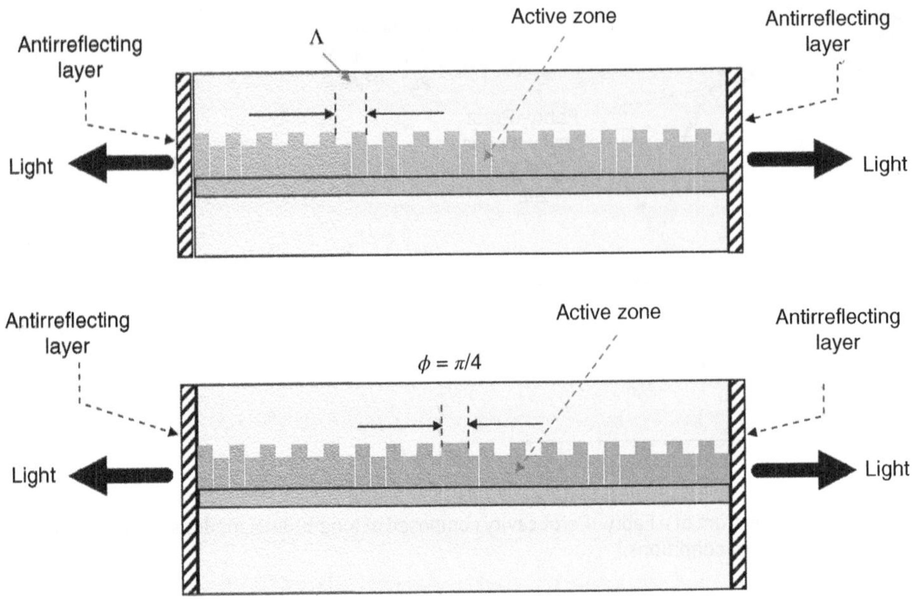

Figure 2.31 Layout of a distributed feedback (DFB) semiconductor laser (top) and a $\lambda/4$ shifted DFB laser.

along the active zone. These modes do not exchange or couple energy between them unless their wavelength is close to the so-called Bragg wavelength given by $\lambda_B = 2\overline{n}\Lambda$. In such a case, the modes continuously exchange energy between them as they propagate in the active zone. The field evolution is described by the *contradirectional coupled mode equations* that are described in detail in Section 2.4:

$$\begin{aligned}\frac{dA}{dz} &= j\kappa B e^{-j(2\beta - K)z} - gA \\ \frac{dB}{dz} &= -j\kappa A e^{j(2\beta - K)z} + gB\end{aligned} \tag{2.91}$$

where B (left to right) and (right to left) A represent the amplitudes of the counterpropagating fields β is the propagation constant of the fundamental mode, κ the mode coupling coefficient and g the material gain that is considered a constant inside the active zone. For practical reasons, it is useful to define a frequency deviation with respect to the Bragg wavelength:

$$\delta = (\beta - K/2) = (\beta - \pi/\Lambda) = (\omega - \omega_B)\overline{n}/c \tag{2.92}$$

So, applying the following transformations

$$\begin{aligned}A(z) &= a(z)e^{-gz} \\ B(z) &= b(z)e^{gz}\end{aligned} \tag{2.93}$$

Equation (2.91) can be recast as [8]:

$$\begin{aligned}\frac{da}{dz} &= j\kappa b e^{-2j(\delta + jg)z} \\ \frac{db}{dz} &= -j\kappa a e^{2j(\delta + jg)z}\end{aligned} \tag{2.94}$$

A reflection coefficient $r(\lambda) = a(0)/b(0)$ can be obtained for the DBF structure by solving (2.94) subject to the following boundary conditions: light is injected in the left part ($z = 0$) of the active zone $b(z = 0) = B_o$ and no light from right to left is injected at the right part $a(z = L) = 0$. Under these conditions, the solutions for (2.94) yield the following reflection coefficient:

$$r(\lambda) = \frac{j\kappa \operatorname{senh}\left[L\sqrt{\kappa^2 + (g - j\delta)^2}\right]}{(g - j\delta)\operatorname{senh}\left[L\sqrt{\kappa^2 + (g - j\delta)^2}\right] - \sqrt{\kappa^2 + (g - j\delta)^2}\cosh\left[L\sqrt{\kappa^2 + (g - j\delta)^2}\right]} \tag{2.95}$$

The *oscillation conditions* for the DBF are obtained when the denominator of (2.95) is zero, in other words, when:

$$(g - j\delta)\operatorname{senh}\left[L\sqrt{\kappa^2 + (g - j\delta)^2}\right] = \sqrt{\kappa^2 + (g - j\delta)^2}\cosh\left[L\sqrt{\kappa^2 + (g - j\delta)^2}\right] \tag{2.96}$$

In general, (2.96) can only be solved numerically. After fixing a value for κ, pairs of values (δ_m, g_m) $m = 0, \pm 1, \pm 2 \pm \ldots$ are obtained that correspond to the oscillation frequencies and their corresponding threshold gains. In the limiting case (which is usually achieved in practice) where the gain is high enough that $g^2 \gg \kappa^2, \delta^2$, the (2.96) can be solved in closed form, yielding:

$$\frac{\kappa^2}{4(g_m - j\delta_m)^2} e^{2L(g_m - j\delta_m)} = -1 \tag{2.97}$$

In (2.97), the phase of the first factor is very close to zero as $g^2 \gg \kappa^2, \delta^2$, hence, the phase of the second factor must verify:

$$-2L\delta_m = (2m + 1)\pi \tag{2.98}$$

And, therefore, from (2.92), the oscillation frequencies of the DFB laser are given by:

$$\omega_m = \omega_B - \left(m + \frac{1}{2}\right)\frac{\pi c}{L\overline{n}}, \quad m = 0, \pm 1, \pm 2, \ldots. \tag{2.99}$$

while their corresponding threshold gains are:

$$\frac{e^{2g_m L}}{g_m^2 + \delta_m^2} = \frac{4}{\kappa^2} \tag{2.100}$$

Figure 2.32 shows a schematic diagram of the oscillation frequency positions for the four lower-order oscillation modes and the required values for their corresponding threshold gains.

As it can be observed, the oscillation frequencies in a DFB laser are located symmetrically around the Bragg frequency. The required value for the threshold gain increases with the difference between the oscillation frequency and the Bragg frequency. A typical threshold gain difference between the two closest modes (below and above) to the Bragg frequency and their adjacents is in the range of 7–50 cm^{-1}, a figure that has to be compared with the 0.5–1 cm^{-1} gain difference that usually exists among competing oscillation modes in a Fabry–Perot laser just above and just below the gain threshold. These high threshold gain differences guarantee the emission of only two longitudinal modes symmetrically placed around the Bragg frequency by a standard DFB structure, even under dynamic (current modulation) conditions. The next modes are emitted with a power level usually around 40 dB below (MSR > 40 dB). However, in most practical cases, a single oscillation frequency is desired and this is achieved by slightly modifying the DBF structure to

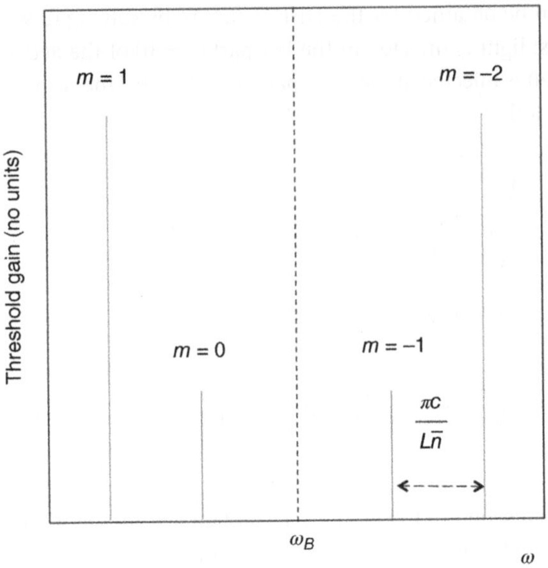

Figure 2.32 Oscillation frequencies and their corresponding threshold gains for the longitudinal modes of a DBF laser.

accommodate an unperturbed central zone such that the product of its constant refractive index times its length provides a phase shift of $\pi/4$. This structure, known as the $\lambda/4$ shifted DFB laser is shown in the bottom part of Fig. 2.32 and features a single longitudinal mode oscillating at ω_B. Commercial DFB lasers are routinely employed in microwave photonic applications, provide modulation bandwidths around 5 GHz and output powers above 10 mW. DFB linewidths in absence of modulation are in the 1–10 MHz range and RIN values can reach −150 dB/Hz.

A second option for the implementation of singlemode semiconductor lasers is the *Distributed Bragg Reflector* or DBR laser. This structure is very similar to a Fabry–Perot laser. However, broadband facet mirrors are replaced by frequency-selective filters implemented using Bragg reflectors [5, 8, 9]. Figure 2.33 represents a DBR laser and, as it can be observed, the Bragg reflectors do not penetrate the active zone. Their role is to select one of the possible oscillation frequencies of the active cavity.

The derivation of the oscillation conditions for a DBR is an identical process to that followed for Fabry–Perot lasers if we assume that $r_1 = r_1(\lambda)$, $r_2 = r_2(\lambda)$, where the spectrum of the reflection coefficients is given by (2.95) with $g = 0$. The oscillation condition is then:

$$r_1(\lambda)r_2(\lambda)e^{[-j2\beta L + (\Gamma g - \alpha_c)l]} = 1 \qquad (2.101)$$

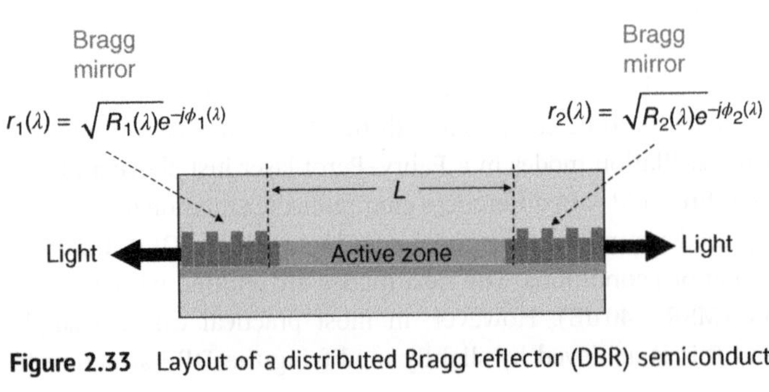

Figure 2.33 Layout of a distributed Bragg reflector (DBR) semiconductor laser.

If we express the complex-valued reflection coefficients in terms of the intensity reflectivities of the Bragg mirrors, then $r_1(\lambda) = \sqrt{R_1(\lambda)}e^{-j\phi_1(\lambda)}, r_2(\lambda) = \sqrt{R_2(\lambda)}e^{-j\phi_2(\lambda)}$, and from (2.101), the following expression is obtained for the threshold gain:

$$g_{th} = \frac{1}{\Gamma}\left(\alpha_c + \frac{1}{2L}\ln\left(\frac{1}{R_1(\lambda)R_2(\lambda)}\right)\right) \tag{2.102}$$

while the phase condition is now:

$$2\beta L + \phi_1 + \phi_2 = q2\pi, \quad q = 1, 2... \tag{2.103}$$

The phase change in the Bragg reflectors can be expressed in terms of an equivalent length L_{effi}, $i = 1, 2$ and their average refractive indices $\overline{n}_i$, $i = 1,2$ so: $\phi_i = 4\pi\overline{n}_i L_{effi}/\lambda$, $i = 1,2$, and, therefore, (2.103) can be written as:

$$\lambda_q = \frac{\overline{n}L + \overline{n}_1 L_{eff1} + \overline{n}_2 L_{eff2}}{q}, \quad q = 1, 2... \tag{2.104}$$

which gives the oscillation frequencies. These are represented in Fig. 2.34 for a typical case, where it can be seen that the structure favors the emission of the longitudinal mode closest to the reflector Bragg wavelength. Adjacent modes, which are deliberately shown with an excessive power in the figure are discriminated by the Bragg reflector. MSR values over 40 dB have been reported for DBR even when subject to direct modulation.

An interesting feature of DBR lasers shown by (2.104) is that once the value of q giving the oscillation frequency is fixed, this value can be tuned by changing the values of $\overline{n}_1$ and/or $\overline{n}_2$, by suitable carrier injection into the Bragg reflector zones. This is the physical principle of *multielectrode lasers* which are shown in Fig. 2.35.

In essence, multielectrode DBR lasers are characterized by the presence of multiple electrodes (2 or 3) that are employed for the injection of independent currents in different zones of the device.

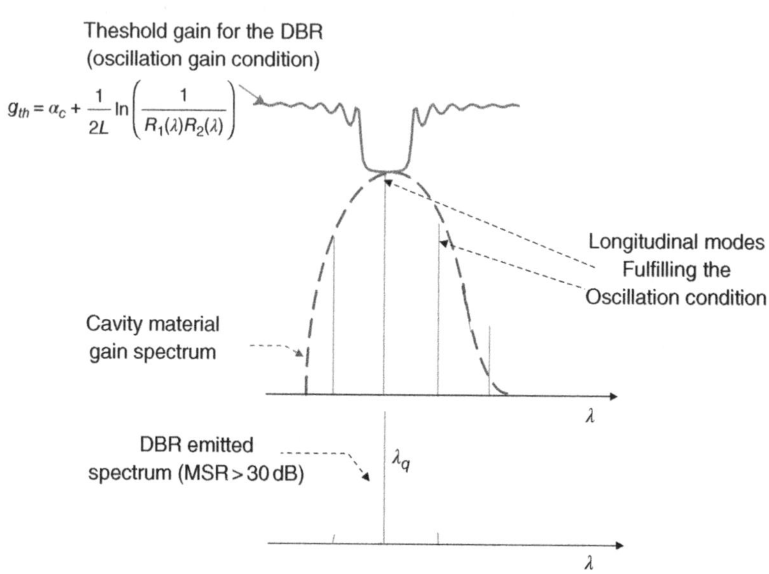

Figure 2.34 Oscillation frequencies and their corresponding threshold gains for the longitudinal modes of a DBR laser.

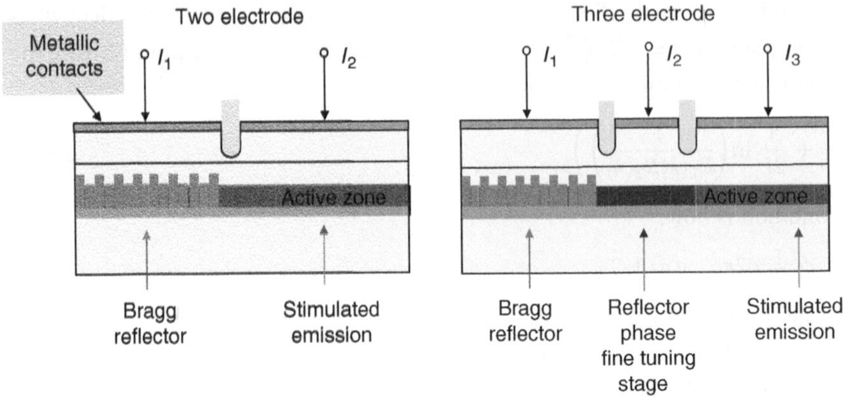

Figure 2.35 Layouts of two and three multielectrode DBR lasers.

The current injected in the active zone provides the bias and therefore the optical gain of the laser and it is also employed for modulation. The currents injected into the adjacent passive zones are employed to fine-tune the central emission wavelength of the laser. As an example, device shown in the right part of Fig. 2.35 is a three-section DBR where the current injected into the Bragg reflector zone ($L \approx 700\ \mu m$) tunes the laser emission wavelength by changing the central wavelength of its reflectivity. The intermediate zone ($L = 70\ \mu m$) is employed to achieve phase matching for the new spectral maximum of the Bragg reflector and therefore avoid mode hopping during the tuning process which can be therefore performed in a continuous way over a given spectral region (typically around a 1–3 nm span). Finally, the active zone ($L = 190\ \mu m$) provides the required stimulated emission. The values for the region lengths in the former example are indicative of typical DBR lasers. Usually, these devices can provide output powers in the 1–10 mW range, MSR > 30 dB featuring linewidths in the 1–70 MHz range and RIN values as low as −150 dB/Hz. These characteristics can be greatly improved by including multi-quantum well (MQW) configurations in the active zone and also by using slightly vernier detuned sampled Bragg grating mirrors at both ends of the cavity.

Both DFB and DBR lasers are lateral emission devices, that is, the emitted light propagates along the active zone parallel to the *pn* junction (z-coordinate) and exits the device at one or both of the lateral interface surfaces (x–y planes). Vertical cavity semiconductor lasers (VCSELs) are surface emission devices where light emission is in a direction perpendicular to the *pn* junction plane as shown in Fig. 2.36.

The design of VCSELs implies several important differences [8]. First, the propagation axis in the cavity is no longer the z- but the y-axis and, thus, cavity lengths are considerably reduced ($L < 10\ \mu m$). This implies that the frequency separation between adjacent longitudinal modes is much higher and, in consequence, only one of them is usually inside the region for which the optical gain is above the lasing threshold. In second place, not all the cavity length L is formed by an active region. This zone characterized by a length d ($d \approx 1\ \mu m$) is usually composed of an MQW structure. The VCSEL resonating cavity is cylindrical with a diameter in the 6–8 μm range. This brings another advantageous feature since the spatial profile of the emitted mode is approximately circular with dimensions compatible with standard singlemode fiber cores. Furthermore, since it is a surface-emission device, its output power can be easily injected into an optical fiber with coupling efficiencies exceeding 90%. The main drawback in the design of VCSELs stems from the fact that since its cavity length is substantially reduced as compared to lateral emission lasers, mirrors with very high reflectivities must

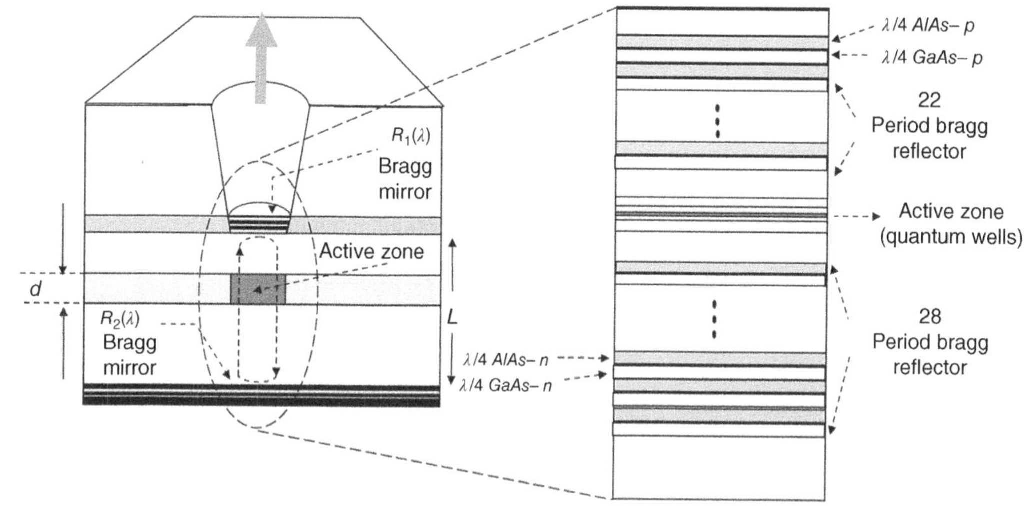

Figure 2.36 Layout of a vertical semiconductor laser (VCSEL).

be employed in order to keep threshold gains within the same order of magnitude. This can be understood assuming that the oscillation conditions for the VCSEL are basically identical to those given by (2.73) and (2.74) for the Fabry–Perot laser. According to (2.73), a reduction in the cavity length increases the value of the threshold gain which can only be compensated by increasing the mirror reflectivities. For this reason, it is customary to employ multi-stack Bragg reflectors formed by alternating layers of two different semiconductors. each layer is characterized by a length d_i, $i = 1, 2$ and a refractive index $\overline{n}_i$ $i = 1, 2$ that must be chosen so the wavelength of the emitted mode λ verifies the Bragg condition:

$$\overline{n}_1 d_1 + \overline{n}_2 d_2 = \lambda / 2\overline{n} \tag{2.105}$$

where $\overline{n}$ is the average refractive index of the reflector. The value of the mirror reflectivity at the Bragg wavelength depends on the number N of stacked periods. For the simplest case, where $d_1 = d_{2.}$ This is given by:

$$R = \left(\frac{1 - (n_1/n_2)^{2N}}{1 + (n_1/n_2)^{2N}} \right)^2 \tag{2.106}$$

Reflectivity values of the order of 99% can be easily achieved by using adequate semiconductor materials and the appropriate number of periods. This means that, overall, the threshold gain of VCSELs can be substantially reduced as compared to DFB and DBR lasers. In fact, devices with threshold currents below 1 mA can be obtained, compared to the standard 25–40 mA value required for DFBs and DBR lasers. A final drawback of VCSELs resides in the fact that due to the high reflectivities, there is a considerable amount of energy inside the cavity. So, internal heating rapidly builds up resulting in a very low differential quantum efficiency (of the order of 4%), which results in a low output power as compared to lateral emission devices. VCSELs are interesting for several microwave photonic applications, in particular, in radio over fiber systems using multimode fibers.

In several applications, it is very important to use very low linewidth (or phase noise) tunable optical sources. These can be obtained by coupling two cavities with slightly different lengths as shown in Fig. 2.37.

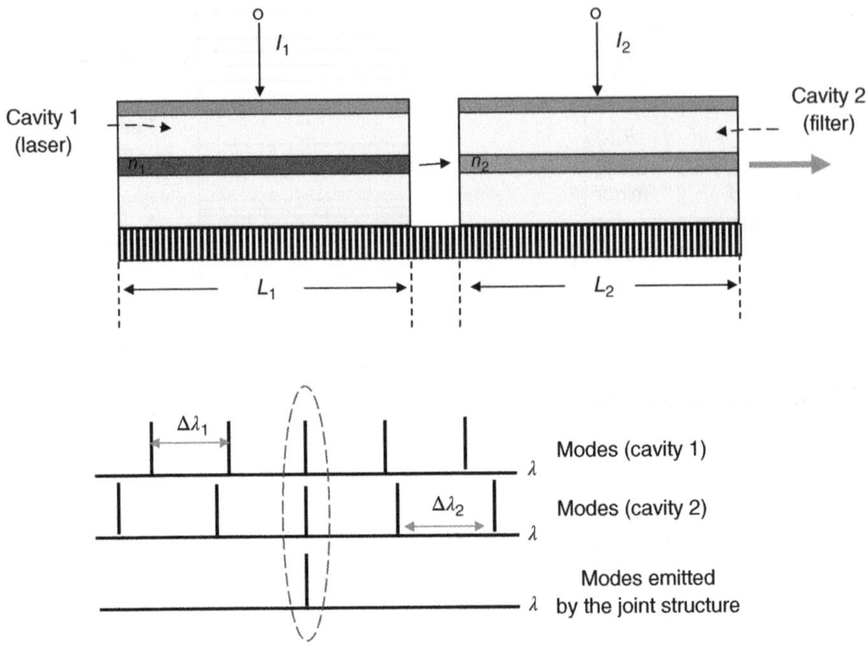

Figure 2.37 Coupled cavity laser structure and operation principle.

In essence, the overall device is composed of two independent Fabry–Perot laser diode cavities placed in series with an intermediate separation (usually air) of a few μm. The Fabry–Perot lasers have, as mentioned previously, slightly different cavity lengths. So, the frequencies of their respective longitudinal modes no not exactly match. Each diode has an independent current injection port, and the device operation can be tuned. The usual operating mode is to have one cavity operating above threshold, that is actually as a laser, while the second is operated below a threshold, acting as a filter where the position of its passbands can be modified by changing the current injection to its active zone, thus changing the value of its refractive index. If $\overline{n}_1$ and $\overline{n}_2$ represent, respectively, the refractive indices of the two cavities and L_1, L_2 their lengths, then the separations between adjacent longitudinal modes are given by:

$$\Delta\lambda_1 = \frac{\lambda^2}{2\overline{n}_1 L_1}, \quad \Delta\lambda_2 = \frac{\lambda^2}{2\overline{n}_2 L_2} \tag{2.107}$$

As it can be observed, the final result, known as *Vernier effect*, is that only those longitudinal modes present simultaneously in both cavities will contribute to laser radiation. The rest will be suppressed by destructive interference. The mode separation of the modes emitted by the overall cavity is much broader than any of the two values given by (2.107):

$$\Delta\lambda_c = \frac{\lambda^2}{2|\overline{n}_1 L_1 - \overline{n}_2 L_2|} \tag{2.108}$$

A special case is the *external cavity laser* (ECL) shown in Fig. 2.38.

Here, the second cavity is formed by the external facet of the first cavity and a spatial diffraction grating that can rotate around a transversal axis. The diffraction grating acts as a very selective mirror such that it only provides feedback to the first cavity within a very limited wavelength region.

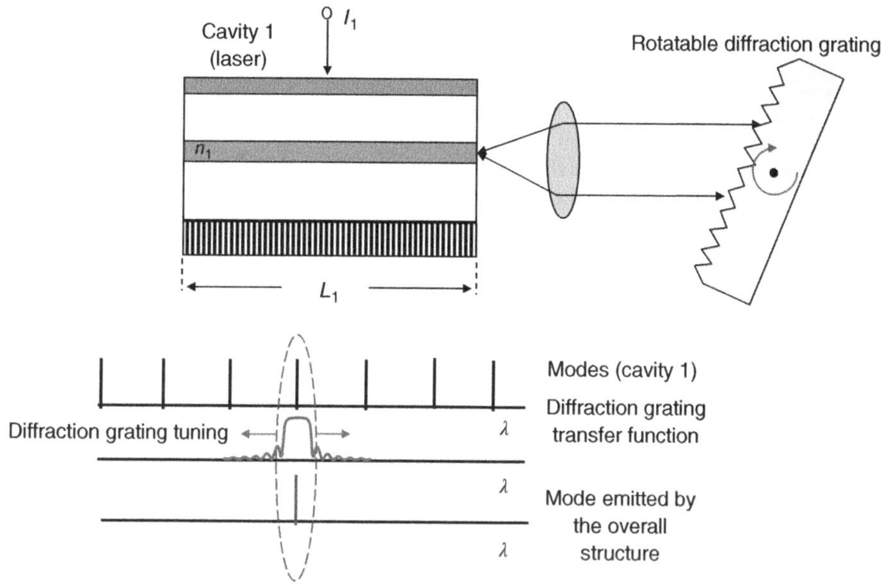

Figure 2.38 Structure and principle of operation of a tunable external cavity semiconductor laser.

Laser tuning is done by rotating the diffraction grating and can be done continuously. Tuning ranges >60 nm and output powers >100 mW can be found in commercial devices, which also exhibit remarkably low linewidth and RIN values (<100 kHz, <−160 dB/Hz).

As mentioned previously, the height (y-coordinate) of active zone in a semiconductor laser is in the range of 0.1–0.5 μm. This value can be decreased to around 10 nm and then quantum effects must be taken into account as the dimensions of the active zone became comparable to the De Broglie wavelength of the electron ($\lambda = h/p$) [8, 9]. One of the most important effects is the quantization of the kinetic energy of electrons propagating in a perpendicular direction to the *pn* junction plane. This means that the kinetic energy of these electrons can only take discrete values, in a similar way as it happens in the problem of a potential well described by Quantum Mechanics. Depending on whether the active zone has one, two, or three dimensions comparable to the De Broglie wavelength for the electron one speaks of *quantum well*, *quantum wire*, and *quantum dot* lasers, respectively. Figure 2.39 shows a scheme of these structures.

Figure 2.40 shows the structure of an active zone for a quantum well laser and the corresponding energy level diagram. As it can be observed in the x and z coordinates, there are no quantum-based restrictions and the relationships between energy and momentum are continuous and parabolic.

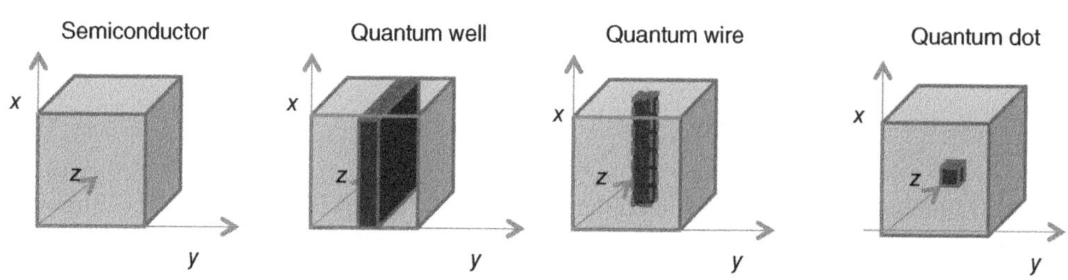

Figure 2.39 Active zone schemes for bulk, quantum well, quantum wire, and quantum dot lasers.

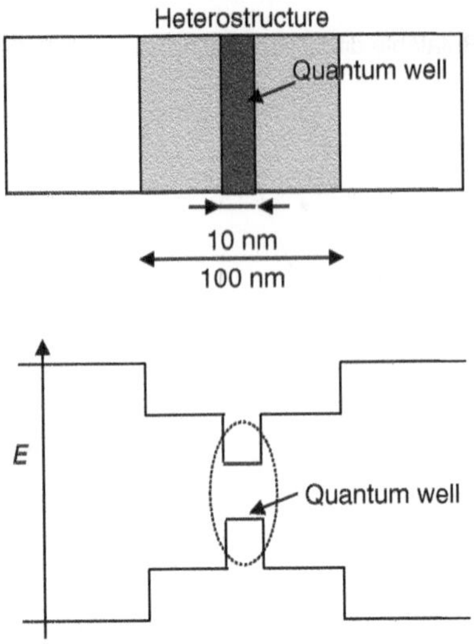

Figure 2.40 Energy levels for a quantum well laser.

$$E = \frac{p^2}{2m} = \frac{\hbar^2 k^2}{2m} = \frac{\hbar^2 \left(k_x^2 + k_z^2\right)}{2m} \quad (2.109)$$

In the y coordinate, however, the dimension L_y of the active zone is comparable to the De Broglie wavelength and thus it features quantum effects. The relationship between energy and momentum in this direction has to be obtained by solving the unidimensional Schrödinger equation for the case of a potential well. Referring to Fig. 2.40, the equations that have to be solved to determine the electron wavefunction Ψ and energy E are:

$$\begin{aligned} &\frac{\hbar^2}{2m}\frac{d^2\Psi}{dy^2} + E\Psi = 0 \quad 0 \le y \le L_y \\ &\frac{\hbar^2}{2m}\frac{d^2\Psi}{dy^2} + (E - V)\Psi = 0 \quad y < 0, y\, y > L_z \end{aligned} \quad (2.110)$$

A solution for the above equations is:

$$\Psi = \begin{cases} Ae^{k_1 y} & y < 0 \\ B\text{sen}(k_2 y + \delta) & 0 \le y \le L_y \\ Ce^{-k_1 y} & y > L_y \end{cases} \quad (2.111)$$

where

$$k_1 = \sqrt{\frac{2m(V-E)}{\hbar^2}}, \quad k_2 = \sqrt{\frac{2mE}{\hbar^2}} \quad (2.112)$$

which are subject to continuity conditions for Ψ and $d\Psi/dy$ in $y = 0$ and $y = L_y$ yield the allowed values for k_1, k_2 :

$$\begin{aligned} &k_1 L_y = k_2 L_y \tan\left(k_2 L_y\right) \\ &k_1^2 + k_2^2 = \frac{2mV}{\hbar^2} \end{aligned} \quad (2.113)$$

Form (2.113) and taking into consideration (2.112), the energy values for the electron kinetic energy in the y coordinate are obtained which are not continuous but a discrete set of values E_n $n = 1, 2, 3....$, both for the valence and the conduction bands. Taking (2.109) and the discretization imposed by (2.112), the energy-momentum relationship for electrons in the quantum well is given by:

$$E(n, k_x, k_z) = E_n + \frac{\hbar^2\left(k_x^2 + k_z^2\right)}{2m_{cn}} \quad (2.114)$$

where m_{cn} is the carrier effective mass (electron or hole) in the energy level E_n. Figure 2.41 shows a typical energy level diagram for the allowed states inside de-conduction and valence bands in a quantum well. Conduction band energy levels (electrons) are labelled by E_{ic} $i = 1, 2, 3....$, while valence band energy levels (holes) are labelled by E_{ihh} $i = 1, 2, 3..$ or E_{ilh} $i = 1, 2, 3...$ depending

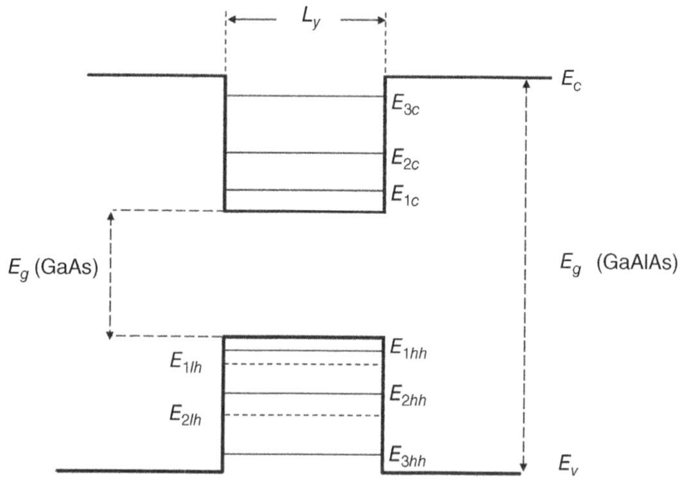

Figure 2.41 Detail of the discretization of the energy levels in the conduction and valence bands of a quantum well.

on whether they refer to heavy or light holes, respectively. As it can be observed, there are no energy levels below E_{1c} in the conduction band.

The energy-level discretization in the conduction and valence bands has important consequences. In particular, it affects the joint density of states ρ_{cv}, which is no longer given by (2.60). Rather, it now remains constant inside the sub-band "i" defined by the range of energies within E_i and E_{i+1}, and is now given by:

$$\rho_{cv,i} = \frac{m_{ci}}{\pi \hbar^2 L_y} \tag{2.115}$$

Figure 2.42 shows the differences and compares the joint density of states for a standard bulk and quantum well, wire, and dot heterojunctions.

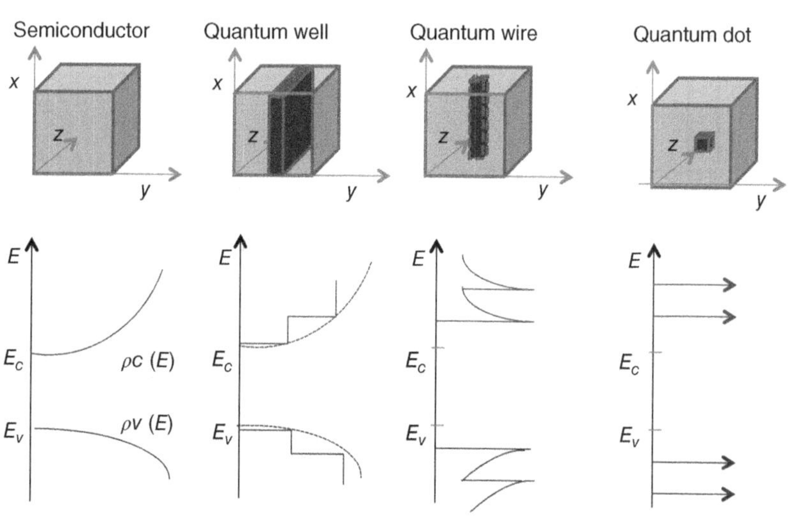

Figure 2.42 Density of states in bulk, quantum well, quantum wire, and quantum dot lasers.

The allowed transitions between the valence and conduction bands must verify the selection rule $\Delta i = 0$. This means, for example, that electrons in level 1 in the conduction band can only recombine with (heavy or light) holes in level 1 of the valence band. The consequence of this is that in the case of stimulated emission, the generated radiation is concentrated on a very narrow wavelength interval and hence its linewidth will be much narrower than that corresponding to the stimulated emission from a standard heterojunction. This, in fact, is one of the advantages of incorporating a quantum well structure in the active zone.

A limitation of quantum well structures is the low confinement factor that results from reducing the dimensions of the active zone. For example, if $L_y = 10$ nm, the confinement factor for an active zone composed of GaAs surrounded by $Al_{0.2}Ga_{0.8}As$ layers is around $\Gamma = 0.0026$. For a standard cavity absorption coefficient of $\alpha_c = 10\,\text{cm}^{-1}$ and mirror losses given by $-(1/2L)\ln(R_1R_2) = 30\,\text{cm}^{-1}$, this results in a threshold gain of $g_{th} = 1.5 \times 10^4\,\text{cm}^{-1}$, which is very high. A solution to overcome this limitation is to use an *MQW* structure composed of alternating *M* quantum well regions with semiconductor layer guiding zones, as shown in Fig. 2.43.

If the quantum wells are separated by a distance enough to uncouple their carrier wavefunctions, then the threshold gain is reduced by g_{th}/M and the laser can be operated at a lower injection density which results in a higher direct modulation bandwidth as compared to standard heterojunction lasers.

The envelope of the field in a singlemode laser is given by:

$$E(t) = \sqrt{S(t)}e^{j\phi(t)} \tag{2.116}$$

where $S(t)$ represents the number of photons in the cavity and $\phi(t)$ represents the field envelope phase. Both quantities depend on time as they can dynamically evolve due either to modulation

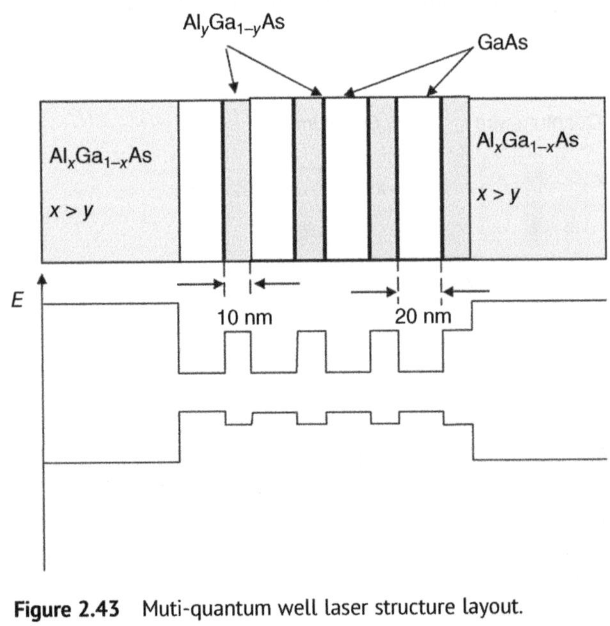

Figure 2.43 Muti-quantum well laser structure layout.

or to noise. This dynamic evolution is governed by the *laser rate equations*. When the laser dynamics are due to a modulation of the injection current $I(t)$, these equations are given by:

$$\begin{aligned}
\frac{dN(t)}{dt} &= \frac{I(t)}{e} - \frac{N(t)}{\tau_c} - GS(t) \\
\frac{dS(t)}{dt} &= S(t)\left(G - \frac{1}{\tau_{ph}}\right) + R_{sp} \\
\frac{d\phi(t)}{dt} &= \frac{\alpha v_g \Gamma \sigma_g}{2V_{act}}(N - N_{th}) \\
G = R_{st} &= x\frac{\sigma_g v_g \Gamma}{V_{act}}(N - N_o) = G_N(N - N_o)
\end{aligned} \tag{2.117}$$

$N(t) = V_{act}n(t)$ represents the number of electrons in the cavity of volume V_{act}, $R_{sp} = n_{sp}R_{st}$ is the spontaneous emission rate with $n_{sp} \geq 1$ being the population inversion factor, N_o, N_{th} account, respectively, for the number of injected carriers required for transparency (optical amplification) and threshold (lasing), e the electron charge, and τ_p the photon lifetime in the cavity, which is given by:

$$\frac{1}{\tau_{ph}} = v_g\left(\alpha_c + \frac{1}{2L}\ln\left(\frac{1}{R_1R_2}\right)\right) = v_g(\alpha_c + \alpha_{mirror}) \tag{2.118}$$

Table 2.1 gives the typical values for two different laser designs operating in the second and third windows.

Table 2.1 Typical parameters including values and units for 1300 and 1550 nm semiconductor lasers.

Parameter	**Units**	**1300 nm heterostructure laser**	**1550 nm DFB laser**
L	μm	250	250
A	μm	2	4
H	μm	0.2	0.5
V_{act}	cm^{-3}	$V_{act} = Lha = 10^{-11}$	5×10^{-11}
Γ	—	0.3	0.24
$\overline{n}$	—	3.4	3.4
$\overline{n}_g$	—	4	4
α	—	−5	−5
α_c	cm^{-1}	40	40
α_{esp}	cm^{-1}	45	60
σ_g	cm^2	2.5×10^{-16}	3×10^{-16}
n_o	cm^{-3}	10^{18}	1.63×10^{18}
R_{sp}	—	2	2
N_{th}	—	2.14×10^{8}	1.6×10^{18}
I_{th}	mA	15.8	12.8
τ_c	nseg	2.2	1
τ_{ph}	pseg	1.6	1.4
ε	—	10^{-7}	5×10^{-6}

The rate equations couple the variations of the number of electrons and photons via a negative feedback mechanism and their solution renders the full variation of the electric field envelope as given by (2.116). They can be applied to the analysis of the laser operation under static, transient, and modulation regimes.

For static operation, $d/dt = 0$ and the solution of (2.117) renders:

$$S = \frac{n_{sp}}{\left(1/\left(G\tau_{ph}\right)\right) - 1} \tag{2.119}$$

The condition for having a significant number of photons generated by stimulated emission in the cavity is $G\tau_{ph} \approx 1$. If $G\tau_{ph} < 1$, stimulated emission is low enough. So, the approximation $S = 0$ can be made, which implies that the device will mainly produce photons by spontaneous emission and will thus operate as a light-emitting diode (LED). If $G\tau_{ph} > 1$, stimulated emission dominates and the laser acts as a laser. The frontier condition between LED and laser operation is thus the so-called *threshold condition* $G\tau_{ph} = 1$, which, combined with $G = G_N(N - N_o)$, gives the number of required injected carriers for threshold and the resulting threshold current:

$$\begin{aligned} N_{th} &= N_o + \frac{1}{G_N \tau_{ph}} \\ I_{th} &= \frac{eN_{th}}{\tau_c} = \frac{e}{\tau_c}\left(N_o + \frac{1}{G_N \tau_{ph}}\right) \end{aligned} \tag{2.120}$$

Typical values for GaAs and InGaAsP laser threshold currents at room temperature are within the 1 and 50 mA range.

Once the number of injected carriers reaches the threshold value, it stabilizes to that value even if the injected current is increased. The reason is that the excess carriers recombine by stimulated emission. Since N remains fixed, the spontaneous emission will also be fixed and the number of generated photons by stimulated emission above the threshold will be directly proportional to $(I - I_{th})$, where I represents the injected current. If the laser bias current is below the threshold, then the number of photons generated in the cavity will be very low as only spontaneous emission contributes to the generation of light. When the threshold condition is achieved, the stimulated emission is the dominant mechanism for light generation and the number of generated photons will substantially increase when the difference between the bias and the threshold current increases. The output optical power from the laser is proportional to the number of photons in the cavity $P_{opt} = \eta_i v_g \alpha_{mirror} \hbar\omega S/2 = KS$ and from the rate equations and the condition for static operation:

$$P_{opt} = \frac{\eta_i v_g \alpha_{mirror} \hbar\omega}{2G}\left(\frac{I}{e} - \frac{N_{th}}{\tau_c}\right) = \eta_d \frac{\hbar\omega}{2e}(I - I_{th}) \tag{2.121}$$

where $\eta_d = \tau_{ph}\eta_i v_g \alpha_{mirror}$ is the *differential quantum efficiency* which is the range of 0.5–0.8. Equation (2.121) represents the well-known static *P–I* curve of a laser that gives a linear relationship of the static output optical power versus injected current of a laser above threshold which is shown in Fig. 2.44.

The solution of the rate equations under transient regime assumes a step function injection current that changes from a value $I_b < I_{th}$ to $I_b + I_m > I_{th}$ in $t = 0$. The output optical signal from the laser, which is illustrated in Fig. 2.45, reveals two important features.

First, the evolution in the number of photons shows a step-like behavior as well, with a *turn-on delay* that can be computed:

$$\tau_d = \tau_c \ln\left(\frac{I_m - I_b}{I_m - I_{th}}\right) \tag{2.122}$$

Second, the number of photons in the cavity fluctuates during a given time before it stabilizes to its final steady-state value $\overline{S}$. These damped oscillations known as *relaxation oscillations* represent a laser internal resonance due to the energy transfer between photons and electrons in the cavity. The damping rate Γ_R can be easily measured and frequency is given by:

$$\Omega_R = \sqrt{\frac{\eta_i G_N (I - I_{th})}{e}} = \sqrt{\frac{\eta_i \Gamma v_g \sigma_g}{eV_{act}}(I - I_{th})} \tag{2.123}$$

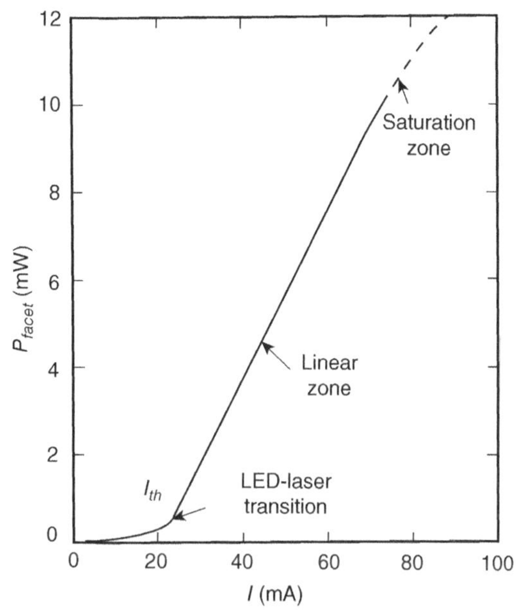

Figure 2.44 Typical static output optical power versus input injected current for a semiconductor laser.

The modulation transfer function of the laser is computed by solving (2.117) under an input sinusoidal current $I(t) = I_b + I_m \sin(\omega_m t)$ that produces a variation in the number of photons given by $S(t) = \overline{S} + \Delta S_m \sin(\omega_m t)$. The result is a second-order low-pass characteristic given by:

$$H(\omega) = \frac{\Delta S_m(\omega)}{\Delta S_m(0)} = \frac{\Delta P_m(\omega)}{\Delta P_m(0)} = \frac{\Omega_R^2 + \Gamma_R^2}{(\Omega_R + \omega - j\Gamma_R)(\Omega_R - \omega + j\Gamma_R)} \tag{2.124}$$

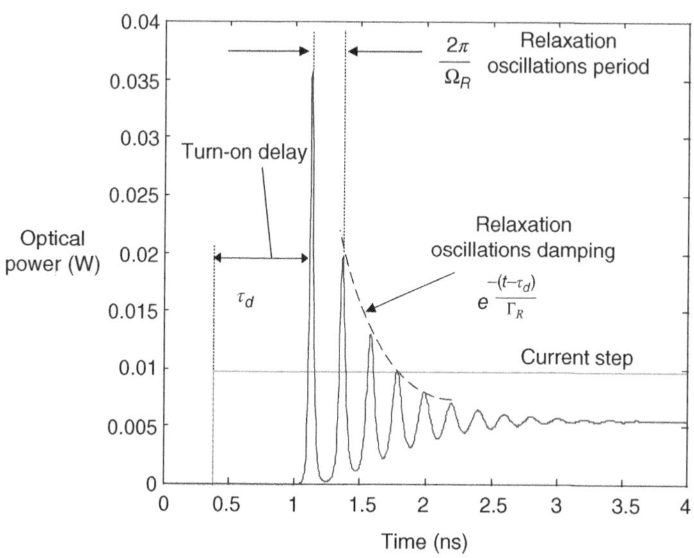

Figure 2.45 Turn-on delay and relaxation oscillations in the laser transient optical response following the application of a step current impulse.

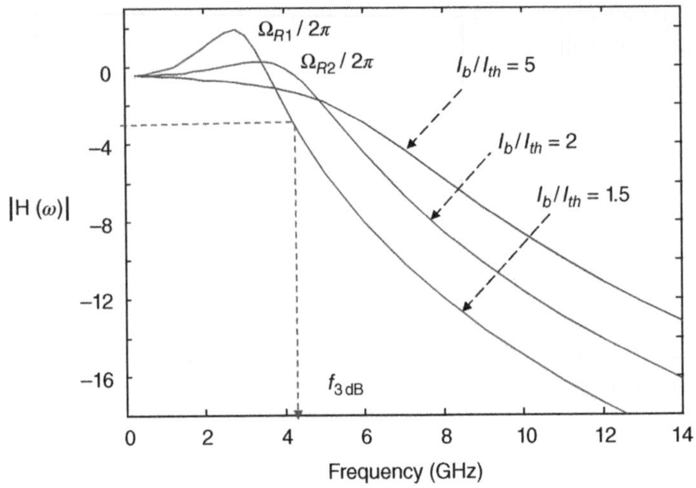

Figure 2.46 Intensity modulation transfer function for a semiconductor laser.

Figure 2.46 shows the modulus of (2.124) for a typical emission laser taking the bias current I_b/I_{th} as a parameter. The transfer function is flat in the region $\omega \ll \Omega_R$, increasing its value when it approaches the frequency of relaxation oscillations until it reaches a maximum at $\omega = \Omega_R$ from which it steadily decays.

The modulation bandwidth of the laser is defined by the 3 dB decay as compared to the DC value and it can be approximated by:

$$f_{3\text{dB}} \approx \frac{\sqrt{3}\Omega_R}{2\pi} \approx \left[\frac{3\eta_i v_g \Gamma \sigma_g}{4\pi^2 e V_{act}}(I_b - I_{th})\right]^{1/2} \tag{2.125}$$

Typical values for the modulation bandwidth of lasers are in the range of 5–10 GHz, which is enough for several applications in MWP. For applications above 10 GHz, directly modulated lasers are not usually employed due to modulation bandwidth and nonlinearity restrictions. Rather, the laser is operated in continuous wave (CW) mode and its output power is then externally modulated. For CW operation, the injection current is constant but nevertheless, the optical power can fluctuate in time due to the effect of intensity and phase noises. These noise sources can be computed using a modified version of (2.117) where white noise and correlated sources are added to the equations describing the evolution of the photon, the carrier number, and the field phase, respectively. The equations are then solved assuming $N(t) = \overline{N} + \Delta N(t)$, $S(t) = \overline{S} + \Delta S(t)$, $\phi(t) = \overline{\phi} + \Delta\phi(t)$ where the time fluctuations are due to noise. Intensity fluctuations are described by the *relative intensity noise (RIN)*, which is defined by the quotient of the noise power spectral density and the squared average number of photons in the cavity $\overline{S}$ and described by a second-order low-pass characteristic.

$$RIN(\omega) = \frac{\Delta\tilde{S}(\omega)}{\overline{S}^2} = K\frac{\Delta\tilde{P}(\omega)}{\overline{P}^2} = \frac{2R_{sp}\left[\left(\Gamma_N^2 + \omega^2\right) + G_N^2\overline{S}^2\left(1 + \overline{N}/\tau_c R_{sp}\overline{S}\right)\right]}{\overline{S}\left[\left(\Omega_R^2 - \omega^2\right)^2 + \left(2\omega\Gamma_R\right)^2\right]} \tag{2.126}$$

where $\Gamma_N = 1/\tau_c + G_N\overline{S}$.

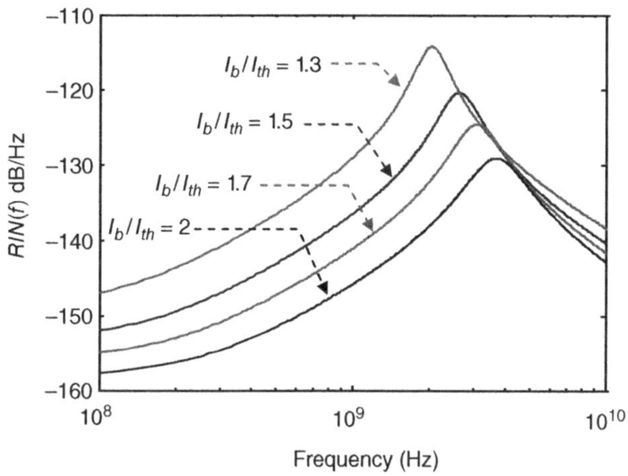

Figure 2.47 Relative intensity noise spectra for an InGaAsP semiconductor laser operating in the second window taking the bias current as a parameter.

Figure 2.47 shows the RIN (in dB/Hz) computed for a second window singlemode InGaAsP laser taking the bias current or alternatively from (2.121), the output optical power as a parameter.

The observation of Fig. 2.47 reveals some aspects of interest. In the first place, the *RIN* value is quite low when $\omega \ll \Omega_R$. It then increases substantially upon approaching the frequency of relaxation oscillations Ω_R. *RIN* then decays after reaching its maximum value. In the second place, fixing the value of the operating frequency, it can be observed that *RIN* decreases when the bias current (i.e. $\overline{S}$ or the output power $\overline{P}$) is increased. In practice, the effect of *RIN* is usually expressed in terms of a frequency-independent (along the modulation bandwidth of the laser) equivalent *RIN* parameter defined as:

$$RIN_{eq}\Delta\omega_{3\text{dB}} = \frac{1}{2\pi}\int_0^\infty RIN(\omega)d\omega \tag{2.127}$$

For microwave photonic applications, RIN_{eq} values as low as possible are desired and a typical range is from −165 to −150 dB/Hz.

The phase fluctuations of the output field generate changes in the instantaneous frequency emitted by the laser which result in a nonzero laser linewidth which is given by:

$$\Delta\nu = \frac{R_{sp}(1+\alpha^2)}{4\pi\overline{S}} = \frac{KR_{sp}(1+\alpha^2)}{4\pi\overline{P}} \tag{2.128}$$

Laser linewidths range from <100 KHz for ECLs to around 1–10 MHz for DFB and DBR lasers.

2.3.4 Optical Detectors

The most commonly employed photodetector in MWP is the *p–i–n* photodiode. Its principle of operation is the same as that of the reversed biased *p–n* junction, of which it is really an improved version. Figure 2.48 shows a schematic diagram of this structure. The applied inverse voltage V_0 produces a carrier-depleted zone around the junction. This is the so-called spatial charge or intrinsic zone, which in the *x*-coordinate, has a width given by $W = W_p + W_n$ that depends on the applied

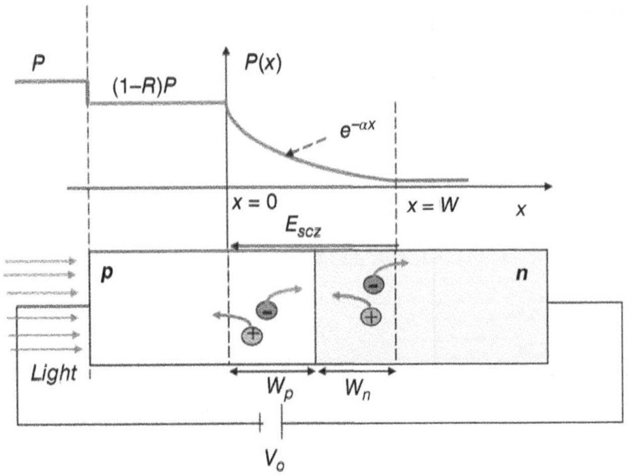

Figure 2.48 Schematic layout of a reversed *pn* junction.

bias voltage. Ionized impurities (positive in the *n* zone and negative in the *p*-zone) produce an electric field E_{scz} that prevents the hole diffusion from the *p* to the *n* zones and the electron diffusion from the *n* to the *p* zone.

The operation of the reversed *pn* junction as a photodetector is fairly simple to explain. Radiation at a given wavelength impinges on one of the external device facets (for example, that corresponding to the *p* zone as shown in Fig. 2.48), which is prepared so the reflection coefficient due to the air or the fiber-semiconductor refractive index discontinuity is small for that wavelength. Most of the incident optical power will then penetrate the device. If the photons from the incident radiation have energies $h\nu > E_g$, where E_g is the gap between the valence and the conduction bands, then they can be absorbed by electrons at the valence band, which will then be promoted to the conduction band generating a hole in the valence band. For the same reasons explained in Section 2.3.2, direct gap semiconductors are preferred in order to maximize the probability of photonic absorption. Hence, the absorption of a photon generates an electron-hole (e-h) pair that can be potentially collected in the form of an external current. In practice, not all the incident photons are absorbed. Some of them are reflected by the input facet. Of the rest, even if absorbed, not all of them will generate e-h pairs that will ultimately incorporate the external current from the device. Finally, even if the photodiode is optimized to absorb all the penetrating photons, it may happen that the e-h pair collection by an external current is frequency dependent leading to bandwidth restrictions. The photodiode quantum conversion efficiency η defined as the rate of generated electrons incorporating the external current divided by the rate of incident optical photons is given by:

$$\eta = \frac{P/(h\nu)(1-R)[1-\exp(-\alpha W)]}{P/(h\nu)} = (1-R)(1-\exp(-\alpha W)) \tag{2.129}$$

where R represents the reflectivity of the semiconductor input facet and α the absorption coefficient of the semiconductor material. To achieve its maximum value $\eta = 1$, one needs $R = 0$ and $\alpha W \to \infty$. A low value for R can be achieved by depositing a multilayer antireflection coating like with DFB lasers. A high value of αW requires the width of the space carrier zone W to be as big as possible. Unfortunately, W in a reversed biased *pn* junction is on the order of a fraction of a micron which means that a very significant part of the incoming radiation will not be absorbed in that zone. This is

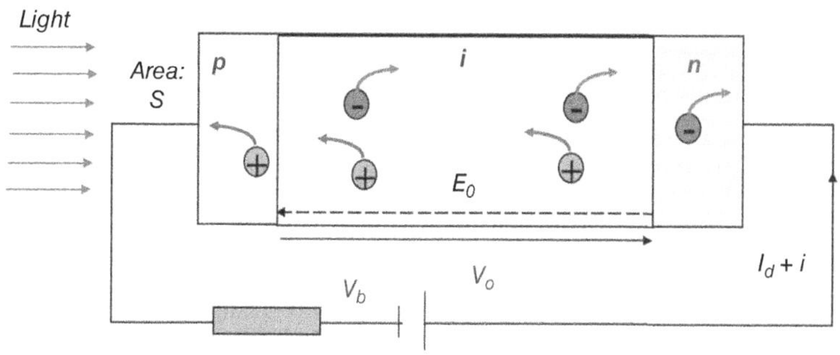

Figure 2.49 Schematic layout of a pin photodiode.

the reason for adding an intermediate intrinsic semiconductor zone in between the *n* and the *p* zones leading to the so-called *p–i–n* structure, which is shown in Fig. 2.49.

Since the intrinsic zone has no impurities, the density of electrons and holes will be very small when no incident radiation is present. This means that compared to the *n* and *p* zones, it can be considered an electric isolator and therefore most of the applied reverse bias will translate to this zone. In this way, a strong electric field, similar to that arising in the *scz* of the standard *pn* junction, is built along its length, but now there is a wide intrinsic region *i* with a length that can be arbitrarily extended (typically up to 10 μm) and therefore there is a noticeable increase in the photon absorption in the device and therefore of its quantum efficiency.

The quantitative study of the *pn* junction and, in particular, its static and dynamic behavior follows a similar path as that described for lasers. In this case, the equivalent role of the rate equations for lasers is played by the *continuity equations*, which describe the interplay between photons, electrons, and holes. For the *p–i–n* photodiode, these are given by:

$$
\begin{aligned}
\frac{\partial n}{\partial t} &= G_n - R_n + \frac{1}{e}\nabla . J_n \\
\frac{\partial p}{\partial t} &= G_p - R_p + \frac{1}{e}\nabla . J_p
\end{aligned}
\tag{2.130}
$$

n and *p* represent the carrier and hole concentrations in the conduction and valence bands, respectively.

The upper equation in (2.130) describes the rate of change of free carriers in a given point of the semiconductor, which is equal to the number of electrons generated at that point per unit of time, G_n, *minus* the number of electrons that disappear at that point due to e-h recombination per unit of time, R_n, *plus* the number of electrons that *arrive* at that point per unit of time *minus* the number of electrons that *leave* that point per unit of time. The last two terms which describe the net flow of incoming and outgoing charges in an infinitesimal volume surrounding the point of interest in the semiconductor are given by the divergence of the surface density current $(1/e)\,\nabla . J_n$. We will actually employ a unidimensional propagation model and, hence, $\nabla = \partial/\partial z$. The interpretation for the rate of change of holes in (2.130) is similar. It should nevertheless be remembered that although the electron and hole flows are in opposite directions, their charge is also opposed in sign and therefore the currents due to both carriers will add.

In general:

$$G_n = G_{tn} + g \tag{2.131}$$

where G_{tn} is the thermal generation rate (i.e. the number of electrons that are thermally generated by a unit of time according to the Fermi–Dirac statistic) and g is the number of carriers generated out of equilibrium by external mechanisms such as light radiation. If, as is the case, the generation is due to the generation of e-h pairs, then $g_n = g_p \equiv g$. For example, in a p-type semiconductor, one has for minority carriers (electrons):

$$G_{tn} - R = -\frac{n - n_{p0}}{\tau_n} \tag{2.132}$$

where n_{po} is the number of thermally generated electrons in the p-zone under equilibrium. So, $n - n_{po}$ represents the excess of electrons. τ_n has a similar meaning to the carrier lifetime in lasers. It represents the average lifetime of electrons in the p zone. Its physical interpretation is therefore as follows. If there is no net current (for example, if the pn junction is left in open circuit configuration), then (2.132) reduces $\partial n/\partial t = -(n - n_{p0})/\tau_n \Rightarrow n(t) = n(0)\exp(-t/\tau_n) + n_{p0}$. So, if at $t = 0$, for any reason, a carrier excess $n(0) - n_{po}$ is generated and the system is left to evolve freely, this excess will decay toward the equilibrium value n_{p0}. In a semiconductor in equilibrium, $\partial/\partial t = 0, J_n = 0$ and $G_n = G_{tn} = R$ since $n = n_{p0}$. In other words, the thermal generation is exactly compensated by carrier recombination.

We now need to know the form of the surface current densities J_n and J_p. It is well known [8] that each current is composed of two parts: the *drift* and the *diffusion* currents.

$$\begin{aligned} J_n &= J_{n,drift} + J_{n,dif} = en\mu_n\varepsilon + eD_n\frac{\partial n}{\partial z} \\ J_p &= J_{p,drift} + J_{p,dif} = en\mu_p\varepsilon + eD_p\frac{\partial p}{\partial z} \end{aligned} \tag{2.133}$$

The drift current is due to the charge movement that under the effect of an electric field ε acquires a certain velocity. If the electric field is not very intense, then the velocity is proportional to the field: $v_n = \mu_n\varepsilon$; $v_p = \mu_p\varepsilon$, where v_n and v_p are, respectively, the electron and hole velocities and μ_n, μ_p their *mobilities*, which depend on the material. Diffusion currents are represented by the second terms in the right-hand side members of (2.133). These are due to the fact that any inhomogeneously distributed charged carriers tend to redistribute homogeneously due to thermal agitation. Diffusion currents are proportional to the gradients of the spatial carrier distributions $n(z)$ and $p(z)$. So, there is no carrier flow if there is no inhomogeneity. D_n and D_p are the diffusion coefficients for electrons and holes, respectively. They depend on the material and the temperature. In the i zone, Eqs. (2.130) and (2.133) reduce to:

$$\begin{aligned} \frac{\partial n}{\partial t} &= g + \frac{1}{e}\frac{\partial J_n^{(i)}}{\partial z} \\ \frac{\partial p}{\partial t} &= g - \frac{1}{e}\frac{\partial J_p^{(i)}}{\partial z} \end{aligned} \tag{2.134}$$

since the high electric field wipes away any free carriers and recombinations can be neglected since generated carriers are drifted in *saturation* regime, that is, at the highest possible velocities v_s, v_p which are no longer proportional to the electric field ε_o ($v_n = v_{sn} < \mu_n\varepsilon_o$ and $v_p = v_{sp} < \mu_p\varepsilon_o$). Thus, drift currents are given here by $J_{n,drift} = env_{sn}$ and $J_{p,drift} = env_{sp}$. Furthermore, g can be expressed as a function of the incident photon flux $\Phi(z, t)$ (photons crossing a unit surface transversal to the propagation direction z per unit of time). Assuming that absorption in the p-zone can be neglected, then:

$$g(z,t) = -\frac{\partial\Phi(z,t)}{\partial z} = \alpha\Phi_0(t)e^{-\alpha z} \tag{2.135}$$

As it has been done with the laser sources, we start considering the operation under a static regime. In this case, the incident optical radiation or the photon flux does not change with time. Hence, $\Phi_0(t) = \Phi_0$ and $\partial/\partial t = 0$, which substituted in (2.135) and (2.134) and taking the boundary conditions $J_n^{(i)}(0) = 0$ and $J_p^{(i)}(W) = 0$ give:

$$\begin{aligned} J_n^{(i)}(z) &= e\Phi_0(1 - e^{-\alpha z}) \\ J_p^{(i)}(z) &= e\Phi_0\left(e^{-\alpha z} - e^{-\alpha W}\right) \end{aligned} \tag{2.136}$$

The total photocurrent generated in the *i* zone will be given then by:

$$J^{(i)} = J_n^{(i)}(z) + J_p^{(i)}(z) = e\Phi_0\left(1 - e^{-\alpha W}\right) = \frac{eP}{Sh\nu}(1 - R)\left(1 - e^{-\alpha W}\right) \tag{2.137}$$

where P represents the incident optical power, S the photodiode cross-section area and ν the frequency of the incident radiation. If we only consider this contribution, the photodetector quantum efficiency is exactly that given by (2.129). In practice, one has to also consider the contribution from the *p* and *n* zones to the current, which results in a slightly modified version of (2.137). The contribution of the *n*-zone can be computed by considering the diffusion of holes to the *i* zone, which results in:

$$\begin{aligned} J^{(n)} &= -e\Phi_0\frac{\alpha L_p}{(1 - \alpha L_p)}e^{-\alpha W} - ep_n\frac{D_p}{L_p} \\ L_p &= \sqrt{D_p\tau_p} \end{aligned} \tag{2.138}$$

where L_p is the hole *diffusion length* that represents the average length traversed by a diffused hole in the *n* zone before it recombines. Note that the second term on the right-hand side of (2.138) is independent of the input radiation. It represents the *dark current* of the photodiode. The total current is now obtained by summing (2.137) and (2.138), resulting in:

$$J = e\Phi_0\left(1 - \frac{e^{-\alpha W}}{1 + \alpha L_p}\right) = \frac{eP}{Sh\nu}(1 - R)\left(1 - \frac{e^{-\alpha W}}{1 + \alpha L_p}\right) \tag{2.139}$$

which states that not all diffused carriers contribute to the photocurrent. From (2.139), the photodetector quantum efficiency is given by:

$$\eta = \eta_i(1 - R)\left(1 - \frac{e^{-\alpha W}}{1 + \alpha L_p}\right) \tag{2.140}$$

where we have introduced an internal quantum efficiency $\eta_i \leq 1$ to account for other mechanisms which prevent the e-h generation from each absorbed photon. Quantum efficiencies for semiconductor *p–i–n* diodes are in the range of 0.5–0.9.

A fundamental parameter of a *p–i–n* photodetector is its responsivity which relates its output electric current to its input optical power:

$$\Re = \frac{e\eta}{h\nu} = \frac{e\eta}{hc}\lambda \tag{2.141}$$

Figure 2.50 represents the typical variation of $\Re(\lambda)$ for the three *p–i–n* photodiodes commonly employed in MWP. Si photodiodes are only appropriate for first window transmission (around 0.8 μm). Germanium photodiodes, on the contrary, can be employed along all the spectral ranges

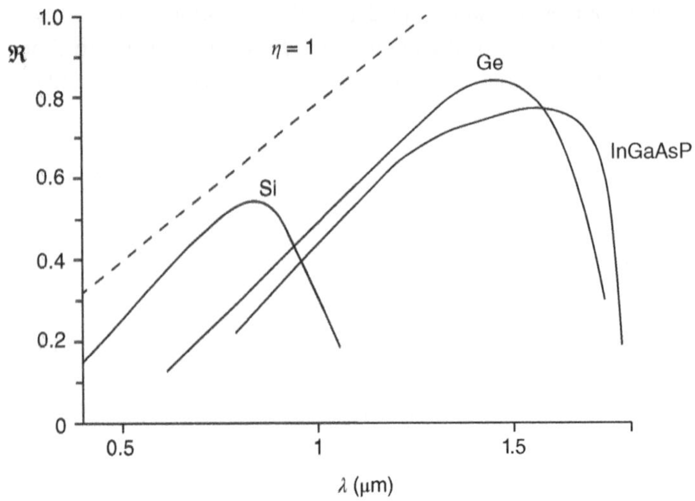

Figure 2.50 Responsivity versus wavelength for Silicon-, Germanium-, and InGaAsP-based photodiodes.

of interest (0.8–1.6 μm). However, they usually feature higher values of dark current and its associated noise. For this reason, Si-based devices are the preferred option if the operating wavelength is ≤1 μm. If the wavelength is >1 μm, then, as happens with optical sources, III–V materials are preferred since their spectral response (energy gap) can be tailored by changing the material composition and they present very low dark currents.

Equation (2.141) describes an apparently linear relationship between $\Re(\lambda)$ and the wavelength. However, as it can be appreciated in Fig. 2.50 for each photodiode, there is one wavelength value from which $\Re$ stops to increase and starts to fall more or less abruptly. This is due to the fact that from that wavelength, the energy of the incident photons is such that their energy $h\nu = hc/\lambda$ is below the semiconductor bandgap E_g and thus cannot be absorbed and generate e-h pairs. On the other hand, the straight zone of the $\Re$ is not exactly linear as the absorption coefficient α depends on the wavelength and, as a consequence, the quantum efficiency is also a wavelength-dependent function $\eta = \eta\,(\lambda)$. Finally, the value of $\Re$ tends to zero for short wavelengths since the absorption coefficient then increases and most of the incident photons are not absorbed in the intrinsic zone but rather in other zones (n or p) where they can recombine and thus decrease the quantum efficiency.

The frequency response of the p–i–n photodiode can be obtained solving (2.130) where now $\partial/\partial t \neq 0$. Fast-response photodiodes require that most of the external current is in the intrinsic zone since, otherwise, the slow diffusion times of minority carriers toward the intrinsic zone will compromise the device response. In this case, the transfer function yields:

$$\begin{aligned}\frac{J^{(i)}(\omega)}{\Phi_0(\omega)} &= \frac{1}{1+j\omega RC}\alpha eW \\ &\left\{\frac{1}{\alpha W - j\omega t_p}\left[\frac{e^{-\alpha W}-1}{\alpha W} + \frac{e^{-\alpha W}\left(e^{j\omega t_p}-1\right)}{j\omega t_p}\right]\frac{1}{\alpha\, W + j\omega t_n}\left[\frac{1-e^{j\omega t_m}}{j\omega t_n} + \frac{1-e^{\alpha W}}{\alpha W}\right]\right\} \\ &= \frac{1}{1+j\omega RC}F(\omega)\end{aligned} \tag{2.142}$$

where $C = \in S/W$ is the capacity of the intrinsic zone, R the photodiode bias circuit resistance, and $t_p \equiv W/v_{sp}$, $t_n \equiv W/v_{sn}$ the *transit times* of holes and electrons in the intrinsic zone, respectively. Equation (2.142) states that the photodiode frequency response is composed of two factors. One, given by the low-pass term $(1 + i\omega\, RC)^{-1}$ is a purely bias circuit contribution. The second factor expresses the real frequency dependence of the current in the intrinsic zone which is influenced by the transit times of electrons and holes. The bandwidth of the second factor can be increased by reducing the transit times which requires to decrease W. This is not a good option since then the quantum efficiency is reduced and the capacity of the intrinsic zone is increased reducing the bandwidth of the first low-pass factor. In practice, it is actually this low-pass factor the dominant term. So, its bandwidth can be increased by reducing the semiconductor cross-section area S.

Optimized p–i–n photodiodes require an increase in the optical absorption and, at the same time, reduce transit time. One possible solution is to use edge-coupling (in the y direction), allowing the optical input to enter the intrinsic region directly and propagate orthogonally to the electric field in the intrinsic zone. The result is that this zone actually becomes a long optical waveguide where a large fraction of the input power is absorbed whilst maintaining low transit times. Using this design, p–i–n photodiodes with bandwidths in excess of 100 GHz have been demonstrated with typical bandwidth–efficiency products of about 55 GHz. The disadvantage of this configuration is that coupling efficiency between the fiber and the intrinsic zone is reduced as the active zone is usually around 1 μm, although this can be mitigated by means of fiber tapers. The second disadvantage is the increase in capacity if the active zone height needs to be increased to absorb more photons.

Many of the above limitations, especially the RC time constant limitation of the photodiode -bandwidth can be overcome using travelling-wave structures which, in addition to offering edge-coupling and optical waveguiding of the incoming photons in the intrinsic zone, incorporate coplanar transmission-line waveguide electric contacts designed to support quasi-TEM mode microwave and RF travelling waves. Since optical absorption occurs in a distributed way along the length (in the y coordinate) of the device, it sets up a travelling wave on the electrodes. In this case, the speed of the device is not limited by the RC constant, but by the mismatch between the optical group velocity and electrical phase velocity. When velocity matching is achieved, long device lengths compared to waveguide photodiodes are possible in principle. Furthermore, the fact that the absorption volume is increased also means that these devices will saturate at a higher power level. Travelling-wave p–i–n photodiodes have been demonstrated, featuring bandwidths in excess of 500 GHz. A similar p–i–n photodiode based on the travelling wave approach but capable of handling high optical powers and offering high electrical output power is the uni-travelling-carrier (UTC) photodiode. UTCs are usually implemented using InP as a base material. The main difference between a UTC and a travelling wave p–i–n is that the optical waveguide and absorption functions are physically separated. Optical waveguiding occurs in the n zone whilst absorption occurs in the p zone. In this way, the photogenerated current consists only of electrons and the main advantage is that since electrons are faster than holes in InP, this leads to leading to potentially high bandwidths as well as high saturation powers.

To conclude with this brief treatment of optical detectors for MWP, we now address the noise sources of a p–i–n photodetector which, strictly speaking, are two: Shot noise due to the signal quantization and the dark noise. The electrical powers due to these sources are given by:

$$\overline{(\Delta i_{shot})^2} = 2e\bar{i}\Delta f \tag{2.143}$$

$$\overline{(\Delta i_{dark})^2} = 2eI_{dark}\Delta f \tag{2.144}$$

where Δf represents the photodiode bandwidth and $\bar{i}$, I_{dark} the average signal and dark currents, respectively. For signal-to-noise calculations, usually a third thermal noise source due to the receiver electronics is added and the RIN from the laser is also considered, hence:

$$\overline{(\Delta i_{th})^2} = \frac{4k_B T \Delta f F_n}{R} \tag{2.145}$$

$$\overline{(\Delta i_{RIN})^2} = RIN\bar{i}^2\Delta f \tag{2.146}$$

In the above expressions, R represents the load resistance seen by the photocurrent, ($F_n > 1$), an equivalent noise factor for the receiver electronics.

2.3.5 Optical Modulators

The direct modulation of semiconductor lasers is not an efficient technique for microwave photonic systems where the frequency of the radiofrequency signals to be transmitted and processed is over 10 GHz. In this case, the laser is operated in CW mode and the radiofrequency signal is impressed into the optical carrier by means of an external modulator. External modulators operate exploiting either the electro-optic or the electroabsorption effect.

By means of the electro-optic or *Pockels effect*, the refractive index in a given direction "i" inside a crystal can be changed by applying a control voltage signal.

$$n_i = n_{oi} \pm k_i V \tag{2.147}$$

where n_{oi} represents the refractive index in the "i" direction when no voltage is applied and k_i depends on the wavelength and a material electro-optic coefficient. Among the material systems that display the Pockels effect, the most widely employed in external modulators are lithium niobate, III–V semiconductors, and polymers. Along a propagation length L orthogonal to the "i" direction, the refractive index variation (2.147) leads to a phase shift given by:

$$\begin{aligned} \phi_i &= \frac{2\pi n_i}{\lambda}L = \frac{2\pi n_{oi}}{\lambda}L \pm \frac{2\pi k_i V}{\lambda}L = \phi_{oi} \pm \pi\left(\frac{V}{V_\pi}\right) \\ V_\pi &= \frac{\lambda}{2k_i L} \end{aligned} \tag{2.148}$$

where V_π is the half-wave voltage; in other words, the voltage required to produce a phase shift of π radians.

An integrated electro-optic phase modulator exploits the effect described by equation (149). The top part of Fig. 2.51 shows a typical configuration.

It is basically composed of a waveguide made of an electro-optic material and length L through which input electric field polarized in the "i" direction propagates. If we assume that the input optical field is given by:

$$E_i = Ae^{j\omega_o t} \tag{2.149}$$

And that the applied voltage to the waveguide is a radiofrequency signal:

$$V(t) = V_b + V_m \sin \Omega t \tag{2.150}$$

where V_b is a DC bias voltage and V_m the amplitude of the time-varying modulation. Then the phase shift experienced by the field (we omit the i subscript from now on) is given by:

$$\begin{aligned} \phi(t) &= \phi_o + \pi\left(\frac{V_b}{V_\pi}\right) \pm \pi\left(\frac{V_m}{V_\pi}\right)\sin \Omega t \\ &= \phi_b \pm m \sin \Omega t = \phi_b \pm \Delta\phi(t) \end{aligned} \tag{2.151}$$

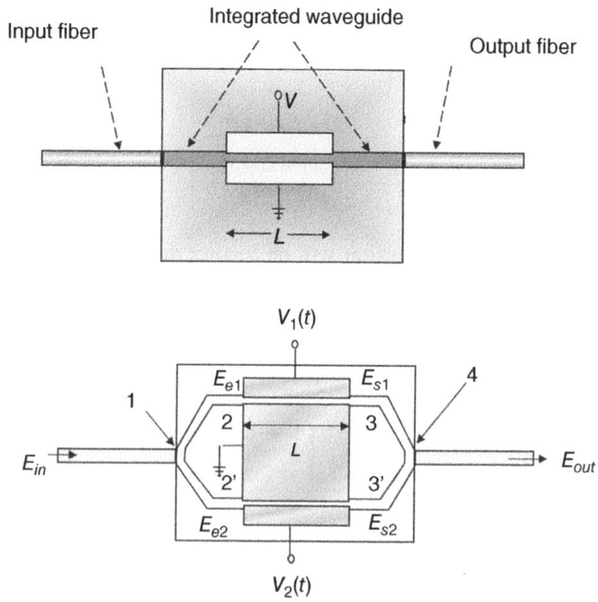

Figure 2.51 (Top) Layout of an integrated electro-optic phase modulator. (Bottom) Layout of a dual drive Mach–Zehnder electro-optic intensity modulator.

where ϕ_b is the bias phase shift and m the modulation index. Introducing (2.151) in (2.149), the output field from the phase modulator is:

$$\begin{aligned} E_o(t) &= Ae^{j[\omega_o t \pm \phi_b \pm m \sin \Omega t]} = Ae^{j[\omega_o t \pm \phi_b]} e^{\pm jm \sin \Omega t} \\ &= Ae^{j[\omega_o t \pm \phi_b]} \sum_{n=-\infty}^{\infty} J_n(\pm m) e^{\pm jn\Omega t} \\ &= Ae^{\pm j\phi_b} \sum_{n=-\infty}^{\infty} J_n(\pm m) e^{\pm j[\omega_o \pm n\Omega]t} \end{aligned} \tag{2.152}$$

where $J_n(x)$ is the Bessel function of the first kind and order n. Note that apart from the desired sidebands at $\pm\Omega$, the phase modulator also produces harmonics at $\pm n\Omega$ $(n > 1)$. The amplitudes of these harmonics can be neglected under the small signal approximation, which requires $m \ll 1$. In such a case:

$$E_o(t) \approx Ae^{\pm j\phi_b} e^{\pm j\omega_o t} \left[J_o(\pm m) - J_1(\pm m) e^{\mp j\Omega t} + J_1(\pm m) e^{\pm j\Omega t} \right] \tag{2.153}$$

An electro-optic intensity modulator can be implemented by embedding one or two integrated electro-optic phase modulators in one (single drive modulator) or the two (dual drive modulator) arms of a Mach–Zehnder interferometer (MZI). In the bottom part of Fig. 2.51, we show the configuration of a typical dual-drive modulator as this configuration has substantial advantages over the single-drive configuration in MWP applications; in particular, it can be employed to produce single sideband modulation (SSB). The input to the MZI is divided into its two arms by a 3 dB Y-branch splitter. In each arm, there is a phase modulator driven by independent voltage $V_1(t)$ for arm 1 and $V_2(t)$ for arm 2. Voltages are applied to electrodes of length L deposited on top of the waveguides made of electro-optic materials. The outputs from the phase modulators are then combined into a single output waveguide by an output 3 dB Y-junction. The output field from the modulator is expressed in terms of the input field as:

$$\begin{aligned} E_o(t) &= \frac{E_i}{2}\left[e^{j[\phi_{b1} \pm \Delta\phi_1(t)]} + e^{j[\phi_{b2} \pm \Delta\phi_2(t)]}\right] \\ &= E_i e^{j(([\phi_{b1} \pm \Delta\phi_1(t) + \phi_{b2} \pm \Delta\phi_2(t)])/2)} \cos\left[\frac{\phi_{b1} - \phi_{b2} \pm (\Delta\phi_1(t) - \Delta\phi_2(t))}{2}\right] \end{aligned} \tag{2.154}$$

The common phase factor represents the modulator chirp that will be further described below. The modulator output-input power transfer characteristic is the given:

$$\begin{aligned} T_{MZM} &= \frac{P_o(t)}{P_i} = \left|\frac{E_o(t)}{E_i}\right|^2 = T_{ff}\cos^2\left[\frac{\phi_{b1} - \phi_{b2} \pm (\Delta\phi_1(t) - \Delta\phi_2(t))}{2}\right] \\ &= T_{ff}\left\{\frac{1 + \cos[\phi_{b1} - \phi_{b2} \pm (\Delta\phi_1(t) - \Delta\phi_2(t))]}{2}\right\} = T_{ff}\left\{\frac{1 + \cos[\phi_T \pm \Delta\phi_T(t)]}{2}\right\} \end{aligned} \tag{2.155}$$

where an optical insertion loss factor T_{ff} has been included. For maximum linearity, the modulator must be biased at a *quadrature point*, in other words:

$$\phi_{TQ} = (2k + 1)\frac{\pi}{2} \quad k = 0, \pm 1, \pm 2, \dots \tag{2.156}$$

And then if $\Delta\phi_T(t) \ll \pi/2$:

$$\begin{aligned} T_{MZM} &\approx \frac{T_{ff}}{2}(1 \mp \Delta\phi_T(t)) \\ \Delta\phi_T &= \frac{\pi}{V_\pi}\Delta V(t) \end{aligned} \tag{2.157}$$

which leads to a modulation efficiency given by:

$$s_M = \frac{\partial I_o}{\partial \Delta V(t)} = \frac{T_{ff}\pi}{2V_\pi}P_i \tag{2.158}$$

Equation (2.158) shows one advantage of using an external modulator, namely that by increasing the CW laser power P_i, we can increase the slope efficiency of the external modulation process. Hence, the laser can be used to control modulator efficiency. Equation (2.158) also shows that one of the key issues of modulator design is the reduction of the voltage V_π. To do so, the microwave-optical interaction length, that is the electrode length, must be increased and this demands reductions in electrode and optical waveguide loss.

The modulator chirp can be eliminated if $\Delta\phi_1(t) = -\Delta\phi_2(t) = \Delta\phi(t)$ since then (2.154) transforms to:

$$E_o(t) = E_i e^{j\frac{[\phi_{b1} + \phi_{b2}]}{2}} \cos\left[\frac{\phi_{b1} - \phi_{b2} \pm 2\Delta\phi(t)}{2}\right] \tag{2.159}$$

And, for a quadrature-biased modulator:

$$T_{MZM} \approx T_{ff}\left(\frac{1}{2} \mp \Delta\phi(t)\right) \tag{2.160}$$

Other operation points of the MZM external modulator of interest for MWP applications are the maximum and minimum transmission bias points (MTBP and mTBP), respectively, that are defined by:

$$\begin{aligned} \phi_T|_{MTTB} &= 2k\pi \quad k = 0, \pm 1, \pm 2, \ldots \\ \phi_T|_{mTTB} &= (2k+1)\pi \quad k = 0, \pm 1, \pm 2, \ldots \end{aligned} \tag{2.161}$$

which result in the following intensity transfer characteristics if $\Delta\phi_T(t) << \pi/2$:

$$\begin{aligned} T_{MZM}|_{MTTB} &\approx T_{ff}\left[1 - \frac{\Delta\phi^2(t)}{4}\right] \\ T_{MZM}|_{mTTB} &\approx T_{ff}\frac{\Delta\phi^2(t)}{4} \end{aligned} \tag{2.162}$$

Figure 2.52 illustrates the typical intensity transmission curve for a MZM including the salient operation points.

As in the case of a detector, travelling wave structures where the velocity of modulating radio-frequency signals is matched to that of the optical signal eliminate the capacitance limiting effect in the frequency response. The electrodes are designed to act as microwave transmission lines, with the microwave signals propagating in the same direction as the optical signals and the optical wave-guides being parallel to the electrodes. Travelling wave lithium niobate MZM external modulators with bandwidths in excess of 40 GHz are commercially available and bandwidths in excess of 100 GHz have been demonstrated in polymer-based devices. Typical MZMs have insertion losses of around 5 dB and operation voltages of around 5–6 V.

In Section 2.3.1, we have reviewed some of the most basic and significant mechanisms of the interaction between radiation and matter, which we have specialized in Section 2.3.2. In particular, absorption is produced when an incident photon with energy greater or equal to the energy gap E_g between the valence and conduction bands is employed to generate an e-h pair which are separated. Really, the minimum value of energy required to absorb a photon corresponds to a transition where the e-h pair is not separated but attracted together by Coulomb electrostatic forces forming an

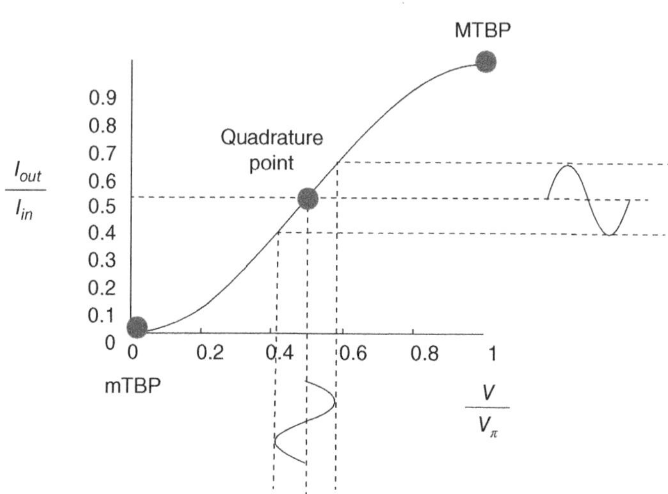

Figure 2.52 Intensity transmission curve of an MZM including the relevant operation bias points.

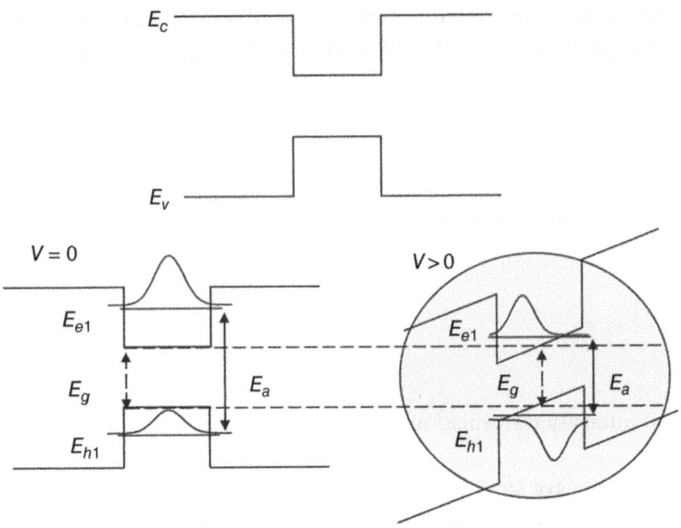

Figure 2.53 Basic operation principle of the electroabsorption modulator.

electrically neutral quasiparticle called *exciton*. If we refer to the bottom left part of Fig. 2.53, we see that the required energy to generate an exciton in a quantum well structure is given by:

$$E_a = E_g + E_{e1} + E_{h1} - B \tag{2.163}$$

where B represents the bonding energy of the exciton. In ordinary semiconductors, this bonding energy is very low. So, in general, the presence of excitons is only observed at very low temperatures. In multiple quantum well semiconductors, however, the bonding energy of the exciton is higher and therefore these are more stable and can be observed at room temperature.

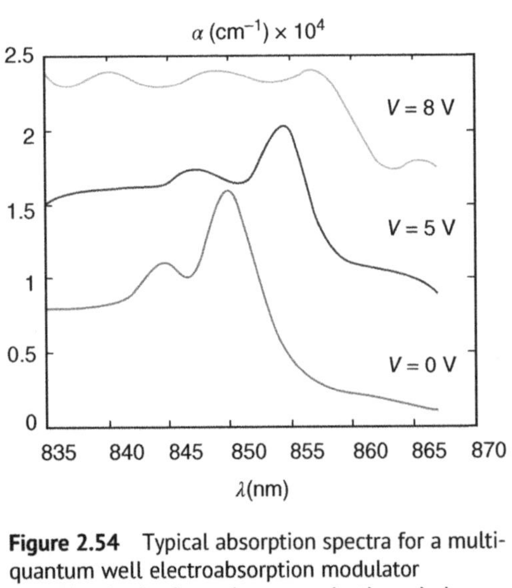

Figure 2.54 Typical absorption spectra for a multi-quantum well electroabsorption modulator operating in the first telecommunication window.

The bottom right part of Fig. 2.53 illustrates the fundamental effect behind the generation of electroabsorption in a semiconductor. When a field is orthogonally applied to the quantum well layer structure, the potential wells that represent the energy bands of the semiconductor are altered with respect to their configuration when no field is applied (left part of Fig. 2.53). The result is that the positions of E_{e1} and E_{h1} are modified by means of the so-called *Quantum-Confined Stark Effect* (QCSE). This effect can be explained by assuming that the applied field actually perturbs the electron and hole wavefunctions with respect to their shapes when no field is applied. The perturbation also decreases the energy values for these wavefunctions. Thus, when a field is applied, the required value for the generation of excitons E_a is reduced with the implications that we will now see. Figure 2.54 shows, for instance, typical absorption spectra for an MQW structure taking the applied voltage in the perpendicular direction to the quantum well layers as a parameter.

When no voltage is applied ($V = 0$), there is no absorption in some regions of the spectrum since $E \geq E_a$ and photons do not have enough energy to promote an electron from the valence to the conduction band. Two different peaks can be observed in the region of the spectrum where absorption is not zero. The first one corresponds to the exciton generation from electrons and heavy hole pairs. The second (of higher energy) corresponds to the exciton generation from electron and light hole pairs. these peaks are centered at different energies since both B and E_{h1} are different in both cases. When a control voltage is applied to the MQW structure, the absorption peaks shift toward lower energy values (i.e. higher wavelengths) since $E \geq E_a$ is reduced. For example, referring to the case shown in Fig. 2.54, for a wavelength of 865 nm, the absorption coefficient when no voltage is applied is approximately 1000 cm^{-1}, while it changes to about 10, 000 cm^{-1}, when a 5 V voltage is applied.

Electroabsorption can be employed to implement high-speed intensity external modulators. Figure 2.55 shows two possible configurations for an electroabsorption modulator (EAM) and illustrates its operation principle by means of the absorption versus frequency curve taking the applied voltage as a parameter.

The configuration shown in the top left part of Fig. 2.55 corresponds to a longitudinal modulator while that shown in the bottom left part corresponds to a transversal modulator. The longitudinal modulator configuration is very similar to that of a heterostructure laser. Light propagates and is modulated in a parallel direction to the interface planes between quantum wells. Moreover, the quantum wells, as it happens with heterostructures, configure an optical waveguide that confines the light as it propagates along the device. A stripe electrode allows the application of the control or modulation voltage. In the case of a transversal modulator, the light propagates in a perpendicular direction to the interface planes between quantum wells. As a consequence, there is no light confinement as it propagates along the device.

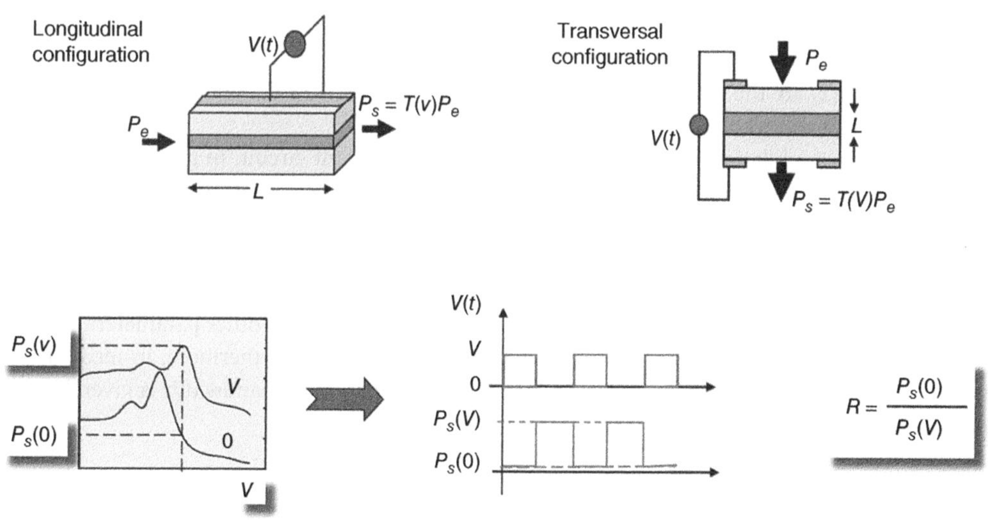

Figure 2.55 (Top) Longitudinal and transversal configurations of an electroabsorption modulator. (Bottom) Optical power versus applied voltage operation curve and modulation example.

The operation principle of the EAM is shown in the top right part of Fig. 2.55. The modulator material is chosen so that when a voltage V is applied to the device, the energy corresponding to the operation wavelength $h\omega_o$ coincides with the exciton energy. When no voltage is applied ($V = 0$) then the energy corresponding to the operation wavelength $h\omega_o$ is below that required for the formation of an exciton and therefore the material is transparent at that wavelength. If we assume that the input optical power to the EAM is given by P_i, then the output optical power from the EAM is given by:

$$P_o(V) = T(V)P_i = e^{-\alpha(V)L}P_i \tag{2.164}$$

Equation (2.164) implicitly includes a time-varying optical power since the voltage applied to the modulator will also be a time-varying function $V(t)$. In fact, a replica of the information signal needs to be modulated onto the CW input power P_i. In the case of a longitudinal modulator, (2.164) has to be slightly modified to account for a confinement factor of the optically guided field. Hence, the absorption coefficient of the semiconductor material $\alpha(V)$ is replaced by $\Gamma\alpha(v)$ where Γ represents the confinement factor of the optical waveguide. The modulator length L depends on its configuration and it is shown in Fig. 2.55 for both configurations. The output field from the EAM can be expressed as:

$$E_o(t) = \sqrt{P_o(t)}e^{j[\omega_o t + \varphi(t)]} = \sqrt{T[V(t)]P_i}e^{j[\omega_o t + \varphi(t)]} \tag{2.165}$$

In the case of the EAM, the instantaneous phase variation leads to a frequency deviation or chirp given by:

$$\Delta\nu(t) = \frac{1}{2\pi}\frac{d\varphi}{dt} = \frac{\alpha}{4\pi}\frac{d(\ln(P_s(t)))}{dt} \tag{2.166}$$

where the chirp parameter is given by:

$$\alpha = \frac{(\partial^2(Ai(u)Bi(u)))/(\partial u^2)}{((\partial^2 Ai^2(u))/(\partial u^2))}$$
$$u = \frac{(E_g - \hbar\omega)}{\hbar\theta} \qquad \theta = \left(\frac{e^2\hbar^2E^2}{2\mu}\right) \tag{2.167}$$

where $Ai(x)$ y $Bi(x)$ are the Airy functions, E represents the applied field, and E_g is the energy gap between the valence and conduction band.

Regarding the modulation bandwidth of an EAM, its equivalent circuit, in general, takes into account the effects of three different parts: an electrical source that generates the modulating signal and is characterized by a generator and a series resistance R, and LC_m network that represents component mounting and a capacity C_{MOD} that represents the semiconductor device.

The frequency response can be computed from the analysis of this circuit provided that one has precise values for the components since it is actually this circuit and not other parameters such as the carrier lifetimes which limits the EAM modulation bandwidth. Furthermore, in most of the cases, the effect of L y C_m can be neglected. So, the EAM modulation bandwidth is given by the simple expression.

$$\Delta f_{3\text{ dB}} = \frac{1}{2\pi RC_{MOD}} \tag{2.168}$$

The modulation bandwidth can be then improved by decreasing the value of C_{MOD} and one option is to reduce the device area. In this case, however, the contact resistance between the device and the source increases and its effect cannot be neglected and eventually reduces the modulation bandwidth. Another alternative is to increase the quantum well layer width but this results again in a tradeoff since more voltage is required to get the same electric field inside the modulator. Travelling wave structures where the electrodes are designed as transmission lines offer a solution again to obtain broadband devices. EAMs have been successfully demonstrated at frequencies as high as 60 GHz. Second- and third-window EAMs are usually fabricated using InP or InGaAsP III–V semiconductors and have the advantage of monolithic integration with laser sources on the same optical chip. EAMS can be operated at low voltages (2 V), and feature typical insertion losses of 10 dB, which can be compensated using semiconductor optical amplifiers (SOAs).

2.4 Fiber Bragg Gratings

An FBG is a periodic or almost a periodic perturbation of the refractive index along the fiber length which is formed by exposure of the core to an intense optical interference pattern. The variation of the refractive index along the length of a FBG can be described as:

$$n(z) = \Delta n_{ac}(z)\cos\left(\frac{2\pi}{\Lambda(z)}z + \theta(z)\right) + n_{dc}(z) \tag{2.169}$$

$\Lambda(z)$ is the grating period, $\theta(z)$ the grating phase, $\Delta n_{ac}(z)$ the amplitude of the perturbation and $n_{dc}(z)$ the average refraction index. The refractive index variations provide small Fresnel reflections, see Fig. 2.56, which create DFB. Light, guiding along the core of an optical fiber, will be scattered by

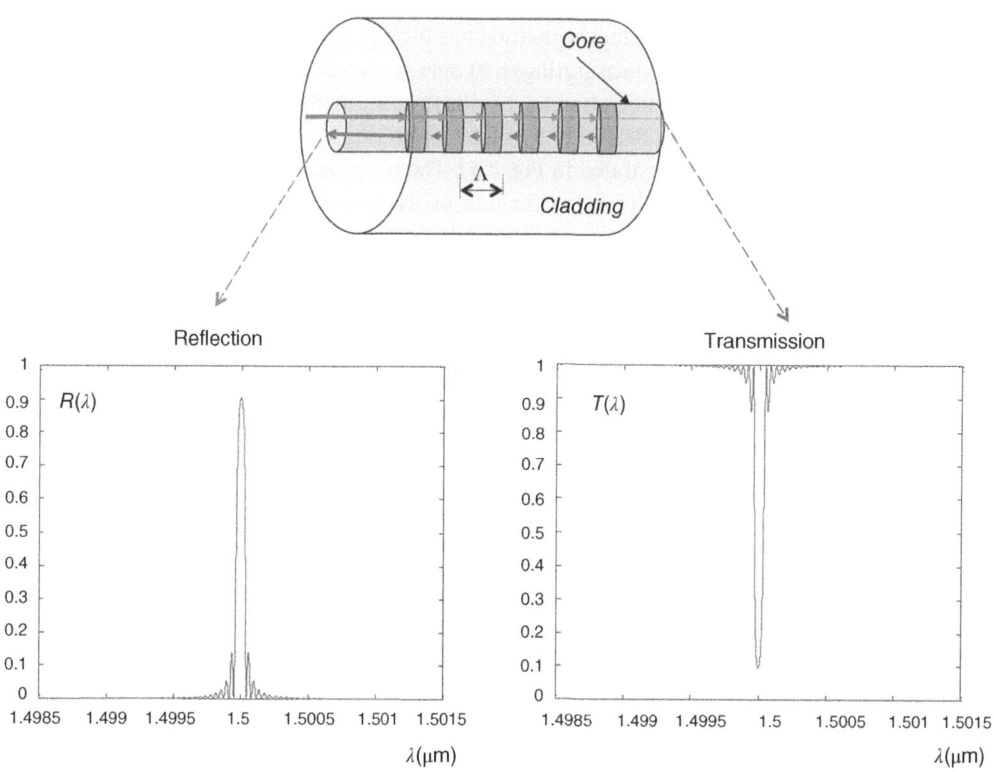

Figure 2.56 (Top) Layout of a Fiber Bragg Grating (FBG) device, (bottom) typical reflection (left), and transmission (right) intensity spectra.

each grating plane. If the Bragg condition is not satisfied, the reflected light from each of the subsequent planes becomes progressively out of phase and will eventually cancel out. If the Bragg condition is satisfied, the contributions of reflected light from each grating plane add constructively in the backward direction to form a back-reflected peak. Several theories have been developed to explain the Bragg condition. An intuitive one supposes the propagation of a forward propagating core-mode and a backward propagating core-mode. The propagation constant is $\beta = k.n_{eff} = 2\pi n_{eff}/\lambda$. Thus, the change of phase in FBG within a constant period is $\Phi = \beta\Lambda$. If the reflected contributions have to add in phase $\Phi = m2\pi$, then the strongest interaction or mode-coupling occurs at the Bragg wavelength given by:

$$\lambda_B = m2n_{eff}\Lambda \cong m2n_{dc}\Lambda \tag{2.170}$$

The FBG acts as a band-rejection filter passing all wavelengths that do not satisfy the Bragg condition. The reflection spectrum is a band-pass filter, simply the complementary of the transmission spectrum. The key characteristics of the FBGs stem from the fact that they are made from optical fibers, i.e. low insertion loss, polarization insensitive, easily controllable characteristics, stability, longevity, ease to connect, and ease to manufacture. Typical lengths are from 1 to 10 cm, some types reach 1 m.

There are several different approaches to writing gratings in fibers. The most common approaches are the two-beam interferometric method and the phase mask method. By exposing the fiber to a UV interference pattern from the side, the pattern is "printed" into the fiber. Only the core is usually doped (for example, with germanium), and, consequently, the grating is only formed in the core and not in the cladding. In the two-beam interferometric method, the laser beam is split into two components that are subsequently recombined at the fiber to produce an interference pattern. The two-beam interferometric method has the advantage that the period of the interference pattern may be tuned to produce gratings that operate over a wide range of wavelengths but requires high stability and alignment, whereas the phase-mask method is relatively insensitive to vibration and alignment but lacks flexibility to select the grating wavelength.

The phase mask technique is illustrated in Fig. 2.57. The phase mask is made from a flat slab of silica glass transparent to ultraviolet light. On one of the surfaces, a one-dimension periodic

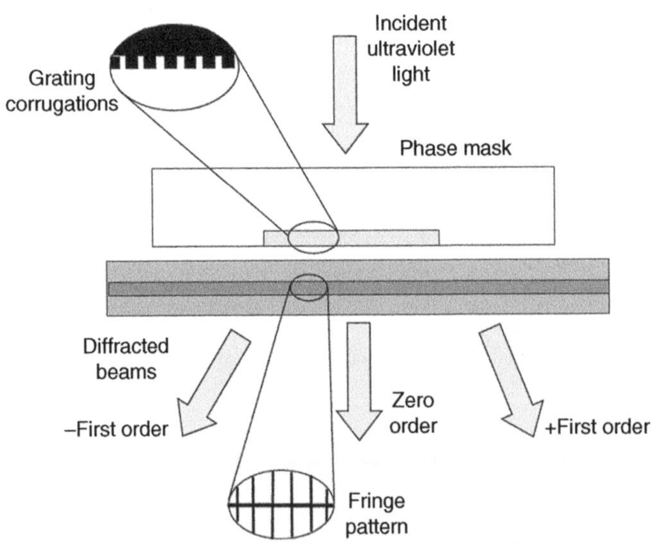

Figure 2.57 FBG fabrication by the phase mask technique.

structure is etched using photolithographic techniques. The shape of the periodic pattern approximates a square wave.

Ultraviolet light, incident normal to the phase mask, passes through and is diffracted by the periodic corrugations of the phase mask. The phase mask is designed to suppress the diffraction into the zero-order and to maximize the diffracted light contained in the ±1 diffraction orders. The two ±1 diffracted order beams interfere to produce a periodic pattern that photoimprints a corresponding grating in the optical fiber. If the period of the phase mask grating is Λ_{mask}, the period of the photoimprinted index grating is $\Lambda_{mask}/2$. In comparison with the holographic technique, the phase mask technique offers easier alignment of the fiber, reduced stability requirements, and lower coherence requirements of the ultraviolet laser beam. A drawback of the mask technique is that a separate phase mask is required for each different Bragg wavelength.

The magnitude of the refractive index change (Δn_{ac}) obtained depends on several different factors such as the irradiation conditions (wavelength, intensity, and the total dosage of irradiating light), the composition of the material forming the fiber core and any processing of the fiber prior to irradiation. Usually, Δn_{ac} is positive in germanium-doped monomode fibers with a magnitude ranging between 10^{-5} and 10^{-3}. Techniques such as "hydrogen loading" or "flame brushing" are available which can be used to process the fiber prior to irradiation in order to enhance the refractive index change. By the use of hydrogen loading, a value as high as 10^{-2} has been obtained.

2.4.1 Theory and Design of Fiber Bragg Grating Filters

Theories of FBG are well-known. The most widely employed methods for their analysis are the coupled-mode theory and the transfer matrix method. The *coupled-mode theory* is straightforward and intuitive. It accurately models most of the optical properties of the FBGs. The FBG is treated as a perturbation on the fiber along the z axis and also weak guiding is assumed. Thus, the total electric field is defined as a superposition of the forward and backward propagating waves,

$$E_z(x,y,z) = b_1(z)\Psi(x,y) + b_{-1}(z)\Psi(x,y) \tag{2.171}$$

The $b_{\pm 1}$ coefficients are dependent on frequency since they include the harmonic propagation factor $\exp(\pm j\beta z)$, being $\beta = \beta(\omega) = n_{eff}\omega/c$ the scalar propagation constant. $\Psi(x,y)$ satisfies the scalar wave of the unperturbed fiber and the total electric field E_z must satisfy the scalar wave equation for the perturbed fiber. The z-dependence of the index perturbation of an FBG is approximately quasi-sinusoidal, Eq. (2.169). Thus, solving the wave equation under the condition of (2.169) yields the couple-mode equations:

$$\begin{aligned} \frac{du}{dz} &= -i\delta u + q(z)v \\ \frac{dv}{dz} &= -i\delta v + q^*(z)u \end{aligned} \tag{2.172}$$

where u and v have been defined as:

$$\begin{aligned} b_1(z) &= u(z)\exp\left(+i\frac{\pi}{\Lambda}z\right)\exp\left(+i\int_0^z \sigma(z')dz'\right) \\ b_{-1}(z) &= v(z)\exp\left(+i\frac{\pi}{\Lambda}z\right)\exp\left(+i\int_0^z \sigma(z')dz'\right) \end{aligned} \tag{2.173}$$

$\sigma(z)$ is a real and slowly varying function that accounts for the DC index variation, $\Delta\varepsilon_{dc}(z)$. Also in (15.22), we have defined the wavenumber detuning $\delta = \beta - \pi/\Lambda$ and the coupling coefficient q of the grating:

$$q(z) = ik(z)\exp(-2i)\int_0^z \sigma(z')dz' \tag{2.174}$$

Any solution $\{u, v\}$ to the coupled-mode equations must satisfy (2.172). The reflection coefficient of an FBG located in $0 \leq z \leq L$ can be computed as $r(z_0) = v(z_0)/u(z_0)$. In two cases, it is possible to find simple closed-form solutions to the coupled-mode equations: *weak gratings* and *uniform gratings* where $q(z)$ = constant. The weak grating is studied under the *first-order Born approximation* which means that the forward propagating wave u is unaffected by the grating, that is $u = \exp(i\delta z)$. Then, we obtain:

$$r(\delta) = -\frac{1}{2}\int_0^\infty q^*\left(\frac{z}{2}\right)\exp(i\delta z)dz \tag{2.175}$$

This weak grating relation is valid when the top reflectivity of the grating is less than approximately 10–40%. Equation (15.25) shows a Fourier transform pair between $r(\delta)$ and $q^*(z)$ which is quite intuitive to be able to understand the behavior of the FBGs.

A uniform grating has a constant coupling coefficient over the grating length. In this situation, the coupled-mode equations can be solved analytically. The resulting reflection coefficient is

$$r(\delta) = \frac{-q^*\sinh(\gamma L)}{\gamma\cosh(\gamma L) - i\delta\sinh(\gamma L)} \tag{2.176}$$

where $\gamma^2 = |q|^2 - \delta^2$, and the transmission coefficient becomes:

$$t(\delta) = \frac{\gamma}{\gamma\cosh(\gamma L) - i\delta\sinh(\gamma L)} \tag{2.177}$$

In most cases, however, the coupled mode Eq. (2.172) have to be solved using numerical solution techniques. There is a variety of methods to compute the reflection and transmission spectra for nonuniform gratings, the Runge-Kutta method being the most commonly used.

The *transfer matrix method* is also extensively used to analyze the FBGs. It divides the grating into smaller sections. The sections are treated as uniform gratings, yielding the overall spectra by transfer matrix multiplication. Let the section length be $\Delta = L/N$, N the number of sections, and L the length of the FBG. By applying the appropriate boundary conditions and solving the coupled-mode equations, the following transfer matrix links the fields at the positions z and $z + \Delta$:

$$\begin{bmatrix} u(z+\Delta) \\ v(z+\Delta) \end{bmatrix} = \begin{bmatrix} \cosh(\gamma\Delta) + i\frac{\delta}{\gamma}\sinh(\gamma\Delta) & \frac{q}{\gamma}\sinh(\gamma\Delta) \\ \frac{q^*}{\gamma}\sinh(\gamma\Delta) & \cosh(\gamma\Delta) - i\frac{\delta}{\gamma}\sinh(\gamma\Delta) \end{bmatrix} \begin{bmatrix} u(z) \\ v(z) \end{bmatrix} \tag{2.178}$$

Hence, the fields at the two ends of the FBG are linked by:

$$\begin{bmatrix} u(L) \\ v(L) \end{bmatrix} = T\begin{bmatrix} u(0) \\ v(0) \end{bmatrix} \quad T = \begin{bmatrix} T_{11} & T_{12} \\ T_{21} & T_{22} \end{bmatrix} \tag{2.179}$$

where $T = T_N.T_{N-1} - T_1$ is the overall transfer matrix and T_j is the transfer matrix of each section. The reflection coefficient and the transmission coefficient are calculated as:

$$r(\delta) = -\frac{T_{21}}{T_{22}}$$
$$t(\delta) = \frac{1}{T_{22}} \tag{2.180}$$

Instead of using (2.178), we can discretize the whole grating in a stack of complex discrete reflectors. Each reflector represents a fraction of the period of the FBG and is characterized by two Transfer matrices $T^{\Delta}.T^{\rho}$, where:

$$T^{\Delta} = \begin{bmatrix} e^{j\beta z} & 0 \\ 0 & e^{-j\beta z} \end{bmatrix}$$
$$T^{\rho} = \frac{1}{1+\rho}\begin{bmatrix} 1 & \rho \\ \rho & 1 \end{bmatrix} \tag{2.181}$$

T^{Δ} is the pure propagation matrix and T^{ρ} is the discrete reflector matrix due to the refractive index changes.

The *synthesis* of FBGs is an important problem that renders from a specified FBG reflection coefficient $r(\delta)$ the required refractive index profile $q(z)$. It has been quite well solved recently. Layer-peeling, or sometimes called the differential inverse scattering method, is the most common algorithm used. The idea of layer-peeling is to calculate the initial part of the FBG from the reflection spectrum. Then, compute the response of the FBG without the initial part of the FBG (by peeling that initial part or layer), taking into account the properties of the transfer matrix and again calculate the initial part of the new FBG, following this procedure till the other end of the FBG. Two types of layer-peeling methods have been developed: the Discrete Layer-Peeling (DLP) and the Continuous Layer-Peeling (CLP). The DLP algorithm can be described as follows:

1) Start with a physically realizable reflection coefficient. It must be a casual filter.
2) Calculate the first coefficient of the impulse response of the filter. This coefficient is the reflection coefficient from the first slab of the FBG. It is straightforward to calculate the refraction index of the first slab from the reflection coefficient.
3) Propagate the fields using the transfer matrices (2.181).
4) Repeat steps 2) and 3) until the entire grating structure is determined.

The CLP method is quite similar to the DLP. The only difference is in the second step. In the DLP, a discrete Fourier transform is used to calculate the reflection coefficient. In the CLP $q(z)$, the coupling coefficient is calculated by evaluating an integral. The DLP method is faster than the CLP method.

2.4.2 Performance Characteristics of FBGs

As seen above, the common reflection response of an FBG is a narrow band-pass filter, λ_B (2.170) being the center wavelength. Any change in fiber properties, such as strain, temperature, or polarization that varies the modal index or grating pitch will change the Bragg wavelength. The tuning of the FBG with strain and temperature can be modelled

$$\Delta\lambda = \lambda_B\{(1-\rho_e)\varepsilon + (\alpha_\Lambda + \alpha_n)\Delta T\} \tag{2.182}$$

where ρ_e is the strain-optic coefficient, ε the applied strain, α_Λ the thermal expansion coefficient, α_n the thermal-optic coefficient, and ΔT the temperature difference. Typical values are: wavelength shift due to strain ranges from 1 and 1.5 pm/μstrain and wavelength shift due to temperature is between 10 and 15 pm/°C. From Eq. (2.176), it is possible to calculate for a uniform FBG the reflectivity at the Bragg wavelength:

$$R(0) = r(0)r^*(0) = \tanh^2(\kappa L) \tag{2.183}$$

Here $\kappa(z) = (\pi/\lambda)\overline{\Delta n}_{ac}(z)\,\eta$ and η is a modal overlap factor. From the former equation, it can be shown that the longer the FBG and/or the higher the value of $\overline{\Delta n}_{ac}(z)$, the higher the value of $R(0)$. This applies also to nonuniform FBGs. A general expression for the approximate full-width-half maximum bandwidth of a grating is given from the equation:

$$\Delta\lambda = \frac{s\lambda^2}{\pi n_g}\sqrt{\kappa^2 + \frac{\pi^2}{L^2}} \tag{2.184}$$

s is a constant which is almost 1 for strong gratings and around 0.5 for weak gratings. Hence, the FWHM is proportional to κ and also inversely proportional to L.

Figure 2.58 shows that the main peak in the reflection spectrum of a finite-length uniform FBG is surrounded by a series of sidelobes due to multiple reflections inside the FBG. It is important in some applications to lower and, if possible, eliminate the reflectivity of these sidelobes, or to apodize the reflection spectrum of the grating. In dense wavelength division multiplexing (DWDM), it is important to have very high rejection of the nonresonant light in order to eliminate crosstalk between information channels.

Nonuniform FBGs where $\Delta n_{ac}(z)$, $n_{dc}(z)$, and $\Lambda(z)$ vary along the propagation distance z have several distinctive spectral characteristics as compared to uniform FBGs. When $\Delta n_{ac}(z)$, $n_{dc}(z)$ and $\Lambda(z) = \Lambda$, the FBG is called apodized. Apodization of the refractive index profile renders a

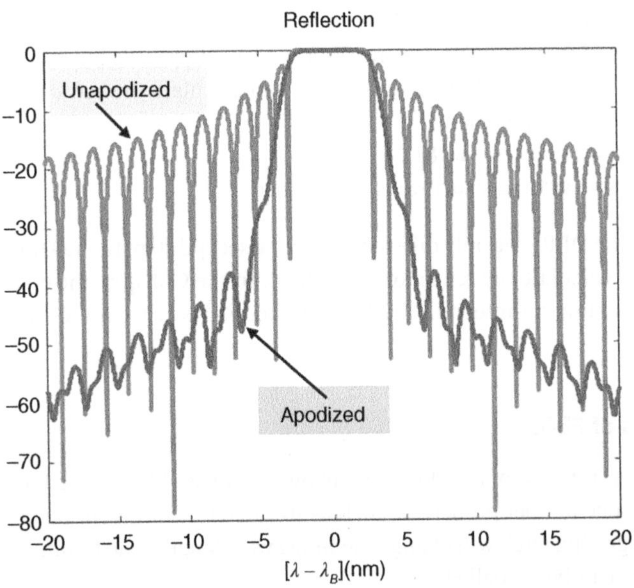

Figure 2.58 Reflection spectra of a uniform and apodized FBG device.

reflection spectrum where sidelobe levels are considerably reduced as compared to an equivalent uniform FBG. $\Delta n_{ac}(z)$ can be changed according to different windowing functions while the average $n_{dc}(z)$ must be also changed accordingly in order to have a fixed corrugation period. Apodized fiber gratings can have very sharp spectral responses, with channel spacing down to 25–100 GHz.

If, in addition, $\Lambda(z)$, then an apodized chirped Bragg grating is obtained. Chirped Bragg gratings can be realized by axially varying the period of the grating and the refractive index of the core. One particular design of interest in MWP in the so-called linearly chirped FBG (LCFBG) where $\Lambda(z) = \Lambda_o + Cz$. In these devices, the Bragg condition changes locally along the device according to:

$$\lambda_B(z) = 2n_{eff}\Lambda(z) = 2n_{eff}(\Lambda_o + Cz) = \lambda_B(0) + \Delta\lambda z \tag{2.185}$$

And, therefore, a multiple wavelength input signal is reflected at various parts of the LCFG with each color suffering a different reflection delay. This effect is employed to design wavelength tunable true time delays and also to implement concentrated dispersive elements.

2.5 Array Waveguide Gratings

The AWG, also known as the Waveguide Grating Router (WGR) or PHASAR is a device that uses an array of integrated optic waveguides to perform the same function as a free space diffraction grating. In the context of MWP, the operation of the AWG as a wavelength division multiplexer/demultiplexer is useful in a number of important applications such as optical beamforming, signal processing, and wave generation. The schematic configuration of the AWG is shown in Fig. 2.59.

The AWG consists of an input waveguide array, two free propagation regions (FPRs), also known as focusing slab waveguides or couplers interconnected by a central array of waveguides, where the length between adjacent waveguides in the array varies by a constant ΔL (that is, the array operates

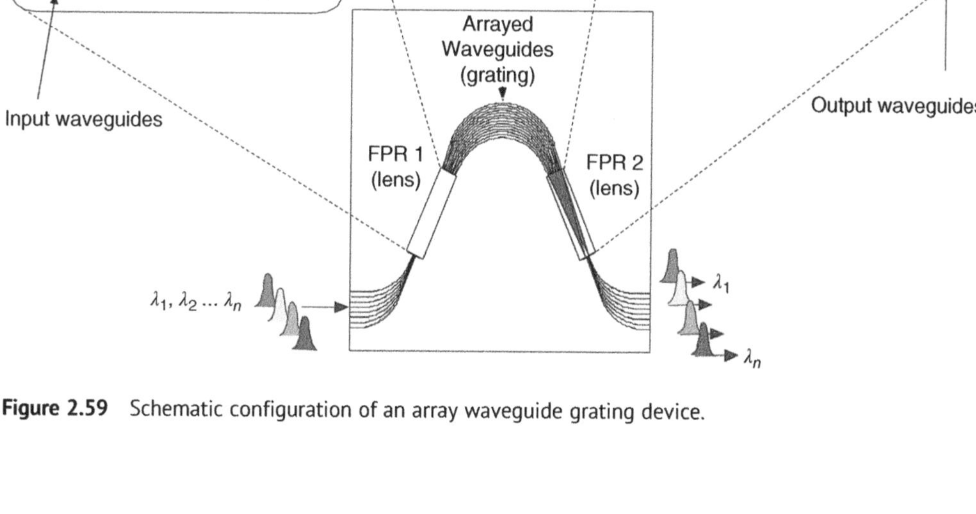

Figure 2.59 Schematic configuration of an array waveguide grating device.

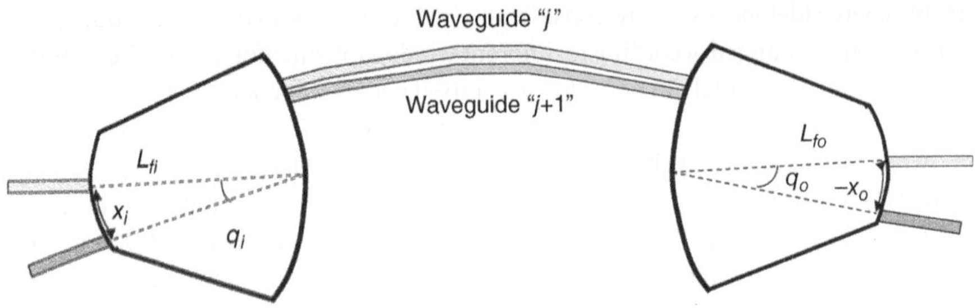

Figure 2.60 Illustration of the phase retardation between the signal travelling through adjacent waveguides in the central part of the AWG.

as the diffraction grating), and finally, an output waveguide array. The device can be operated as a $1 \times N$ device if it has a single input and multiple outputs, a $K \times 1$ device or, most generally, a $K \times N$ device having multiple inputs and outputs.

The operation of the AWG can be explained in simple terms referring to Fig. 2.60 as follows. We consider the first FPR (input lens) where the input waveguide separation is given by D_i, the arrayed waveguide separation is d_{ai} and the focal length of the lens is L_{fi}. We will consider that the waveguide parameters in the two FPRs (lenses) can be different (although, in practice, most of the times are equal). Therefore, in the second slab region, the output waveguide separation is D_o, the arrayed waveguide separation is d_{ao}, and the focal length is given by L_{fo}. The input light at position x_i (measured positive in a counter-clockwise direction) is radiated to the first slab and excites the arrayed waveguides. Within each waveguide in the array, the field amplitude profile has usually a Gaussian distribution. After travelling through the array light, beams interfere constructively at one focal point x_o (measured positive in counterclockwise direction) at the output surface of the second FPR.

The grating equation for this device can be derived as follows. Let us consider the phase retardation for the two light beams passing through the "j" and "$j+1$" arrayed waveguides as shown in Fig. 2.60. If n_c represents the refractive index in the arrayed waveguides and n_s the refractive index at the FPRs, then the grating equation is given by the condition that the difference between the total phase retardation experienced by the two light beams originated at x_i and passing through waveguide "j" and waveguide "$j+1$" must be an integer of 2π in order to achieve constructive interference at x_o.

In other words:

$$m\lambda = n_c \Delta L + n_s d_{ai} \sin\theta_i - n_s d_{ao} \sin\theta_o \tag{2.186}$$

Since $\sin\theta_i \approx x_i/L_{fi}$, $\sin\theta_o \approx x_0/L_{f_o}$, the former equation can be expressed as:

$$m\lambda = n_c \Delta L + \frac{n_s d_{ai} x_i}{L_{fi}} - \frac{n_s d_{ao} x_o}{L_{fo}} \tag{2.187}$$

Let us assume that the device is designed in such a way that for a given wavelength λ_o, the constructive interference for a signal injected at $x_i = 0$ is achieved at $x_o = 0$. In other words, $n_c \Delta L = m\lambda_o$. Then, according to (2.187), if this wavelength is input at position x_i, the output position x_o will be given by the following expression:

$$\frac{n_s d_{ai} x_i}{L_{fi}} = \frac{n_s d_{ao} x_o}{L_{fo}} \tag{2.188}$$

Usually, the waveguide parameters for the first and the second FPRs are identical as pointed out previously. In this case, (2.188) transforms into $x_o = x_i$ and in consequence, the input and output distances are the same. This principle can be generalized to show the wavelength routing operation of the device. Consider now another wavelength λ_b for which the constructive interference for a signal injected at $x_i = 0$ is achieved at $x_o = rD_o$ with r being an integer. In this case:

$$m\lambda_b = n_c\Delta L - \frac{n_s d_{ao} r D_o}{L_{fo}} \tag{2.189}$$

If the input signal is now injected at $x_i = lD_i$ with l integer substitution in (2.187) with $\lambda = \lambda_b$, and assuming equal value for the parameters in the two FPRs and the same waveguide separation at the input and output planes ($D_i = D_o = D$) yields:

$$x_o = (l + r)D \tag{2.190}$$

Equation (2.190) illustrates an important property of AWGs. If the input point for a signal of wavelength λ_b is shifted by an integer multiple of the waveguide separation D, then output focal point is shifted by the same discrete amount. Figure 2.61 shows a specific example of a 16 × 16 AWG. Here the chosen wavelength is such that if it is injected in the central port $x_i = 0$, then $x_o = 2D$ (i.e. $r = 2$). If the same wavelength is now injected through the port located at $x_i = 4D$ ($l = 4$), then according to (2.190), it will be focused at port $x_o = 6D$.

Another parameter of importance, as in the case of free space gratings, is the dispersion of the focal position x_o with respect to the wavelength λ for a fixed input position x_i. This can be computed by direct differentiation of Eq. (2.187), yielding:

$$D_{x_o} = \frac{dx_o}{d\lambda} = \frac{N_c L_f \Delta L}{n_s d\lambda_o} = m\left(\frac{N_c L_f}{n_s n_c d}\right) \tag{2.191}$$

In the above equation, N_c represents the group index corresponding to n_c, that is $N_c = n_c - \lambda dn_c/d\lambda$. Note that the dispersion increases linearly with the value of m. A correct device design for its application as a WDM Mux/Demux requires that for a wavelength separation $\Delta\lambda$ corresponding to the

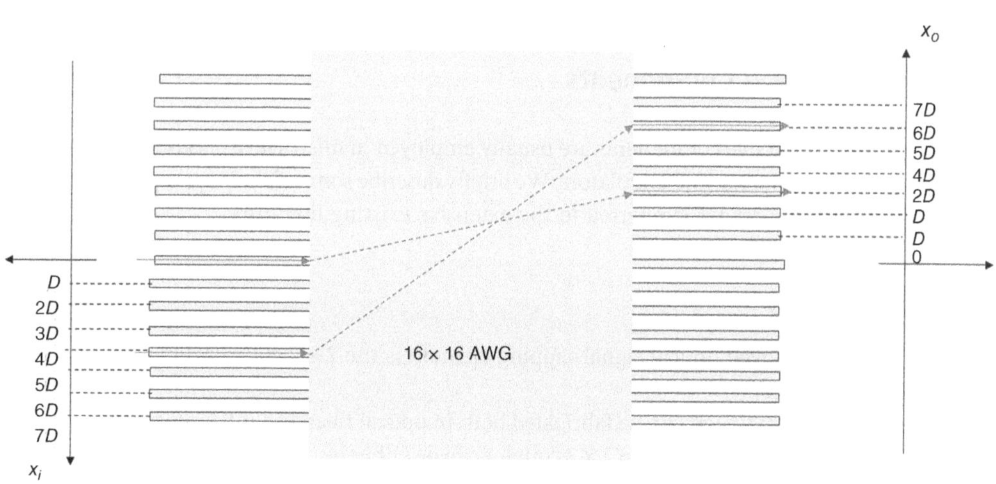

Figure 2.61 Signal routing example in an AWG for a fixed wavelength.

fixed ITU channel separation, the spatial dispersion in the focusing plane corresponds to the separation D between adjacent output waveguides. This implies the fulfilment of the following condition:

$$\Delta x_0 = D = D_{x_o}\Delta\lambda = \frac{N_c L_f \Delta L \Delta\lambda}{n_s d\lambda_o} \Rightarrow \Delta\lambda = \frac{n_s D\lambda_o}{N_c L_f \Delta L} \tag{2.192}$$

Because the input signal expanded by the input FPR is spatially sampled by the grating array, the diffraction pattern of a given wavelength follows a periodic angular pattern in the output plane of the second FPR, each period corresponding to a given order m of diffraction. Therefore, there is a limited spatial region called the *spatial free spectral range* FSR_x within the output plane where wavelengths can be univocally resolved which is given by the spatial separation between the mth and the $m+1$th focused beams for the same wavelength. From (2.187), it follows immediately that:

$$FSR_x = \frac{\lambda_o L_f}{n_s d} = \frac{n_c \Delta L L_f}{n_s d m} \tag{2.193}$$

It is interesting to note that like in the case of the free space diffraction grating, the value of FSR_x is inversely proportional to m. Thus, increasing the number of the diffraction order used to implement the wavelength separation will decrease the value of the wavelength region that can be actually demultiplexed by the device.

Finally, the frequency response for a given input-output configuration is periodic. The value of the period or frequency Free Spectral Range, FSR_f is given by:

$$FSR_f = \frac{c}{n_c \Delta L} \tag{2.194}$$

The model described so far is based on the propagation of plane waves and provides useful insight into many of the important properties of AWG. However, this model is only approximate and does not take into account the guided wave nature of the propagation inside the waveguides of the array, the power coupling from the field at the FPR, and the waveguides in the array and, finally, the coupling between the far field at the second FPR and the output waveguides. Such a description requires an electromagnetic treatment.

2.6 Other Passive Components

A number of optical passive components are usually employed in microwave photonic applications including couplers, isolators, and circulators. We briefly describe some of their operation principles. For further details, the reader is referred to the extensive existing literature.

2.6.1 2 × 2 Couplers

The most widely employed optical signal-tapping element is the 2 × 2 coupler. Figure 2.62 shows the layout of the device.

It is a four-port device which can be fabricated both in optical fiber and integrated optic technology. The physical principle behind the 2 × 2 coupler can be either by evanescent field coupling or by modal interference. In the first case, the evanescent part of the fundamental mode in one of the waveguides excites the fundamental mode of the other and vice versa while, in the second case,

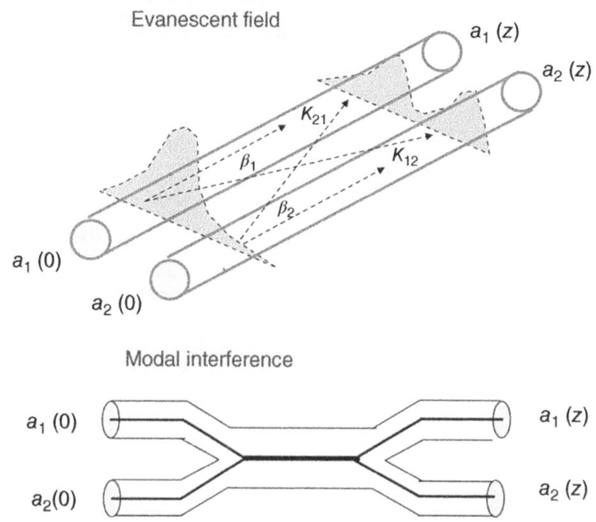

Figure 2.62 Diagram of an evanescent (top) and a modal interference (bottom) 2 × 2 waveguide coupler.

a bimodal zone is created in the coupling region and signal coupling is obtained by energy transfer between the two possible modes as they propagate along the coupling region.

The operation of the 2 × 2 coupler can be analyzed using the coupled mode theory. In this case, two co-directional modes are involved in the equations rather than two contradirectional modes as it is the case of FBGs. The overall performance for a z-length device is given by its transfer matrix:

$$\begin{pmatrix} a_1(z) \\ a_2(z) \end{pmatrix} = \sqrt{1-\gamma}\begin{pmatrix} \sqrt{1-k(z)} & j\sqrt{k(z)} \\ j\sqrt{k(z)} & \sqrt{1-k(z)} \end{pmatrix} \cdot \begin{pmatrix} a_1(0) \\ a_2(0) \end{pmatrix} \tag{2.195}$$

Here $a_1(z)$ and $a_2(z)$ represent the output fields from the device in waveguides 1 and 2, respectively, while $a_1(0)$ and $a_2(0)$ are the input fields to the device in waveguides 1 and 2, respectively. γ accounts for the device's excess losses and $k(z) = \sin^2(cz)$ is the power coupling constant that describes the power coupled from one waveguide to the other. k depends through the coefficient c on the wavelength and the waveguide geometry. In general, the coupler can be designed so the coupling constant can be wavelength insensitive within a given transmission window. Commercial devices have excess loss values <0.1 dB and feature k values between 0 and 1. A special case is the so-called 3 dB coupler where $k(z) = 1/2$.

2.6.2 Isolators

Figure 2.63 shows the configuration of an optical isolator. Its function is to allow the lightwave propagation in a given direction (for example, from left to right in the figure) with minimum insertion losses while avoiding the lightwave propagation in the opposite direction (from right to left in the figure).

The isolator is a two-port component that can be described by a 2 × 2 scattering matrix:

$$S_a = \begin{pmatrix} S_{11} & S_{12} \\ S_{21} & S_{22} \end{pmatrix} \tag{2.196}$$

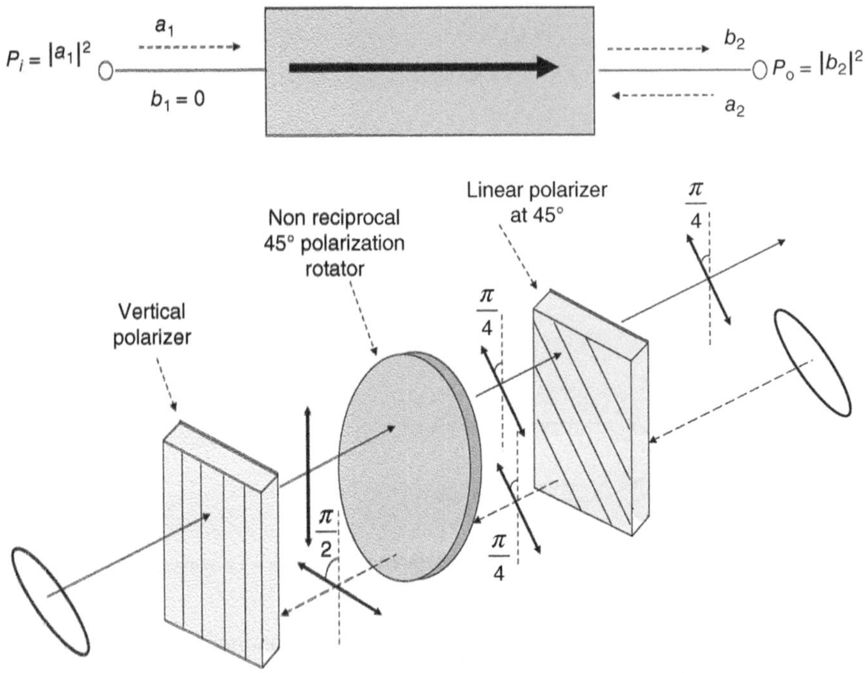

Figure 2.63 (Top) Scheme of an optical isolator and (bottom) layout of the nonreciprocal polarization rotator required for its implementation.

The performance of this device is usually specified by the manufacturer through a couple of parameters expressed in dB. These are the insertion losses L and the isolation ratio I. These relate to the optical power at the device input and output which depend, in turn, on the input and output fields and therefore on the scattering matrix parameters. The dependence is as follows:

$$\begin{aligned} L &= 10\log\left(\frac{P_i}{P_o}\right) = 10\log\left(\frac{1-|S_{11}|^2}{|S_{21}|^2}\right) \\ I &= 10\log\left(\frac{|a_2|^2}{|b_1|^2}\right) = 10\log\left(\frac{1}{|S_{12}|^2}\right) \end{aligned} \tag{2.197}$$

Here P_i and P_o represent the input and output optical powers in the privileged direction of propagation while a_i, b_i represent the input and output fields in port i. Hence, $|a_2|^2$, $|b_1|^2$ represent the input and output optical powers in the forbidden direction of propagation. An ideal isolator is defined by $L = 0$ dB and $I = \infty$. isolators are mainly applied to prevent undesired optical feedback from the optical components that compose the MWP system to the optical laser. The nonreciprocal propagation nature of the isolator requires the use of polarization-sensitive components and Faraday effect-based polarization rotators in their construction. Typical devices for the third telecommunication window feature insertion losses below 0.4 dB and isolation ratios above 60 dB.

2.6.3 Circulators

A circulator is an N-port device. Figure 2.64 shows a schematic of a three-port device that includes the allowed signal flow.

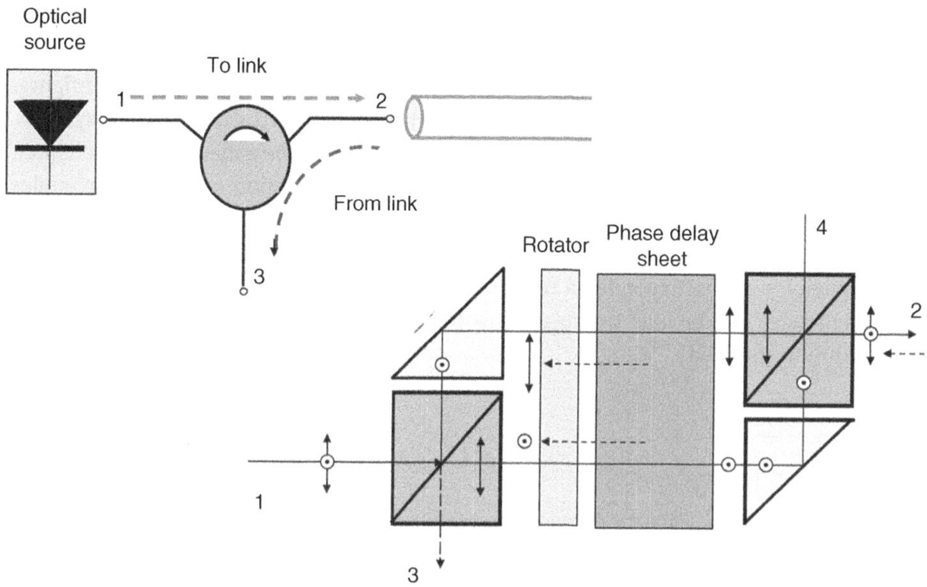

Figure 2.64 (Top) operation diagram of an optical isolator and (bottom) internal configuration using polarization-based devices.

For an ideal circulator, the input signal to port 1 must exit the device only from port 2. The input signal to port 2 must exit the device only from port 3 and the input signal to port 3 must exit the device only from port 1. Therefore, since there are no signal reflections in an ideal circulator, the scattering matrix can be expressed by:

$$[S]_{ideal} = \begin{pmatrix} 0 & 0 & 1 \\ 1 & 0 & 0 \\ 0 & 1 & 0 \end{pmatrix} \tag{2.198}$$

In practice, however, there are signal reflections and therefore the scattering matrix given by (2.198) must be modified. If t_i, $i = 1, 2, 3$ and r_i, $i = 1, 2, 3$ represent the field transmission and reflection coefficients for each port, then:

$$[S]_{real} = \begin{pmatrix} 0 & r_2 & t_3 \\ t_1 & 0 & r_3 \\ r_1 & t_2 & 0 \end{pmatrix} \tag{2.199}$$

In practice, $|t_i|^2$ must be as close to unity as possible and $|r_i|^2$ as close to zero as possible. For each port, the device performance is measured as in the case of isolators by means of its individual insertion losses and isolation ratios as given by (2.197). As in the case of isolators, their implementation requires the use of polarization-sensitive components and Faraday effect-based polarization rotators. Commercial devices have insertion losses of around 1 dB per port and isolation ratios >45 dB per port. Optical circulators have several applications in MWP. Perhaps one of the most important is for the extraction of reflected signals in LCFBG devices used as tunable delay lines.

2.7 Chapter Summary

In this chapter, we have provided a summary and description of the basic principles and configurations of those photonic components that are most widely employed in microwave photonic systems. This treatment is necessarily limited and no attempt has been made to be exhaustive. The chapter, however, provides the information necessary to understand the operation of the most complex MWP subsystems and systems that will be described in the book. Of course, more detailed and comprehensive coverage of these and other photonic components can be found in the literature and for the sake of simplicity, we have appended a reference section that is composed exclusively of textbooks where the reader will be able to acquire this information without having to surf the very extensive literature in the field.

References

1 Yariv, A. and Yeh, P. (2006). *Photonics: Optical Electronics in Modern Communications*, Oxford Series in Electrical and Computer Engineering. Oxford University Press.

2 Saleh, B.E.A. and Tech, M.C. (2007). *Fundamentals of Photonics*, 2e. Wiley.

3 Agrawal, G.P. (2004). *Lightwave Technology: Components and Devices*. Wiley.

4 Agrawal, G.P. (1995). *Nonlinear Fiber Optics*, 2e. Academic Press.

5 Agrawal, G.P. and Dutta, N.K. (1993). *Semiconductor Lasers*. Academic Press.

6 Coldren, L.A., Corzine, S., and Mashanovitch, M.L. (2012). *Diode Lasers and Photonic Integrated Circuits*. Wiley.

7 Petermann, K. (1991). *Laser Diode Modulation and Noise*. Kluwer Academic Press.

8 Kashyap, R. (2009). *Fiber Bragg Gratings*, 2e. Academic Press.

9 Othonos, A. and Kalli, K. (1999). *Fiber Bragg Gratings: Fundamentals and Applications in Telecommunications and Sensing*. Artech House.

10 Okamoto, K. (2005). *Fundamentals of Optical Waveguides*, 2e. Academic Press.

11 Dutta, A.K. and Dutta, N.K. (2003). *WDM Technologies: Passive Optical Components*. Academic Press.

12 Kashima, N. (1995). *Passive Optical Components for Optical Fiber Transmission*. Artech House.

13 Desurvire, E. (1994). *Erbium-doped Fiber Amplifiers: Principles and Applications*. Wiley.

14 Desurvire, E. and Bayart, D. (2002). *Erbium-doped Fiber Amplifiers: Device and System Developments*. Wiley.

15 Dutta, N.K. and Wang, Q. (2013). *Semiconductor Optical Amplifiers*, 2e. World Scientific.

3

Photonic Generation of Microwave Signals

3.1 Introduction

A low-phase noise and frequency-tunable microwave source is desirable for many applications such as radar, wireless communications, software-defined radio, and modern instrumentation. Conventionally, a microwave signal is generated using an electronic oscillator at a relatively low frequency, with the frequency increased by many stages of frequency multiplication to achieve the desired high frequency. The system is complicated and costly. In addition, many stages of multiplication will increase the phase noise of the signal. Furthermore, for many applications, the generated microwave signal should be distributed to a remote site. The distribution of a microwave signal in the electrical domain is not practical due to the high loss associated with electrical distribution lines, such as a coaxial cable. Due to the extremely broad bandwidth and low loss of state-of-the-art optical fibers, the distribution of a microwave signal over an optical fiber is an ideal solution to fulfill this task. Therefore, the ability to generate a microwave signal in the optical domain would allow the distribution of the signal via optical fiber from a central office to a remote site without the need for an additional electrical to optical conversion, greatly simplifying the equipment requirement.

Usually, a microwave signal can be generated in the optical domain based on optical heterodyning in which two optical waves with different wavelengths beat at a photodetector, as shown in Fig. 3.1. An electrical beat signal is then generated at the output of the photodetector with a frequency corresponding to the wavelength spacing of the two optical waves [1]. Assume that we have two optical waves given by

$$E_1(t) = E_{01} \cos(\omega_1 t + \phi_1) \tag{3.1}$$

and

$$E_2(t) = E_{02} \cos(\omega_2 t + \phi_2) \tag{3.2}$$

where E_{01}, E_{02} are the amplitude terms and ϕ_1, ϕ_2 are the phase terms of the two optical waves.

Considering the limited bandwidth of the photodetector, the current at the output of the photodetector is given by

$$i_{RF} = A \cos[(\omega_1 - \omega_2)t + (\phi_1 - \phi_2)] \tag{3.3}$$

where A is a constant which is determined by E_{01}, E_{02}, and the responsivity R of the photodetector.

As can be seen from (3.3), an electrical signal with a frequency equal to the frequency difference between the two optical waves is generated. This technique is capable of generating an electrical

Microwave Photonics, First Edition. Jianping Yao, José Capmany, and Ming Li.

Companion website: www.wiley.com/go/yao/microwavephotonics

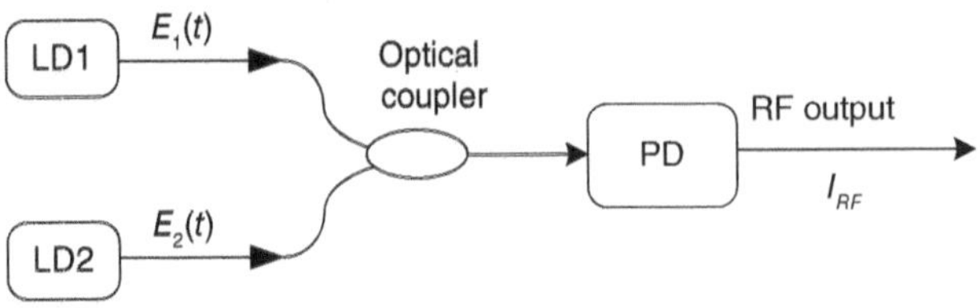

Figure 3.1 Optical heterodyning of two optical waves at a photodetector to generate a microwave signal. LD, laser diode; PD, photodetector.

signal with a frequency up to the THz band, limited only by the bandwidth of the photodetector. However, beating two optical waves from two free-running laser sources would generate a microwave signal with a high phase noise since the phase terms of the two optical waves are not correlated, and the beating process will make the phase noise terms from the two optical waves transferred directly to the generated microwave signal. Numerous techniques have been proposed and demonstrated in the past few years to generate low-phase-noise microwave signals with the phase terms of the two optical waves being locked. These techniques can be generally classified into five categories: (i) Optical injection locking, (ii) Optical phase-locked loop (OPLL), (iii) Microwave generation using external modulation, (iv) Dual-wavelength laser source, and (v) Optoelectronic oscillator (OEO).

3.2 Optical Injection Locking

To generate a high-quality microwave signal, the phase terms of the two optical waves used for heterodyning must be highly correlated. The phase correlation of the two laser sources can be realized by using optical injection locking [2]. Figure 3.2 shows an optical injection-locking system that consists of one master laser and two slave lasers. As can be seen, a low-frequency microwave reference signal is applied to the master laser, which is usually a laser diode (LD). Due to frequency modulation (FM) at the master laser, an optical carrier and different orders of optical sidebands are generated at the output of the master laser. The optical signal at the output of the master laser is then injected via an isolator into the two slave lasers. The two slave lasers are selected such that their free-running wavelengths are close to two sidebands, say the +2nd and −2nd order sidebands in Fig. 3.2. Therefore, the wavelengths of the two slave lasers are locked to the +2nd-order and −2nd-order sidebands and optical injection locking is thus achieved [2]. Since the two wavelengths from the two slave lasers are phase correlated, the beating between the two wavelengths at a photodetector would generate a beat signal with low phase noise. In addition, depending on the design, the frequency of the beat signal is equal to an integer multiple of the frequency of the microwave reference signal applied to the master laser.

In a simplified version [3], the two slave lasers could be replaced by a single multi-longitudinal-mode slave laser, which could be a Fabry-Perot (FP) laser. Again, the master laser is frequency-modulated by a microwave reference signal at a lower frequency. Two longitudinal modes of the slave laser are thus locked, say, by the +3rd-order and −3rd-order sidebands of the frequency-modulated signal. A beat signal at a frequency that is six times the frequency of the microwave reference signal is generated. In [3], the frequency of the microwave reference signal is 5.846 GHz, and the frequency of the generated microwave signal is about 35 GHz.

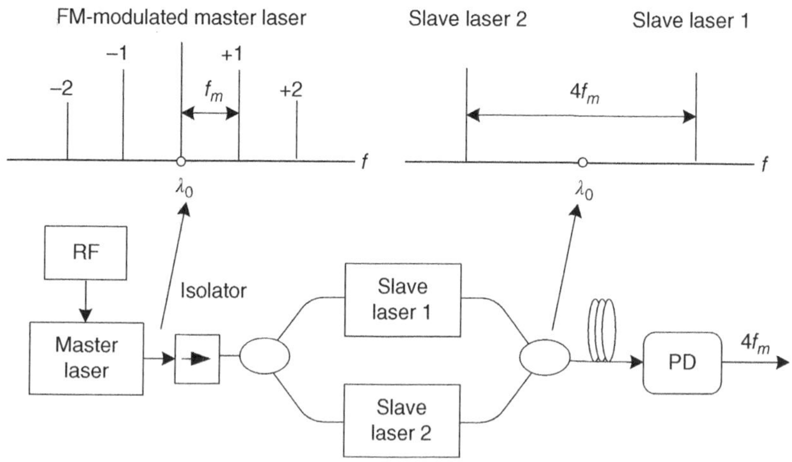

Figure 3.2 Optical injection locking of two slave lasers. The master laser is directly modulated by a low-frequency microwave reference with its output injected into the two slave lasers. The slave lasers are wavelength-locked by the +2nd-order and −2nd-order sidebands from the master laser.

3.3 Optical Phase-Locked Loop

Another approach to achieving phase correlation between two optical waves is to use an optical phase-locked loop (OPLL), in which the phase of one laser source is actively locked to that of a second laser source through the OPLL, as shown in Fig. 3.3. This technique has been explored extensively in the past few years [4–10]. To achieve effective phase locking, the two laser sources should be selected to have narrow linewidths and therefore have phase fluctuations only at low frequencies, which would ease significantly the requirement for a very short feedback loop.

As shown in Fig. 3.3, a beat signal is generated at the output of the photodetector. The phase of the beat signal is compared with that of a microwave reference from a microwave generator at an electronic mixer followed by a low-pass loop filter. The module in the dotted box is an electrical phase detector, with the output current or voltage being proportional to the phase difference between the beat signal and the microwave reference signal, which is an error current or voltage that is fed back to control the phase of one of the laser sources by changing the laser cavity length or

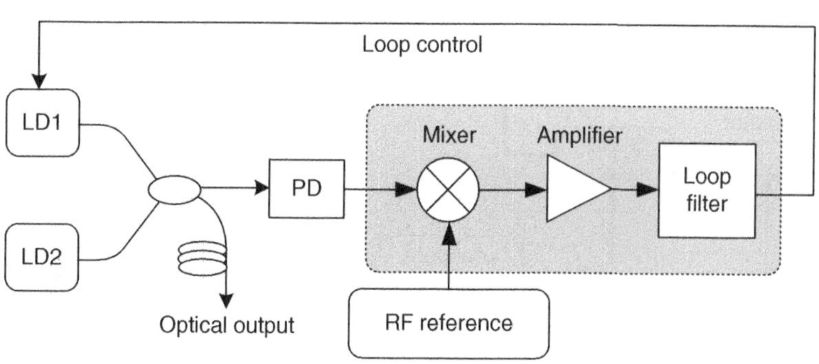

Figure 3.3 Schematic of an optical phase-locked loop. LD, laser diode; PD, photodetector.

the injection current. With a proper feedback loop gain and response time, the relative phase fluctuations between the two laser sources are significantly reduced, and the phase of the beat signal is locked to that of the reference signal.

It was reported that an OPLL using two Nd: YAG lasers could generate a microwave signal continuously tunable from 6 to 34 GHz with a linewidth less than 1 mHz [8].

To increase the frequency acquisition capability, a modified OPLL incorporating a frequency discriminator was reported [9]. It was demonstrated with the incorporation of the frequency discriminator, a pull-in range as large as 300 MHz was realized.

To reduce the feedback frequency, an OPLL incorporating a frequency down-conversion module was proposed and demonstrated [10]. The use of the frequency down-conversion module allows the use of lower-frequency components in the phase control module, which would reduce significantly the system cost. In addition, in a discriminator-aided OPLL, the use of the frequency down-conversion module would also allow the use of lower-frequency components in the frequency control module to reduce the system cost [10].

To use a low-frequency reference to generate a high-frequency microwave signal, one may use an electronic harmonic generator to generate higher-order harmonics, and the beat microwave signal is mixed with one of the high-order harmonics. Once locked, the beat signal would be at a frequency that is a high-order harmonic of the reference signal. Of course, the phase noise of the high-order harmonic is poorer and the generated beat signal also has a poorer phase noise performance.

3.4 Optical Injection Phase Locking

To further improve the signal quality, it was proposed in [7] that the two techniques of optical injection locking and optical phase locking can be combined in a single optical phase-locking system. The diagram of an optical injection-locking and phase-locking system is shown in Fig. 3.4. The system consists of a master laser source, a slave laser source, an external modulator which is a Mach–Zehnder modulator, a photodetector, and a phase detection module (dotted box).

As can be seen from Fig. 3.4, a light wave from the master laser is divided into two channels, with one channel coupled into the Mach–Zehnder modulator before being injected into the slave laser source. The slave laser source is locked to one sideband of the modulated signal for optical injection locking. The other channel of the light from the master laser source is combined with the output of the slave laser source and beat at the photodetector. The beat signal is mixed with a microwave

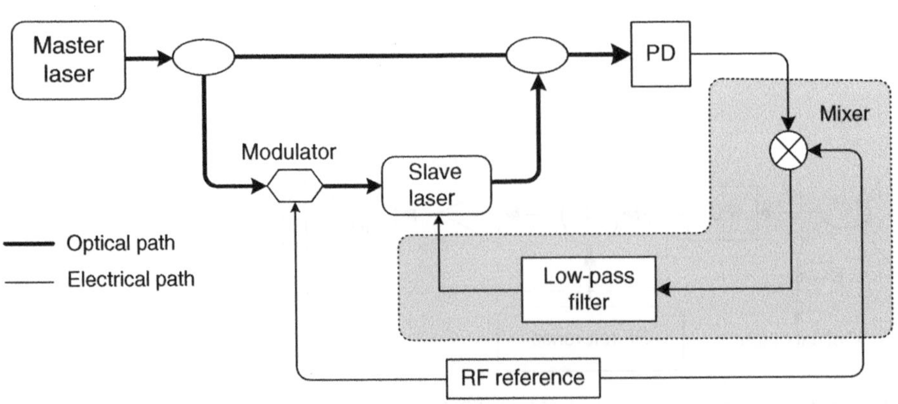

Figure 3.4 Schematic of an optical injection-locking and phase-locking system.

reference signal, with the mixing output filtered by a low-pass filter to achieve phase-locking of the wavelength from the master laser source. It was demonstrated experimentally that the proposed system offers lower phase noise compared with the techniques using an injection-locking-only or OPLL-only system.

3.5 Microwave Generation Based on External Modulation

In addition to the techniques using optical injection locking and OPLL, high-quality microwave signals can also be generated based on external modulation [11–16]. A method to generate a microwave signal using an external optical modulation technique was first proposed in 1992 [11]. A frequency-doubled electrical signal was optically generated by biasing a Mach–Zehnder modulator to suppress the even-order optical sidebands. A 36-GHz millimeter-wave signal was generated when the Mach–Zehnder modulator was driven by an 18-GHz microwave signal. Such a system was employed for the remote delivery of video services [12]. In 1994, another method was proposed to generate a frequency-quadrupled electrical signal. Instead of biasing the Mach–Zehnder modulator to suppress the even-order optical sidebands, the method [13] was based on the quadratic response of a Mach–Zehnder modulator. The optical carrier and the 1st- and 3rd-order optical sidebands were suppressed by adjusting the power level of the microwave drive signal. A 60-GHz millimeter-wave signal was generated when a 15-GHz drive signal was applied to the Mach–Zehnder modulator. However, to ensure a clean spectrum at the output of a photodetector, an imbalanced Mach–Zehnder filter with a free spectral range (FSR) equal to the spacing of the two second-order optical sidebands is used to select the two 2nd-order sidebands and to suppress the other unwanted optical components.

In addition to the use of a Mach–Zehnder modulator, an optical phase modulator can also be used to generate a microwave signal [14]. The advantage of using a phase modulator is that a phase modulator is not biased, which would eliminate the bias drift problem existing in a Mach–Zehnder modulator-based system. In [14], a frequency-quadrupled microwave signal was generated using a phase modulator. In the system, an FP filter was used to select the two second-order optical sidebands. A microwave signal that has four times the frequency of the microwave drive signal was generated by beating the two second-order sidebands at a photodetector.

A key advantage of the approaches reported in [13, 14] is that an optical modulator with a maximum operating frequency of f_m can generate a microwave signal with a frequency of $4f_m$. However, since both approaches rely on the use of an optical filter to select the two optical sidebands, to generate tunable microwave signals, a tunable optical filter must be used, which would increase significantly the complexity and cost of the system.

For system applications with frequency reconfigurability, such as wideband surveillance radar and spread-spectrum or software-defined radio, a continuously tunable microwave signal is highly desirable. In [15, 16], two approaches to generating frequency-tunable microwave signals using a wavelength-fixed optical filter were demonstrated.

3.5.1 Intensity Modulator-Based Approach

Figure 3.5 shows a system to generate a continuously tunable microwave signal based on external modulation using a Mach–Zehnder modulator and a wavelength-fixed optical filter [15]. The significance of the technique is that no tunable optical filter is required, which simplifies the system implementation significantly.

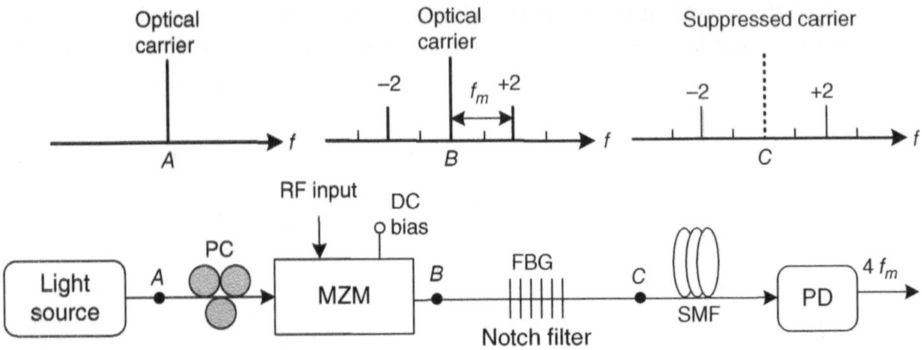

Figure 3.5 Microwave signal generation based on external modulation using a Mach–Zehnder modulator and a wavelength-fixed optical filter (which can be an FBG). FBG: fiber Bragg grating; MZM, Mach–Zehnder modulator; PD, photodetector.

As can be seen from Fig. 3.5, the system consists of a Mach–Zehnder modulator that is biased at the maximum transmission point of the transfer function to suppress the odd-order optical sidebands. A fiber Bragg grating (FBG) serving as a wavelength-fixed notch filter is then used to filter out the optical carrier. A stable, low-phase noise microwave signal that has four times the frequency of the microwave drive signal is generated at the output of the photodetector. In the experimental demonstration, a 32–50 GHz microwave signal was observed on an electrical spectrum analyzer when the microwave drive signal was tuned from 8 to 12.5 GHz. The quality of the generated microwave signal was maintained after transmission over a 25-km standard single-mode fiber (SMF).

3.5.2 Phase-Modulator-Based Approach

The approach using a Mach–Zehnder modulator to microwave generation, as discussed above, can produce a high-quality frequency-tunable microwave signal with a simple system structure. To suppress the odd- or even-order optical sidebands, however, the Mach–Zehnder modulator should be biased at the minimum or maximum transmission point of the transfer function, which would cause the bias-drifting problem, leading to poor system stability or a sophisticated bias-control circuit must be employed to minimize the bias drift.

A simple solution to this problem is to replace the Mach–Zehnder modulator with an optical phase modulator. The key advantage of using an optical phase modulator is that no DC bias is required, which eliminates the bias drifting problem [16]. An experimental setup for the generation of a microwave signal using an optical phase modulator is illustrated in Fig. 3.6.

A light wave from a tunable laser source is sent to the optical phase modulator through a polarization controller (PC). It is different from the use of a Mach–Zehnder modulator, which can be biased to suppress the odd- or even-order sidebands, the use of an optical phase modulator will generate all sidebands including the optical carrier. Therefore, a narrow-band optical notch filter is used to filter out the optical carrier. In Fig. 3.6, the optical notch filter is an FBG. In the experiment, the wavelength of the optical carrier was tuned to match the maximum attenuation wavelength of the FBG. The remaining sidebands at the output of the FBG were amplified by an erbium-doped fiber amplifier (EDFA), and then transmitted over a SMF. The beat of these optical sidebands at a photodetector generates the required microwave signal. It should be noted that the optical sidebands transmitted over the SMF will suffer from the chromatic dispersion of the

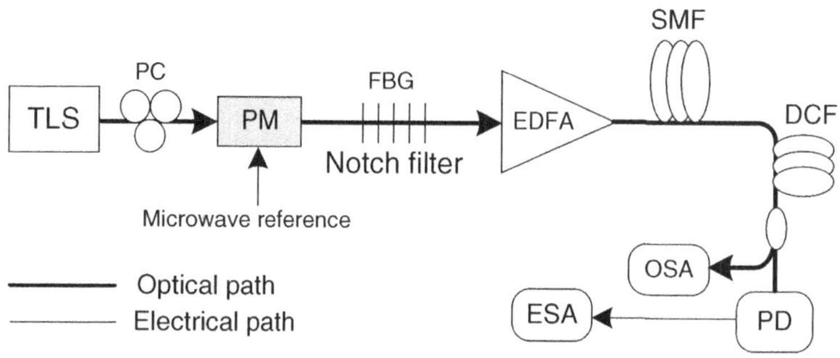

Figure 3.6 Microwave signal generation using a phase modulator. DCF, Dispersion compensating fiber; ESA, electrical spectrum analyzer; FBG, fiber Bragg grating; OSA, optical spectrum analyzer; PC, Polarization controller; PD, Photodetector; PM, Phase modulator; TLS, Tunable laser source.

fiber, which will alter the phase relationship of the sidebands. To maintain the same phase relationship, dispersion-compensation is required to eliminate the power fluctuation of the generated signal and to maintain the suppression of the odd-order electrical harmonics when the optical signal is distributed over the SMF. It was experimentally demonstrated that when the RF drive signal was tuned from 18.8 to 25 GHz, two bands of mm-wave signals from 37.6 to 50 GHz and from 75.2 to 100 GHz with high spectral purity were generated locally and remotely.

3.6 Microwave Generation Using a Dual-Wavelength Laser

Microwave signals can also be generated using a dual-wavelength laser source with the two wavelengths separated at a desired frequency [17, 18]. It is different from the techniques of optical injection locking and the OPLL, the two wavelengths from a dual-wavelength laser source are not locked in phase. However, due to the fact that the two wavelengths are generated from the same cavity, the phase correlation between the two wavelengths is better than the use of two physically separated, free-running laser sources. The advantage of using a dual-wavelength laser source to generate a microwave signal is that the system does not need a microwave reference source, which can reduce the system cost significantly.

Figure 3.7(a) shows a dual-wavelength fiber ring laser. To ensure that the two wavelengths are in single-longitudinal mode, a dual-band filter with ultra-narrow passbands must be used to limit the number of longitudinal modes to one in each passband. In the experimental demonstration, the ultra-narrow dual-band filter was a dual-wavelength ultra-narrow transmission band FBG with two ultra-narrow transmission bands, which was designed and fabricated based on the equivalent phase shift (EPS) technique [18]. As can be seen from Fig. 3.7(b), the two ultra-narrow transmission bands of FBG1 are selected by the two reflection bands of FBG2, with the entire spectral response of the two cascaded FBGs shown in the bottom figure of Fig. 3.7(b). Instead of using an EDFA in the ring cavity, a semiconductor optical amplifier (SOA) was employed to provide the gain. An SOA has less homogeneous line broadening at room temperature than an EDFA, the use of which would reduce significantly the mode competition of the two lasing wavelengths.

The key device was the ultra-narrow band FBG, which was designed and fabricated based on the EPS technique [18]. It is different from the technique with a true phase shift. An EPS is introduced by changing the sampling period of a sampled FBG. In the fabrication of an EPS FBG, the fiber and

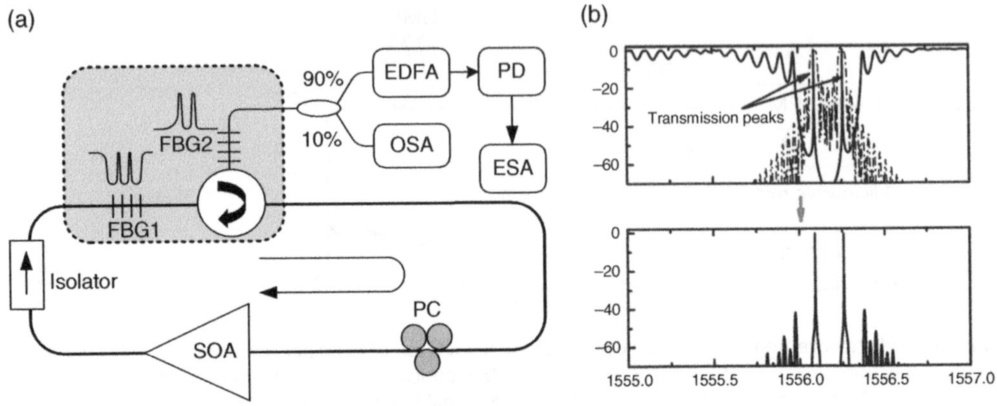

Figure 3.7 A dual-wavelength single-longitudinal-mode fiber ring laser for microwave generation. (a) Schematic of the laser, and (b) the spectral responses of the two cascaded FBGs.

the phase mask are both fixed, and there are less phase fluctuations compared to the true phase shift method in which the fiber or the phase mask must be laterally shiftable. In addition, the EPS can be controlled more precisely because it only requires micrometer precision instead of nanometer precision for a true phase shift during the FBG fabrication. Thus, an FBG with a more precise phase shift leading to a narrower transmission bandwidth is possible.

Since the two lasing wavelengths share the same gain cavity, the relative phase fluctuations between the two wavelengths are low. In a demonstration, three dual-wavelength ultranarrow transmission-band FBGs with wavelength spacings of 0.148, 0.33, and 0.053 nm were incorporated into the laser cavity. Microwave signals at 18.68, 40.95, and 6.95 GHz were generated, respectively. The spectral width of the generated microwave signals was as small as 80 kHz with frequency stability better than 1 MHz in the free-running mode at room temperature.

The homogeneous line broadening of an EDFA that causes strong gain competition at room temperature [19] is a serious issue for achieving highly stable microwave generation. The two wavelengths in the laser cavity are competing, and consequently, the generated microwave signal becomes unstable. One solution is to cool the erbium-doped fiber down to a cryogenic temperature in liquid nitrogen [20]. However, this technique is not easy to fulfill for practical applications. Another solution is to jointly use an EDFA and an SOA, or simply use an SOA as the gain medium. The SOA is biased at its low-gain regime. Since an SOA has much weaker homogeneous line broadening as compared with an EDFA, the incorporation of an SOA in the laser cavity would greatly reduce the gain competition of the two modes [21]. The major limitation of using an SOA as the sole gain medium is the high optical noise generated by the SOA, which will be translated to the generated microwave signal. In addition, the heat caused by the pump current may make the gain fluctuate, which would deteriorate the stability of the generated microwave signal. Therefore, microwave generation using either an EDFA- or SOA-based dual-wavelength SLM fiber laser has some clear drawbacks such as poor stability, small frequency tunability, and high phase noise.

The stimulated Brillouin scattering (SBS) effect can be used to implement a dual-wavelength fiber laser. The SBS gain spectrum is intrinsically narrow, only several or tens of MHz [22]. Figure 3.8 shows a schematic of a typical SBS dual-wavelength fiber laser. As can be seen, the fiber loop in the ring is pumped by two CW laser sources through an optical circulator. The two pump waves provide two SBS gains with different central wavelengths determined by the wavelengths of the pump waves. Due to the narrow bandwidths of the SBS gain spectra, two longitudinal modes will be

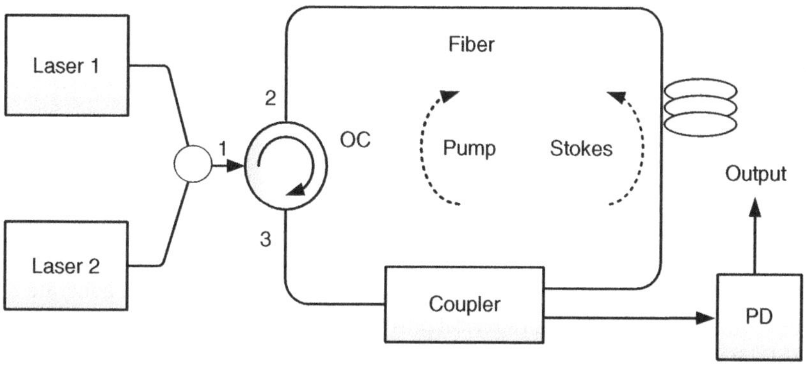

Figure 3.8 Schematic of a typical SBS-based dual-wavelength fiber laser.

generated in the loop. A key advantage of using an SBS-gain-based dual-wavelength SLM fiber laser for microwave signal generation is that the frequency of the generated microwave or the mm-wave signal can be easily controlled by changing the wavelengths of the two pump waves. A drawback of this method is the frequency drift of the generated microwave signal, since the wavelength drifts of the pump waves will be directly reflected as the frequency drifts of the two lasing modes.

The methods in [17, 18] can generate a microwave signal with a frequency that corresponds to the wavelength spacing of the two wavelengths which are determined by the two transmission bands of the narrow-band optical filter. Thus, the frequency is not tunable since, once the dual-passband filter is fabricated, the band spacing cannot be tuned. The use of an SBS-based dual-wavelength laser source can simply achieve frequency tuning, but the wavelength drift from any of the two pumped CW sources will be translated directly to the generated microwave signal [22]. To generate a microwave signal having a largely tunable frequency and high-frequency accuracy, a solution is to use the birefringence of the laser cavity to generate two orthogonally polarized light waves [23, 24]. The wavelength spacing is determined by the birefringence, which can be tuned by applying a voltage to change the birefringence. Figure 3.9 shows a dual-wavelength solid-state laser source incorporating a birefringence crystal in the cavity to achieve dual-wavelength lasing with continuously tunable wavelength spacing. The laser cavity consists of two mirrors (M1 and M2) in which an active medium and a crystal are incorporated. This birefringent crystal allows the oscillation of two spatially separated eigenmodes, an ordinary one with a linear polarization along the y axis

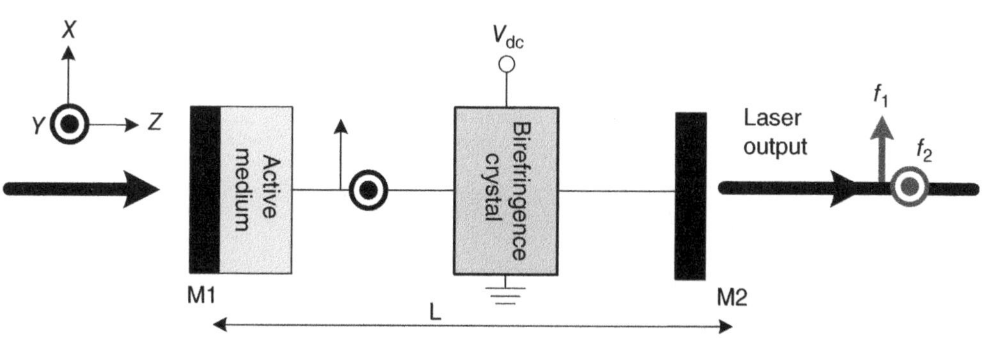

Figure 3.9 A dual-wavelength solid-state laser source incorporating a birefringence crystal in the cavity to achieve dual-wavelength lasing with continuous wavelength spacing tuning.

and an extraordinary one with a linear polarization along the x axis. The tuning of the wavelength spacing is done by applying a voltage to the birefringent crystal. It was demonstrated that a microwave signal tunable from 2 to 130 GHz could be generated. The quality of the microwave source is good and meets the requirements for radar LO distribution.

3.7 Microwave Generation Using an Optoelectronic Oscillator

In addition to the techniques of heterodyning two optical waves at a photodetector to generate a microwave signal, a microwave signal with high spectral purity can also be generated using an OEO [21, 25–27]. Compared with the techniques using injection locking, phase-lock loop, or external modulation, the use of an OEO can generate a microwave signal without the need for a microwave reference source. A typical configuration of an OEO is shown in Fig. 3.10. It consists of a CW laser source, a modulator (usually a Mach–Zehnder modulator), a length of SMF, a photodetector, an electrical bandpass filter, and an electronic amplifier (EA). The length of the SMF can be long to increase the Q factor of the OEO loop, thus reducing the phase noise of the generated microwave signal.

A light wave from the CW laser source is introduced into the Mach–Zehnder modulator and modulated by the noise signal from the EA. The signal at the output of the modulator is sent through a length of SMF to the photodetector. At the photodetector, the optical signal is converted to an electrical signal. After being filtered by the electronic bandpass filter, it is amplified by the EA and fed back to the modulator, to close the OEO loop. For a given frequency, if the loop gain is high enough to exceed the loss in the loop, the signal at this particular frequency will sustain and oscillation will start. The phase noise performance of the generated signal is determined by the Q factor of the OEO loop. Due to the low loss of an SMF, a long SMF loop up to tens of km would make the OEO have a high Q factor. Thus, a microwave signal with an ultra-low phase noise can be generated.

However, a long SMF would make the system have poor stability. The change in environmental temperature will cause the change in the length of the fiber, which will cause a drift of the oscillation frequency [28]. The use of two combined fibers, with one having a positive thermal coefficient and the other having a negative thermal coefficient may reduce the thermal effect, but the size is still large. In addition, a long OEO loop will cause a large number of closely spaced eigenmodes. To ensure a single-frequency oscillation without mode hopping, a solution is to use an OEO with

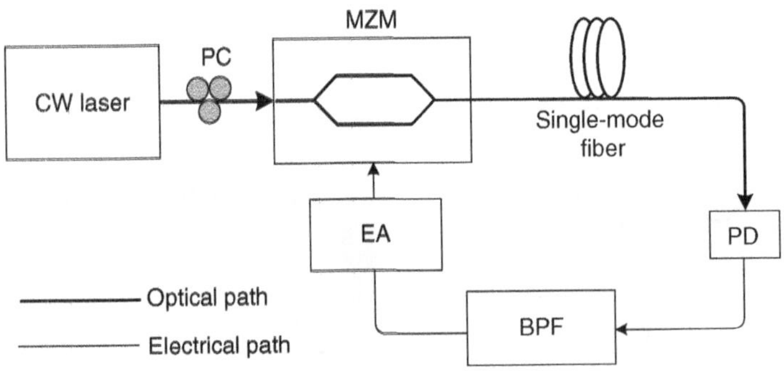

Figure 3.10 A typical configuration of an optoelectronic oscillator. BPF, bandpass filter; CW, continuous-wave; EA, electronic amplifier; PD, photodetector; MZM, Mach–Zehnder modulator.

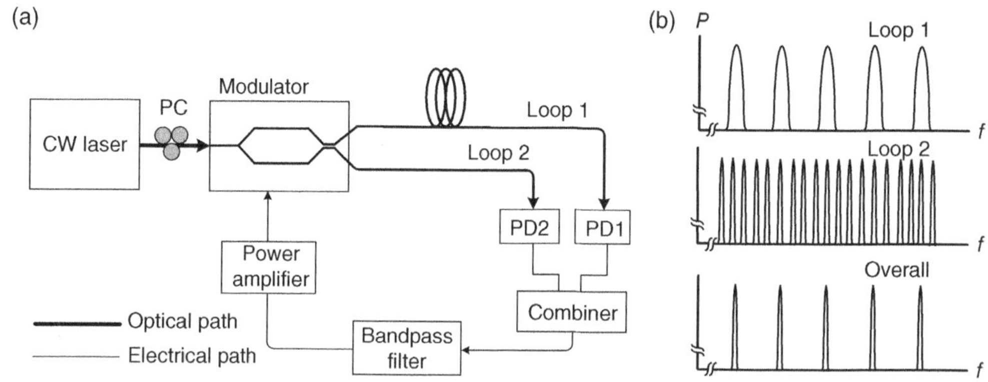

Figure 3.11 A dual-loop optoelectronic oscillator. (a) The configuration and (b) the spectral response of the two loops and the combination of the two.

two or multiple loops [29–34]. Due to the Vernier effect, the effective mode spacing can be significantly increased. For example, for a two-loop OEO, a long loop is used to ensure a high Q factor and a short loop is used to ensure a large FSR. The combination of the two loops corresponds to a resonator with a high Q factor and a large FSR. Thus, a single-frequency oscillation without mode hopping is realized. Figure 3.11(a) shows the operation of a dual loop OEO and the relationship between the spectral responses of the long and the short loop and the combined spectral response are shown in Fig. 3.11(b).

Another solution to ensure a single-frequency oscillation is to use an injection-locked dual-OEO [35–38]. In an injection-locked dual-OEO, a master OEO with a long loop with multiple modes is used to injection-lock a slave OEO with a short loop with a signal mode. At the output of the slave OEO, a single-frequency microwave signal with a significantly reduced phase noise can be achieved. The major limitation of implementing an OEO with two or multiple loops, or based on injection locking is the complexity of the system. In addition, the tuning of the microwave frequency is not feasible since the frequency is fixed once the loop lengths are determined. To ensure a single-frequency oscillation without using multiple loops, a simpler solution is to use an electrical bandpass filter with an ultra-narrow passband (~kHz), but again, the frequency tuning is not possible or very limited.

To achieve a high-Q OEO without using a long fiber or an electronic filter, a solution is to use a high-Q optical bandpass filter, such as a whispering gallery mode (WGM) resonator [40–43], a ring resonator [44], or a phase-shifted fiber Bragg grating (PS-FBG) [45]. Due to the ultra-high Q factor offered by a WGM resonator or a PS-FBG, a microwave signal with ultra-low phase noise can be generated. The small size of a WGM is particularly useful for the implementation of an OEO with a smaller size. Figure 3.12 shows a miniaturized OEO developed by OEwaves in which a lithium-niobate WGM resonator is employed [39, 46].

To achieve signal-frequency oscillation with a largely tunable frequency, a tunable electronic bandpass filter is needed. However, a purely electronic bandpass filter is not tunable or with very limited tunability. A simple solution to demonstrate an OEO with a large frequency tuning range is to use a tunable photonic microwave bandpass filter, which can be realized based on a phase modulation and phase-modulation to intensity-modulation conversion in a high-Q optical filter, such as a WGM resonator [40–43], a ring resonator [44], or a phase-shifted FBG [45]. Figure 3.13(a) shows an OEO using a photonic microwave bandpass filter based on phase

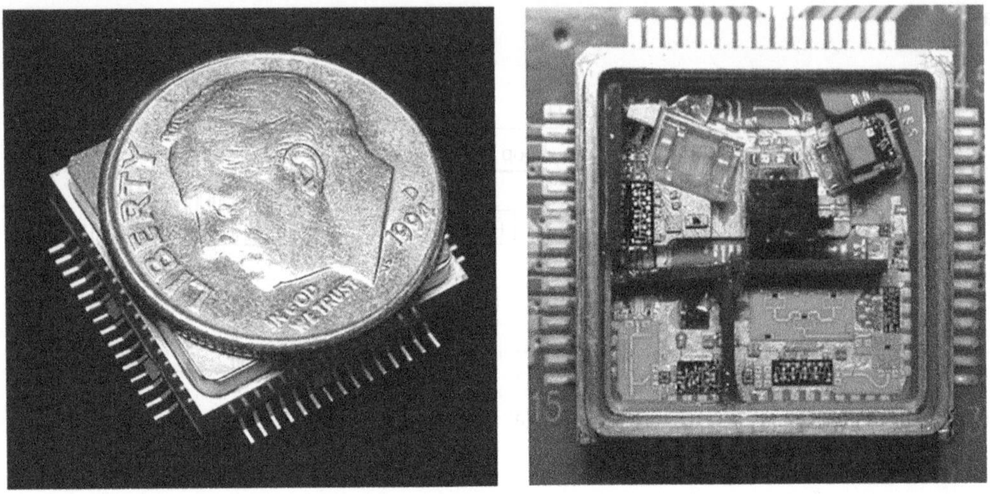

Figure 3.12 Miniature OEO incorporating a lithium-niobate WGM resonator. *Source:* Maleki [39]/Reproduced from Springer Nature.

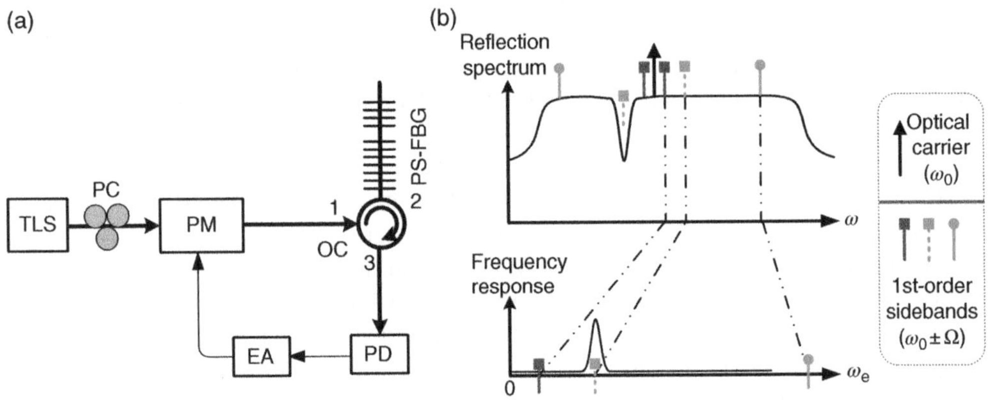

Figure 3.13 (a) Schematic of an OEO incorporating a microwave photonic bandpass filter based on phase modulation and phase-modulation to intensity-modulation conversion in a phase-shifted FBG.
(b) The operation of the phase-modulation to intensity-modulation conversion to form a microwave photonic bandpass filter. *Source:* Adapted from Li and Yao [45].

modulation and phase-modulation to intensity-modulation conversion in a phase-shifted FBG. Figure 3.13(b) shows the operation of the phase-modulation to intensity-modulation conversion to form a microwave photonic bandpass filter.

A tunable photonic microwave bandpass filter can also be implemented using an FP-LD that is biased below the lasing threshold, to selectively amplify a sideband of an intensity-modulated signal at a specific frequency. The operation corresponds to a microwave bandpass filter. The frequency tuning can be achieved by either tuning the FP-LD through thermal tuning or the tuning of the carrier wavelength [47]. An OEO with a tunable frequency from 6.41 to 10.85 GHz was demonstrated [47]. A tunable OEO can also be realized by incorporating a high-Q microwave photonic delay-line filter in the OEO loop [48]. The photonic microwave transversal filter is implemented using a sliced broadband optical source and a dispersive element to form a delay-line filter. If the number of the taps is large, a large Q factor is achieved. The central frequency of the photonic microwave filter is tuned by tuning the wavelength spacing of the sliced optical source or the chromatic dispersion of the dispersive element. An OEO with a frequency-tunable range of 9.7 GHz was demonstrated. The generated microwave signal exhibited a good phase noise performance with a phase noise of −120 dBc/Hz at an offset of 10 kHz [48].

To increase the frequency of a generated microwave signal using relatively low-frequency components in an OEO loop, one may employ frequency multiplication technique to allow the OEO to oscillate at a lower frequency for a wavelength in the loop and the lower frequency is multiplied through frequency multiplication for a different wavelength [49–51]. For example, in [49] two CW laser sources operating at 1550 and 1310 nm are used. Due to the wavelength dependence of the half-wave voltage of the EO modulator in the OEO loop, the modulator is biased at the quadrature transmission point for the wavelength at 1310 nm to sustain oscillation at this wavelength to produce a low-frequency microwave signal, and at the minimum transmission point for the wavelength at 1550 nm to generate a frequency-doubled signal. In [50], a special phase modulator that can support both TE and TM modes with opposite phase modulation indices is employed to realize frequency doubling. The special phase modulator is also known as a polarization modulator, whose bias status is dependent on the polarization status of the light wave. Therefore, by controlling the polarization statuses of the two light waves from the same laser source, two bias statuses of the modulator can be achieved. One light wave travels in the OEO loop in which the modulator is biased at the quadrature transmission point to generate a low-frequency microwave signal, while the other goes through the modulator which is biased at the minimum transmission point to generate a frequency-doubled microwave signal.

3.8 Performance Comparison of the Techniques for Photonic Microwave Generation

The performance in terms of the frequency, frequency tunability, frequency stability, and system complexity of all the methods discussed above is summarized in Table 3.1.

From Table 3.1, we can see that different methods have their own advantages and disadvantages. As far as frequency tunability and stability are concerned, microwave generation based on external modulation and optoelectronic oscillation are two solutions with better performance.

The phase noise performance for the microwave generation methods is shown in Fig. 3.14 [39, 52–55]. The phase noise performance of the generated microwave signal using an EDFA-based and SOA-based dual-wavelength SLM fiber laser is not included since the generated microwave is so unstable that the phase noise measurement is hard to be realized. From Fig. 3.14, the combination of OIL and OPLL has the poorest phase noise, and an OEO employing 16 km SMF provides the best phase noise performance except for many undesired frequency peaks at offset frequencies higher than 10 kHz, which are the densely spaced oscillation modes due to the long SMF in the loop. The electrical techniques still outperform the photonic techniques but with very limited tunability.

The phase noise performance of a photonically generated microwave signal can be improved by tens of dBc/Hz using a technique called photonic microwave frequency division [56, 57]. Such a frequency division is accomplished by phase-locking a passively locked femtosecond laser to an ultra-stable continuous-wave optical reference source. The phase locking could transfer the phase

Table 3.1 Comparison of the methods for photonic microwave generation.

Method		Frequency	Tunability	Stability	Complexity
Optical injection locking		<100 GHz	Fair	Fair	Medium
Optical phase, lock loop		<100 GHz	Fair	Fair	Medium
Dual-wavelength laser source	Semiconductor laser	~THz	Fair	Good	Low
	Fiber laser (using EDFA or SOA as gain medium)	~THz	Poor	Poor	High
	Fiber laser (based on SBS-gain)	~THz	Good	Fair	Medium
External modulation		<100 GHz (no optical nonlinear effects)	Good	Excellent	Low
		<1 THz (assisted with optical nonlinear effects)	Fair	Excellent	High
Optoelectronic oscillator		<100 GHz	Fair	Good	Low

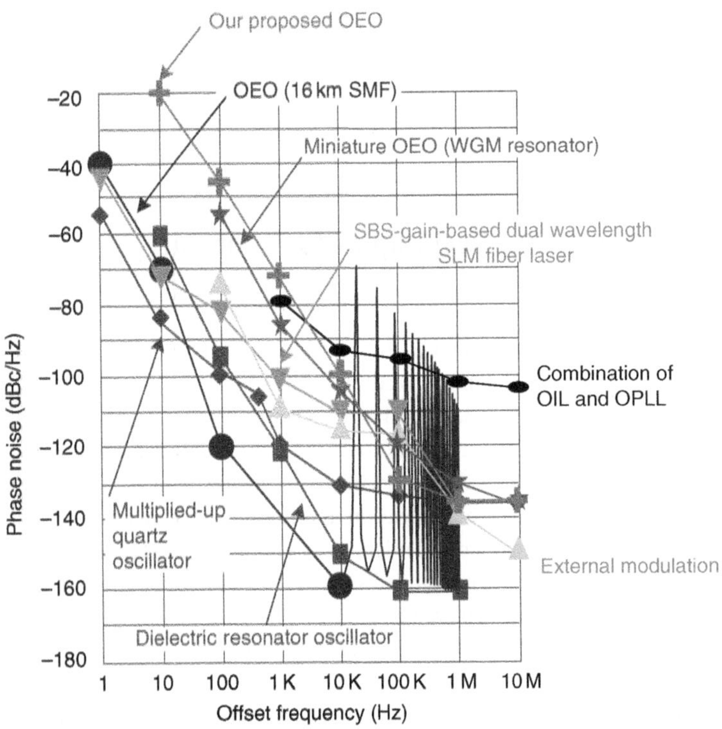

Figure 3.14 Phase noise performance of photonic and electrical techniques for microwave generation. All methods operate at 10 GHz except the combination of OIL and OPLL that operate at 36 GHz, and the miniature OEO at 30 GHz.

noise of the ultra-stable optical reference to the timing between the femtosecond laser pulses, and hence to a microwave signal acquired by beating two adjacent comb lines from the phase-locked femtosecond at a photodetector. Due to the phase locking, the phase noise of the generated microwave could be reduced by a factor of a ratio of the frequency of the optical reference source to that of the microwave signal. A few demonstrations have been reported [58–62]. However, this technique is extremely complicated and cannot be widely used for applications where the size should be small and the cost must be low.

3.9 Summary

In this chapter, we have reviewed different techniques proposed and demonstrated in the last few decades for the generation of microwave signals based on photonics. The key advantage of using photonics for microwave generation is the low phase noise. In addition, the ease of transmitting a microwave signal over a fiber for microwave distribution is another advantage that makes photonic-assisted microwave generation techniques attractive. Among the different techniques, microwave generation using an OEO is considered the best solution since the system is simple and the quality of the generated microwave signal is high.

References

1 Gliese, U., Nielsen, T.N., Nørskov, S., and Stubkjaer, K.E. (1998). Multifunctional fiber-optic microwave links based on remote heterodyne detection. *IEEE Trans. Microw. Theory Tech.* 46 (5): 458–468.

2 Goldberg, L., Taylor, H.F., Weller, J.F., and Bloom, D.M. (1983). Microwave signal generation with injection locked laser diodes. *Electron. Lett.* 19 (13): 491–493.

3 Goldberg, L., Yurek, A., Taylor, H.F., and Weller, J.F. (1985). 35 GHz microwave signal generation with injection locked laser diode. *Electron. Lett.* 21 (18): 714–715.

4 Harrison, J. and Mooradian, A. (1989). Linewidth and offset frequency locking of external cavity GaAlAs lasers. *IEEE J. Quantum Electron.* 25 (6): 1252–1255.

5 Ramos, R.T. and Seeds, A.J. (1992). Fast heterodyne optical phase-lock loop using double quantum well laser diodes. *Electron. Lett.* 28 (1): 82–83.

6 Gliese, U., Nielsen, T.N., Bruun, M. et al. (1992). A wideband heterodyne optical phase-locked loop for generation of 3–18 GHz microwave carriers. *IEEE Photonics Technol. Lett.* 4 (8): 936–938.

7 Bordonalli, A.C., Walton, C., and Seeds, A.J. (1999). High-performance phase locking of wide linewidth semiconductor lasers by combined use of optical injection locking and optical phase-lock loop. *J. Lightw. Technol.* 17 (2): 328–342.

8 Williams, K.J. (1989). 6–34 GHz offset phase locking of Nd: YAG 1319 nm nonplanar ring lasers. *Electron. Lett.* 25 (18): 1242–1243.

9 Fan, Z.F. and Dagenais, M. (1997). Optical generation of a mHz-linewidth microwave signal using semiconductor lasers and a discriminator-aided phase-locked loop. *IEEE Trans. Microw. Theory Tech.* 45 (8): 1296–1300.

10 Rideout, H., Seregelyi, J., Paquet, S., and J.Y. (2006). Discriminator-aided optical phase-lock loop incorporating a frequency down-conversion module. *IEEE Photonics Technol. Lett.* 18 (22): 2344–2346.

11 O'Reilly, J.J., Lane, P.M., Heidemann, R., and Hofstetter, R. (1992). Optical generation of very narrow linewidth millimeter wave signals. *Electron. Lett.* 28 (25): 2309–2311.

12 O'Reilly, J.J. and Lane, P.M. (1994). Remote delivery of video services using mm-wave and optics. *J. Lightw. Technol.* 12 (2): 369–375.

13 O'Reilly, J.J. and Lane, P.M. (1994). Fiber-supported optical generation and delivery of 60 GHz signals. *Electron. Lett.* 30 (16): 1329–1330.

14 Shen, P., Gomes, N.J., Davies, P.A. et al. High-purity millimeter-wave photonic local oscillator generation and delivery. *Proc. Int. Microwave Photonics Topical Meeting* (10–12 September 2003), Budapest, Hungary, pp. 189–192.

15 Qi, G., J.Y., Seregelyi, J. et al. (2005). Generation and distribution of a wide-band continuously tunable mm-wave signal with an optical external modulation technique. *IEEE Trans. Microw. Theory Tech.* 53 (10): 3090–3097.

16 Qi, G., J.Y., Seregelyi, J. et al. (2005). Optical generation and distribution of continuously tunable millimeter-wave signals using an optical phase modulator. *J. Lightw. Technol.* 23 (9): 2687–2695.

17 Chen, X., Deng, Z., and J.Y. (2006). Photonic generation of microwave signal using a dual-wavelength single-longitudinal-mode fiber ring laser. *IEEE Trans. Microw. Theory Tech.* 54 (2): 804–809.

18 Chen, X., J.Y., and Deng, Z. (2005). Ultranarrow dual-transmission-band fiber Bragg grating filter and its application in a dual-wavelength single-longitudinal-mode fiber ring laser. *Opt. Lett.* 30 (16): 2068–2070.

19 Desurvire, E. (1994). *Erbium-Doped Fiber Amplifiers*. Wiley.

20 Yamashita, S. and Hotate, K. (1996). Multiwavelength erbium-doped fiber laser using intracavity etalon and cooled by liquid nitrogen. *Electron. Lett.* 32 (14): 1298–1299.

21 Neyer, A. and Voges, E. (1982). Hybrid electro-optical multivibrator operating by finite feedback delay. *Electron. Lett.* 18 (2): 59–60.

22 Agrawal, G.P. (2001). *Nonlinear Fiber Optics*. Academic.

23 Baxter, G.W., Dawes, J.M., Dekker, P., and Knowles, D.S. (1996). Dual-polarization frequency-modulated laser source. *IEEE Photonics Technol. Lett.* 8 (8): 1015–1017.

24 Brunel, M., Bretenaker, F., and Le Floch, A. (1997). Tunable optical microwave source using spatially resolved laser eigenstates. *Opt. Lett.* 22 (6): 384–386.

25 Yao, X.S. and Maleki, L. (1996). Optoelectronic microwave oscillator. *J. Opt. Soc. Am. B* 13 (8): 1725–1735.

26 Yao, X.S. and Maleki, L. (1996). Optoelectronic oscillator for photonic systems. *IEEE J. Quantum Electron.* 32 (7): 1141–1149.

27 Li, W. and J.Y. (2010). An optically tunable optoelectronic oscillator. *J. Lightw. Technol.* 28 (18): 2640–2645.

28 Zhang, L., Madhavan, V., Patel, R.P. et al. (2013). Ultra low FM noise in passively temperature compensated microwave opto-electronic oscillators. *2013 IEEE MTT-S International Microwave and RF Conference* (14–16 December 2013), New Delhi, India, pp. 14–16.

29 Yao, X.S. and Maleki, L. (2000). Multiloop optoelectronic oscillator. *IEEE J. Quantum Electron.* 36 (1): 79–84.

30 Jiang, Y., Yu, J., Wang, Y. et al. (2007). An optical domain combined dual-loop optoelectronic oscillator. *IEEE Photonics Technol. Lett.* 19 (11): 807–809.

31 Shumakher, E. and Eisenstein, G. (2008). A novel multiloop optoelectronic oscillator. *IEEE Photonics Technol. Lett.* 20 (22): 1881–1883.

32 Shumakher, E. (2009). The optoelectronic oscillator: review and recent advances. *COMCAS 2009. IEEE Int. Conf. on Microwaves, Communications, Antennas and Electronics Systems, 2009* (9–11 November 2009), Tel-Aviv, Israel, pp. 1–5.

33 Tseng, W.H. and Feng, K.M. (2012). Enhancing long-term stability of the optoelectronic oscillator with a probe-injected fiber delay monitoring mechanism. *Opt. Express* 20 (2): 1597–1607.

34 Zhou, W. and Blasche, G. (2005). Injection-locked dual opto-electronic oscillator with ultra-low phase noise and ultra-low spurious level. *IEEE Trans. Microw. Theory Tech.* 53 (3): 929–933.

35 Li, W., Kong, F., and J.Y. (2014). Stable and frequency-hopping-free microwave generation based on a mutually injection-locked optoelectronic oscillator and a dual-wavelength single-longitudinal-mode fiber laser. *J. Lightw. Technol.* 32 (22): 3572–3577.

36 Lee, K.H., Kim, J.Y., and Choi, W.Y. (2008). Injection-locked hybrid optoelectronic oscillators for single-mode oscillation. *IEEE Photonics Technol. Lett.* 20: 1645–1647.

37 Okusaga, O., Zhou, W., Levy, E.C. et al. (2009). Experimental and simulation study of dual injection-locked OEOs. *IEEE Freq. Cont. Symp.* (20–24 April 2009), Besancon, France, pp. 875–879.

38 Okusaga, O., Adles, E.J., Levy, E.C. et al. (2011). Spurious mode reduction in dual injection-locked optoelectronic oscillators. *Opt. Express* 19 (7): 5839–5854.

39 Maleki, L. (2011). Sources: the optoelectronic oscillator. *Nat. Photonics* 5 (12): 728–730.

40 Ilchenko, V.S. and Matsko, A.B. (2006). Optical resonators with whispering-gallery modes—Part II: applications. *IEEE J. Sel. Top. Quantum Electron.* 12 (1): 15–32.

41 Matsko, A.B., Maleki, L., Savchenkov, A.A., and Ilchenko, V.S. (2003). Whispering gallery mode based optoelectronic microwave oscillator. *J. Mod. Opt.* 50 (15–17): 2523–2542.

42 Volyanskiy, K., Salzenstein, P., Tavernier, H. et al. (2010). Compact optoelectronic microwave oscillators using ultra-high *Q* whispering gallery mode disk-resonators and phase modulation. *Opt. Express* 18 (21): 22358–22363.

43 Savchenkov, A.A., Ilchenko, V.S., Byrd, J. et al. (2011). Whispering-gallery mode based optoelectronic oscillators. *IEEE Freq. Cont. Symp.* (2–5 May 2011), San Fransisco, CA, USA, pp. 554–557.

44 Saleh, K., Merrer, P., Llopis, O., and Cibiel, G. (2012). Optical scattering noise in high *Q* fiber ring resonators and its effect on optoelectronic oscillator phase noise. *Opt. Lett.* 37 (4): 518–520.

45 Li, W. and J.Y. (2012). Wideband frequency-tunable optoelectronic oscillator incorporating a tunable microwave-photonic filter based on phase-modulation to intensity-modulation conversion using a phase-shifted fiber-Bragg grating. *IEEE Trans. Microw. Theory Tech.* 60 (6): 1735–1742.

46 Savchenkov, A.A. et al. (2010). Whispering-gallery mode based opto-electronic oscillators. *2010 IEEE International Frequency Control Symposium*, Newport Beach, CA, USA, pp. 554–557. http://www.oewaves.com.

47 Pan, S. and J.Y. (2010). Wideband and frequency-tunable microwave generation using an optoelectronic oscillator incorporating a Fabry–Perot laser diode with external optical injection. *Opt. Lett.* 35 (11): 1911–1913.

48 Li, M., Li, W., and J.Y. (2012). Tunable optoelectronic oscillators incorporating a high-Q spectrum-sliced photonic microwave transversal filter. *IEEE Photonics Technol. Lett.* 24 (14): 1251–1253.

49 Shin, M., Grigoryan, V.S., and Kumar, P. (2007). Frequency-doubling optoelectronic oscillator for generating high-frequency microwave signals with low phase noise. *Electron. Lett.* 43 (4): 242–244.

50 Pan, S. and J.Y. (2009). A frequency-doubling optoelectronic oscillator using a polarization modulator. *IEEE Photonics Technol. Lett.* 21 (13): 929–931.

51 Zhu, D., Pan, S., and Ben, D. (2012). Tunable frequency-quadrupling dual-loop optoelectronic oscillator. *IEEE Photonics Technol. Lett.* 24 (3): 194–196.

52 Howe, D.A. and Hati, A. (2005). Low-noise X-band oscillator and amplifier technologies: comparison and status. *Proceedings of the Joint IEEE Int. Frequency Control Symp. and Precise Time and Time Interval (PTTI) Systems and Applications Meeting* (29–31 August 2005), Vancouver, Canada (August 2005), pp. 481–487.

53 Johansson, L.A. and Seeds, A.J. (2000). Millimeter-wave modulated optical signal generation with high spectral purity and wide locking bandwidth using a fiber-integrated optical phase-lock loop. *IEEE Photonics Technol. Lett.* 12 (6): 690–693.

54 Li, W. and J.Y. (2010). Investigation of photonically assisted microwave frequency multiplication based on external modulation. *IEEE Trans. Microw. Theory Tech.* 58 (11): 3259–3268.

55 Callahan, P.T., Gross, M.C., and Dennis, M.L. (2010). Phase noise measurements of a dual-wavelength Brillouin fiber laser. *IEEE Topical Meet. Microwave Photonics* (5–9 October 2020), Montréal, QC, Canada, pp. 155–158.

56 Ramond, T.M., Diddams, S.A., Hollberg, L., and Bartels, A. (2002). Phase-coherent link from optical to microwave frequencies by means of the broadband continuum from a 1-GHz Ti:sapphire femtosecond oscillator. *Opt. Lett.* 27 (20): 1842–1844.

57 Bartels, A., Diddams, S.A., Oates, C.W. et al. (2005). Femtosecond-laser-based synthesis of ultrastable microwave signals from optical frequency references. *Opt. Lett.* 30 (6): 667–669.

58 Zhang, W., Xu, Z., Lours, M. et al. (2011). Advanced noise reduction techniques for ultra-low phase noise optical-to-microwave division with femtosecond fiber combs. *IEEE Trans. Ultrason. Ferroelectr. Freq. Control* 58 (5): 900–908.

59 Millo, J., Boudot, R., Lours, M. et al. (2009). Ultra-low-noise microwave extraction from fiber-based optical frequency comb. *Opt. Lett.* 34 (23): 3707–3709.

60 Swann, W.C., Baumann, E., Giorgetta, F.R., and Newbury, N.R. (2011). Microwave generation with low residual phase noise from a femtosecond fiber laser with an intracavity electro-optic modulator. *Opt. Express* 19 (24): 24387–24395.

61 Fortier, T.M., Kirchner, M.S., Quinlan, F. et al. (2011). Generation of ultrastable microwaves via optical frequency division. *Nat. Photonics* 5 (7): 425–429.

62 Quinlan, F., Fortier, T.M., Kirchner, M.S. et al. (2011). Ultralow phase noise microwave generation with an Er:fiber-based optical frequency divider. *Opt. Lett.* 36 (16): 3260–3262.

4

Photonic-Assisted Microwave Signal Processing

4.1 Introduction

Microwave signals can be processed in the electrical domain based on analog or digital electronics. Thanks to the advancement of digital electronics, microwave signal processing using digital electronics has been widely used. However, for broadband and high-frequency microwave signals, the currently available digital electronic circuits may not be able to operate at such a high frequency and wideband width, limited mainly by analog-to-digital conversion (ADC) and digital-to-analog conversion (DAC). One solution is to process microwave signals in the optical domain by taking advantage of the inherent broad width and high speed of modern photonics. Among the numerous signal processing functions, microwave filtering and frequency up or down conversion are two important functions. In this chapter, the discussion will be focused on the design and implementation of microwave filters and microwave up or down converters based on photonics.

4.2 Microwave Photonic Filters

Microwave photonic filters can be implemented either in the incoherent or in the coherent regime. For a microwave photonic filter in the incoherent regime, a delay-line configuration is usually used with a finite impulse response (FIR). To avoid optical interferences, an incoherent light source or a laser array is used. The tuning and reconfiguration of the filter are achieved by changing the time delay and the tap coefficients. For a microwave photonic filter in the coherent regime, however, a single wavelength is needed, and the spectral response of the filter is directly translated from the spectral response of an optical filter. Thus, in a coherent microwave filter, a well-defined optical filter with a precisely controlled spectral response is needed. In this chapter, techniques to implement both incoherent and coherent tunable and reconfigurable microwave photonic filters are discussed.

It was proposed by Wilner and Van den Heuvel [1] that optical fibers could be utilized as delay lines for signal processing since optical fibers have low loss and large bandwidth. In the last few years, extensive efforts have been directed to the design and implementation of microwave photonic filters with different configurations to fulfill different functionalities. Figure 4.1 shows a diagram of a generic microwave photonic delay-line filter with an FIR. It consists of a light source, a modulator, which can be a Mach–Zehnder modulator or a phase modulator (PM), a delay-line

Microwave Photonics, First Edition. Jianping Yao, José Capmany, and Ming Li.

Companion website: www.wiley.com/go/yao/microwavephotonics

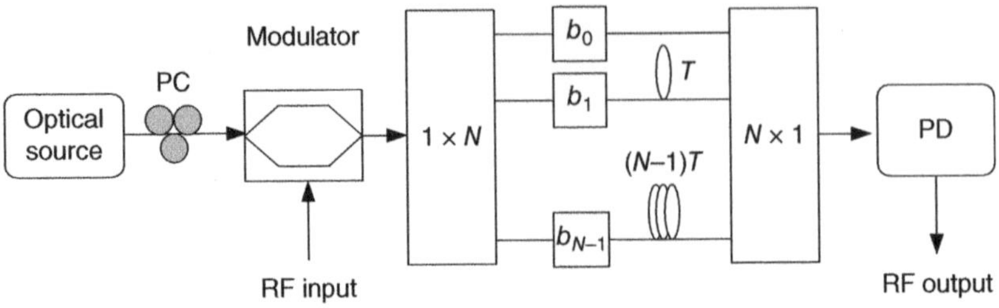

Figure 4.1 A diagram showing a generic photonic microwave delay-line filter with a finite impulse response.

module, and a photodetector (PD). The key device in a microwave photonic filter is the optical delay-line module, which can be implemented using optical couplers [2–4], fiber Bragg gratings (FBGs) [5–13], arrayed waveguide gratings (AWGs) [14, 15], Mach–Zehnder lattices [16], or simply a chirped FBG or a dispersive fiber [17, 18].

For an N-tap microwave photonic filter, the RF signal at the output of the PD is given by

$$y(t) = b_0x(t) + b_1x(t-T) + b_2x(t-2T) + \cdots + b_{N-1}x(t-(N-1)T) \tag{4.1}$$

where b_0, b_1, b_2, ..., b_{N-1} are the coefficients of the N taps, and T is the time delay between two adjacent taps. By applying Fourier transform to (4.1), we have

$$Y(j\omega) = b_0X(j\omega) + b_1\,e^{j\omega T}X(j\omega) + b_2\,e^{j\omega 2T}X(j\omega) + \cdots + b_{N-1}\,e^{j\omega(N-1)T}X(j\omega) \tag{4.2}$$

The transfer function of the microwave photonic filter is

$$H(j\omega) = \frac{Y(j\omega)}{X(j\omega)} = b_0 + b_1\,e^{j\omega T} + b_2\,e^{j\omega 2T} + \cdots + b_{N-1}\,e^{j\omega(N-1)T} \tag{4.3}$$

Figure 4.2(a) shows the spectral response of a 4-tap microwave photonic filter with four coefficients of (1, 1, 1, 1). Since the coefficients are all positive, a baseband resonance at zero frequency is resulted. The filter is a low-pass filter. The time delay between two adjacent taps is 125 ps,

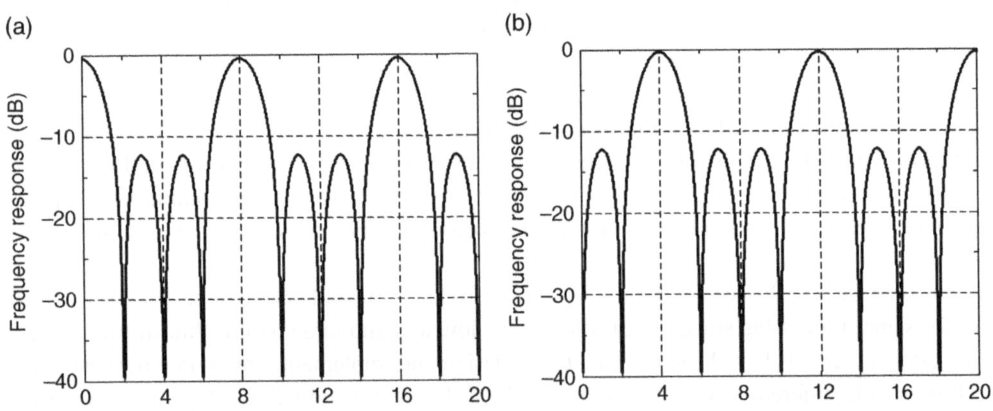

Figure 4.2 Spectral response of (a) a 4-tap microwave photonic filter with four coefficients of (1, 1, 1, 1), and (b) a 4-tap microwave photonic filter with four coefficients of (1, −1, 1, −1). The time delay between two adjacent taps is 125 ps, corresponding to a free spectral range of 8 GHz.

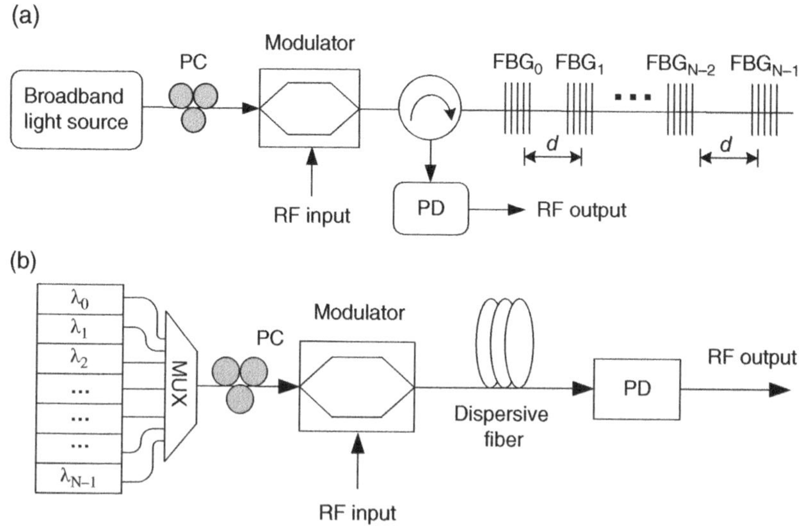

Figure 4.3 Microwave photonic filters operating in the incoherent regime based on (a) a broadband light source and (b) a laser array.

corresponding to a free spectral range (*FSR* = 1/T) of 8 GHz. Figure 4.2(b) shows the spectral response of a 4-tap microwave photonic filter with four coefficients of (1, −1, 1, −1). Two coefficients are negative. The filter is a bandpass filter with its central frequency of the passband at 4 GHz.

To avoid optical interference, most of the proposed microwave photonic filters are operating in the incoherent regime. Figure 4.3(a), (b) show two microwave photonic filters operating in the incoherent regime based on a broadband light source and a laser array, respectively. In Fig. 4.3(a), a broadband light source is modulated by an RF signal at a modulator and sent to an array of FBGs. The modulated light is then sliced by the FBGs in the array and reflected via an optical circulator (OC) to a PD. The physical spacing between two adjacent FBGs determines the time delay and the reflectivities of the FBGs determine the coefficients of the taps. In Fig. 4.3(b), N wavelengths from a laser array are multiplexed at a multiplexer and modulated by an RF signal at a modulator, the modulated signal is then sent to a length of dispersive fiber. If the dispersion parameter of the dispersive fiber is D (ps/nm km), and the length of the dispersive fiber is L, and the wavelength spacing between two adjacent wavelengths is $\Delta\lambda$, then the time delay between two adjacent taps is $T = \Delta\lambda DL$.

It is known a photonic microwave delay-line filter operating in the incoherent regime would have all-positive tap coefficients. Based on signal processing theory, an all-positive-coefficient microwave delay-line filter would only operate as a low-pass filter. To overcome this limitation, considerable efforts have been taken to design and implement a photonic microwave delay-line filter with negative or complex tap coefficients, to achieve bandpass filtering functionality in the incoherent regime. Comprehensive overviews of microwave photonic filters have been published in [19–22]. In the next section, the discussion will be focused only on techniques for the implementation of photonic microwave delay-line filters with negative and complex coefficients.

4.2.1 Photonic Microwave Delay-Line Filters with Negative Coefficients

A straightforward way to generate a negative coefficient is to use differential detection [3, 23]. As shown in Figure 4.4, a light wave from a laser source is modulated by an RF signal at a modulator.

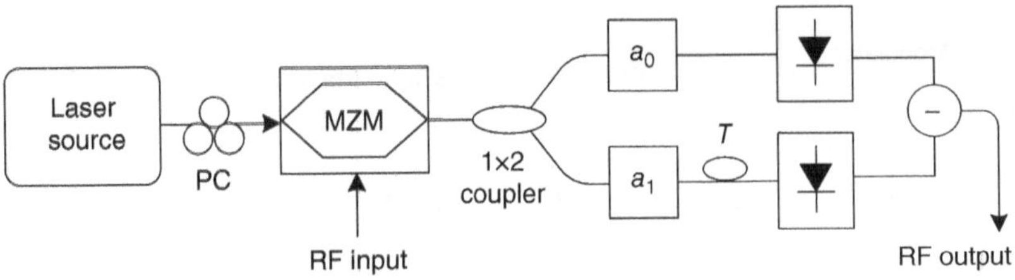

Figure 4.4 A photonic microwave delay-line filter with a negative coefficient using differential detection. MZM, Mach–Zehnder modulator.

The modulated signal is time-delayed by optical fiber delay lines with a time delay difference between two adjacent taps of T. The output signals from the fiber delay lines are fed to a differential PD, which consists of two matched PDs with the detected microwave signals combined and subtracted electrically, leading to the generation of a positive and a negative coefficient. In this approach, the negative coefficient was not generated directly in the optical domain, the filter is not all-optical, but hybrid. The two-tap photonic microwave delay-line filter shown in Fig. 4.4 can be extended to have multiple taps if the single-wavelength source is replaced by a multiwavelength source and the 3-dB coupler is replaced by a WDM demultiplexer.

A few techniques have been proposed to implement an all-optical photonic microwave delay-line filter with negative coefficients. One approach to generating negative coefficients is to use wavelength conversion based on cross-gain modulation in a semiconductor optical amplifier (SOA) [24]. As shown in Fig. 4.5, a tunable laser source (TLS) operating at λ_1 is modulated by an input microwave signal, and then split into two parts. One part goes through a length of optical fiber, to introduce a time delay. The other part is combined with a continuous-wave (CW) light wave from a DFB laser diode operating at a different wavelength λ_2, and then fed into an SOA. Due to the cross-gain modulation in the SOA [25], the CW beam λ_2 is also modulated by the input RF signal, but with a π phase inversion compared with the microwave signal carried by λ_1, which leads to the generation of a negative coefficient. An optical bandpass filter is used to filter out the residual signal at λ_1. Then, the time-delayed microwave signal carried by λ_1 in the upper channel and the π-phase-inverted RF signal carried by λ_2 in the lower channel are combined and detected at a PD. A two-tap photonic microwave bandpass filter with one negative coefficient is thus realized. Since the two

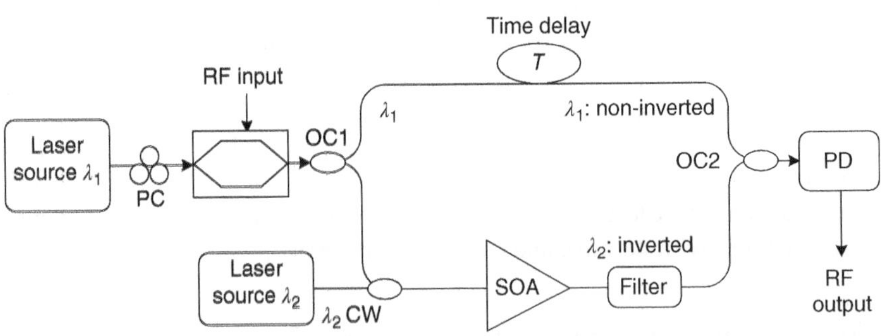

Figure 4.5 A photonic microwave delay-line filter with a negative coefficient based on cross-gain modulation in an SOA.

wavelengths are generated by two independent laser sources, the detection at the PD is incoherent. To avoid the beat signal between the two wavelengths falling in the passband of the filter, a large wavelength spacing should be chosen. Again, the filter can be extended to be a multi-tap microwave delay-line filter if the light sources and the two couplers are replaced by multiwavelength sources, a WDM demultiplexer, and a multiplexer.

A similar approach to generating negative coefficients using an injection-locked Fabry–Perot (FP) laser diode and an FBG array was demonstrated [26]. The filter structure is shown in Fig. 4.6. Again, the modulated microwave signal is split into two channels. The signal in the upper channel is applied to a PD through a delay line. In the lower channel, the signal is injected into an FP laser diode. The FP laser diode is operating with multiple longitudinal modes. One longitudinal mode is locked by the injected optical signal and the other modes that are not locked will experience a cross-intensity modulation. It is similar to the cross-gain modulation in an SOA. The signal modulated on the unlocked modes is with a π phase inversion. As a result, negative coefficients are generated by the unlocked modes. The time delay difference between adjacent unlocked modes is introduced by an FBG array. The major disadvantage of the technique is the mode competition in the lasing output, which may cause instability of the system. In addition, the mode spacing should be large enough to avoid the beat note between two adjacent modes falling in the passband of the filter.

The use of the carrier depletion effect in a DFB laser diode to generate a negative coefficient was also proposed [27], as shown in Fig. 4.7. Instead of using a multi-longitudinal mode FP laser

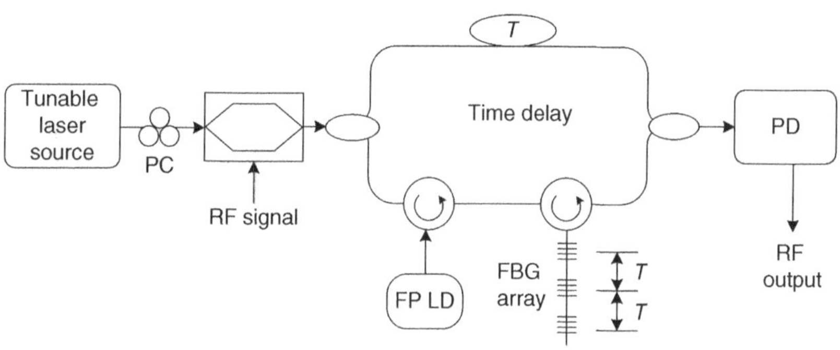

Figure 4.6 A photonic microwave delay-line filter with a negative coefficient based on injection-locking of an FP laser diode.

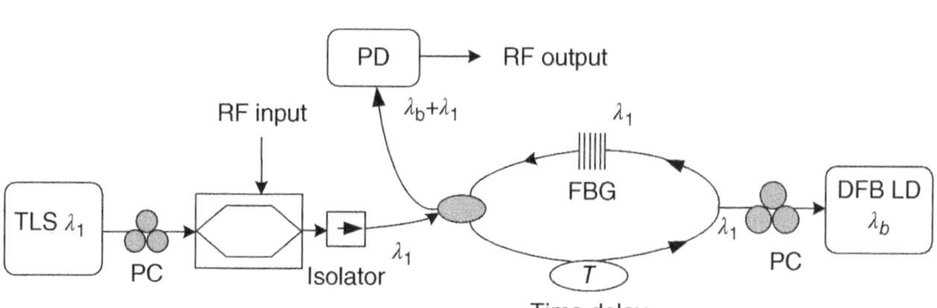

Figure 4.7 A photonic microwave delay-line filter with negative coefficients based on carrier depletion effect in a DFB laser diode.

diode, a DFB laser diode operating in a single-longitudinal mode is used. The lasing wavelength of the DFB laser diode is injection-locked by the injected optical signal. Due to the carrier depletion effect, the microwave signal modulated on the injection carrier is transferred to the lasing wavelength with a π phase inversion, leading to the generation of a negative coefficient. Note that the lasing wavelength of the DFB laser diode should be slightly different from the injection wavelength sent to the PD through an FBG.

Negative coefficients can also be generated by using an erbium-doped fiber amplifier (EDFA)-amplified spontaneous emission (ASE) source cascaded with uniform FBGs [28]. As shown in Fig. 4.8, the transmission spectrum at the output of FBGs is altered which is used to generate negative taps. Positive taps are generated by using another multiwavelength source, the output of which is combined with the filtered ASE source. A two-tap delay-line filter with one negative coefficient is shown in Fig. 4.8.

More recently, a new technique to generate negative coefficients based on two Mach–Zehnder modulators that are biased at the complementary transmission slopes was reported [29]. The operation of the microwave phase inversion is shown in Fig. 4.9. As can be seen, the two Mach–Zehnder modulators are biased at the linear regions of the left and the right slopes of the transfer functions. When a microwave signal is applied to the two modulators, the envelopes of the modulated optical signals are complementary. At the output of a PD, two complementary microwave signals are generated, leading to the generation of a negative coefficient. The time delay difference between two adjacent taps is generated due to the chromatic dispersion of the dispersive device. To implement a microwave delay-line filter with multiple taps, a multiwavelength laser source or a laser diode array is needed. For those taps with positive and negative coefficients, the corresponding wavelengths must be sent separately to the two Mach–Zehnder modulators. A similar technique using only a single Mach–Zehnder modulator was proposed [30]. Considering the wavelength dependence of the transfer function of a Mach–Zehnder modulator, a proper DC bias would make the modulator operate at the complementary slopes of the transfer functions when the optical wavelengths are set at 1550 and 1310 nm windows.

All the filters discussed above are implemented based on the use of one or multiple Mach–Zehnder modulators. Recently, a novel approach to implementing a photonic microwave bandpass filter with negative coefficients based on an optical PM was proposed [31]. The negative coefficients are generated based on phase modulation to intensity modulation (PM-IM) conversion in

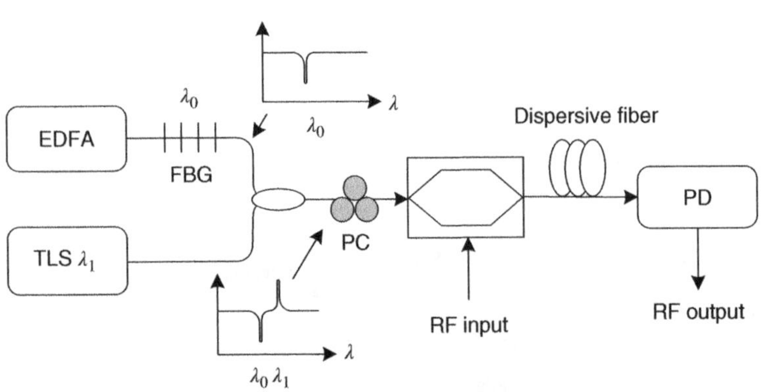

Figure 4.8 A photonic microwave delay-line filter implemented using a broadband ASE source cascaded with uniform FBGs to produce negative coefficients.

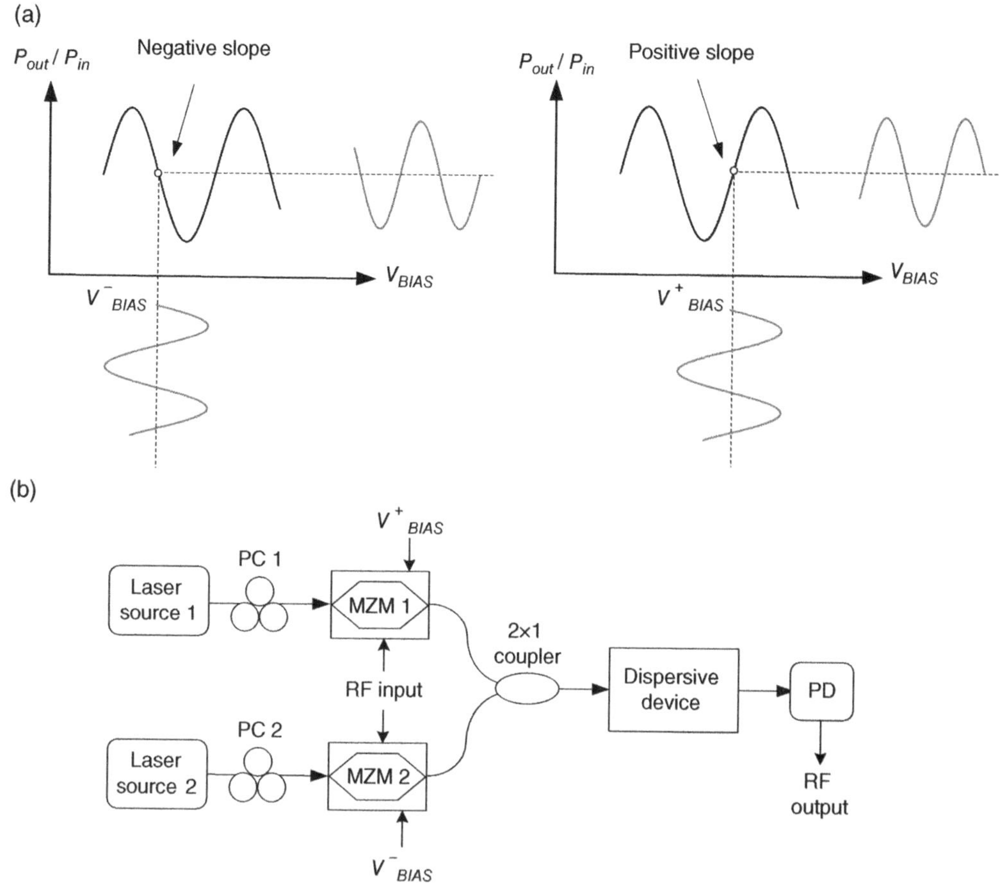

Figure 4.9 Photonic microwave delay-line filter with negative coefficients based on phase inversion using complementarily biased intensity modulators. (a) The operation of phase inversion, (b) the schematic of the filter.

dispersive elements with complementary dispersions, such as linearly chirped FBGs (LCFBGs) with complementary chirps, by reflecting the phase-modulated optical signals from the LCFBGs with positive or negative chirps, microwave signals without or with π phase inversion are generated at the PD. An added advantage of using an optical PM is that a PM is not biased, which eliminates the bias drifting problem existing in a Mach–Zehnder modulator. The fundamental concept of the filter operation is shown in Fig. 4.10(a). An RF signal is applied to the optical PM via the RF port to phase-modulate the multiple wavelengths applied to the PM via the optical port. Since a PD functions as an envelope detector, if a phase-modulated signal is directly applied to a PD, no modulating signal will be recovered except a DC. This conclusion can also be explained based on the spectrum of a phase-modulated signal. As shown in Fig. 4.10(a), a small-signal phase-modulated signal has a spectrum with the +1st- and −1st-order sidebands out of phase. The beating between the optical carrier and the +1st-order sideband exactly cancels the beating between the optical carrier and the −1st-order sideband. However, if the phase-modulated optical signal passes through a dispersive element, the phase relationship between the two sidebands and the optical carrier will be changed, leading to the conversion from PM-IM. In addition, depending on the sign of the chromatic

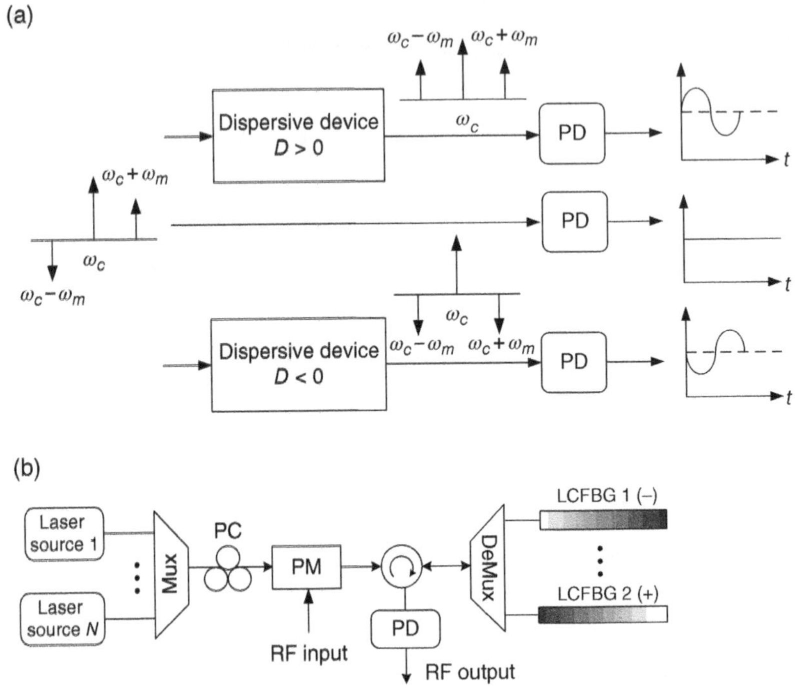

Figure 4.10 (a) RF phase inversion based on PM-IM conversion through opposite dispersions. (b) A photonic microwave delay-line filter with a negative coefficient based on PM-IM conversion in two LCFBGs with opposite dispersions.

dispersion, a recovered RF signal with or without a π phase inversion would be obtained, leading to the generation of negative coefficients. The system configuration of the filter is shown in Fig. 4.10(b).

Instead of using LCFBGs, the PM-IM conversion can also be realized using an optical frequency discriminator, such as a Sagnac loop filter, as shown in Fig. 4.11 [32]. By locating the optical carriers from the laser sources at the positive or negative slopes of the optical filter spectral response, PM-IM conversion with recovered microwave signals that are in phase or out of phase would be generated, leading to the generation of positive or negative tap coefficients.

Recently, a simple approach to generating negative coefficients in a photonic microwave delay-line filter using a polarization modulator (PolM) was proposed [33, 34]. A PolM is a special optical PM that can support both TE and TM modes but with opposite phase modulation indices. The operation principle is shown in Fig. 4.12(a). The light wave from a TLS is sent to a PolM via a polarization controller with its polarization direction aligned to have a 45° relative to one principal axis of the PolM, which is modulated by an input RF signal. Thanks to the polarization modulation at the PolM, two out-of-phase RF signals carried by two optical carriers with identical wavelengths but orthogonal polarizations are achieved at the output of the PolM. The optical microwave signals are fed into a section or two sections of polarization-maintaining fiber (PMF) to produce a delay line with two- or four-time delays. A photonic microwave bandpass filter of two or four taps with one or two negative coefficients was experimentally demonstrated [33].

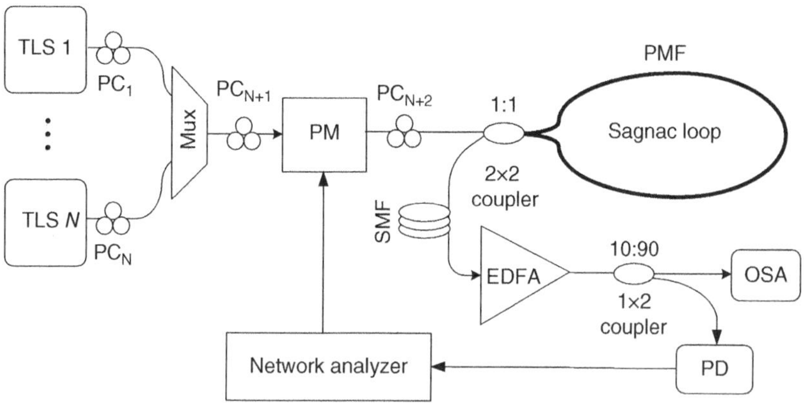

Figure 4.11 Photonic microwave delay-line filter with negative coefficients based on PM-IM conversion using optical frequency discriminators.

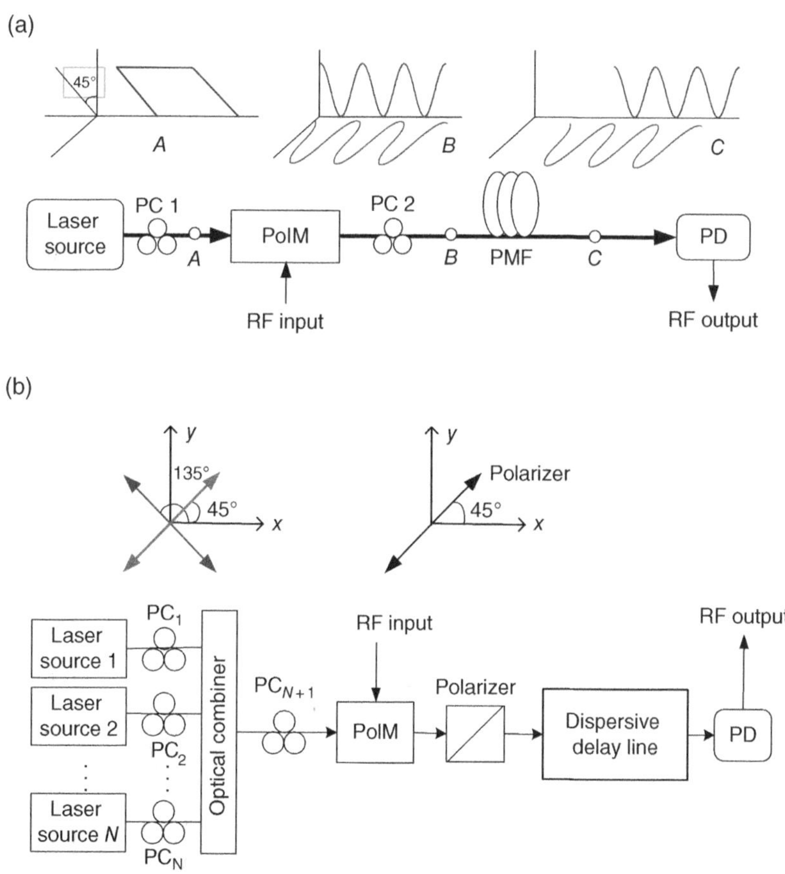

Figure 4.12 Photonic microwave delay-line filter with negative coefficients based on a polarization modulator. (a) A photonic microwave delay-line filter based on a PolM using a single wavelength with time delays generated by a section or two sections of PMF. (b) A photonic microwave delay-line filter based on a PolM using N wavelengths with the arbitrary number of taps and arbitrary tap coefficients.

To implement a photonic microwave delay-line filter with arbitrary number of taps and arbitrary tap coefficients, a modified structure based on a PolM was proposed, as shown in Fig. 4.12(b) [34]. Instead of using a single-wavelength light source, to achieve a microwave delay-line filter with N taps, a light source with N wavelengths or a laser array with N laser diodes is needed. An optical polarizer is connected at the output of the PolM with its transmission axis aligned at an angle of 45° to one principal axis of the PolM. By adjusting the polarization directions of the input light waves to be 45° or 135° to one principal axis of the PolM, an in-phase or out-of-phase intensity-modulated optical microwave signal is obtained at the output of the optical polarizer, leading to the generation of negative or positive coefficients. A time-delay difference between adjacent taps is generated by using a wavelength-dependent delay line, such as a dispersive fiber or an LCFBG.

4.2.2 Photonic Microwave Delay-Line Filters with Complex Coefficients

The tunability of a photonic microwave delay-line filter is usually achieved by adjusting the time-delay difference. However, the change in the time-delay difference would lead to a change in the FSR, which results in the change of the 3-dB bandwidth as well as the entire shape of the filter frequency response. For many applications, it is highly desirable that only the center frequency of the passband or stop-band is changed while maintaining the shape of the frequency response unchanged. A solution to this problem is to design a photonic microwave delay-line filter with complex coefficients.

An N-tap microwave delay-line filter with complex coefficients should have a transfer function given by

$$\begin{aligned} H(j\omega) &= a_0 + a_1 e^{-j\theta} \cdot e^{-j\omega T} + \cdots + a_{N-1} e^{-j(N-1)\theta} \cdot e^{-j\omega(N-1)T} \\ &= \sum_{n=0}^{N-1} a_n e^{-jn\theta} \cdot e^{-j\omega nT} \end{aligned} \tag{4.4}$$

where T is the time delay difference between two adjacent taps. To tune the filter while maintaining the shape of the frequency response unchanged, the phase shifts of all the taps should maintain a fixed relationship, as can be seen from (4.4). Therefore, the phase shift of each tap should be tuned independently.

Three photonic microwave delay-line filter architectures with complex coefficients have been reported [35–37]. In [35], a two-tap photonic microwave delay-line filter with one complex coefficient was implemented using a system consisting of three optical attenuators and two microwave couplers. Specifically, we expect the filter to have a transfer function given by

$$H(j\omega) = \cos(\omega T + \varphi) \tag{4.5}$$

where T is a constant. By tuning the phase φ, the filter transfer function will be shifted laterally along the horizontal direction without changing the shape. Eq. (4.5) can be rewritten as

$$\begin{aligned} H(j\omega) &= \frac{a}{2}\left(e^{j\omega T} + e^{-j\omega T}\right) - \frac{b}{2j}\left(e^{j\omega T} - e^{-j\omega T}\right) \\ &= e^{-j\omega T}\left\{\frac{a}{2}\left(e^{j\omega 2T} + 1\right) - \frac{b}{2j}\left(e^{j\omega 2T} - 1\right)\right\} \end{aligned} \tag{4.6}$$

where $a = \cos(\varphi)$, $b = \sin(\varphi)$. As can be seen, the filter has one complex coefficient $-b/2j$. The term $e^{-j\omega T}$ only introduces a linear phase, which will not affect the shape of the filter spectral response.

As can be seen from (4.6), the filter can be realized by a system shown in Fig. 4.13. By changing the values of a and b through tuning the attenuations of the variable attenuators, the frequency response of the filter is laterally shifted, while the FSR and the 3-dB bandwidth are maintained unchanged, as can be seen from Fig. 4.14. Since the values of a and b are changed by using the optical attenuators, the tuning range of a and b is from 0 to 1, which limits the RF phase shift from 0 to 180° or a tunable range of a half FSR. We should note that the complex coefficient is generated in the electrical domain after the PDs, the system is not all-optical, but hybrid.

An all-optical tunable photonic microwave delay-line filter with a complex coefficient was proposed and demonstrated [36]. The complex coefficient was generated by changing the phase of the RF signal, which was realized based on a combined use of optical single-sideband (SSB) modulation and stimulated Brillouin scattering (SBS). The experimental setup is shown in Fig. 4.15. It was demonstrated that the phase of a microwave signal carried by an optical carrier will experience microwave phase shift if the spectrum of the optical carrier or the sideband is falling in the SBS gain spectrum when passing through an optical fiber in which an SBS is resulted [38].

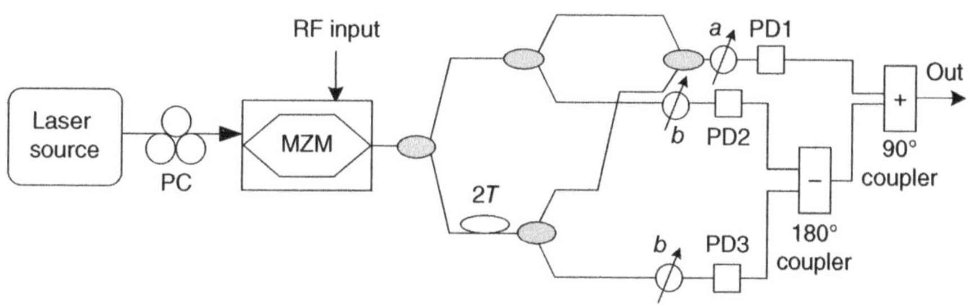

Figure 4.13 A two-tap photonic microwave filter with a complex coefficient using three optical attenuators and two microwave couplers.

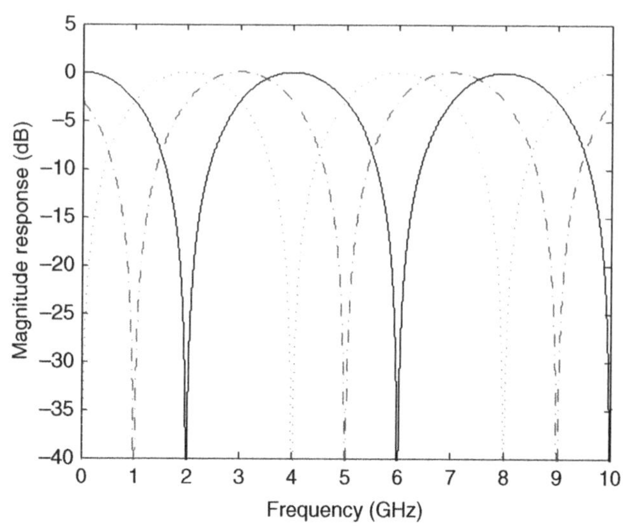

Figure 4.14 Frequency response of a photonic microwave delay-line filter with a complex coefficient. Solid: $a = 1$, $b = 0$; dotted: $a = 0$, $b = 1$; dash-dot: $a = b = 0.71$.

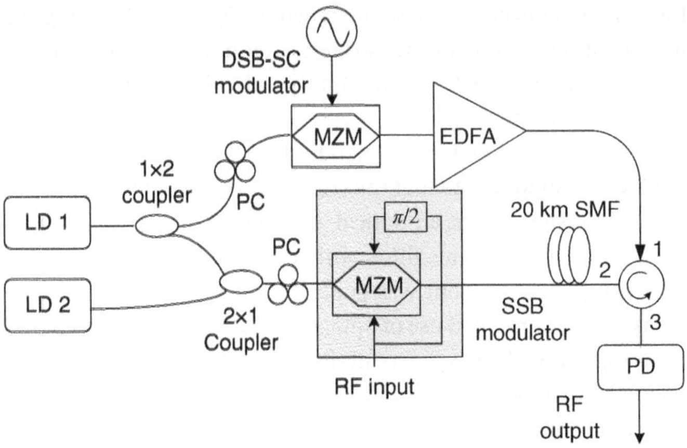

Figure 4.15 A two-tap photonic microwave delay-line filter with a complex coefficient based on SSB modulation and SBS.

Since the generation of a complex coefficient involves the use of an EDFA, an additional Mach–Zehnder modulator, and a long fiber (20 km in the experiment), to implement a multi-tap filter, the system will be extremely complicated, highly power-consuming, and costly. A simpler system architecture for a tunable photonic microwave delay-line filter with complex coefficients was demonstrated [37]. The complex coefficient is generated using a wideband tunable optical RF phase shifter that consists of two electro-optic Mach–Zehnder modulators, as shown in Fig. 4.16(a). The phase of the RF signal is shifted by simply adjusting the bias voltages applied to the two electro-optic Mach–Zehnder modulators, and the phase shift remains constant over the microwave spectral region of interest. Figure 4.16(b) shows the measured phase shifts for different bias voltages over a large microwave frequency band. As can be seen, the phase shifts are independent of the microwave frequency.

4.2.3 Nonuniformly Spaced Photonic Microwave Delay-Line Filters

The filters in [35–37] with complex coefficients have the potential to be extended to have multiple taps, but the filter complexity would be significantly increased. To design a photonic microwave delay-line filter with complex coefficients having a simple structure. A new concept by generating complex coefficients based on a delay-line filter architecture with nonuniformly spaced taps was reported. It was demonstrated the complex coefficients can be equivalently generated by introducing additional time delays to the taps [39].

It is known that a regular, uniformly spaced microwave delay-line filter has an impulse response $h_R(t)$ given by

$$h_R(t) = \sum_{k=0}^{N-1} a_k \delta(t - kT) \tag{4.7}$$

where N is the number of taps, a_k is the filter coefficient of the kth tap, $T = 2\pi/\Omega$ is the time delay difference between two adjacent taps, and Ω is the FSR of the filter. Apply the Fourier transform to (4.7), we have the frequency response of the filter,

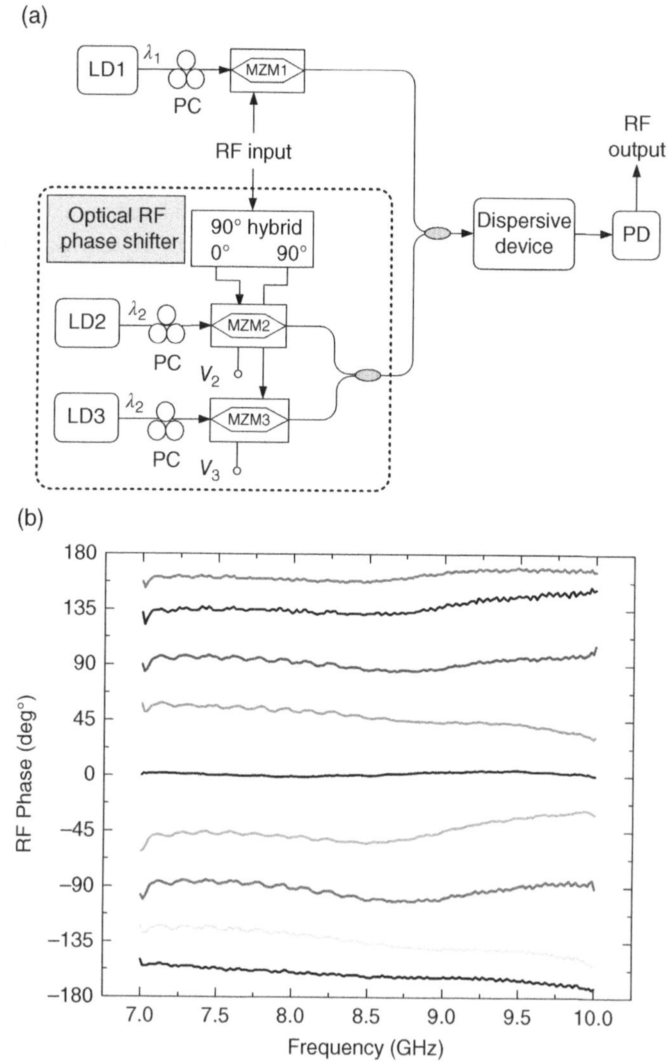

Figure 4.16 A photonic microwave delay-line filter with a complex coefficient generated based on an optical RF phase shifter. (a) The system architecture. (b) The measured phase shifts for different bias voltages over a large microwave frequency band. The phase shifts are independent of the microwave frequency.

$$H_R(\omega) = \sum_{k=0}^{N-1} \alpha_k \exp\left(-jk\frac{2\pi}{\Omega}\omega\right) \tag{4.8}$$

It is known that $H_R(\omega)$ has a multi-channel frequency response with adjacent channels separated by an FSR, with the mth channel located at $\omega = m\Omega$. Note that except for the different central frequencies, the frequency responses for all the channels are exactly identical.

In a regular photonic microwave delay-line filter based on incoherent detection, the coefficients are usually all positive, or special designs have to be incorporated to generate negative or complex coefficients, as discussed earlier in this section. However, a phase term can be introduced to a specific coefficient by adding an additional time delay at the specific tap, which is termed time-delay-based phase shift [39]. For example, at $\omega = m\Omega$, a time delay shift of $\Delta\tau$ will generate a phase shift given by $\Delta\varphi = -\Delta\tau \times m\Omega$. Note that such a phase shift is frequency-dependent, which is accurate only for the frequency at $m\Omega$, but approximately accurate for a narrow frequency band at around $m\Omega$. For most of applications, the filter is designed to have a very narrow frequency band. Therefore, for the frequency band of interest, the phase shift can be considered constant over the entire bandwidth. As a result, if the mth bandpass response, where $m \neq 0$, is considered, one can then achieve the desired phase shift at the kth tap by adjusting the time delay shift by $\Delta\tau_k$. Considering the time delay shift of $\Delta\tau_k$, one can get the frequency response of the nonuniformly spaced delay-line filter at around $\omega = m\Omega$,

$$
\begin{aligned}
H_N(\omega) &= \sum_{k=0}^{N-1} \alpha_k \exp\left[-j\left(k\frac{2\pi}{\Omega} + \Delta\tau_k\right)\omega\right] \\
&= \sum_{k=0}^{N-1} \alpha_k \exp(-j\omega\Delta\tau_k) \times \exp\left(-jk\frac{2\pi}{\Omega}\omega\right) \\
&\approx \sum_{k=0}^{N-1} [\alpha_k \exp(-jm\Omega\Delta\tau_k)] \times \exp\left(-jk\frac{2\pi}{\Omega}\omega\right)
\end{aligned}
\tag{4.9}
$$

As can be seen from (4.9), one can get an equivalent phase shift (EPS) for each tap coefficient. Specifically, if the desired phase shift for the kth tap is φ_k, the total time delay τ_k for the kth tap is $\tau_k = kT - \varphi_k/m\Omega$. As a result, if the time delay of each tap is adjusted, the filter coefficients would have the required phase shifts to generate the required passband with the desired bandpass characteristics.

A seven-tap photonic microwave delay-line filter with nonuniformly spaced taps to produce a flat-top bandpass frequency response was designed and experimentally demonstrated [39]. The experimental setup is shown in Fig. 4.17(a). Assuming that the passband of interest is at $m = 1$ and the frequency response of the bandpass has the shape of a rectangle, the corresponding impulse response should be a *Sinc* function, which has both positive and negative values extending to infinity along the horizontal axis. For practical implementation, the calculated impulse response should be cut off to enable a physically realizable filter. If a regular photonic microwave delay-line filter is employed to produce the frequency response, the filter coefficients can be selected to be [−0.12, 0, 0.64, 1, 0.64, 0, −0.12]. The frequency response is shown as a dotted line in Fig. 4.17(b), where $T = 82.6$ ps, which corresponds to an FSR of 12.1 GHz. The 3-dB bandwidth of the filter is 5.0 GHz with a central frequency of 12.1 GHz.

With the proposed technique, the same bandpass characteristics at $m = 1$ can be generated by using nonuniform spacing. The filter was designed to have all-positive coefficients of [0.12, 0, 0.64, 1, 0.64, 0, 0.12]. The time delays for the seven taps are then [−2.5T, −2T, -T, 0, T, 2T, 2.5T], with the taps nonuniformly spaced. Since at $m = 0$, there are no phase shifts introduced to the taps, there is always a baseband resonance which was eliminated by using an optical PM in the system. It is known the PM-IM conversion in a dispersive fiber would produce a frequency response with a notch at DC [40–42]. The frequency response of the PM-IM conversion is designed to make its first peak be located at $f = 12.1$ GHz, which is shown as the dash-dot line in Fig. 4.17(b). The overall frequency response of the nonuniformly spaced filter is calculated and shown as solid line in Fig. 4.17(b). It is clearly seen a flap-top frequency response is achieved in a photonic microwave delay-line filter with all-positive coefficients. The 3-dB bandwidth was 4.9 GHz and the central frequency was 12.1 GHz. The frequency response of the passband is close to that generated by a regular photonic microwave delay-line filter with true negative taps.

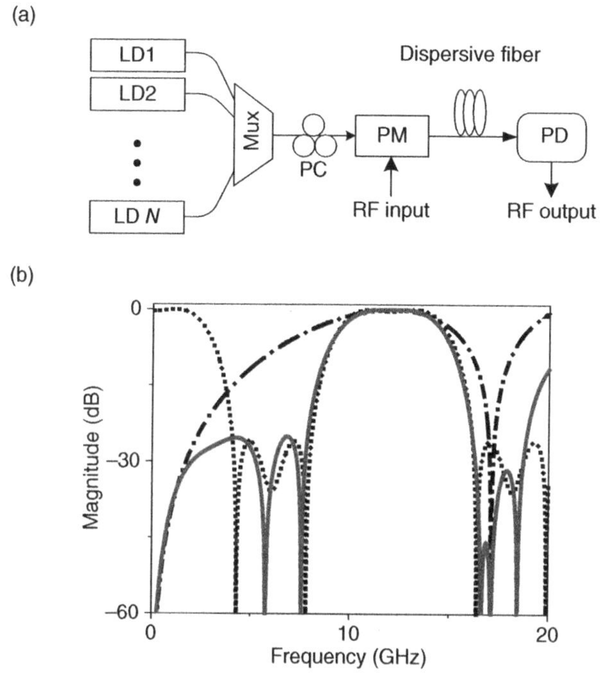

Figure 4.17 (a) A nonuniformly spaced photonic microwave delay-line filter. (b) Dotted line: the frequency response of a regular photonic microwave delay-line filter with true positive and negative coefficients. Dash-dot line: the frequency response of the PM-IM conversion. Solid line: the frequency response of the nonuniformly spaced photonic microwave delay-line filter.

As we mentioned earlier, the time-delay-based phase shift is frequency-dependent. As a result, the phase shifts due to the additional time delays would be different for different passbands. Therefore, in a nonuniformly spaced photonic microwave delay-line filter, the frequency responses for different spectral channels are different. In addition, within the mth passband, the time-delay-based phase shift is accurate only for the frequency at $m\Omega$, and approximately accurate for a narrow band at around $m\Omega$. The maximum error of the phase shift is determined by the maximum bandwidth of the bandpass concerned and the central frequency. For many applications, it is usually required that the bandwidth of the passband is very narrow, and then the errors due to the frequency-dependent phase shift are small and negligible, which ensures the effectiveness of the proposed technique.

The proposed technique to design a nonuniformly spaced microwave delay-line filter is particularly useful for applications such as arbitrary microwave waveform generation [43]. Recently, the same concept to introduce phase shifts to an RF pulse to implement RF pulse phase encoding was reported [43, 44]. A nonuniformly spaced microwave delay-line filter with a quadratic phase response for chirped microwave pulse generation was also reported [43, 45].

4.3 Optical Mixing of Microwave Signals

In an optical microwave signal processor, in addition to the functionality of microwave filtering, optical microwave mixing is another important functionality which can find applications in radio-over-fiber systems and other microwave systems for microwave frequency up- or down-conversion.

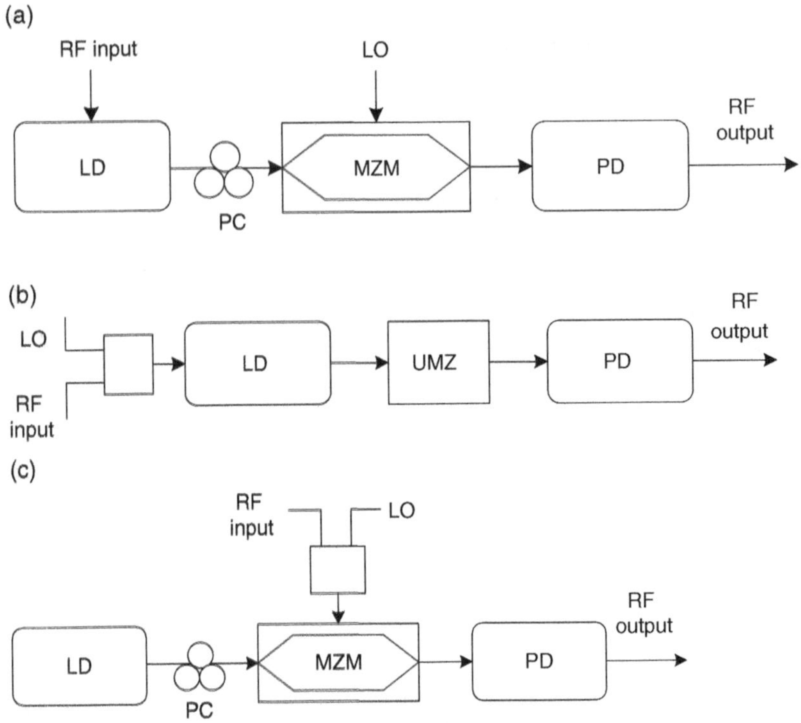

Figure 4.18 Architectures for all-optical microwave mixing. (a) Optical mixing that combines direct modulation and external modulation, (b) Optical microwave mixing based on direct modulation at a laser diode with an unbalanced interferometer to perform FM-IM conversion, (c) Optical microwave mixing based on external modulation.

Different mixing architectures have been summarized in [46, 47]. In this subsection, we will give a brief introduction to optical microwave mixing and then a few examples demonstrated recently will be provided.

Figure 4.18 shows three different optical microwave mixing architectures. The architecture shown in Fig. 4.18(a) combines direct modulation of an RF signal at a laser diode with an external modulation of a local oscillator (LO) signal at a Mach–Zehnder modulator. If the LO frequency is not too high, the system can be simplified, as shown in Fig. 4.18(b), in which both the RF signal and the LO are combined and sent to a laser diode. Due to the chirping at the laser diode, the output signal is frequency-modulated. The use of an unbalanced Mach–Zehnder (UMZ) interferometer, by locating the optical carrier at one slope of the spectral response, frequency-modulation to intensity-modulation (FM-IM) conversion is realized. The UMZ interferometer can also be replaced by an FBG. To increase the dynamic range, the FBG should be designed to maximize the linear region of the slope. In addition, by locating the optical carrier at a point closer to the bottom of the FBG spectral response, the dynamic range can also be improved since the optical carrier-induced noise at the PD can be significantly reduced, leading to an increased spurious-free dynamic range (SFDR) [48, 49]. For some applications, both the RF signal frequency and the LO frequency are high. In this case, the mixing has to be implemented at an external modulator, by combining the RF signal and the LO using a microwave combiner and then apply to the external modulator. One concern in an optical microwave mixing system is the existence of many unwanted frequency components, which are usually eliminated in the electrical domain using a microwave filter. To implement an all-optical microwave mixer, both the mixing and bandpass filtering have to be realized in the optical domain. An approach was reported which functions simultaneously as an all-optical microwave mixer and an all-optical microwave bandpass filter [50]. The schematic of the system is shown in Fig. 4.19.

The architecture is similar to that shown in Fig. 4.18(c), except that the modulator is an optical PM and the light source is a multiwavelength source. The microwave mixing and filtering functions are simultaneously implemented. The mixing operation is done at the PM, to which a large LO signal is applied to make the PM operate in the nonlinear region, leading to the generation of different mixing spectral components. The filtering operation is realized thanks to the use of a multiwavelength source. If the number of wavelengths is N, the system is an N-tap microwave delay-line filter. Considering that the PM-IM conversion in a dispersive fiber will generate a notch at DC, the overall frequency response of the system is a bandpass filter with baseband resonance eliminated by the notch at DC [40, 41]. Therefore, the entire system functions as a microwave mixer and a bandpass filter. By properly designing the filter to make the central frequency of the passband locate at the frequency of interest, an up- or down-converted microwave signal free of other frequency components is obtained.

An experiment was performed to achieve microwave frequency up-conversion from 3 to 11.8 GHz [50]. In the experimental setup, a fiber ring laser with 30 wavelengths and a wavelength spacing of 0.2 nm was used as the multiwavelength source [51]. The dispersive fiber was a single-mode fiber. Considering a wavelength spacing of 0.2 nm, a length of 25 km standard single-mode fiber would lead to a frequency response with the central frequency of the passband located at 11.8 GHz. Therefore, the up-covered microwave component at 11.8 GHz is selected. Figure 4.20 shows the experimental results of the mixing output, with and without filtering operation. Figure 4.20(a) shows the spectrum of the mixing components when the light source was a single wavelength source, with no bandpass filtering operation. All frequency components were observed. Figure 4.20(b) shows the spectrum when the 30-wavelength laser source was used. The system was operating as a bandpass filter. The up-converted frequency component at 11.8 GHz was selected with other components being suppressed.

The use of a PM for optical microwave mixing to study the system performance with data transmission has been investigated in [52, 53].

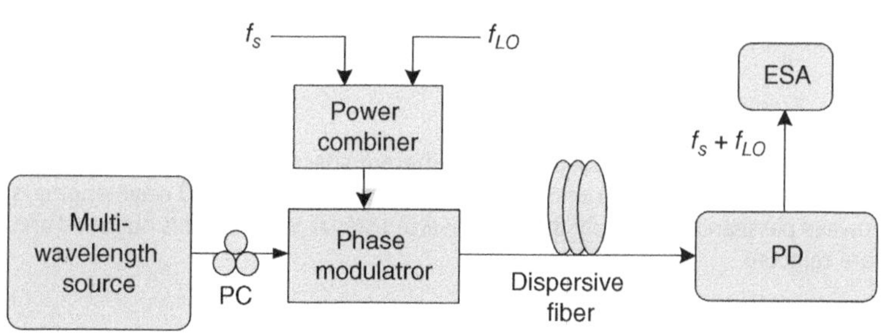

Figure 4.19 An all-optical microwave signal processor that functions simultaneously as an all-optical mixer and a bandpass filter.

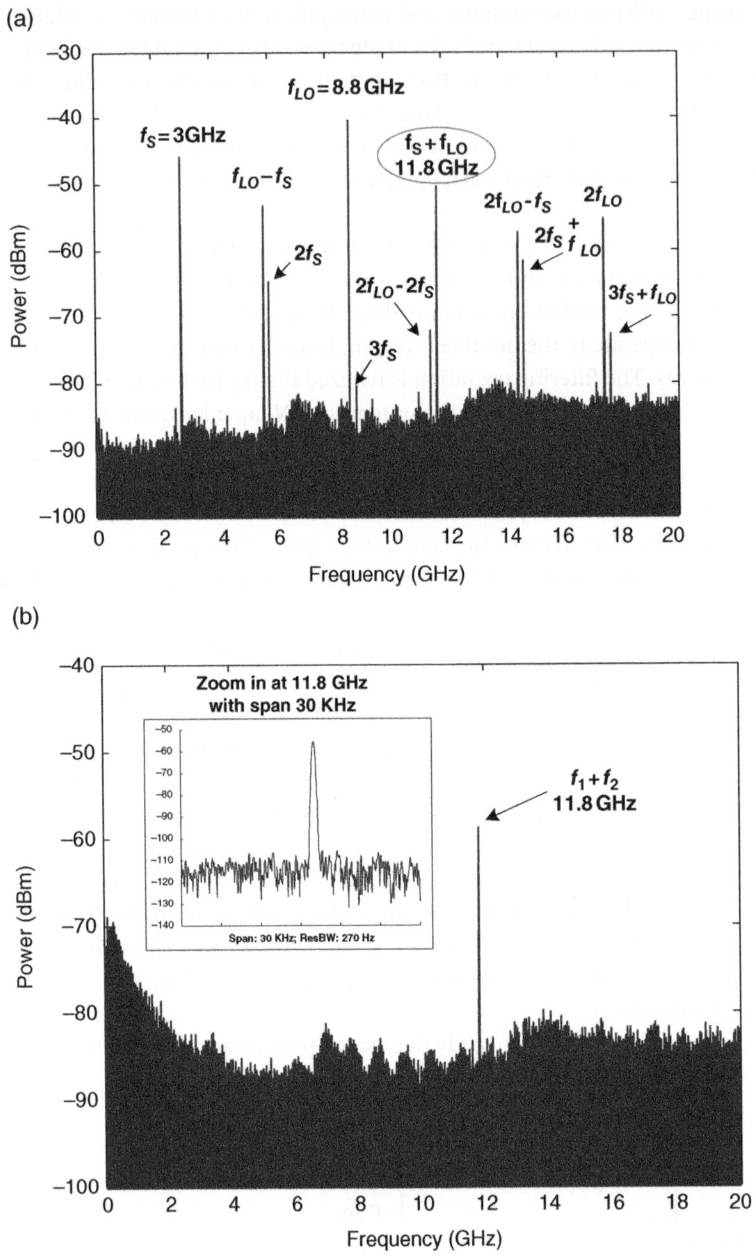

Figure 4.20 Power spectra at the output of the photodetector. (a) Different spectral components are observed if a single wavelength source is used. (b) When a multiwavelength source with over 30 wavelengths is used, the system is also a microwave passband filter. Only the up-converted signal at 11.8 GHz is obtained and other frequency components are rejected.

4.4 Coherent Microwave Photonic Filters

The delay-line microwave photonic filters discussed above belong to incoherent microwave photonic filters, where optical interferences are eliminated by using either a broadband incoherent light source or a laser array. In fact, a microwave photonic filter can also be implemented in the coherent regime using a single wavelength light source. Since a coherent microwave photonic filter does not have a delay line configuration, optical interferences are not an issue that would affect the stable operation of the filter.

A general structure of a coherent microwave photonic filter is shown in Fig. 4.21. A narrow linewidth light wave from a laser source is sent to a PM. Assuming small signal modulation, at the output of the PM, an optical carrier and two sidebands are generated. Note that for phase modulation, the two sidebands are out of phase. Thus, the detection of a phase-modulated signal directly at a PD will not generate an RF signal except a DC since the beating between the optical carrier and the lower sideband will cancel completely the beating between the optical carrier and the upper sideband. However, if one of the sidebands is removed by an optical notch filter in transmission or using a dual passband filter in reflection [54], such as an FBG or two cascaded FBGs, then the phase-modulated signal is converted to a SSB intensity-modulated signal and the detection of the SSB intensity-modulated signal at a PD will generate an RF signal. Figure 4.21 shows a coherent microwave photonic filter, in which an optical notch filter is used to filter out one sideband of a phase-modulated signal, thus achieving PM-IM conversion. As can be seen, the entire operation is equivalent to a microwave filter with the bandwidth determined by the bandwidth of the notch of the optical notch filter (an FBG). The center frequency is determined by the wavelength spacing between the optical carrier and the center wavelength of the notch. Thus, by simply tuning the center frequency of the notch filter or the wavelength of the laser source, the center frequency of the microwave bandpass filter can be tuned. When tuning, the spectral shape is maintained unchanged. This is an advantage as compared with an incoherent microwave photonic filter, where the spectral shape changes when the filter is tuned, or complex coefficients must be used to avoid such a spectral shape change.

In [54], instead of using a single FBG in transmission, two cascaded FBGs are used in reflection, by using one FBG to select the optical carrier and the other FBG to select one sideband, a phase-modulated signal is converted to a SSB intensity-modulated signal. A microwave photonic filter with the spectral shape identical to the notch of the second FBG is generated. The major limitation of the technique in [54] is the large passband since a uniform FBG was employed. The passband of

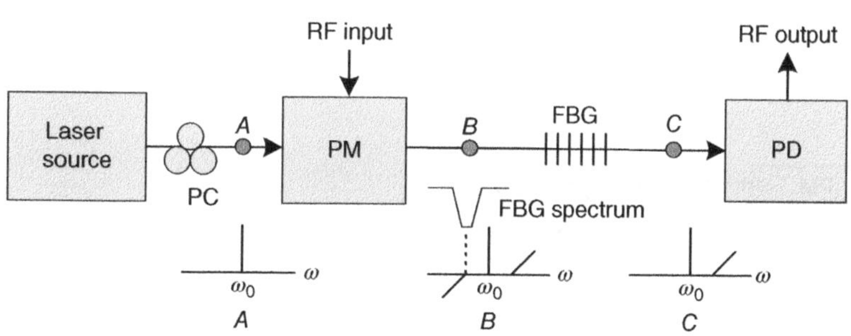

Figure 4.21 Coherent microwave photonic filter, in which an optical notch filter is used to filter out one sideband of a phase-modulated signal, thus achieving PM-IM conversion.

the microwave photonic filter is determined by the bandwidth of the FBG to select the sideband. By using a ring resonator [55], the passband can be smaller, but still too large for most applications.

Due to the ultra-narrow bandwidth of an SBS gain or loss spectrum, a microwave photonic filter with a narrow passband can be implemented based on phase modulation and PM-IM conversion in an SBS gain or loss spectrum. Figure 4.22 shows a configuration to implement a narrowband microwave filter based on SBS. A light wave from a laser source at ω_0 is split into two paths. At the upper path, the light is modulated by an LO signal at a dual-parallel MZM (DP-MZM) to generate a SSB with a suppressed carrier (SSB-SC) signal. Assume that the frequency of the LO signal is ω_{LO}, the generated lower sideband is at $\omega_0 - \omega_{LO}$. The SSB-SC signal is sent to an EDFA to increase its power and then launched into an SMF to stimulate the SBS effect. The SBS gain and loss are located at $\omega_O - \omega_{LO} - \omega_B$ and $\omega_O - \omega_{LO} + \omega_B$, where ω_B is the SBS frequency shift. If one sideband of a phase-modulated signal is amplified by the SBS gain, then the phase-modulated signal is converted to an intensity-modulated signal, and the detection of the intensity-modulated signal at a PD will generate a microwave signal at $\omega_{LO} - \omega_B$, as shown in Fig. 4.22. The entire operation of the system is equivalent to a microwave bandpass filter with the passband determined by the SBS gain spectrum. The advantage of this approach is that the filter can be tunable by tuning the LO frequency and the passband is very narrow. However, the system is quite complicated especially a high-power EDFA and a long SMF are needed.

A simple solution is to use an optical filter with an ultra-narrow passband. Figure 4.23 shows the implementation of a coherent microwave photonic filter using a phase-shifted FBG (PS-FBG) to perform PM-IM conversion [56]. A PS-FBG is a special FBG with a phase shift introduced to the FBG during the inscription process [56]. It is known for a uniform FBG, if a phase shift is introduced to the FBG, an ultra-narrow notch with a phase jump in the notch would be generated in the reflection band [57]. If a phase-modulated optical signal is injected into the PS-FBG that is employed to modify the magnitude and the phase of one sideband, PM-IM conversion would be achieved, which would lead to the implementation of an ultra-narrow passband, as shown in Fig. 4.24. Figure 4.25 shows the frequency response of a coherent microwave photonic filter using a PS-FBG. The wavelength of the optical carrier is tuned such that the lower sideband falls in the notch of the PS-FBG when the microwave frequency is equal to the difference between the frequency of the optical carrier and the center frequency of the notch. By increasing the wavelength of the optical carrier, the center frequency of the microwave photonic filter is accordingly increased.

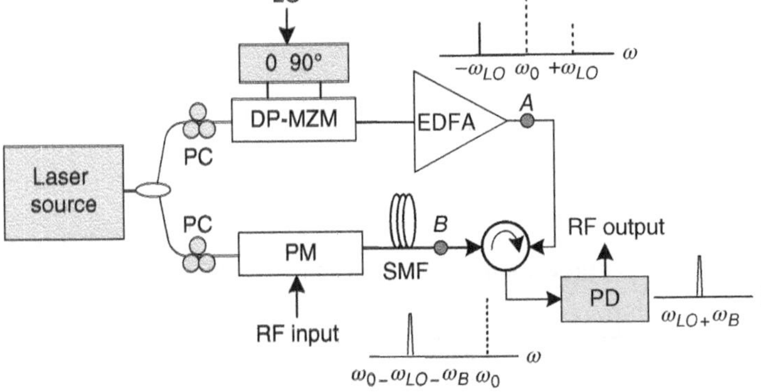

Figure 4.22 A coherent microwave photonic filter implemented based on phase modulation and PM-IM using the SBS gain.

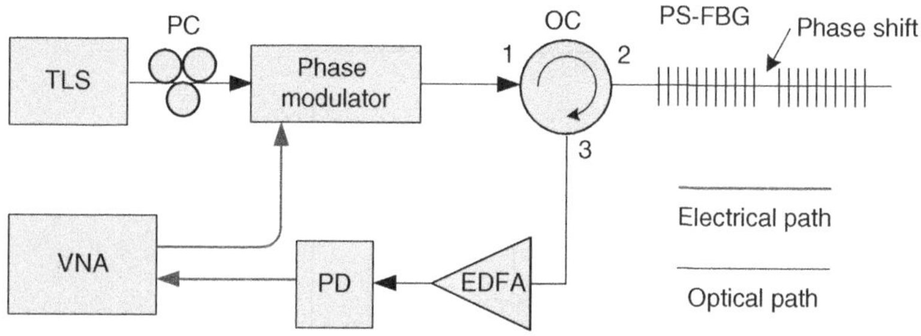

Figure 4.23 A coherent microwave photonic filter implemented based on phase modulation and PM-IM conversion using a PS-FBG.

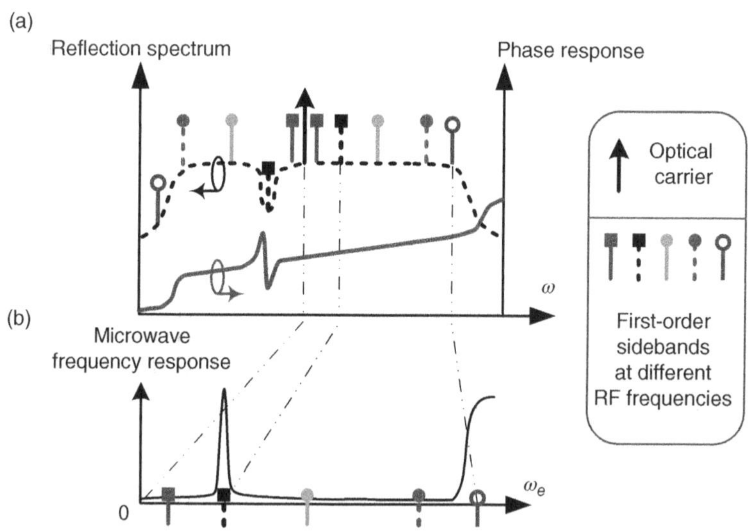

Figure 4.24 The operation of a coherent microwave photonic filter. (a) The reflection spectrum (dashed line) and phase response (solid line) of the PS-FBG. (b) The frequency response of the coherent microwave photonic filter.

Figure 4.25(a) shows the superimposed frequency responses of the microwave photonic filter with a tunable central frequency covering a frequency range of about 15 GHz with a tuning step of 1.45 GHz. From Fig. 4.25(a) we can also see that the ratio of the transmission peak to the sidelobe can be as large as 40 dB. In Fig. 4.25(b), a zoom-in view of the measured frequency response at about 6.9 GHz is shown. The 3-dB bandwidth is about 60 MHz. The bandwidth can be further decreased by using a PS-FBG with a narrower notch.

The same concept to implement a coherent microwave photonic filter based on PM-IM conversion to translate the spectral response of an optical filter to that of a microwave photonic filter can be used to implement a microwave photonic filter with a flat top [58]. Instead of using SBS or a PS-FBG, a specially designed superstructured FBG (SFBG) is used. An SFBG is a special FBG that is spatially modulated by a periodic sampling function. Due to the spatial sampling, the spectral

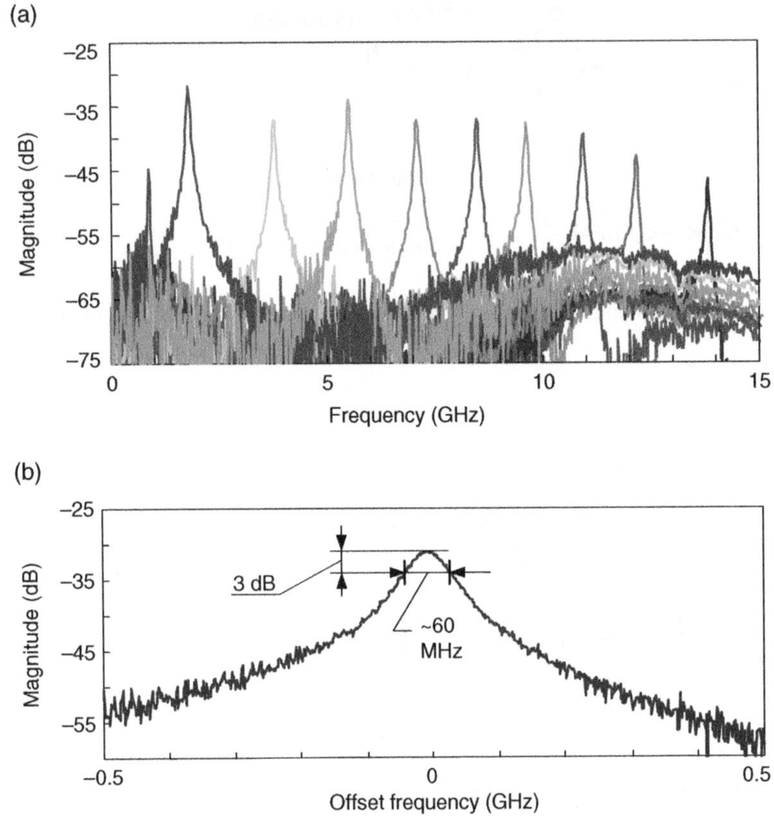

Figure 4.25 (a) Measured frequency responses of the coherent microwave photonic filter when a PS-FBG is employed. (b) The zoom-in view of the frequency response when the center frequency is tuned at 6.9 GHz.

response has multiple channels. If one period of the sampling function is intentionally increased by a half period, a π phase shift is introduced equivalently to the grating, leading to a narrow notch. The technique is called EPS technique [59]. To have a flat bottom in the notch, the SFBG is designed to have two closely spaced notches which are achieved by introducing two EPSs to the structure. Each phase shift would produce a Lorentz-shaped notch and the combination of the two closely spaced Lorentz-shaped notches would lead to a notch with a flat bottom. When the SFBG with a flat-bottom notch is incorporated to perform PM-IM conversion, a passband with a flat top is achieved. The reflection bandwidth and notch width of the SFBG, which determine the frequency tunable range and bandwidth of the passband, can be controlled by controlling the length and maximum index modulation (MIM) of the SFBG. Figure 4.26(a) shows a simulated reflection spectrum of a PS-FBG with a single phase shift and Fig. 4.26(b) shows a simulated reflection spectrum of an SFBG with two phase shifts. Figure 4.26(c) shows the structure of an SFBG in which two EPSs are incorporated.

Figure 4.27(a) shows the magnitude and phase response in reflection of an SFBG with a length of 32.4 mm. Due to the limited resolution of the optical vector analyzer, the notch depth and shape cannot be precisely shown. Figure 4.27(b) shows the frequency response of the microwave photonic filter. As can be seen, it has a narrow passband with a flat top. The flatness of the passband is within

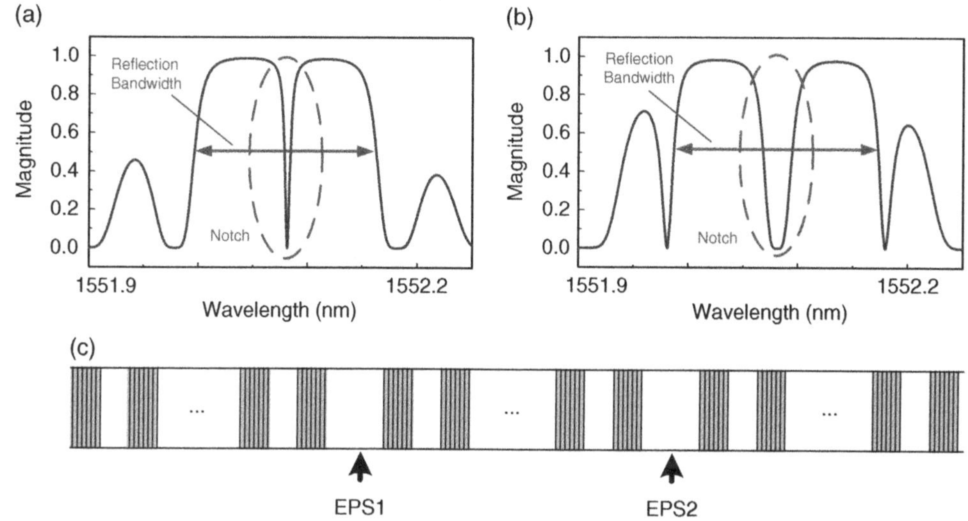

Figure 4.26 (a) Simulated reflection spectrum of a PS-FBG with a single phase shift. (b) Simulated reflection spectrum of an SFBG with two phase shifts. (c) Structure of an SFBG.

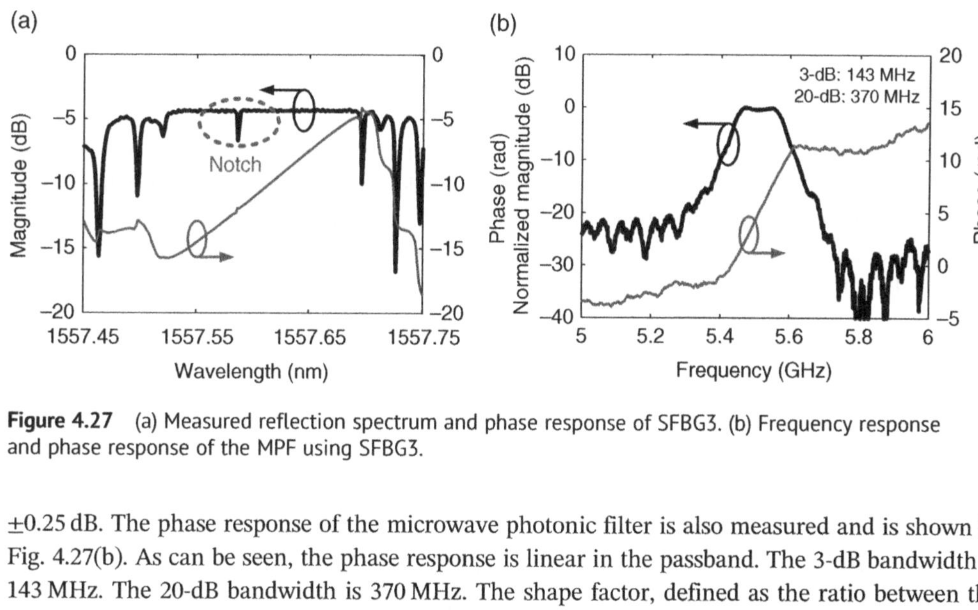

Figure 4.27 (a) Measured reflection spectrum and phase response of SFBG3. (b) Frequency response and phase response of the MPF using SFBG3.

±0.25 dB. The phase response of the microwave photonic filter is also measured and is shown in Fig. 4.27(b). As can be seen, the phase response is linear in the passband. The 3-dB bandwidth is 143 MHz. The 20-dB bandwidth is 370 MHz. The shape factor, defined as the ratio between the 20-dB and 3-dB bandwidths, is calculated to be 2.6, which is much smaller than those reported in [56]. A smaller shape factor represents a better selectivity. By tuning the wavelength of the optical carrier, the central frequency of the passband is tuned from 0.4–6.4 GHz, as shown in Fig. 4.28. The magnitude and bandwidth of the passband keep almost unchanged during the tuning, a feature that is important for applications where the loss and bandwidth are required to be constant. The insertion loss is measured, which is about 28 dB. The use of a high power-handling PD may reduce the loss [56]. The SFDR of the MPF is also measured, which is about 83 dB $Hz^{2/3}$.

For certain applications, microwave filters with dual passbands are needed due to the increasing demand for multiband/multifunctional microwave systems that support various modern services.

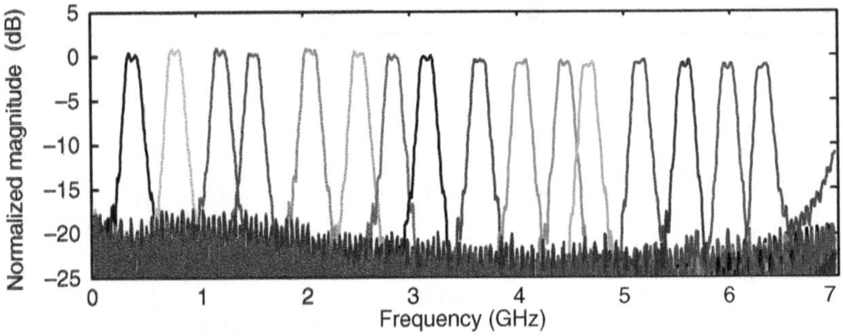

Figure 4.28 Measured frequency response of the narrow and flat-top MPF with a tunable passband from 0.4 to 6.4 GHz.

Such systems require microwave circuits and components that can handle several different frequency bands. The same concept can also be used to implement a microwave photonic filter with two passbands [42]. The key component in a dual-band microwave photonic filter is an EPS-FBG, which is designed and fabricated by using the EPS technique [59]. The EPS-FBG has multiple channels due to the spatial sampling of the grating structure. EPSs introduced to the ±1st channels are realized through changing the sampling function. Therefore, the fabrication of an EPS-FBG is significantly simplified since the control of the spatial sampling is on a micrometer scale while the control of the phase shift in the fabrication of a true PS-FBG is on a nanometer scale. In the design, two π phase shifts are introduced to both of the ±1st channels to produce an ultra-narrow and flat-bottom notch in each channel. Figure 4.29 shows a simulated reflection spectrum of an EPS-FBG with two π phase shifts. As can be seen, the notches are flat due to the use of two-phase shifts. A bandpass filter is achieved due to the PM-IM conversion by filtering out one sideband of a phase-modulated signal. In a dual-band microwave photonic filter, the PM-IM conversion is performed at both the ±1st channels, thus two independently tunable passbands are generated, as shown in Fig. 4.30. Because two π phase shifts are introduced to achieve flat-bottom notches,

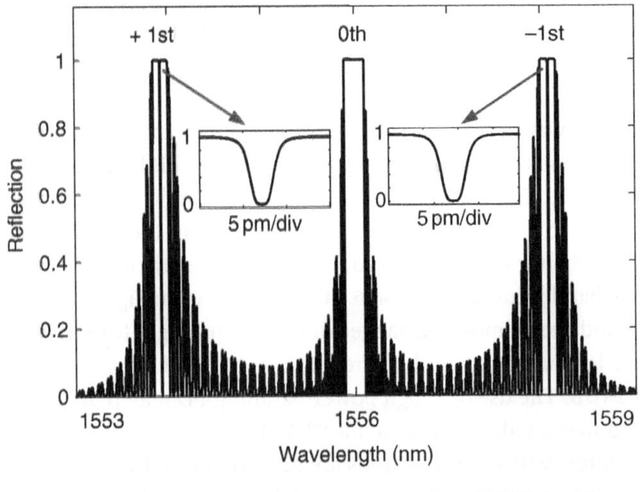

Figure 4.29 Simulated reflection spectrum of an EPS-FBG with two-phase shifts.

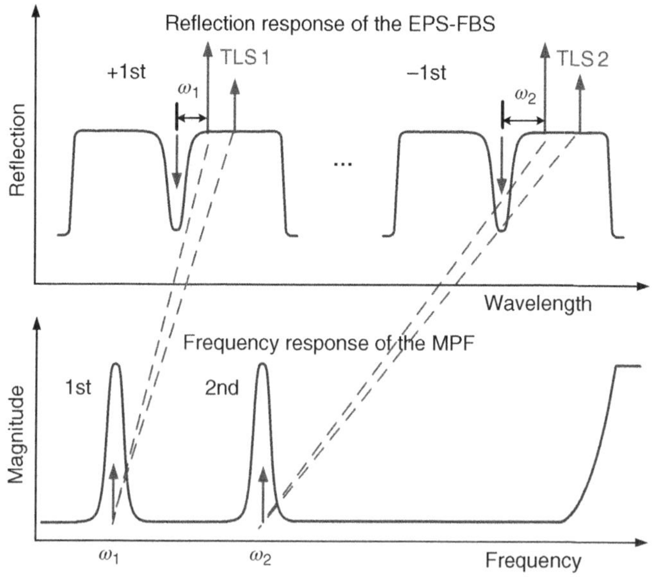

Figure 4.30 Illustration of the generation of the two passbands.

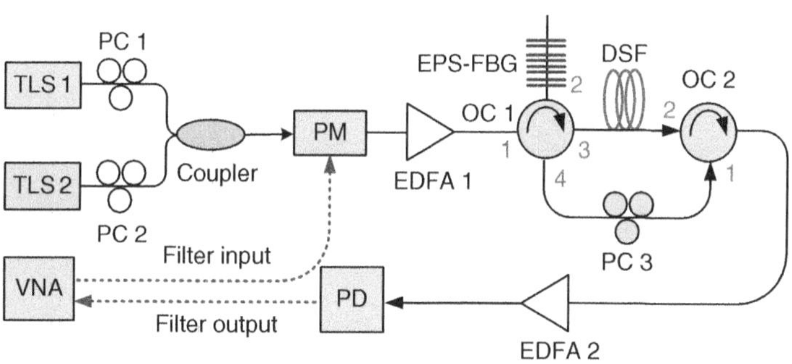

Figure 4.31 Schematic of a dual-passband MPF. DSF, dispersion-shifted fiber; EDFA, erbium-doped fiber amplifier; EPS-FBG, equivalent phase-shifted fiber Bragg grating; OC, optical circulator; PC, polarization controller; PD, photodetector; PM, phase modulator; TLS, tunable laser source; VNA, vector network analyzer.

the passbands have small shape factors. Here, the shape factor is again defined as the ratio between the 20 and 3-dB bandwidths.

To produce two independent passbands, two optical carriers from two TLSs (TLS1 and TLS2) are tuned to locate at the ±1st channels of the EPS-FBG, as shown in Fig. 4.31. When one sideband of a phase-modulated signal is suppressed by the notch in the +1st channel, a passband with a central frequency of ω_1, the frequency difference between the optical carrier and the notch in the +1st channel, will be produced due to the PM-IM conversion. Similarly, when one sideband of a phase-modulated signal is suppressed by the notch in the −1st channel, a passband with a central frequency of ω_2, the frequency difference between the optical carrier and the notch in the −1st channel, will also be produced. Therefore, there will be two passbands with the central frequencies of ω_1 and ω_2. The central frequencies of the passbands can be tuned by shifting the wavelengths of TLS1 and TLS2 independently. Figure 4.32 shows the spectral response of the dual passband filter. The

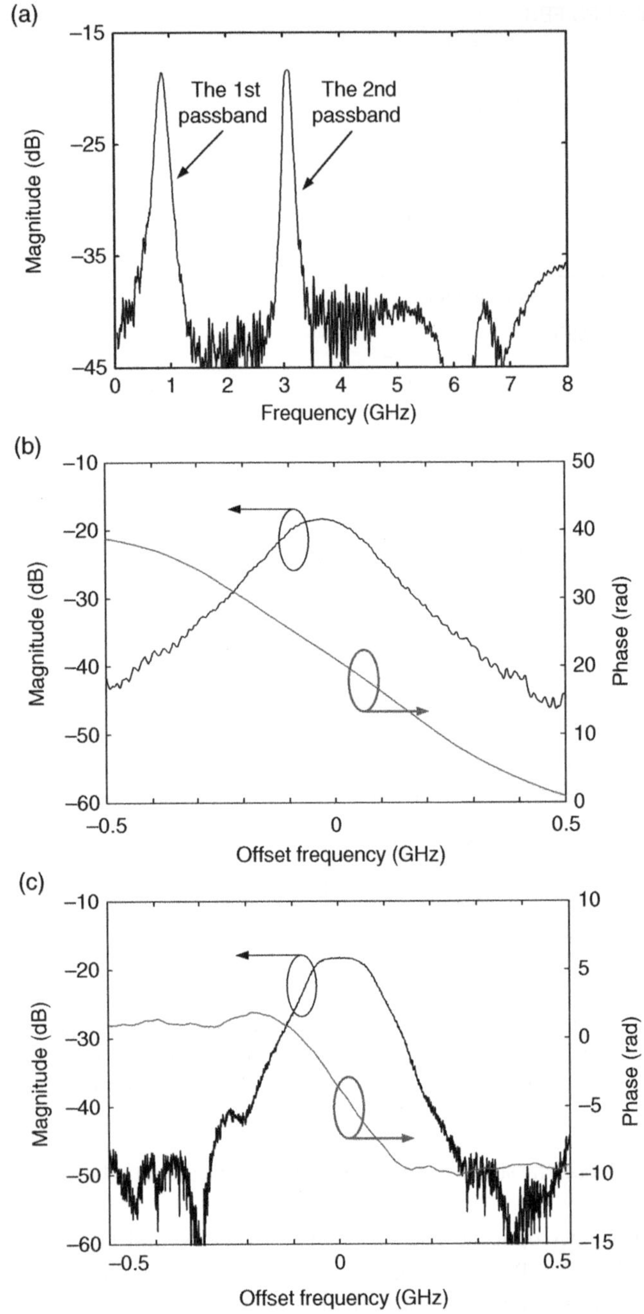

Figure 4.32 (a) Frequency response of the dual-passband MPF. (b) Frequency response and phase response of the 1st passband. (c) Frequency response and phase response of the 2nd passband.

bandwidth and shape factor of the 1st passband are 167.3 MHz and 3.8, respectively, and those of the 2nd passband are 143.4 MHz and 3.3, respectively. The 1st and 2nd passbands have frequency tunable ranges of 5.4 and 7.4 GHz, respectively. The central frequencies of both passbands can be continuously tuned. Again, the shape of the spectral response of the two passbands is maintained unchanged during tuning. The variations of the magnitude over the tunable ranges are maintained within ±0.5 dB.

The SFDRs for the two passbands and the noise figures (NFs) of the filter are two important parameters that define the performance of the filter. The SFDRs and the NFs can be improved if the optical carriers are partially suppressed while maintaining the powers to the PD unchanged. This operation is equivalent to increasing the filter gains. This can be done using an SBS-assisted filter, as shown in Fig. 4.31. The SBS-assisted filter is implemented using a dispersion-shifted fiber (DSF). When the light waves pass through the DSF, and the powers of the optical carriers are above the SBS threshold, Stokes waves are generated, which are traveling along the DSF in an opposite direction of the optical carriers. Note that, under small-signal modulation conditions, the powers of the sidebands are below the SBS threshold. Thus, SBS occurs only at the optical carriers and the sidebands are kept unchanged. The Stokes waves go through PC3, which is used to adjust the polarization state of the Stokes waves and then they are launched into the DSF via OC2. As the Stokes waves consecutively circulate in the ring, the optical carriers are suppressed. When the optical power at the input of the PD is constant, the suppression of the optical carriers will increase the gain of the system. The SFDRs for the two passbands are 82.9 dB $Hz^{2/3}$ and 83.2 dB $Hz^{2/3}$, respectively, when the SBS-assisted filter is not incorporated into the system. After connecting PC3 with port 1 of OC2, the optical carriers are partially suppressed, and when the optical power at the input of the PD is kept identical to what is without carrier suppression, the gain of the system is improved by about 10 dB. The measured SFDRs of the 1st and 2nd passbands are then 89.8 dB $Hz^{2/3}$ and 90.6 dB $Hz^{2/3}$, respectively, which are improved by about 7 dB, and the NFs of the two passband filters are decreased by 10 dB.

4.5 Dynamic Range of a Microwave Photonic Filter

An incoherent microwave photonic filter usually has a spectral response that is periodic (the period is equal to the FSR). To achieve a microwave photonic filter with a single passband, one may use a broadband optical source sliced by a Mach–Zehnder interferometer (MZI) [60]. Due to the continuous nature of the optical spectrum, an MPF with a single passband can be realized. However, the NF of the MPF is very large and the dynamic range is small due to the strong relative intensity noise (RIN) of the broadband optical source. On the other hand, a single bandpass MPF can be implemented in the coherent regime using a PM and an optical filter with an ultra-narrow notch, where a TLS with a low RIN is utilized. By using the notch of the optical notch filter to remove one sideband of the phase-modulated signal, the phase-modulated signal is converted to an SSB intensity-modulated signal. At the output of a PD, a microwave signal is detected. The entire operation is equivalent to a single passband MPF with the center frequency tunable by tuning the wavelength of the optical carrier. The optical notch filter can be a whispering gallery mode resonator [61], a fiber or integrated ring resonator [62] or a phase-shifted fiber Bragg grating (PS-FBG) [56]. Although a TLS is used in the MPF, the SFDR is still limited due to the low gain of the MPF. In [63], a single bandpass MPF with an improved dynamic range using an SBS-assisted adaptive filter was reported. By partially suppressing the optical carrier, the gain of the MPF is improved, which leads to the improvement of the SFDR. The major limitation of the approach in [63] is the large size. To trigger the SBS effect, a long fiber loop in the SBS-based optical filter is needed. In addition, the input optical power is also very high, which increases the complexity of the MPF.

A single bandpass MPF with an improved dynamic range can be implemented using a dual-parallel Mach–Zehnder modulator (DP-MZM) and a PS-FBG [64]. A DP-MZM has an MZI structure with two MZMs in the two arms and a phase shifter in the lower arm. The DP-MZM is utilized to generate an equivalent phase-modulated (EPM) signal with an adjustable optical carrier power by controlling the DC bias voltage applied to the internal MZM in the upper arm. The EPM signal with a partially suppressed optical carrier at the output of the DP-MZM is amplified by an EDFA and sent to the PS-FBG via an OC, where one sideband of the EPM signal is removed and the EPM signal is converted to an SSB intensity-modulated signal. The operation is known as PM-IM conversion. Upon detection at a PD, a microwave signal is generated and the entire operation is equivalent to a single bandpass MPF with the central frequency of the passband determined by wavelength difference between the optical carrier and notch wavelength. Due to the partial suppression of the optical carrier, the carrier-to-sideband ratio (CSR) is decreased. If the optical power at the input of the PD is maintained unchanged by amplifying the EPM signal using the EDFA, the gain of the MPF is enhanced, which results in the improvement of the SFDR. An experiment is performed. A single bandpass MPF with a passband width of 150 MHz and a frequency-tunable range of about 5.5 GHz with an improved SFDR by 11 dB is demonstrated.

4.6 Conclusion

An overview of microwave photonic filters implemented in the incoherent and coherent regime has been provided. For microwave photonic filters implemented in the incoherent regime, an incoherent light source, usually a broadband light source, such as an LED source, an ASE source, or a laser array, is needed. The problem with using a broadband light source is the low optical power and high noise, which may make the microwave photonic filters have very poor NF. The use of a laser array can avoid the problem, but the cost is high. In addition, the tuning of the filter is usually performed by tuning the wavelengths of the laser array, which is very complicated (the wavelength spacing must be tuned). In addition, the implementation of an incoherent microwave photonic filter with an arbitrary spectral response requires the filter to have complex coefficients, which are hard to realize. The use of the nonuniform tap spacing concept can simplify the implementation, but the complex coefficients are realized via changing the tap spacing which is frequency-dependent and only accurate for a specific frequency. Thus, the frequency response is precise for narrowband operation. On the other hand, the implementation of microwave photonic filters in the coherent regime can have a much simpler configuration. The filter spectral response is, in fact, the spectral response of an optical filter that is translated to the microwave domain. Thus, the key to implementing a coherent microwave photonic filter is to design an optical filter with a well-defined spectral response. The tuning of a coherent microwave photonic filter can be simply done by tuning the wavelength of the coherent light source.

Although both incoherent and coherent microwave photonic filters offer advantages of high operating frequency and large frequency tuning, the filters usually have poor NF and small dynamic range. The use of high-power handling PDs may increase the noise performance, but the dynamic range can only be improved by using techniques such as pre- or post-linearization which should be done in the electronic domain, making the filters complicated and costly. An effective solution to increase the dynamic range is to cancel the third-order intermodulation terms – a technique that has been used for microwave photonic links [65].

References

1 Wilner, K. and Van den Heuvel, A.P. (1976). Fiber-optic delay lines for microwave signal processing. *Proc. IEEE* 64 (5): 805–807.

2 Heyde, E.C. and Minasian, R.A. (1994). A solution to the synthesis problem of recirculating optical delay line filter. *IEEE Photonics Technol. Lett.* 6 (7): 833–835.

3 Capmany, J., Cascon, J., Martin, J.L. et al. (1995). Synthesis of fiber-optic delay line filters. *J. Lightw. Technol.* 13 (10): 2003–2012.

4 Capmany, J. and Martin, J. (1995). Solutions to the synthesis problem of optical delay line filters. *Opt. Lett.* 20 (23): 2438–2440.

5 Hunter, D.B. and Minasian, R. (1995). Reflectivity tapped fiber optic transversal filter using in-fiber Bragg gratings. *Electron. Lett.* 31 (12): 1010–1012.

6 Hunter, D.B. and Minasian, R. (1996). Microwave optical filters using in-fiber Bragg grating arrays. *IEEE Microw. Guid. Wave Lett.* 6 (2): 103–105.

7 Hunter, D.B. and Minasian, R. (1997). Photonic signal processing of microwave signals using an active-fiber Bragg-grating-pair structure. *IEEE Trans. Microw. Theory Tech.* 45 (8): 1463–1466.

8 Marti, J., Ramos, F., and Laming, R.I. (1998). Photonic microwave filter employing multimode optical sources and wideband chirped fiber gratings. *Electron. Lett.* 34 (18): 1760–1761.

9 Capmany, J., Pastor, D., and Ortega, B. (1999). New and flexible fiber-optic delay-line filters using chirped Bragg gratings and laser arrays. *IEEE Trans. Microw. Theory Technol.* 47 (7): 1321–1326.

10 Marti, J., Polo, V., Ramos, F., and Moodie, D. (1999). Photonics tunable microwave filters employing electro-absorption modulators and wideband chirped fiber gratings. *Electron. Lett.* 35 (4): 305–306.

11 Yu, G., Zhang, W., and Williams, J.A.R. (2000). High-performance microwave transversal filter using fiber Bragg grating arrays. *IEEE Photonics Technol. Lett.* 12 (9): 1183–1185.

12 Pastor, D., Capmany, J., and Ortega, B. (2001). Broad-band tunable microwave transversal notch filter based on tunable uniform fiber Bragg gratings as slicing filters. *IEEE Photonics Technol. Lett.* 13 (7): 726–728.

13 Zhang, W., Williams, J.A.R., and Bennion, I. (2001). Polarization synthesized optical transversal filter employing high birefringence fiber gratings. *IEEE Photonics Technol. Lett.* 13 (5): 523–525.

14 Pastor, D., Capmany, J., and Ortega, B. (2000). Experimental demonstration of parallel fiber-optic-based RF filtering using WDM techniques. *IEEE Photonics Technol. Lett.* 12 (1): 77–78.

15 Polo, V., Vidal, B., Corral, J.L., and Marti, J. (2003). Novel tunable photonics microwave filter based on laser arrays and N × N AWG-based delay lines. *IEEE Photonics Technol. Lett.* 15 (4): 584–586.

16 Cusick, T.A., Iezekiel, S., Miles, R.E. et al. (1997). Synthesis of all-optical microwave filters using Mach–Zehnder lattices. *IEEE Trans. Microw. Theory Tech.* 45 (8): 1458–1462.

17 Norton, D., Johns, S., Keefer, C., and Soref, R. (1994). Tunable microwave filtering using high dispersion fiber time delays. *IEEE Photonics Technol. Lett.* 6 (7): 831–832.

18 Lee, K.H., Choi, W.Y., Choi, S., and Oh, K. (2003). A novel tunable fiber-optic microwave filter using multimode DCF. *IEEE Photonics Technol. Lett.* 15 (7): 969–971.

19 Capmany, J., Ortega, B., and Pastor, D. (2006). A tutorial on microwave photonic filters. *J. Lightw. Technol.* 24 (1): 201–229.

20 Minasian, R.A. (2006). Photonic signal processing of microwave signals. *IEEE Trans. Microw. Theory Tech.* 54 (2): 832–846.

21 Yao, J. (2015). Photonics to the rescue: a fresh look at microwave photonic filters. *IEEE Microw. Mag.* 16 (8): 46–60.

22 Capmany, J., Mora, J., Gasulla, I. et al. (2013). Microwave photonic signal processing. *J. Lightw. Technol.* 31 (4): 571–586.

23 Sales, S., Capmany, J., Marti, J., and Pastor, D. (1995). Experimental demonstration of fiber-optic delay line filters with negative coefficients. *Electron. Lett.* 31 (13): 1095–1096.

24 Coppinger, F., Yegnanarayanan, S., Trinh, P.D., and Jalali, B. (1997). All-optical RF filter using amplitude inversion in a semiconductor optical amplifier. *IEEE Trans. Microw. Theory Tech.* 45 (8): 1473–1477.

25 Mukai, T., Inoue, K., and Saitoh, T. (1987). Homogeneous gain saturation in 1.5 m InGaAsP traveling-wave semiconductor laser amplifiers. *Appl. Phys. Lett.* 51 (6): 381–383.

26 Wang, X. and Chan, K.T. (2000). Tunable all-optical incoherent bipolar delay-line filter using injection-locked Fabry–Perot laser and fiber Bragg gratings. *Electron. Lett.* 36 (24): 2001–2002.

27 Li, S., Chiang, K.S., Gambling, W.A. et al. (2000). A novel tunable all-optical incoherent negative-tap fiber-optical transversal filter based on a DFB laser diode and fiber Bragg gratings. *IEEE Photonics Technol. Lett.* 12 (9): 1207–1209.

28 Mora, J., Andres, M.V., Cruz, J.L. et al. (2003). Tunable all-optical negative multitap microwave filters based on uniform fiber Bragg gratings. *Opt. Lett.* 28 (15): 1308–1310.

29 Capmany, J., Pastor, D., Martinez, A. et al. (2003). Microwave photonics filter with negative coefficients based on phase inversion in an electro-optic modulator. *Opt. Lett.* 28 (16): 1415–1417.

30 Vidal, B., Corral, J.L., and Marti, J. (2006). All-optical WDM multi-tap microwave filter with flat bandpass. *Opt. Express* 14 (2): 581–586.

31 Zeng, F., Wang, J., and J.Y. (2005). All-optical microwave bandpass filter with negative coefficients based on a phase modulator and linearly chirped fiber Bragg gratings. *Opt. Lett.* 30 (17): 2203–2205.

32 Wang, J., Zeng, F., and J.Y. (2005). All-optical microwave bandpass filter with negative coefficients based on PM-IM conversion. *IEEE Photonics Technol. Lett.* 17 (10): 2176–2178.

33 J.Y. and Wang, Q. (2007). Photonic microwave bandpass filter with negative coefficients using a polarization modulator. *IEEE Photonics Technol. Lett.* 19 (9): 644–646.

34 Wang, Q. and J.Y. (2008). Multi-tap photonic microwave filters with arbitrary positive and negative coefficients using a polarization modulator and an optical polarizer. *IEEE Photonics Technol. Lett.* 20 (2): 78–80.

35 You, N. and Minasian, R.A. (2001). A novel tunable microwave optical notch filter. *IEEE Trans. Microw. Theory Tech.* 49 (10, pt. 2): 2002–2005.

36 Loayssa, A., Capmany, J., Sagues, M., and Mora, J. (2006). Demonstration of incoherent microwave photonic filters with all-optical complex coefficients. *IEEE Photonics Technol. Lett.* 18 (16): 1744–1746.

37 Yan, Y. and J.Y. (2007). A tunable photonic microwave filter with a complex coefficient using an optical RF phase shifter. *IEEE Photonics Technol. Lett.* 19 (19): 1472–1474.

38 Loayssa, A., Hernandez, R., Benito, D., and Galech, S. (2004). Characterization of stimulated Brillouin scattering spectra by use of optical single-sideband modulation. *Opt. Lett.* 29 (6): 638–640.

39 Dai, Y. and J.Y. (2008). Nonuniformly-spaced photonic microwave delay-line filter. *Opt. Express* 16 (7): 4713–4718.

40 Zeng, F. and J.Y. (2004). All-optical bandpass microwave filter based on an electro-optic phase modulator. *Opt. Express* 12 (16): 3814–3819.

41 Zeng, F. and J.Y. (2005). Investigation of phase modulator based all-optical bandpass microwave filter. *J. Lightw. Technol.* 23 (4): 1721–1728.

42 Gao, L., Zhang, J., Chen, X., and Yao, J. (2014). Microwave photonic filter with two independently tunable passbands using a phase modulator and an equivalent phase-shifted fiber Bragg grating. *IEEE Trans. Microw. Theory. Tech.* 62 (2): 380–387.

43 Dai, Y. and Yao, J. (2010). Nonuniformly spaced photonic microwave delay-line filters and applications. *IEEE Trans. Microw. Theory. Tech.* 58 (11): 3279–3289.

44 Dai, Y. and J.Y. (2007). Microwave pulse phase encoding using a photonic microwave delay-line filter. *Opt. Lett.* 32 (24): 3486–3488.

45 Dai, Y. and J.Y. (2009). Chirped RF pulse generation using a photonic microwave filter with a quadratic phase response. *IEEE Photonics Technol. Lett.* 21 (9): 569–571.

46 Cabon, B., Le Guennec, Y., Lourdiane, M., and Maury, G. (2006). Photonic mixing in RF modulated optical links. *LEOS* 2006: 408–409.

47 Tang, Z.Z., Li, Y.F., Yao, J.P., and Pan, S.L. (2020). Photonics-based microwave frequency mixing: methodology and application. *Laser Photonics Rev.* 14: 1800350-1-25.

48 Yan, Y. and J.Y. (2008). Photonic microwave bandpass filter with improved dynamic range. *Opt. Lett.* 33 (15): 1756–1758.

49 Zhang, J., and Darcie, T.E. (2006), Low-biased microwave-photonic link using optical frequency or phase modulation and fiber-Bragg-grating discriminator. *Optical Fiber Communication Conference* (5–10 March 2006), Anaheim, Calif., paper OWG1.

50 Zeng, F. and J.Y. (2005). All-optical microwave mixing and bandpass filtering in a radio-over-fiber link. *IEEE Photonics Technol. Lett.* 17 (4): 899–901.

51 Yao, J., J.Y., Deng, Z., and Liu, J. (2005). Investigation of room-temperature multiwavelength fiber-ring laser that incorporates an SOA-based phase modulator in the laser cavity. *J. Lightw. Technol.* 23 (8): 2484–2490.

52 J.Y., Maury, G., Guennec, Y.L., and Cabon, B. (2005). All-optical subcarrier frequency conversion using an electrooptic phase modulator. *IEEE Photonics Technol. Lett.* 17 (11): 2427–2429.

53 Le Guennec, Y., Maury, G., J.Y., and Cabon, B. (2006). New optical microwave up-conversion solution in radio-over-fiber networks for 60 GHz wireless applications. *J. Lightw. Technol.* 24 (3): 1277–1282.

54 Yi, X. and Minasian, R.A. (2009). Microwave photonic filter with single bandpass response. *Electron. Lett.* 45 (7): 362–363.

55 Palací, J., Villanueva, G.E., Galán, J.V. et al. (2010). Single bandpass photonic microwave filter based on a notch ring resonator. *IEEE Photonics Technol. Lett.* 22 (17): 1276–1278.

56 Li, W., Li, M., and J.Y. (2012). A narrow-passband and frequency-tunable micro-wave photonic filter based on phase-modulation to intensity-modulation conversion using a phase-shifted fiber Bragg grating. *IEEE Trans. Microw. Theory Tech.* 60 (5): 1287–1296.

57 Erdogan, T. (1997). Fiber grating spectra. *J. Lightw. Technol.* 15 (8): 277–1294.

58 Gao, L., Chen, X., and J.Y. (2013). Microwave photonic filter with a narrow and flat-top passband. *IEEE Microw. Wirel. Compon. Lett.* 23 (7): 362–364.

59 Dai, Y., Chen, X., Jiang, D. et al. (2004). Equivalent phase shift in a fiber Bragg grating achieved by changing the sampling period. *IEEE Photonics Technol. Lett.* 16 (10): 2284–2286.

60 Mora, J., Ortega, B., Diez, A. et al. (2006). Photonic microwave tunable single-bandpass filter based on a Mach–Zehnder interferometer. *J. Lightwave Technol.* 24 (7): 2500–2509.

61 Savchenkov, A.A., Matsko, A.B., Strekalov, D. et al. (2005). Mode filtering in optical whispering gallery resonators. *Electron. Lett.* 41 (8): 495–497.

62 Palací, J., Villanueva, G.E., Galán, J.V. et al. (2010). Single bandpass photonic microwave filter based on a notch ring resonator. *IEEE Photon. Technol. Lett.* 22 (17): 1276–1278.

63 Li, W. and Yao, J. (2013). A narrow-passband frequency-tunable microwave photonic filter with an improved dynamic range. *Optical Fiber Communication Conference/National Fiber Optic Engineers Conference 2013*, OSA Technical Digest (online), Optica Publishing Group, paper OTu2H.3.

64 Han, X. and Yao, J. (2016). Tunable single bandpass microwave photonic filter with an improved dynamic range. *IEEE Photon. Technol. Lett.* 28 (1): 11–14.

65 Li, W., Kong, F., and Yao, J. (2013). Dynamic range improvement of a microwave photonic link based on bi-directional use of a polarization modulator in a Sagnac loop. *Opt. Express* 21 (13): 15692–15697.

5

Photonic Distribution of Microwave Signals

5.1 Introduction

The photonic distribution of radiofrequency (RF) signals lies at the heart of microwave photonics (MWP) [1–4]. This chapter deals with this topic in certain detail dividing its contents into two main parts. The first part focuses on the MWP link (MWPL), which is the cornerstone for the implementation of RF signal distribution systems and networks by optical means. A brief introduction to MWPLs given in Sections 5.2 and 5.3 covers the important topic of simple MWPLs where the optical source and detector are separated by a lossy and dispersionless optical link. These simple MWPLs provide an excellent framework to introduce the fundamental concepts related to the sources of degradation and the figures of merit (FOMs), which are employed to evaluate their impact. Section 5.4 covers the more general topic of filtered MWP links that embrace all kinds of possible configurations. A full mathematical model is provided that leads to formal expressions for the computation of the relevant FOMs both in intensity and phase-modulated links.

The second part of the chapter deals with the main application of MWPLs, which is in the implementation of fiber-wireless (FW) telecommunication systems and networks. After a brief introduction to these systems in Section 5.5, we proceed to describe in Section 5.6 the main techniques that are currently employed for the optical transport of wireless signals. The specialization of the degradation mechanisms in these systems is discussed in Section 5.7 together with the main techniques reported in the literature for overcoming them. Section 5.8 discusses two topics, which are relevant for the incorporation of FW systems into the existing wavelength division multiplexing (WDM) access network infrastructure: first, the design of specialized optical layer subsystems such as optical add-drop multiplexers (OADMs), multichannel multiplexers, and demultiplexers required at the central office (CO) and remote nodes (RNs) and, second, the design of optoelectronic frontends for the base stations (BSs). The chapter is finalized in Section 5.9 where we touch upon two important application fields: converged access networks where digital and radio services must share the same fiber infrastructure and in-home and in-building networks.

5.2 Introduction to Microwave Photonics Links

Microwave photonics links (MWPLs) are the building blocks of any system or network in charge of the distribution of RF signals by optical means [5–11]. A general layout of an MWPL is shown in Fig. 5.1 for reference and it will serve us to introduce some basic definitions and concepts.

Microwave Photonics, First Edition. Jianping Yao, José Capmany, and Ming Li.

Companion website: www.wiley.com/go/yao/microwavephotonics

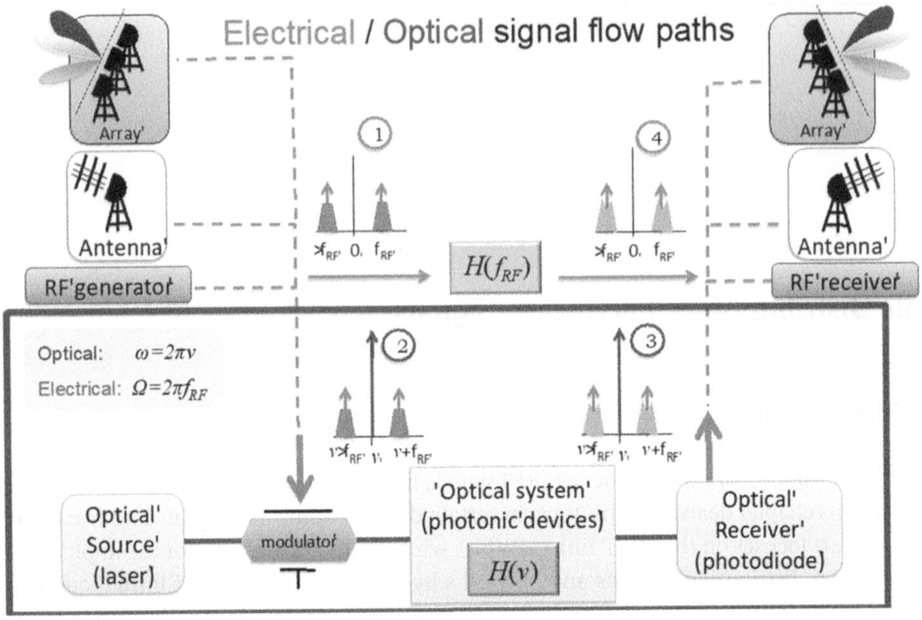

Figure 5.1 General layout of a microwave photonic link.

The starting point is an input RF signal with spectrum sidebands centered at frequency $\pm f_{RF}$, as shown in point 1. This signal, which can have its origin either from an RF generator or from the detection by means of a single or an array of antennas, modulates the output of an optical source to upconvert its spectrum to the optical region of the spectrum (point 2), such that the sidebands are now centered at $\nu \pm f_{RF}$, where ν represents the central frequency of the optical source. The combined optical signal is then processed by an optical system composed of several photonic devices (including the possibility of an optical fiber link) and characterized by an overall lumped optical field transfer function $H(\nu)$. The MWPL modifies the spectral characteristics of the sidebands as illustrated in point 3. Finally, an optical detector is employed to down-convert the sidebands again to the RF part of the spectrum by suitable beating with the optical carrier. So, the recovered RF signal, processed by the MWP link (as shown in point 4), is ready to be sent to an RF receiver or to be re-radiated. The overall electrical transfer function $H(f_{RF})$ is also shown and must not be confused with the optical field transfer function of the auxiliary optical system.

MWP links can range from very simple to very complex configurations depending, fundamentally, on the nature of the auxiliary optical system transfer function $H(\nu)$. The simplest configuration, which historically was the first to be considered, corresponds to the case where the optical system is a dispersionless optical fiber link. Though limited in its applications, this configuration is quite adequate for introducing the concepts related to the operation and FOMs.

5.3 Figures of Merit of a Simple Microwave Photonic Link

Figure 5.2 shows the diagram of a simple MWPL [5], where the optical transmitter and the optical receiver are connected by means of a lossy and dispersionless L km optical fiber link. Fiber losses are given by an attenuation coefficient α in dB/km.

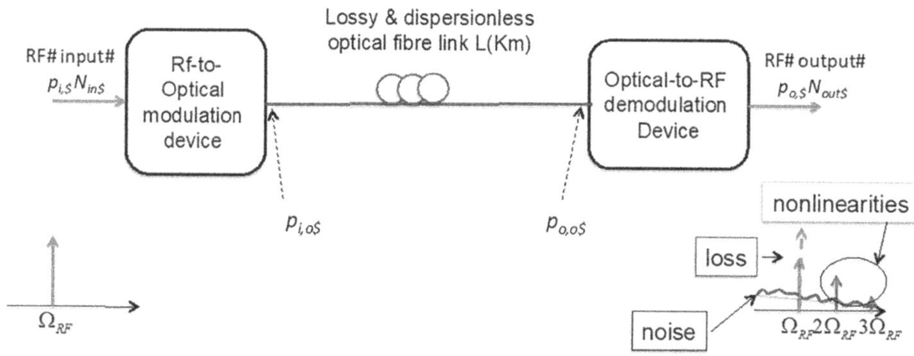

Figure 5.2 Diagram of a simple MWP link showing the relevant signal nomenclature at different points of the link and the input (bottom left-hand side)/output (bottom right-hand side) signal spectra for an input RF tone.

If an input RF tone at frequency Ω_{RF} is injected into the link as shown in the bottom left-hand side inset, the output RF signal, shown in the bottom right-hand side inset, is generally composed of three contributions: the desired input tone itself with a decreased amplitude due to the system losses, noise spread over a given spectral region, and nonlinear harmonic distortion contributions appearing at $n\Omega_{RF}$ $n = 2, 3, \ldots$. These three effects characterize the quality of the MWP in terms of three FOMs (figures of merit): the RF link gain G_{RF}, the noise figure *NF*, and the dynamic or spurious-free dynamic range (SFDR). In very simple terms, G_{RF} accounts for the overall end-to-end system losses, *NF* accounts for the overall noise level introduced by the system components, and the *SFDR* qualifies the impact of nonlinearities. In the following subsections, we will provide a simple definition and description of these three magnitudes, deferring a deeper description of them to Section 5.4.

5.3.1 RF Gain

Referring to the MWP configuration of Fig. 5.2, let p_i represent the RF power of the input tone at frequency Ω_{RF} injected to the MWP link. In the same way, p_o represents the RF power of the output tone at frequency Ω_{RF} extracted from the MWP link. The *RF Link Gain* in natural units is defined as [5–7, 9]:

$$G_{RF} = \frac{p_o}{p_i} \tag{5.1}$$

It is convenient to re-express (5.1) in terms of some intermediate auxiliary values of the optical power at the input and the output of the optical fiber link. If we define $p_{i,o}$ as the rms power magnitude of the time-varying optical power at the link input and $p_{o,o}$ as the rms power magnitude of the time-varying optical power at the link output, then:

$$p_{o,o} = 10^{-((\alpha L)/(10))} p_{i,o} = T_{M-D} p_{i,o} \tag{5.2}$$

where $T_{M\text{-}D}$ represents the optical transmittance function from the optical transmitter to the optical detector. Using the relationship given by (5.2) in (5.1), we get [6]:

$$G_{RF} = \left(\frac{p_o}{p_{o,o}^2}\right) \cdot T_{M-D}^2 \cdot \left(\frac{p_{i,o}^2}{p_i}\right) \tag{5.3}$$

or in decibel units:

$$G_{RF}(\text{dB}) = 10\log\left(\frac{p_o}{p_{o,o}^2}\right) + 20\log T_{M\text{-}D} + 10\log\left(\frac{p_{i,o}^2}{p_i}\right) \tag{5.4}$$

Equation (5.3) seems in principle an unnecessary complication of the simple version given in (5.1). However, as we shall see, it turns out to be extremely useful in practice, as it is the product of three factors that separately characterize the performance of the optical receiver, the optical link, and the optical transmitter, respectively.

The $p_{i,o}^2/p_i$ factor represents the equivalent input-output RF power transmission at the optical transmitter. It depends on whether direct modulation of the laser current or external modulation of the output power of a continuous wave (CW) optical source is considered. It also depends on the electrical circuit configuration employed to match the resistance R_s of the current/voltage RF source to the input resistance R_{in} of the laser/external modulator. Figure 5.3 illustrates the case when an external Mach–Zehnder intensity modulator (MZM) is employed.

The optical power transmission versus input modulation voltage of the MZM has been derived in Chapter 2:

$$T_{MZM} = \frac{p_{i,o}}{P_o} = \frac{\alpha_{MZM}}{2}\left[1 + \cos\left(\frac{\pi V_M}{V_\pi}\right)\right] \tag{5.5}$$

where P_o represents the output power from the CW laser (input power to the modulator), α_{MZM} is the MZM optical loss, and the modulation voltage is expressed as $V_T = V_M + v_M(t)$, where V_M is the bias voltage and $v_M(t)$ the time-varying RF signal. The modulation efficiency in terms of output

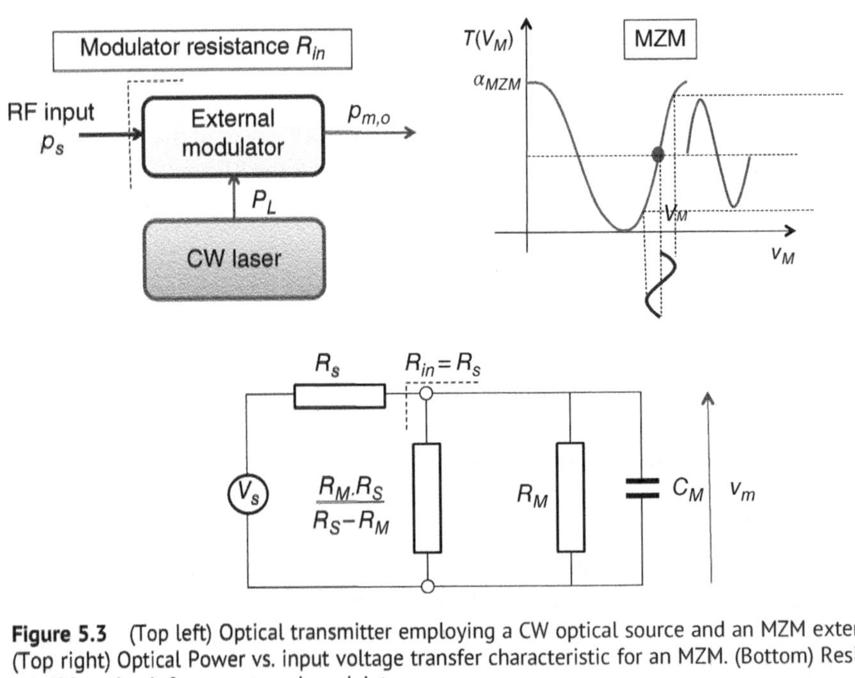

Figure 5.3 (Top left) Optical transmitter employing a CW optical source and an MZM external modulator. (Top right) Optical Power vs. input voltage transfer characteristic for an MZM. (Bottom) Resistive impedance matching circuit for an external modulator.

optical power vs. input electrical current for an MZM biased at the quadrature point ($V_M = V_\pi/2$) is given by:

$$s_m \triangleq \left(\frac{dp_{i,o}}{di_M}\right)_{I_M} = R_{in}\left(\frac{dp_{i,o}}{dv_M}\right)_{V_M} = R_{in}P_o\left(\frac{dT(v_M)}{dv_M}\right)_{V_M} = \frac{\pi\alpha_{MZM}P_oR_{in}}{2V_\pi} \tag{5.6}$$

In the case of using a resistively matched network between the voltage source and the modulator as shown in the bottom part of Fig. 5.3, it is straightforward to show by means of circuit analysis that:

$$\frac{p_{i,o}^2}{p_i} \approx \frac{s_m^2}{R_{in}} = \left(\frac{\pi\alpha_{MZM}P_o}{2V_\pi}\right)^2 R_{in} \tag{5.7}$$

Similar analyses for direct modulation and external modulation configurations with simple and more complex matching electrical circuits can be found in [5].

The $p_o/p_{o,o}^2$ factor represents the equivalent input–output RF power transmission at the optical receiver. Again, it is advisable to employ electrical circuits to match the output or load resistance to that of the optical receiver. In that case, it is possible to make the following approximation [5, 6]:

$$\frac{p_o}{p_{o,o}^2} \approx \Re^2 R_{out} \tag{5.8}$$

where $\Re$ represents the detector responsivity.

Introducing (5.2), (5.7), and (5.8) into (5.3), one gets:

$$G_{RF} = \left(\frac{\pi I_{dc}}{V_\pi}\right)^2 R_{in}R_{out} \tag{5.9}$$

where the average received photocurrent is given by:

$$I_{dc} = \frac{\alpha_{MZM}\Re P_o 10^{-((\alpha L)/(10))}}{2} \tag{5.10}$$

5.3.2 Noise

Figure 5.2 also shows the noise power at the input N_{in} and output N_{out} of the MWP link. The link noise figure NF is defined as the quotient between the signal-to-noise ratio at the input and the signal-to-noise ratio at the output:

$$NF = 10\log\left(\frac{p_i/N_{in}}{p_o/N_{out}}\right) = 10\log\left(\frac{N_{out}}{G_{RF}N_{in}}\right) \tag{5.11}$$

where we have made use of (5.1).

There are two important consequences derived from (5.11). First, if no noise is added by the link $NF = 0$ dB, then $N_{out} = G_{RF}N_{in}$. Secondly, NF does not depend on the RF signal power at the link input. In practice, however, noise is added by several electronic and photonic components in the link (laser, detector, electronic components, etc.), as will be shown in detail in Section 5.4. This implies that, in general, $N_{out} = G_{RF}N_{in} + N_{add}$, with N_{add} representing the power of additional noise sources at the system output. Hence:

$$NF = 10\log\left(\frac{G_{RF}N_{in} + N_{add}}{G_{RF}N_{in}}\right) = 10\log\left(1 + \frac{N_{add}}{G_{RF}N_{in}}\right) \tag{5.12}$$

Note that the effect of the additional noise can be referred to the system input as an additional noise source with power given by N_{add}/G_{RF}. Furthermore, (5.11) can be generalized to include noise sources appearing at intermediate points in the system. In this case, if $N_{add,i}$ represents the noise power generated at the system output generated at point i and $G_{RF,I}$, the RF gain between the system input and point i, then the overall contribution to the noise figure is $N_{add,i}/G_{RF,i}$. Hence, [7, 9]:

$$NF = 10\log\left(1 + \frac{\sum_i (N_{add,i}/G_{RF,i})}{N_{in}}\right) \tag{5.13}$$

Depending on the system complexity, more or less noise sources will have to be considered. However, the most important ones, i.e. those that have to be considered in any system, are the thermal and shot noises at the receiver and the laser relative intensity noise (RIN). For a system with RF bandwidth given by B, the thermal noise power is given by:

$$N_{o,th} = 4kTF_nB = RIN_{o,th}I_{dc}^2R_{out}B \tag{5.14}$$

where k is the Boltzmann constant and F_n is a noise factor that takes into account the contribution of all electronic components at the receiver. In (5.14), the noise power is defined in terms of an equivalent thermal noise RIN parameter given by:

$$RIN_{o,th} = \frac{4kTF_n}{R_{out}I_{dc}^2} \tag{5.15}$$

The shot noise power is given by:

$$N_{shot} = 2eI_{dc}R_{out}B = RIN_{shot}I_{dc}^2BR_{out} \tag{5.16}$$

where e represents the charge of the electron. As with the thermal noise, the shot noise power is defined as well in terms of an equivalent shot noise RIN parameter given by:

$$RIN_{shot} = \frac{2e}{I_{dc}} \tag{5.17}$$

Finally, the laser noise is directly defined through its RIN parameter:

$$N_{laser} = RIN_{laser}I_{dc}^2R_{out}B \tag{5.18}$$

The overall noise contribution can be represented by:

$$N_{out} = N_{o,th} + N_{shot} + N_{laser} = RIN_{tot}I_{dc}^2R_{out}B \tag{5.19}$$

where $RIN_{tot} = RIN_{o,th}+RIN_{shot}+RIN_{laser}$. In general, as we will see in Section 5.4, more contributions to the noise will have to be considered for a more complete description.

5.3.3 Dynamic Range

For a given RF gain, Eq. (5.1) implies that a higher output RF power will be achieved by increasing the input RF power. Unfortunately, the input RF power cannot be increased at will since nonlinear terms start to appear in the output RF spectrum that can distort it. The dynamic range gives a measure of how much RF power can be injected into the MWP link before the effects of nonlinear terms become comparable to the link noise. Figure 5.4 reproduces again the configuration of a simple MWP link but this time the input and output RF signals are outlined and an external modulation configuration is considered (the discussion for a directly modulated link is similar).

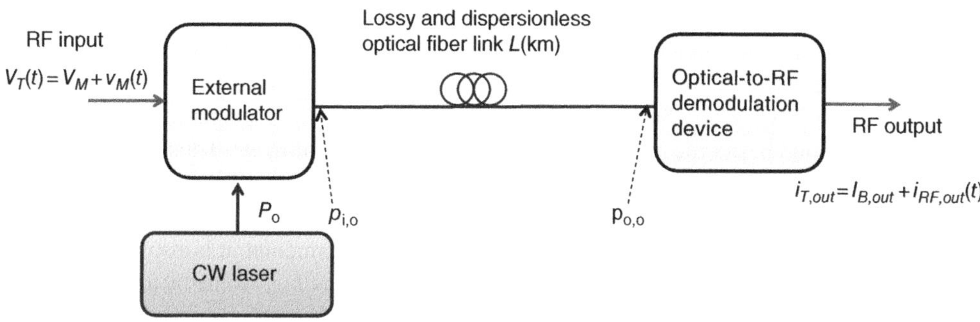

Figure 5.4 Simple MWP link configuration showing explicitly the input and output electronic signals.

Let us assume that it is possible to establish a relationship between the output current and the input voltage by means of a function f [6]:

$$\begin{aligned} i_{T,out} &= f[V] = f[V_M + v_M(t)] \\ &= f(V_M) + \left(\frac{df}{dV}\right)_{V_M} v_M(t) + \frac{1}{2!}\left(\frac{d^2f}{dV^2}\right)_{V_M} v_M^2(t) + \frac{1}{3!}\left(\frac{d^3f}{dV^3}\right)_{V_M} v_M^3(t) + \cdots \\ &= I_{B,out} + f_1 v_M(t) + f_2 v_M^2(t) + f_3 v_M^3(t) + \cdots \end{aligned} \tag{5.20}$$

In general, the f_i coefficients can be both time- and frequency-dependent but for the sake of simplicity, we will however assume no such dependence in the following discussion. An analysis of the harmonic distortion due to nonlinearities can be performed if we assume a sinusoidal modulation for the input RF voltage where the amplitude v_s represents the rms value delivered by the source:

$$v_M(t) = \sqrt{2}v_s \cos \Omega_{RF} t \tag{5.21}$$

Substitution of (5.21) in (5.20) yields the following nonlinear harmonic distortion terms:

$$\begin{aligned} v_M^2(t) &= \left(\sqrt{2}v_s \cos \Omega_{RF} t\right)^2 = \frac{1}{2}\left(\sqrt{2}v_s\right)^2 [1 + \cos(2\Omega_{RF} t)] \\ v_M^3(t) &= \left(\sqrt{2}v_s \cos \Omega_{RF} t\right)^3 = \frac{1}{4}\left(\sqrt{2}v_s\right)^3 [3\cos(\Omega_{RF} t) + \cos(3\Omega_{RF} t)] \\ v_M^4(t) &= \left(\sqrt{2}v_s \cos \Omega_{RF} t\right)^4 = \frac{1}{8}\left(\sqrt{2}v_s\right)^4 [3 + 4\cos(2\Omega_{RF} t) + \cos(4\Omega_{RF} t)] \\ &\cdots\cdots\cdots \end{aligned} \tag{5.22}$$

Note from (5.22) and (5.20) that the f_k term contributes to the kth harmonic distortion tern at $k\Omega_{RF}t$ but it also contributes terms with frequencies ranging from $(k-2)\Omega_{RF}t$ to $(k-2m)\Omega_{RF}t$ terms with m integer and $k - 2m > 0$. Substituting (5.22) into (5.20), we get:

$$\begin{aligned} i_{T,out} &= I_{B,out} + \frac{f_2}{2}\left(\sqrt{2}v_s\right)^2 + \frac{6f_4}{16}\left(\sqrt{2}v_s\right)^4 + \cdots \\ &+ \left[f_1\left(\sqrt{2}v_s\right) + \frac{3f_3}{4}\left(\sqrt{2}v_s\right)^3 + \frac{10f_5}{16}\left(\sqrt{2}v_s\right)^5 + \cdots\right]\cos(\Omega_{RF} t) \\ &+ \left[\frac{f_2}{2}\left(\sqrt{2}v_s\right)^2 + \frac{4f_4}{8}\left(\sqrt{2}v_s\right)^4 + \cdots\right]\cos(2\Omega_{RF} t) \\ &+ \left[\frac{f_3}{4}\left(\sqrt{2}v_s\right)^3 + \frac{5f_5}{16}\left(\sqrt{2}v_s\right)^5 + \cdots\right]\cos(3\Omega_{RF} t) + \cdots \end{aligned} \tag{5.23}$$

The inspection of (5.23) reveals several interesting features. In the first place, the dependence between the output RF current $i_{T,out}$ and the input RF voltage v_s is only linear if the contributions from f_i, $i > 1$ can be neglected, otherwise, harmonic distortion terms must be considered. However, in practice, usually $f_1 > f_2 > f_3$ .... So, within a reasonable range of v_s, the fundamental tone is much larger than the second harmonic, which, in turn, is larger than the third, etc. Thus, in practice, for the majority of systems, we need only to consider f_1, f_2, and f_3 in the series expansion given in (5.22). Note that the largest coefficient for the nth harmonic contains a v_s^n term. Hence, the RF power of the nth harmonic increases at least n times faster than that of the fundamental. It is also interesting to point out that the output current DC component is shifted from the $I_{B,out}$ value by a term proportional to f_2.

The total output RF power can be computed from (5.23) as [5, 6]:

$$
\begin{aligned}
p_o &= \frac{\Omega_{RF}}{2\pi}\int_0^{2\pi/\Omega_{RF}} i_{T,out}^2 R_{out}\,dt = \sum_k p_k \\
p_1 &= 4f_1^2 R_{out} R_{in} p_i \\
p_2 &= 8f_2^2 R_{out} R_{in}^2 p_i^2 \\
p_3 &= 16f_3^2 R_{out} R_{in}^3 p_i^3 \\
&\ldots \\
p_k &= 2^{k+1} f_k^2 R_{out} R_{in}^k p_i^k
\end{aligned}
\tag{5.24}
$$

That is, the total RF output power is the sum of the RF powers of the individual terms in (5.23) since the cross-terms yield zero average:

$$
\frac{\Omega_{RF}}{2\pi}\int_0^{2\pi/\Omega_{RF}} \cos(m\Omega_{RF}t)\cos(n\Omega_{RF}t)dt = \delta_{m,n} \tag{5.25}
$$

For the following discussion, it is useful to express the relationships given in (5.24) in logarithmic units. Thus:

$$
\begin{aligned}
P_1(\text{dBm}) &= P_i(\text{dBm}) + 10\log\left(4f_1^2 R_{out} R_{in}\right) \\
P_2(\text{dBm}) &= 2P_i(\text{dBm}) + 10\log\left(8f_2^2 R_{out} R_{in}^2\right) \\
P_3(\text{dBm}) &= 3P_i(\text{dBm}) + 10\log\left(16f_3^2 R_{out} R_{in}^3\right) \\
&\ldots \\
P_k(\text{dBm}) &= kP_i(\text{dBm}) + 10\log\left(2^{k+1} f_k^2 R_{out} R_i^k\right)
\end{aligned}
\tag{5.26}
$$

A similar procedure can be followed for the analysis of the intermodulation distortion. In this case, one has to assume at least a two-tone RF input to the system:

$$
v_M(t) = \sqrt{2}v_s[\cos\Omega_{RF,1}t + \cos\Omega_{RF,2}t] \tag{5.27}
$$

which substituted in (5.20) and taking up to the third-order term shows the appearance of second-order terms in: $2\Omega_1$, $2\Omega_2$, $\Omega_1+\Omega_2$, $\Omega_1-\Omega_2$, and third-order terms in: $3\Omega_1$, $3\Omega_2$, $2\Omega_1+\Omega_2$, $2\Omega_1-\Omega_2$, $2\Omega_2+\Omega_1$, $2\Omega_2-\Omega_1$. In the so-called sub-octave applications, where the bandwidth is less than one octave, second-order distortion terms fell out of band and can be filtered out. Of the remaining terms, the sum of terms can be usually neglected. So, the limitation in dynamic range comes from the terms in $2\Omega_1 - \Omega_2$ and $2\Omega_2 - \Omega_1$ which are usually known as third-order intermodulation (IM3) distortion terms. A straightforward but lengthy calculation renders for the amplitude of the cos $(2\Omega_2 - \Omega_1)t$ term the value $3f_3\left(\sqrt{2}v_s\right)^3/4$ which is three times bigger than the corresponding third-order harmonic distortion (HD3) term given by (5.23). Thus, the power of the IMD3 terms is 9 times higher than that of the HD3 terms.

In general, the calculation of an analytic expression of $f(V)$ for an end-to-end link is not possible, but a simplified analysis can be carried out by identifying the component that causes most of the nonlinear distortion and treating the rest of the link as linear. In many cases, this component is the external modulator. Alternatively, a numerical calculation for the end-to-end function $f(V)$ can be employed.

In any case, it is possible to represent the values of the output RF power values for the signal, harmonic, and intermodulation distortion terms as a function of the RF power of the input signal to the system. This is usually done in logarithmic units, following the form given by (5.26). Figure 5.5 shows a typical representation that we will take as a reference in order to explain several important parameters related to the dynamic range. In particular, we have represented the straight lines corresponding to the RF signal or fundamental, the second-order harmonic distortion (HD2), and the third-order intermodulation (IM3) and harmonic distortion (HD3). The lines show a different slope as expected from (5.25) with a value of one for the fundamental term, two for HD2, and three for IM3 and HD3. We have also plotted the noise floor $\eta = RIN_{out}I_{dc}{}^2R_{out}$ (W.Hz^{-1}) due to the contribution of all the relevant noise sources.

For the region of input powers comprised between points 1 and 2, the relevant distortion source is noise and the system performance is given in terms of the signal-to-noise dynamic range or SNDR. As it can be immediately observed, this value is obtained at any point in this region as the difference between the output RF power level corresponding to the fundamental and the noise floor level. SNDR is given in dB.Hz as the noise floor is given per Hz of bandwidth. From this value and

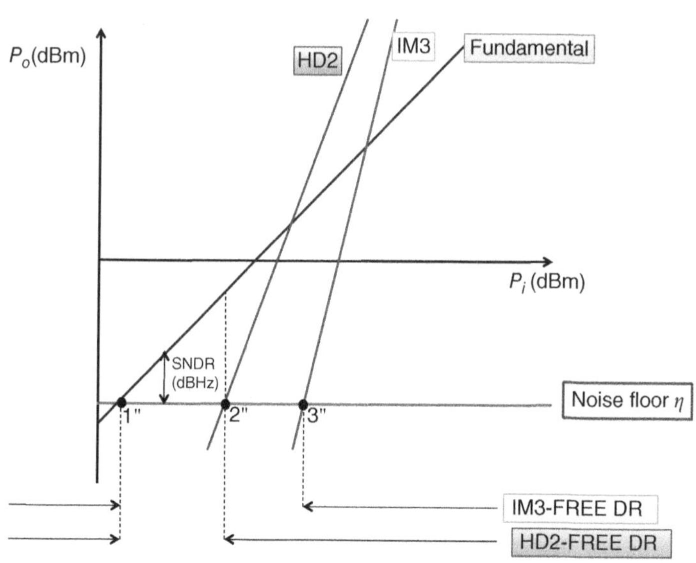

Figure 5.5 Logarithmic representation of the output RF power versus input RF power for the signal (fundamental), and second- and third-order distortion terms in an MWP link.

the knowledge of the system RF bandwidth B, it is straightforward to obtain the real SNR limited by noise:

$$SNR(\text{dB}) = SNDR(\text{dB}) - 10\log([B(\text{Hz})]) \tag{5.28}$$

When the input RF power reaches the value given by point 2, second-order nonlinear distortion becomes the dominant factor over noise. The input power margin between points 1 and 2 generates an output signal free from second-order nonlinearities. This input power range is known as *second-order SFDR* (SFDR_2). In a similar way, the input power margin between points 1 and 3 generates an output signal free from third-order nonlinearities. Hence, this power range is known as *third-order SFDR* (SFDR_3).

The dependence of $SFDR_2$ on the system bandwidth can be computed as follows referring to Fig. 5.5. From (5.26), we know that:

$$\begin{aligned} P_1(\text{dBm}) &= P_i(\text{dBm}) + 10\log\left(4f_1^2 R_{out} R_{in}\right) \rightarrow p_1 = c_1 p_i \\ P_2(\text{dBm}) &= 2P_i(\text{dBm}) + 10\log\left(8f_2^2 R_{out} R_{in}^2\right) \rightarrow p_2 = c_2 p_i^2 \end{aligned} \tag{5.29}$$

But in point 2, the RF power for HD2 equals that of the noise per unit bandwidth. So:

$$p_2 = c_2 p_i^2 = \eta \rightarrow p_i = \sqrt{\frac{\eta}{c_2}} \tag{5.30}$$

which, substituted into the first expression in (5.29), yields:

$$p_1 = c_1 p_i = \sqrt{\frac{c_1^2 \eta}{c_2}} \rightarrow SFDR_2 = \frac{p_1}{\eta} = \sqrt{\frac{c_1^2}{c_2 \eta}} \tag{5.31}$$

Since p_1 is in watts, the units of $SFDR_2$ are $\text{Hz}^{1/2}$ or $\text{dB.Hz}^{1/2}$ if we express it in logarithmic units. To distinguish the fact that this quantity is for a 1 Hz bandwidth, we will name it $SFDR_2^{1\text{Hz}}$. If second-order nonlinearities are the limiting factor, then the $SFDR_2$ for a system bandwidth of B Hz can be computed using:

$$SFDR_2(\text{dB}) = SFDR_2^{1\text{Hz}}(\text{dB}) - 5\log([B(\text{Hz})]) \tag{5.32}$$

A similar procedure can be employed to obtain the system dependence of $SFDR_2^{1\text{Hz}}$ on the system bandwidth. Again, from (5.26), we know that[1]:

$$\begin{aligned} p_1 &= c_1 p_i \\ p_3 &= c_3 p_i^3 \end{aligned} \tag{5.33}$$

But in point 3, the RF power for IM3 equals that of the noise per unit bandwidth. So:

$$p_3 = c_3 p_i^3 = \eta \rightarrow p_i = \sqrt[3]{\frac{\eta}{c_3}} \tag{5.34}$$

which, substituted into the first expression in (5.33), yields:

$$p_1 = c_1 p_i = \sqrt[3]{\frac{c_1^3 \eta}{c_3}} \rightarrow SFDR_3^{1\text{Hz}} = \frac{p_1}{\eta} = \sqrt[3]{\frac{c_1^3}{c_3 \eta^2}} \tag{5.35}$$

1 Note that according to (5.1) $c_1 = G_{RF}$.

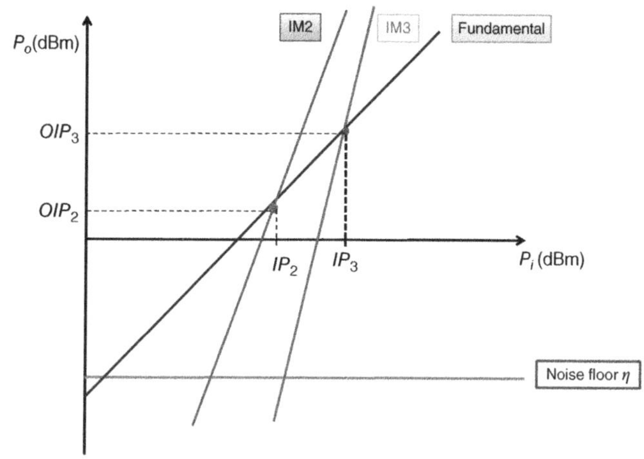

Figure 5.6 Second- and third-order optical interception points for the distortion terms in an MWP link.

Since p_I is in watt, the units of $SFDR_2^{1Hz}$ are $Hz^{2/3}$ or dB. $Hz^{2/3}$ if we express it in logarithmic units. If third-order nonlinearities are the limiting factor (for example, in a suboctave system), then the $SFDR_3$ for a system bandwidth B can be computed using:

$$SFDR_3(\text{dB}) = SFDR_3^{1\text{Hz}}(\text{dB}) - \frac{20}{3}\log([B(\text{Hz})]) \tag{5.36}$$

In practice, the values of $SFDR_2$ and $SFDR_3$ are usually computed through the so-called interception points, which are shown in Fig. 5.6.

The interception points: (IIP_2, OIP_2) and (IIP_3, OIP_3) are the intersection between the lines representing the fundamental and those representing the second- and third-order distortions, respectively. These points can be easily obtained after experimental measurements. Referring to Fig. 5.6 and (5.26), it is

$$\left.\begin{aligned} OIP_2 &= c_1 IP_2 \\ OIP_2 &= c_2 IP_2^2 \end{aligned}\right\} \Rightarrow IP_2 = \frac{c_1}{c_2}$$
$$\left.\begin{aligned} OIP_3 &= c_1 IP_3 \\ OIP_3 &= c_3 IP_3^3 \end{aligned}\right\} \Rightarrow IP_3 = \sqrt{\frac{c_1}{c_3}} \tag{5.37}$$

Which, substituted in (5.31) and (5.35), yield:

$$SFDR_2^{1\text{Hz}} = \sqrt{\frac{c_1^2}{\eta c_2}} = \sqrt{IP_2\left(\frac{c_1}{\eta}\right)} = \sqrt{\frac{G_{RF}IP_2}{\eta}} = \sqrt{\frac{OIP_2}{\eta}}$$
$$SFDR_2 = \sqrt{\frac{G_{RF}IP_2}{\eta B}} = \sqrt{\frac{OIP_2}{\eta B}} = \sqrt{\frac{OIP_2}{N_{out}}} \tag{5.38}$$

$$SFDR_3^{1\text{Hz}} = \sqrt[3]{\frac{c_1^3}{\eta^2 c_3}} = \sqrt[3]{\frac{c_1^2}{\eta^2} IP_3^2} = \sqrt[3]{\frac{G_{RF}^2 IP_3^2}{\eta^2}} = \left(\frac{OIP_3}{\eta}\right)^{2/3}$$
$$SFDR_3 = \sqrt[3]{\frac{G_{RF}^2 IP_3^2}{\eta^2 B^2}} = \left(\frac{OIP_3}{\eta B}\right)^{2/3} = \left(\frac{OIP_3}{N_{out}}\right)^{2/3} \tag{5.39}$$

or, in logarithmic units:

$$SDFR_2(\text{dB}) = \frac{IP_2(\text{dBm}) + G_{RF}(\text{dB}) - N_{out}(\text{dBm})}{2} = \frac{OIP_2(\text{dBm}) - N_{out}(\text{dBm})}{2} \tag{5.40}$$

$$SDFR_3(\text{dB}) = \frac{2}{3}[IP_3(\text{dBm}) + G_{RF}(\text{dB}) - N_{out}(\text{dBm})] = \frac{2}{3}[OIP_3(\text{dBm}) - N_{out}(\text{dBm})] \tag{5.41}$$

5.4 Figures of Merit of a Filtered Microwave Photonic Link

In the prior section, we have considered a simple link in order to introduce and define the basic FOMs that characterize MWP systems. We have restricted ourselves to considering the most basic configuration composed of an optical transmitter, a lossy and dispersionless optical fiber link, and an optical receiver. Furthermore, we have considered intensity modulation exclusively. We have followed this approach for two main reasons: on one hand, the simple MWP link configuration lends itself to an easy introduction and physical explanation of the FOMs. Secondly, basic MWP links were historically the first to be implemented.

With the development of new photonic technologies and components, more complex MWP links including new optical devices, modulation, and detection techniques have started to be considered [12–20]. For instance, IM-DD MWP links including optical amplification have been proposed to provide low RF loss or even gain [12] and, recently, links combining phase modulation and differential delay detection have been considered to enhance the dynamic range [13]. Furthermore, it is envisaged that this trend will not stop here but rather will be expanded to accommodate more advanced features, such as, for instance, multiport MWP links, whereby a common signal from a central station is sent to different BSs or as in the so-called FW networks considered in Section 5.5, where different RF signals are sent to different BSs by means of WDM. Other examples focus on frequency modulation direct-detection (FM-DD) links [14] as well as on the use of frequency [18] or phase [20] modulation combined with a photonic frequency discriminator to enhance the dynamic range.

All the former examples and the simple MWP links considered in the previous section can be represented by the general model shown in Fig. 5.7 for a single-port system, where either intensity or phase modulation (or both simultaneously) can be applied. The effect of all intermediate optical components placed between the electro-optic (EO) and the optical-to-electronic (OE) conversion stages can be lumped into an optical transfer function $H(\omega)$ connecting the input to the output of the system. We use the term filtered MWP links to address these systems. Reference [14] similarly considers the inclusion of an arbitrary optical filter, acting as an FM discriminator, for the particular case of directly modulated FM-DD links.

In this section, we are going to develop the expressions leading to the computation of the main FOMs, that is: G_{RF}, NF, and $SFDR$, in general, externally modulated filtered MWP links [21]. The model works under very general conditions for the coherence properties of the optical source, only restricted by the assumption that the source fluctuations can be regarded as a stationary random process. Direct modulated filtered MWP links where the laser is frequency modulated by current injection follow a similar treatment and the interested reader can find the main results in [14].

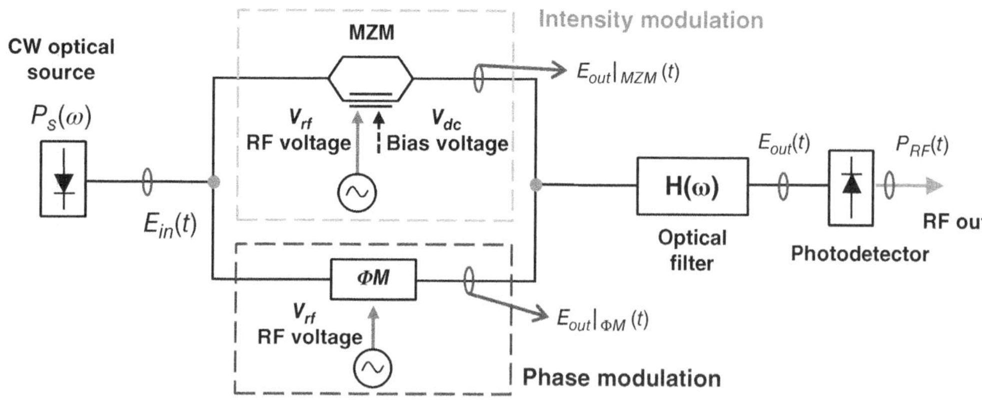

Figure 5.7 Schematic of a general single-port filtered MWP link.

The linear end-to-end RF gain (or loss) of a representative filtered MWP link, as the one illustrated in Fig. 5.7, has been defined in the previous section as the ratio of the RF power (at the modulating angular frequency Ω) delivered to a matched load at the photodetector output, $P_{RF}|_{out}$, to the available RF power at the input (at the modulating angular frequency Ω), $P_{RF}|_{in}$, delivered to the modulation device: $G_{RF} = P_{RF}|_{out}(\Omega)/P_{RF}|_{in}(\Omega)$. For both modulation processes, we will have $P_{RF}|_{in} = V_{rf}^2/(2R_{in})$, where V_{rf} is the amplitude of the voltage signal applied to the external modulator and R_{in} is the input resistance.

The evaluation of the degradation experienced by the microwave signal due to the existing noise sources will be carried in this section by evaluating the noise figure parameter defined in the previous section as $NF = N_{out}/(G_{RF}\, N_{in})$, being N_{out} and N_{in}, respectively, the total output and input noise spectral densities. The NF will be referred to as the total output RIN parameter (RIN_{tot}) [12], defined as $RIN_{tot} = N_{out}/(I|_{dc}^2\, R_{out})$, where $I|_{dc}$ represents the DC photocurrent and R_{out} is the output resistance. The RIN_{tot} is derived taking into account the input and output thermal noises, the shot noise and laser noise contributions: $RIN_{tot} = RIN_{i,th} + RIN_{o,th} + RIN_{shot} + RIN_{laser}$.

For the dynamic range evaluation, we employ the SFDR defined in the previous section as the carrier-to-noise ratio when the noise floor in the signal bandwidth equals the power of a given order intermodulation product. As has been shown in Section 5.3.3, the SFDR of a link limited by second- (IMD_2) or third-order (IMD_3) intermodulation distortion can be computed, respectively, from $SFDR_2 = (OIP_2/N_{tot})^{1/2}$ or $SFDR_3 = (OIP_3/N_{tot})^{2/3}$, where OIP_2 and OIP_3 are the optical interception points, also defined in Section 5.3.3.

5.4.1 Filtered Intensity Modulated Direct Detection Links

We consider a general filtered IM-DD link, as that shown in Fig. 5.7, assuming that the light source emits a signal whose amplitude, separating the rapid sinusoidal oscillations at the central angular frequency ω_0 from the slower source fluctuations $A(t)$ [22], can be described by f $E_{in}(t) = (2P_0)^{1/2} e^{j\omega_0 t} A(t)$, where P_0 is the optical power at ω_0.

When the modulating signal is composed of an RF tone characterized by an angular frequency Ω, the optical field at the output of one arm of the push-pull MZM is given by

$$E_{out}|_{MZM}(t) = j\sqrt{\alpha_{MZM}} E_{in}(t) \sin\left[\phi_{dc}/2 + \phi_{rf}/2 \sin(\Omega t)\right] \tag{5.42}$$

where α_{MZM} is the MZM optical loss while $\phi_{rf} = \pi V_{rf}/V_\pi$ and $\phi_{dc} = \pi V_{dc}/V_\pi$ for the bias voltage V_{dc}. After expansion of Eq. (5.42) in terms of Bessel functions of the first kind, the amplitude spectrum of the modulated signal is obtained as

$$E_{out}|_{MZM}(\omega) = 2\pi\sqrt{\alpha_{MZM}} \sum_{n=-\infty}^{\infty} B_n J_n\left(\phi_{rf}/2\right) E_{in}(\omega - n\Omega) \tag{5.43}$$

with

$$B_n = (-1)^n j^{|n|+1} \sin(\phi_{dc}/2 + |n|\pi/2) \tag{5.44}$$

where, at the modulator quadrature bias point, one must set $\phi_{dc} = \pi/2$.

We will need the autocorrelation function of the spectrum, given from (5.43) as

$$\begin{aligned}&\left\langle E_{out}|_{MZM}(\omega) E^*_{out}|_{MZM}(\omega')\right\rangle \\ &= (2\pi)^2 \alpha_{MZM} 2P_0 \sum_{n=-\infty}^{\infty} \sum_{m=-\infty}^{\infty} B_n B^*_m J_n\left(\phi_{rf}/2\right) J_m\left(\phi_{rf}/2\right) P_s(\omega - \omega_0 + n\Omega)\delta[\omega' - (\omega + (n-m)\Omega)]\end{aligned} \tag{5.45}$$

where the symbol < > indicates an ensemble average. $P_s(\omega)$ is the spectral density function of the optical source, usually modeled as a Gaussian, [22, 23], or Lorentzian function, which is defined as the Fourier transform of the temporal autocorrelation function $R(u)$:

$$P_s(\omega) = \int_{-\infty}^{\infty} R(u) e^{-i\omega u} du \tag{5.46}$$

being the source fluctuations regarded as a stationary random process so that their temporal autocorrelation function can be written as $\langle A(t)\, A^*(t')\rangle = R(t-t')$.

The ensemble average power of light after propagation through a general optical filter, characterized by its optical transfer function $H(\omega)$, can be expressed in terms of the autocorrelation function of the spectrum at the filter output as follows:

$$\begin{aligned}\langle P(t)\rangle &= \frac{1}{(2\pi)^2} \int_{-\infty}^{\infty} \int_{-\infty}^{\infty} \langle E_{out}(\omega) E^*_{out}(\omega')\rangle e^{j(\omega-\omega')t} d\omega d\omega' \\ &= \alpha_{MZM} 2P_0 \sum_{n=-\infty}^{\infty} \sum_{m=-\infty}^{\infty} B_n B^*_m J_n\left(\phi_{rf}/2\right) J_m\left(\phi_{rf}/2\right) e^{-j(n-m)\Omega t} \\ &\quad \int_{-\infty}^{\infty} P_s(\omega - \omega_0 + n\Omega) H(\omega) H^*[\omega + (n-m)\Omega] d\omega\end{aligned} \tag{5.47}$$

Assuming a small-signal approximation, the RF photodetected power for the signal contribution is developed as:

$$P^I_{RF}|_{out}(\Omega) = 2I_{dc}{}^2 R_{out}\left(\phi_{rf}/4\right)^2 \sin^2(\phi_{dc}) |A^I_\Omega|^2 \tag{5.48}$$

where I_{dc} is defined as $I_{dc} = \Re \alpha_{MZM} P_0/2$ for a photodetector responsivity $\Re$, as in [13], and

$$A^I_\Omega = \sum_{n=\{0,1\}} \int_{-\infty}^{\infty} P_s(\omega - \omega_0 - n\Omega) H(\omega) H^*(\omega - \Omega) d\omega. \tag{5.49}$$

In consequence, Eq. (5.48) implies an end-to-end RF link gain given by:

$$G_{RF}^{I}(\Omega) = (I_{dc}\pi/V_{\pi})^2 \sin^2(\phi_{dc}) R_{in} R_{out} \left|A_{\Omega}^{I}/2\right|^2 \tag{5.50}$$

In relation to the link noise characterization and taking into account the previous definition of RIN_{tot}, we obtain the following expressions for the *NF*, as well as for the *RIN* terms referenced to the output due to thermal noise at the input, thermal noise at the output, and shot noise:

$$NF^{I} = 4\frac{RIN_{tot} V_{\pi}^{2}}{\pi^2 R_{in} k_B T} \frac{[1-\cos(\phi_{dc})]^2}{\sin^2(\phi_{dc})} \frac{\left|A_{DC}^{I}\right|^2}{\left|A_{\Omega}^{I}\right|^2} \tag{5.51}$$

$$RIN_{i,th}^{I} = \frac{k_B T \pi^2 R_{in}}{4V_{\pi}^{2}} \frac{\sin^2(\phi_{dc})}{[1-\cos(\phi_{dc})]^2} \frac{\left|A_{\Omega}^{I}\right|^2}{\left|A_{DC}^{I}\right|^2} \tag{5.52}$$

$$RIN_{o,th}^{I} = \frac{k_B T}{I_{dc}^{2} R_{out}} \frac{1}{[1-\cos(\phi_{dc})]^2 \left|A_{DC}^{I}\right|^2} \tag{5.53}$$

$$RIN_{shot}^{I} = \frac{2e}{I_{dc}} \cdot \frac{1}{[1-\cos(\phi_{dc})]\left|A_{DC}^{I}\right|} \tag{5.54}$$

being k_B the Boltzmann constant, T the temperature, e the electronic charge constant, and

$$A_{DC}^{I} = \int_{-\infty}^{\infty} P_s(\omega-\omega_0)|H(\omega)|^2 d\omega \tag{5.55}$$

For the dynamic range evaluation, we assume, as it was defined in Section 5.3, a modulating signal composed of two RF tones, characterized by the same amplitude V_{rf} at angular frequencies Ω_1 and Ω_2, respectively. In principle, one is primarily concerned with the $SFDR_3$ related to the products placed at $2\Omega_1 - \Omega_2$ or $2\Omega_2 - \Omega_1$, as these frequencies appear within the system bandwidth in any MWP link, either suboctave or broadband. On the other hand, the $SFDR_2$ becomes also important for systems whose bandwidth is more than one octave since the IMD_2 products at $\Omega_1 \pm \Omega_2$ fell within the bandpass of interest, while they can be neglected in suboctave links.

The expression for the photodetected RF power relative to the IM_2 term placed at angular frequency $\Omega_1 \pm \Omega_2$ is obtained as

$$P_{RF}^{I}\Big|_{out}(\Omega_1 \pm \Omega_2) = 2 I_{dc}^{2} R_{out} \left(\phi_{rf}/4\right)^4 \left|A_{\Omega_1 \pm \Omega_2}^{I}\right|^2 \tag{5.56}$$

where, similarly to Eq. (5.49), we define

$$A_{\Omega_1 \pm \Omega_2}^{I} = \pm \sum_{\substack{n=\{0,1\} \\ m=\{-1,0\}}} \left[(-1)^{n+m} - \cos(\phi_{dc})\right] \int_{-\infty}^{\infty} P_s(\omega-\omega_0-n\Omega_1 \pm m\Omega_2) H(\omega) H^*(\omega-\Omega_1 \mp \Omega_2) d\omega. \tag{5.57}$$

Given Eqs. (5.48) and (5.56), which lead to the calculation of the OIP_2 value, we can express the $SFDR_2$ in terms of the RIN_{tot} parameter as:

$$SFDR_2^{I} = \sqrt{\frac{2}{RIN_{tot}} \frac{\sin^2(\phi_{dc})}{|1-\cos(\phi_{dc})|} \frac{\left|A_{\Omega_1}^{I}\right|^2}{\left|A_{\Omega_1 \pm \Omega_2}^{I}\right|\left|A_{DC}^{I}\right|}} \tag{5.58}$$

And, in a similar way, for the third-order intermodulation distortion term at $2\Omega_1 - \Omega_2$, we have

$$P_{RF}^{I}\big|_{out}(2\Omega_1 - \Omega_2) = 2I_{dc}{}^2R_{out}\left(\phi_{rf}/4\right)^6 \sin^2(\phi_{dc})\left|A_{2\Omega_1-\Omega_2}^{I}\right|^2 \tag{5.59}$$

with

$$A_{2\Omega_1-\Omega_2}^{I} = \sum_{\substack{n=\{0,1,2\}\\ m=\{-1,0\}}} (1/2)^{|n-1|} \int_{-\infty}^{\infty} P_s(\omega - \omega_0 - n\Omega_1 - m\Omega_2)H(\omega)H^*(\omega - 2\Omega_1 + \Omega_2)d\omega \tag{5.60}$$

which results in:

$$SFDR_3^I = \left[\frac{2}{RIN_{tot}}\frac{\sin^2(\phi_{dc})}{|1-\cos(\phi_{dc})|}\frac{|A_{\Omega_1}^{I}|^3}{|A_{2\Omega_1-\Omega_2}^{I}||A_{DC}^{I}|^2}\right]^{2/3} \tag{5.61}$$

Note that if the optical source can be assumed as monochromatic and, subsequently, the spectral density function of the source $P_s(\omega)$ approximated by a delta function, then (5.49), (5.55), (5.57), and (5.60) can be simplified to:

$$A_{\Omega}^{I}\big|_{\delta} = \sum_{n=\{0,1\}} H(\omega_0 + n\Omega)H^*[\omega_0 + (n-1)\Omega] \tag{5.62}$$

$$A_{DC}^{I}\big|_{\delta} = |H(\omega_0)|^2 \tag{5.63}$$

$$A_{\Omega_1 \pm \Omega_2}^{I}\big|_{\delta} = \pm \sum_{\substack{n=\{0,1\}\\ m=\{-1,0\}}} \left[(-1)^{n+m} - \cos(\phi_{dc})\right]H(\omega_0 + n\Omega_1 \mp m\Omega_2)H^* [\omega_0 + (n-1)\Omega_1 \mp (m+1)\Omega_2] \tag{5.64}$$

and

$$A_{2\Omega_1-\Omega_2}^{I}\big|_{\delta} = \sum_{\substack{n=\{0,1,2\}\\ m=\{-1,0\}}} (1/2)^{|n-1|}H(\omega_0 + n\Omega_1 + m\Omega_2)H^*[\omega_0 + (n-2)\Omega_1 + (m+1)\Omega_2] \tag{5.65}$$

It must be noted that if we consider a back-to-back configuration where we can omit the presence of the optical filter by setting $H(\omega) = 1$ for every frequency, our IM-DD model provides exactly the same expressions for G_{RF}, but different *RIN* contributions, *NF* and *SFDR* for third-order intermodulation with those previously derived in Section 5.3 for a simple MWP link under external intensity modulation of a monochromatic optical source.

5.4.2 Filtered Phase Modulated Links

The procedure followed for the performance analysis of filtered phase modulated (PM) links is similar to that previously employed for filtered IM-DD links where now the output of the phase modulator is

$$E_{out}\big|_{\Phi M}(t) = j\sqrt{\alpha_{\Phi M}}E_{in}(t)e^{j\phi_{rf}\sin(\Omega t)} \tag{5.66}$$

being $\alpha_{\Phi M}$ the phase modulator insertion loss.

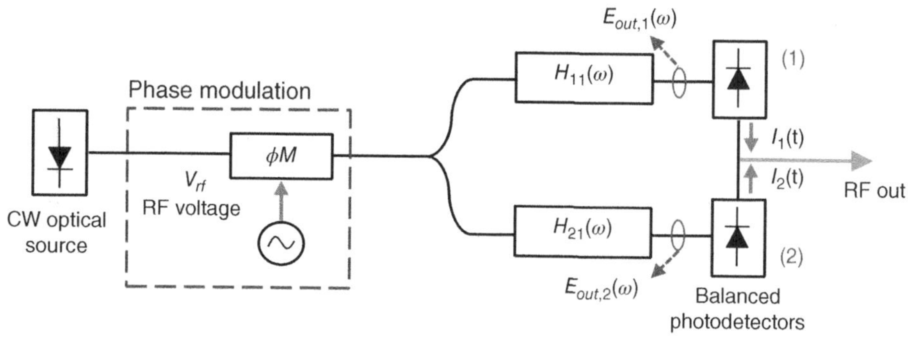

Figure 5.8 Schematic of a filtered MWP link applying phase modulation with a general balanced detection scheme.

Figure 5.8 represents the layout of a balanced photodetection PM MWP link where each detector is preceded by a different optical filter, $H_{11}(\omega)$ and $H_{21}(\omega)$, and where the photodetected total current is obtained by subtracting the current obtained from the second detector from the one corresponding to the first: $I(t) = I_1(t) - I_2(t)$. For the particular case of PM single-port links, where only a single photodetector is used in the detection stage, all the results derived in this section can be employed, provided that we set $H_{21}(\omega) = 0$.

The RF photodetected power and link gain are then respectively derived as

$$P_{RF}^{\Phi}\Big|_{out}(\Omega) = 2I_{dc}{}^2 R_{out}\phi_{rf}{}^2 \left|A_{\Omega}^{\Phi}\right|^2 \tag{5.67}$$

where:

$$A_{\Omega}^{\Phi} = \sum_{n=\{0,1\}} (-1)^n \int_{-\infty}^{\infty} P_s(\omega - \omega_0 - n\Omega)\left[H_{11}(\omega)H_{11}^*(\omega - \Omega) - H_{21}(\omega)H_{21}^*(\omega - \Omega)\right] d\omega \tag{5.68}$$

and

$$G_{RF}^{\Phi}(\Omega) = 4(I_{dc}\pi/V_{\pi})^2 R_{in}R_{out}\left|A_{\Omega}^{\Phi}\right|^2 \tag{5.69}$$

For modelling the noise metrics in the balanced detection scheme under evaluation, we will take into account that the noise current contributions add incoherently in the balanced detection process, and subsequently refer them to the total RIN as $RIN_{tot} = N_{out}/(I|_{dc}{}^2 R_{out})$, where $I|_{dc}$ is defined for convenience as the sum of the DC photocurrents provided by both detectors: $I|_{dc} = I_1|_{dc} + I_2|_{dc} = 2\, I_{dc}A_{DC}^{\Phi}$, with:

$$A_{DC}^{\Phi} = \int_{-\infty}^{\infty} P_s(\omega - \omega_0)\left[|H_{11}(\omega)|^2 + |H_{21}(\omega)|^2\right] d\omega \tag{5.70}$$

obtaining:

$$NF^{\Phi} = \frac{RIN_{tot}V_{\pi}{}^2}{\pi^2 R_{in}k_B T}\frac{\left|A_{DC}^{\Phi}\right|^2}{\left|A_{\Omega}^{\Phi}\right|^2} \tag{5.71}$$

$$RIN_{o,th}^{\Phi} = \frac{k_B T}{4I_{dc}{}^2 R_{out}} \cdot \frac{1}{\left|A_{DC}^{\Phi}\right|^2} \tag{5.72}$$

$$RIN_{i,th}^{\Phi} = \frac{k_B T \pi^2 R_{in}}{V_\pi^{\ 2}} \cdot \frac{\left|A_\Omega^\Phi\right|^2}{\left|A_{DC}^\Phi\right|^2} \tag{5.73}$$

$$RIN_{shot}^{\Phi} = \frac{e}{I_{dc}} \cdot \frac{1}{\left|A_{DC}^\Phi\right|} \tag{5.74}$$

The IMD_2 output power and $SFDR_2$ can be computed, yielding:

$$P_{RF}^{\Phi}\Big|_{out}(\Omega_1 \pm \Omega_2) = \frac{I_{dc}^2 R_{out}}{2} \phi_{rf}^4 \left|A_{\Omega_1 \pm \Omega_2}^\Phi\right|^2 \tag{5.75}$$

where

$$A_{\Omega_1 \pm \Omega_2}^{\Phi} = \pm \sum_{\substack{n=\{0,1\} \\ m=\{-1,0\}}} (-1)^{n+m} \int_{-\infty}^{\infty} P_s(\omega - \omega_0 - n\Omega_1 \pm m\Omega_2)\left[H_{11}(\omega)H_{11}^*(\omega - \Omega_1 \mp \Omega_2) - H_{21}(\omega)H_{21}^*(\omega - \Omega_1 \mp \Omega_2)\right]d\omega \tag{5.76}$$

and

$$SFDR_2^{\Phi} = \sqrt{\frac{2}{RIN_{tot}} \frac{\left|A_{\Omega_1}^\Phi\right|^2}{\left|A_{\Omega_1 \pm \Omega_2}^\Phi\right|\left|A_{DC}^\Phi\right|}} \tag{5.77}$$

Finally, for third-order intermodulation distortion, we obtain:

$$P_{RF}^{\Phi}\Big|_{out}(2\Omega_1 - \Omega_2) = \frac{I_{dc}^2 R_{out}}{8} \phi_{rf}^6 \left|A_{2\Omega_1 - \Omega_2}^\Phi\right|^2 \tag{5.78}$$

where:

$$A_{2\Omega_1 - \Omega_2}^{\Phi} = \sum_{\substack{n=\{0,1,2\} \\ m=\{-1,0\}}} (1/2)^{|n-1|}(-1)^{m+n} \int_{-\infty}^{\infty} P_s(\omega - \omega_0 - n\Omega_1 - m\Omega_2)\left[H_{11}(\omega)H_{11}^*(\omega - 2\Omega_1 + \Omega_2) - H_{21}(\omega)H_{21}^*(\omega - 2\Omega_1 + \Omega_2)\right]d\omega \tag{5.79}$$

and, finally:

$$SFDR_3^{\Phi} = \left[\frac{4}{RIN_{tot}} \frac{\left|A_{\Omega_1}^\Phi\right|^3}{\left|A_{2\Omega_1 - \Omega_2}^\Phi\right|\left|A_{DC}^\Phi\right|^2}\right]^{2/3} \tag{5.80}$$

Note again that if the spectral density function of the source $P_s(\omega)$ can be approximated by a delta function, then (5.68), (5.70), (5.76), and (5.79) simplify to:

$$A_\Omega^\Phi\Big|_\delta = \sum_{n=\{0,1\}} (-1)^n\left[H_{11}(\omega_0 + n\Omega)H_{11}^*(\omega_0 + (n-1)\Omega) - H_{21}(\omega_0 + n\Omega)H_{21}^*(\omega_0 + (n-1)\Omega)\right], \tag{5.81}$$

$$A_{DC}^{\Phi}\big|_{\delta} = |H_{11}(\omega_0)|^2 + |H_{21}(\omega_0)|^2 \tag{5.82}$$

$$A_{\Omega_1 \pm \Omega_2}^{\Phi}\big|_{\delta} = \pm \sum_{\substack{n=\{0,1\}\\ m=\{-1,0\}}} (-1)^{n+m}\big[H_{11}(\omega_0 + n\Omega_1 \mp m\Omega_2)H_{11}^{*}(\omega_0 + (n-1)\Omega_1 \mp (m+1)\Omega_2)$$
$$-H_{21}(\omega_0 + n\Omega_1 \mp m\Omega_2)H_{21}^{*}(\omega_0 + (n-1)\Omega_1 \mp (m+1)\Omega_2)\big] \tag{5.83}$$

and:

$$A_{2\Omega_1 - \Omega_2}^{\Phi}\big|_{\delta} = \sum_{\substack{n=\{0,1,2\}\\ m=\{-1,0\}}} (1/2)^{|n-1|}(-1)^{m+n}\{H_{11}(\omega_0 + n\Omega_1 + m\Omega_2)\big[H_{11}^{*}\omega_0 + (n-2)\Omega_1 + (m+1)\Omega_2\big]$$
$$-H_{21}(\omega_0 + n\Omega_1 + m\Omega_2) - H_{21}^{*}[\omega_0 + (n-2)\Omega_1 + (m+1)\Omega_2]\} \tag{5.84}$$

5.4.3 Application Examples

The principal advantage of the filtered MWP link model resides in the global character of the developed FOMs, which allows the potential application to an extensive variety of MWP transmission systems. One emerging area of application is, for instance, multiport MWP links where an electrical common signal is distributed to a set of different base stations (BS_m). Another very important area of application is WDM FW systems and networks, which will be considered in detail in the next sections. Fig. 5.9 depicts a schematic of WDM transmission in a FW optical distribution network which can be considered a bank of parallel optical filters, where $H_{mn}(\omega)$ represents the optical transfer function between output port (base station) m (output signal S_{outm}) and input port (central station) n (input signal S_{inn}).

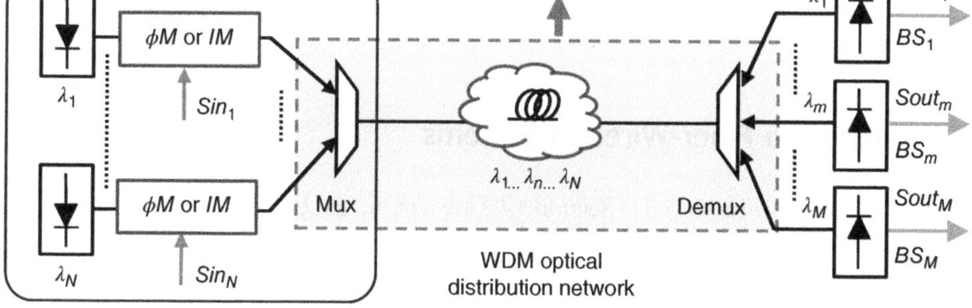

Figure 5.9 Multiport filtered MWP link implementing WDM distribution.

In Section 5.3, we considered the simple MWP where the optical fiber link was considered lossy but dispersionless. The effect of chromatic dispersion, which is key to the description of FW and Radio over Fiber (RoF) systems, can be readily incorporated into the description of the figures of merit by considering its effect as equivalent to having a phase filter in the general model for filtered MWP links.

For second-order dispersive optical fiber links, the propagation constant $\beta(\omega)$ can be expanded in Taylor's series around the central angular frequency ω_0 as

$$\beta(\omega)\approx\beta(\omega_0)+\left.\frac{d\beta(\omega)}{d\omega}\right|_{\omega=\omega_0}(\omega-\omega_0)+\left.\frac{1}{2}\frac{d\beta^2(\omega)}{d\omega^2}\right|_{\omega=\omega_0}(\omega-\omega_0)^2=\beta_o+\beta_1(\omega-\omega_0)+\frac{1}{2}\beta_2(\omega-\omega_0)^2. \tag{5.85}$$

Moreover, if we consider an optical source represented by a Gaussian power spectrum $P_s(\omega)$ with a root mean square (RMS) linewidth ΔW related to the source RMS coherence time as $\Delta W = 1/(\sqrt{2}\sigma_c)$:

$$P_s(\omega) = \frac{P_0}{2\Delta W\sqrt{\pi}}e^{((-\omega^2)/((2\Delta W)^2))}. \tag{5.86}$$

For a singlemode fiber (SMF) link of length L, for which we assume that the power optical loss $\alpha(\omega)$ is independent of ω, i.e. $\alpha(\omega)\approx\alpha(\omega_0)=\alpha^0$, we must employ the following optical transfer function in the general model:

$$H|_{SMF}(\omega) = e^{-\alpha_0 L/2}e^{-j\left[\beta_o+\beta_1(\omega-\omega_0)+\beta_2/2(\omega-\omega_0)^2\right]L} \tag{5.87}$$

to directly obtain the characterization of the FOMs of interest. For instance, the substitution of Eqs. (5.86) and (5.87) in the closed-form expressions for the RF gain both in intensity, Eq. (5.50), and phase, Eq. (5.69), modulated MWP links with single-detection, give the following closed-form equations:

$$G_{RF}^{I}(\Omega)\big|_{SMF} = (I_{dc}\pi/V_\pi)^2\sin^2(\phi_{dc})R_{in}R_{out}e^{-2(\beta_2 L\,\Omega\,\Delta W)^2}\cos^2(\beta_2\Omega^2L/2)e^{-2\alpha^0 L} \tag{5.88}$$

$$G_{RF}^{\Phi}(\Omega)\big|_{SMF} = (4I_{dc}\pi/V_\pi)^2R_{in}R_{out}e^{-2(\beta_2 L\,\Omega\,\Delta W)^2}\sin^2(\beta_2\Omega^2L/2)e^{-2\alpha^0 L} \tag{5.89}$$

These results are in concordance with previous work related to the characterization of fiber dispersion in intensity [24] and PM [19] RoF systems. As it can be observed in both expressions, the first exponential factor is a low-pass frequency response term which depends on β'' and the spectral source linewidth ΔW. The intermediate sinusoidal factors depending on $\beta''\Omega^2L/2$ are the well-known carrier suppression effect (CSE) terms that arise from the phase offset between the upper and lower RF modulation sidebands of an optical carrier due to chromatic dispersion. This effect will be further considered and physically explained in Section 5.7.1.

5.5 Introduction to Fiber-Wireless Systems

FW systems (FWSs), briefly introduced in Section 5.4.3 are a particular but important application of MWP links [25–28]. The building block of FWSs is the MWP link where analog radio signals giving support to one or several telecommunication services are transmitted from a CO to the antenna of a BS and vice versa. A Fiber-Wireless Network (FWN) is, in turn, composed of multiple FWSs, integrated as an overlay in the existing infrastructure of an access network, and more particularly, in

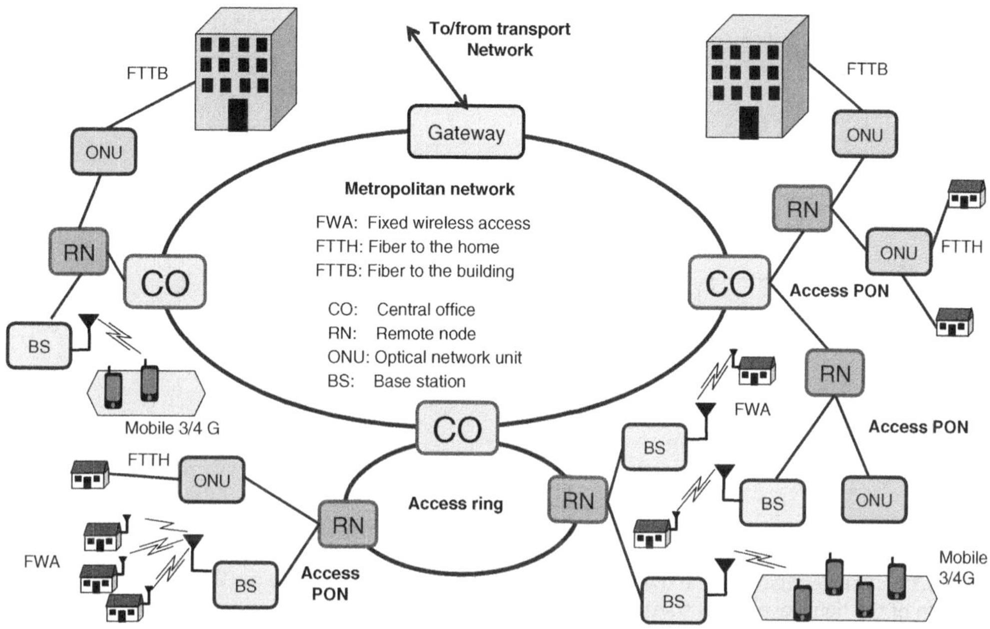

Figure 5.10 Fiber-wireless network integrated in a PON access infrastructure.

passive optical networks (PONs), as shown in Fig. 5.10. While the PON infrastructure provides the end-to-end digital baseband connectivity from the CO to the end users and vice versa, the FWN provides the end-to-end wireless connectivity from the CO to the end users, supporting a considerable and increasing variety of emerging telecommunication services and standards like 3G/4G and LTE for cellular communications, fixed-wireless access services, and UMTS. The entry points for the end users of the FWN are the BS that are connected like the PON optical network units (ONUs) via RNs to the rest of the network.

The first step in the design of FWN is the transport of the wireless signals over the fiber-based FWSs that constitute it. There are a variety of alternatives that can be employed for this purpose which are considered in detail in the next section.

5.6 Optical Transport of Wireless Signals

There are three options for the transport of wireless signals modulating an optical carrier from the CO to the BS and vice versa: RoF, intermediate frequency over fiber (IFoF) and baseband over fiber (BBoF). These have been reported and well documented in the literature [25, 26]. Each one of these techniques has its advantages and limitations that are briefly outlined in the following subsections.

5.6.1 Radio Over Fiber

Figure 5.11 displays an FWS from the CO to a BS and vice versa. In essence, the wireless signal at f_{RF} is externally modulated onto the optical carrier f_c, usually resulting in a double sideband (DSB) signal. Sidebands are thus separated from the optical carrier by f_{RF} as shown in the spectrum displayed

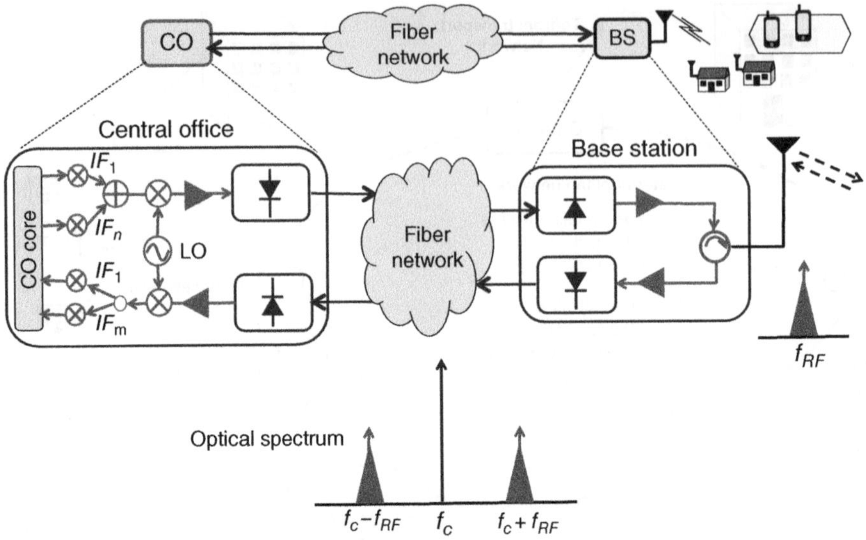

Figure 5.11 RoF approach for the transport of wireless signals from the CO to the BS and vice versa (top). Spectrum of the lightwave signal transported (bottom).

on the bottom part of Fig. 5.11. The wireless signal is thus directly transported along the fiber link. Upon detection at the BS, the wireless signal is recovered from the beating of the sidebands and the optical carrier. Since no frequency translation is required, this means that the BS design is very simple and thus this configuration benefits from the centralized control. It also enables multiwireless band operation since each band is recovered after beating with the optical carrier.

The main disadvantages of the RoF approach are that it requires high-speed and broadband modulation/detection schemes and that, as we will see, it is very sensitive to the effect of chromatic dispersion.

5.6.2 Intermediate Frequency Over Fiber

A second alternative shown in Fig. 5.12 is known as IFoF. Here the wireless signal at a lower intermediate frequency f_{IF} is externally modulated onto the optical carrier f_c, usually resulting in a DSB signal. Sidebands are thus separated from the optical carrier by f_{IF} as shown in the bottom part of Fig. 5.12. Upon detection at the BS, the wireless signal is recovered at f_{IF} from the beating of the sidebands and the optical carrier. Now, however, an upconversion stage is required before radiation. This is achieved by mixing the recovered signal with a local oscillator of frequency f_{LO}, such that $f_{RF} = f_{IF} + f_{LO}$. This complicates the design of the BS as a stable LO (this component might be saved by remotely delivering the LO from the CO) and a mixer is required. However, as we will see, the signal transmission at f_{IF} is much more immune to the effect of the fiber chromatic dispersion. Another advantage is due to the fact that smaller bandwidths are required for both the modulator and the optical receiver as they need to span only until intermediate frequency region. The BS configuration may be less flexible to network upgrades requiring the inclusion of further wireless channels or alterations in the wireless frequencies.

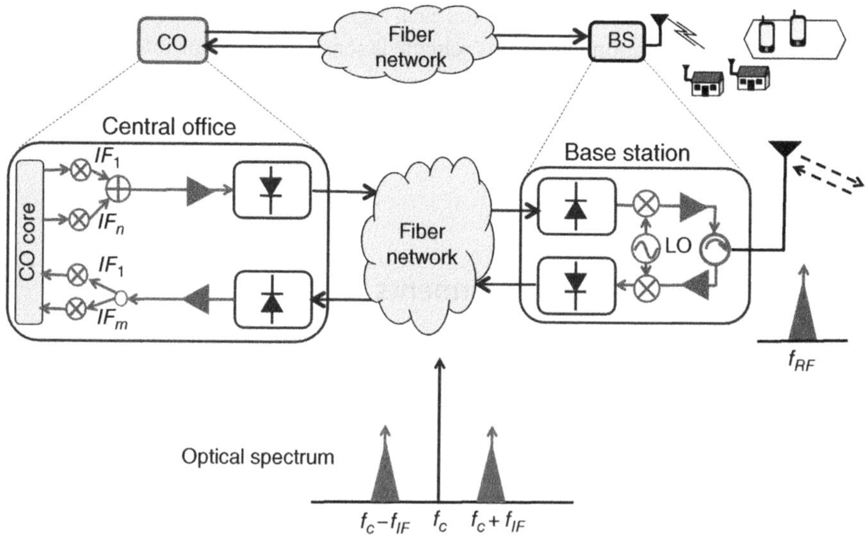

Figure 5.12 IFoF approach for the transport of wireless signals from the CO to the BS and vice versa (top). Spectrum of the lightwave signal transported (bottom).

5.6.3 Baseband Over Fiber

In this approach, which is shown in Fig. 5.13, the wireless signal is transported as baseband over the optical fiber link. Upon reaching the BS, it is detected and upconverted to RF before being radiated. Thus, a stable RF oscillator which is more expensive than an IF oscillator is required (although it can be remotely delivered) and high-frequency mixers will be required as well, which are also expensive. On the other hand, the detected baseband signal can be efficiently processed by mature electronic signal processors before being upconverted. Furthermore, low-bandwidth optoelectronic

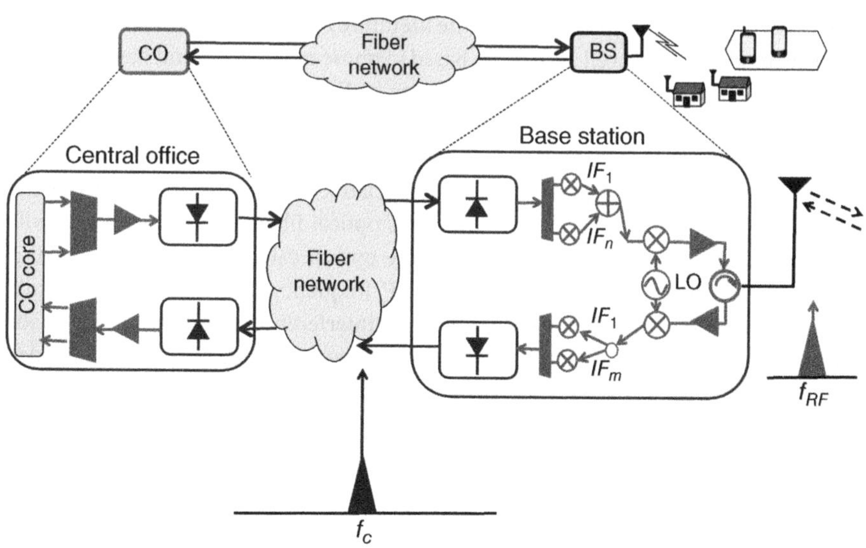

Figure 5.13 BBoF approach for the transport of wireless signals from the CO to the BS and vice versa (top). Spectrum of the lightwave signal transported (bottom).

devices can be employed. An additional advantage is the even higher immunity to the effect of chromatic dispersion. A significant limitation is that this scheme is sensitive to the air interface which means that the BS must be able to process the incoming wireless signals before sending them in baseband format to the CO. This might require, among others, frequency measurement and tracking and increases the BS complexity. Thus, a low-cost transceiver solution may require the use of integrated optoelectronic chips.

5.7 Sources of Degradation and Impairments

The point-to-point link connecting the CO and BS or vice versa is a particular case of an MWP link and it is thus subject to several sources of impairment which impact its FOMs [25–28]. For example, fiber losses, optical amplification, and chromatic dispersion impact the RF link gain, while the noise figure is affected by optical amplification and the quality of the optical source and the detector. Finally, transmitter and receiver nonlinearities impact the dynamic range. Several sources of impairment can be completely overcome as, for example, the effect of chromatic dispersion. For others like the system nonlinearities, there exist a variety of techniques that can help to substantially improve the link performance. These span from linearization to tailoring the modulation indexes of the RF signals.

Further to improving the performance of single channel links, there is also a need for improving the spectral efficiency of FWSs, in particular, of DSB modulation systems. This is another limiting factor that can severely bound the number of transportable wireless signals in the available optical spectrum.

In this section, we consider these degradation effects separately and we describe some of the most efficient solutions to overcome them reported in the literature.

5.7.1 Chromatic Dispersion

Chromatic dispersion affects FWSs principally by producing notches or fading regions in the spectrum of their RF gain [29]. This fading, also known in the literature as CSE, appears both in systems based on DSB intensity modulation as well as in those based in phase modulation and has its origins in the sinusoidal dependence of their RF gain spectra as given by Eqs. (5.88) and (5.89). In intensity (phase) modulation systems, CSE happens because an optical double-sideband-with-carrier (ODSB) modulation format results where the sidebands are located at the RF frequency on either side of the optical carrier with the same (opposed) phase relative to the carrier. When the RF wave-modulated optical signal propagates along a dispersive optical fiber, the two sidebands will experience a different amount of phase shifts relative to the optical carrier. Upon detection, the sidebands beat with the carrier and interfere with the RF frequency. The interference result depends on the phase shift. If this is 90° (0°), destructive interference or fading is obtained. Figure 5.14 shows as an example the case corresponding to intensity modulation where the sidebands have a field modulation index m with respect to the optical carrier.

From (5.88) and Fig. 5.14, it is clear that the received RF power follows the law [24, 29]:

$$
\begin{aligned}
p_o &\propto \cos^2\left(\frac{\phi(f_{RF}) + \phi(-f_{RF})}{2}\right) \\
\phi(f_{RF}) &= \frac{\beta_2 L \Omega_{RF}^2}{2}
\end{aligned}
\tag{5.90}
$$

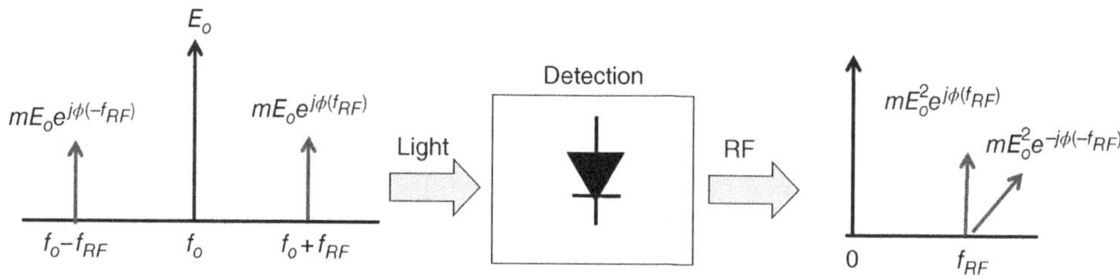

Figure 5.14 Explanation for the origin of dispersion-induced power fading in a Fiber-Wireless system.

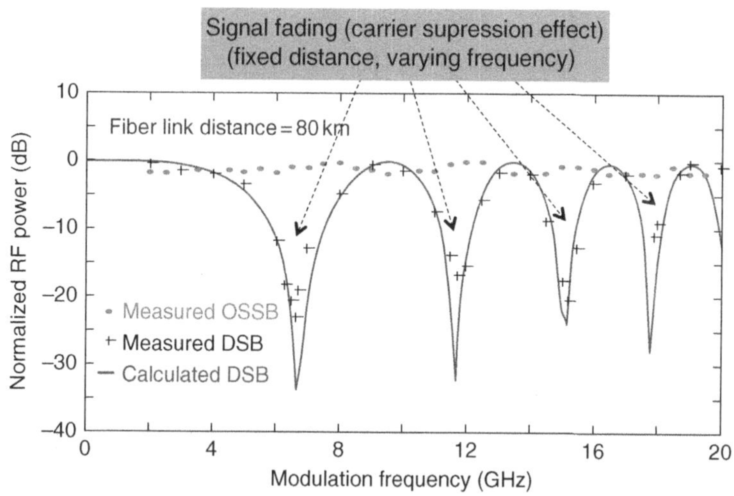

Figure 5.15 Output RF power versus RF frequency for an 80 km standard singlemode fiber link shows the regions where signal fading appears.

Figure 5.15 shows, for instance, the variation of the output RF power versus the RF frequency for an 80 km standard SMF link.

As it can be observed both in the calculated and measured curves corresponding to the ODSB case, there are spectral regions where the output RF power vanishes and no output signal is delivered to the link output. Within these regions, given by the condition:

$$\Omega_{RF} = \sqrt{\frac{(2k+1)\pi}{2\beta_2 L}} \quad k = 0, 1, 2... \tag{5.91}$$

Transmission is unfeasible due to chromatic dispersion. This is a distinctive effect of the deleterious impact of dispersion in analog photonic systems as opposed to the case of digital communication systems where dispersion produces pulse spreading and overlap of consecutive bits.

Despite the fact that dispersion can be overcome in FWSs by means of lumped photonic passive components such as linearly chirped fiber Bragg grating filters and dispersion compensating fibers, in the case of analog systems, the standard and easiest approach is to use optical single sideband plus carrier (OSSB + C) modulation at the transmitter. As is shown in the top part of Fig. 5.16, this avoids having two interference terms arising from the beating of the carrier and the sidebands and,

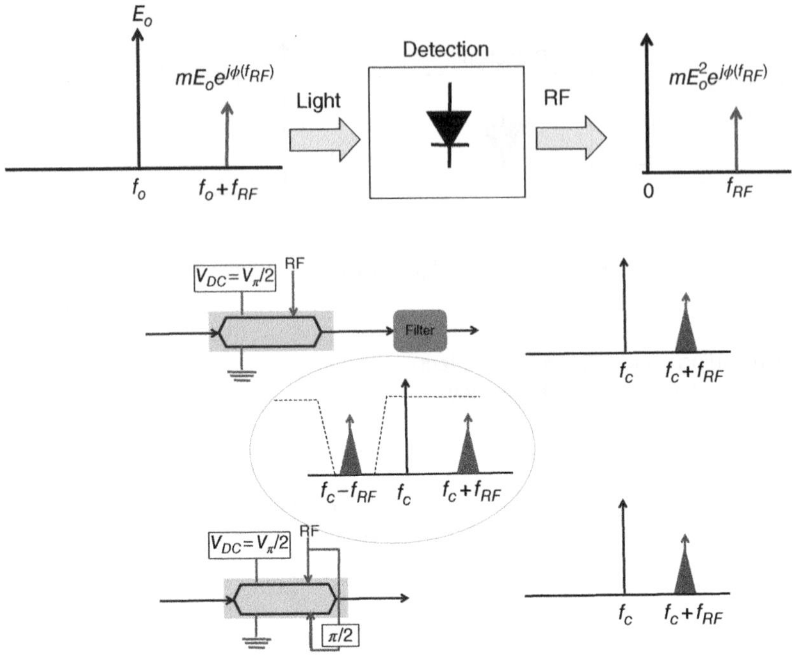

Figure 5.16 (Top) OSSB + C transmission with no interference after optical detection. (Intermediate) OSSB + C signal generation using optical sideband filtering after ODSB modulation. (Bottom) OSSB + C signal generation using 90° phase-shifted dual-drive intensity modulation.

therefore, the possibility of fading due to destructive interference. The bulleted trace in Fig. 5.15 shows how power fading is overcome by means of using OSSB+C modulation in the same 80 km dispersive fiber link.

There are two main possibilities for obtaining an OSSB + C signal that are shown in the intermediate and bottom parts of Fig. 5.16. The first alternative consists of generating the ODSB signal by intensity modulation and then filtering out one of the sidebands by means of a passive optical filter [30, 31]. This technique is potentially broadband and can be simultaneously applied to different sidebands but requires the wavelength locking between the filter central frequency and the output signal from the modulator. The second alternative, which is more compact but more restrictive in terms of bandwidth involves the use of a dual-drive external modulator where the RF signal is injected at the two modulation ports with a 90° phase difference [32].

The optical carrier suppression technique is another method to combat dispersion effects in FW links as shown in Fig. 5.17. Biasing a single-electrode MZM at the minimum transmission point of the optical power versus the input voltage transfer function, the optical carrier will be suppressed to generate a double-sideband-suppressed-carrier (DSB-SC) optical signal. This can be readily checked by making $V_{dc} = 0$ ($\phi_{dc} = 0$), which renders $B_0 = 0$ in (5.44) and results in a carrierless output signal from the modulator according to (5.43). This implementation requires only half the desired modulating frequency to drive the MZM. The mixing of the two optical carriers at a high-speed photodetector generates a single beat component at twice the drive frequency that is not affected by dispersion-induced RF power penalties. Despite its simplicity, this technique requires a large RF drive power to obtain a desirable modulation depth as the modulator is biased in the nonlinear region.

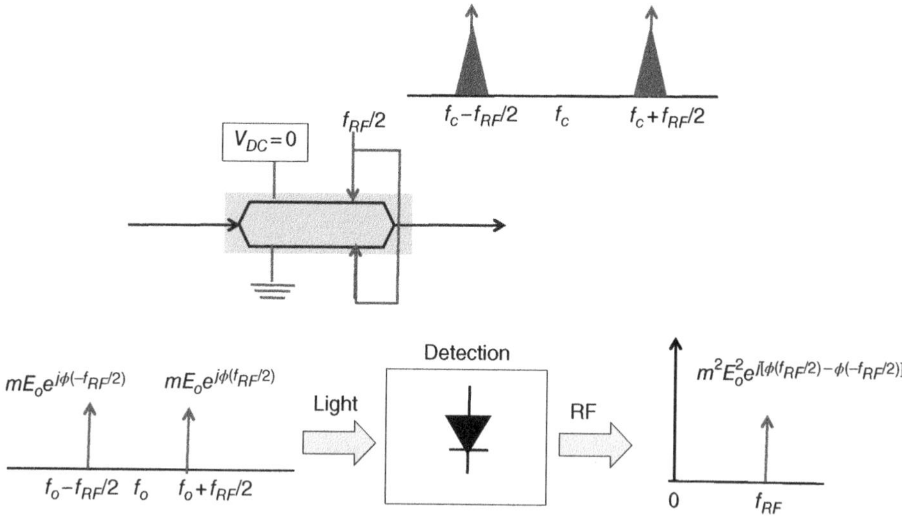

Figure 5.17 (Top) DSB + SC signal generation using a dual-drive intensity modulation in the minimum biasing point. (Bottom) DSB + SC signal detection scheme and recovered sideband.

5.7.2 Optical Nonlinearities

As has been seen in Sections 5.3 and 5.4, nonlinearities limit the dynamic range and hence the performance of MWP systems in general and in particular of FWSs. Nonlinearities occur at the optical frontends and also in the fiber transmission medium. Several techniques have been proposed to combat nonlinearities including the use of optical feedforward configurations, gain modulation, predistortion circuits, and parallel modulator configurations among others [33–37]. These can be categorized into four groups: (i) predistortion techniques, (ii) postcompensation techniques, (iii) optical techniques and (iv) digitized RF signal transmission techniques.

Predistortion techniques focus on the mitigation of IMD within the electro-optic frontend and require a predistorter at the source that combats IMD by generating frequency components of the same amplitude but opposite in phase. Post-compensation techniques are applied at the receiver end and use estimation and equalization to mitigate IMD via signal processing. Optical techniques are applied in the optical section of the MWP link. Optical feed-forward is probably the best-known approach in this category and it is employed to reduce both third-order IMD and also the impact of laser RIN. It is based on using two sections with one determining the error and the other cancelling the error, relying on the careful tuning of the amplitudes and phases of the signals to ensure perfect cancellation. Another approach is based on the use of efficient modulator configurations where two or more external amplitude or phase modulators are employed to reduce the impact of third-order IMD.

Digitized RF transmission techniques [26, 38] are based on bandpass sampling and are attractive since most of the wireless applications use signal bandwidths that are a small fraction of their carrier frequencies. It uses a much lower sampling rate (i.e. comparable to the wireless information bandwidth rather than the wireless carrier frequency). Figure 5.18 depicts the basic principle of bandpass sampling.

If we assume an RF signal $X(f)$ of bandwidth B, bounded by a low-frequency $f_L = 2B$ and a high-frequency $f_H = 3B$, then by sampling at the Nyquist rate given by $Fs = 2B$, one obtains:

$$x_s(t) = x(t) \sum_{n=-\infty}^{\infty} \delta\left(t - \frac{n}{F_s}\right) \Leftrightarrow X_s(f) = F_s \sum_{n=-\infty}^{\infty} X(f - nF_s) \tag{5.92}$$

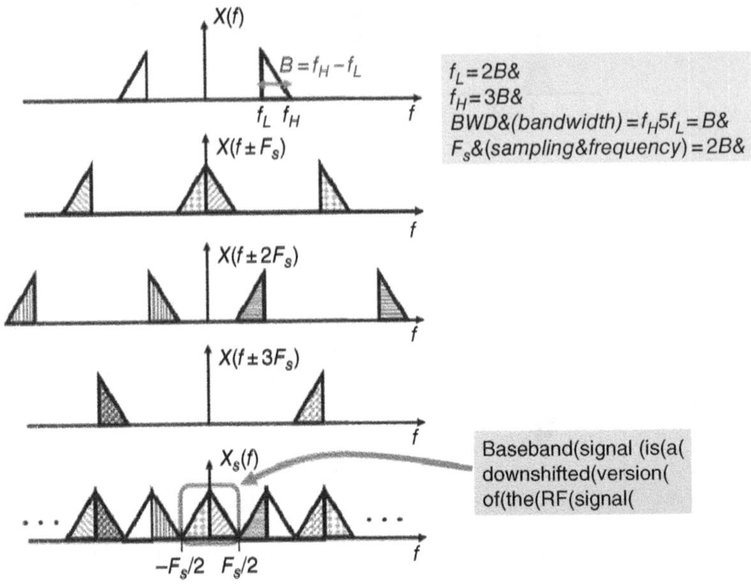

Figure 5.18 The principle of baseband sampling, applied to the digitization of radiofrequency signals.

The contribution of $n = 0, \pm 1, \pm 2, \pm 3...$ is shown by the first four traces in Fig. 5.18, while the top-down trace shows the overall spectrum obtained by summing all the contributions. It is straightforward to appreciate that the original RF spectrum is downshifted to the baseband region and thus it can be recovered by proper filtering. Bandpass sampling hence uses a much lower sampling rate (i.e. comparable to the wireless information bandwidth rather than the wireless carrier frequency). Figure 5.19 shows a proposed layout to implement digitized RF transmission.

The digitization of the wireless signal produces a sampled digital data stream in a serial format that can directly modulate an optical source. This approach enables the use of digital photonic links to transport the wireless signals and corresponds to the BBoF alternative previously described. As can be observed in Fig. 5.19, only a minimal set of frontend components (i.e. analog-to-digital

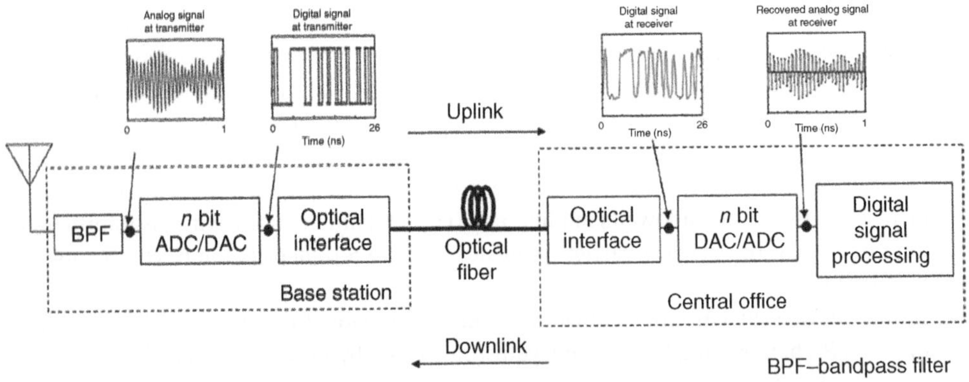

Figure 5.19 System layout of the digitized RF technique. *Source:* Lim et al. [26]/IEEE.

converter (ADC) and digital-to-analog converter (DAC)) are needed in the antenna BS leaving all signal processing functions to be located in the CO.

Given that IMD mainly arises from the nonlinear electrical-to-optical conversion in analog photonic links, this effect can now be completely avoided. In other words, an optical link employing digitized RoF transmission can maintain its dynamic range independent of the fiber transmission distance until when the received signal goes below the link sensitivity. The main disadvantage of digitized radio systems is that although they are very robust in terms of dynamic range and the use of bandpass sampling enables the use of a lower sampling frequency, the ADC requires the analog bandwidth to be at least equivalent to the wireless carrier frequency.

5.8 Fiber-Wireless Networks

The capacity of FW networks can be drastically increased by employing dense WDM (DWDM) [25, 26, 39], [27]. Furthermore, in order to be economically feasible, they have to be compatible and coexist with already existing fiber access network infrastructures and technologies as shown by the network layout of Fig. 5.10. The integration of FW networks into existing PON architectures requires the solution of two important problems: the first is related to the spectral inefficiency of FW systems where a considerable part of the spectrum of a wavelength channel is unused. The second is related to the implementation of cost-effective optical subsystems installed either at RNs or BSs that can route and process the optical channels carrying wireless signals directly in the optical domain. In this section, we address these issues and also illustrate by means of examples some of the application scenarios of FW networks.

5.8.1 Spectral Efficiency

FW RoF systems are in principle inefficient in terms of spectral usage because there is a large portion of unused spectrum between the carrier and the sidebands/sideband in the ODSB/OSSB + C modulation approach. While the OSSB+C modulation approach overcomes the limitation imposed by fiber chromatic dispersion and improves the optical spectral usage by at least 50% compared to the ODSB case, the transportation of modulated radio signals, even in this modulation format, still leads to the inefficient use of the optical spectrum [26]. This spectral inefficiency has a special impact in a WDM environment where the actual information bandwidth of the radio signals modulated onto an optical WDM channel (typically at 50 or 100 GHz spacing) is <2 GHz.

To overcome this limitation, several techniques based on interleaving multiple RoF channels using the spectral band between the optical carriers and sidebands in an OSSB + C signal and that between the two sidebands in the optical carrier suppression technique have been proposed [39–41]. Figure 5.20 shows, for instance, an ingenious approach developed to interleave wavelength division multiplexed OSSB + C RoF channels where the optical carrier separation Δf between adjacent channels is lower than the RF sideband to carrier separation f_{RF} within a given channel.

This technique has been shown to improve, for instance, the overall capacity within the standard 1550 nm Erbium-doped fiber amplifier (EDFA) gain window by a factor of 3 for a 37.5 GHz FW link. Similar approaches have been reported for millimeter-wave systems operating in the 60 GHz band [42–45], one of which is shown in Fig. 5.21.

Here, the wavelength channel separation grid of 100 GHz as defined by the International Telecommunication Union (ITU) is divided into four segments and adjacent optical carriers are

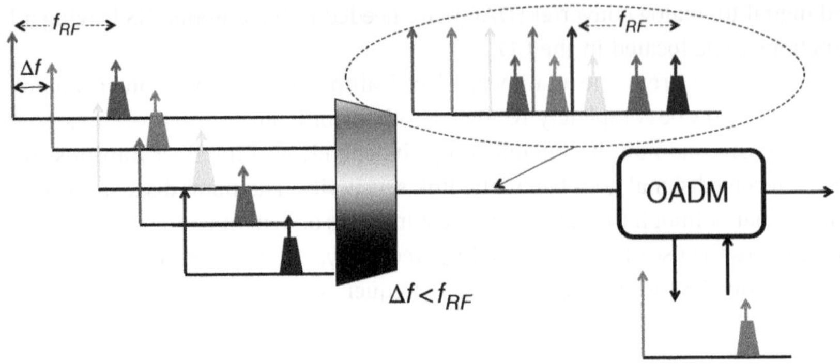

Figure 5.20 Wavelength interleaving scheme for improving the spectral efficiency of WDM OSSB + C FW networks.

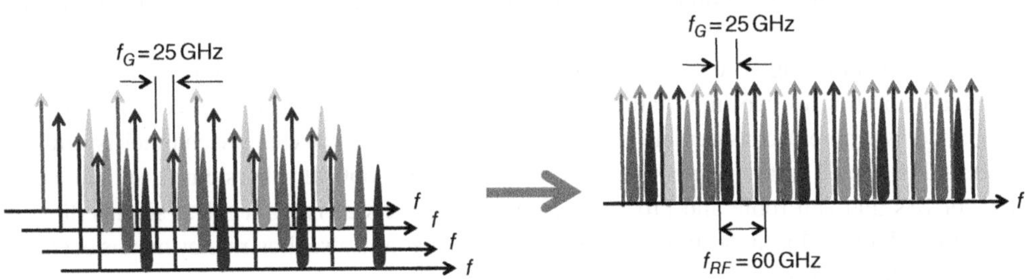

Figure 5.21 Wavelength interleaving for a millimeter-wave DWDM system with adjacent carrier separation of 25 GHz. Each optical carrier is subject to single-sideband modulation by a 60 GHz subcarrier. *Source:* Kuri et al. [45]/IEEE.

interleaved with a 25 GHz separation. Each optical carrier is a single sideband modulated by a 60 GHz RF subcarrier.

The interleaving technique requires that all the FW channels are packed together in adjacent locations occupying a compact spectral region different from that reserved for standard high-speed digital communications but overall it allows for the integration of both types of communication channels over the same infrastructure. However, the routing and processing of FW channels require a specific design for optical layer components such as OADMs as will be shown in the next section.

5.8.2 Optical Subsystems for Fiber-Wireless Networks

One of the most important subsystems for the coexistence of digital and FW services over the same access network infrastructure under WDM is the availability of specific OADM and multiplexing/demultiplexing devices capable of inserting and extracting FW channels leaving unaltered the standard digital channels. As it is shown schematically in Fig. 5.20, the OADM or, in its case, a demultiplexer must extract both the carrier and the sideband, taking into account that the intermediate spectral region between them is not empty due to the FW channel interleaving. OADMs and demultiplexers can be necessary within the CO, RNs, and BSs. As shown in Fig. 5.10, these can be part of a ring-type or a tree-type subnetwork. In each case, the design constraints are different and several architectures have been reported for each network topology scenario. For instance, in ring architectures, the OADM is the preferred solution for RNs. Figure 5.22 shows a layout for an OADM based on a 7-port optical circulator (OC) in conjunction with one double-notch and one single-notch fiber Bragg gratings [46, 47]. This configuration provides wavelength add, drop, and reuse functionalities.

The device operation is described in the following lines. As can be observed, it consists of 5 ports: "IN," "OUT," "DOWNLINK DROP," "REUSE," and "ADD." The experiment was conducted for a system composed of three RF channels each one separated by 37.5 GHz from its optical carrier. The OSSB + C channels were interleaved such that the spacing between an optical carrier and the sideband corresponding to the adjacent channel (which entered the optical interface via circulator port 1 IN) was 12.5 GHz. The double notch filter at port 2 reflected 100% of the desired downlink (DL) optical carrier (in this case) and its corresponding sideband, while it transmitted the "thru" channels to port 6 of the circulator, to be routed out from the interface via port 7 (OUT).

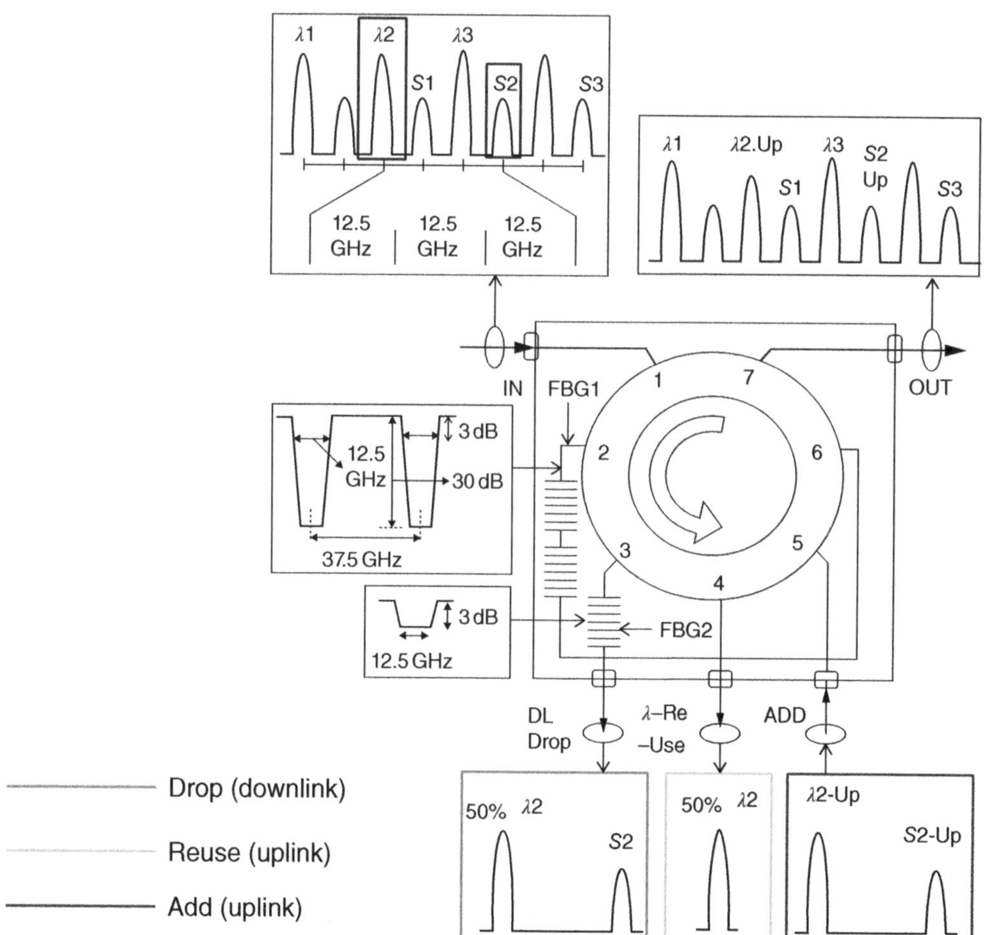

Figure 5.22 ROADM for a remote node in ring-type FW network based on a 7 port circulator and fiber Bragg grating filters. *Source:* Bakaul et al. [46]/IEEE.

The FBG at port 3 was designed to reflect 50% of the optical carrier while the remaining 50% of the optical carrier and the corresponding sideband were dropped at port 3 (the DOWNLINK DROP). The 50% of the reflected optical carrier was recovered at port 4 (REUSE) and can be reused as the uplink (UL) optical carrier through being modulated by the upstream mm-wave radio signals. The UL optical signal can be added back to the mainstream via port 5 (ADD) and combined with the "thru" channels and exits via port 7 (OUT).

In star-tee architectures, an optical interface with simultaneous demultiplexing and multiplexing of several channels is preferred in the CO and RNs of the FW network to avoid the complexity resulting from cascading OADMs. It is essential to combine these functionalities required in both the CO and the RNs into a single device to obtain low complexity and cost-effective architectures. It is also important that passive WDM components in the COs and RNs are transparent to the UL channels generated by reusing the DL optical carrier, as this allows (as we shall see later in this section) to simplify the structure of the BS by removing the light source from the UL path. Again, several approaches have been reported in the literature [42, 48–50], which fulfill these requirements to a lower or higher degree. For instance, Fig. 5.23 shows a solution which is based on the use of a $(2N+2) \times (2N+2)$ arrayed waveguide grating (AWG) device.

The subsystem shown in Fig. 5.23 is designed to effectively multiplex and demultiplex WI-FW channels and it ensures the transparency of the CO and the RN to UL channels generated by reusing the DL optical carriers, which enables a simple, compact, and low-cost BS through the complete removal of the UL light source. The basic multiplexing configuration is shown in the top part of Fig. 5.23. Note that the optical carriers (C_1, C_2, ... C_N) are separated by $2\Delta f$, while the separation between the carrier and its associated RF sideband (Ci-Si) is $3\Delta f$. In this way, using AWG devices with channel spacing given by 12.5 GHz allows for the multiplexing of 25 GHz-spaced optical channels, each one carrying an RF signal at 37.5 GHz (i.e. in millimeter-wave part of the spectrum). Furthermore, the spectral-spatial diversity of the AWG is fully exploited as all interleaved channels (optical/RF) are separated by Δf.

The MUX/DEMUX, shown in the bottom right part of Fig. 5.23, comprises a $(2N+2) \times (2N+2)$ AWG with a channel bandwidth $<\Delta f$ and a channel spacing of Δf, in conjunction with multiple OCs and optical isolators (OIs). The input (A) and output (B) ports of the AWG, reciprocal in nature, are numbered from 1 to $2N+2$. The characteristic routing matrix of the AWG that governs the distribution of different channels at various ports is shown in the bottom left part of Fig. 5.23. For the sake of clarity, the proposed scheme is considered to be located at an RN where the UL channels are multiplexed and the DL channels are demultiplexed simultaneously.

Channels from the CO via the feeder network enter the RN, are split by a 3 dB coupler, and pass through the left-hand side circulators before entering the AWG via ports A_1 and A_4. These input ports are selected in such a way that the optical carriers and their respective modulation sidebands are demultiplexed together and exit the AWG via the odd-numbered output ports B_1, B_3, B_5 ... followed by the right-hand side circulators which separate DL from UL channels.

In the UL direction, the modulated optical mm-wave channels are generated by either using the optical carriers that correspond to wavelengths spaced at multiples of the free spectral range (FSR) of the AWG from the DL optical carriers, or by reusing the DL optical carriers recovered by applying a wavelength reuse technique are applied to the AWG via the ports B_1, B_3, B_5.... Due to the reciprocal and cyclic characteristics of the AWG, the UL optical carriers and their respective modulation sidebands combine at ports A_4 and A_1, respectively. The composite UL carriers at are then passed through and looped back to the AWG through port B_2 that redistributes the carriers respectively to the odd-numbered ports, starting with A_3.

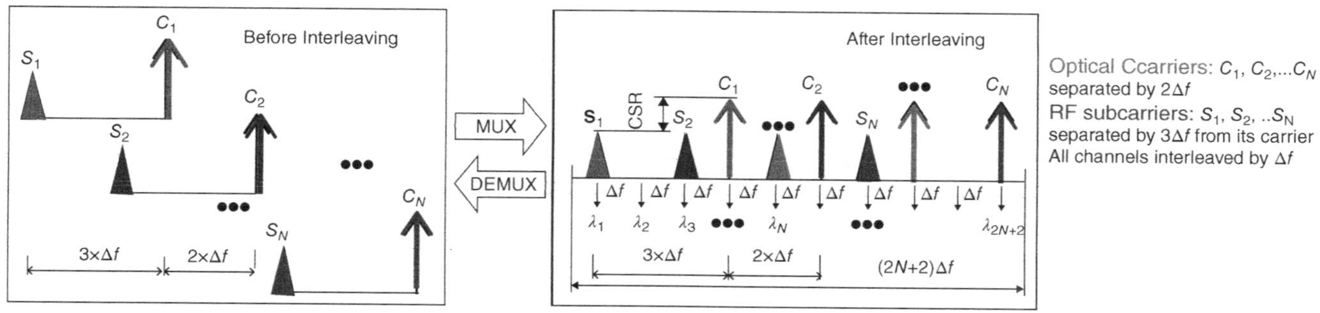

I \ O	B_1	B_2	B_3	•••	B_{N-1}	B_N	B_{N+1}	•••	B_{2N}	B_{2N+1}	B_{2N+2}
A_1	λ_1	λ_2	λ_3		λ_{N-1}	λ_N	λ_{N+1}		λ_{2N}	λ_{2N+1}	λ_{2N+2}
A_2	λ_2	λ_3	λ_4		λ_N	λ_{N+1}	λ_{N+2}		λ_{2N+1}	λ_{2N+2}	λ_1
A_3	λ_3	λ_4	λ_5		λ_{N+1}	λ_{N+2}	λ_{N+3}		λ_{2N+2}	λ_1	λ_2
A_4	λ_4		[illegible]								
A_{N-1}	λ_{N-1}	λ_N	λ_{N+1}		λ_{2N-3}	λ_{2N-2}	λ_{2N-1}		λ_{N-4}	λ_{N-3}	λ_{N-2}
A_N	λ_N	λ_{N+1}	λ_{N+2}		λ_{2N-2}	λ_{2N-1}	λ_{2N}		λ_{N-3}	λ_{N-2}	λ_{N-1}
A_{N+1}	λ_{N+1}	λ_{N+2}	λ_{N+3}		λ_{2N-1}	λ_{2N}	λ_{2N+1}		λ_{N-2}	λ_{N-1}	λ_N
•••											
A_{2N}	λ_{2N}	λ_{2N+1}	λ_{2N+2}		λ_{N-4}	λ_{N-3}	λ_{N-2}		λ_{2N-3}	λ_{2N-2}	λ_{2N-1}
A_{2N+1}	λ_{2N+1}	λ_{2N+2}	λ_1		λ_{N-3}	λ_{N-2}	λ_{N-1}		λ_{2N-2}	λ_{2N-1}	λ_{2N}
A_{2N+2}	λ_{2N+2}	λ_1	λ_2		λ_{N-2}	λ_{N-1}	λ_N		λ_{2N-1}	λ_{2N}	λ_{2N+1}

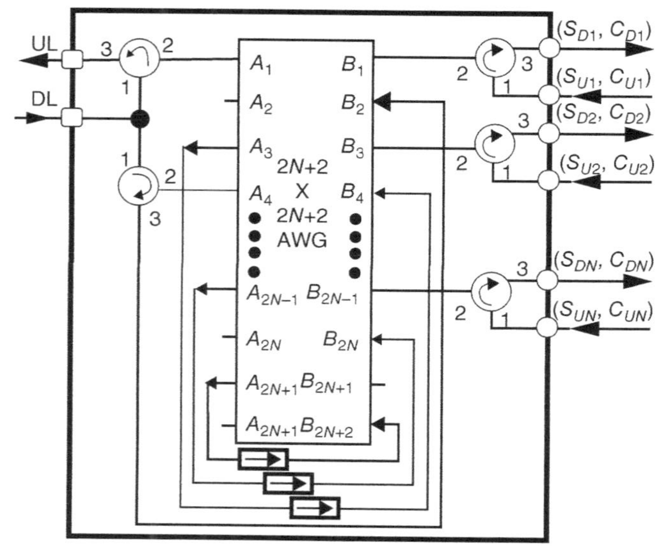

Figure 5.23 Node configuration for simultaneous multiplexing and demultiplexing of wavelength interleaved channels in a star-type network. (Top left) Interleaving spectral scheme. (Bottom left) The input–output AWG routing matrix. (Bottom right) MUX/DEMUX configuration using a $(2N+2) \times (2N+2)$ AWG. *Source:* Adapted from Bakaul et al. [50].

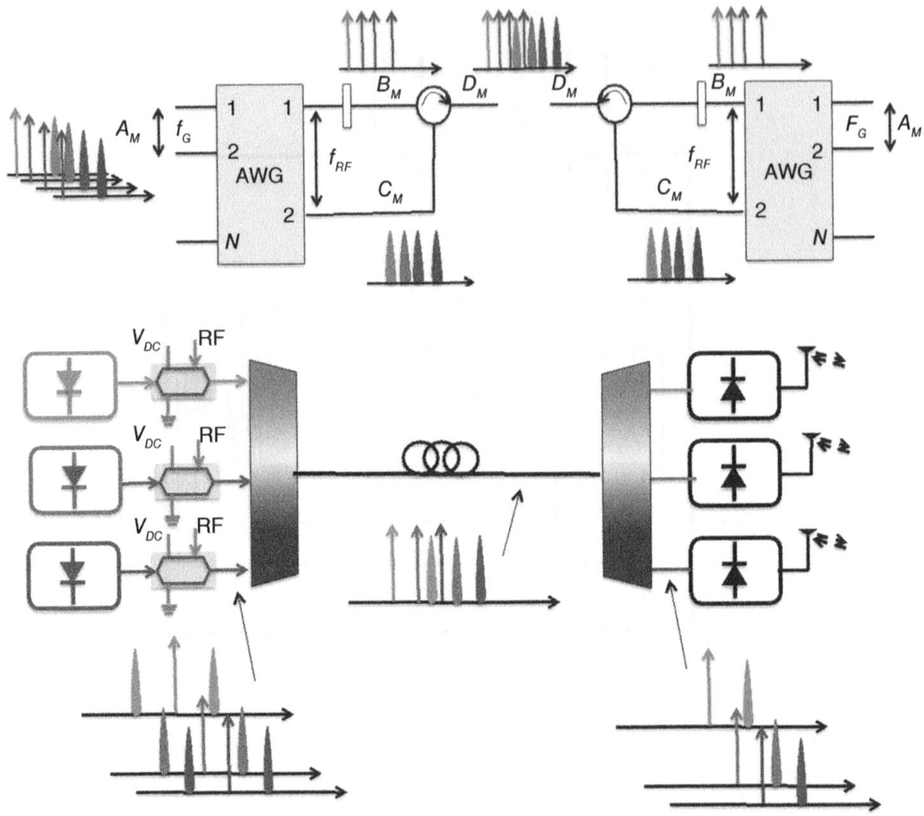

Figure 5.24 Multiplexer (top left) and demultiplexer (top right) and RoF system configurations (bottom) for the millimeter-wave DWDM interleaved system proposed in [42]. *Source:* Adapted from Toda et al. [42].

The configuration described in Fig. 5.23 is rather complex and simpler configurations are feasible as well at the cost of duplicating the number of AWG devices. One example is shown in Fig. 5.24 where the multiplexer, demultiplexer and FW system configurations for the millimeter-wave interleaved system described in Fig. 5.21 are shown [42]. For the multiplexer, the main building block here is an $N \times 2$ AWG device that routes the carrier and sidebands injected through ports 1, 2 ... N to two different output ports. Carriers are directed to the upper port where they are subsequently passed by a 25 GHz free spectral range Fabry–Perot filter, where the filter resonances coincide with the optical carrier spectral position. Subcarriers are directed to the lower output port and sent via an OC to the Fabry–Perot filter where they are reflected and combined with the comb of optical carriers. The demultiplexer operates exactly in the reverse way.

A second critical issue for FWN operation is that optical links have to support full-duplex transmission from the CO to the BS and vice versa. In principle, this implies that optical interfaces within the antenna BS must include optical sources which can be modulated by the UL wireless signals and furthermore feature narrow linewidths to minimize phase noise degradation. This option however is not attractive for UL transmission as ultra-stable, low-cost, and narrow-linewidth optical sources are difficult to realize and therefore expensive. A possible solution to overcome these problems is to completely remove the need for an optical source in the antenna BS. This also translates

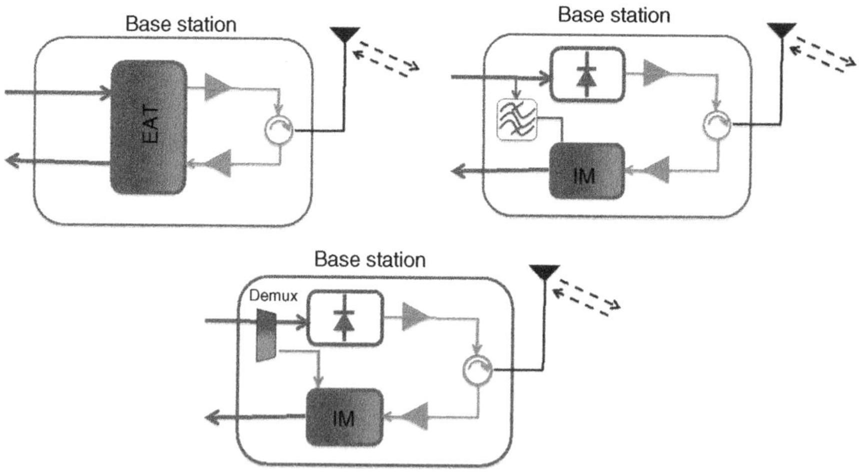

Figure 5.25 BS frontend configurations which do not require the presence of an optical source to deliver upstream traffic to the CO. (Top left) Electroabsorption transceiver frontend. *Source:* Stohr et al. [51]. (Top right) Wavelength reuse frontend. *Source:* Nirmalathas et al. [47]. (Bottom) Remote carrier delivery and selection frontend. *Source:* Chen et al. [52].

the wavelength assignment and source monitoring functionalities to the CO. This amounts to further relaxing the stringent requirements on the antenna BS hardware that is very cost-sensitive, especially in applications where pico and femtocells are required to provide area coverage. Figure 5.25 shows three possible solutions [47, 51, 52] for the BS frontend design, which do not require the physical presence of an optical source.

The layout shown in the top left part of Fig. 5.25 corresponds to the electroabsorption transceiver (EAT) frontend [51], where an electroabsorption modulator is used as a detector (for the DL carrier) and a modulator (for the UL carrier). The unmodulated UL carrier is remotely delivered from the CO being transparent to detection at the EAT. The second option, shown in the top right part of Fig. 5.25, corresponds to the wavelength reuse frontend [47] where an optical bandpass filter is employed to extract part of the DL optical carrier and reinject it to the optical input of the UL intensity modulator. Finally, the bottom part of Fig. 5.25 shows the layout corresponding to the remote carrier delivery and selection frontend [52]. Here, the UL carrier is remotely provided by the CO. The optical source in the BS is replaced by a demultiplexer, which separates the unmodulated UL carrier from the DL signals. The upstream wireless signals are modulated onto this carrier using an external modulator.

5.8.3 Application Scenarios

FW networks can be employed in two main application scenarios. The first one, currently under development, corresponds to the access segment of telecommunications network while the second, subject to current research and expected to be developed within the next years, are *in-home* networks.

Regarding the application of FW networks in access networks, we have previously discussed the convenience of integrating them into the already installed fiber infrastructure. This integration leads to the concept of *converged networks* [53, 54] where traditional "wired" digital (baseband) optical access technologies co-exist with RoF wireless access technologies in the same fiber infrastructure (usually a PON). Research has focused on using subcarrier multiplexing to isolate the wired and wireless signals and several wireless standards including 3G, WiFi, and WiMAX have been demonstrated by suitable overlay over the PON infrastructure.

The implementation of a fully converged access network requires however going one step further in the wireless delivery, which implies the development of an integrated optical access infrastructure to enable the simultaneous distribution of multiple signal bands (BB, IF, and RF) per optical channel. In particular, this requires a further refinement of the routing and demultiplexing subsystems initially devised for single RF channel delivery per optical channel, some of which have been presented in the previous section.

The most flexible approach is therefore to provision passive WDM components in the RNs to support this functionality. A first necessary step is considering a frequency allocation plan that can efficiently pack in the available region of the spectrum all the targeted services in each optical channel. Figure 5.26 illustrates one proposal reported in [54] in which N optical super-channels, each including BB, IF, and RF channels (the last including carrier and sideband). Internal separation is given by $\Delta\lambda$, while the RF sideband optical carrier separation is given by $n\Delta\lambda$ ($n = 3$). Equivalent services in adjacent superchannels are separated by $4\Delta\lambda$, while the total multiplex spans a region of $4(N+1)\Delta\lambda$.

Figure 5.27 shows a multiplexer device compatible with this frequency plan, which is based on an AWG device.

The multiplexer is designed in such a way that it features $(4N+1)$ input and $(4N+1)$ ports and the bandpass bandwidth is equal to the adjacent channel spacing of the desired multiplexing scheme, in this case $\Delta\lambda$. The input (A) and output (B) ports of the AWG are numbered from 1 to $4N+1$. The optically modulated RF, BB, and IF input signals enter the AWG starting with port A_2 to A_{4N}, leaving the 5th, 9th, 13th, $(4N+1)$th ports unused. The RF channels are modulated in OSSB+C modulation format. The AWG combines all the modulation sidebands of the RF channels ($S_1, S_2, \ldots S_N$) as well as the BB and IF channels at the output port B_1. Due to the cyclic characteristics of the AWG, described in Chapter 2, the optical carriers of the RF channels ($C_1, C_2, \ldots C_N$) also exit as a composite signal via the output port B_4. The composite carriers are then looped back to the AWG

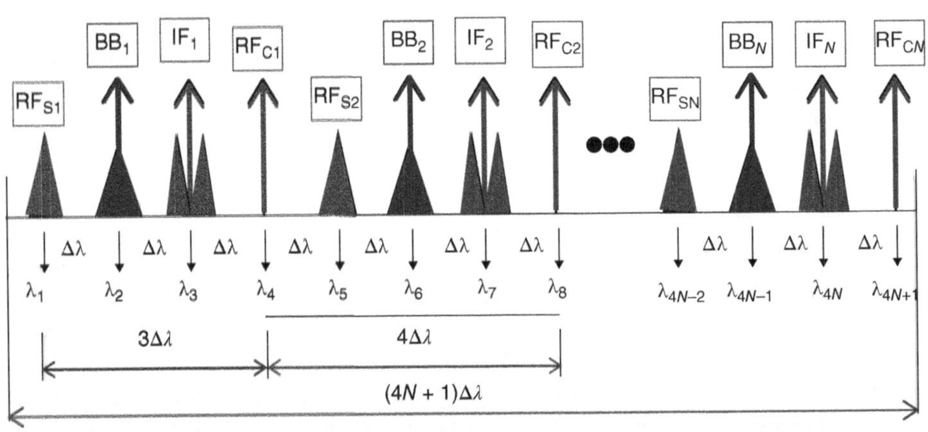

Figure 5.26 Frequency plan for multiservice converged access FW networks compatible with DWDM. *Source:* Adapted from Bakaul et al. [54].

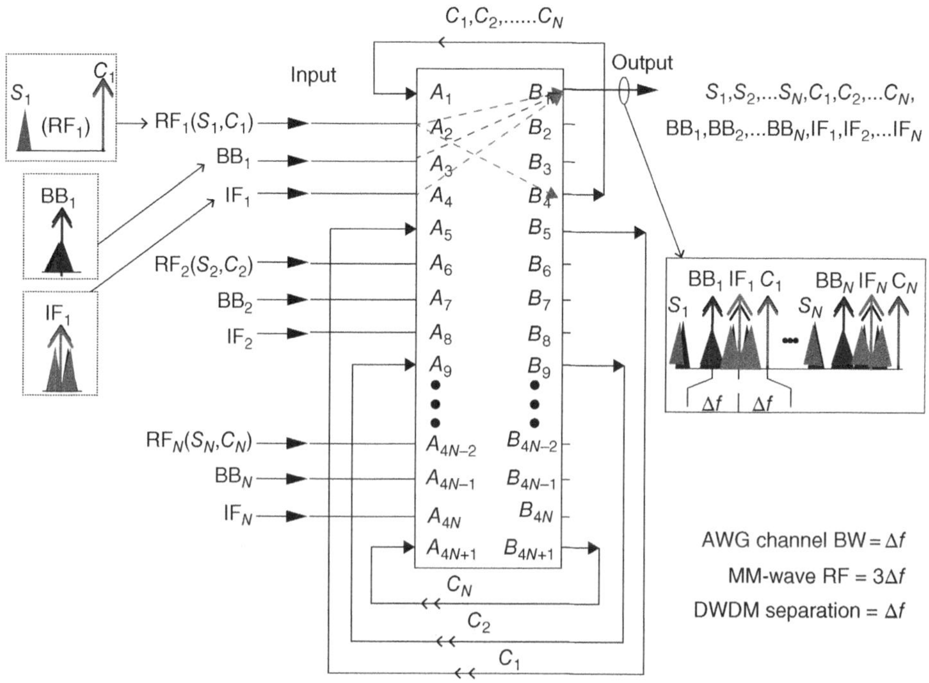

I/O		Output ports										
		B_1	B_2	B_3	B_4	B_{N-1}	B_N	B_{N+1}		B_{4N+1}	B_{4N}	B_{4N+1}
Input ports	A_1	λ_1	λ_2	λ_3	λ_4	λ_{N-1}	λ_N	λ_{N+1}		λ_{4N-1}	λ_{4N}	λ_{4N+1}
	A_2	λ_2	λ_3	λ_4		λ_N	λ_{N+1}	λ_{N+2}		λ_{4N}	λ_{4N-1}	λ_1
	A_3	λ_3	λ_4	λ_5		λ_{N+1}	λ_{N+2}	λ_{N+3}		λ_{4N+1}	λ_1	λ_2
	A_{N-1}	λ_{N-1}	λ_N	λ_{N+1}	λ_{N+2}	λ_{4N-4}	λ_{4N-3}	λ_{4N-2}		λ_{N-4}	λ_{N-3}	λ_{N-2}
	A_N	λ_N	λ_{N+1}	λ_{N+2}		λ_{4N-3}	λ_{4N-2}	λ_{4N-1}		λ_{N-3}	λ_{N-2}	λ_{N-1}
	A_{N+1}	λ_{N+1}	λ_{N+2}	λ_{N+3}		λ_{4N-2}	λ_{4N-1}	λ_{4N}		λ_{N-2}	λ_{N-1}	λ_N
	A_{4N-1}	λ_{4N-1}	λ_{4N}	λ_{4N-1}	λ_{4N+2}	λ_{N-4}	λ_{N-3}	λ_{N-2}		λ_{4N-4}	λ_{4N-3}	λ_{4N-2}
	A_{4N}	λ_{4N}	λ_{4N+1}	λ_1		λ_{N-3}	λ_{N-2}	λ_{N-1}		λ_{4N-3}	λ_{4N-2}	λ_{4N-1}
	A_{4N+1}	λ_{4N-1}	λ_1	λ_2		λ_{N-2}	λ_{N-1}	λ_N		λ_{4N-2}	λ_{4N-1}	λ_{4N}

Figure 5.27 Flexible multiplexer/demultiplexer based on an AWG device for a multiservice DWDM FW network using the frequency plan of Fig. 5.26 (left). Routing table of the AWG (right). *Source:* Adapted from Bakaul et al. [54].

through the input port that redistributes the carriers to the 5th, 9th, 13th, and $(4N + 1)$th output ports. To realize the desired multiplexing, the distributed carriers are again looped back to the AWG via the unused input ports and the resultant output at port B_1 is the RF, BB, and IF signals multiplexed with the BB and IF channels interleaved between the optical carrier and the modulation sideband of the RF channels.

With fiber to the home becoming a reality in a growing number of cities, offering broadband capacity up to the home's doorstep, the current bottleneck for broadband communication is the

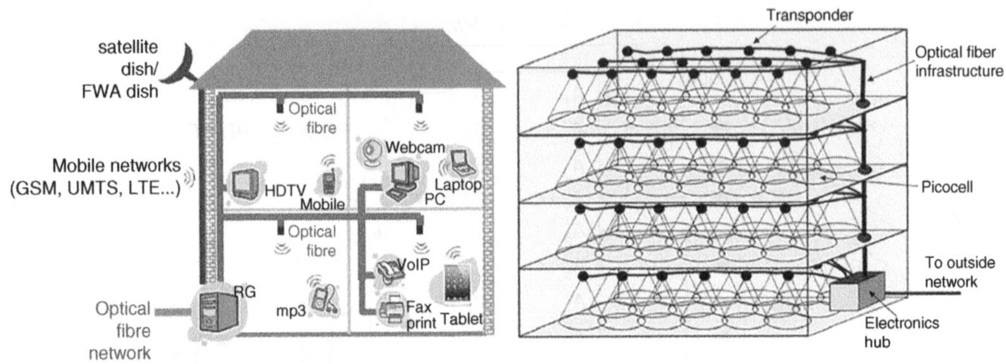

Figure 5.28 Fiber-wireless network application scenarios for in-home [55] (left) and in-building (right) networks. *Source:* Adapted from Sauer et al. [56].

extension of the fiber's high capacity into the home, by means of the so-called *in-home* and *in-building* networks [55, 56]. While presently there is usually a mixture of wired and wireless networks inside homes and buildings, a single network, which can support all these services is recommended in order to achieve greater than the currently available bit rates. An interesting option is to employ a fiber backbone to distribute digital services in combination with radio signals, which can provide support to pico-cellular networks covered by very small antennas. By reducing the cell size and limiting the number of users per cell, this scheme is able to support very high data rate transmission per user especially in a dense in-building scenario. Figure 5.28 shows two examples corresponding to an in-home and an in-building scenario, respectively.

In the first case, an in-home optical fiber backbone physically separated from the access network by means of a residential gateway (RG) provides the infrastructure to deliver the wired and wireless services simultaneously. Although in terms of bandwidth, the best solution is SMF, in terms of cost and handling multimode, and even large core area plastic fibers represent a serious competing alternative [55]. As for in-building networks, due to the relatively short distances in a building so far most of the optical distribution for in-building architectures are based on multimode fibers. Here as shown in the right part of Fig. 5.28, a large number of densely packed pico-cells are connected via an optical backbone of multimode fiber to a CO [56].

5.9 Chapter Summary

In this chapter, we have considered in detail the photonic distribution of RF signals dividing its contents into two main parts. In the first, covering Section 5.2–5.4, we have focused on microwave photonic links introducing their basic operation principles, limiting factors, and FOMs. In the second part of the chapter, which covers Sections 5.5–5.8, we have dealt with the main application of MWPLs which is in the implementation of FW telecommunication systems and networks covering the main relevant topics related to system operation, techniques currently employed for the optical transport of wireless signals, specialization of the degradation mechanisms, and the main techniques reported in the literature to combat them, optical layer subsystem design, and application scenarios.

Problems

5.1 Derive Eq. (5.3) for the RF Gain of a lossy and dispersionless MWP link.

5.2 The RF Gain for a lossy and dispersionless MWP link that employs an external modulator is given by Eq. (5.9) in terms of the relevant parameters of the modulator, the optical source, and the detector. In this exercise, we will derive the steps leading to that expression. First of all, consider the external modulator and derive an expression for its equivalent input–output RF power transmission factor $p_{i,o}^2/p_i$. Assume that there is perfect impedance matching between the voltage generator and the modulator electrical input and that the modulator is biased at quadrature. In the second place, derive the expression for the TM-D factor characterizing the propagation losses from the transmitter to the receiver. In the third place, derive the equivalent input-output RF power transmission at the optical receiver $p_o/p_{o,o}^2$ now assuming impedance matching between the photodetector output impedance and the load impedance. Finally, use Eq. (5.3) to derive the RF gain expression given by Eqs. (5.9) and (5.10).

5.3 Derive Eq. (5.23) that yields the values of the harmonic distortion terms in a nonlinear system with response given by (5.20) when the input signal is:

$$v_M(t) = \sqrt{2}v_s \cos \Omega_{RF} t$$

5.4 Show that the spurious-free dynamic range for a microwave photonics link limited by nth order intermodulation distortion can be expressed as:

$$SFDR = \sqrt[n]{\frac{c_1^n}{c_n \overline{n}^{n-1}}}$$

where the RF powers of the fundamental and the nth order distortion are given as a function of the input RF power to the system P_{RF} by: $P_1 = c_1 P_{RFin}$, $P_n = c_n P_{RFin}^n$ and N_{out} represents the noise power. Assume that you know the interception point OIP_N for this intermodulation order. Show then that:

$$SFDR(\text{dB}) = \frac{(n-1)}{n}[OIP_N(\text{dBm}) - N_{out}(\text{dBm})]$$

5.5 An analog photonic link is defined by OIP3 of 95 dB and a noise level of −80 dB/Hz. Which is the spurious-free dynamic range for a 10 MHz bandwidth?

5.6 The dynamic range of an MWP link limited by third-order intermodulation distortion can be written as:

$$SFDR\left(\text{dB Hz}^{2/3}\right) = \frac{2}{3}[OIP3 - N_{out}]$$

where $OIP3(\text{dBm}) = IIP3(\text{dBm}) + G_{RF}(\text{dB})$ is the third-order optical interception point, which depends on the RF gain $G_{RF}(\text{dB})$and the input power interception point $IIP3(\text{dBm})$ and $N_{out}(\text{dBm/Hz}) = 10\log[\eta_{out}(\text{W/Hz})/10^{-3}]$ represents the noise spectral density in logarithmic units at the link output.

Knowing that the noise figure for an MWP link is defined as:

$$nf = \frac{(P_{RF,in}/k_B T)}{(P_{RF,out}/N_{out})}$$
$$N_{out} = 10\log\left[N_{out}(\mathrm{W/Hz})/10^{-3}\,\mathrm{W}\right]$$
$$NF(\mathrm{dB}) = 10\log(nf)$$

where B is the system bandwidth and k_B is the Boltzmann constant ($k_B = 1.38 \times 10^{-23}$ J°/K), showing that the dynamic range can be expressed as:

$$SFDR\left(\mathrm{dB.Hz}^{2/3}\right) = \frac{2}{3}[IIP3 - NF - 10\log(k_B T)]$$

and specialize this expression for $T = 300$ °K.

5.7 Derive Eqs. (5.88) and (5.89) that yield the RF gain of a dispersive MWP link for amplitude and phase modulation, respectively.

5.8 Most of the links in the access network are below 10 km with an average value of 2 km. Consider this last figure as representative case for an access network link length. Compute the values of the fading frequencies for both intensity and phase modulation in the spectral region ranging from the L band (1–2 GHz) to the V band (50–75 GHz) when employing a standard SMF featuring a dispersion parameter of $\beta_2 = -21$ ps^2/km @1550 nm.

References

1 Yao, J. (2009). Microwave photonics. *IEEE J. Lightw. Technol.* 27: 314–335.
2 Capmany, J. and Novak, D. (2007). Microwave photonics combines two worlds. *Nat. Photonics* 1: 316–330.
3 Seeds, A. (2002). Microwave photonics. *IEEE Trans. Microw. Theory Tech.* 50 (3): 877–887.
4 Seeds, A.J. and Williams, K.J. (2006). Microwave photonics. *IEEE J. Lightw. Technol.* 24: 4628–4641.
5 Cox, C.H. III (2004). *Analog Photonic Links: Theory and Practice*. Cambridge: Cambridge University Press.
6 Chang, W.S.C. (ed.) (2007). *RF Photonic Technology in Optical Fiber Links*. Cambridge University Press.
7 Iezekiel, S. (ed.) (2009). *Microwave Photonics*. Wiley.
8 Vilcot, A., Cabon, B., and Chazelas, J. (2003). *Microwave-Photonics: From Components to Applications and Systems*. Springer.
9 Rumerhald, C., Algani, C., and Billabert, A.L. (2013). *Microwaves Photonic Links: Components and Circuits*. Wiley.
10 Cox, C.H. III, Ackerman, E.I., Betts, G.E., and Prince, J.L. (2006). Limits on the performance of RF-over-fiber links and their impact on device design. *IEEE Trans. Microw. Theory Tech.* 54: 906–920.
11 Roussell, H.V., Regan, M.D., Prince, J.L. et al. (2007). Gain, noise figure and bandwidth-limited dynamic range of a low-biased external modulation link. *2007 Interntional Topical Meeting on Microwave Photonics*, Victoria, Canada, pp. 84–87. http://doi.org/10.1109/MWP.2007.4378142.
12 Urick, V.J., Rogge, M.S., Bucholtz, F., and Williams, K.J. (2006). The performance of analog photonic links employing highly-compressed erbium-doped fiber amplifiers. *IEEE Trans. Microw. Theory Tech.* 54: 3141–3145.

13 Urick, V.J., Bucholtz, F., Devgan, P.S. et al. (2007). Phase modulation with interferometric detection as an alternative to intensity modulation with direct detection for analog-photonic links. *IEEE Trans. Microw. Theory Tech.* 55: 1978–1985.

14 Wyrwas, J.M. and Wu, M.C. (2009). Dynamic range of frequency modulated direct-detection analog fiber optic links. *IEEE J. Lightw. Technol.* 27: 5552–5562.

15 Darcie, T.E. and Driessen, P.F. (2006). Class-AB techniques for high-dynamic-range microwave-photonic links. *IEEE Photonics Technol. Lett.* 18: 929–931.

16 McKinney, J.D. and Williams, K.J. (2009). Sampled analog optical links. *IEEE Trans. Microw. Theory Tech.* 57: 2093–2099.

17 Urick, V.J., Godinez, M., Devgan, P.S. et al. (2009). Analysis of an analog fiber-optic link employing a low-biased Mach–Zehnder modulator followed by an erbium-doped fiber amplifier. *IEEE J. Lightw. Technol.* 27: 2013–2019.

18 Darcie, T.E., Zhang, J., Driessen, P.F., and Eun, J.-J. (2007). Class-B microwave-photonic link using optical frequency modulation and linear frequency discriminators. *IEEE J. Lightw. Technol.* 25: 157–164.

19 Chi, H., Zou, X., and Yao, J. (2009). Analytical models for phase-modulation-based microwave photonic systems with phase modulation to intensity modulation conversion using a dispersive device. *IEEE J. Lightw. Technol.* 27: 511–521.

20 Marpaung, D., Roeloffzen, C., Leinse, A., and Hoekman, M. (2010). A photonic chip based frequency discriminator for a high performance microwave photonic link. *Opt. Express* 18: 27359–27370.

21 Gasulla, I. and Capmany, J. (2011). Analytical model and figures of merit for filtered microwave photonic links. *Opt. Express* 19: 19758–19774.

22 Marcuse, D. (1980). Pulse distortion in single-mode fibers. *Appl. Opt.* 19: 1653–1660.

23 Moslehi, B. (1986). Analysis of optical phase noise in fiber-optic systems employing a laser source with arbitrary coherence time. *IEEE J. Lightw. Technol.* 4: 1334–1351.

24 Devaux, F., Sorel, Y., and Kerdiles, J.F. (1993). Simple measurement of fiber dispersion and of chirp parameter of intensity modulated light emitter. *IEEE J. Lightw. Technol.* 11: 1937–1940.

25 Kitayama, K. (2000). Architectural considerations on fiber radio millimeter-wave wireless access systems. *J. Fiber Integr. Opt.* 19: 167–186.

26 Lim, C., Nirmalathas, A., Bakaul, M. et al. (2010). Fiber-wireless networks and subsystem technologies. *IEEE J. Lightw. Technol.* 28 (4): 390–405.

27 Chang, G.-K., Chowdhury, A., Jia, Z. et al. (2009). Key technologies of WDM-PON for future converged optical broadband access networks (invited paper). *OSA J. Opt. Commun. Netw.* 1 (4): C35–C50.

28 Olmos, J.J.V., Kuri, T., and Kitayama, K. (2007). Dynamic reconfigurable WDM 60-GHz millimeter-waveband radio-over-fiber access network: architectural considerations and experiment. *IEEE J. Lightw. Technol.* 25 (11): 3374–3380.

29 Gliese, U., Norskov, S., and Nielsen, T.N. (1996). Chromatic dispersion in fiber-optic microwave and millimeter-wave links. *IEEE Trans. Microw. Theory Tech.* 44 (10): 1716–1724.

30 Park, J., Sorin, W.V., and Lau, K.Y. (1997). Elimination of the fibre chromatic dispersion penalty on 1550 nm millimetre-wave optical transmission. *Electron. Lett.* 33 (6): 512–513.

31 J. Capmany, B. Ortega, A. Martinez, D. Pastor, M. Popov and P. Y. Fonjallaz, "Multiwavelength single sideband modulation for WDM radio-over-fiber systems using a fiber grating array tandem device," in *IEEE Photonics Technology Letters*, vol. 17, no. 2, pp. 471–473, Feb. 2005, https://doi.org/10.1109/LPT.2004.840017.

32 Smith, G.H., Novak, D., and Ahmed, Z. (1997). Technique for optical SSB generation to overcome dispersion penalties in fibre-radio systems. *Electron. Lett.* 33 (1): 74–75.

33 Park, S.-H. and Choi, Y.-W. (2005). Significant suppression of the third intermodulation distortion in transmission system with optical feedback forward linearized transmitter. *IEEE Photonics Technol. Lett.* 17 (6): 1280–1282.

34 Ismail, T., Liu, C.-P., Mitchell, J.E., and Seeds, A.J. (2007). High-dynamic-range wireless-over-fiber link using feedforward linearization. *IEEE J. Lightw. Technol.* 25 (11): 3274–3282.

35 Lee, S.H., Kang, J.M., Won, Y.Y. et al. (2005). Linearization of RoF optical source by using light-injected gain modulation. International Topical Meeting on Microwave Photonics, MWP 2005 (10–12 october 2005), Seoul, Korea, Republic.

36 Roselli, L., Borgioni, V., Zepparelli, F. et al. (2003). Analog laser predistortion for multiservice radio-over-fiber systems. *IEEE J. Lightw. Technol.* 21 (5): 1211–1223.

37 Shah, A.R. and Jalali, B. (2005). Adaptive equalisation for broadband predistortion linearisation of optical transmitters. *IEE Proc. Optoelectron.* 152 (1): 16–32.

38 Gamage, P.A., Nirmalathas, A., Lim, C. et al. (2008). Experimental demonstration of digitized RF transport over optical fiber links. *2008 International Topical Meeting on Microwave Photonics jointly held with the 2008 Asia-Pacific Microwave Photonics Conference*, Gold Coast, QLD, Australia, pp. 15–18. http://doi.org/10.1109/MWP.2008.4666623.

39 Toda, H., Yamashita, T., Kitayama, K.-I., and Kuri, T. (2001). A DWDM mm-wave fiber-radio system by optical frequency interleaving for high spectral efficiency. *2001 International Topical Meeting on Microwave Photonics. Technical Digest. MWP'01 (Cat. No.01EX476)*, Long Beach, CA, USA, pp. 85–88. http://doi.org/10.1109/MWP.2002.981802

40 Lim, C., Nirmalathas, A., Novak, D. et al. (2001). Technique for increasing optical spectral efficiency in millimeter-wave WDM fiber-radio. *Electron. Lett.* 37 (16): 1043–1045.

41 Lim, C., Nirmalathas, A., Novak, D., and Waterhouse, R.B. (2003). Capacity analysis for WDM fiber-radio backbones with star-tree and ring architecture incorporating wavelength interleaving. *IEEE J. Lightw. Technol.* 21 (12): 3308–3315.

42 Toda, H., Yamashita, T., Kuri, T., and Kitayama, K. (2003). Demultiplexing using an arrayed-waveguide grating for frequency-interleaved DWDM millimeter-wave radio-on-fiber systems. *IEEE J. Lightw. Technol.* 21 (8): 1735–1741.

43 Toda, H., Nakasyotani, T., Kuri, T., and Kitayama, K. (2004). WDM mm-wave-band radio-on-fiber system using single supercontinuum light source in cooperation with photonic up-conversion. *2004 IEEE International Topical Meeting on Microwave Photonics (IEEE Cat. No.04EX859)*, Ogunquit, ME, USA, pp. 161–164. http://doi.org/10.1109/MWP.2004.1396864.

44 Kuri, T., Toda, H., and Kitayama, K. (2007). Novel demultiplexer for dense wavelength division mutliplexed millimeter-wave-band radio-over-fiber systems with optical frequency interleaving technique. *IEEE Photonics Technol. Lett.* 19 (24): 2018–2020.

45 Kuri, T., Toba, H., Vegas-Olmos, J.J., and Kitayama, K.-I. (2010). Reconfigurable dense wavelength-division-multiplexing millimeter-waveband radio-over-fiber access system technologies. *IEEE J. Lightw. Technol.* 28: 2247–2257.

46 Bakaul, M., Nirmalathas, A., and Lim, C. (2005). Multifunctional WDM optical interface for millimeter-wave fiber-radio antenna base station. *IEEE J. Lightw. Technol.* 23 (3): 1210–1218.

47 Nirmalathas, A., Novak, D., Lim, C., and Waterhouse, R.B. (2001). Wavelength re-use in the WDM optical interface of a millimeter-wave fiber wireless antenna base-station. *IEEE Trans. Microw. Theory Tech.* 49 (10): 2006–2012.

48 Toda, H., Yamashita, T., Kuri, T., and Kitayama, K. (2003). 25-GHz channel spacing DWDM multiplexing using an arrayed waveguide grating for 60-GHz band radio-on-fiber systems. *MWP 2003 Proceedings. International Topical Meeting on Microwave Photonics, 2003.*, Budapest, Hungary, pp. 287–290. http://doi.org/10.1109/MWP.2003.1422901.

49 Bakaul, M., Nirmalathas, A., Lim, C. et al. (2005). Simplified multiplexing scheme for wavelength-interleaved DWDM millimetre-wave fiber-radio systems. *2005 31st European Conference on Optical Communication, ECOC 2005*, Glasgow, vol. 4, pp. 809–810.

50 Bakaul, M., Nirmalathas, A., Lim, C. et al. (2005). Efficient multiplexing scheme for wavelength-interleaved DWDM millimeter-wave fiber-radio systems. *IEEE Photonics Technol. Lett.* 17 (12): 2718–2720.

51 Stohr, A., Kitayama, K., and Jager, D. (1999). Full-duplex fiber-optic RF subcarrier transmission using a dual-function modulator/photodetector. *IEEE Trans. Microw. Theory Tech.* 47 (7): 1338–1341.

52 Chen, L., Wen, H., and Wen, S. (2006). A radio-over-fiber system with a novel scheme for millimeter-wave generation and wavelength reuse for up-link connection. *IEEE Photonics Technol. Lett.* 18 (19): 2056–2058.

53 Lim, C., Nirmalathas, A., Attygalle, M. et al. (2003). On the merging of millimeter-wave fiber-radio backbone with 25 GHz WDM ring networks. *IEEE J. Lightw. Technol.* 21 (10): 2203–2210.

54 Bakaul, M., Nirmalathas, A., Lim, C. et al. (2006). Hybrid multiplexing of multiband optical access technologies towards an integrated DWDM network. *IEEE Photonics Technol. Lett.* 18 (21): 2311–2313.

55 Shi, Y. (2013). Optical broadband in-home networks for converged service delivery. PhD Thesis. Technical University of Eindhoven.

56 Sauer, M., Kobyakov, A., and George, J. (2007). Radio over fiber for picocellular network architectures. *IEEE J. Lightw. Technol.* 25 (11): 3301–3320.

6

Photonic Generation of Ultra-Wideband Signals

6.1 Introduction

Ultra-wideband (UWB), regulated by the Federal Communications Commission (FCC) for indoor UWB systems operating in the frequency range from 3.1 to 10.6 GHz [1], has been considered a solution for short-range high-throughput wireless communications and sensor networks thanks to their intrinsic properties, such as the immunity to multipath fading, extremely short time duration, carrier-free, low duty cycle, wide bandwidth, and low power consumption [1–8]. Based on the FCC definition, a UWB signal should have a spectral bandwidth greater than 500 MHz or a fractional bandwidth greater than 20% [1]. There are, in general, two types of UWB, the direct sequence UWB [7] and the multiband UWB [8]. Direct-sequence impulse radio is one of the most attractive techniques for UWB communications since it is carrier-free. Therefore, there is no need for frequency mixers and local oscillators to down- or up-convert the carrier frequency. The selection of the impulse signal types is one of the fundamental considerations in the design of an impulse UWB radio system because the impulse types will determine the performance of the system [9]. For UWB communications, Gaussian pulses are the most widely used waveforms thanks to advantageous features such as simplicity and achievability. However, a Gaussian pulse has a spectrum that is centered at the zero frequency, which cannot be directly used for UWB systems since the frequency range of a UWB system must comply with the FCC-defined bandwidth from 3.1 to 10.6 GHz. Thus, a Gaussian waveform should be shaped to have a spectrum that is in the UWB band. Basically, by passing a Gaussian pulse through a bandpass filter that acts in a manner similar to a first- or second-order frequency differentiator [2], the spectrum will be shifted to a higher frequency band. This can be implemented using electronic circuits, but it is rather difficult to generate such pulses with a fractional bandwidth greater than 100% at the central frequency of around 7 GHz [10–12].

On the other hand, by wireless transmission, UWB communications systems can only operate within a limited distance of a few to tens of meters. Such short-range wireless networks can operate mainly in an indoor environment in a standalone mode, with a nearly nonexistent integration into fixed wired networks or wireless wide-area infrastructures. To offer availability of undisrupted service across different networks and eventually to achieve high-data rate access at any time and from any place, UWB combined with fiber transmission, a technology called UWB over fiber, may provide an effective solution. Fig. 6.1 shows a conceptual illustration of a UWB-over-fiber system for broadband indoor wireless access. In the system, UWB pulses are generated and encoded in the central office (CO) and distributed to the access points (APs) via optical fibers.

Microwave Photonics, First Edition. Jianping Yao, José Capmany, and Ming Li.

Companion website: www.wiley.com/go/yao/microwavephotonics

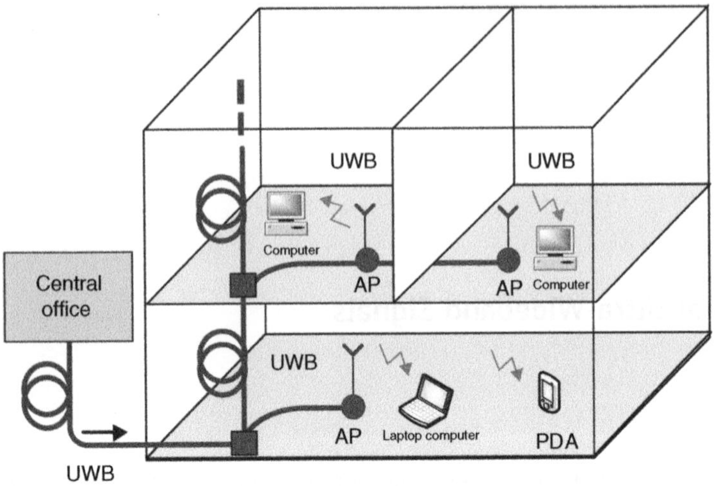

Figure 6.1 UWB over fiber for achieving broadband indoor wireless access. AP, access point.

UWB pulses can be generated electrically using an electronic circuit, such as a microwave ring filter [10], a microwave resonator-based bandpass filter [11], or a microwave differentiator [12]. To distribute UWB signals over optical fiber, it is highly desirable that the UWB signals can be generated directly in the optical domain without the need for extra electrical to optical conversions. In addition, the use of optical techniques to generate UWB pulses has many other advantages, such as light weight, small size, large tunability, and immunity to electromagnetic interference (EMI).

In this chapter, techniques to generate UWB pulses in the optical domain will be discussed. These techniques are classified into three categories: (i) UWB pulse generation based on phase-modulation to intensity-modulation (PM-IM) conversion. The PM-IM conversion can be implemented in the optical domain by using either a dispersive device or an optical frequency discriminator. (ii) UWB pulse generation using a photonic microwave delay-line filter. To generate a UWB monocycle, a two-tap microwave delay-line filter with one negative coefficient is needed, while to generate a UWB doublet, a three-tap microwave delay-line filter with one negative coefficient is needed. The negative coefficient in the microwave delay-line filters can be generated based on cross-gain modulation (XGM) in a semiconductor optical amplifier (SOA) or polarization modulation (XPolM) in a polarization modulator (PolM). (iii) UWB pulse generation based on optical spectral shaping and frequency-to-time mapping. The optical spectral shaping can be implemented using a spatial light modulator (SLM)-based or an optical-filter-based spectrum shaper, to make the shaped spectrum have a shape corresponding to a UWB monocycle or doublet. The frequency-to-time mapping is realized using a dispersive device, such as a dispersive fiber. A comparison of these techniques is provided. The use of the techniques for UWB pulse coding is also briefly discussed.

6.2 UWB Pulse Generation Based on PM-IM Conversion

UWB pulse generation can be implemented in the optical domain based on optical PM-IM conversion. In this section, the principle of optical phase modulation will be discussed. Then, two different techniques to achieve PM-IM conversions, using either a dispersive device or an optical frequency discriminator, will be discussed. Optical implementations of these techniques will also be presented.

6.2.1 Optical Phase Modulation

The electrical field of a phase-modulated optical carrier $e_{PM}(t)$ can be expressed as

$$e_{PM}(t) = \cos[\omega_o t + \Delta\varphi(t)] \tag{6.1}$$

where ω_o is the angular frequency of the optical carrier, and $\Delta\varphi(t)$ is the phase change induced by the modulation signal. Without loss of generality, $\Delta\varphi(t)$ can be expressed as

$$\Delta\varphi(t) = \beta_{PM} \times f(t) \tag{6.2}$$

where β_{PM} is the phase modulation index, defined as the phase change of the optical carrier when a unit voltage is applied (rad/volt), and $f(t)$ is the electrical modulating signal. If $f(t)$ is a single-frequency sinusoidal signal with zero initial phase, given by

$$f(t) = V_e \cos\omega_m t \tag{6.3}$$

where V_e and ω_m are, respectively, the amplitude and angular frequency of the modulating signal, then (6.1) can be expanded in terms of Bessel functions of the first kind,

$$e_{PM}(t) = \sum_{n=-\infty}^{+\infty} J_n(\beta_{PM} V_e) \cos\left[(\omega_o + n\omega_m)t + \frac{1}{2}n\pi\right] \tag{6.4}$$

where $J_n(\cdot)$ denotes the nth-order Bessel function of the first kind. For simplicity, the argument $\beta_{PM}V_e$ will be omitted in the remainder of the chapter. From (6.4), we can see that the phase modulation generates a series of sidebands with their amplitude coefficients given by the Bessel functions. For small signal modulation, only the upper and lower first-order sidebands need to be considered and the higher-order sidebands are very small and can be ignored.

With small-signal assumption, (6.4) can be further simplified as

$$e_{PM}(t) = \left\{J_o \cos\omega_o t + J_1 \cos\left[(\omega_o + \omega_m)t + \frac{\pi}{2}\right] + J_{-1} \cos\left[(\omega_o - \omega_m)t - \frac{\pi}{2}\right]\right\} \tag{6.5}$$

For Bessel functions, we have

$$J_n = -J_{-n}, \text{when } n \text{ is odd} \tag{6.6}$$

Applying the Fourier transform to the two sides of (6.5), we have

$$\begin{aligned} E_{PM}(\omega) \approx\ & \pi J_o[\delta(\omega - \omega_o) + \delta(\omega + \omega_o)] - j\pi J_1[\delta(\omega + \omega_o + \omega_m) - \delta(\omega - \omega_o - \omega_m)] \\ & - j\pi e_o J_1[\delta(\omega + \omega_o - \omega_m) - \delta(\omega - \omega_o + \omega_m)] \end{aligned} \tag{6.7}$$

Figure 6.2 illustrates the spectrum of a modulating electrical signal and that of the corresponding optical phase-modulated signal. As can be seen, the lower and upper sidebands are exactly out of phase. If the phase-modulated signal is directly detected at a photodetector (PD), no signal except a DC will be generated since the beating between the lower sideband with the optical carrier will exactly cancel the beating between the upper sideband with the optical carrier.

6.2.2 PM-IM Conversion

To recover the information encoded in optical phase, coherent detection based on heterodyne or homodyne scheme should be used, in which the phase-modulated optical signal is mixed with the light from a local laser source. However, it is a challenge to build a local oscillator light source with its phase locked to that of the input light source. In addition, the temperature and mechanical

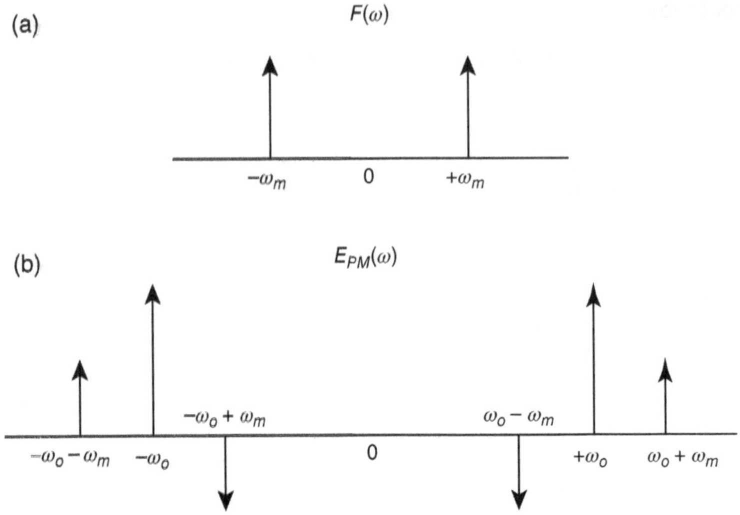

Figure 6.2 The spectrum of (a) a single-frequency sinusoidal modulating signal, and (b) that of the corresponding phase-modulated optical signal.

vibrations in the transmission line will impose phase and polarization fluctuations on the transmitted optical light, which would appear as a noise after photodetection. Two simple incoherent methods may be used to convert a phase-modulated signal to an intensity-modulated signal [13–15]. In the first method [13, 14], a dispersive device is used to change the phase relationship between the two first-order sidebands from out of phase to partially or fully in phase. In the second method [15], an optical filter is used to act as an optical frequency discriminator, to perform PM-IM conversion. The operation of the latter method can be explained by the magnitude relationships among the two sidebands and the optical carrier are changed, leading to PM-IM conversion. After the PM-IM conversion, a PD is used to detect the intensity-modulated signal.

Figure 6.3 shows the principle of PM-IM conversion using a dispersive device. As can be seen, a phase-modulated signal propagates through a dispersive device and then is fed to a PD.

Assume that the dispersive device has a unity magnitude response (which is true if the dispersive device is a dispersive fiber with a very low loss) but a quadratic phase response. Then, the optical signal at the output of the dispersive device can be expressed as

$$e_{DD}(t) = \left\{ J_o \cos(\omega_o t + \theta_o) + J_1 \cos\left[(\omega_o + \omega_m)t + \frac{\pi}{2} + \theta_{+1}\right] - J_1 \cos\left[(\omega_o - \omega_m)t - \frac{\pi}{2} + \theta_{-1}\right]\right\} \tag{6.8}$$

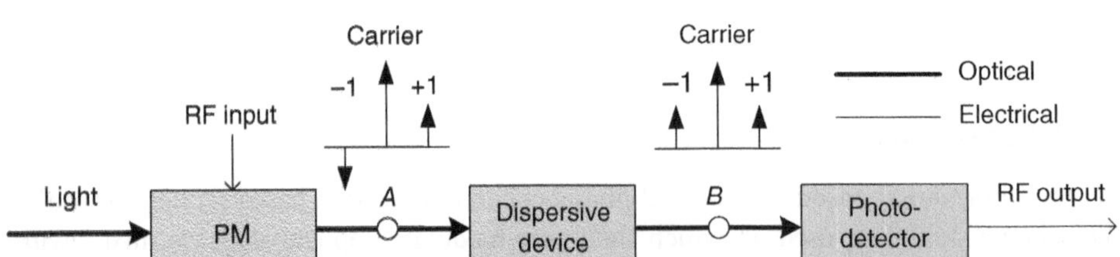

Figure 6.3 Chromatic-dispersion-based PM-IM conversion and direct photodetection.

where θ_o, θ_{-1}, and θ_{+1} are the phase shifts experienced by the optical carrier, the lower sideband, and the upper sideband, respectively.

Generally, a phase shift introduced by a dispersive device can be expressed as $\theta = \beta z$, where β is the propagation constant and z is the traveled distance. Expanding β in Taylor series, we have

$$\theta(\omega) = z\beta(\omega_o) + z\beta'(\omega_o)(\omega - \omega_o) + \frac{1}{2}z\beta''(\omega_o)(\omega - \omega_o)^2 + \cdots \tag{6.9}$$

where the β' and β'' are the first- and second-order derivatives of β with respect to the optical angular frequency.

We know that a group delay $\tau(\omega)$ is defined as $d\theta/d\omega$,

$$\tau(\omega) \equiv \frac{d\theta}{d\omega} = z\beta'(\omega_o) + z\beta''(\omega_o)(\omega - \omega_o) + \cdots \tag{6.10}$$

For a dispersive device with a quadratic phase response, its group delay response is linear. If the third- and higher-order derivatives of β in (6.10) are ignored, (6.10) is simplified as

$$\tau(\omega) \approx \underbrace{z\beta'(\omega_o)}_{\tau_0} + z\beta''(\omega_o)(\omega - \omega_o) \tag{6.11}$$

where the first term (τ_0) is the group delay experienced by the optical carrier at frequency ω_o, and the second term describes the group delay variation as a linear function with respect to the angular frequency ω.

The first-order chromatic dispersion is defined as the first-order derivative of the group delay with respect to the angular frequency ω, which is a constant in this case,

$$D_\omega = z\beta''(\omega_o) \tag{6.12}$$

Evaluating θ at the optical carrier of ω_o and the sidebands of $\omega_o \pm \omega_m$, we have

$$\begin{cases} \theta_o = z\beta(\omega_o) \\ \theta_{-1} = z\beta(\omega_o) - \tau_0\omega_m + \frac{1}{2}D_\omega\omega_m^2 \\ \theta_{+1} = z\beta(\omega_o) + \tau_0\omega_m + \frac{1}{2}D_\omega\omega_m^2 \end{cases} \tag{6.13}$$

When the light wave passes through a dispersive device and is applied to a PD, a photocurrent is generated at the output of the PD. Taking only the microwave signal centered at the modulating frequency ω_m, and ignoring the DC and the higher-order harmonics, we have

$$i_{PD}(t) \propto \sin\left(\frac{\theta_{+1} + \theta_{-1}}{2} - \theta_o\right) \cdot \cos\left(\omega_m t + \frac{\theta_{+1} - \theta_{-1}}{2}\right) = \underbrace{\sin\left(\frac{1}{2}D_\omega\omega_m^2\right)}_{H_{PM-IM}(\omega_m)} \cdot \cos\left[\omega_m(t - \tau_0)\right] \tag{6.14}$$

From (6.14), we can see that the chromatic-dispersion-based PM-IM conversion with direct detection has a frequency response given by

$$H_{PM-IM}(\omega_m) = \sin\left(\frac{1}{2}D_\omega\omega_m^2\right) \tag{6.15}$$

The frequency response is shown in Fig. 6.4, from which a quasi-periodic function with a notch at DC is observed. The first peak and the second notch are obtained by letting $D_\omega\omega_m^2/2 = \pi/2$ and π,

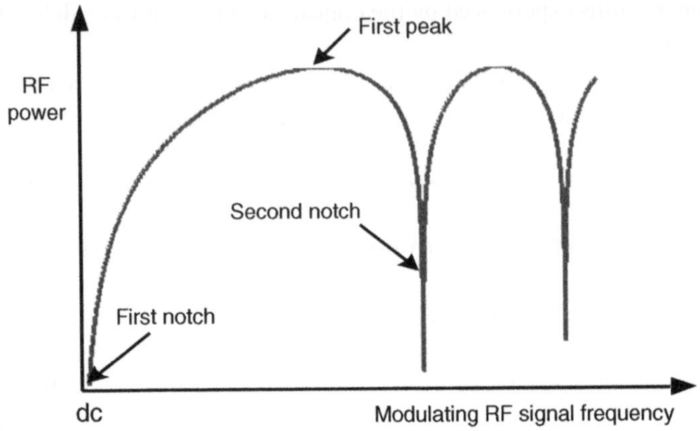

Figure 6.4 Frequency response corresponding to the chromatic-dispersion-based PM-IM conversion.

respectively. The frequency response between the first two notches forms a passband, which can be directly used to shape the spectrum of the modulating signal. If the input modulating signal is a Gaussian pulse, the spectrum at lower frequency will be filtered out, a spectrum corresponding to a Gaussian monocycle or doublet may be generated. In addition, the feature of PM-IM conversion with a notch at dc can be used to implement a photonic microwave filter with a bandpass [13–18]. The baseband resonance of a microwave photonic delay-line filter with all-positive coefficients can be eliminated by the DC notch, and a passband at $\omega_m = \sqrt{(\pi)/(2D_\omega)}$ would be generated.

Optical PM-IM conversion can also be implemented using an optical frequency discriminator. Based on the theoretical analysis in Section 6.2.1, a phase-modulated signal has two sidebands that are out of phase. If the phase relationship between the two sidebands is changed to partially or totally in phase, a phase-modulated signal is converted to an intensity-modulated signal, and PM-IM conversion is realized. This is the chromatic-dispersion-based PM-IM conversion. On the other hand, if the magnitude relationship between the optical carrier and the sidebands is changed, PM-IM conversion can also be realized. For instance, using an optical filter to filter out one sideband would generate a single-sideband intensity-modulated signal. In the following, a general discussion on optical-filter-based PM-IM conversion is presented.

Figure 6.5 shows the frequency response of an ideal optical filter. It has two linear and complementary slopes. By locating the optical carrier of the phase-modulated signal at one of the two slopes, PM-IM conversion can be realized.

We assume again the input to the phase modulator is an electrical Gaussian pulse $f(t)$. As a result, the phase of the input light wave is modulated. The electrical field of the phase-modulated light wave can be expressed as [16]

$$e_{PM}(t) = \exp[j\omega_o t + j\beta_{PM} s(t)] \tag{6.16}$$

where β_{PM} is again the phase modulation index, and ω_o is the angular frequency of the optical carrier. The phase-modulated light wave is then fed to an optical frequency discriminator, which is usually an optical bandpass filter with a transfer function having two opposite linear slopes. When the optical carrier is located at one of the two opposite linear slopes, PM-IM conversion is realized.

The frequency response of the optical filter for the left slope, shown in Fig. 6.5, can be written as

$$H_d(\omega) = K(\omega - \omega_1) \tag{6.17}$$

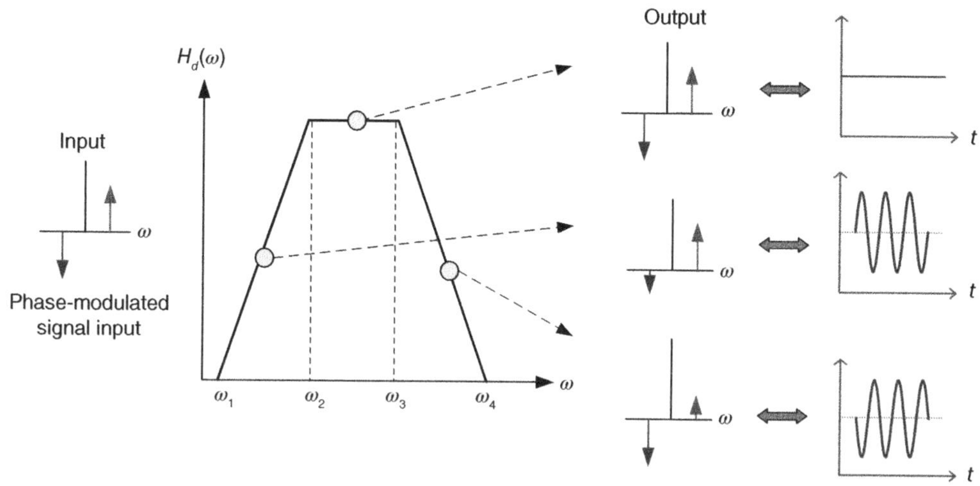

Figure 6.5 Frequency response of an optical filter with two linear slopes and a flat top.

where K is the left slope of the frequency response and ω_1 is an angular frequency at which $H_d(\omega) = 0$. When the phase-modulated light is sent to the optical filter that serves as a linear frequency discriminator, the output signal in the frequency domain is given by

$$e_{out}(\omega) = K(\omega - \omega_1)e_{PM}(\omega) \tag{6.18}$$

where $e(\omega)$ is the Fourier transform of $e_{PM}(t)$. Equation (6.18) is obtained based on the convolution property of a linear time-invariant system. For a linear time-invariant system, in the frequency domain, the output is a multiplication of the Fourier transform of the input signal and the frequency response of the system.

Applying the inverse Fourier transform to (6.18), we obtain the signal in the time domain

$$A_{out}(t) = [K(\omega_c - \omega_o) + K\beta_{PM}s'(t)]A(t) \tag{6.19}$$

where $s'(t)$ is the first-order derivative of $s(t)$. After photo-detection at the PD, we obtain the photocurrent, given by

$$\begin{aligned} i_{PD}(t) = K^2(\omega_c - \omega_o)^2 + K^2\beta_{PM}^2[s'(t)]^2 \\ + 2K^2(\omega_c - \omega_o)\beta_{PM}s'(t) \end{aligned} \tag{6.20}$$

The first term on the right-hand side in Eq. (6.20) is a DC term. Usually, for a small phase modulation index, we have $2K^2(\omega_c - \omega_o)\beta_{PM}s'(t) \gg K^2\beta_{PM}^2[s'(t)]^2$, so the term $K^2\beta_{PM}^2[f'(t)]^2$ can be neglected and the ac term of the photocurrent at the output of the PD is

$$i_{sig}(t) = 2K^2(\omega_o - \omega_1)\beta_{PM}s'(t) \tag{6.21}$$

From Eq. (6.21), we can see that the detected signal after the PD is proportional to the first-order derivative of the applied modulating signal. Therefore, if the phase modulating signal is a Gaussian pulse, a Gaussian monocycle would be generated, as shown in Fig. 6.5. In addition, the sign of $i_{sig}(t)$ in (6.21) is determined by$(\omega_c - \omega_o)$. If $(\omega_c - \omega_o) > 0$, $i_{sig}(t)$ has the same waveform as $s'(t)$, while if $(\omega_c - \omega_o) < 0$, $i_{sig}(t)$ is an inverted version of $s'(t)$.

If the optical carrier is located at the left slope of the transfer function, as shown in Fig. 6.5, we have $(\omega_c - \omega_o) > 0$. If it is located at the right slope, we have $(\omega_c - \omega_o) < 0$. Therefore, by locating the optical carrier at either slope of the optical bandpass filter, UWB pulses with inverted polarities can be generated, as shown in Fig. 6.5, where K is the slope of the filter, and ω is the optical frequency. From Fig. 6.5, we can see that the frequency response of the optical filter consists of three linear sections: the left slope, the center, and the right slope. To simplify the analysis, we assume that within each section, the phase response is linear. This assumption was verified to be acceptable if the filter is a fiber Bragg grating (FBG) with proper apodization. In addition, if the phase-modulated optical signal has a narrow bandwidth and the optical carrier is properly selected to make it located at the center of each spectral section of the frequency response of the optical filter, as shown in Fig. 6.5, the impulse response of the optical filter $h_d(t)$ can be approximated as

$$h_d(t) \approx \begin{cases} -K(\omega_1 - \Delta\omega)\delta(t) - jK\delta'(t); & \text{left slope} \\ K\Delta\omega\delta(t); & \text{center} \\ K(\omega_2 + \Delta\omega)\delta(t) + jK\delta'(t); & \text{right slope} \end{cases} \tag{6.22}$$

where the group delay derived from the linear phase response is neglected, $\delta(t)$ is the unit impulse, and $\delta'(t)$ is the first-order derivative of the unit impulse. Note that $\delta(t) \overset{FT}{\longleftrightarrow} 1$ and $d/dt(x(t)) \overset{FT}{\longleftrightarrow} j\omega \cdot X(\omega)$. After the phase-modulated optical signal passing through the optical filter, we obtain the optical field, given by

$$\begin{aligned} e_d(t) &= e_{PM}(t) * h_d(t) \\ &\approx \begin{cases} e_{PM}(t)[K(\omega_o - \omega_1 + \Delta\omega) + K\beta_{PM}f'(t)]; & \text{left slope} \\ e_{PM}(t)K\Delta\omega; & \text{center} \\ e_{PM}(t)[K(\omega_2 + \Delta\omega - \omega_o) - K\beta_{PM}f'(t)]; & \text{right slope} \end{cases} \end{aligned} \tag{6.23}$$

where $*$ denotes the convolution operation, and $f'(t)$ is the first-order derivative of the modulating signal $f(t)$.

Again, the photocurrent at the output of the PD is given by

$$i_{PD}(t) \sim \begin{cases} K^2\left\{(\omega_o - \omega_1 + \Delta\omega)^2 + [\beta_{PM}f'(t)]^2 + 2(\omega_o - \omega_1 + \Delta\omega)\beta_{PM}f'(t)\right\}; & \text{left slope} \\ K^2\Delta\omega^2; & \text{center} \\ K^2\left\{(\omega_2 + \Delta\omega - \omega_o)^2 + [\beta_{PM}f'(t)]^2 - 2(\omega_2 + \Delta\omega - \omega_o)\beta_{PM}f'(t)\right\}; & \text{right slope} \end{cases} \tag{6.24}$$

The first term on the right-hand side for each case, which equals to $|H_d(\omega)|_{\omega=\omega_o}$, represents a DC component and can be eliminated by using a DC blocker. When the optical carrier ω_o is located at the left or the right slope with an assumption of small signal modulation, the second term is much smaller than the third term and can be neglected. Finally, we obtain the recovered RF signal,

$$r(t) \sim \begin{cases} 2K^2(\omega_o - \omega_1 + \Delta\omega)\beta_{PM}f'(t); & \text{left slope} \\ 0; & \text{center} \\ -2K^2(\omega_2 + \Delta\omega - \omega_o)\beta_{PM}f'(t); & \text{right slope} \end{cases} \tag{6.25}$$

From (6.25), we can conclude that (i) no signal can be recovered if the optical carrier is located at the center passband of the optical filter, (ii) the recovered signal is the first-order derivative of the modulating signal when the optical carrier is located at either slope of the optical filter, and (iii) the detected signals have different signs when the carriers are located at the opposite slopes, which is a very important feature and would find many interesting applications in all-optical microwave signal processing [19, 20].

6.2.3 UWB Pulse Generation Based on PM-IM Conversion in a Dispersive Device

As discussed in Section 6.2.2, PM-IM conversion in a dispersive device has a frequency response corresponding to a bandpass filter. If the filter is properly designed by taking into consideration the frequency responses of the phase modulator and the PD, then the spectrum of a Gaussian pulse can be shaped to have a spectrum that satisfies the FCC spectrum mask. Figure 6.6 shows a system to generate and distribute UWB signals over standard single mode fiber (SMF) [21]. As can be seen, an AP is connected to a central station (CS). At the CS, a light wave from a laser diode (LD) is fed to an optical phase modulator which is driven by a data sequence to be transmitted. The phase-modulated optical signal is then transmitted to the AP via a length of SMF, which serves as a transmission medium as well as a dispersive device. Thanks to the chromatic dispersion induced by the SMF, at the AP, the phase-modulated signal is converted to an intensity-modulated signal, and the electrical signal is obtained at the output of a PD, which is ready to radiate to the free space via a UWB antenna.

Under small-signal condition, the frequency response of the system including the frequency responses of the phase modulator and the PD is given by

$$H(\omega) = \sin\left(\frac{1}{2}D_\omega \omega_m^2\right) \times H_{PM}(\omega) \times H_{PD}(\omega) \tag{6.26}$$

where the first term on the right-hand side is the frequency response of the PM-IM conversion, which is given in Eq. (6.15), $H_{PM}(\omega)$ and $H_{PD}(\omega)$ are, respectively, the frequency responses of the phase modulator and the PD. Since the frequency responses of the phase modulator and the PD are usually bandwidth-limited, the overall frequency response of the system can be limited to a passband close to DC. In addition, since the frequency response of the PM-IM conversion depends on the optical carrier wavelength and the total dispersion of the transmission medium, the bandwidth and shape of the bandpass filter can be tailored by changing the carrier wavelength and the total dispersion to optimize the spectrum of the generated UWB signals. Figure 6.7 shows an experimental frequency response of the system, with the LD operating at 1550 nm and the SMF length of 25 km. The SMF has a chromatic dispersion of 17 ps/nm km at 1550 nm, with 25 km of this fiber having a total dispersion of 425 ps/nm. From Fig. 6.7 we can see that a notch at the DC is observed. The passband peak and the lower and higher −10 dB cutoff frequencies are 10.5, 4.1, and 15.9 GHz, respectively.

Then, a Gaussian-like pulse with a full width at half maximum (FWHM) of about 63 ps is applied to the phase modulator via its RF port. The temporal shape and the spectrum of the input pulse are shown in Fig. 6.8(a) and (b), respectively. After passing through the 25 km SMF link, the

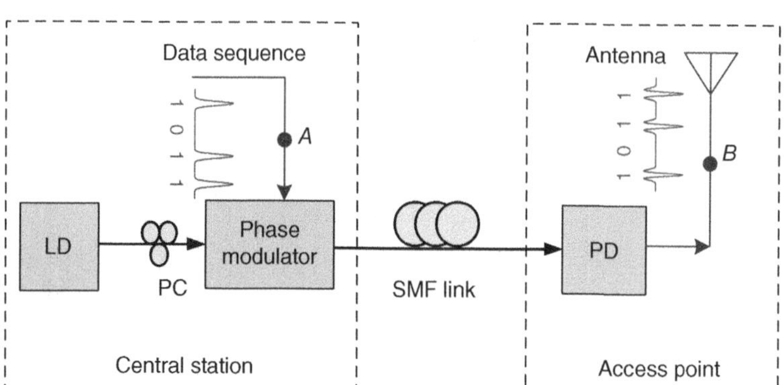

Figure 6.6 A UWB generation system based on PM-IM conversion in a dispersive device. The dispersion device is a length of SMF. LD, DC; PC, polarization controller; PD, photodetector.

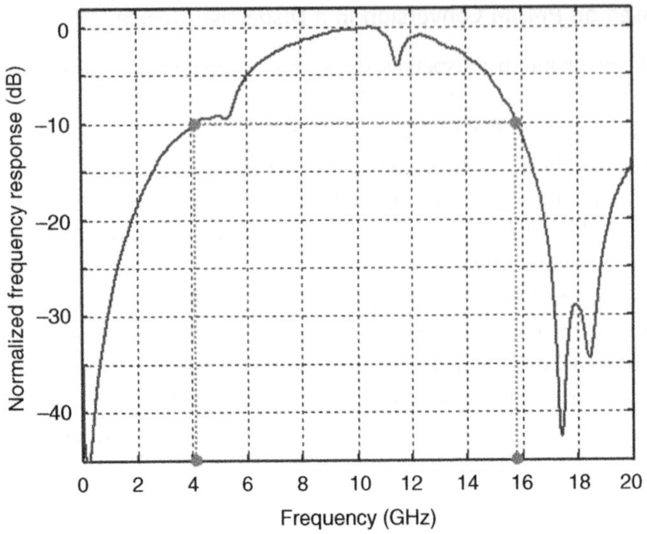

Figure 6.7 Frequency response of the UWB pulse generation system based on PM-IM conversion in a dispersive device.

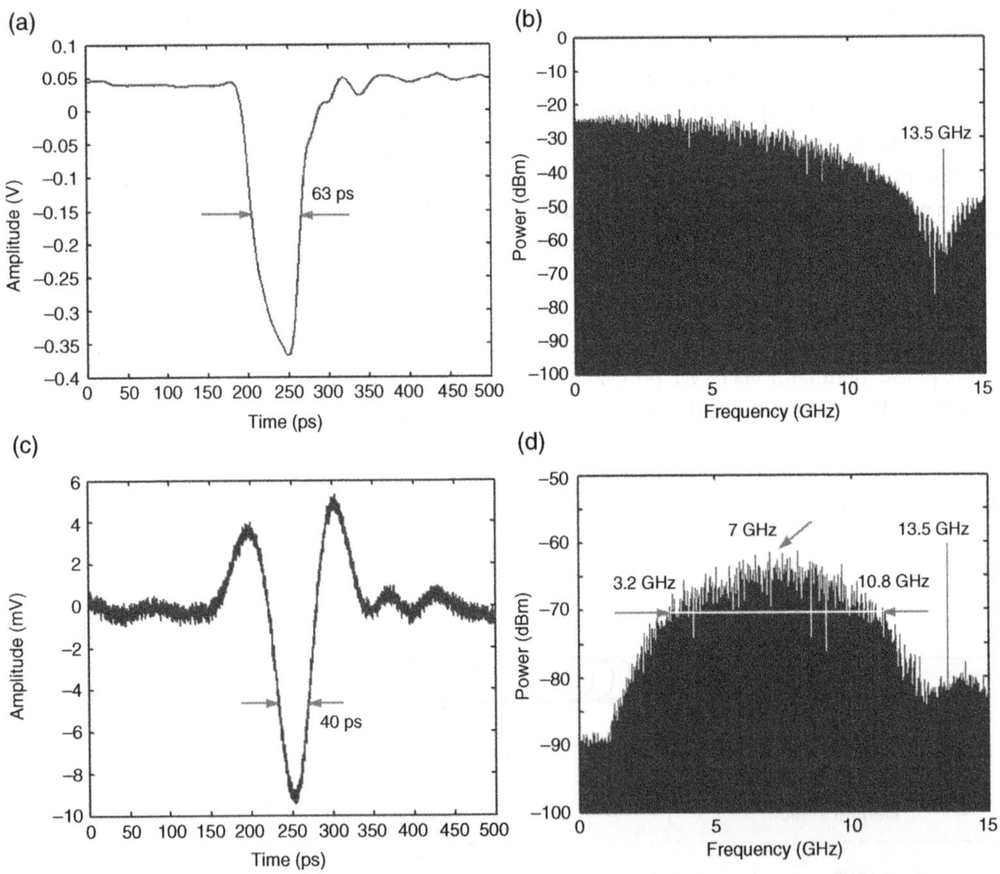

Figure 6.8 (a) The waveform of a Gaussian-like input pulse, and (b) the power spectrum of the Gaussian-like pulse. (c) The generated UWB doublet, and (d) the power spectrum of the doublet.

phase-modulated optical signal is converted to an intensity-modulated signal, which is then detected by the PD at the AP. Figure 6.8(c) shows the temporal waveform of the generated pulse, which is a Gaussian doublet with an FWHM of about 40 ps. The spectrum of the doublet is shown in Fig. 6.8(d). It can be seen that the spectrum has a central frequency of about 7 GHz, the lower and higher frequencies at −10 dB points are 3.2 and 10.8 GHz, respectively. The generated UWB doublet has a fractional bandwidth of about 109%.

6.2.4 UWB Pulse Generation Based on PM-IM Conversion in an Optical Frequency Discriminator

The technique to generate UWB pulses discussed in Section 6.2.3 requires the use of a dispersive element, which was 25 km SMF in the experiment. There are two limitations associated with the use of optical fiber as a dispersive device. First, the fiber length is long. For most of the applications, the communication distance is much less than 25 km; therefore, the system is bulky. Second, the fiber should have a fixed length to achieve a desirable frequency response. If the communication distance is changed with a different fiber length, the frequency response of the system is changed, leading to a distortion to the generated UWB pulses. A method to avoid using a long optical fiber is to use an optical filter-based frequency discriminator, to perform PM-IM conversion [22]. The theoretical details of an optical frequency discriminator have been discussed in Section 6.2.2. If a phase-modulated Gaussian pulse is sent to an optical frequency discriminator, depending on the location of the optical carrier at the linear or quadrature slope, a Gaussian monocycle or doublet would be generated. In addition, the location of the optical carrier at the positive or negative slope would lead to the generation of UWB pulses with opposite polarity. Figure 6.9 shows the generation of a UWB monocycle using a linear optical frequency discriminator.

For real implementation, the optical frequency discriminator can be an optical filter, such as an FBG [22]. Figure 6.10 shows the reflection spectrum of a 10-mm apodized FBG with a central wavelength at 1536.12 nm, a 3-dB bandwidth of 0.23 nm, and a reflectivity of 90%. By locating the optical carrier with phase modulation of a Gaussian pulse at the linear slope or the quadrature of the FBG reflection spectrum, a UWB monocycle or doublet pulse is obtained at the output of a PD.

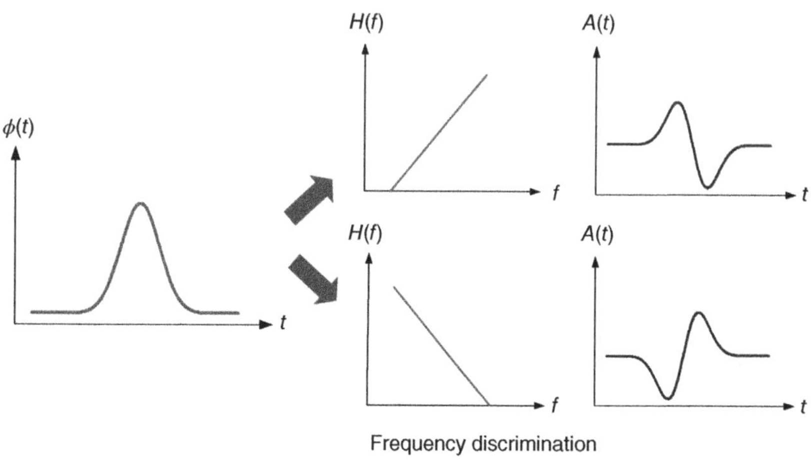

Figure 6.9 UWB monocycle generation using a linear optical frequency discriminator.

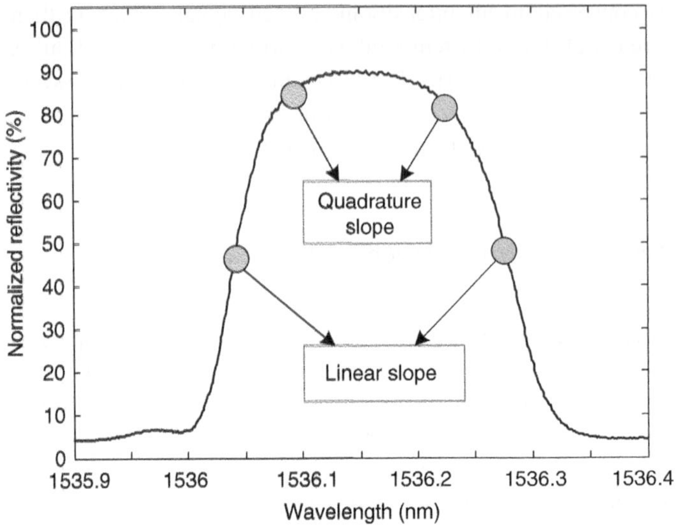

Figure 6.10 Reflection spectrum of an apodized FBG.

In addition, UWB pulses with opposite polarities can be generated by locating the optical carrier at the right or the left slope of the FBG reflection spectrum. This property provides the possibility to implement two different UWB pulse modulation schemes by simply shifting the optical carrier: the pulse shape modulation (PSM) (monocycle-doublet) and the pulse polarity modulation (PPM).

Figure 6.11 shows an experimental setup to generate UWB pulses based on PM-IM conversion in an optical frequency discriminator. The lightwave from an LD is fed to an optical phase modulator which is driven by an electrical Gaussian pulse. The phase-modulated optical signal is then applied to an FBG via an optical circulator. The PM-IM conversion is achieved by using the FBG serving as a frequency discriminator. The PM-IM converted signal is then detected at a PD.

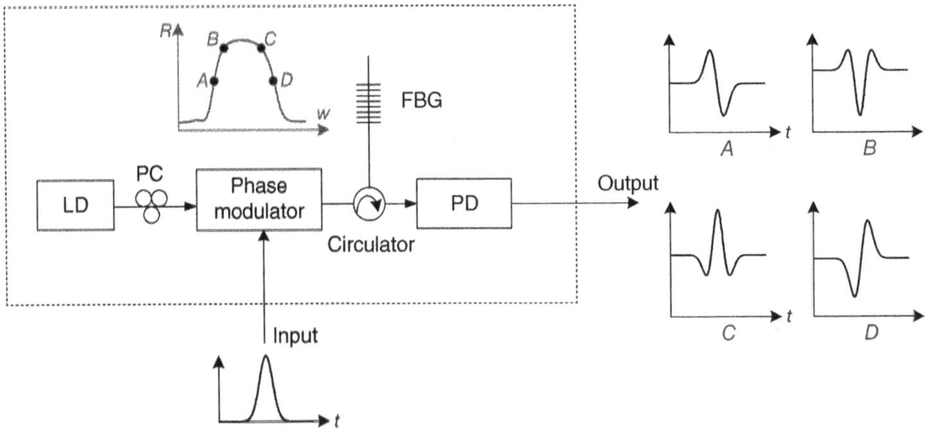

Figure 6.11 Experimental setup for UWB pulse generator based on PM-IM conversion in an optical frequency discriminator.

Based on the configuration shown in Fig. 6.11, when the phase-modulated light is located at the linear region of the FBG reflection slopes, as shown in Fig. 6.11 at *A*, the *AC* part of the recovered signal at the output of the PD can be written as

$$r(t) \sim \Re P\beta_{PM}Ks'(t) \tag{6.27}$$

where $\Re$ is the responsivity of the PD, *P* is the optical power reflected from the FBG, *K* is the slope of the FBG power spectrum, and $s'(t)$ is the first-order derivative of the modulating signal $s(t)$. A UWB monocycle is thus generated.

When the optical carrier is located at the opposite slope of the FBG reflection spectrum, as shown in Fig. 6.11 at *D*, the output pulses will be a monocycle with an opposite polarity. This feature enables the realization of PPM when two optical carriers corresponding to these two complementary pulses are employed and switched by a data sequence to be transmitted. In addition, if the optical carrier is located at the quadrature slopes of the FBG reflection spectrum, as shown in Fig. 6.11 at *B* and *C*, UWB doublets are generated. Therefore, by locating the optical carrier at different locations, UWB pulses with different shapes can be generated in the same configuration, and eventually, the implementation of PSM is possible.

When the wavelength of the optical carrier is tuned to be reflected at the right turning corner and the right linear slopes of the FBG spectrum, the generated pulses, shown in Fig. 6.12, are the inverted versions of those shown in Fig. 6.13(a) and (c).

Experimental results are shown in Fig. 6.13. When the carrier wavelength is tuned to 1536.032 nm, which is located at the left linear slope of the FBG spectrum shown in Fig. 6.10, a Gaussian monocycle pulse is generated, as shown in Fig. 6.13(a). The spectrum of the generated Gaussian monocycle is shown in Fig. 6.13(b), which has a central frequency of about 3.45 GHz, and a 10-dB bandwidth of about 7.94 GHz. When the carrier wavelength is tuned at 1536.098 nm, which is located at the left turning corner (quadrature slope) of the FBG spectrum, the generated pulse is a Gaussian doublet, as shown in Fig. 6.13(c). From its power spectrum shown in Fig. 6.13(d), we can see that the central frequency is increased to 7.14 GHz and the 10-dB bandwidth is about 8.8 GHz. These results are expected according to the mathematical definition of a Gaussian pulse with different orders of derivatives [23].

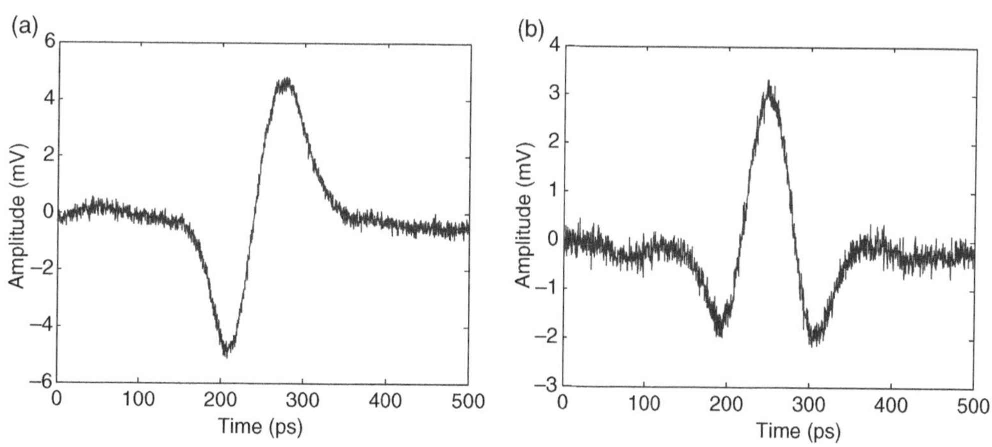

Figure 6.12 The waveforms of the Gaussian monocycle and doublet when the optical carrier is located at the opposite slopes of the FBG.

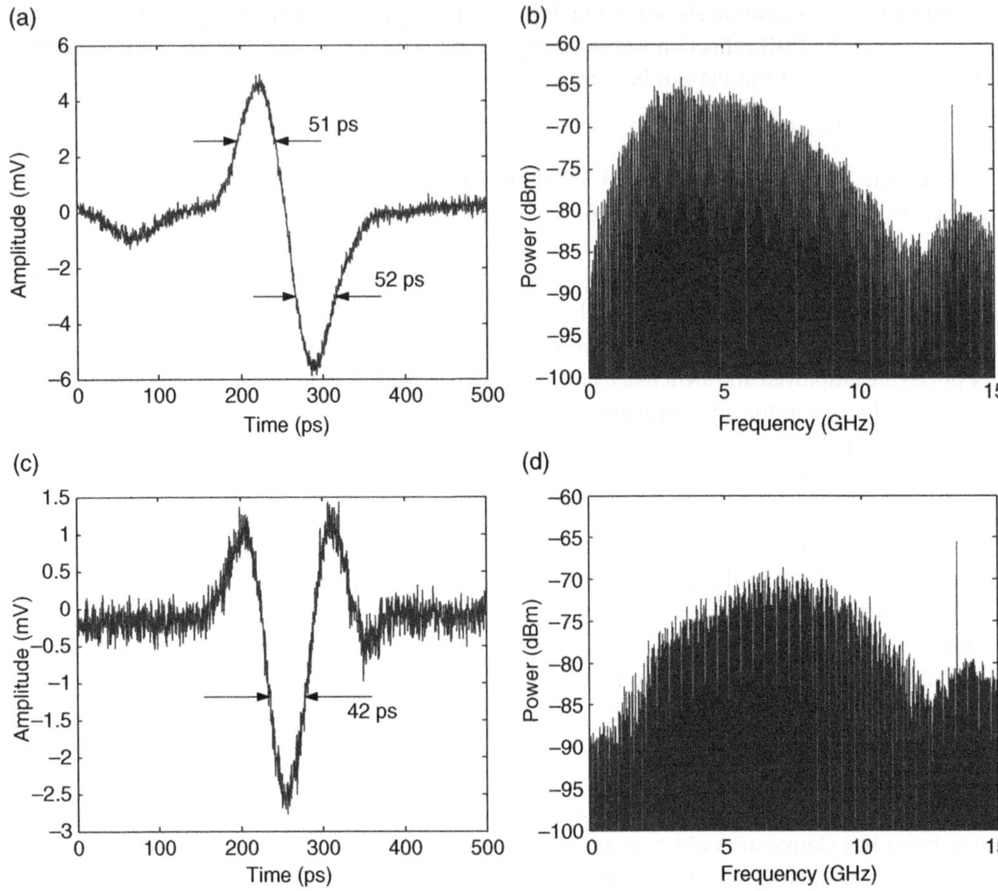

Figure 6.13 (a) The waveform of the generated Gaussian monocycle, and (b) the corresponding power spectrum. (c) The waveform of the generated Gaussian doublet, and (d) the corresponding power spectrum.

We should note that in the systems shown in Figs. 6.6 and 6.11, the input Gaussian pulse is generated electrically using a pulse generator, and the phase modulation is implemented using an optical phase modulator. Therefore, the two systems are not all-optical, but hybrid. To generate UWB pulses using an all-optical system, the Gaussian pulse should be generated in the optical domain. One possible solution is to generate the Gaussian pulse using a mode-locked laser source. In addition, the optical phase modulation should also be implemented in the optical domain, which can be realized based on optical cross-phase modulation (XPM). Optical XPM can be implemented in an optical nonlinear device, such as a length of nonlinear fiber.

Figure 6.14 shows an all-optical UWB pulse generation system [24]. It is different from the system shown in Fig. 6.11 where the Gaussian pulse is generated using an electrical pulse generator; in the system shown in Fig. 6.14, the optical Gaussian pulse is generated using a femtosecond pulse laser (FSPL) and the phase modulation is implemented based on XPM in a length of nonlinear fiber. To control the pulse width, a tunable bandpass filter (TBF) is incorporated after the FSPL. The generated optical pulse is then injected with a CW probe into a length of dispersion-shifted fiber (DSF) to achieve optical XPM. The phase-modulated signal carried by the probe is then converted to

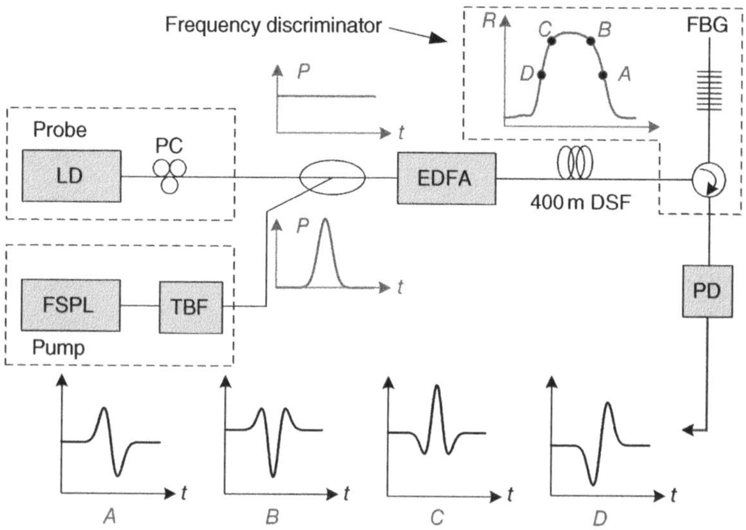

Figure 6.14 An all-optical UWB pulse generator. The Gaussian pulse is generated by the femtosecond pulse laser with proper spectrum slicing to control the pulse width. XPM is realized in the 400-m DSF.

intensity-modulated signal at an FBG (serving as a frequency discriminator). The FBG also serves as an optical bandpass filter to remove the residual pump and the amplified spontaneous emission (ASE) noise from the EDFA. Depending on the location of the probe at the FBG reflection spectrum, a UWB monocycle or doublet is generated.

6.3 UWB Pulse Generation Based on a Photonic Microwave Delay Line Filter

It is known that a Gaussian monocycle or doublet is the first- or second-order derivative of a Gaussian pulse [23]. The first- or second-order derivative can be implemented in the optical domain using an optical frequency discriminator [22], as discussed in Section 6.2.4. The approach discussed in Section 6.2.3 that uses a dispersive device to perform PM-IM conversion can also be considered a frequency discriminator [21]. Mathematically, the first- and second-order derivatives can be approximated by the first- and second-order differences, which can be implemented using a photonic microwave delay-line filter, with two or three taps, both with one negative tap. It is known that to design a photonic microwave delay-line filter, to avoid optical interference, the filter should operate in the incoherent regime using incoherent detection. A photonic microwave delay-line filter with incoherent detection can usually have positive coefficients only [25–27]. Therefore, to generate a negative coefficient while maintaining incoherent detection, special designs must be incorporated. In this section, two methods to generate a negative coefficient will be discussed. In the first method, a negative coefficient is generated based on XGM in an SOA, by injecting two wavelengths, one saving as the pump and the other as the probe. In the second method, a PolM is used. The negative coefficient is generated based on polarization modulation in a PolM.

6.3.1 Photonic Microwave Delay-Line Filters for UWB Pulse Generation

Assume $x(t)$ is an input Gaussian pulse which is applied to a two-tap microwave delay-line filter with one positive and one negative coefficient of [1 −1], as shown in Fig. 6.15. The output $y(t)$ is given

$$y(t) = x(t) - x(t - \tau) \tag{6.28}$$

where τ is the time delay difference. Applying the Fourier transform to the two sides of (6.28), we have

$$Y(\omega) = X(\omega) - e^{-j\omega\tau}X(\omega) = X(\omega)\left(1 - e^{-j\omega\tau}\right), \tag{6.29}$$

where $Y(\omega)$ and $X(\omega)$ are, respectively, the Fourier transforms of $y(t)$ and $x(t)$. The magnitude frequency response of the filter is given

$$|H(\omega)| = \left|\frac{Y(\omega)}{X(\omega)}\right| = \sqrt{2 - 2\cos(\omega\tau)} \tag{6.30}$$

Figure 6.16 shows the frequency response of the two-tap microwave delay-line filter with a time delay difference of $\tau = 70$. As can be seen, the filter is a bandpass filter with the center frequency of the first passband located at $\omega_o = \pi/\tau$ (≈7 GHz). For practical applications, considering the limited bandwidth of the intensity modulator and the PD, the resonances at higher frequencies can be ignored. Therefore, by properly selecting the time delay difference τ, a filter with a frequency response that can shape the spectrum of the input Gaussian pulse to have a spectrum corresponding to a UWB monocycle would be obtained.

A UWB doublet can be generated by implementing the second-order derivative of a Gaussian pulse, which can be approximated by the second-order difference,

$$y(t) = [x(t) - x(t - \tau)] - [x(t - \tau) - x(t - 2\tau)] = x(t) - 2x(t - \tau) + x(t - 2\tau) \tag{6.31}$$

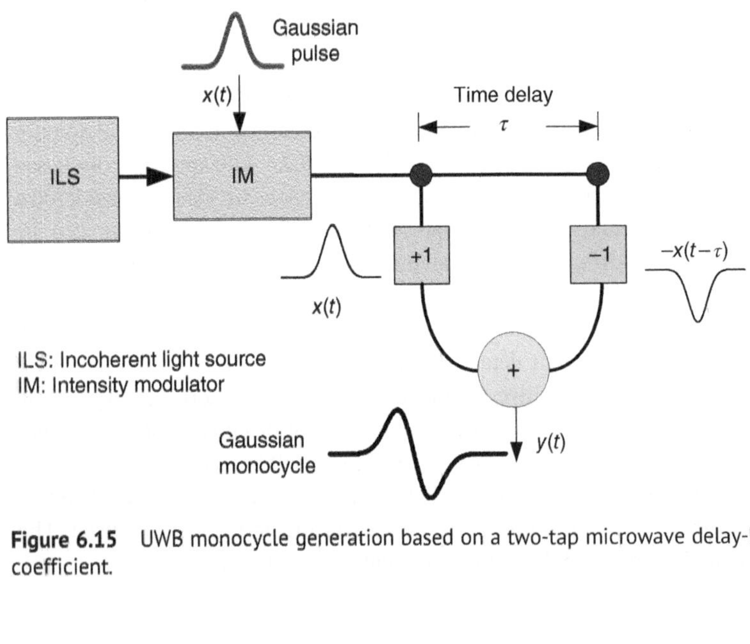

Figure 6.15 UWB monocycle generation based on a two-tap microwave delay-line filter with one negative coefficient.

Figure 6.16 Magnitude frequency response of a two-tap microwave delay-line filter with coefficients [1 −1] and a time delay difference of τ = 70 ps.

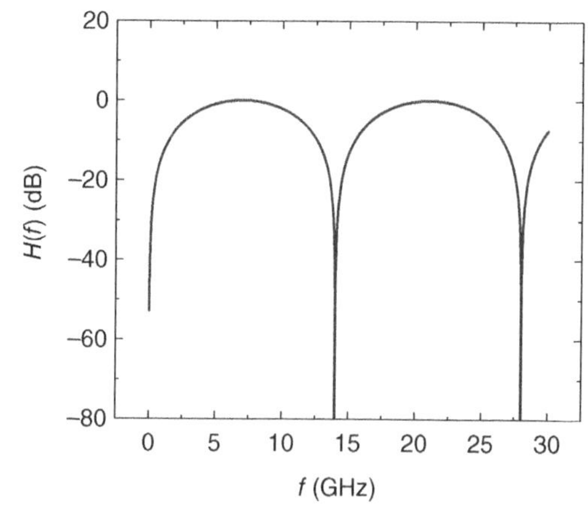

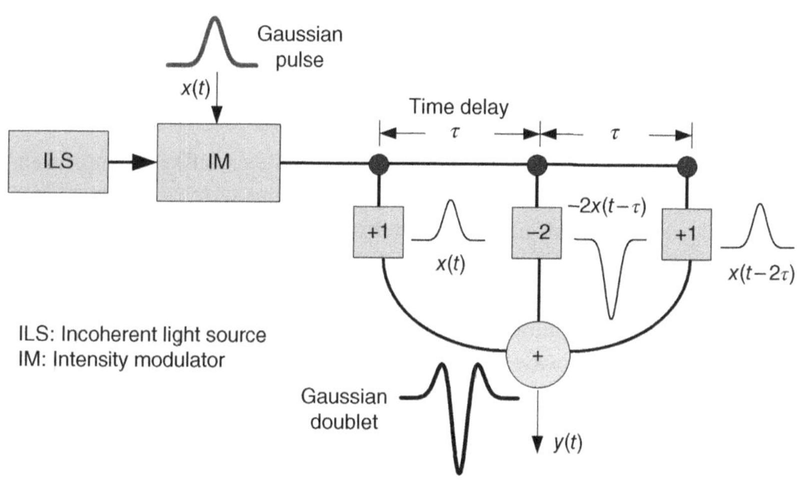

Figure 6.17 UWB doublet generation based on a three-tap delay-line filter with coefficients [1 −2 1] and a time delay difference of τ = 70 ps.

From (6.31), we can see that a UWB doublet can be generated using a microwave delay-line filter with three taps with coefficients of [1 −2 1], as shown in Fig. 6.17. The frequency response of the three-tap microwave delay-line filter is given,

$$H(\omega) = \frac{Y(\omega)}{X(\omega)} = 1 - 2e^{-j\omega\tau} + e^{-2j\omega\tau} \tag{6.32}$$

The magnitude frequency response is shown in Fig. 6.18. It is a bandpass filter with a narrower passband compared to that of the two-tap microwave delay-line filter shown in Fig. 6.16. Because of the narrower passband, the Gaussian pulse is shaped to have a spectrum corresponding to a Gaussian doublet.

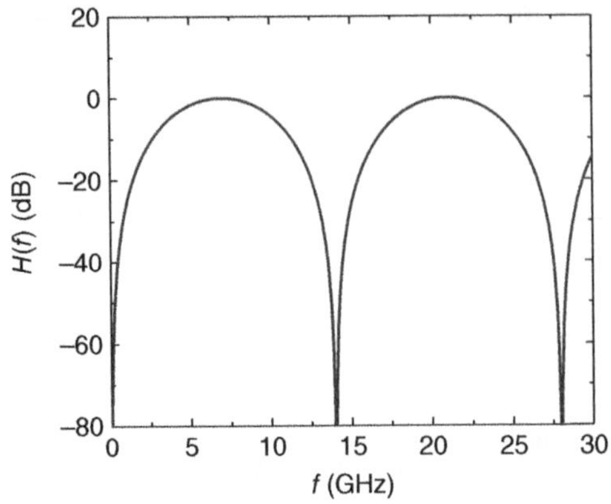

Figure 6.18 Magnitude frequency response of a three-tap microwave delay line filter with coefficients [1 −2 1] and a time delay difference of τ = 70 ps.

6.3.2 UWB Monocycle Generation with a Two-Tap Microwave Delay-Lines Filter

In this subsection, two demonstrations of two-tap microwave delay-line filters with coefficient of [1 −1] for Gaussian monocycle generation will be discussed. In the first filter, the negative coefficient is generated based on XGM in an SOA. In the second filter, a PolM is used to generate the negative coefficient.

Figure 6.19 shows a two-tap microwave delay-line filter with one negative coefficient implemented using an SOA for UWB monocycle generation [28]. The function of the SOA in the system is to generate a negative coefficient based on XGM. As can be seen from Fig. 6.19, two light waves, one a high-power pulsed light wave (the pump, modulated by a Gaussian pulse) and the other a low-power CW light wave (the probe), are injected into the SOA. Thanks to the XGM, the power of the probe would vary inversely with the pump power; a pair of complementary optical pulses are thus generated, with one at the pump wavelength and the other at the probe wavelength. If a proper time delay difference is introduced between the two pulses, a monocycle is generated. The pulse width of the monocycle can be controlled by altering the time delay difference to make its spectrum meet the FCC spectrum mask.

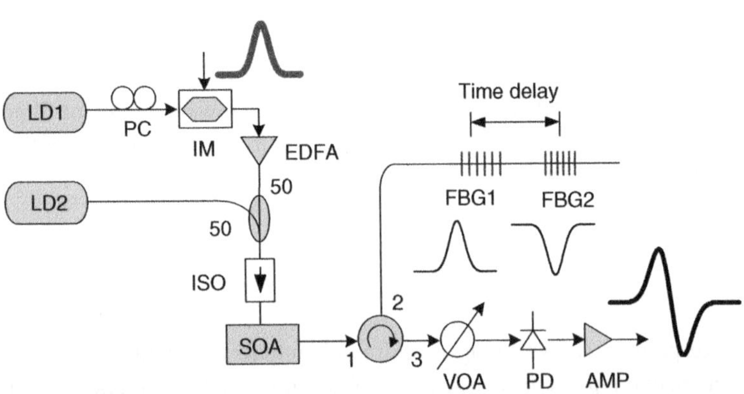

Figure 6.19 UWB monocycle generation based on a two-tap photonic microwave delay-line filter using an SOA. PC, polarization controller; IM, intensity modulator; EDFA, erbium-doped fiber amplifier; ISO, isolator; SOA, semiconductor optical amplifier; VOA, variable optical attenuator; AMP, amplifier.

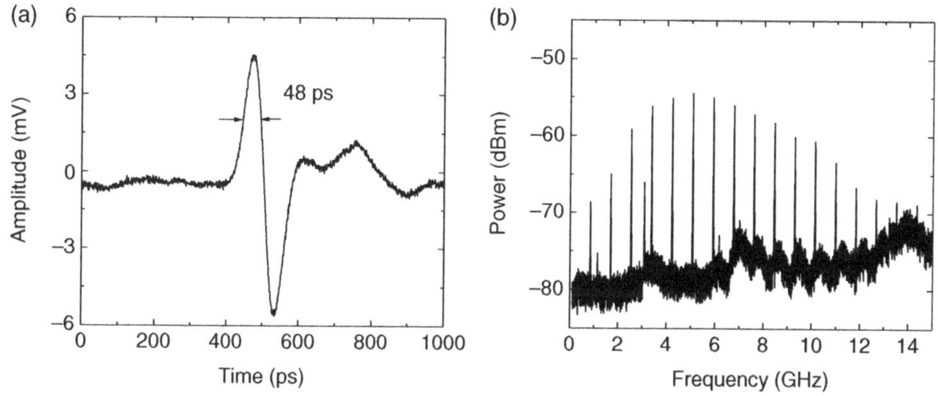

Figure 6.20 Generated UWB monocycle pulse; (a) the temporal waveform, and (b) its spectrum.

Figure 6.20 shows the generated monocycle and its spectrum. In the experiment, a Gaussian pulse with an FWHM of about 72 ps is applied to the intensity modulator. The output from the intensity modulator, amplified by an EDFA, is injected into the SOA with a probe generated by a second LD. Thanks to the XGM, a pair of polarity-reversed pulses are generated at the output of the SOA. The time delay difference between the two pulses is introduced by two FBGs. The two FBGs are fabricated in a hydrogen-loaded SMF with a physical spacing of $L = 5$ mm between the two FBGs. Since the pump power is higher than the probe power, the amplitudes of the two pulses are different. To have a good Gaussian monocycle, FBG1 is made with a much lower reflectivity than that of FBG2. In addition to the introduction of the time delay difference, the two FBGs also function as an optical bandpass filter to filter out the ASE noise generated by the SOA. The FWHM of the monocycle pulse is about 48 ps, which is narrower than that of the original Gaussian pulse. This is because the time delay difference between the pump and a probe pulse is smaller than the width of the Gaussian pulse. The envelope of the discrete spectrum lines corresponds to the spectrum of a single monocycle pulse, which has a center frequency at about 5.0 GHz and a −10-dB bandwidth of about 9.4 GHz, with a fractional bandwidth of about 188%. It is different from the UWB generation schemes discussed in Sections 6.2.3 and 6.2.4, where a single wavelength is used. The system here employs two wavelengths. When the UWB pulse is distributed over fiber, the fiber chromatic dispersion effect would have a higher impact compared with the approaches using a single wavelength, as the wavelength spacing between the two wavelengths is usually larger than the spectrum bandwidth of the UWB pulse.

Note that the two FBGs in the system can be replaced by a dispersive device, such as a dispersive fiber or a chirped FBG. Thanks to the chromatic dispersion, the two pulses will travel in the dispersive device at different velocities, and a time delay difference will be generated. The advantage of using a dispersive device is that the time delay difference can be easily tuned by tuning the wavelength spacing between the pump and the probe wavelengths.

To avoid using two wavelengths, we may use a different architecture in which the negative coefficient is generated using a PolM. It was reported that a two-tap photonic microwave bandpass filter with one negative coefficient can be implemented using a PolM [29]. The same architecture can be employed for UWB monocycle generation if the input modulating signal is a Gaussian pulse. Figure 6.21 shows a PolM-based UWB monocycle generation system. As can be seen, a single wavelength generated by an LD is sent to the PolM via a polarization controller (PC) to align its polarization direction with an angle of 45° to one principal axis of the PolM. A Gaussian pulse

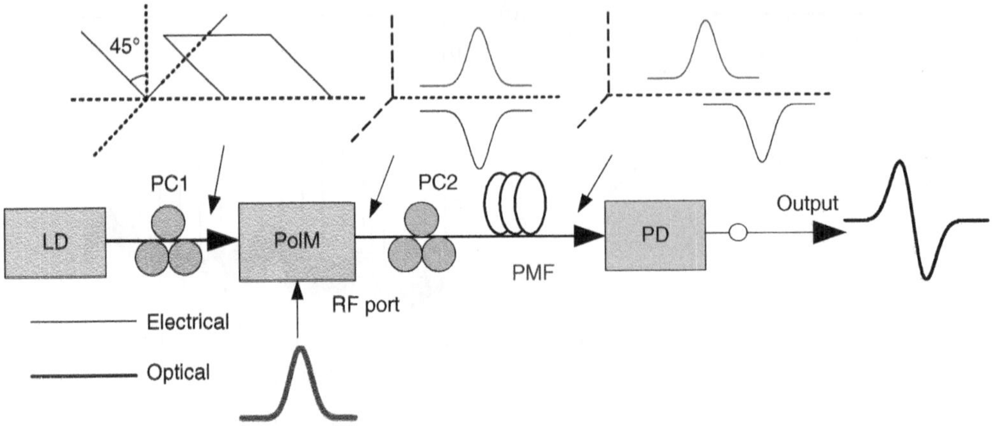

Figure 6.21 UWB monocycle generation using a two-tap microwave delay-line filter based on a PolM.

is applied to the PolM via its RF port. Thanks to the polarization modulation, two complementary Gaussian pulses modulated on two orthogonally polarized optical carriers are generated. The optical signals are then sent to a polarization-maintaining fiber (PMF)-based delay line through a second PC (PC2), to align its polarization directions with the principal axes of the PMF, to introduce a time delay difference. The time delay difference is determined by the birefringence of the PMF and the length of the PMF. The time-delayed Gaussian pulses are then detected at a PD.

6.3.3 UWB Doublet Generation with a Three-Tap Microwave Delay-Line Filter

The two-tap microwave delay-line filter shown in Figs. 6.19 and 6.21 can generate UWB monocycle using an SOA or a PolM. The filter architectures can be modified to generate a Gaussian doublet. As discussed in Section 6.3.1, to generate a UWB doublet, a three-tap microwave delay-line filter with coefficients of [1 −2 1] is needed. Figures 6.22 and 6.23 show two architectures, one based on an SOA, and the other based on a PolM, for Gaussian doublet generation.

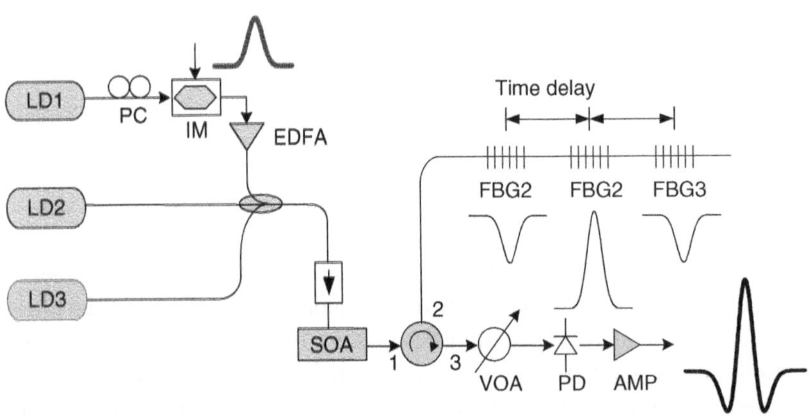

Figure 6.22 UWB doublet generation using a three-tap photonic microwave delay-line filter with one negative coefficient based on XGM in an SOA.

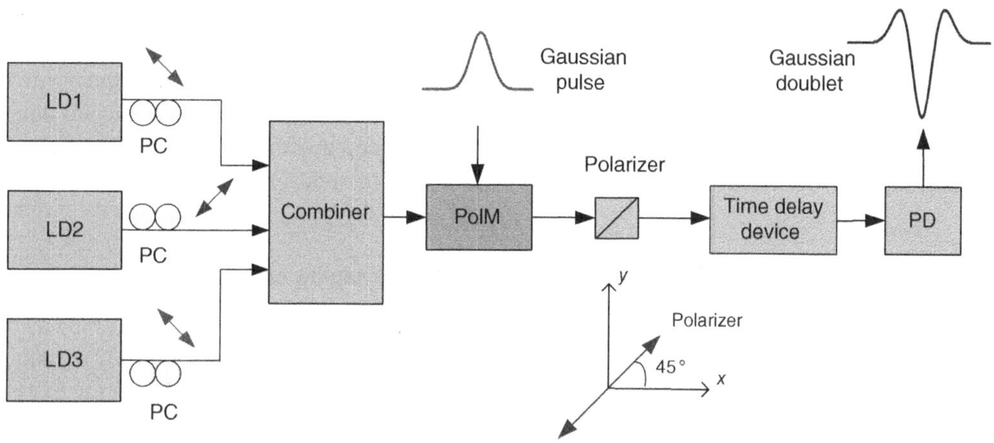

Figure 6.23 UWB doublet generation using a three-tap photonic microwave delay-line filter with one negative coefficient based on XPolM in a PolM.

The difference between the filters in Figs. 6.19 and 6.22 is that in Fig. 6.22, two CW probes are used. Again, thanks to the XGM in the SOA, the pulses carried by the two probes are both inverted versions of the one carried by the pump. If the three pulses are reflected by three FBGs with a proper time delay difference and a reflectivity ratio of 1 : 2 : 1, a UWB doublet is generated.

In Fig. 6.23, three wavelengths generated by three LDs are sent to the PolM. The polarization directions of the three wavelengths should be controlled using three PCs to have angles of 135°, 45°, and 135°, with respect to one principal axis of the PolM. At the output of the PolM, a polarizer with its polarization axis adjusted to have an angle of 45° with the same principal axis of the PolM. Thanks to the polarization modulation, three intensity-modulated Gaussian pulses with or without polarity inversion are generated. By introducing a time delay difference using a time delay device (three cascaded FBGs with a reflectivity ratio of 1 : 2 : 1), a three-tap microwave delay-line filter with coefficients of [1 −2 1] is being realized.

Note that the negative coefficient has a magnitude that is twice that of the positive coefficients. Therefore, in designing the time delay device, the reflection for the wavelength from LD2 should be twice that for LD1 and LD3, which can be realized by controlling the reflectivities of the three cascaded FBGs. In addition, the three FBGs should be cascaded with a physical spacing between two adjacent FBGs corresponding to the required time delay difference.

6.4 UWB Pulse Generation based on Spectral Shaping and Frequency-to-Time Mapping

Spectral shaping and frequency-to-time mapping for ultra-fast electrical pulse shaping or generation have been intensively investigated in the past few years [30]. Thanks to the well-known duality between paraxial diffraction in space and dispersion in time, a dispersive device, such as a dispersive fiber or a linearly chirped FBG (LCFBG), can serve as a real-time Fourier transformer to perform the frequency-to-time conversion [30, 31]. Dispersion-induced frequency-to-time mapping based on the space-time duality was analyzed in detail in [30]. In [32], an approach proposed to generating broadband arbitrary RF waveforms based on frequency-to-time mapping was implemented using free-space optics. In the approach in [32], a desirable spectrum waveform was obtained by space Fourier transform of an optical pulse using a liquid crystal spatial light modulator (SLM) and a Fourier lens. The temporal UWB pulses were then generated based on frequency-to-time conversion in a dispersion medium. Optical spectral shaping and dispersive frequency-to-time mapping were also implemented to generate adaptive broadband microwave arbitrary waveforms [33] and measure optical fiber dispersion, and also evaluate an optical source spectrum [34].

6.4.1 UWB Pulse Generation Based on Optical Spectral Shaping and Frequency-to-Time Mapping

Recently, UWB pulse generation based on SLM-based spectrum shaper and frequency-to-time mapping was proposed and demonstrated [32, 35–38]. The major difficulty associated with the SLM-based approaches for UWB pulse generation is that the pulse shaping process is implemented in free space, which makes the system bulky and complicated.

UWB pulses can also be generated based on spectral shaping and frequency-to-time conversion using pure fiber-optic components [39]. The use of fiber-optic components instead of free-space optics has the advantage of smaller size, lightweight, and the potential for photonic integration using photonic integrated circuit (PIC) technique.

Figure 6.24(a) shows a UWB pulse generation system based on spectral shaping and frequency-to-time mapping using all-fiber components. In the system, the optical power spectrum of a femtosecond pulse from a mode-locked fiber laser (MLFL) is shaped by two optical filters, to obtain a spectral shape corresponding to a UWB monocycle or a doublet. A length of SMF acting as a dispersive device is used to perform frequency-to-time mapping, and at the same time, as a transmission medium to distribute the UWB pulse to a remote site. A UWB monocycle or doublet pulse is then obtained at the output of a high-speed PD. The UWB pulse has a shape that is a scaled version of the user-designed power spectrum. The pulse width is determined by the total dispersion of the SMF. As shown in Fig. 6.24(b), an ultrashort pulse from the MLFL source is divided into two

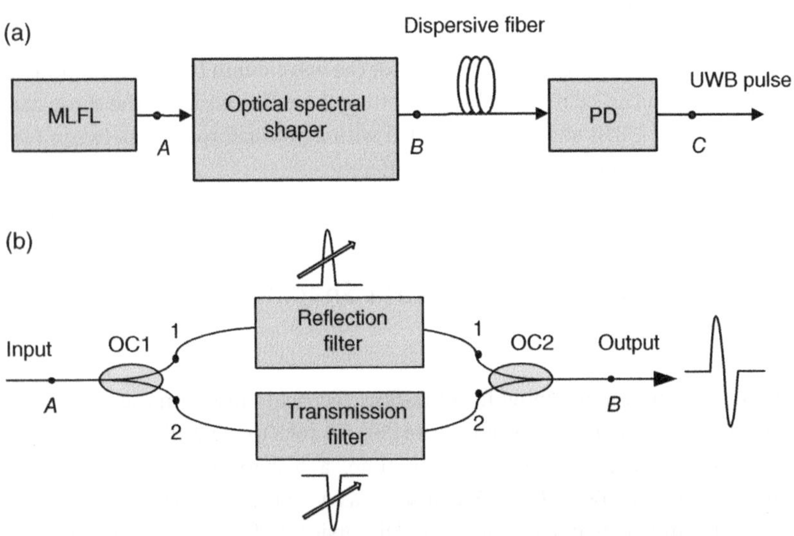

Figure 6.24 (a) UWB pulse generation based on spectral shaping and frequency-to-time mapping using all-fiber components. (b) All-fiber spectrum shaper configuration.

channels by an optical coupler. The input pulse spectrum from one port is shaped by a reflection filter and the input pulse from the other port is spectrally shaped by a transmission filter. The spectra shapes of the two filters are complementary, which ensures that the time-domain pulses are polarity-reversed.

The spectrum-shaped pulse is then sent to a length of dispersive fiber to perform frequency-to-time mapping. If the duration of the filtered pulse, denoted as Δt_0, and the chromatic dispersion χ (ps/nm) of the dispersive fiber satisfy the following relation

$$\left|\frac{\Delta {t_0}^2 c}{2\pi\lambda_0^2\chi}\right| \ll 1 \tag{6.33}$$

where λ_0 is the center wavelength of the MLFL source and c is the light velocity in vacuum, then the average optical power of the output pulse is proportional to the energy spectrum of the input pulse envelope [31],

$$|y(t)|^2 \propto | X(\omega) | ^2{}_{\omega = (2\pi ct)/\left({\lambda_0}^2\chi\right)} \tag{6.34}$$

where $y(t)$ is the complex envelope of output pulse, and $X(\omega)$ is the Fourier transform of the input pulse envelope. The mapping relationship from the spectrum bandwidth $\Delta\lambda$ to the temporal pulse width Δt is determined by

$$\Delta t = \chi\Delta\lambda \tag{6.35}$$

Therefore, a properly determined chromatic dispersion according to (6.35) would yield a pulse with appropriate temporal width to ensure the UWB pulse spectrum meets the FCC regulations. As a result, after the optical to electrical conversion in a high-speed PD, we can obtain a UWB monocycle or doublet pulse.

6.4.2 Implementation of All-Fiber UWB Pulse Generation based on Spectral Shaping and Frequency-to-Time Mapping

The UWB pulse generation system, as shown in Fig. 6.24(a), is experimentally implemented [39]. The optical spectrum shaper can be configured to generate UWB monocycle or doublet pulse by adjusting the parameters (the spectral width and the center wavelength) of the two optical filters. The input ultrashort pulse is split into two channels by an optical coupler (OC1). To generate a UWB monocycle, the input pulse spectrum from port 1 of OC1 is shaped by a reflection filter with its center wavelength tuned at 1557.71 nm, with a 3-dB bandwidth of 0.2 nm. The input spectrum from port 2 of OC1 is shaped by a transmission filter (an FBG) with a 3-dB bandwidth of 0.25 nm and a center wavelength of 1558.2 nm. The spectrum-shaped pulses are then combined at a second optical coupler (OC2). To ensure the positive and negative spectral peaks have an identical magnitude, the coupling ratio of OC1 can be properly selected. In the experiment, the coupling ratio is 70 : 30 since the reflection filter has higher insertion loss. The optical lengths of the two branches are carefully controlled to guarantee a temporal synchronization of optical pulses from different paths.

Figure 6.25(a) shows the optical spectrum at the output of the spectrum shaper. The shaped optical spectrum exhibits a monocycle pulse shape but is superimposed on a broader Gaussian-like pedestal, which is the spectrum of the pulse from the MLFL, as shown in Fig. 6.25(b). The spectrally shaped optical pulse is then applied to a 10-km SMF serving as the dispersive medium to perform frequency-to-time mapping. The total chromatic dispersion of the SMF is about 170 ps/nm. The UWB pulse obtained at the output of the PD is shown in Fig. 6.25(c). The pulse has the same shape

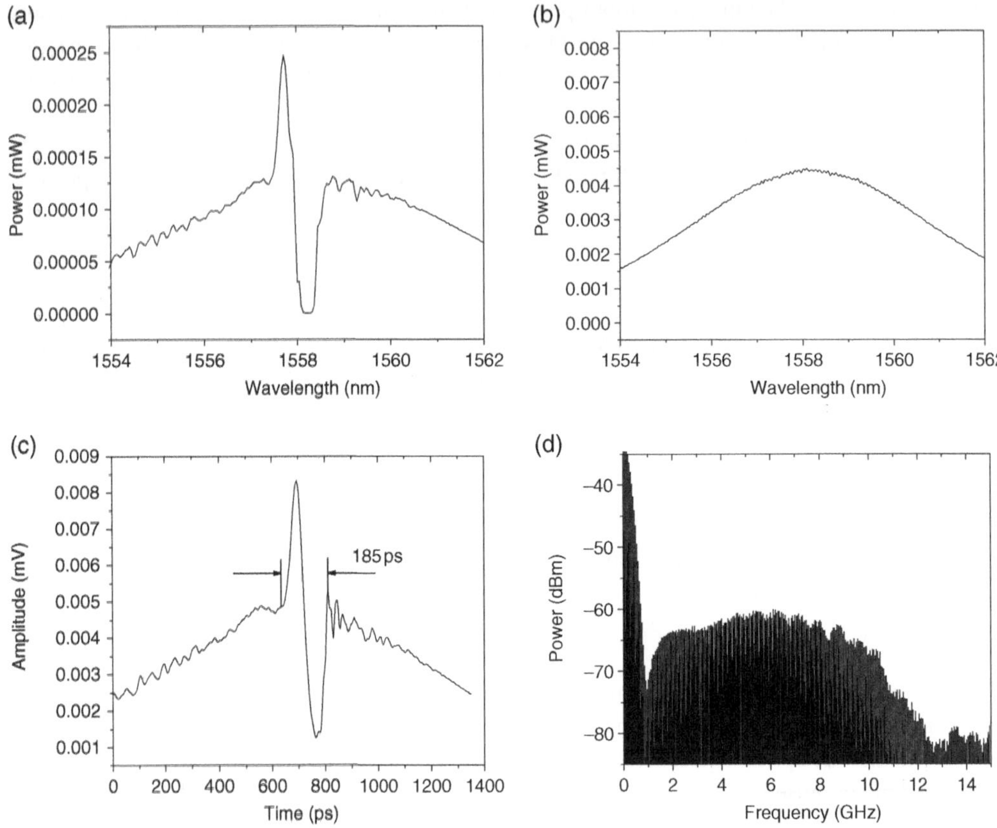

Figure 6.25 All-fiber UWB monocycle pulse generation based on spectral shaping and frequency-to-time mapping. (a) Spectrum after spectral shaping. (b) Spectrum of the incident ultrashort pulse. (c) Generated UWB monocycle pulse. (d) Power spectrum of the generated monocycle pulse.

as the optical power spectrum before mapping. The pulse width of the monocycle is about 185 ps, which is in good agreement with the theoretical value of 187 ps, calculated based on (6.35) using the total fiber dispersion and the monocycle pulse spectrum width. The spectrum of the generated monocycle pulse is shown in Fig. 6.25(d). As can be seen, the spectrum has a central frequency of 6 GHz with a 10-dB bandwidth of 9 GHz, from 1.5 to 10.5 GHz. The fractional bandwidth is about 150%. There is a spectral component at the baseband with a bandwidth of less than 1 GHz, as shown in Fig. 6.25(d), which results from the Gaussian-like pedestal. To reduce or eliminate this spectral component, we may use an MLFL with a much narrower pulse spectrum or use a DC block at the output of the PD.

To generate a UWB doublet, the spectrum shaper has to be reconfigured by using a transmission filter (an FBG) with a broader spectral width. The input pulse from port 1 of OC1 is spectrally shaped by the reflection filter, with the center wavelength set at 1559.45 nm. In the other path, the input pulse spectrum is shaped by an FBG with a broader bandwidth of 0.45 nm centered at 1559.4 nm. The spectrum-shaped optical pulses from the two arms are re-combined by OC2. Figure 6.26(a) shows the optical spectrum after the spectrum shaper. It has a shape corresponding to a doublet. After distribution over 10-km SMF, a UWB doublet pulse is generated thanks to the dispersion-induced frequency-to-time mapping. As shown in Fig. 6.26(b), the generated UWB

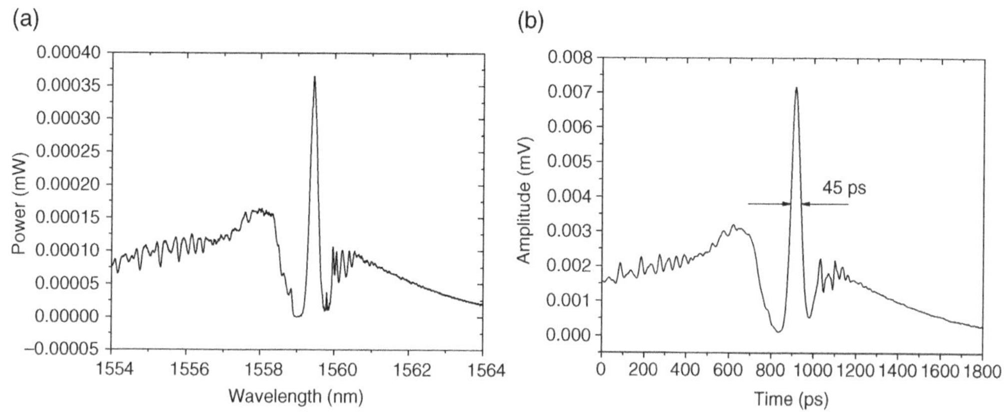

Figure 6.26 UWB doublet pulse generation. (a) Spectrum after spectrum shaping. (b) Generated UWB doublet pulse.

doublet pulse has an FWHM of about 45 ps, which is again superimposed on a broad and small Gaussian-like pedestal. The generated UWB doublet pulse has a 10-dB bandwidth of about 9.5 GHz, from 1.5 to 11 GHz.

6.5 Discussion and Conclusion

In this chapter, techniques to generate UWB pulses in the optical domain have been discussed. These techniques were divided into three categories: (i) PM-IM conversion in a dispersive device or an optical frequency discriminator, (ii) microwave filtering using a photonic microwave delay-line filter having two or three taps with one negative tap, and (iii) optical spectral shaping and frequency-to-time mapping in a dispersive device. The key feature of the techniques in all three categories is that all could be implemented using all fiber-optic components, which provide the potential for integration using integrated photonic circuits. Compared to the techniques in the second and third categories, the techniques in the first category have a simpler structure with only a single LD required. The approach based on spectral shaping and frequency-to-time mapping in the third category has more flexibility in generating UWB pulses with arbitrary shapes. The use of an optical pulsed source may make the system more complicated and costly.

In addition to the techniques discussed, there are a few other techniques developed recently for UWB pulse generation, such as UWB generation using a frequency shift keying (FSK) modulator [40], a nonlinearly biased Mach–Zehnder (MZM) modulator [41], dual-parallel MZM [42], and dual-drive MZM [43]. The key advantage of using optics to generate UWB pulses is that the UWB pulses generated directly in the optical domain can be distributed over an optical fiber without the need for an extra electrical-to-optical conversion. In addition, the extremely broad bandwidth offered by optics enables the generation of UWB pulses with large fractional bandwidth which is difficult to realize using electronic circuits.

In a UWB over fiber network, a central office is connected via optical fiber to base stations or APs. Optical generation of UWB pulses is only one of the many functions in a UWB over fiber network. To demonstrate a practical UWB over fiber network, other functions should also be implemented, such as UWB pulse encoding at the central office. The pulse encoding can be implemented in the systems based on PM-IM conversion by simply switching the optical wavelength to realize PPM or PSM. A major limitation of these techniques is the limited speed of wavelength switching, which may not be suitable for a UWB communication system operating at a high data rate. In a UWB communication system, to implement the PPM or PSM, it is highly desirable that the UWB pulses can be switched at a speed higher than 100 MHz [2]. A technique to generate UWB pulses that is able to perform PPS and PSM at a speed as high as 40 Gb/s has been recently demonstrated [44].

Other topics that need to be investigated for UWB over fiber applications include fiber dispersions and nonlinearity on the performance of the UWB over fiber systems. The UWB antenna design would also be a topic for investigation.

The key challenge that limits the practical applications of UWB over fiber technologies for broadband wireless indoor networking is the high cost of optoelectronic components. The use of PIC technology would be a solution to reduce the cost.

References

1 Federal Communications Commission (2002). Revision of part 15 of the commission's rules regarding ultra-wideband transmission systems. Tech. Rep., ET-Docket 98–153, FCC02–48 (April 2002).

2 Ghavami, M., Michael, L.B., and Kohno, R. (2004). *Ultra Wideband Signals and Systems in Communication Engineering*. West Sussex: Wiley.

3 Porcine, D., Research, P., and Hirt, W. (2003). Ultra-wideband radio technology: potential and challenges ahead. *IEEE Commun. Mag.* 41 (7): 66–74.

4 Aiello, G.R. and Rogerson, G.D. (2003). Ultra-wideband wireless systems. *IEEE Microw. Mag.* 4 (2): 36–47.

5 Siwiak, K. and McKeown, D. (2004). *Ultra-Wideband Radio Technology*. Chichester: Wiley.

6 Yang, L.Q. and Giannakis, G.B. (2004). Ultra-wideband communications: an idea whose time has come. *IEEE Signal Process. Mag.* 21: 26–54.

7 Nassar, C.R., Zhu, F., and Wu, Z. (2003). Direct sequence spreading UWB systems: frequency domain processing for enhanced performance and throughput. *IEEE Int. Conf. Commun.* 3: 2180–2186.

8 Balakrishnan, J., Batra, A., Dabak, A. (2003). A multi-band OFDM system for UWB communication. *2003 IEEE Conference on Ultra Wideband Systems and Technologies* (16–19 November 2003), Reston, VA, USA, pp. 354–358.

9 Wu, X., Tian, Z., Davidson, T.N., and Giannakis, G.B. (2006). Optimum waveform design for UWB radios. *IEEE Trans. Sign. Process.* 54 (6): 2009–2021.

10 Ishida, H. and Araki, K. (2004). Design and analysis of UWB bandpass filter with ring filter. *IEEE MTT-S Int. Dig.* 3: 1307–1310.

11 Zhu, L., Sun, S., and Menzel, W. (2005). Ultra-wideband (UWB) bandpass filters using multiple-mode resonator. *IEEE Microw. Wirel. Compon. Lett.* 15 (11): 796–798.

12 Lin, W.P. and Chen, J.Y. (2005). Implementation of a new ultrawide-band impulse system. *IEEE Photonics Technol. Lett.* 17 (11): 2418–2420.

13 Zeng, F. and J.Y. (2004). All-optical bandpass microwave filter based on an electro-optic phase modulator. *Opt. Express* 12 (16): 3814–3819.

14 Zeng, F. and J.Y. (2005). Investigation of phase modulator based all-optical bandpass filter. *J. Lightw. Technol.* 23 (4): 1721–1728.

15 Wang, J., Zeng, F., and J.Y. (2005). All-optical microwave bandpass filters implemented in a radio-over-fiber link. *IEEE Photonics Technol. Lett.* 17 (8): 1737–1739.

16 Zeng, F. and J.Y. (2005). All-optical microwave mixing and bandpass filtering in a radio-over-fiber link. *IEEE Photonics Technol. Lett.* 17 (4): 899–901.
17 J.Y., Maury, G., Guennec, Y.L., and Cabon, B. (2005). All-optical subcarrier frequency conversion using an electrooptic phase modulator. *IEEE Photonics Technol. Lett.* 17 (11): 2427–2429.
18 Le Guennec, Y., Maury, G., J.Y., and Cabon, B. (2006). New optical microwave up-conversion solution in radio-over-fiber networks for 60 GHz wireless applications. *J. Lightw. Technol.* 24 (3): 1277–1282.
19 Wang, J., Zeng, F., and J.Y. (2005). All-optical microwave bandpass filters with negative coefficients based on PM-IM conversion. *IEEE Photonics Technol. Lett.* 17 (10): 2176–21780.
20 Zeng, F., Wang, J., and J.Y. (2005). All-optical microwave bandpass filter with negative coefficients based on an electro-optic phase modulator and linearly chirped fiber Bragg gratings. *Opt. Lett.* 30 (17): 2203–2205.
21 Zeng, F. and J.Y. (2006). An approach to ultra-wideband pulse generation and distribution over optical fiber. *IEEE Photonics Technol. Lett.* 18 (7): 823–825.
22 Zeng, F. and J.Y. (2006). Ultrawideband signal generation using a high-speed electrooptic phase modulator and an FBG-based frequency discriminator. *IEEE Photonics Technol. Lett.* 18 (19): 2062–2064.
23 Chen, X. and Kiaei, S. (2002). Monocycle shapes for ultra wide-band system. *IEEE Int. Symp. Circ. Syst.* 1: 26–29.
24 Zeng, F., Wang, Q., and J.Y. (2007). All-optical UWB impulse generation based on cross phase modulation and frequency discrimination. *Electron. Lett.* 43 (2): 119–121.
25 Seeds, A.J. (2002). Microwave photonics. *IEEE Trans. Microw. Theory Tech.* 50 (3): 877–887.
26 Minasian, R.A. (2006). Photonic signal processing of microwave signals. *IEEE Trans. Microw. Theory Tech.* 54 (2): 832–846.
27 Capmany, J., Ortega, B., and Pastor, D. (2006). A tutorial on microwave photonic filters. *J. Lightw. Technol.* 24 (1): 201–229.
28 Wang, Q., Zeng, F., Blais, S., and J.Y. (2006). Optical UWB monocycle pulse generation based on cross-gain modulation in a semiconductor optical amplifier. *Opt. Lett.* 31 (21): 3083–3085.
29 J.Y. and Wang, Q. (2007). Photonic microwave bandpass filter with negative coefficients using a polarization modulator. *IEEE Photonics Technol. Lett.* 19 (9): 644–646.
30 Weiner, A.M. (2000). Femtosecond pulse shaping using spatial light modulators. *Rev. Sci. Instrum.* 71 (15): 1929–1960.
31 Muriel, M., Azaña, J., and Carballar, A. (1999). Real-time Fourier transformer based on fiber gratings. *Opt. Lett.* 24 (1): 1–3.
32 Lin, I., McKinney, J.D., and Weiner, A.M. (2005). Photonic synthesis of broadband microwave arbitrary waveforms applicable to ultra-wideband communication. *IEEE Microw. Wirel. Compon. Lett.* 15 (4): 226–228.
33 Chou, J., Han, Y., and Jalali, B. (2003). Adaptive RF-photonic arbitrary waveform generator. *IEEE Photonics Technol. Lett.* 15 (4): 581–583.
34 Tong, Y.C., Chan, L.Y., and Tsang, H.K. (1997). Fiber dispersion or pulse spectrum measurement using a sampling oscilloscope. *Electron. Lett.* 33 (11): 983–985.
35 Jalali, B., Chou, J., and Yan, H. (2004). Optically sculpt UWB waveforms. *Microw. RF* 43 (8): 54–62.
36 McKinney, J.D., Lin, I.S., and Weiner, A.M. (2006). Shaping the power spectrum of ultra-wideband radio-frequency signals. *IEEE Trans. Microw. Theory Tech.* 54 (12, Part 1): 4247–4255.
37 Xiao, S. and Weiner, A.M. (2006). Programmable photonic microwave filters with arbitrary ultra-wideband phase response. *IEEE Trans. Microw. Theory Tech.* 54 (11): 4002–4008.

38 McKinney, J.D. and Weiner, A.M. (2006). Compensation of the effects of antenna dispersion on UWB waveforms via optical pulse-shaping techniques. *IEEE Trans. Microw. Theory Tech.* 54 (4, Part 2): 1681–1686.

39 Wang, C., Zeng, F., and J.Y. (2007). All-fiber ultra wideband pulse generation based on spectral shaping and dispersion-induced frequency-to-time conversion. *IEEE Photonics Technol. Lett.* 19 (3): 137–139.

40 Kawanishi, T., Sakamoto, T., and Izutsu, M. (2004). Ultra-wide-band signal generation using high-speed optical frequency-shift-keying technique. *IEEE International Topical Meeting on Microwave Photonics – Technical Digest, MWP'04* (4–6 October 2004), Ogunquit, ME, USA, pp. 48–51.

41 Wang, Q. and J.Y. (2006). UWB doublet generation using a nonlinearly-biased electro-optic intensity modulator. *IEE Electron. Lett.* 42 (22): 1304–1305.

42 Zhang, F. and Pan, S. (2013). Background-free millimeter-wave ultra-wideband signal generation based on a dual-parallel Mach–Zehnder modulator. *Opt. Express* 21 (22): 27017–27022.

43 Li, W., Wang, W., Sun, W. et al. (2014). Photonic generation of background-free millimeter-wave ultra-wideband pulses based on a single dual-drive Mach–Zehnder modulator. *Opt. Lett.* 39 (5): 1201–1203.

44 Wang, Q. and J.Y. An electrically switchable optical ultra-wideband pulse generator. *J. Lightw. Technol.* 25: 3626–3633, Nov. 2007.

7

Photonic Generation of Microwave Arbitrary Waveforms

7.1 Introduction

Microwave arbitrary waveforms are widely used in radar, communications, medical imaging, and modern instrumentation systems. Microwave arbitrary waveforms are usually generated electrically using digital electronics. Due to the limited sampling rate, the generation of a microwave arbitrary waveform based on digital electronics is limited to a low frequency and small bandwidth. For many applications, however, high frequency and large bandwidth waveforms are needed. A solution is to generate microwave arbitrary waveforms using photonics, to take advantage of the high speed and broad bandwidth offered by modern photonics. In general, photonic-assisted microwave waveform generation can be classified into four categories: (i) direct space-to-time (DST) pulse shaping, (ii) spectral-shaping and wavelength-to-time mapping, (iii) temporal pulse shaping (TPS), and (iv) microwave pulse generation based on photonic microwave delay-line filtering. These techniques can be implemented in free space where a spatial light modulator (SLM) is usually employed to perform temporal or spectral shaping. The key advantage of using an SLM for microwave arbitrary waveform generation is its flexibility. The two-dimensional response (both magnitude and phase) of an SLM can be updated in real time, making the system reconfigurable. However, a pulse shaping system based on an SLM is usually implemented in free space, making the system bulky and costly. In addition, the coupling between fiber and free space and free space to fiber makes the system lossy and sensitive to environmental changes. Microwave arbitrary waveform generation can also be implemented using pure fiber-optic devices. Considering the low loss and small size, a microwave waveform generation system using fiber-optic devices is considered a promising alternative to that implemented based on free space optics. In this chapter, techniques to use fiber-optic devices to implement arbitrary microwave waveform generation are reviewed. All four different techniques that are implemented using fiber-optic devices are discussed. The use of the techniques to generate frequency-chirped and phase-coded microwave waveforms is discussed. The challenges in the implementation of the systems for practical applications [1] are also discussed.

7.2 Direct Space-to-Time Pulse Shaping

Arbitrary waveform generation can be realized based on DST mapping [2–11] in which an arbitrary optical pulse sequence is generated in the optical domain and then applied to a high-speed optical-to-electrical converter to generate a microwave waveform. By this technique, reprogrammable

Microwave Photonics, First Edition. Jianping Yao, José Capmany, and Ming Li.

Companion website: www.wiley.com/go/yao/microwavephotonics

cycle-by-cycle synthesis of an arbitrary microwave waveform such as a frequency-chirped microwave waveform could be implemented. In such a system, a bandwidth-limited optical-to-electrical converter was usually used to convert an optical pulse burst consisting of isolated optical pulses into a smooth microwave waveform. Figure 7.1 shows a DST mapping system in which an ultrashort optical pulse from a pulsed laser source is sent to an optical pulse shaper to generate a pulse burst, with the pulse spacing increasing temporally. The pulse burst is then applied to an optical-to-electrical converter or a photodetector (PD). Due to the bandwidth-limited nature of the PD, the high-frequency components are eliminated and a smooth frequency-chirped microwave waveform is generated.

The key device in the DST mapping system is the optical pulse shaper. It could be implemented using free-space optical components [2–8], but with large size, high loss, and poor stability. A simpler but more effective solution is to generate a pulse burst using a photonic integrated device, such as an arrayed waveguide grating (AWG) [9–11], as shown in Fig. 7.2. An ultrashort pulse is launched into the AWG through an amplitude mask. Due to the different time delays resulting from the different physical lengths of the waveguides in the AWG, a pulse burst with temporally spaced pulses is generated. To generate a pulse burst with the desired temporal pattern, the input amplitude mask can be configured to block the light from going into some of the optical waveguides. The use of the AWG-based pulse shaper to generate a pulse burst of more than 30 pulses as an ultrafast optical data packet over approximately an 80-ps temporal window was demonstrated [9]. The generation of high-repetition-rate femtosecond WDM pulses was also demonstrated [10].

The theory behind the generation of a microwave arbitrary waveform using a DST mapping system is that the waveform to be generated can be obtained by filtering the pulse burst using a band-limited filter [12]. It was demonstrated that a temporally spaced pulse burst, such as a pulse burst

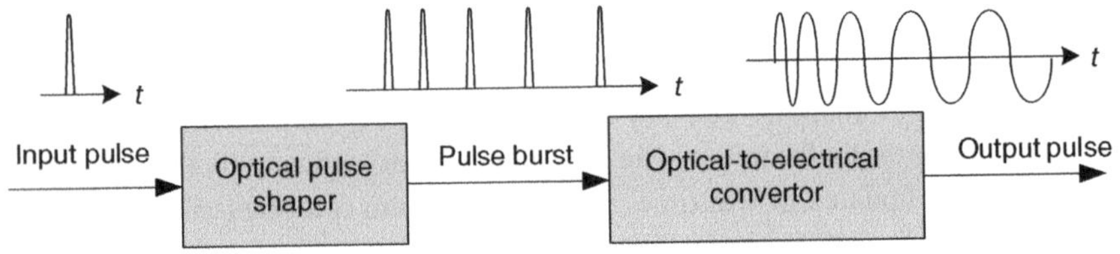

Figure 7.1 Arbitrary waveform generation based on direct space-to-time mapping.

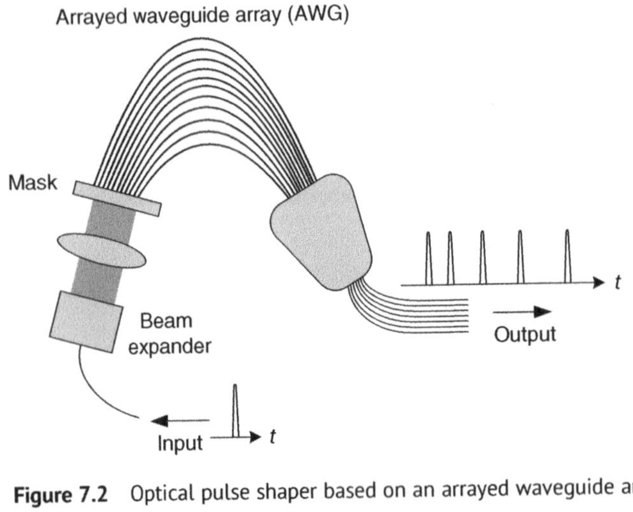

Figure 7.2 Optical pulse shaper based on an arrayed waveguide array.

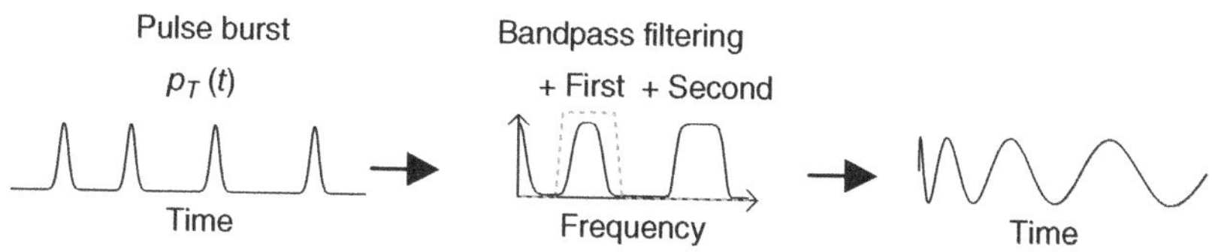

Figure 7.3 Generation of a chirped microwave waveform from a pulse burst with increasing pulse spacing through bandpass filtering.

with increasing or decreasing temporal spacing, would have a multi-channel spectral response, with one channel having a spectrum that corresponds to the spectrum of the waveform to be generated [12]. Figure 7.3 shows the generation of a linearly chirped microwave waveform from a pulse burst with increasing pulse spacing through bandpass filtering. As can be seen, the +1st-order channel has a spectral response that is equal to that of a linearly chirped microwave waveform. By employing a bandpass filter to select the +1st-order channel, a linearly chirped microwave waveform would be generated. In the following, a brief mathematical derivation is provided to show the generation of an arbitrary microwave waveform through pulse position modulation [13] and bandpass filtering.

A uniformly spaced intensity-modulated optical pulse burst $p_T(t)$, in which the kth pulse has a time delay of $\tau_k = kT$, where T is the time-delay difference between two adjacent pulses, can be expressed as

$$p_T(t) = \sum_{k=1}^{N} \alpha_k g(t - kT) \tag{7.1}$$

where $g(t)$ is a single short pulse, α_k is the coefficient weighted on the kth pulse, and N is the number of pulses in the pulse burst. The pulse burst can be expressed in another form,

$$p_T(t) = g(t) * \sum_{k=1}^{N} \alpha_k \delta(t - kT) = g(t) * [a(t) \times s(t)] \tag{7.2}$$

where $a(t)$ is the weight profile which is given as $a(kT) = \alpha_k$ for $1 \le k \le N$, otherwise $a(t) = 0$; $s(t)$ is a unit impulse train given by $s(t) = \sum_k \delta(t - kT)$, and $*$ denotes the convolution operation.

The spectrum of the pulse burst, $P_T(\omega)$, can be calculated by the Fourier transform,

$$P_T(\omega) = G(\omega) \times \left[\frac{1}{2\pi} A(\omega) * \sum_m \frac{2\pi}{T} \delta(\omega - m\Omega)\right] = \sum_m \frac{1}{T} G(\omega) A(\omega - m\Omega) \tag{7.3}$$

where $\Omega = 2\pi/T$, $A(\omega)$ is the Fourier transform of $a(t)$, and $G(\omega)$ is the spectrum of the short pulse $g(t)$. Since the pulse $g(t)$ is usually ultrashort, its spectrum $G(\omega)$ changes much slower compared with $A(\omega - m\Omega)$ within the bandwidth at Ω. Therefore, the change of $G(\omega)$ within the mth channel could be ignored, and $G(\omega)$ can be approximated as $G(m\Omega)$. Thus, the pulse burst has a multi-channel spectral response, with all channels having the same spectral profile $A(\omega)$ and the mth channel being located at $m\Omega$.

If a microwave bandpass filter with its bandpass located at $\omega = m\Omega$ is used in the system, the spectrum of the mth channel of the pulse burst is selected. As a result, the signal at the output of the microwave bandpass filter is given by

$$P(\omega) \approx \frac{1}{T} G(m\Omega) A(\omega - m\Omega) \tag{7.4}$$

In the time domain, the output RF signal, $p(t)$, is the inverse Fourier transform of (7.4), which is given by

$$\begin{aligned} p(t) &= \frac{1}{T} G(m\Omega) a(t) \times \exp(jm\Omega t) \\ &= \frac{1}{T} G(m\Omega) |a(t)| \times \exp\{j[\varphi(t) + m\Omega t]\} \end{aligned} \tag{7.5}$$

where $\varphi(t)$ is the phase response of $a(t)$. We can clearly see that the output signal is a microwave signal with a central frequency located at $m\Omega$. The generated phase modulation is just the phase of the weight profile, $\varphi(t)$.

Based on (7.5), we can see that if an arbitrary microwave waveform is generated then $a(t)$ should be a function having an arbitrary phase response. While in a DST mapping system, since only the power of the individual pulse is detected at the optical-to-electrical converter (which is a PD), the coefficients, α_k, are always positive. Therefore, the required phase response $\varphi(t)$ cannot be introduced. A solution to this problem is to introduce a phase shift through varying the spacing of the pulse burst, which is also called pulse position modulation. For a specific frequency, a time shift corresponds to a phase shift, and the inclusion of the phase shift would make the weight profile complex-valued, leading to the generation of an arbitrary microwave waveform.

Assume a function $f(t)$ is introduced to describe the pulse position modulation, that is, $s(t + f(t)) = \sum_k \delta(t + f(t) - kT)$, a new optical pulse burst is given by

$$\tilde{p}_T(t) = g(t) * \{a(t) \times s[t + f(t)]\} \tag{7.6}$$

With the introduction of $f(t)$, the time spacing of the pulse burst is no longer uniform. Based on Fourier series expansion, $s(t)$ can be expressed by its Fourier series, $s(t) = \sum_m (1/T) \exp(jm\Omega t)$. By variable substitution, (7.6) can be written as

$$\tilde{p}_T(t) = g(t) * \sum_m \frac{1}{T} a(t) \exp[jm\Omega f(t)] \times \exp(jm\Omega t) \tag{7.7}$$

Note that the above equation is obtained by variable substitution, it is no longer a Fourier series expansion (an aperiodic signal does not have a Fourier series expansion).

Since $g(t)$ is ultrashort, we may model it as an unit impulse, that is, $g(t) = \delta(t)$, where $\delta(t)$ is the Dirac delta function. Thus, (7.7) can be approximated as

$$\tilde{p}_T(t) \approx \sum_m \frac{1}{T} a(t) \exp[jm\Omega f(t)] \times \exp(jm\Omega t) \tag{7.8}$$

It is clearly seen from (7.8) that the nonuniformly spaced pulse burst is expressed as the sum of multiple bandpass microwave signals with different central frequencies. If T is sufficiently small such that the mth channel is not interfered by its adjacent channels, the spectral component at $m\Omega$ can be filtered out by a microwave bandpass filter with its central frequency at $m\Omega$. Then the output RF signal is now given by

$$\tilde{p}(t) = \frac{1}{T} \{a(t) \exp[jm\Omega f(t)]\} \times \exp(jm\Omega t) \tag{7.9}$$

Comparing (7.9) with (7.5), we can see that an additional phase modulation is introduced to the microwave signal due to the pulse position modulation introduced by $f(t)$. Therefore, although the weight coefficients are all positive, by using a specially designed pulse burst with pulse position modulation, an arbitrary microwave waveform can be obtained.

For example, to generate a phase-coded microwave waveform, if the phase modulation function is $\varphi(t)$, then the relationship between the pulse position modulation function $f(t)$ and the desired phase modulation can be obtained by letting $\varphi(t) = m\Omega f(t)$ or

$$f(t) = \frac{\varphi(t)}{m\Omega} \tag{7.10}$$

Substituting (7.10) into (7.6), we get the time delay of each pulse in the pulse burst,

$$\tau_k + \frac{\varphi(\tau_k)}{m\Omega} = kT \tag{7.11}$$

Considering the property of the Dirac function, the coefficients can also be obtained, which are given by

$$\alpha_k = \left| \frac{a(\tau_k)}{1 + \varphi'(\tau_k)/m\Omega} \right| \tag{7.12}$$

Based on the analysis, we conclude that if a pulse burst with pulse position modulation described by (7.11) and with the coefficients given by (7.12), a microwave waveform with a central frequency located at $m\Omega$ and a phase modulation of $a(t)\exp[j\varphi(t)]$ is then obtained at the output of a microwave bandpass filter with its central frequency located at $m\Omega$.

7.3 Spectral-Shaping and Wavelength-to-Time Mapping

Arbitrary waveform generation can be realized based on spectral-shaping and wavelength-to-time mapping. The fundamental principle of the technique is shown in Fig. 7.4(a). The system consists of a pulsed source, a spectral shaper, and a dispersive element. The spectral shaper is used to modify the spectrum emitted from the pulsed laser source, which can be a passively or actively mode-locked laser source. The shaped spectrum then undergoes wavelength-to-time mapping in a dispersive device, which can be a length of dispersive fiber or a linearly chirped fiber Bragg grating. A microwave waveform is generated in the electrical domain at a high-speed PD. If the dispersive element is a length of fiber with a value of dispersion of $\ddot{\Phi}$, for an input pulse $g(t)$ with a temporal width of Δt_0, the signal at the output of the dispersive element is given by

$$\begin{aligned} y(t) &= g(t) * \exp\left(j\frac{t^2}{2\ddot{\Phi}}\right) = \int_{-\infty}^{\infty} g(\tau) \times \exp\left[j\frac{(t-\tau)^2}{2\ddot{\Phi}}\right] d\tau \\ &= \exp\left(j\frac{t^2}{2\ddot{\Phi}}\right) \times \int_{-\infty}^{\infty} g(\tau) \times \exp\left(j\frac{\tau^2}{2\ddot{\Phi}}\right) \times \exp\left[-j\left(\frac{t}{\ddot{\Phi}}\right)\tau\right] d\tau \\ &\approx \exp\left(j\frac{t^2}{2\ddot{\Phi}}\right) \times \int_{-\infty}^{\infty} g(\tau) \times \exp\left[-j\left(\frac{t}{\ddot{\Phi}}\right)\tau\right] d\tau \\ &= \exp\left(j\frac{t^2}{2\ddot{\Phi}}\right) \times G(\omega)\big|_{\omega = \frac{t}{\ddot{\Phi}}} \end{aligned} \tag{7.13}$$

where $G(\omega)$ is the Fourier transform of $g(t)$.

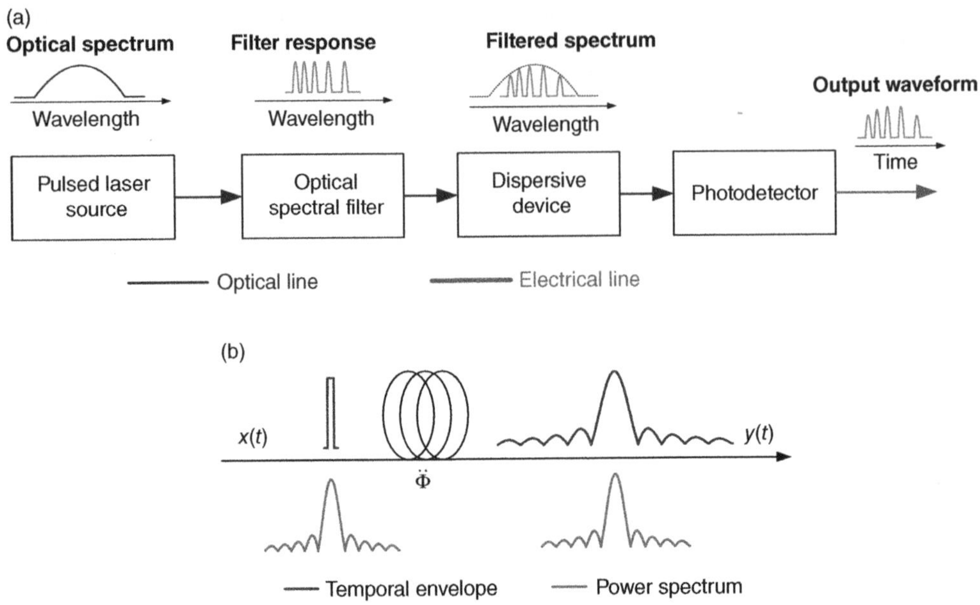

Figure 7.4 Arbitrary waveform generation based on spectral-shaping and wavelength-to-time mapping. (a) Schematic of a microwave waveform generation system based on spectral-shaping and wavelength-to-time mapping. (b) Illustration of wavelength-to-time mapping in a dispersive element.

As can be seen, the output signal envelope is proportional to the Fourier transform of the input signal envelope. Note that (7.13) is obtained if the duration of the input ultrashort pulse, Δt_0, and the second-order dispersion $\ddot{\Phi}$ of the dispersive element satisfy the following condition,

$$\left|\frac{\Delta t_0{}^2}{2\ddot{\Phi}}\right| \ll 1 \tag{7.14}$$

which means the phase term $(\tau^2)/(2\ddot{\Phi})$ in (7.13) satisfies $((\tau^2)/(2\ddot{\Phi})) \leq ((\Delta t_0{}^2)/(2\ddot{\Phi})) \ll 1$, thus we have $\exp(j((\tau^2)/(2\ddot{\Phi}))) \approx 1$ [14].

The wavelength-to-time mapping is illustrated in Fig. 7.4(b). If the input to the dispersive device is a rectangular pulse, then the output temporal waveform should be a sinc function. As can be seen, the key device in the arbitrary microwave waveform generator is the spectral shaper, which should be designed to have a magnitude response that can make the shaped spectrum have the same shape as the microwave waveform to be generated.

Based on this concept, a few approaches to generating arbitrary microwave waveforms [15–23] were demonstrated. The main effort in these approaches is to design an optical spectral shaper that has a magnitude response with its shape identical to that of the microwave waveform to be generated. Figure 7.5(a) shows a spectral-shaping and wavelength-to-time mapping system for the generation of a chirped microwave waveform [17]. An ultrashort pulse from a mode-locked laser is sent to an optical spectral shaper. For chirped microwave waveform generation, the magnitude response of the optical spectral shaper should have an increasing or decreasing free spectral range (FSR), which is termed chirped FSR in [17]. The spectrum-shaped pulse is then sent to a dispersive element, which is a length of single mode fiber (SMF), as shown in Fig. 7.5(a). To obtain a spectral response with a chirped FSR, the spectral shaper is designed by superimposing two chirped

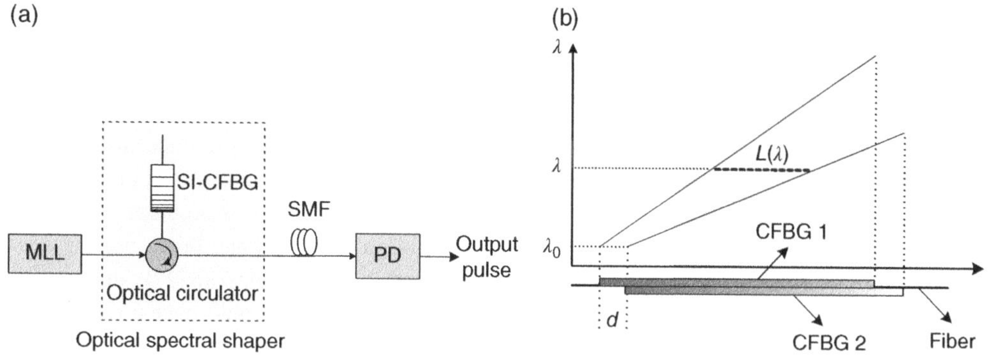

Figure 7.5 Chirped microwave waveform generation based on spectral-shaping and wavelength-to-time mapping. (a) Schematic of the chirped microwave pulse generation system. (b) Optical spectral shaper consisting of two superimposed chirped fiber Bragg gratings with different chirp rates and a small longitudinal offset. MLL, mode-locked laser; PD, photodetector; SI-CFBG, superimposed chirped fiber Bragg grating; SMF, singlemode fiber.

fiber Bragg gratings with different chirp rates into the same fiber with a small longitudinal offset, as shown in Fig. 7.5(b). Distributed Fabry–Perot interference is then produced in the fiber due to the reflections between the two chirped fiber Bragg gratings. This arrangement would create an in-fiber optical filter with an FSR inversely proportional to the cavity length L. Since the two chirped fiber Bragg gratings have different linear chirp rates, the equivalent cavity length L varies linearly with respect to optical wavelength λ. As a result, the FSR is not constant but is increasing or decreasing with respect to optical wavelength.

From Fig. 7.5(b), we can see the equivalent cavity length L is linearly proportional to the wavelength λ within the filter bandwidth,

$$L(\lambda) = d + \frac{C_1 - C_2}{C_1 C_2}(\lambda - \lambda_0) \tag{7.15}$$

where C_1 and C_2 are the chirp rates of the two chirped fiber Bragg gratings (in nm/mm), d is the longitudinal offset, and λ_0 is the start wavelength.

The FSR of the distributed Fabry–Perot filter is given by

$$FSR \cong \frac{\lambda_0{}^2}{2n_{eff}L(\lambda)} \tag{7.16}$$

where n_{eff} is the effective refractive index of the fiber. After the dispersion-induced linear frequency-to-time mapping, the FSR is mapped to the temporal period of the generated chirped pulse, namely $\Delta\tau$, with a mapping relationship $\lambda \to t/\chi$, where χ (ps/nm) is the total dispersion of the SMF.

For simplicity, the instantaneous microwave carrier frequency of the generated temporal pulse, f_{RF}, can be approximated by the reciprocal of the temporal period $\Delta\tau$, given by

$$f_{RF}(t) = \frac{1}{\Delta\tau} = 2n_{eff}\left[\frac{C_1 - C_2}{C_1 C_2} \times \left(\frac{t}{\lambda_0{}^2\chi^2} - \frac{1}{\lambda_0\chi}\right) + \frac{d}{\lambda_0{}^2\chi}\right] \tag{7.17}$$

It can be seen that the frequency of the microwave waveform is linearly proportional to time t. Therefore, the generated microwave waveform is linearly chirped. For the SMF with a given length, the central carrier frequency of the generated chirped pulse at $t = 0$ is dependent only upon the longitudinal offset d. The chirp rate of the generated pulse is determined by the chirp rates of the two chirped fiber Bragg gratings. Therefore, by choosing the longitudinal offset and the chirp rates of the chirped fiber Bragg gratings, a linearly chirped microwave waveform with a high central frequency and a large chirp rate can be generated.

The major limitation of the approach in [17] is that the optical spectral shaper, once fabricated, is not reconfigurable. For many applications, however, it is expected that the chirp rate and the center frequency of the microwave chirped waveform can be tunable. To solve this problem, an optical spectral shaper that is a Sagnac-loop mirror incorporating a chirped fiber Bragg grating was proposed [18]. Similar to the superimposed chirped fiber Bragg gratings in [17], due to the incorporation of a chirped fiber Bragg grating, the Sagnac-loop mirror would have a distributed Fabry–Perot interference due to the reflections from the two opposite directions of the chirped fiber Bragg grating. Thus, an in-fiber optical filter with an FSR that is inversely proportional to the length difference is formed. Since the reflection point is wavelength dependent, the equivalent length difference varies linearly with respect to optical wavelength λ. As a result, the FSR is not constant but is increasing or decreasing when tuning the optical wavelength.

Figure 7.6 shows the optical spectral shaper based on an all-fiber Sagnac-loop mirror incorporating a chirped fiber Bragg grating [18]. The Sagnac-loop mirror is constructed from a fused 3-dB fiber coupler spliced to the terminals of the chirped fiber Bragg grating, which is located approximately at the middle point of the fiber loop. A tunable delay-line (TDL) is located in the fiber loop to finely tune the time-delay difference between two fiber lengths L_1 and L_2. A polarization controller (PC) is also placed in the loop to optimize the visibility of the interference pattern at the output of the loop mirror. A three-port optical circulator is used to direct the ultrashort pulse into the loop mirror and to output the spectrum-shaped pulse for wavelength-to-time mapping.

Mathematically, the Sagnac-loop mirror incorporating a chirped fiber Bragg grating can be modeled as a two-tap delay-line filter. The transfer function of the Sagnac-loop mirror is expressed as

$$T(\lambda) = \frac{1}{2}W(\lambda)\left[1 + \cos\left(\frac{2\pi n_{eff}}{\lambda}2\Delta L\right)\right], \left(|\lambda - \lambda_0| \leq \frac{B_\lambda}{2}\right) \tag{7.18}$$

where $W(\lambda)$ is the intensity reflection spectrum of the chirped fiber Bragg grating with a bandwidth B_λ, and n_{eff} is the effective refractive index of the fiber core. $\Delta L = L_1 - L_2$ is the fiber length difference, with L_1 and L_2 measured from the center of the chirped fiber Bragg grating to the fiber coupler along the clockwise and counterclockwise paths, as shown in Fig. 7.6. The fiber length difference

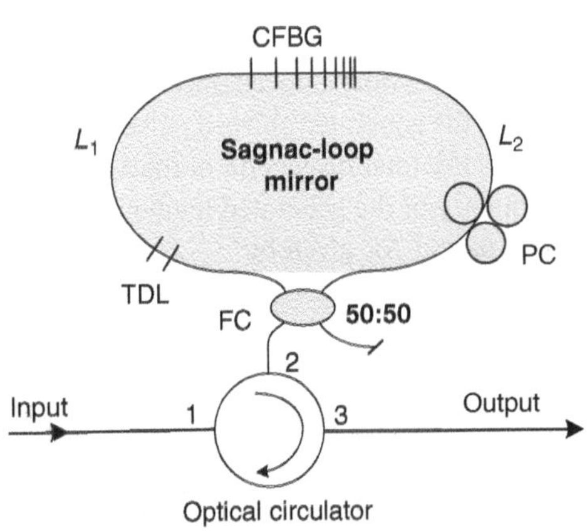

Figure 7.6 An all-fiber optical spectral shaper consisting of Sagnac-loop mirror incorporating a chirped fiber Bragg grating. CFBG, chirped fiber Bragg grating, FC, fiber coupler; PC, polarization controller; TDL, tunable delay-line.

ΔL comes from two sources: the wavelength-independent path difference, ΔL_0, and the wavelength-dependent fiber length difference introduced by the chirp of the chirped fiber Bragg grating, $\Delta L(\lambda)$. Note that ΔL_0 can be controlled to be either a positive or a negative value by tuning the TDL in the fiber loop. $\Delta L(\lambda)$ is determined by the bandwidth and the chirp parameter of the chirped fiber Bragg grating, and can be calculated using $\Delta L(\lambda) = \delta\lambda/C$, where $\delta\lambda$ (nm) is the wavelength detuning from the center wavelength λ_0, and C (nm/cm) is the chirp parameter of the chirped fiber Bragg grating. Then the filter transfer function $T(\lambda)$ can be rewritten as

$$T(\lambda) = \frac{1}{2}W(\lambda)\left\{1 + \cos\left[\frac{4\pi n_{eff}}{\lambda_0^2}\lambda\left(\Delta L_0 + \frac{\delta\lambda}{C}\right)\right]\right\} \tag{7.19}$$

Since the linearly chirped fiber Bragg grating is located in the fiber loop, an optical signal with different wavelengths will be reflected from a different position in the chirped fiber Bragg grating. As a result, an optical spectral filter with a wavelength-dependent FSR is formed. The FSR of the optical spectral filter response is a function of the wavelength and can be expressed as

$$FSR = \frac{\lambda^2}{2n_{eff}|\Delta L|} = \frac{\lambda_0^2}{2n_{eff}|((\delta\lambda)/C) + \Delta L_0|} \tag{7.20}$$

According to (7.20), by properly choosing the parameters of the chirped fiber Bragg grating and by controlling the TDL in the fiber loop, the FSR of the Sagnac-loop mirror can be controlled.

After the spectrum-shaped optical pulse propagates through the dispersive element and is detected by the high-speed PD, the shaped spectrum is mapped into a temporal microwave pulse as $T(\lambda) \rightarrow y(t)$ thanks to the dispersion-induced linear wavelength-to-time mapping. Assume the input ultrashort optical pulse is a unit impulse, according to the mapping relationship $\lambda \rightarrow t/\chi$, the time-domain waveform is given by

$$y(t) = \frac{1}{2}W\left(\frac{t}{\chi}\right)\left\{1 + \cos\left[\frac{4\pi n_{eff}}{\lambda_0^2\chi}t\left(\Delta L_0 + \frac{\delta t}{C\chi}\right)\right]\right\} \tag{7.21}$$

where δt is the time detuning from the center of the temporal waveform, which is given by the mapping relationship $\delta\lambda \rightarrow \delta t/\chi$. The time-domain pulse duration ΔT of the generated microwave pulse is determined by the window function $W(t/\chi)$, and is calculated by $\Delta T = B_\lambda\chi$. Considering that the pulse width of the input ultrashort optical pulse is not zero, the detected pulse envelope should be modified by adding an envelope $r(t)$,

$$y(t) = \frac{1}{2}r(t)W\left(\frac{t}{\chi}\right)\left\{1 + \cos\left[\frac{4\pi n_{eff}}{\lambda_0^2\chi}t\left(\Delta L_0 + \frac{\delta t}{C\chi}\right)\right]\right\} \tag{7.22}$$

where $r(t)$ is the pulse envelope after the input pulse passing through the dispersive element. Assuming that the input ultrashort optical pulse has a Gaussian envelope as $g(t) \propto \exp(-t^2/\Delta t_0^2)$, where Δt_0 is the half pulse-width at $1/e$ maximum, then the envelope of output pulse from a dispersive element will maintain the Gaussian shape, but with a broadened pulse width of $|\ddot{\Phi}|/\Delta t_0$. Actually, the envelope $r(t)$ is a scaled version of the spectrum envelope of the input pulse, which is mapped to the time domain thanks to the dispersion-induced frequency-to-time mapping in the chirped fiber Bragg grating.

The instantaneous radio-frequency (RF) carrier frequency of the generated waveform can be obtained from the phase term of (7.22), which is expressed as

$$f_{RF}(\delta t) = \frac{1}{2\pi} \times \frac{d\Psi}{dt} = \frac{2n_{eff}}{\lambda_0^2\chi}\left(|\Delta L_0| \pm \frac{\delta t}{C\chi}\right) \tag{7.23}$$

It is shown that the generated microwave waveform is linearly chirped. For a given dispersive element, the central RF carrier frequency of the generated chirped microwave pulse is only determined by the absolute value of wavelength-independent fiber length difference ΔL_0. Therefore, the central frequency of the generated microwave chirped waveform can be tuned by simply tuning ΔL_0. The chirp rate of the generated microwave waveform is given by

$$CR_{RF} = df_{RF}(\delta t)/d\delta t = \pm 2n_{eff}/(C\lambda_0{}^2\chi^2), \tag{7.24}$$

is only dependent on the dispersion of the chirped fiber Bragg grating. The sign of the chirp rate corresponds to the positive and negative value of ΔL_0. Therefore, by appropriately controlling the TDL and choosing the dispersion of the chirped fiber Bragg grating, a linearly chirped microwave pulse with a high central frequency and a tunable chirp rate can be generated. Figure 7.7 shows the spectra of a Gaussian pulse after spectral shaping by the Sagnac-loop mirror for different ΔL_0. The corresponding temporal domain waveforms after wavelength-to-time mapping are also shown.

In [17, 18], the spectral shaper for spectrum shaping and the dispersive element for wavelength-to-time mapping are two separate components. In fact, the two components can be a single component if the magnitude response and the group delay response can be individually controlled to perform simultaneously spectrum shaping and wavelength-to-time mapping. In [19], a single chirped fiber Bragg grating was employed to perform the two functions, as shown in Fig. 7.8(a).

The key component in the system is the linearly chirped fiber Bragg grating, which should be designed to have a magnitude as well as a group delay response that can fulfill the requirements for both spectral shaping and wavelength-to-time mapping. Different techniques have been proposed to synthesize an FBG, such as the well-known discrete layer-peeling (DLP) algorithm [24–26] and the Gelfand-Levitan-Marchenko (GLM) inverse scattering algorithm [27].

Here a simplified approach is employed to synthesize the linearly chirped fiber Bragg grating [19]. Figure 7.8(b) shows the magnitude and the group delay response of a linearly chirped fiber Bragg grating. The magnitude response has an increasing FSR and the group delay response is linear. Due to the inherent linear group delay response, a linearly chirped fiber Bragg grating can always act as a linear wavelength-to-time mapper. Therefore, the focus of the work is to synthesize the grating refractive index modulation profile from the target grating magnitude response. The synthesis is performed based on an accurate mapping of grating reflection response to the refractive index modulation [19]. We can first set up the mapping relationship by applying a linearly increasing index modulation function to a test grating with the use of a linearly chirped phase mask and then measuring the grating reflection spectrum. A linear index modulation function is first constructed, which is expressed as

$$\Delta n_L(z) = \Delta n_{\max}\frac{z}{L}\exp\left(j\frac{2\pi z}{\Lambda_0}\right)\exp\left[-j\frac{2\pi C(z-L/2)^2}{\Lambda_0{}^2}\right] \tag{7.25}$$

where Λ_0 is the grating period at the center of the linearly chirped fiber Bragg grating, L is the length of the linearly chirped fiber Bragg grating, and C is the grating chirp rate. We can imprint the index modulation function in (7.25) into the test grating using a given linearly chirped phase mask. Then the reflection spectrum $R_{test}(\lambda)$ of the fabricated test grating is measured. By equally dividing the grating into N consecutive segments with positions z_i ($1 \le i \le N$), we can get the sampled reflection spectrum $R_{test}(\lambda_i)$ thanks to the unique mapping relationship between $\Delta n_L(z_i)$ and $R_{test}(\lambda_i)$.

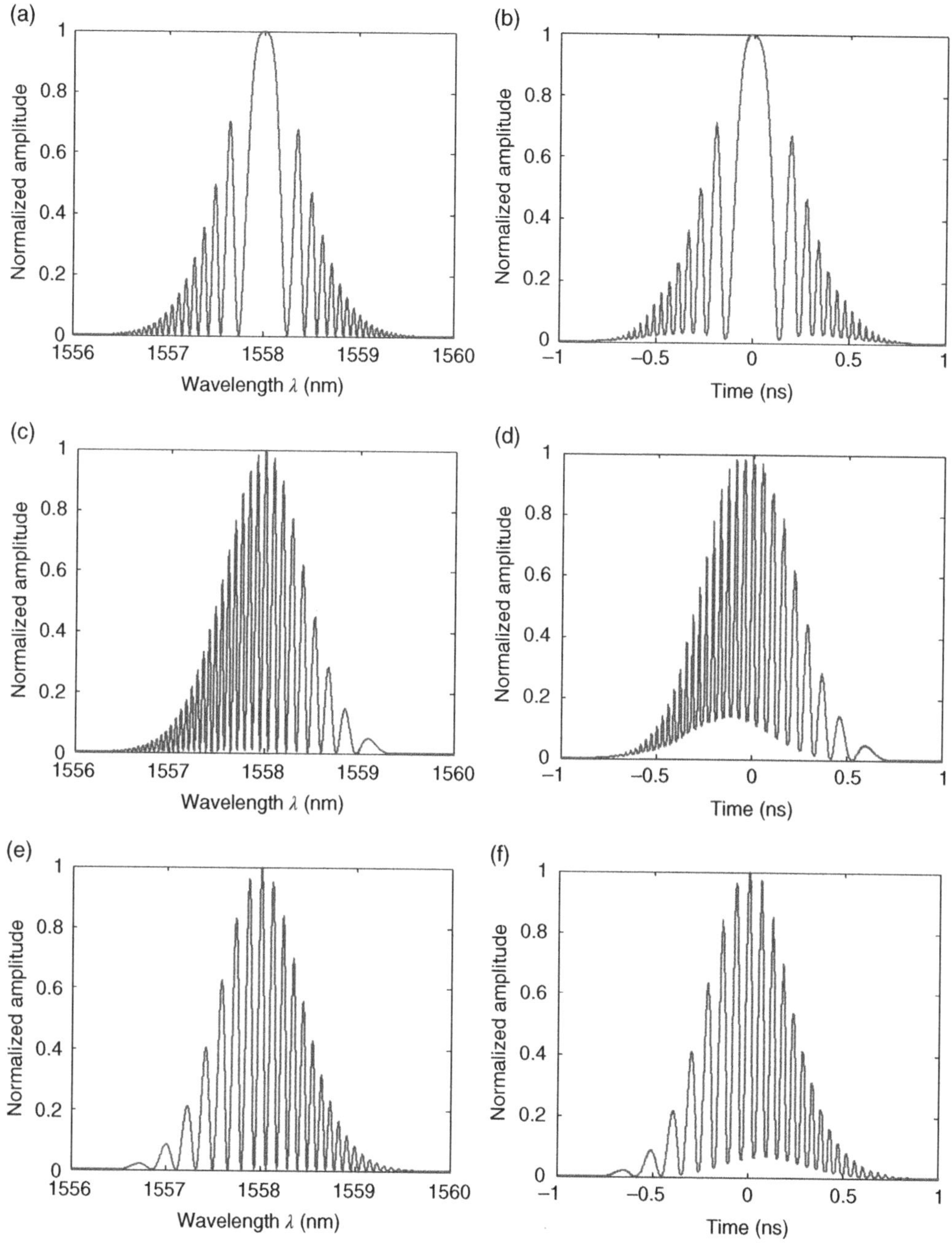

Figure 7.7 The spectra of an optical Gaussian pulse after spectral shaping by the Sagnac-loop mirror and the corresponding temporal waveforms. (a) A symmetrical FSR ($\Delta L_0 = 0$), (c) an increasing FSR ($\Delta L_0 = -9.7$ mm), and (e) a decreasing FSR ($\Delta L_0 = 6.9$ mm). The generated time-domain waveforms with (b) a symmetrical chirp rate and a zero central frequency, (d) a negative chirp rate and a central frequency of 22.3 GHz, and (f) a positive chirp rate and a central frequency of 16.1 GHz.

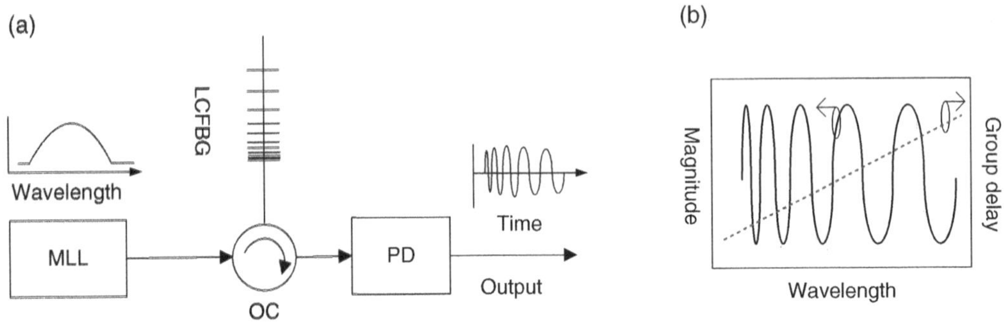

Figure 7.8 (a) A microwave arbitrary waveform generator based on spectral-shaping and wavelength-to-time mapping using a single linearly chirped fiber Bragg grating. (b) The magnitude and phase responses of the linearly chirped fiber Bragg grating. LCFBG, Linearly chirped fiber Bragg grating; MLL, mode-locked laser; PD, photodetector.

For a grating with a target reflection spectrum $R_{tgt}(\lambda)$, we can compare it with the test grating response $R_{test}(\lambda_i)$ wavelength by wavelength and then determine the desired index modulation function $\Delta n_D(z_i)$ by querying the linear index modulation function $\Delta n_L(z_i)$ segment by segment. Therefore, by applying the amplitude-only index modulation $\Delta n_D(z_i)$ under the same experimental condition, a desired linearly chirped fiber Bragg grating with the target reflection spectrum can be easily fabricated with the current FBG fabrication technology.

To improve the reconfigurability, recently an integrated arbitrary microwave waveform generator that incorporates a fully programmable spectral shaper fabricated on a silicon photonics platform was demonstrated [20]. The spectral shaper consists of a cascade of multi-channel microring resonators on a silicon photonics platform that is compatible with electronic integrated circuit technology. The system reconfigurability is achieved by thermally tuning both the resonant frequencies and the coupling strengths of the microring resonators. Two generations of the microring spectral shaper were developed. In the first generation system, the resonators have a ring structure with the central wavelength of each resonator independently tuned by a microheater placed above the microring, and have a tuning speed in millisecond to microsecond range. Since the heating of the ring has little impact on the coupling efficiency, which is crucial in controlling the magnitude profile of the spectral shaper, a coupler with a Mach–Zehnder structure in the input port of each microring was added in the second generation system. By thermally tuning the phase shift between the two arms, the coupling coefficient into a ring can be adjusted. It was demonstrated that for each resonant frequency, full tuning from the on state (no dip) to the off state can be achieved. By incorporating the spectral shaper into a photonic arbitrary microwave waveform generation system, a variety of different waveforms are generated, including those with an apodized amplitude profile, multiple π phase shifts, two-tone waveforms and frequency-chirped waveforms [20].

More recently, a more flexible approach to the generation of microwave arbitrary waveforms using a single spatially discrete chirped fiber Bragg grating (SD-CFBG) was reported [21]. The SD-CFBG functions to perform simultaneously spectral slicing, frequency-to-time mapping, and temporal shifting of the input optical pulse, which leads to the generation of an optical pulse burst with the individual pulses in the burst temporally spaced by the time delays determined by the SD-CFBG. With the help of a bandpass filter, a smooth microwave waveform is obtained. The SD-CFBG is fabricated using a linearly chirped phase mask by axially shifting the photosensitive fiber to introduce a spatial spacing between two adjacent sub-gratings during the fabrication process. By properly designing the fiber shifting function, a large time-bandwidth-product microwave arbitrary waveform with the desired frequency chirping or phase coding can be generated. The photonic generation of large TBWP microwave waveforms with linear, nonlinear, and stepped frequency chirping was experimentally demonstrated.

Instead of using a dispersive element with only the second-order dispersion for linear wavelength-to-time mapping, a dispersive element with both the second- and third-order dispersion can be used to achieve nonlinear wavelength-to-time mapping. For the case of chirped microwave waveform generation, for example, if a dispersive element with both the second- and third-order dispersion is employed, a chirped microwave waveform can be generated using an optical spectral shaper with a uniform FSR, which would simplify the implementation [22, 23].

If the dispersion up to the third order is considered, then a new wavelength-to-time mapping function would be used, which is given by [28]

$$\omega = \frac{t}{\ddot{\Phi}} - \frac{\dddot{\Phi}\, t^2}{2\ddot{\Phi}^3}, \tag{7.26}$$

where $\dddot{\Phi}$ is the third-order dispersion. Assuming that the spectral shaper is a Sagnac-loop filter, the transfer function of the two-tap Sagnac-loop filter is given by

$$H(\omega) = \frac{1}{2}[1 + \cos(\omega\tau_0)] \tag{7.27}$$

where τ_0 is the time-delay difference between the two taps. Based on wavelength-to-time mapping, a microwave waveform to be generated is given

$$i(t) = r(t) \times \frac{1}{2}\left[1 + \cos\left(\frac{t}{\ddot{\Phi}} - \frac{\dddot{\Phi}\, t^2}{2\ddot{\Phi}^3}\right)\tau_0\right] \tag{7.28}$$

where $r(t)$ is again the pulse envelope. If the input short pulse from the mode-locked laser is a Gaussian pulse, the output pulse envelope $r(t)$ can be analytically expressed using the Airy function [22].

The instantaneous microwave carrier frequency of the obtained waveform can be written as

$$\omega_{RF}(t) = \frac{d}{dt}\left(\frac{t}{\ddot{\Phi}} - \frac{\dddot{\Phi}\, t^2}{2\ddot{\Phi}^3}\right)\tau_0 = \frac{\tau_0}{\ddot{\Phi}} - \frac{\dddot{\Phi}\,\tau_0}{\ddot{\Phi}^3}t \tag{7.29}$$

As can be seen, the frequency of the microwave carrier is not constant, but a function of time. The microwave waveform is linearly chirped.

7.4 Temporal Pulse Shaping

Microwave waveforms can also be generated based on TPS [29–32]. A TPS system usually consists of a model-locked laser source, a pair of complementary dispersive elements and a modulator. The modulator can be a Mach–Zehnder modulator (MZM) [30, 31] or a phase modulator [32]. Figure 7.9 shows a schematic of a TPS system in which the modulator is an MZM.

Again, if $|\Delta t_0{}^2/\ddot{\Phi}| \ll 1$, where Δt_0 is the temporal width of the input short pulse, the electrical field of the short pulse $g(t)$ after propagating through the first dispersive element with a dispersion of $-\ddot{\Phi}$ can be expressed as [14]

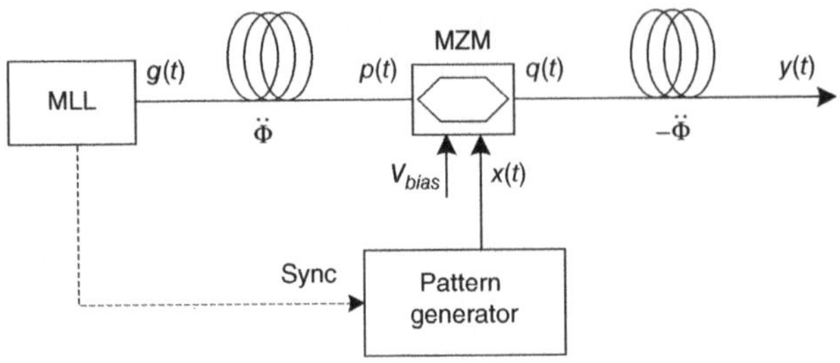

Figure 7.9 Schematic of a TPS system for arbitrary waveform generation.

$$
\begin{aligned}
p(t) &= g(t) \times \exp\left(j\frac{t^2}{2\ddot{\Phi}}\right) \\
&= \int_{-\infty}^{\infty} g(\tau) \times \exp\left[j\frac{(t-\tau)^2}{2\ddot{\Phi}}\right] d\tau \\
&= \exp\left(j\frac{t^2}{2\ddot{\Phi}}\right) \int_{-\infty}^{\infty} g(\tau) \times \exp\left(j\frac{t^2}{2\ddot{\Phi}}\right) \times \exp\left[j\left(-\frac{t}{\ddot{\Phi}}\right)\tau\right] d\tau \\
&\approx \exp\left(j\frac{t^2}{2\ddot{\Phi}}\right) \int_{-\infty}^{\infty} g(\tau) \times \exp\left[j\left(-\frac{t}{\ddot{\Phi}}\right)\tau\right] d\tau \\
&= \exp\left(j\frac{t^2}{2\ddot{\Phi}}\right) G(\omega)\big|_{\omega=\frac{t}{\ddot{\Phi}}}
\end{aligned}
\tag{7.30}
$$

where $G(\omega)$ is the Fourier transform of $g(t)$. The signal at the output of the MZM,

$$
q(t) = p(t) \times x(t) = \exp\left(j\frac{t^2}{2\ddot{\Phi}}\right) \times G\left(\frac{t}{\ddot{\Phi}}\right) \times x(t) \tag{7.31}
$$

where $x(t)$ is the input microwave signal to the MZM.

After propagating through the second dispersive element that has an opposite chromatic dispersion $\ddot{\Phi}$, we obtain the output temporal signal as the convolution of $q(t)$ with the impulse response of the dispersive element $\exp(-jt^2/2\ddot{\Phi})$,

$$
\begin{aligned}
y(t) &= q(t) \times \exp\left(-j\frac{t^2}{2\ddot{\Phi}}\right) = \int_{-\infty}^{\infty} q(\tau) \times \exp\left[-j\frac{(t-\tau)^2}{2\ddot{\Phi}}\right] d\tau \\
&= \exp\left(-j\frac{t^2}{2\ddot{\Phi}}\right) \times \int_{-\infty}^{\infty} q(\tau) \times \exp\left(-j\frac{\tau^2}{2\ddot{\Phi}}\right) \times \exp\left[j\left(\frac{t}{\ddot{\Phi}}\right)\tau\right] d\tau \\
&= \exp\left(-j\frac{t^2}{2\ddot{\Phi}}\right) \times \int_{-\infty}^{\infty} \exp\left(j\frac{\tau^2}{2\ddot{\Phi}}\right) \times G\left(\frac{\tau}{\ddot{\Phi}}\right) \times x(\tau) \times \exp\left(-j\frac{\tau^2}{2\ddot{\Phi}}\right) \times \exp\left[j\left(\frac{t}{\ddot{\Phi}}\right)\tau\right] d\tau \\
&= \exp\left(-j\frac{t^2}{2\ddot{\Phi}}\right) \times \int_{-\infty}^{\infty} \left[G\left(\frac{\tau}{\ddot{\Phi}}\right) \times x(\tau)\right] \times \exp\left[-j\left(-\frac{t}{\ddot{\Phi}}\right)\tau\right] d\tau \\
&= \exp\left(-j\frac{t^2}{2\ddot{\Phi}}\right) \times F\left[G\left(\frac{t}{\ddot{\Phi}}\right) \times x(t)\right]\Big|_{\omega=-\frac{t}{\ddot{\Phi}}} \\
&= 2\pi|\ddot{\Phi}| \exp\left(-j\frac{t^2}{2\ddot{\Phi}}\right) \times \left[g(-t) * X\left(-\frac{t}{\ddot{\Phi}}\right)\right]
\end{aligned}
\tag{7.32}
$$

where F denotes the Fourier transform operation, $*$ denotes the convolution operation, and $X(\omega)$ is the Fourier transform of $x(t)$.

As can be seen from (7.32), the output waveform is a convolution between the input optical pulse and the Fourier transform of the input modulation signal. If the input optical pulse is ultrashort, say, a unit impulse function, and the convolution of a function with a unit impulse is the function itself, then the generated waveform is simply the Fourier transform of the modulation signal. Based on the property of the Fourier transform, a slow waveform would lead to the generation of a fast waveform with a narrow temporal width.

The major challenge in implementing a TPS system is the complexity in the modulation stage since the modulation signal is usually complex-valued. For example, to generate a nonsymmetrical waveform, based on Fourier transform property, the modulation signal is complex-valued. Therefore, in the modulation stage, an amplitude modulator and a phase modulator must be employed, and the magnitude and phase information must be precisely synchronized, which makes the system extremely complicated. Considering the fact that the Fourier transform of a real and symmetrical waveform is still real and symmetrical, it is possible to use an MZM that is biased at the minimum transmission point to perform temporal spectrum shaping with a real signal that has both positive and negative values [33]. With this concept, waveforms such as square waves, rectangular waves, triangular waves, or doublets can be easily generated.

The TPS systems in [30–33] were employed for the generation of optical waveforms. In fact, the TPS technique can be extended for the generation of high-frequency microwave waveforms. Recently, the generation of a microwave waveform with a continuously tunable frequency by use of an unbalanced TPS system was proposed [34]. As can be seen from (7.32), in a conventional TPS system, the output waveform is the Fourier transformation of the input modulation signal. If a second Fourier transform is applied to the output waveform of a conventional TPS system, the finally generated waveform would be a scaled version of the input modulation signal. The second Fourier transformation can be performed by adding a third dispersive element, the entire system is then called an unbalanced TPS system. The schematic of the system is illustrated in Fig. 7.10.

In general, the values of the third-order dispersion of the two dispersive elements are small and negligible, and only the second-order dispersion or group velocity dispersion (GVD) is considered. The dispersive elements can then be characterized by the transfer function given by $H_i(\omega) = \exp(-j\ddot{\Phi}_i\omega^2/2)$, $(i = 1, 2)$, where $\ddot{\Phi}_1$ and $\ddot{\Phi}_2$ (ps^2) are the dispersion of the two

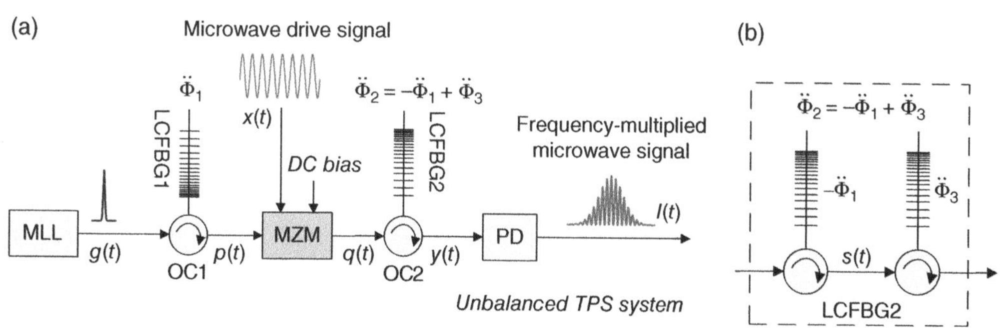

Figure 7.10 (a) An unbalanced TPS system for the generation of a microwave waveform with tunable carrier frequency. (b) The second linearly chirped fiber Bragg grating can be considered a cascade of two linearly chirped fiber Bragg gratings with the first one having the opposite dispersion to LCFBG1 and the second one having a residual dispersion.

dispersive elements. In the unbalanced TPS system, the dispersion values should satisfy $\ddot{\Phi}_1\ddot{\Phi}_2 < 0$, and $|\ddot{\Phi}_1| \neq |\ddot{\Phi}_2|$. Therefore, the entire unbalanced TPS system can be modeled as a typical TPS system with a pair of complementary dispersive elements, followed by a residual dispersive element with a transfer function of $H_3(\omega) = \exp(-j\ddot{\Phi}_3\omega^2/2)$, where $\ddot{\Phi}_3 = \ddot{\Phi}_1 + \ddot{\Phi}_2$ is defined as the residual dispersion.

Mathematically, when a continuous-wave $x(t) = \exp(j\omega_m t)$ with an angular frequency of ω_m is applied to an MZM, the modulated signal $e_{IM}(t)$ at the output of the MZM is given by $e_{IM}(t) = \exp(j\omega_0 t) \times \{\exp[j\beta x(t)] + \exp[-j\beta x(t) + j\phi_0]\}$. Based on Taylor expansion, we have

$$e_{IM}(t) = \exp(j\omega_0 t) \times \left\{ \sum_{n=0}^{\infty} \frac{[j\beta x(t)]^n}{n!} + e^{j\phi_0} \sum_{n=0}^{\infty} \frac{[-j\beta x(t)]^n}{n!} \right\} \tag{7.33}$$

where β is the phase modulation index, and ϕ_0 is a phase shift introduced by the DC bias. To make the MZM operate at the minimum transmission point for the suppression of the optical carrier, a DC voltage is applied to the MZM to introduce a π phase shift (i.e. $\phi_0 = \pi$) between the two arms of the MZM. If the modulation index β is sufficiently small, $e_{IM}(t)$ can be approximated to be $e_{IM}(t) \approx \exp(j\omega_0 t) \times [2j\beta x(t)]$. Therefore, the modulation function of the MZM biased at the minimum transmission point is $2j\beta x(t)$.

It is known that if the first dispersive element has an adequate dispersion, i.e. $|\Delta t_0{}^2/\ddot{\Phi}| \ll 1$, Δt_0 is the pulse width of the input optical pulse $g(t)$, the output signal of the typical TPS system, $s(t)$, shown in Fig. 7.10(b), is the convolution between the input signal and the Fourier transform of the modulation function,

$$s(t) \propto g(t) * E_{IM}(\omega)|_{\omega = t/\ddot{\Phi}_1} = J_1(\beta)[g(t - T_1) + g(t + T_1)] \tag{7.34}$$

where $E_{IM}(\omega)$ is the Fourier transform of $e_{IM}(t)$, $*$ denotes the convolution operation, and $T_1 = |\omega_m \ddot{\Phi}_1|/2\pi$. Therefore, two time-delayed replicas of the input pulse are generated at the output of the typical TPS system, which correspond to the two optical sidebands at the output of the DSB-SC modulator.

The electrical field at the output of the entire unbalanced TPS system, $y(t)$, is obtained by propagating $s(t)$ through the residual dispersive element. If $|T_1{}^2/2\ddot{\Phi}_3| \ll 1$ is satisfied, then $y(t)$ can be approximated by the real-time FT of $s(t)$ in the residual dispersive element [34],

$$\begin{aligned} y(t) &\approx \exp\left[jt^2/(2\ddot{\Phi}_3)\right] S(\omega)|_{\omega = t/\ddot{\Phi}_3} \\ &= \exp\left[jt^2/(2\ddot{\Phi}_3)\right] J_1(\beta) G(t/\ddot{\Phi}_3) \cos(tT_1/\ddot{\Phi}_3) \end{aligned} \tag{7.35}$$

where $G(\omega)$ is the Fourier transform of the input pulse $g(t)$.

The current at the output of the PD is proportional to the intensity of the input electrical field, which is given by

$$I(t) = R|y(t)|^2 = K \exp\left(-\frac{t^2}{\tau^2}\right)\left[1 + \cos\left(2\pi \frac{2T_1}{\ddot{\Phi}_3} t\right)\right] \tag{7.36}$$

where R is the responsivity of the PD, $K = RJ_1{}^2(\beta)\pi\Delta t_0{}^2/2$ is a time-independent constant, and $\tau = \sqrt{2}\Delta\ddot{\Phi}/\Delta t_0$ is the output pulse width. A frequency-multiplied microwave signal is generated, which is a pulsed microwave signal with a Gaussian envelope. The new carrier frequency is

$$\omega_{RF} = 2\pi|2T_1/\Delta\ddot{\Phi}| = \omega_m|2\ddot{\Phi}_1/\ddot{\Phi}_3| \tag{7.37}$$

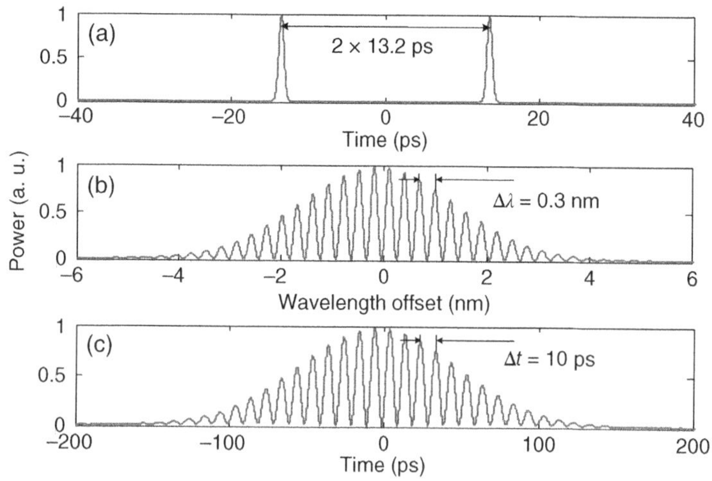

Figure 7.11 Simulation results. (a) Signal at the output of the typical TPS system. (b) The optical spectrum of the signal in (a). (c) The frequency-multiplied microwave signal at the output of the entire system.

From (7.37), we can conclude that the frequency multiplication factor $M = \omega_{RF}/\omega_m = 2|\ddot{\Phi}_1/\ddot{\Phi}_3|$ is determined by both the stretching dispersion $\ddot{\Phi}_1$ and the residual dispersion $\ddot{\Phi}_3$. Figure 7.11 shows the simulation results of an unbalanced TPS system for the generation of a microwave waveform with tunable carrier frequency. Note that the MZM in the unbalanced TPS system is biased at the minimum transmission point to suppress the optical carrier. If the MZM is biased at the quadrature point, double-sideband modulation would result. The use of double-sideband modulation would cause the dispersion-induced power cancellation, leading to a reduced fringe visibility, as shown in Fig. 7.12.

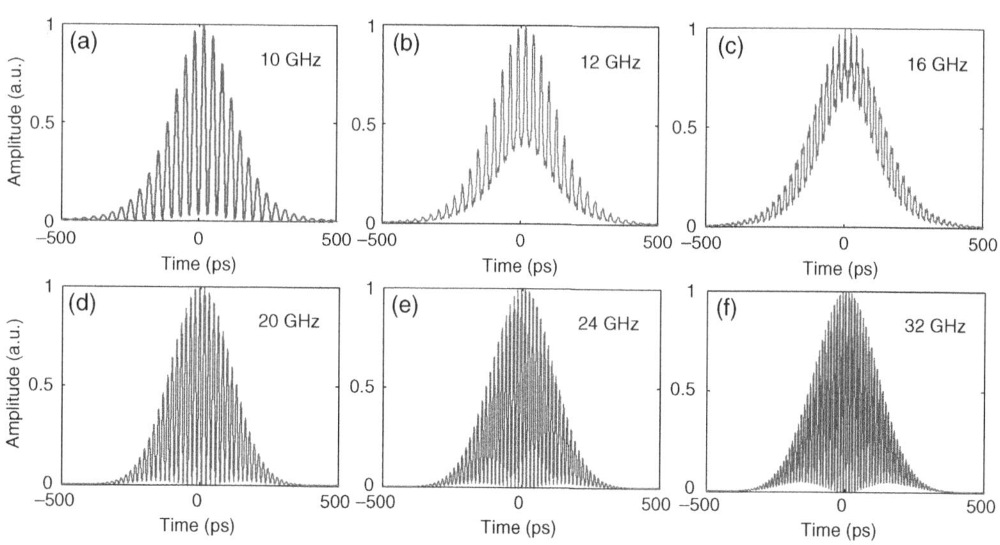

Figure 7.12 Simulated output microwave waveforms based on double-sideband modulation with a carrier frequency of (a) 10 GHz, (b) 12 GHz, and (c) 16 GHz, and DSB-SC modulation with a carrier frequency of (d) 20 GHz, (e) 24 GHz, and (f) 32 GHz.

If the dispersion up to the third order is considered, then a new wavelength-to-time mapping function would be used, which is given in (7.26). Then, the signal $q(t)$ at the output of the MZM is given by

$$q(t) = e_{\mathrm{IM}}(t) \times G(\omega)\big|_{\omega = \frac{t}{\ddot{\Phi}_1} - \frac{\dddot{\Phi}_1 t^2}{2\ddot{\Phi}_1^3}} = [2j\beta x(t)] \times G(\omega)\big|_{\omega = \frac{t}{\ddot{\Phi}_1} - \frac{\dddot{\Phi}_1 t^2}{2\ddot{\Phi}_1^3}} \tag{7.38}$$

Since the first and second dispersive elements have conjugate dispersion and the third-order dispersion of the dispersive element is very small, the signal $s(t)$ at the output of the second dispersive element, shown in Fig. 7.10(b), is given by

$$s(t) = F[q(t)]\Big|_{\omega = \frac{t}{\ddot{\Phi}_2} - \frac{\dddot{\Phi}_2 t^2}{2\ddot{\Phi}_2^3}} = \frac{2j\beta}{|\ddot{\Phi}_2|} \times g(t) * X(\omega)\Big|_{\omega = \frac{t}{\ddot{\Phi}_2} - \frac{\dddot{\Phi}_2 t^2}{2\ddot{\Phi}_2^3}} \tag{7.39}$$

with its spectrum $S(\omega)$ given by

$$S(\omega) = 2j\beta G(\omega) \times x\left(\ddot{\Phi}_2 \omega + \frac{\dddot{\Phi}_2 \omega^2}{2}\right) \tag{7.40}$$

The output signal in the frequency domain is given by

$$Y(\omega) = P(\omega) \times H_3(\omega) = P(\omega) \times \exp\left[-j\left(\frac{\ddot{\Phi}_3 \omega^2}{2} + \frac{\dddot{\Phi}_3 \omega^3}{6}\right)\right] \tag{7.41}$$

where $H_3(\omega)$ is the transfer function of the residual dispersive element, $\ddot{\Phi}_3$ and $\dddot{\Phi}_3$ are the second-order and the third-order dispersion of the residual dispersive element, respectively.

By using the nonlinear wavelength-to-time mapping function given in (7.26), the output signal in the time domain is given by

$$y(t) = F[s(t)]\big|_{\omega = \frac{t}{\ddot{\Phi}_3} - \frac{\dddot{\Phi}_3 t^2}{2\ddot{\Phi}_3^3}} \propto G(\omega) x\left(\ddot{\Phi}_2 \omega + \frac{\dddot{\Phi}_2 \omega^2}{2}\right)\Big|_{\omega = \frac{t}{\ddot{\Phi}_3} - \frac{\dddot{\Phi}_3 t^2}{2\ddot{\Phi}_3^3}} \tag{7.42}$$

If the optical signal at the output of the third dispersive element is applied to a PD, we have the output current, given by

$$\begin{aligned} i(t) &= R|y(t)|^2 \propto [G(\omega)]^2 \left[x\left(\ddot{\Phi}_2 \omega + \frac{\dddot{\Phi}_2 \omega^2}{2}\right)\right]^2 \Bigg|_{\omega = \frac{t}{\ddot{\Phi}_3} - \frac{\dddot{\Phi}_3 t^2}{2\ddot{\Phi}_3^3}} \\ &= \pi\tau_0^2 \exp\left[-\frac{\tau_0^2}{4}\left(\frac{t}{\ddot{\Phi}_3} - \frac{\dddot{\Phi}_3 t^2}{2\ddot{\Phi}_3^3}\right)^2\right] \times \left[x\left(\frac{\ddot{\Phi}_2 t}{\ddot{\Phi}_3} - \frac{\ddot{\Phi}_2 \dddot{\Phi}_3 t^2}{2\ddot{\Phi}_3^3} + \frac{\dddot{\Phi}_2}{2}\left(\frac{t^2}{\ddot{\Phi}_3^2} - \frac{\dddot{\Phi}_3 t^3}{\ddot{\Phi}_3^4} + \frac{\dddot{\Phi}_3^2 t^4}{4\ddot{\Phi}_3^6}\right)\right)\right]^2 \end{aligned} \tag{7.43}$$

where R is again the responsivity of the PD. Usually, the third-order dispersion, $\dddot{\Phi}$, is much smaller than the second-order dispersion $\ddot{\Phi}$, and the condition $|T_1^2/2\ddot{\Phi}_3| \ll 1$ should be satisfied in the system where $T_1 = |\omega_m \ddot{\Phi}_1|/2\pi$, we also have $\left((t_R^2)/(\ddot{\Phi}_3^2)\right) \gg \left((\dddot{\Phi}_3 t_R^3)/(\ddot{\Phi}_3^4)\right)$ and $\left((t_R^2)/(\ddot{\Phi}_3^2)\right) \gg \left((\dddot{\Phi}_3^2 t_R^4)/(4\ddot{\Phi}_3^6)\right)$, Eq. (7.43) is then simplified to

$$i(t) \propto \exp\left[-\frac{\tau_0^2}{4}\left(\frac{t}{\ddot{\Phi}_3} - \frac{\dddot{\Phi}_3 t^2}{2\ddot{\Phi}_3^3}\right)^2\right] \times \left\{x\left[\frac{\ddot{\Phi}_2}{\ddot{\Phi}_3} t + \left(\frac{\dddot{\Phi}_2 \ddot{\Phi}_3 - \ddot{\Phi}_2 \dddot{\Phi}_3}{2\ddot{\Phi}_3^3}\right) t^2\right]\right\}^2 \tag{7.44}$$

To generate a chirped microwave pulse, we assume that the microwave modulation signal applied to the MZM is also a sinusoidal waveform with a frequency of ω_m, and the third-order dispersion of the three dispersive elements cannot be ignored. It can be seen from (7.44) that the output signal is also a product of a Gaussian-like function with a chirped microwave waveform. Since the angular frequency should be kept positive, the instantaneous angular frequency of the chirped microwave waveform is approximately given as

$$\omega_{RF} \cong 2\omega_m \left| \frac{\ddot{\Phi}_2}{\ddot{\Phi}_3} + \left(\frac{\dddot{\Phi}_2 \ddot{\Phi}_3 - \ddot{\Phi}_2 \dddot{\Phi}_3}{\ddot{\Phi}_3^3} \right) t \right| \tag{7.45}$$

Equation (7.45) shows that a chirped microwave waveform can be generated if both the second and the third dispersive elements have nonzero third-order dispersion. The chirp rate is given by

$$CR_{RF} = 2\omega_m \left(\frac{\dddot{\Phi}_2 \ddot{\Phi}_3 - \ddot{\Phi}_2 \dddot{\Phi}_3}{\ddot{\Phi}_3^3} \right) \times \operatorname{sgn}\left(\frac{\ddot{\Phi}_2}{\ddot{\Phi}_3} \right) \tag{7.46}$$

where sgn is the sign function. From (7.46), it can be seen that the chirp rate is determined by the values of the third-order dispersion $\dddot{\Phi}_2$ and $\dddot{\Phi}_3$. When $t = 0$, the frequency is the central frequency of the generated microwave waveform which is also equal to the angular frequency ω_{RF} in (7.37). Therefore, a microwave waveform with a tunable central frequency and tunable chirp rate can be generated by tuning the second-order dispersion and third-order dispersion of the second and third dispersive elements. By varying the third-order dispersion of a nonlinearly chirped FBG (NLC-FBG) based on a strain-gradient beam tuning technique, the chirp rate can be continuously tunable. On the other hand, if the frequency chirp in the generated microwave waveform is not desired, the frequency chirp can be made zero by tuning the dispersion to make $(\dddot{\Phi}_2 \ddot{\Phi}_3 - \ddot{\Phi}_2 \dddot{\Phi}_3)$ equal to zero.

Figure 7.13 shows the experimental results of an unbalanced TPS system for the generation of microwave waveform without and with chirp [35]. In the experiment, the frequency of the input

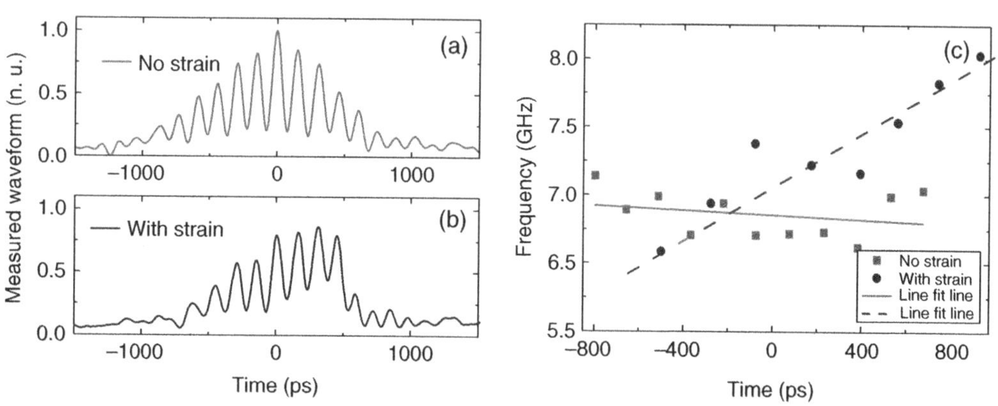

Figure 7.13 Measured microwave waveforms for an unbalanced system using a chirped FBG that is glued on a cantilever beam (a) with no strain, (b) with strain, and (c) the corresponding frequency chirps with and without mechanical force applied to the free end of the cantilever beam.

microwave waveform is 1.3 GHz and the first dispersive element is a dispersion compensating fiber (DCF) with a dispersion value of $-485.07\ \text{ps}^2$. A chirped FBG glued on a cantilever beam was employed in the system connected to the output of the MZM. When no force is applied to the cantilever beam, the total second-order dispersion and third-order dispersion after the MZM are $-485.07\ \text{ps}^2$ and $3.34\ \text{ps}^3$, respectively. Since the unbalanced TPS system is a linear time-invariant system, the values of the second-order dispersion of the three dispersive elements in the entire system are $\ddot{\Phi}_1 = 772.62\ \text{ps}^2$, $\ddot{\Phi}_2 = -772.62\ \text{ps}^2$, and $\ddot{\Phi}_3 = 287.54\ \text{ps}^2$. Based on $M = \omega_{RF}/\omega_m = 2\left|\ddot{\Phi}_1/\Delta\ddot{\Phi}\right|$, the multiplication factor is calculated to be 5.38. Since the frequency of the input microwave waveform is 1.3 GHz, the central frequency of the generated microwave waveform should be 6.99 GHz. In addition, the values of the third-order dispersion of the three dispersive elements in the entire system are $\dddot{\Phi}_1 = -4.44\ \text{ps}^3$, $\dddot{\Phi}_2 = 4.44\ \text{ps}^3$, and $\dddot{\Phi}_3 = -1.09\ \text{ps}^3$. The generated microwave waveform is shown in Fig. 7.13(a). The central frequency is measured to be 6.75 GHz, as shown in Fig. 7.13(c), which agrees well with the theoretically calculated value of 6.99 GHz. Since the third-order dispersion after the MZM is $-1.09\ \text{ps}^3$, according to (7.46), the chirp rate is calculated to be -4.75×10^{-2} GHz/ns, which is close to the chirp rate of -5.35×10^{-2} GHz/ns estimated from the generated waveform shown in Fig. 7.13(a).

When a force is applied to the cantilever beam, the values of the second-order dispersion and the third-order dispersion after the MZM are $-517.02\ \text{ps}^2$ and $-2.81\ \text{ps}^3$, respectively. The values of the second-order dispersion of the three dispersive elements in the entire system are $\ddot{\Phi}_1 = 772.62\ \text{ps}^2$, $\ddot{\Phi}_2 = -772.62\ \text{ps}^2$,$\ddot{\Phi}_3 = 255.61\ \text{ps}^2$. Again, the multiplication factor is calculated to be 6.04. Since the frequency of the input microwave waveform is 1.3 GHz, a central frequency of the generated microwave waveform should be 7.85 GHz. The central frequency of the generated waveform estimated from the waveform shown in Fig. 7.13(b) is 7.5 GHz, which is close to the theoretical value of 7.85 GHz. The values of the third-order dispersion of the three dispersive elements in the entire system are $\dddot{\Phi}_1 = -4.44\ \text{ps}^3$, $\dddot{\Phi}_2 = 4.44\ \text{ps}^3$, and $\dddot{\Phi}_3 = 5.75\ \text{ps}^3$. The chirp rate calculated based on (7.46) is 0.868 GHz/ns, which is close to the chirp rate of 0.715 GHz/ns obtained from the generated chirped waveform shown in Fig. 7.13(b). In addition, the chirp rate of the experimentally generated microwave waveform is also reversed due to the change of the sign of the third-order dispersion, as shown in Fig. 7.13(c).

7.5 Microwave Waveform Generation Based on a Photonic Microwave Delay-Line Filter

The implementation of microwave delay-line filters has been a topic of interest in the last few years [36, 37]. To avoid optical interferences which are extremely sensitive to environmental changes, a photonic microwave delay-line filter is usually implemented in the incoherent regime based on incoherent detection. The major limitation of an incoherent microwave delay-line filter is that the tap coefficients are all positive, which would limit the filter to operate as a low-pass filter only. Numerous techniques have been proposed to implement a microwave delay-line filter with negative coefficients, including the use of differential detection [38], biasing a pair of MZMs at the opposite slopes [39], cross-gain modulation in a semiconductor optical amplifier [1], polarization modulation [40], and phase-modulation to intensity-modulation conversion [41]. A comprehensive overview of these techniques can be found in [42]. The use of a microwave delay-line filter for the generation of ultra-wideband (UWB) impulse waveform has been demonstrated [43–46].

However, all the filters implemented based on the techniques in [1, 38–41], including the filters for the generation of UWB waveforms [43–46], have linear group delay response. For microwave waveform generation, such as the generation of a chirped microwave waveform, a group delay response that is nonlinear is required. It is known for a delay-line filter with nonlinear group delay response, the filter should have tap coefficients that are complex, which is difficult to implement, especially for a delay-line filter with a large number of taps [47–49]. Recently, a new concept to implement a microwave delay-line filter with arbitrary group delay response based on a delay-line structure with nonuniformly spaced taps was reported [50]. It was theoretically proved that a delay-line filter with an arbitrary frequency response in a specific resonance band can be achieved by introducing additional time delays to the taps, making the taps nonuniformly spaced.

It is known that a regular, uniformly spaced microwave delay-line filter has an impulse response $h_0(t)$ given by

$$h_0(t) = \sum_{k=0}^{N-1} \alpha_k \delta(t - kT) \tag{7.47}$$

where N is the tap number, α_k is the tap coefficient of the kth tap, $T = 2\pi/\Omega$ is the time-delay difference between two adjacent taps, and Ω is the FSR of the filter. The expression of $h_0(t)$ can be expressed in another form

$$h_0(t) = a(t) s_0(t) \tag{7.48}$$

where $a(t)$ is the coefficient profile which can be complex-valued,

$$\alpha_k = a(kT) \text{ and } a(t) = 0 \text{ if } t < 0 \text{ or } t \geq NT \tag{7.49}$$

and $s(t)$ is the sampling function given by

$$s(t) = \sum_{k} \delta(t - kT) \tag{7.50}$$

Applying the Fourier Transfer to (7.47), we have the frequency response of the filter,

$$H_0(\omega) = \sum_{k=0}^{N-1} \alpha_k \exp\left(-jk\frac{2\pi}{\Omega}\omega\right) \tag{7.51}$$

It is known that $H_0(\omega)$ has a multi-channel frequency response with adjacent channels separated by an FSR, with the mth channel located at $\omega = m\Omega$.

In a regular photonic microwave delay-line filter based on incoherent detection, the coefficients are usually all positive, or special designs have to be incorporated to generate negative or complex coefficients [42]. However, a phase term can be introduced to a specific coefficient by adding an additional time delay at the specific tap, which is termed time-delay-based phase shift [50]. For example, at $\omega = m\Omega$, a time-delay shift of $\Delta\tau$ will generate a phase shift given by $\Delta\varphi = -\Delta\tau \times m\Omega$. Note that such a phase shift is frequency-dependent, which is accurate only for the frequency at $m\Omega$, but approximately accurate for a narrow frequency band at around $m\Omega$. For most of the applications, the filter is designed to have a very narrow frequency band. Therefore, for the frequency band of interest, the phase shift can be considered constant over the entire bandwidth. As a result, if the mth bandpass response, where $m \neq 0$, is considered, one can then achieve the desired phase shift at the kth tap by adjusting the time-delay shift by $\Delta\tau_k$. Considering the time-delay shift of $\Delta\tau_k$, one can get the frequency response of the nonuniformly spaced delay-line filter at around $\omega = m\Omega$,

$$\begin{aligned} H_N(\omega) &= \sum_{k=0}^{N-1} \alpha_k \exp\left[-j\left(k\frac{2\pi}{\Omega} + \Delta\tau_k\right)\omega\right] \\ &= \sum_{k=0}^{N-1} \alpha_k \exp(-j\omega\Delta\tau_k) \times \exp\left(-jk\frac{2\pi}{\Omega}\omega\right) \\ &\approx \sum_{k=0}^{N-1} [\alpha_k \exp(-jm\Omega\Delta\tau_k)] \times \exp\left(-jk\frac{2\pi}{\Omega}\omega\right) \end{aligned} \tag{7.52}$$

As can be seen from (7.52), one can get an equivalent phase shift for each tap coefficient. Specifically, if the desired phase shift for the kth tap is φ_k, the total time delay τ_k for the kth tap is $\tau_k = kT - \varphi_k/m\Omega$. As a result, if the time delay of each tap is adjusted, the filter coefficients would have the required phase shifts to generate the required passband with the desired bandpass characteristics.

For a nonuniformly spaced delay-line filter with a spectral response at the $m\Omega$ that is identical to that of a regular delay-line filter with a coefficient profile given by $a(t)$, the time delays and the coefficients can be calculated by [51]

$$\tau_k + \frac{\varphi(\tau_k)}{m\Omega} = kT \tag{7.53a}$$

$$\beta_k = \left|\frac{a(\tau_k)}{1 + \varphi'(\tau_k)/m\Omega}\right| \tag{7.53b}$$

where $\phi(t)$ is the phase term of $a(t)$. As can be seen to realize the desired bandpass response, if a regular delay-line filter is used, the tap coefficients are determined by (7.49), which may be negative or complex-valued; however, if a nonuniformly spaced delay-line filter is used, the time delays and the tap coefficients are determined by (7.53a) and (7.53b), and the filter has positive-only coefficients.

Since the filter can be designed to have an arbitrary spectral response, the employment of the filter for the generation of an arbitrary microwave waveform can be realized.

Figure 7.14 shows a delay-line filter for the generation of phase-coded microwave waveform [51, 52]. To generate the required phase code using a delay-line filter with all positive coefficients, the filter should have nonuniformly spaced taps. The desired phase coding is implemented by adjusting the time-delay differences between the adjacent taps. According to (7.53a), we have

$$\tau_k = T \times \left(k - \frac{\varphi_k}{2\pi m}\right) = \left(km - \frac{\varphi_k}{2\pi}\right) T_0, \quad k = 0, 1, 2, \ldots, N-1 \tag{7.54}$$

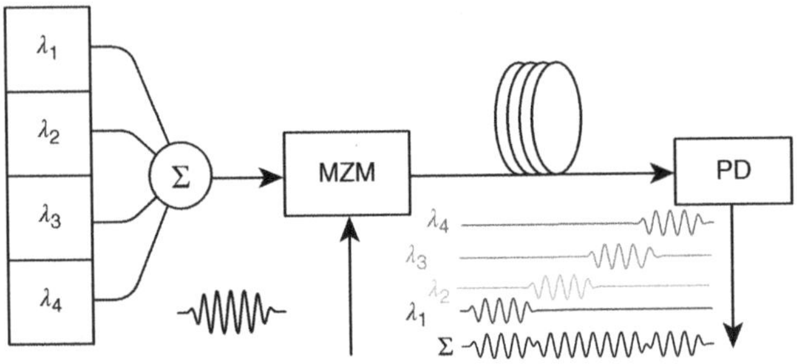

Figure 7.14 A photonic microwave delay-line filter for microwave phase coding.

where T is the time-delay difference between two adjacent taps for a uniformly spaced filter, and T_0 is the period of the microwave carrier.

For example, for a four-tap filter, we have $N = 4$. If $m = 4$ is selected, for a code pattern of $\{0, \pi, \pi, 0\}$, the time delays should be $\{0, 7/8T, 15/8T, 3T\}$.

As a design example, a binary phase-coded microwave signal with a 13-chip Barker code, [+1, +1, +1, +1, +1, −1, −1, +1, +1, −1, +1, −1, +1], is generated [51]. The Barker codes are usually used in direct-sequence spread-spectrum communication systems and pulse compression radar systems thanks to the excellent correlation performance. In the design, the carrier frequency is 40 GHz and $m = 4$, i.e. the chip rate is 10 GHz, and T_0 is 25 ps. Based on (7.54), the time delays of all the taps are calculated, which are given [0, 4, 8, 12, 16, 19.5, 23.5, 28, 32, 35.5, 40, 43.5, 48] × 25 ps. If the input microwave signal is a super-Gaussian pulse with a full width at half maximum (FWHM) of 100 ps, the output of the filter is calculated which is shown in Fig. 7.15.

A chirped microwave waveform can also be generated by using a nonuniformly spaced delay-line filter [51, 53]. If a broadband chirp-free microwave pulse is passing through a microwave delay line filter with a quadratic phase response or equivalently, a linear group-delay response, a pulse burst with increasing or decreasing pulse spacing is generated. As it was proved in [12] that a pulse-position-modulated pulse burst would have a multi-channel spectral response, with one channel having a spectrum corresponding to the desired phase modulated microwave signal. By using a microwave bandpass filter to select the channel of interest, a frequency-chirped microwave signal would be generated. Figure 7.16 shows a delay-line filter that can be designed to have nonuniformly

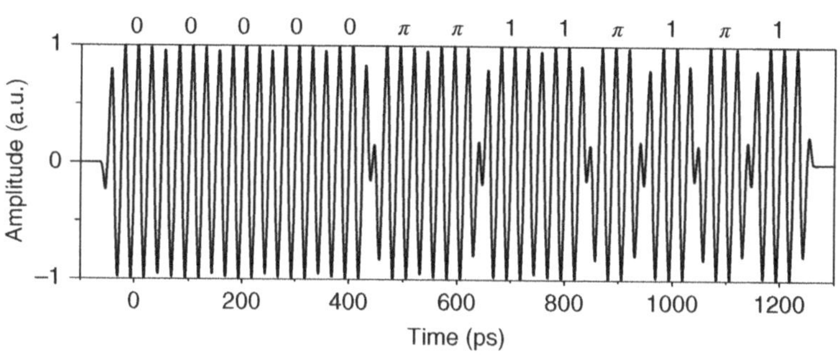

Figure 7.15 Phase-coded microwave pulse with a 13-chip Barker code generated by a nonuniformly spaced delay-line filter.

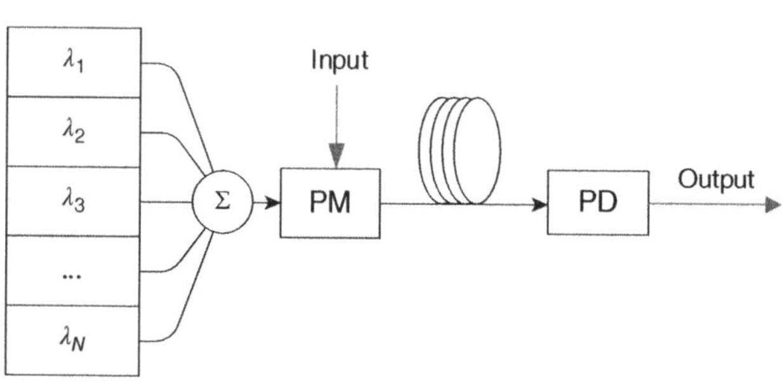

Figure 7.16 A nonuniformly spaced delay-line filter for the generation of a chirped microwave waveform. The baseband resonance is eliminated by the notch at DC due to the PM-IM conversion. PM: phase modulator. PD: photodetector.

spaced taps to produce a quadratic phase response. Note that instead of using an MZM, a phase modulator is employed. It was demonstrated that the phase-modulation to intensity-modulation conversion in a dispersive element will generate a notch at the DC, which is used to eliminate the resonance at the DC. The nonuniform tap spacing can be achieved in the system by tuning the wavelength spacing. For example, if the total dispersion of the dispersive element is χ(ps/nm), then the wavelengths of the multiple wavelength sources can be calculated by

$$\lambda_k = \lambda_0 + \frac{\tau_k - \tau_0}{\chi} \tag{7.55}$$

where λ_0 is the wavelength for the 0th tap.

The generation of a chirped microwave pulse using a five-tap nonuniformly spaced microwave FIR filter was experimentally demonstrated. The desired pulse has an FWHM of 550 ps, a mean period of 110 ps, and a chirp rate of 13.2 GHz/ns, as shown in the dotted line in Fig. 7.17(a). The input pulse is a chirp-free Gaussian pulse generated in the experiment by a pattern generator with an FWHM of 65 ps. The five-tap delay line filter is designed based on (7.53). The time delays of the taps, calculated based on (7.54), are $T \times [-2.31, -1.08, 0.00, 0.81, 1.51]$, where $T = 110$ ps. The wavelengths from the multi-wavelength source are then calculated based on (7.55). In the experiment, λ_0 is 1543.30 nm, and the five wavelengths are set as [1542.70, 1543.02, 1543.30, 1543.51, 1543.69] nm. In the experiment, the output power of each laser is controlled such that the tap coefficients, α_k, are optimized to minimize the errors between the desired and the generated chirped pulses. The tap coefficients are calculated to be [1.0, 1.0, 1.0, 1.5, 1.4]. The dispersive fiber is a length of 25 km standard SMF with a total dispersion of about 425 ps/nm. The generated pulse is then measured by a high-speed oscilloscope, which is shown as the solid line in Fig. 7.17(a). A chirped RF pulse having a shape close to the theoretical microwave chirped pulse is generated. A slight distortion in the pulse shape is due to the small tap number. It is known that a chirped microwave pulse can be compressed at a receiver using a microwave matched filter. To demonstrate the pulse compression performance, the correlation between the measured and the reference chirped microwave pulses (the two curves in Fig. 7.17(a)) is calculated, and the result is shown in Fig. 7.17(b). It is clearly seen that the microwave pulse is compressed, which confirms that the generated microwave pulse is chirped.

If the input pulse to the delay-line filter has a spectral response that is complex conjugate to that of the delay-line filter, then the delay-line filter becomes a matched filter, which can be used for pulse compression for applications in radar and microwave imagining systems. The use of the

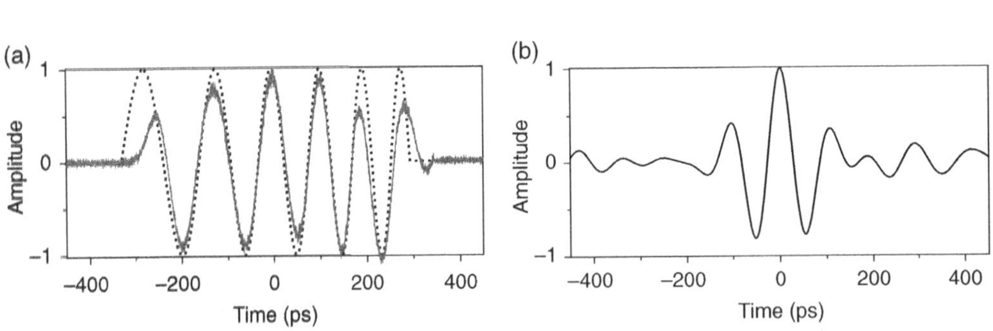

Figure 7.17 (a) Solid line: the generated chirped RF pulse. Dotted line: the desired chirped RF pulse. (b) The correlation between the measured and the reference pulses.

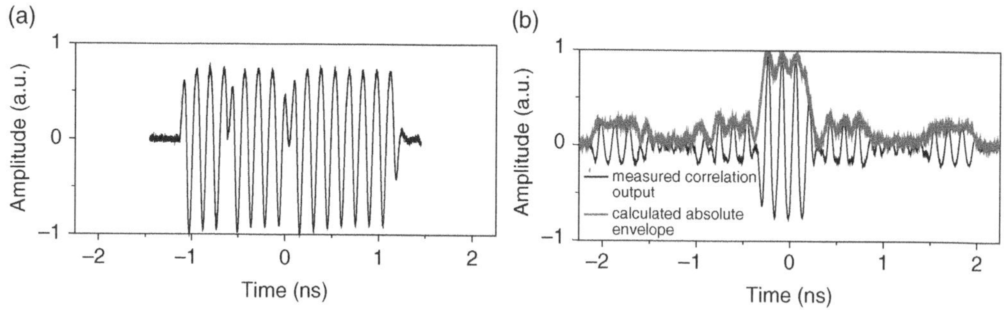

Figure 7.18 (a) The input phase-coded microwave signal. (b) The measured correlation output with the calculated absolute envelope.

delay-line filter for the compression of a phase-coded microwave pulse was demonstrated [51, 54]. Again, the system setup shown in Fig. 7.16 can be used. The dispersive fiber is a 25-km standard SMF. The output is measured by a high-speed oscilloscope. The input signal has a 6.75 GHz carrier, which is phase coded by a 4-chip Barker code [+1, 1, +1, +1]. In our design, $m = 4$ so that the chip rate is 1.6875 GHz. Then, based on (7.54), the time delays for the four taps are calculated which are given as [0.5926, 1.1852, 1.8519, 2.3704] ns. The wavelengths of the four lasers are then calculated according to (7.55), which are given [1543.10, 1544.49, 1546.06, 1547.28] nm. The original input phase-coded microwave signal is shown in Fig. 7.18(a). When the phase-coded signal is sent to the nonuniformly spaced delay-line filter, an auto-correlation is obtained at the output of the filter, which is shown in Fig. 7.18(b). It is clearly seen that the phase-coded signal is correctly decoded.

7.6 Conclusion

Arbitrary microwave waveforms can find numerous applications such as in radar and communication systems. In most of the applications, microwave arbitrary waveforms are generated using digital electronics, which provides advantages such as high flexibility and low cost. Due to the limited sampling speed of digital electronics, the generation of high-frequency and broadband microwave waveforms is still a challenge. A solution is to use photonics. The high frequency and broad bandwidth offered by modern optics allow the generation of microwave waveforms with much higher frequency and wider bandwidth. In this chapter, the techniques to generate arbitrary microwave waveforms based on fiber-optics were reviewed. Four different techniques were discussed: direct space-to-time pulse shaping; spectral-shaping and wavelength-to-time mapping; TPS; and microwave pulse generation based on photonic microwave delay-line filters. The advantages of these techniques compared with free-space-optics-based solutions include simplicity, low cost, and small size. In addition, fiber-optics-based systems have a high potential for photonic integration. The key limitation of these techniques is the poor reconfigurability. Although some of the techniques could provide limited waveform tunability [18, 34], in general, the reconfigurability of the systems is low. For practical applications, a system with large reconfigurability should be employed. The use of photonic integrated circuits (PICs), such as the silicon-photonic chip-based spectral shaper, as was demonstrated [20, 55–58] and reconfigurable InP-based PICs [59], provides a potential solution for low-cost and large reconfigurable microwave waveform generation.

References

1 Coppinger, F., Yegnanarayanan, S., Trinh, P.D., and Jalali, B. (1997). All-optical RF filter using amplitude inversion in a semiconductor optical amplifier. *IEEE Trans. Microw. Theory Tech.* 45: 1473–1477.

2 Leaird, D.E. and Weiner, A.M. (1999). Femtosecond optical packet generation via a direct space-to-time pulse shaper. *Opt. Lett.* 24: 853–855.

3 Leaird, D.E. and Weiner, A.M. (2001). Femtosecond direct space-to-time pulse shaping. *IEEE J. Quantum Electron.* 37: 494–504.

4 Leaird, D.E. and Weiner, A.M. (2000). Chirp control in the direct space-to-time pulse shaper. *Opt. Lett.* 25: 850–852.

5 McKinney, J.D., Leaird, D.E., and Weiner, A.M. (2002). Millimeter-wave arbitrary waveform generation with a direct space-to-time pulse shaper. *Opt. Lett.* 27: 1345–1347.

6 McKinney, J.D., Seo, D.S., and Weiner, A.M. (2003). Photonically assisted generation of continuous arbitrary millimeter electromagnetic waveforms. *Electron. Lett.* 39: 309–310.

7 McKinney, J.D., Seo, D., Leaird, D.E., and Weiner, A.M. (2003). Photonically assisted generation of arbitrary millimeter-wave and microwave electromagnetic waveforms via direct space-to-time optical pulse shaping. *J. Lightw. Technol.* 21: 3020–3028.

8 Xiao, S., McKinney, J.D., and Weiner, A.M. (2004). Photonic microwave arbitrary waveform generation using a virtually-imaged phase-array (VIPA) direct space-to-time pulse shaper. *IEEE Photonics Technol. Lett.* 16: 1936–1938.

9 Leaird, D.E. and Weiner, A.M. (2004). Femtosecond direct space-to-time pulse shaping in an integrated-optic configuration. *Opt. Lett.* 29: 1551–1553.

10 Leaird, D.E., Weiner, A.M., Shen, S. et al. (2001). High repetition rate femtosecond WDM pulse generation using direct space-to-time pulse shapers and arrayed waveguide gratings. *Opt. Quant. Electron.* 13: 811–826.

11 Vega, A., Leaird, D.E., and Weiner, A.M. (2010). High-speed direct space-to-time pulse shaping with 1 ns reconfiguration. *Opt. Lett.* 35: 1554–1556.

12 Dai, Y. and J.Y. (2008). Arbitrary phase-modulated RF signal generation based on optical pulse position modulation. *J. Lightw. Technol.* 26: 3329–3336.

13 Carlson, A.B. and Crilly, P.B. (2010). *Communication Systems: An Introduction to Signals and Noise in Electrical Communication*, 5e. McGraw-Hill.

14 Muriel, M.A., Azaña, J., and Carballar, A. (1999). Real-time Fourier transformer based on fiber gratings. *Opt. Lett.* 24: 1–3.

15 Chou, J., Han, Y., and Jalali, B. (2003). Adaptive RF-photonic arbitrary waveform generator. *IEEE Photonics Technol. Lett.* 15: 581–583.

16 Lin, I.S., McKinney, J.D., and Weiner, A.M. (2005). Photonic synthesis of broadband microwave arbitrary waveforms applicable to ultra-wideband communication. *IEEE Microw. Wireless Compon. Lett.* 15: 226–228.

17 Wang, C. and J.Y. (2008). Photonic generation of chirped microwave pulses using superimposed chirped fiber Bragg gratings. *IEEE Photonics Technol. Lett.* 20: 882–884.

18 Wang, C. and J.Y. (2009). Chirped microwave pulse generation based on optical spectral shaping and wavelength-to-time mapping using a Sagnac-loop mirror incorporating a chirped fiber Bragg grating. *J. Lightw. Technol.* 27: 3336–3341.

19 Wang, C. and J.Y. (2009). Simultaneous optical spectral shaping and wavelength-to-time mapping for photonic microwave arbitrary waveform generation. *IEEE Photonics Technol. Lett.* 21: 793–795.

20 Khan, M.H., Shen, H., Xuan, Y. et al. (2010). Ultrabroad-bandwidth arbitrary radiofrequency waveform generation with a silicon photonic chip-based spectral shaper. *Nature Photonics* 4: 117–122.

21 Wang, C. and J.Y. (2010). Large time-bandwidth product microwave arbitrary waveform generation using a spatially discrete chirped fiber Bragg grating. *J. Lightw. Technol.* 28: 1652–1660.

22 Chi, H. and J.Y. (2008). Chirped RF pulse generation based on optical spectral shaping and wavelength-to-time mapping using a nonlinearly chirped fiber Bragg grating. *J. Lightw. Technol.* 26: 1282–1287.

23 Wang, C. and J.Y. (2008). Photonic generation of chirped millimeter-wave pulses based on nonlinear frequency-to-time mapping in a nonlinearly chirped fiber Bragg grating. *IEEE Trans. Microwav. Theory Tech.* 56: 542–553.

24 Feced, R., Zervas, M.N., and Muriel, M.A. (1999). An efficient inverse scattering algorithm for the design of nonuniform fiber Bragg gratings. *IEEE J. Quantum Electron.* 35: 1105–1115.

25 Belai, O.V., Frumin, L.L., Podivilov, E.V., and Shapiro, D.A. (2007). Efficient numerical method of the fiber Bragg grating synthesis. *J. Opt. Soc. Am. B* 24: 1451–1457.

26 Skaar, J., Wang, L.G., and Erdogan, T. (2001). On the synthesis of fiber Bragg gratings by layer peeling. *IEEE J. Quantum Electron.* 37: 165–173.

27 Peral, E., Capmany, J., and Marti, J. (1996). Iterative solution to the Gel'Fand-Levitan-Marchenko coupled equations and application to synthesis of fiber gratings. *IEEE J. Quantum Electron.* 32: 2078–2084.

28 Xia, H. and J.Y. (2009). Characterization of sub-picosecond pulses based on temporal interferometry with real-time tracking of all-order dispersion and optical time delay. *J. Lightw. Technol.* 27: 5029–5037.

29 Weiner, A.M. (2000). Femtosecond pulse shaping using spatial light modulators. *Rev. Sci. Instrum.* 71: 1929–1960.

30 Heritage, J.P. and Weiner, A.M. (1990). Optical systems and methods based upon temporal stretching, modulation and recompression of ultrashort pulses. US Patent 4928316.

31 Saperstien, R.E., Alic, N., Pasasenko, D. et al. (2005). Time-domain waveform processing by chromatic dispersion for temporal shaping of optical pulses. *J. Opt. Soc. Am. B* 22: 2427–2436.

32 Azana, J., Berger, N.K., Levit, B., and Fischer, B. (2005). Reconfigurable generation of high-repetition-rate optical pulse sequences based on time domain phase-only filtering. *Opt. Lett.* 30: 3228–3230.

33 Chi, H. and J.Y. (2007). Symmetrical waveform generation based on temporal pulse shaping using an amplitude-only modulator. *Electron. Lett.* 43: 415–417.

34 Wang, C., Li, M., and J.Y. (2010). Continuously tunable photonic microwave frequency multiplication by use of an unbalanced temporal pulse shaping system. *IEEE Photonics Technol. Lett.* 22: 1285–1287.

35 Li, M., Wang, C., Li, W., and J.Y. (2010). An unbalanced temporal pulse shaping system for chirped microwave waveform generation. *IEEE Trans. Microw. Theory Tech.* 58: 2968–2975.

36 Capmany, J., Ortega, B., and Pastor, D. (2006). A tutorial on microwave photonic filters. *J. Lightw. Technol.* 24: 201–229.

37 Minasian, R.A. (2006). Photonic signal processing of microwave signals. *IEEE Trans. Microw. Theory Tech.* 54: 832–846.

38 Sales, S., Capmany, J., Marti, J., and Pastor, D. (1995). Experimental demonstration of fiber-optic delay line filters with negative coefficients. *Electron. Lett.* 31: 1095–1096.

39 Capmany, J., Pastor, D., Martinez, A. et al. (2003). Microwave photonics filter with negative coefficients based on phase inversion in an electro-optic modulator. *Opt. Lett.* 28: 1415–1417.

40 J.Y. and Wang, Q. (2007). Photonic microwave bandpass filter with negative coefficients using a polarization modulator. *IEEE Photonics Technol. Lett.* 19: 644–646.

41 Zeng, F., Wang, J., and J.Y. (2005). All-optical microwave bandpass filter with negative coefficients based on a phase modulator and linearly chirped fiber Bragg gratings. *Opt. Lett.* 30: 2203–2205.

42 J.Y. (2009). Microwave photonics. *J. Lightw. Technol.* 27: 314–335.

43 J.Y., Zeng, F., and Wang, Q. (2007). Photonic generation of ultra-wideband signals. *IEEE/OSA J. Lightw. Technol.* 25: 3219–3235.

44 Wang, Q. and J.Y. (2007). Optically switchable UWB monocycle and doublet generation using a reconfigurable photonic microwave delay-line filter. *Opt. Express* 15: 14667–14672.

45 Bolea, M., Mora, J., Ortega, B., and Capmany, J. (2009). Optical UWB pulse generator using an N tap microwave photonic filter and phase inversion adaptable to different pulse modulation formats. *Opt. Express* 17: 5023–5032.

46 Mu, H. and J.Y. (2009). Polarity- and shape-switchable UWB pulse generation based on a photonic microwave delay-line filter with a negative tap coefficient. *IEEE Photonics Technol. Lett.* 21: 253–1255.

47 You, N. and Minasian, R.A. (2001). A novel tunable microwave optical notch filter. *IEEE Trans. Microw. Theory Tech.* 49: 2002–2005.

48 Loayssa, A., Capmany, J., Sagues, M., and Mora, J. (2006). Demonstration of incoherent microwave photonic filters with all-optical complex coefficients. *IEEE Photonics Technol. Lett.* 18: 1744–1746.

49 Yan, Y. and J.Y. (2007). A tunable photonic microwave filter with a complex coefficient using an optical RF phase shifter. *IEEE Photonics Technol. Lett.* 19: 1472–1474.

50 Dai, Y. and J.Y. (2008). Nonuniformly-spaced photonic microwave delay-line filter. *Opt. Express* 16: 4713–4718.

51 Dai, Y. and J.Y. (2010). Nonuniformly spaced photonic microwave delay-line filters and applications. *IEEE Trans. Microw. Theory Tech.* 58: 3279–3289.

52 Dai, Y. and J.Y. (2007). Microwave pulse phase encoding using a photonic microwave delay-line filter. *Opt. Lett.* 32: 3486–3488.

53 Dai, Y. and J.Y. (2009). Chirped microwave pulse generation using a photonic microwave delay-line filter with a quadratic phase response. *IEEE Photonics Technol. Lett.* 21: 569–571.

54 Dai, Y. and J.Y. (2009). Microwave correlator based on a nonuniformly spaced photonic microwave delay-line filter. *IEEE Photonics Technol. Lett.* 21: 969–971.

55 Zhang, W., Zhang, J., and Yao, J. (2014). Largely chirped microwave waveform generation using a silicon-based on-chip optical spectral shaper. *Proc. MWP*, Sapporo, Japan (October 2014).

56 Zhang, W. and Yao, J. (2015). Photonic generation of linearly chirped microwave waveforms using a silicon-based on-chip spectral shaper incorporating two linearly chirped waveguide Bragg gratings. *J. Lightw. Technol.* 33: 5047–5054.

57 Wang, J., Shen, H., Fan, L. et al. (2015). Reconfigurable radio-frequency arbitrary waveforms synthesized in a silicon photonic chip. *Nat. Commun.* 6: 5957.

58 Zhang, W. and Yao, J. (2016). Silicon-based on-chip electrically-tunable spectral shaper for continuously tunable linearly chirped microwave waveform generation. *J. Lightw. Technol.* 34: 4664–4672.

59 Coldren, L.A. (2010). Photonic integrated circuits for microwave photonics. *2010 International Topical Meeting on Microwave Photonics, WE1-1*, Montreal, Canada (October 5–9, 2010).

8

Microwave Photonics Beamforming Networks for Phased Array Antennas

8.1 Introduction

Phased array antennas (PAAs) are radiating structures composed of multiple antenna elements usually disposed in one- or two-dimension configurations, which are individually and coherently fed by means of either variable phase or true time delay (TTD) control elements, in order to scan a complex radiation pattern to a given angular direction in space [1, 2]. Furthermore, the shape of this pattern can also be independently reconfigured by including variable amplitude control elements. Figure 8.1 shows two examples of 1D and 2D PAA configurations and also introduces the axis and angle conventions that we will use throughout this chapter and the element nomenclature.

As can be observed, a PAA is composed of three main parts: the RF source or receiver from/to which the signals are generated/directed, and a beamforming network connecting the RF source/receiver to the third part, the radiating elements or antennas. PAAs can operate in two modes: as a radiating element whereby the signal from the RF source is sent and processed by the beamforming network in such a way that a constructive interference signal is generated by the antennas in a given spatial direction or as a receiving element, whereby the incoming radiation from a given direction is constructively added by the beamforming network and directed to the RF receiver. The key element of the PAA is the beamforming network which is composed of several independent physical paths connecting the radiating elements and the RF source/detector. The transmission characteristics of each path and thus the privileged direction of radiation of the PAA can be changed if phase/delay and amplitude control elements are included. Furthermore, the former are usually dynamic, meaning that they can be controlled by means of a time-varying external signal.

PAAs are mainly employed in radar and radiofrequency communication systems where there is a need to produce a directive beam that can be scanned in response to an input control signal. A further advantage is related to the fact that due to the presence of multiple radiating elements, PAAs allow more precise control and shaping of the radiation pattern, and also the possibility of achieving lower sidelobes.

Microwave Photonics (MWP) subsystems and techniques can be advantageously employed in the implementation of the beamforming network that connects the input (output) to the antenna elements in an emitting (receiving) PAA [3]. To the usual advantages of low weight, low loss, and immunity to electromagnetic interference, MWP techniques bring other important added values such as fast angular scanning and radiation pattern shaping. Two features are, respectively, very similar to the tuning and reconfiguration of MWP filters discussed in Chapter 4. In the case of a photonic beamforming network, the techniques reported in the literature can be classified into two categories: beamforming based on optical fibers and integrated waveguides and beamforming based on free space optics. In this chapter, we only focus on the first group as it has attracted more

Microwave Photonics, First Edition. Jianping Yao, José Capmany, and Ming Li.

Companion website: www.wiley.com/go/yao/microwavephotonics

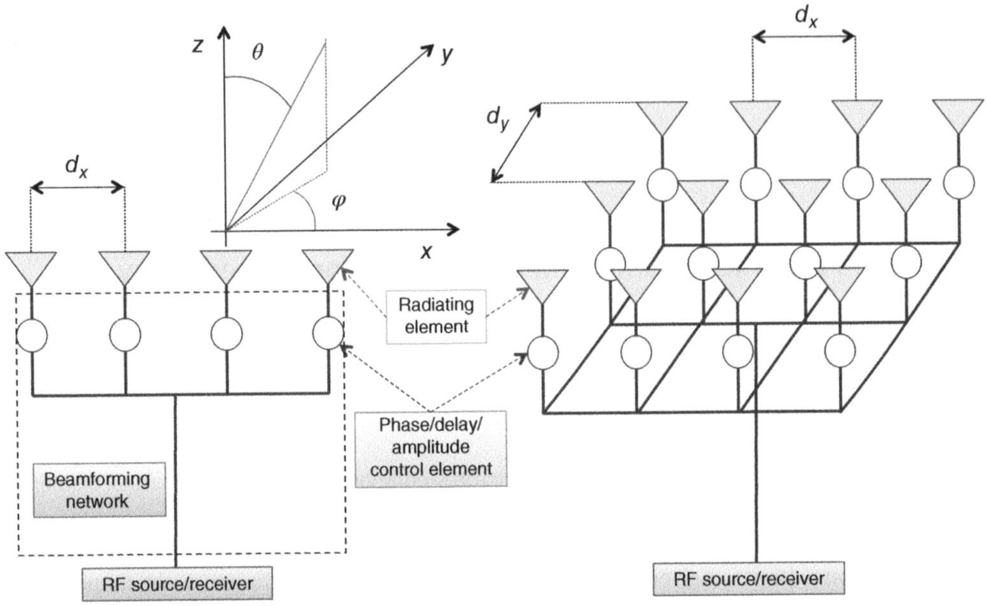

Figure 8.1 Two examples of 1D (left) and 2D (right) PAA configurations.

interest among researchers during the last two decades due to the advantages in terms of compact design, lower losses, and potential integration. The interested reader is nevertheless referred to the existing literature on free space approaches [4–11].

The objective of this chapter is to provide a comprehensive treatment of the use of MWP subsystems and technologies in the implementation of the beamforming network in PAA structures. We present some general and technology-independent concepts on PAAs in Section 8.2, which the reader, depending on his/her background, needs to refresh or learn in order to understand their basic system principles. In Section 8.3, we deal with broadband optical beamforming networks where the control elements are implemented by optical TTDs. We will describe some of the most important approaches so far reported in the literature including those based on switched delay lines, dispersive optical fiber links, and linearly chirped fiber Bragg gratings (LCFBGs). The advantages and limitations of each configuration are described in detail. Section 8.4 deals with narrowband optical beamforming networks where the control elements are implemented by means of variable phase shifters (VPS). We describe the work reported based on several technologies such as integrated ring resonators, semiconductor optical amplifiers (SOAs), and stimulated Brillouin scattering (SBS) in optical fibers. The chapter is finally summarized in Section 8.5.

8.2 Basic Concepts on Phased Array Antennas

8.2.1 Principles of Operation

A detailed mathematical analysis of PAAs involves extremely complex electromagnetic methods and is beyond the scope of this chapter [1, 2]. Here we will employ a simpler yet powerful approach based on the so-called array theory that provides a very reasonable approximation to the behavior of PAAs while keeping the mathematical treatment affordable.

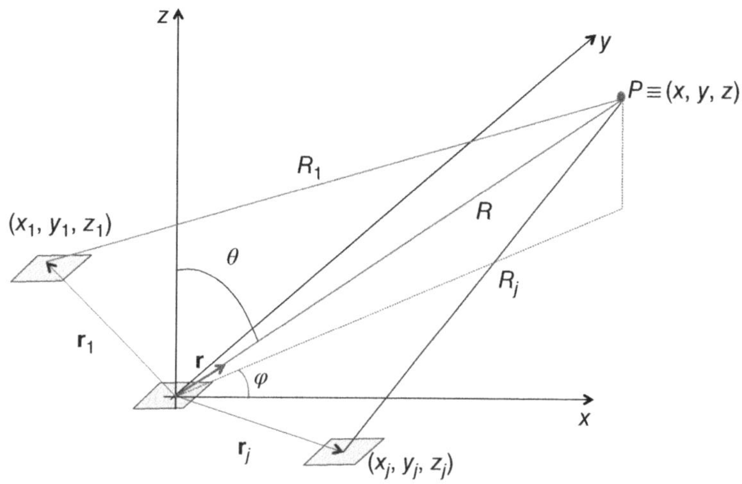

Figure 8.2 Geometry, axes, and arbitrary distribution of elements in a PAA.

Figure 8.2 shows an arbitrary distribution of array elements. In particular, the figure displays three elements, the first element in the array ($s = 1$), an arbitrary element $s = j$, and a central element which is located at the center of coordinates and will be used as a reference. Each element is shown as a small radiating surface, the shape of which is of no concern here. Also the axes, angles, and relevant coordinates and vectors are displayed for reference. The arbitrary point of the space $P \equiv (x, y, z)$ in which we wish to compute the radiated field from the PAA is also shown.

Although each element radiates a vector directional pattern that has both angle and radial dependence near it, we will assume as a first approximation that $P \equiv (x, y, z)$ is very far from the element. This far-field approximation allows us to express the field from a given element s as the product of a spherical wave factor $\exp[-jkR_s]/R_s$ and a vector function of the angle $\mathbf{f}_s(\theta, \varphi)$ called the element radiation pattern. Here $k = 2\pi/\lambda$ is the free space wave number and R_s is the radial distance between the radiating element and $P \equiv (x, y, z)$.

Thus, the far field from the element s of the array at $P \equiv (x, y, z)$ is given by:

$$\mathbf{E}_s(r, \theta, \varphi) = \frac{\mathbf{f}_s(\theta, \varphi) e^{-jkR_s}}{R_s} \tag{8.1}$$

where:

$$R_s = \sqrt{(x - x_s)^2 + (y - y_s)^2 + (z - z_s)^2} \tag{8.2}$$

At a distance very far away from the array, Eq. (8.2) can be approximated by:

$$R_j \approx R - \hat{\mathbf{r}} \cdot \mathbf{r}_j \tag{8.3}$$

where R is the distance measured from the central element of the array in Fig. 8.2 and $P \equiv (x, y, z)$, $\mathbf{r}_s$ is the position vector of the element s, and $\hat{\mathbf{r}}$ is a unit vector in the direction connecting the origin of coordinates and $P \equiv (x, y, z)$. With this approximation, (8.1) is now given by:

$$\mathbf{E}_s(r, \theta, \varphi) = \left(\frac{e^{-jkR}}{R}\right) \mathbf{f}_s(\theta, \varphi) e^{jk\hat{\mathbf{r}} \cdot \mathbf{r}_s} \tag{8.4}$$

It is customary to write the vectors $\mathbf{r}_s$ and $\hat{\mathbf{r}}$ as:

$$\begin{aligned}\mathbf{r}_s &= x_s\hat{\mathbf{x}} + y_s\hat{\mathbf{y}} + z_s\hat{\mathbf{z}} \\ \hat{\mathbf{r}} &= u\hat{\mathbf{x}} + v\hat{\mathbf{y}} + \cos\theta\hat{\mathbf{z}}\end{aligned} \tag{8.5}$$

where $u = \sin\theta\cos\varphi$ and $v = \sin\theta\sin\varphi$ are the direction cosines and $R > 2L^2/\lambda$ with L representing the largest array dimension for the far field approximation to be correct.

The overall field from a PAA at a given point can be obtained by superposition from the fields of each radiating element:

$$\mathbf{E}(\mathbf{r}) = \frac{e^{-jkR}}{R}\sum_{s=1}^{N} a_s\mathbf{f}_s(\theta,\phi)e^{jk\mathbf{r}_s\cdot\hat{\mathbf{r}}} \tag{8.6}$$

In (8.6), the coefficients a_s represent the complex-valued *weights* or *tapers* applied to the incident signals by means of phase, delay, and amplitude control elements placed in each independent path of the beamforming network.

Although, in principle, the element radiation patterns are different for each element in the array, we will consider them identical, that is $\mathbf{f}_s(\theta, \varphi) = \mathbf{f}(\theta, \varphi), \forall s$. Thus:

$$\mathbf{E}(\mathbf{r}) = \frac{e^{-jkR}\mathbf{f}(\theta,\phi)}{R}\sum_{s=1}^{N} a_s e^{jk\mathbf{r}_s\cdot\hat{\mathbf{r}}} \tag{8.7}$$

The spherical wave factor $\exp[-jkR]/R$ is usually removed as the overall field is usually described on a sphere of constant radius and, therefore, the overall field can be expressed as the product of the element radiation pattern and a scalar quantity known as the *array factor*, $AF(\theta, \phi)$:

$$\begin{aligned}\mathbf{E}(\mathbf{r}) &= \mathbf{f}(\theta,\phi)AF(\theta,\phi) \\ AF(\theta,\phi) &= \sum_{s=1}^{N} a_s e^{jk\mathbf{r}_s\cdot\hat{\mathbf{r}}}\end{aligned} \tag{8.8}$$

The array scanning can be achieved by changing the array factor which, in turn, implies the changing of the complex-valued weights a_s. There are two different options in which this can be implemented. The first one is known as TTD scanning and requires weights in the form:

$$\begin{aligned}a_s &= |a_s|e^{-jk\mathbf{r}_s\cdot\hat{\mathbf{r}}_o} \\ \hat{\mathbf{r}}_o &= u_o\hat{\mathbf{x}} + v_o\hat{\mathbf{y}} + \cos\theta_o\hat{\mathbf{z}}\end{aligned} \tag{8.9}$$

which introduced in (8.8) yields:

$$AF(\theta,\phi) = \sum_{s=1}^{N}|a_s|e^{jk\mathbf{r}_s\cdot(\hat{\mathbf{r}}-\hat{\mathbf{r}}_o)} \tag{8.10}$$

The array factor is maximum for $\theta = \theta_o$, $\phi = \phi_o$ since at this pointing direction $\hat{\mathbf{r}} = \hat{\mathbf{r}}_o$, all the exponential terms in (8.10) become unity. Furthermore, this pointing direction is stationary for all frequencies. Note that to implement this technique, the required exponential dependence has a linear phase relationship with frequency, which corresponds to inserting TTD control elements since:

$$\begin{aligned}\varphi_s &= \arg[a_s] = \frac{\mathbf{r}_s\cdot\hat{\mathbf{r}}_o}{c}\omega \\ \tau_s &= \frac{\partial\varphi_s}{\partial\omega} = \frac{\mathbf{r}_s\cdot\hat{\mathbf{r}}_o}{c}\end{aligned} \tag{8.11}$$

These time delays are chosen so the path length differences of the different elements in the array are compensated in order to make the signals from all of them arrive simultaneously at some distant point (for radiating PAAs) or at the receiver (for receiving PAAs). Thus, changing the pointing direction implies switching time delays.

The second option to control the beamsteering is by using tunable phase shifters. In this case, the weights have the form:

$$\begin{aligned} a_s &= |a_s| e^{-jk_o \mathbf{r_s} \cdot \hat{\mathbf{r}}_\mathbf{o}} \\ k_o &= \frac{\omega_o}{c} \end{aligned} \tag{8.12}$$

which, substituted in (8.8), yields:

$$AF(\theta, \phi) = \sum_{s=1}^{N} |a_s| e^{jk_o \mathbf{r}_s \cdot ((\omega/\omega_o)\hat{\mathbf{r}} - \hat{\mathbf{r}}_\mathbf{o})} \tag{8.13}$$

In this case, as it can be immediately observed, the array pattern has its peak at a direction $\hat{\mathbf{r}} = (\omega_o/\omega)\hat{\mathbf{r}}_\mathbf{o}$, which depends on the frequency. This feature is behind the so-called *beam squint* phenomenon that is explained later in this section.

8.2.2 Design Parameters

The important parameters of PAAs which are *gain*, *beamwidth*, *sidelobe level*, and *bandwidth* will be introduced and explained by means of the simple one-dimensional configuration shown in the left part of Figure 8.1. We thus consider an array of N elements (with equal radiating profile) arranged in a line where the radiating elements are located at positions $x_n = nd_x$. The normalized array radiation pattern in the far field is:

$$\mathbf{E}(\mathbf{r}) = \mathbf{f}(\theta, \phi) \sum_{s=1}^{N} a_s e^{jksd_x u} \tag{8.14}$$

which, at a given frequency f_o, is given by:

$$\mathbf{E}(\mathbf{r}) = \mathbf{f}(\theta, \phi) \sum_{s=1}^{N} a_s e^{jk_o sd_x u} \tag{8.15}$$

If the complex-valued coefficients are given by $a_s = |a_s| \exp(-jk_o sd_x u_o)$, with $u = \sin\theta_o$, then the array factor becomes:

$$AF(\theta, \phi) = \sum_{s=1}^{N} |a_s| e^{jk_o sd_x (u - u_o)} \tag{8.16}$$

which is maximum if $u - u_o = \sin\theta \cos\varphi - \sin\theta_o = 0$, in other words, in the direction $(\theta_o, 0)$. Note that the array factor given by (8.16) is only a function of $u - u_o$ which means that if the array is scanned to any given angle, then the pattern remains unchanged except for a translation in the u coordinate.

A case of special interest is a PAA where the amplitude of the excitations is equal for all the branches ($|a_s| = 1, \forall s$). This is the so-called uniformly illuminated array. In this case, (8.16) can be summed in closed-form resulting in the following normalized value:

$$AF(\theta, \phi) = \frac{\sin[(N\pi d_x f_o (u - u_o))/c]}{N \sin[(\pi d_x f_o (u - u_o))/c]} \tag{8.17}$$

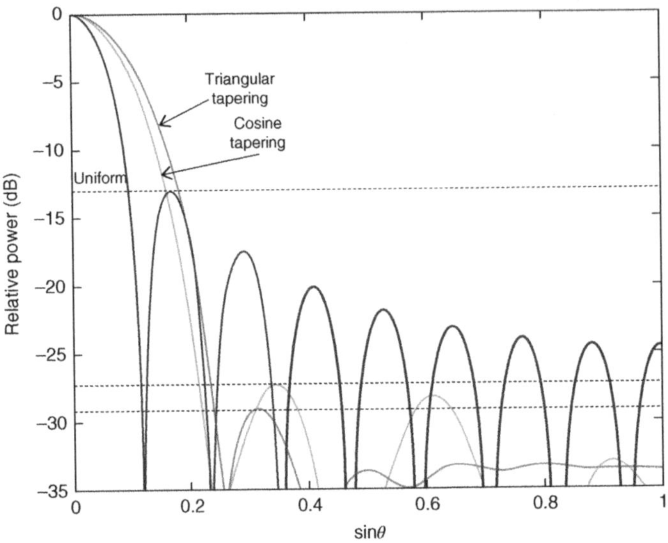

Figure 8.3 Broadside array factor for a uniform and tapered illuminated PAA composed of 16 elements spaced half a wavelength apart.

Figure 8.3 shows the array factor corresponding to this case for a PAA composed of 16 elements spaced half a wavelength apart. We consider the so-called *broadside pattern* where $u_o = 0$ and the independent variable is $u = \sin\theta$. The figure also represents the radiation characteristics of other PAAs subject to coefficient weighting or tapering.

Several interesting features, much of them closely related to that of MWP filters, can be obtained from its observation. First, note that the 3-dB beamwidth of the uniformly illuminated array is the narrowest of any illumination profile. In fact, for a broadside pattern, one has:

$$\theta_{3\text{-dB}} = \frac{0.866 B_b \lambda}{N d_x} \tag{8.18}$$

where B_b the beam broadening factor is $B_b = 1$ for the uniformly illuminated array and $B_b > 1$ for the rest of apodization profiles. Furthermore, when the beam is scanned to an angle θ_o and the pattern is displaced from the broadside pattern, the beam width is not constant in angle space (although it is in the u space) as it broadens according to:

$$\theta_{3\text{-dB}} = \sin^{-1}\left(u_o + \frac{0.433 B_b \lambda}{N d_x}\right) - \sin^{-1}\left(u_o - \frac{0.433 B_b \lambda}{N d_x}\right) \tag{8.19}$$

The second feature that deserves attention in the radiation characteristics shown in Fig. 8.3 is the existence of sidelobes and, in particular, the level of the first sidelobes. These secondary peaks appear whenever the exponent in the summation (8.16) is a multiple p of 2π. In other words, if:

$$\begin{aligned} &\frac{2\pi d_x}{\lambda}(u - u_o) = \frac{2\pi d_x}{\lambda}(\sin\theta - \sin\theta_o) = 2p\pi \\ &\sin\theta_p = \sin\theta_o + \frac{p\lambda}{d_x}; \quad p = \pm 1, \pm 2 \ldots\ldots \end{aligned} \tag{8.20}$$

And p is constrained to values for which $|\sin\theta_p| \leq 1$. In the case of the uniformly illuminated array, the level of the first sidelobes is the highest (−13 dB) but this value can be greatly reduced to figures below −40 dB by suitable coefficient tapering, much in the same way as sidelobes in transversal microwave filters are reduced.

The bandwidth of a PAA can be limited by many factors, including own bandwidth of the radiating elements in the array. This occurs in particular when TTD elements are employed as, in this case, the beam peak is obtained at (θ_o, φ_o) for all frequencies. Often, however, the most restrictive limitation is caused by the use of phase shifters to scan the array instead of using TTDs. In this case, the peak is scanned to the desired angle (θ_o, φ_o) only at the central frequency f_o. For other frequencies, the beam is scanned in other directions as implied by Eq. (8.13). In particular, for the case of one-dimensional PAAs, one has that the value of u corresponding to the beam peak is given by:

$$u = u_o \left(\frac{f_o}{f} \right) \tag{8.21}$$

This result, known as *beam squint* indicates that the beam peak angle is reduced for frequencies above the design frequency f_o and increased for frequencies below it. Figure 8.4 shows qualitatively this effect. The limiting effect due to beam squint can be computed by noting that if we define the bandwidth by the frequency limits at which the gain is reduced to half power (see right part of Fig. 8.4), then it follows from (8.21) that:

$$\frac{\Delta f}{f_o} = \frac{\Delta u}{u_o} = \frac{\theta_{3\text{-dB}}}{\sin\theta_o} = 0.866 B_b \left(\frac{\lambda_o}{N d_x \sin\theta_o} \right) \tag{8.22}$$

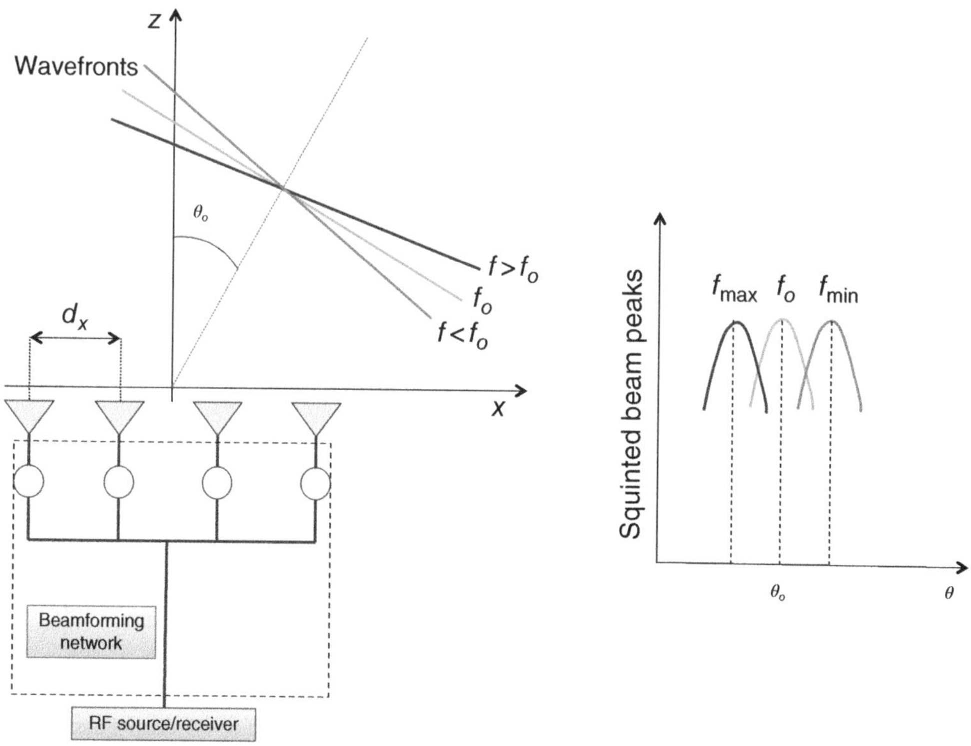

Figure 8.4 Illustration of the beamsquint phenomenon.

It is clear that the bandwidth becomes smaller as the array is made larger or as the scan angle is increased.

To avoid beamsquint and therefore to guarantee broadband operation of the PAA is necessary to resort to TTD feeding schemes as has been previously pointed. For instance, Fig. 8.5 shows the far-field radiation pattern corresponding to a PAA composed of six elements separated by a distance of 1 cm ($d = \lambda/2$) operating at a central frequency placed at the K-band ($f = 15$ GHz). The bottom part of Fig. 8.5 shows the evolution of the far-field radiation pattern when phase shifters are employed to steer the radiation pattern and the central frequency is swept from 10 to 20 GHz. It can be clearly appreciated how the orientation of the main lobe varies with the feed signal frequency. When a TTD scheme is employed, the radiation pattern remains constant for the same frequency interval as shown in the bottom right part of Fig. 8.5.

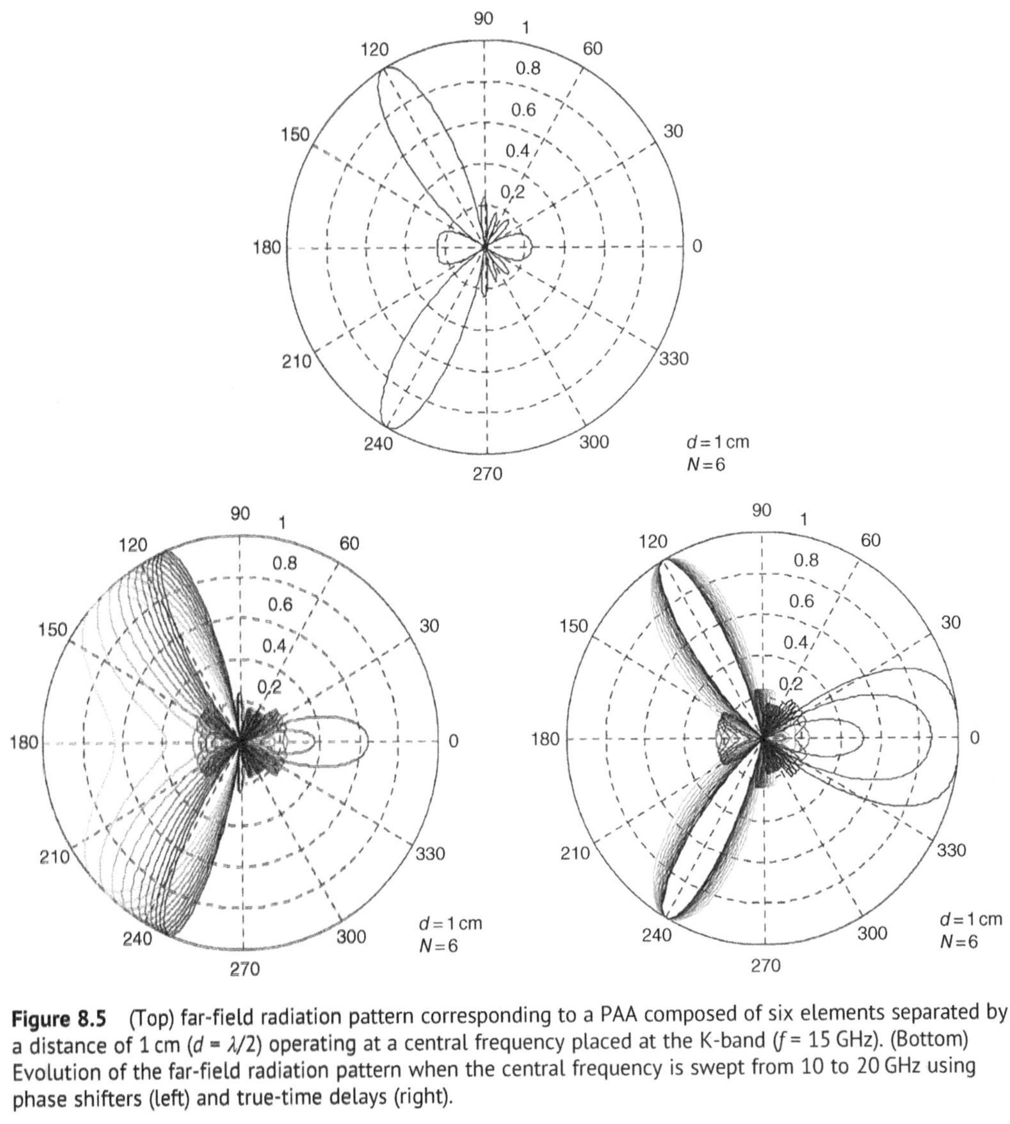

Figure 8.5 (Top) far-field radiation pattern corresponding to a PAA composed of six elements separated by a distance of 1 cm ($d = \lambda/2$) operating at a central frequency placed at the K-band ($f = 15$ GHz). (Bottom) Evolution of the far-field radiation pattern when the central frequency is swept from 10 to 20 GHz using phase shifters (left) and true-time delays (right).

8.2.3 PAA Feed Architectures

The PAA configurations shown in Fig. 8.1 are examples of what is known in the literature as *constrained feeds* which are based on power dividers and transmission lines to bring the signal to each radiating element. When the transmission lines have equal length, the configurations are known as *corporate feeds* and are used to provide equal phase signal distribution for wideband arrays. There are, of course, other alternatives, including space-fed lens array beamforming networks where the array is fed by means of a single antenna that illuminates its back face. There a lens providing phase control for each radiating element is placed. Both reflection and transmission configurations are possible. For example, the top left part of Fig. 8.6 shows a transmission configuration.

Most of the PAA configurations are single beams, that is, they only produce one (fixed or scannable) radiation beam. However, for certain applications, such as satellite communications, it is interesting to consider *multiple beam* PAAs where they can be employed to produce switched individual beams or clusters of beams to cover particular areas on the Earth globe. The feed network in this case has a multiport input configuration where each input port excites an independent beam in space as shown schematically in the top right part of Fig. 8.6. Multiple beam array feeds can be implemented in a variety of ways. For instance, the bottom part of Fig. 8.6 shows a layout which is based on a Rotman lens, a familiar configuration which is also employed in the implementation of the coupling regions in arrayed waveguide grating (AWG) photonic devices. These devices have the property of focusing different plane wavefront angular directions into different points of the output surface. Another interesting option is the use of *Butler matrices*, which mimic the implementation of the Fast Fourier algorithm radiating orthogonal sets of beams with uniform aperture illumination.

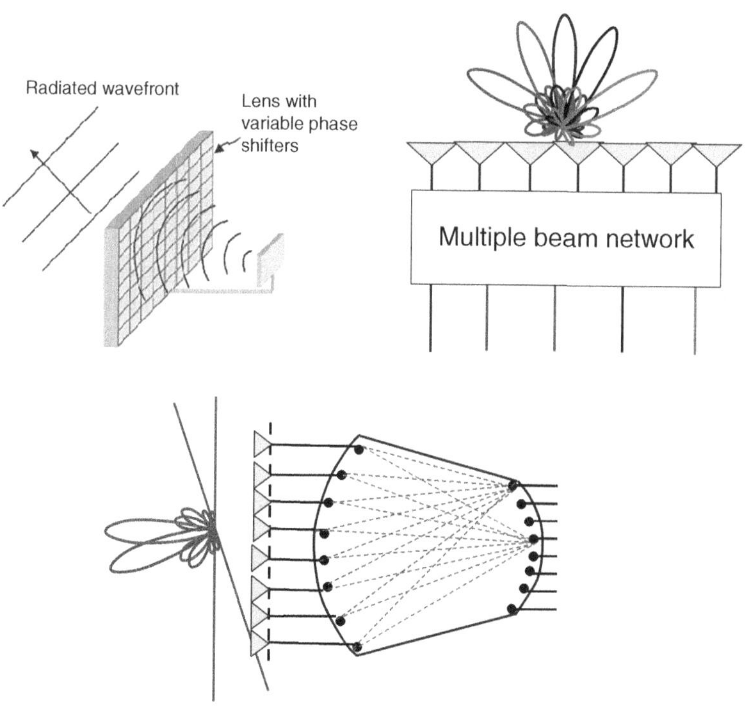

Figure 8.6 Different configurations for PAAs. Transmission space-fed lens array beamforming network (top left), multibeam PAA (top right), and PAA based on a Rotman lens (bottom).

8.3 True Time Delay Optical Beamforming Networks

A considerable number of contributions have addressed the topic of TTD optical beamforming network design over the last three decades [12–66]. An early review [12] provides an extensive coverage of some of the first developed techniques.

Different guided-wave TTD optical beamforming networks have been proposed which rely either on the use of switched delay line configurations or on exploiting chromatic dispersion in optical fibers. In the first case, each radiating element is fed by a bank of optical fibers with lengths adjusted to provide different pointing angles. A particular direction is selected by switching a given fiber in the bank. This approach only allows for discrete scanning and can become complex and bulky when the number of elements in the PAA is increased. The second approach is based on dispersive optical delay lines (ODLs) where the delay experienced by a microwave subcarrier modulating a lightwave carrier can be changed just by tuning its wavelength.

An initial landmark in switched guided-wave techniques was reported in [13], where a first experimental demonstration of an optically steered TTD-PAA was presented in the microwave region of the spectrum. The system that could operate both in the L (1–2.6 GHz) and the X (8–12 GHz) bands was implemented using switched optical fiber delay lines. In particular, the radiating elements in both bands were divided into four subarrays. Each subarray was time-steered by a shared 3-bit switched delay line element composed of eight parallel analog fiber links of different lengths cut to provide a prespecified set of differential time delays. Each fiber was fed by a different laser and finished by a different receiver. During steering, one of the eight delay lines, as specified by the steering angle, was selected by switching on its input laser by means of activating its bias current. The radiation pattern of each subarray was, in turn, steered by means of RF phase shifters. Thus, the overall radiation pattern was the product of the array patterns due to TDD feeding of the four subarrays and the radiation pattern of one subarray. This simple system was free from beamsquint and could achieve angle steering from −28° to 28° at 2 and 9 GHz with 4° increment steps. A similar approach was reported by Goutzolis and co-workers [14], based on binary fiber-optic delay lines in which series of integrated optic space switches were used to route the optical signal through the proper path to achieve the desired delay. In [15], an integrated 6-bit true-time ODL was proposed for application to a 3–6 GHz radar beamformer. The proposed designs were based on lithium niobate 2×2, 4×4, and 8×8 switch matrices as shown in Fig. 8.7. The experimental demonstration was based on the case of 4×4 switches, the structure containing four cascaded devices and 3×4 ODLs of different delays (3 delay lines with 0 delay plus individual lines featuring delays given by τ, 2τ, 3τ, 4τ, 8τ, 12τ, 16τ, 32τ, and 48τ that provided discrete switchable delays from 0 to 64τ, i.e. from 0 to 30 ns for $\tau = 470$ ps.

The first contributions proposing the use of chromatic dispersion as a means to implement optical TTD were reported by Esman and co-workers [16] and Soref [17] almost simultaneously. The ODL reported in [16] is shown in the top left part of Fig. 8.8.

A tunable laser externally modulated by a microwave signal is sent through a dispersive optical fiber link of length L km. If λ_o represents the central operation wavelength and D the dispersion parameter, the group delay at any other wavelength can be approximated by:

$$\tau(\lambda) \approx \frac{L}{v_{go}} + D(\lambda - \lambda_o)L = \frac{L}{v_{go}} + D\Delta\lambda L \tag{8.23}$$

where V_{go} represents the group velocity at λ_o. Thus, the relative delay with respect to that corresponding to this reference wavelength $\tau(\lambda_o) = v_{go}L$ is given by $\Delta\tau(\lambda) \approx D\Delta\lambda L$ that can be continuously changed by tuning the value of the laser wavelength and, in turn, $\Delta\lambda$.

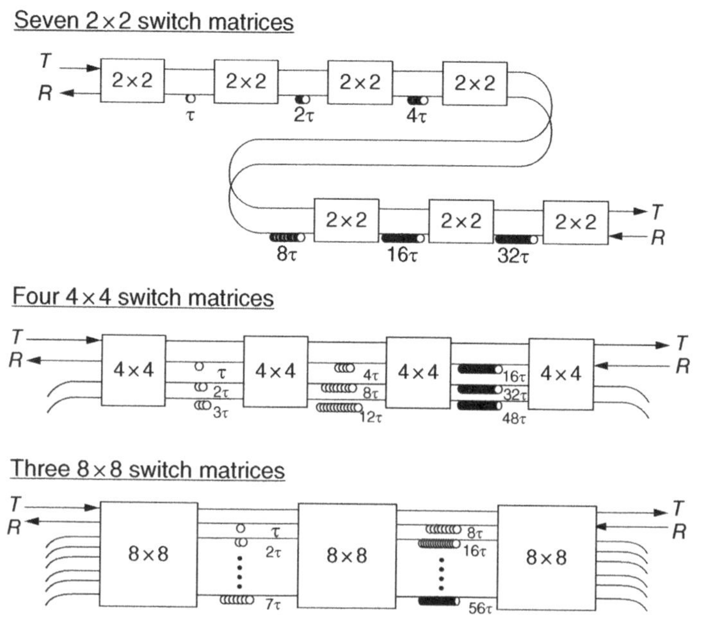

Figure 8.7 Integrated 6-bit true-time optical delay line proposed for application to a 3–6 GHz radar beamformer based on lithium niobate 2 × 2, 4 × 4, and 8 × 8 switch matrices. *Source:* Ackerman et al. [15]/IEEE.

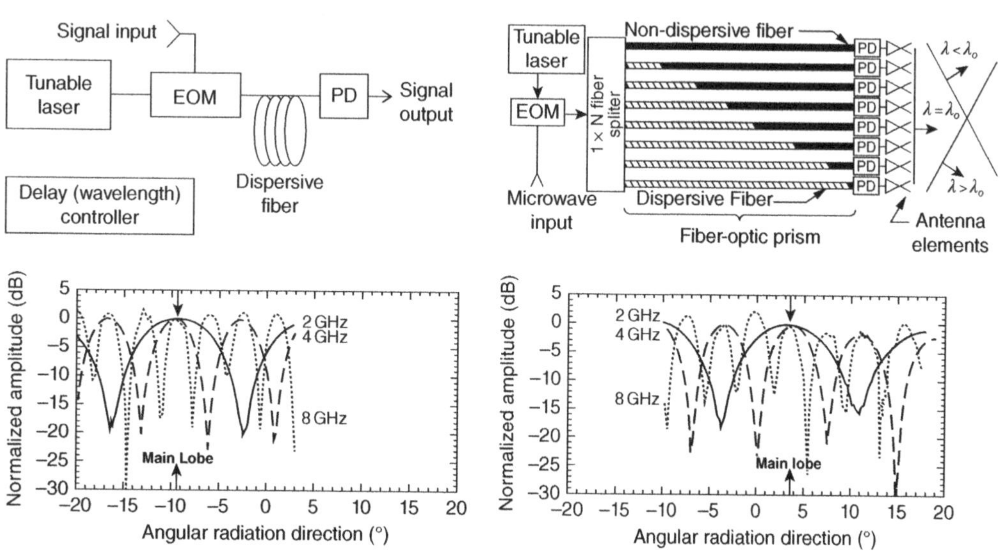

Figure 8.8 (Top left) Configuration proposed in [16] based on chromatic dispersion, as a means to implement optical TTD. *Source:* Adapted from Esman et al. [16]. (Top right) Fiber-optic prism delay line proposed in [17] and demonstrated radiation patterns (bottom). *Source:* Adapted from Soref [17].

Esman and co-workers [16] demonstrated the wideband (2–20 GHz) TTD operation of this dispersive delay line using a tunable σ laser controlled by voltage signals. Up to 1.7 ns relative delay was achieved using a 1.8 km link of Corning SMF-28 fiber over a wavelength range of 40 nm around 1532 nm. The extension of this technique to an antenna array of N elements was first discussed in [17], where a parallel battery of identical dispersive fiber-optic links, each containing an independently tunable laser, is considered.

An improvement over this configuration was subsequently reported by Esman and co-workers [18] based on a so-called fiber-optic prism fed by a single laser. The prism, shown in the top right part of Fig. 8.8 is composed of N dispersive optical links fed by the same laser via a $1 \times N$ star coupler. Each link has nominally the same length L, but it is composed by connecting varying amounts of high dispersion $\overline{D}$and nondispersive $D = 0$ (at λ_o) fiber. If V_{go} represents the fiber group delay at λ_o assumed equal both for the dispersive and nondispersive fibers, and the link length corresponding to the dispersive section is given by $(k-1)L'$, $k = 1, 2...N$, then the overall group delay provided by the kth link is given by:

$$\tau_k(\Delta\lambda) \approx \frac{L}{v_{go}} + (k-1)\overline{D}\Delta\lambda L' \tag{8.24}$$

This corresponds to an incremental delay of $\Delta\tau \approx \overline{D}\Delta\lambda L'$ per branch and accordingly to a pointing direction:

$$\sin[\theta(\Delta\lambda)] = \frac{\Delta L}{d_x} = \frac{c\Delta\tau(\Delta\lambda)}{d_x} = \frac{c\overline{D}L'}{d_x}\Delta\lambda \tag{8.25}$$

which can be continuously tuned by changing $\Delta\lambda$. For instance, broadside emission is achieved for $\Delta\lambda = 0$, $\theta < 0$ if $\Delta\lambda < 0$ and $\theta > 0$ if $\Delta\lambda > 0$. An $N = 2$ setup was assembled and its beamsquint-free operation was demonstrated in the 2–8 GHz frequency range for two pointing directions ($\theta = -9.4°, \theta = 3.4°$) as shown in the bottom part of Fig. 8.8. Further advantages provided by this configuration include the possibility of multibeam and multifrequency operation by means of adding several independent modulated and tuned input optical sources. Also, PAA apodization can be implemented by means of inserting variable attenuators. Finally, this configuration can be extended to two-dimensional arrays.

A more simple and compact approach to implement switched delay lines was proposed by Ball, Glenn and Morey in 1994 [19], where wavelength instead of space switching is employed. The key element upon which this variable delay line is based is a series of in FBGs fabricated along a single length of optical fiber with spatial separation between adjacent devices given by d_g. The configuration shown in the top part of Fig. 8.9 is completed by a feeding tunable laser subject to RF external modulation and a coupler or circulator to extract the delayed signals that arise as reflections from the FBGs. Each FBG has a different central wavelength and individual gratings are addressed simply by tuning the central wavelength of the optical carrier. Thus, the FBG addressing is equivalent to the selection of a specific length of fiber traversed by the RF-modulated optical carrier and thus of a specific discrete delay. In particular, a configuration composed of N FBGs provides the following set of discrete delays: 0, $\Delta\tau_g$, $2\Delta\tau_g \cdots (N-1)\Delta\tau_g$ or, in other words, a $\log_2(N)$ bit delay line, where $\Delta\tau_g = 2nd_g/c$, being $\Delta\tau_g$ the fiber refractive index.

In [19], a set of 5 FBGs was used as a phase-shifter for a 17 MHz radiofrequency signal. A maximum delay of 50 ns, with programmable variation of 10 nsec increments was reported. Molony et al. [20] subsequently developed this concept for the specific application case of PAAs providing the first experimental demonstration of a 3-bit TTD of radiofrequency signals in the range of

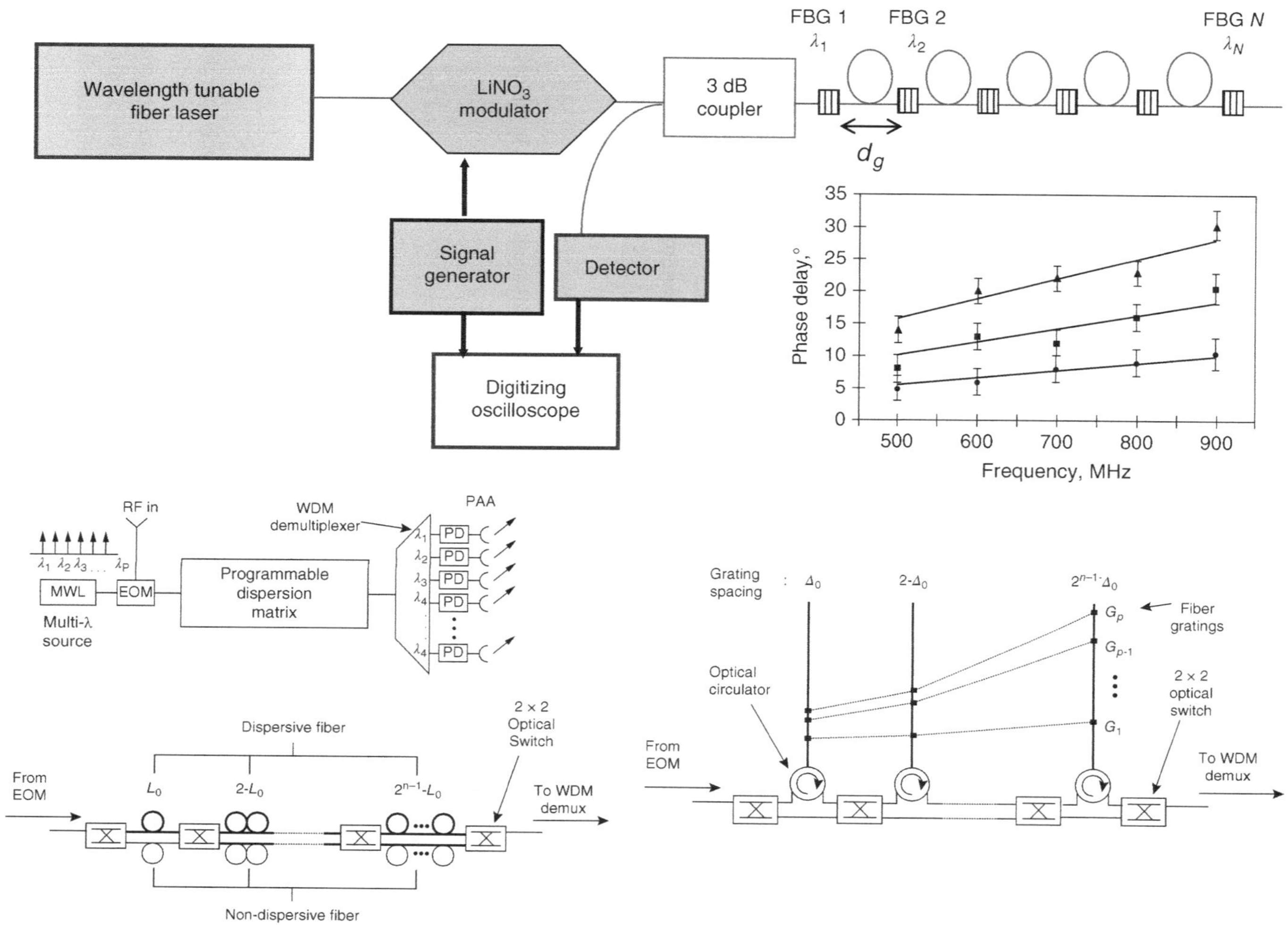

Figure 8.9 (Top) Wavelength switched fiber delay line configuration and measured phased delay versus frequency results. *Source:* Adapted from Ball et al. [19]. (Bottom) Beamforming network based on a programmable dispersion matrix. *Source:* Adapted from Tong and Wu [21, 22].

500–900 MHz as shown in the top right part of Fig. 8.9. FBG lengths were 2 mm and the distance between adjacent elements 4 mm, providing a minimum delay step of 31 ps. Two of these delay lines were integrated in a novel architecture proposed by Tong and Wu [21, 22] to fabricate the first 2-bit beamforming network based on a programmable dispersion matrix (PDM). A generic PDM with *n-bit* resolution [21] comprises *n* dispersive elements with exponentially increasing dispersion (i.e. D_o, $2D_o$, ... $2^{n-1}D_o$ ps/nm) and $n \times n$ 2×2 optical switches to program the total dispersion as shown in the bottom left part of Fig. 8.9. The total dispersion provided by the PDM is given by:

$$D_{TOT} = \sum_{i=1}^{n} 2^{i-1} D_o S_i \tag{8.26}$$

where $S_i = \{0, 1\}$ represents the state of the *i*th switch. By programming the optical switches, the total dispersion of the PDM can range from 0 to $(2^n - 1)D_0$ ps/km in increments of D_o. Thus, the PDM creates a wavelength-dependent relative time delay among different wavelength channels. So, when multiple wavelengths are simultaneously modulated by an external modulator and then sent through the PDM for TTD processing, the resulting time delay between adjacent optical channels is:

$$\Delta T = D_{TOT} \cdot \Delta\lambda \tag{8.27}$$

where $\Delta\lambda$ is the channel spacing. The dispersive elements can be made of any wavelength-dependent time-delay units. In the layout reported in [22], they were made of serially spaced FBGs as shown in the bottom right part of Fig. 8.9.

Further progress in serially spaced FBG TTDs resulted in the reduction of the minimum delay step down to 10 ps [23] to demonstrate the operation of a 3-bit ODL suitable for beamforming up to frequencies of 3 GHz.

Referring to the layout displayed in the top part of Fig. 8.9, the maximum number of delay elements that can be fabricated in each delay line is determined by the tuning range of the optical source $\Delta\Lambda$ and the optical bandwidth of each grating $\Delta\lambda$:

$$N_{\max} \approx \frac{\Delta\Lambda}{\Delta\lambda} \tag{8.28}$$

$\Delta\lambda$ determines the minimum possible wavelength spacing between adjacent gratings in the delay line. For uniform FBG, one has:

$$\Delta\lambda = \frac{\lambda^2}{2nL_g} \tag{8.29}$$

where $\Delta\lambda$ is the length of the FBG device.

On the other hand, the minimum achievable time delay unit is given by:

$$\Delta\tau_{\min} = \frac{2nd_g}{c} \tag{8.30}$$

Equation (8.30) shows a major drawback of the discrete FBG variable delay line, namely that they supply a phase distribution to the ports of the PAA given by the distance between adjacent gratings d_g. Since this length cannot be modified in a continuous manner, the orientation of the principal lobe of the PAA can only be switched in a discrete set of angles. Furthermore, although the pointing direction does not change with frequency using a TTD line feed, this is not the case with phase shift resolution due to discretization, which decreases with increasing RF frequency. In order to achieve a reasonable resolution for high frequencies, it should be possible to decrease the value of d_g. However, there is a physical limit to the minimum value of d_g achievable with this scheme since $d_g \geq L_g$.

A first solution to overcome this limitation and achieve continuous beam-scanning over a broadband region of the spectrum with reasonable phase resolution is to substitute the discrete set of FBGs with a linearly chirped FBG (LCFBG). The first proposal was due to Cruz and co-workers [24] followed by other research groups during the same year [23–27]. The LCFBG is a highly dispersive reflector providing a time delay $\tau(\lambda)$ that strongly depends on the optical wavelength. The basic principle behind the operation of an LCFBG as an RF phase shifter is shown in the top part of Fig. 8.10.

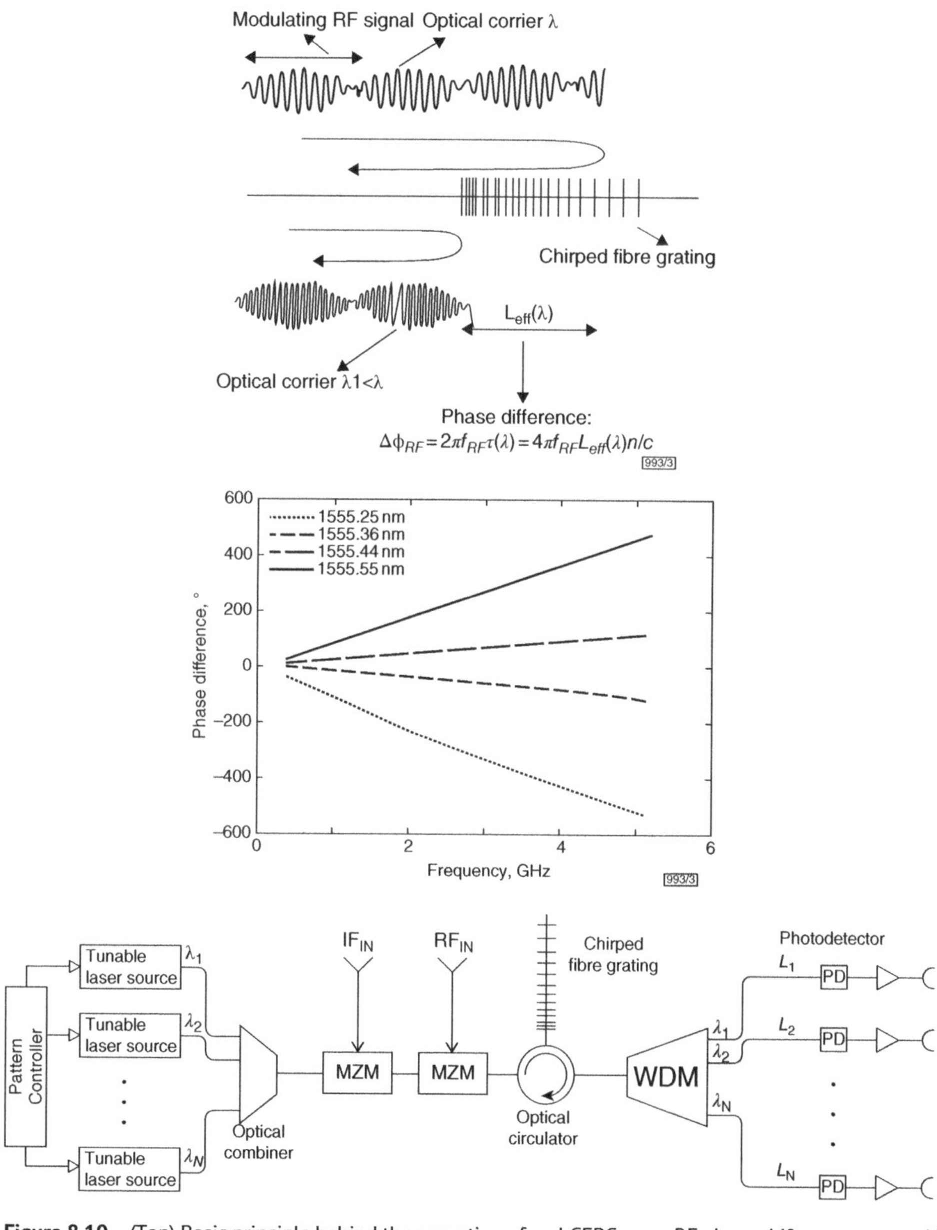

Figure 8.10 (Top) Basic principle behind the operation of an LCFBG as an RF phase shifter as proposed in [24]. *Source:* Cruz et al. [24]/The Institution of Electrical Engineers. (Middle) Measured phase delays versus frequency using the configuration. (Bottom) WDM beamforming network based on the same concept.

Assuming that each wavelength from an optical carrier is reflected from a single point of the LCFBG, then different optical wavelengths travel different distances $L_{eff}(\lambda)$ in the device. When the optical carrier is modulated, the modulating signal suffers a delay which is dependent on the depth reached by the lightwave in the grating $\tau(\lambda) = 2nL_{eff}(\lambda)/c$, where n represents the average refractive index along the grating and c is the speed of light. If an optical carrier is modulated by an RF signal of frequency f_{RF}, then the RF signal experiences a phase delay in the device given by $\Delta\Phi_{RF}(\lambda) = 2\pi f_{RF}\tau(\lambda)$. In other words, the LCFBG produces a linear phase shift in the modulating RF signal the slope of which can be continuously varied by changing the wavelength of the optical carrier. In their initial experiment, Cruz and co-workers achieved time delay variations up to 556 ps within an RF range from 390 MHz to 5.2 GHz using a 6 cm long, 0.4 nm bandwidth LCFBG as shown in the bottom left part of Fig. 8.10. Corral et al. [26] combined the use of a broadband 4 mn, 40 cm long LCFBG and a bank of tunable lasers, as shown in the bottom right part of Fig. 8.10, to drive a four-element PAA in the L and S bands (2–6 GHz) with a 5.3 ps time resolution step as well as Minasian and Alameh [27] extended the concept to implement a beamforming network capable of providing a large number of simultaneous and independent beams.

The extension of the concept of fiber-optic prism for PAA applications based on FBG delay lines was first proposed by Soref in 1996 [28] and experimentally demonstrated by Zmuda and co-workers [29]. A layout for an $N = 4$ element array and $M = 5$ beamsteering angular positions is shown in the top part of Fig. 8.11.

The basic incremental delay between radiators is selected by changing the wavelength of the tunable laser. In this particular configuration, λ_3 corresponds to the broadside radiation pattern ($\theta = 0°$) and for a given wavelength λ_i, the steering angle is given by:

$$\sin\theta_i = c\frac{\Delta t_i}{d_x} = \frac{2\left(d_{i,j} - d_{i,j+1}\right)n}{d_x} \tag{8.31}$$

where $d_{i,j}$ represents the spacing between the center of the ith grating and the center of the broadside grating in fiber j, and n is the average fiber refractive index. On the other hand, the steering angle resolution can be expressed as:

$$\theta_{\min} = \arcsin\left(\frac{4nf_{RF,\max}\,d_{\min}}{d_x}\right) \tag{8.32}$$

where $f_{RF,\max}$ is the maximum RF frequency to be processed by the PAA and $d_{\min}$ represents the minimum distance between adjacent gratings in the setup (in the case of that shown in Fig. 8.11 is the distance between adjacent gratings in the uppermost array). Note that Eq. (8.8.32) implies either a minimum resolution when the frequency is fixed or, alternatively, a maximum operation frequency for a fixed resolution.

The configuration allows both transmission and reception modes as explained in detail in [28]. The experimental demonstration featured the operation of an $N = 3$, $M = 4$ PAA system with steering angles given by $\theta = 0°, 30°, 59°, 60°$ over a 3.5 GHz bandwidth. The bottom part of Fig. 8.11 displays nine nonzero resolvable delays corresponding to $\theta = 30°, 59°, 60°$.

Multibeam operation of a fiber-optic PAA beamforming network was first demonstrated by Frankel and Esman [30]. Their system was based on the fiber-optic prism reported in [16] and consisted of eight broadband spiral elements in a sparsely populated array. Fully independent dual-beam, dual-frequency, ultrawideband (6–16 GHz) antenna transmitter operation was demonstrated featuring squint-free operation over >70° azimuth. A later experiment [31] reported a wideband (2–18 GHz) fiber-optic TTD array receiver capable of multiple independent beam

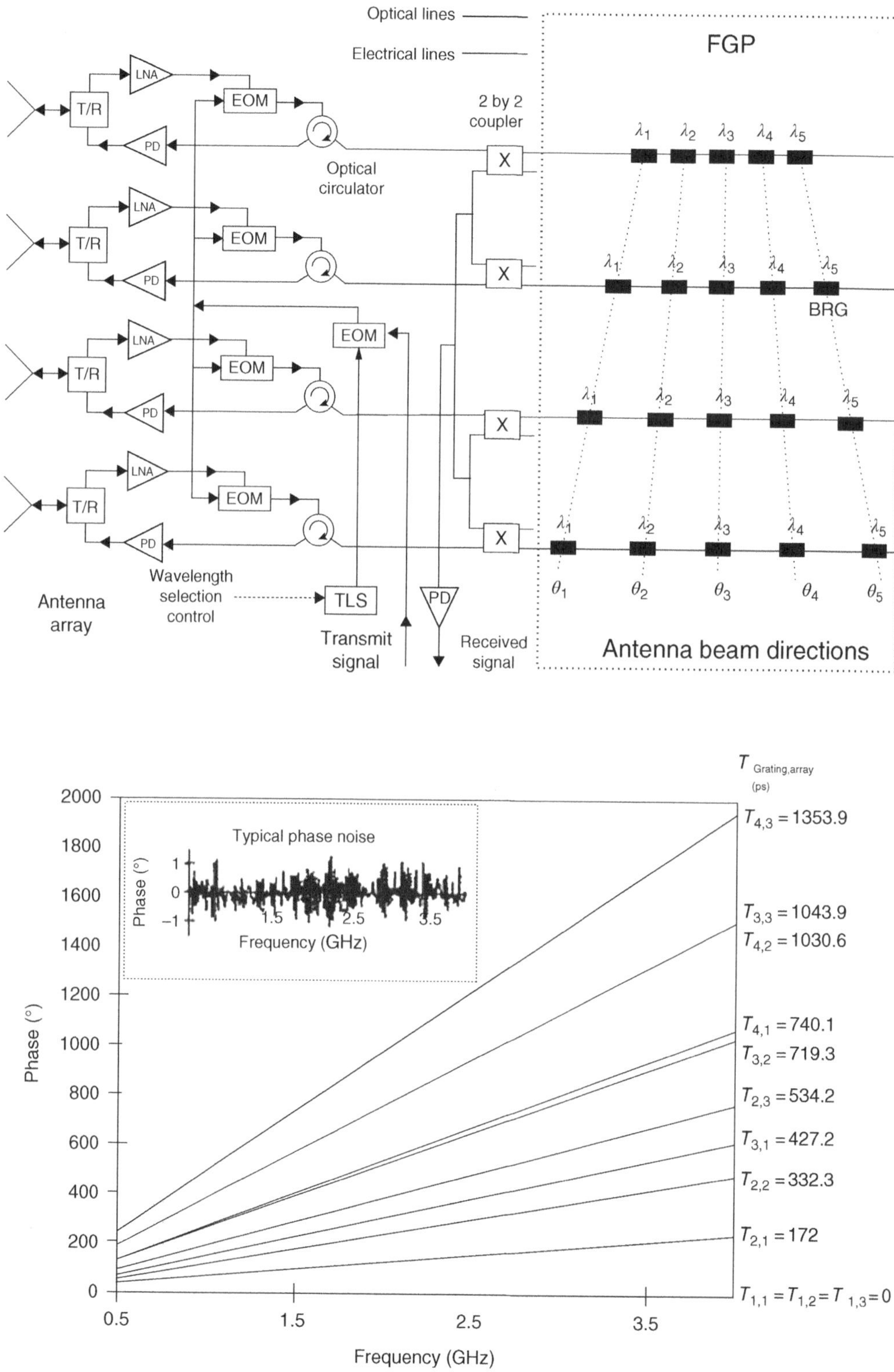

Figure 8.11 (Top) Layout for an $N = 4$ element array and $M = 5$ beamsteering angular positions fiber-optic prism PAA based on FBG delay lines. (Bottom) Nine nonzero resolvable delays corresponding to $\theta = 30°$, 59°, 60° implemented with the configuration. *Source:* Adapted from Soref [28] and Zmuda et al. [29].

simultaneous reception. The particular experimental configuration demonstrated the independent reception of two squint-free beams using two tunable lasers in a ±60° azimuthal range.

For this configuration, the fiber-optic prism design was changed featuring a binary-tree-type architecture using wavelength flattened 50/50 couplers which had already been tested by the authors [32].

Two-dimensional guided-wave photonic beamforming architectures for PAA where first proposed by Frankel and co-workers [32] also based on the fiber-optic prism concept [18]. The basic configuration, which is shown in the top part of Fig. 8.12 was based on independent dispersive prisms for separate azimuth and elevation control stages. The azimuth dispersive prism stage included an amount of dispersion in each link proportional to the corresponding column position in the array. Here microwave signals, properly time delayed for azimuth steering, were amplified and input to identical dispersive prisms feeding the elements in each column. The elevation dispersion stages included another amount of dispersion in each link which was proportional to the corresponding row position in the array. Using this configuration, they were able to demonstrate independent ±30° azimuth and ±30° elevation steering with no beamsquint over a bandwidth of 6–18 GHz. In another experiment [33], Tong and Wu extended their concept of PDM to implement a 4 × 2 2D array with 2-bit × 2-bit resolution. This was achieved by means of a combination of wavelength-dependent and wavelength-independent time delays as shown in the bottom part of Fig. 8.12.

One of the most serious limitations of using dispersive elements to implement TTDs is their inherent band limitation due to the carrier suppression effect that arises from the signal cancellation when the beating of both RF sidebands and the optical carrier are equal in magnitude but opposite in sign. This limitation occurs when employing double sideband (DSB) modulation and was anticipated in [34, 35]. This limitation can be overcome using single-sideband (SSB) modulation [36, 37]. For instance, in [37], Ortega and co-workers demonstrated the feasibility of this approach when using an LCFBG and multiple wavelengths to implement TTDs as shown in the top part of Fig. 8.13. For an equivalent 32-antenna element configuration, beampointing direction was maintained over a microwave frequency range of 2–7 GHz when using DSB as shown in the bottom left part of Fig. 8.13 while this range was extended to over 18 GHz when using SSB (bottom right part of Fig. 8.13). Furthermore, it was also shown that DSB technique is more restrictive in relation to the required grating characteristics since the operation up to a certain maximum microwave frequency imposes a limitation on the time-delay slope and therefore the spectral region where the chirp is active must be large enough to provide the allocation of a fixed number of elements. For instance, it was shown that while a 10 nm chirp bandwidth is required to feed a 32-antenna element from 0 to 18 GHz using DSB, this figure drops down to 2.5 nm if SSB is employed.

The continuous beamsteering schemes based on LCFBGs can be easily extended to implement a simultaneous upconversion of the microwave signal either by placing a second electro-optic modulator in tandem with the original one [38, 39] or by using a single device biased at the minimum or maximum transmission point [40].

A significant advantage of using a single LCFBG as a TTD element for the whole PAA system is that it avoids the time inaccuracies induced by the position mismatches when using many gratings in serial configuration. However, the resulting system can be certainly complex as in principle one needs to use multiwavelength tunable laser sources where the emission wavelengths must be supervised by a programmable wavelength controller [26, 35]. Furthermore, the tunable bandpass filters required at the front of each photodetector should also be tuned synchronously with the wavelengths of the tunable laser sources. In [41], Ortega and co-workers proposed a variable time delay line where strain perturbation is applied at different positions of a uniform FBG to get different time delays. This enables the use of a single fixed optical wavelength to operate the system. In essence,

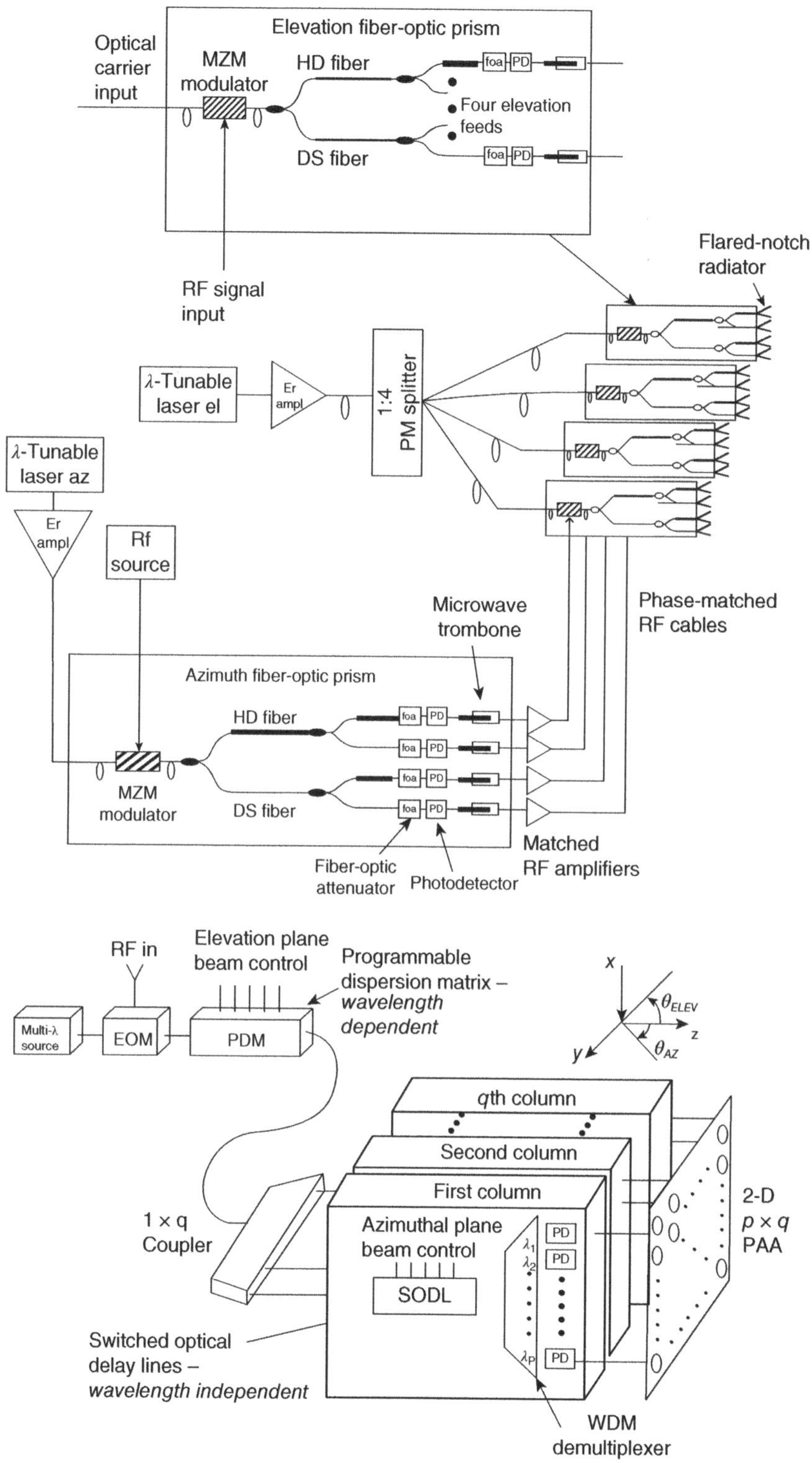

Figure 8.12 Configurations for 2D guided-wave PAAS. (Top) Independent dispersive prisms for separate azimuth and elevation control stages. *Source:* Frankel et al. [32]/IEEE. (Bottom) 2D array with 2-bit × 2-bit resolution based on the concept of programmable dispersion matrix. *Source:* Tong et al. [33]/IEEE.

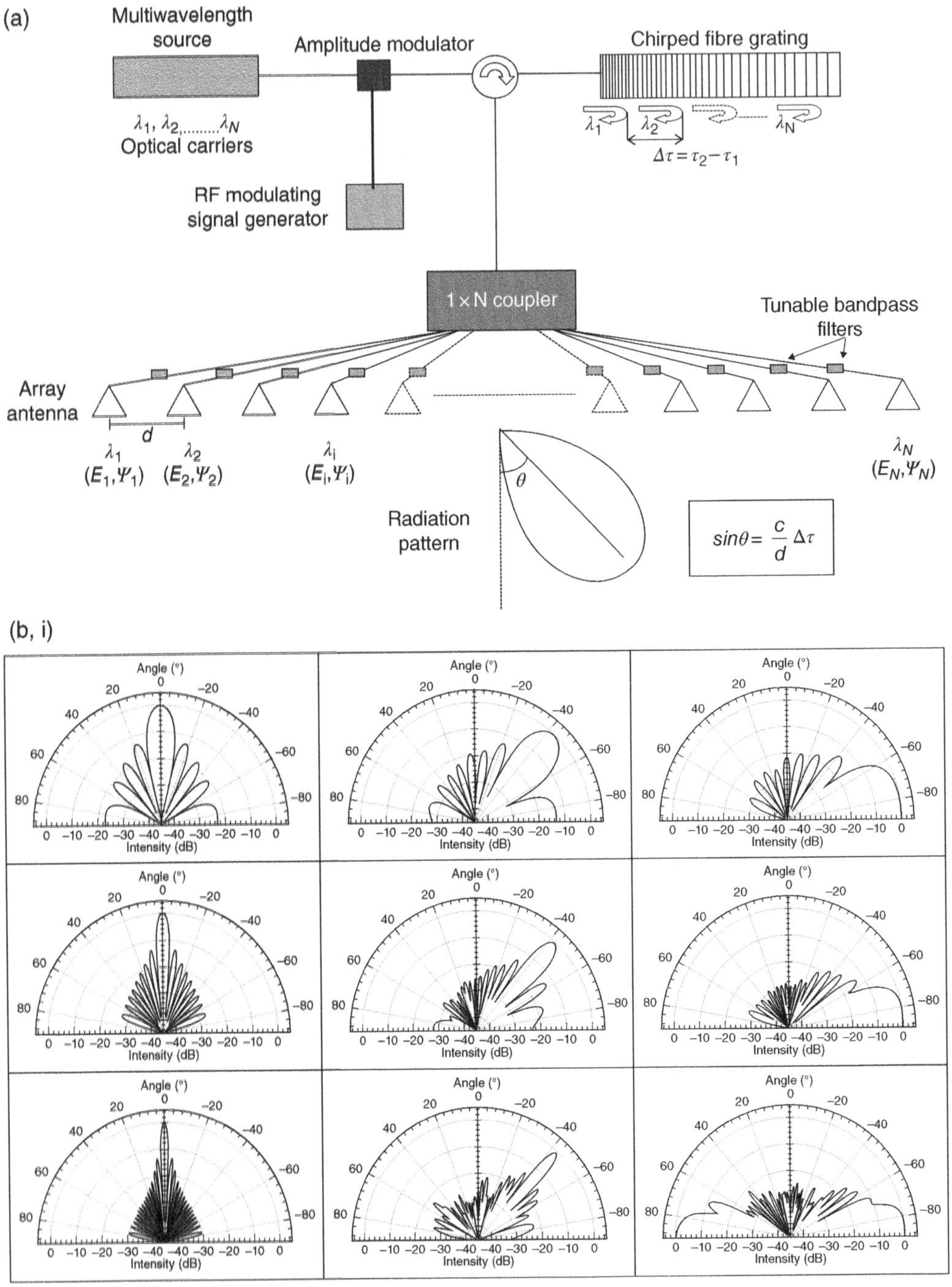

Figure 8.13 (a) Broadband PAA WDM architecture based on single-sideband modulation and an LCFBG delay line. *Source:* Ortega et al. [37]/IEEE. (b) Radiation patterns for a 32-element antenna using double-sideband modulation (i) and single-sideband modulation (ii).

(b, ii)

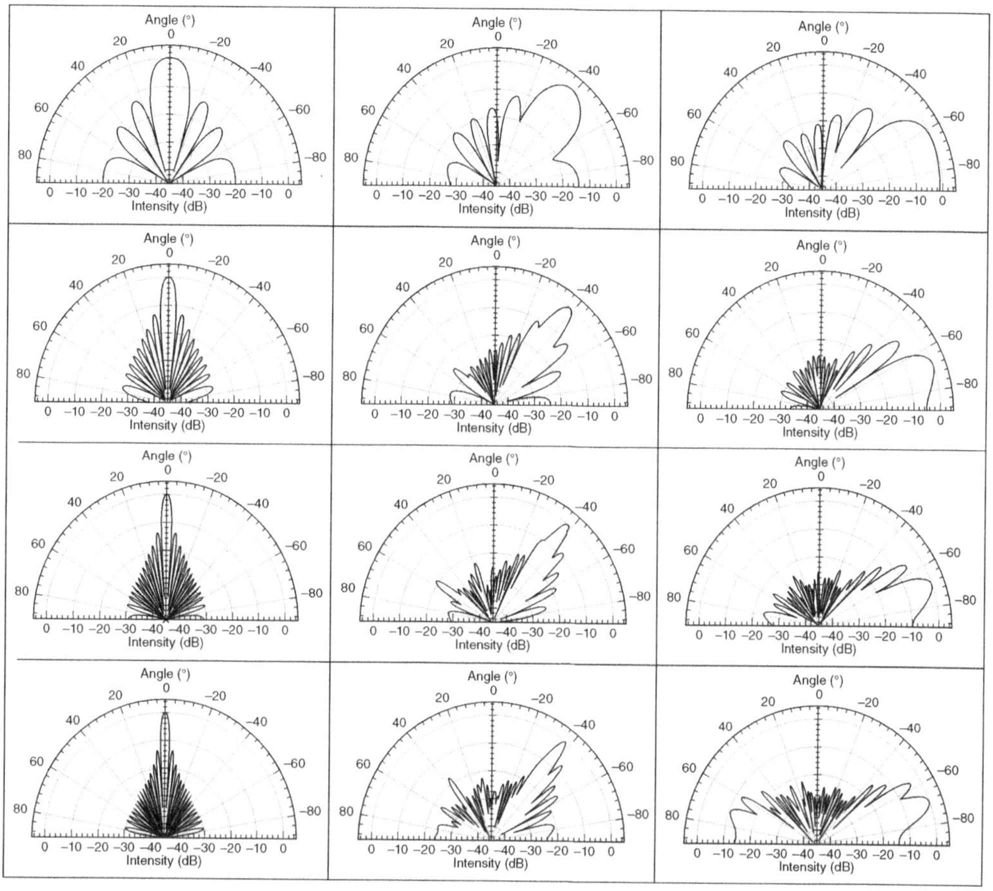

Figure 8.13 (Continued)

the configuration, which is shown in the top part of Fig. 8.14, is based on a uniform FBG with a section whose period is modified by an external actuator. The optical carrier conveying the RF modulation is reflected by the perturbed section of the grating and the time delay is modified by changing the position of the perturbation along the grating.

By using a 5 cm long grating, continuous time delay operation was demonstrated up to 14 GHz with a maximum value of 330-ps time delay obtained. The bottom left part of Fig. 8.14 shows the reflectivity and group delay obtained when the perturbed area was located at two different positions along the grating (5 and 40 mm respectively). The right part of the spectrum shows the resonance due to the perturbed section. The bottom right part of the figure shows the theoretical and measured values of the phase shift experienced by the RF carrier versus the frequency taking the position of the perturbed area as a parameter.

Another alternative is to use a multiwavelength tunable source that can simultaneously generate multiwavelength lasing with equally increased or decreased wavelength spacing. A simple and efficient approach based on a modification of a multiwavelength fiber laser using a cascade of sampled FBGs (SFBG) [42] is reported in [43] where the SFBGs are replaced by cascaded FBGs and tuning is realized by stretching or compressing the FBGs cascaded in the laser cavity as shown in the top left part of Fig. 8.15. To make the wavelength spacing increase or decrease equally, the gratings were mounted onto an organic plate with angles between adjacent gratings selected in such a way that

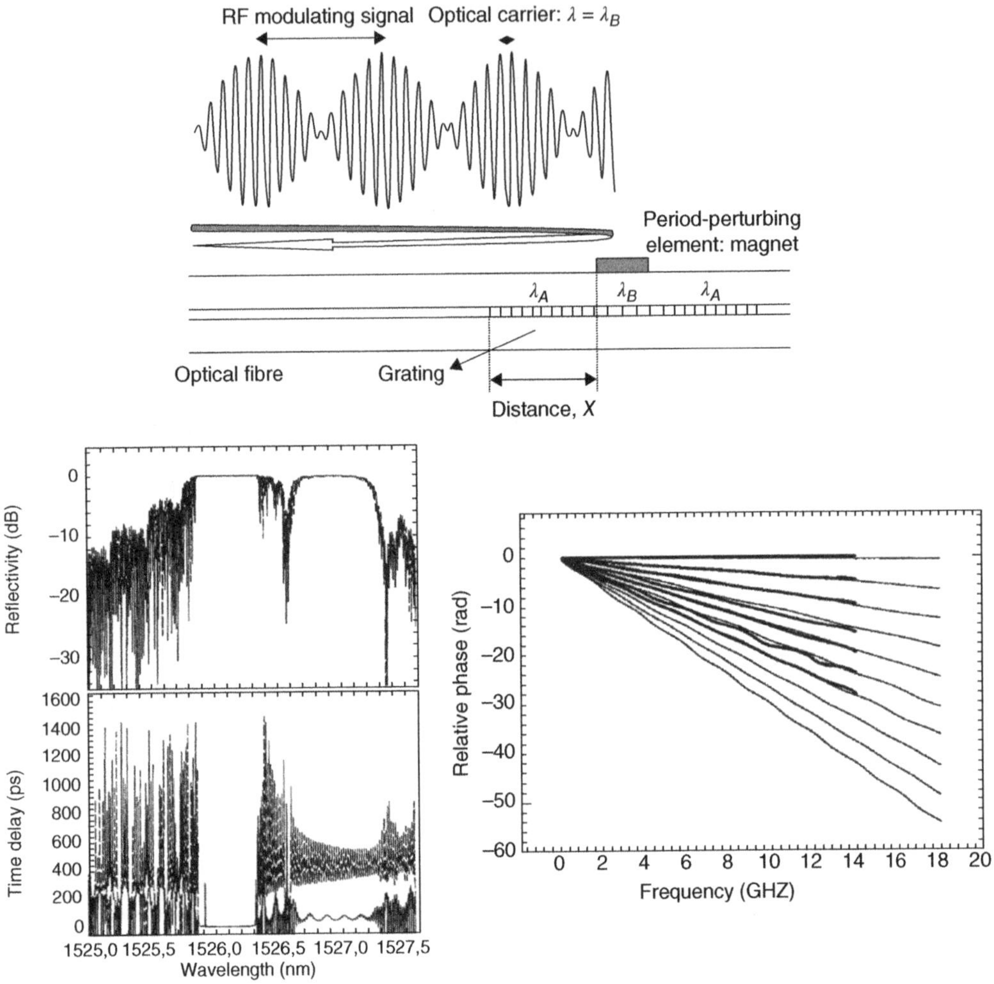

Figure 8.14 (Top) Tunable time-delay line where strain perturbation is applied at different positions of a uniform FBG to get different time delays. *Source:* Ortega et al. [41]/IEEE. (Bottom) Reflectivity and group delay characteristics (left). Theoretical and measured values of the phase shift experienced by the RF carrier versus the frequency taking the position of the perturbed area as a parameter (right).

the forces applied to the gratings generated equally spaced increased or decreased wavelength spacing as shown in the top middle part of Fig. 8.15. In particular, the angles were selected such that the compressing or stretching forces applied to the gratings were 0, ΔF, $2\Delta F$, $3\Delta F$, $\rightleftharpoons$ which resulted in wavelength shiftings in the gratings given by 0, $\Delta\lambda$, $2\Delta\lambda$, $3\Delta\lambda$, $\rightleftharpoons$ as shown in the diagram depicted in the top right part of Fig. 8.15. Using this configuration, a four-chirped-grating-based TTD system was experimentally demonstrated operating at 4 GHz. The experimental configuration, shown in the bottom left part of Fig. 8.15, required tunable filters synchronized with the multiwavelength laser but was able to produce an excellent linear time delay vs. wavelength spacing characteristic as shown in the bottom right part of the same figure.

A simpler approach based on tunable LCFBGs, fixed optical filters and a fixed multiwavelength laser was reported by Liu and co-workers [44]. The layout, shown in the top left part of Fig. 8.16 avoids the complicated wavelength tuning and the synchronous control of the tunable lasers as well as the tunable bandpass filters.

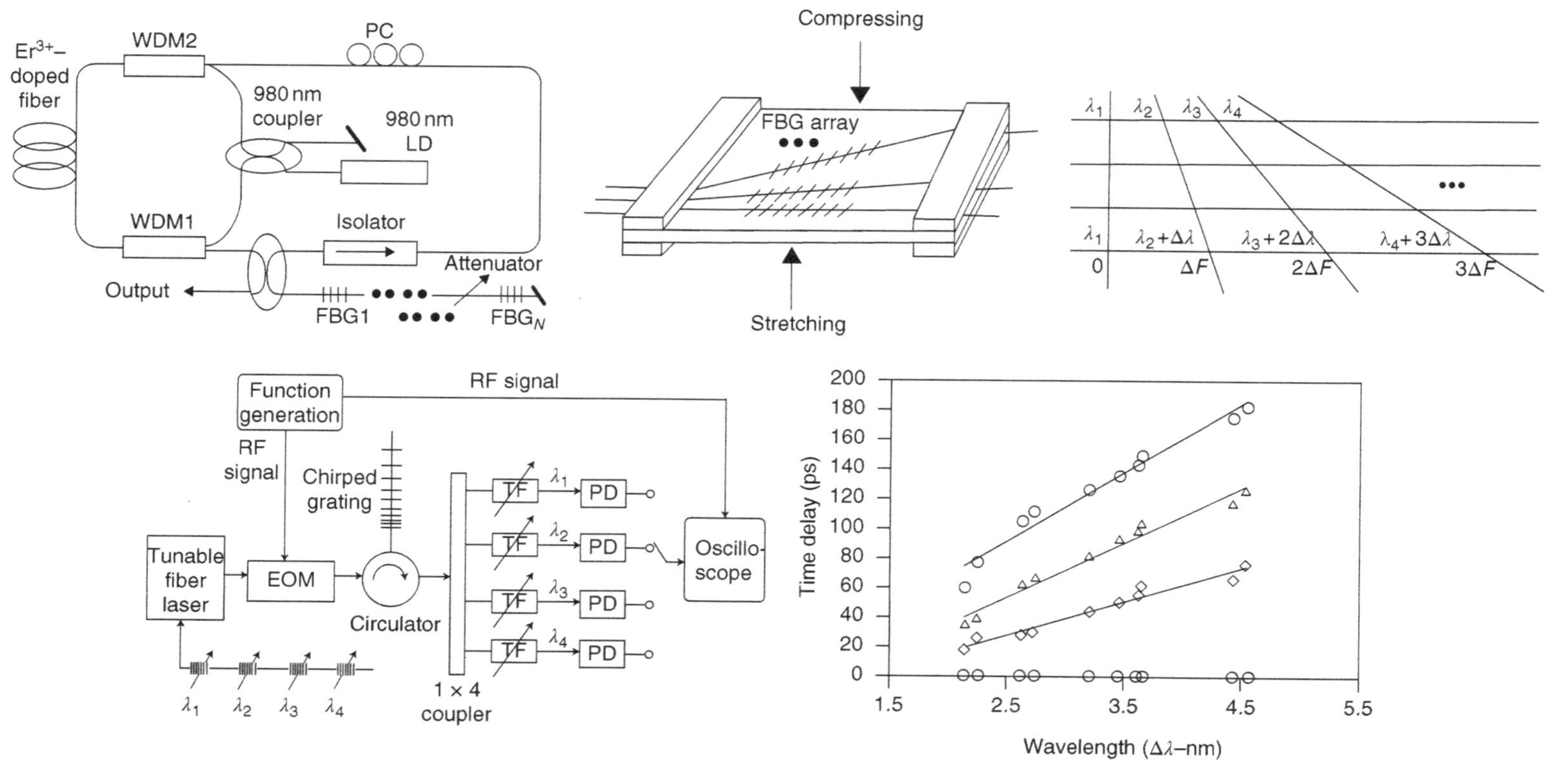

Figure 8.15 (Top) Tunable multiwavelength fiber laser source based on an FBG array (left) mounted on an organic plate (Middle) to achieve equal wavelength separation in response to a given stretching or compressing force (right). (Bottom) PAA configuration based on a fixed LCFBG and the tunable multiwavelength source (left). Time delay versus wavelength spacing characteristic for the PAA at 4 GHz. *Source:* Yao et al. [43]/IEEE.

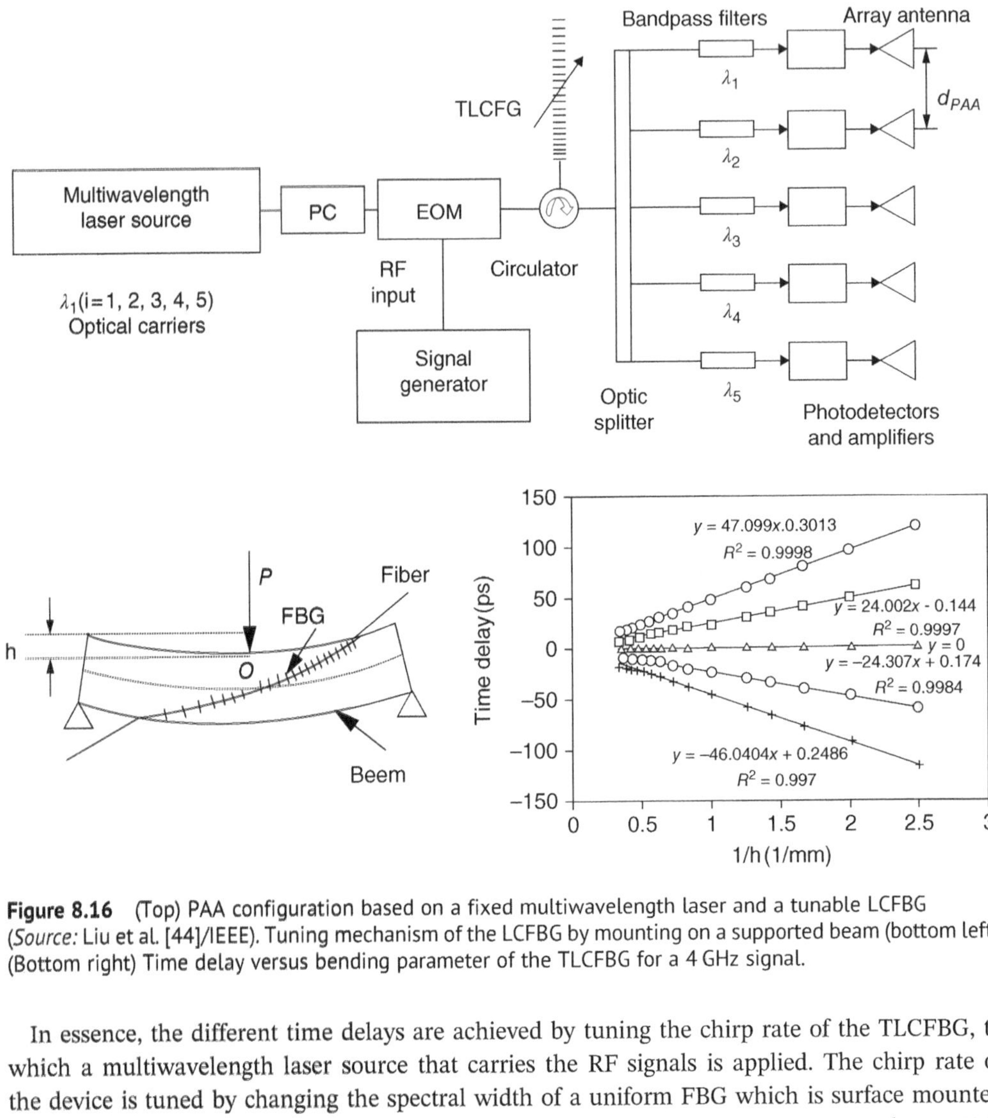

Figure 8.16 (Top) PAA configuration based on a fixed multiwavelength laser and a tunable LCFBG (*Source:* Liu et al. [44]/IEEE). Tuning mechanism of the LCFBG by mounting on a supported beam (bottom left). (Bottom right) Time delay versus bending parameter of the TLCFBG for a 4 GHz signal.

In essence, the different time delays are achieved by tuning the chirp rate of the TLCFBG, to which a multiwavelength laser source that carries the RF signals is applied. The chirp rate of the device is tuned by changing the spectral width of a uniform FBG which is surface mounted on a simply supported beam as shown in the top right part of Fig. 8.16. Referring to the TLFBG, if $\Delta\lambda$ represents the wavelength separation between the optical carriers produced by the multiwavelength source, then the time delay between adjacent carriers can be expressed as:

$$\Delta t = C\Delta\lambda \tag{8.33}$$

where C is the chirp parameter of the TLFBG. For the simple beam configuration setup depicted in Fig. 8.16, the spectral width of the chirped grating can be expressed as $\Delta\lambda_G = kh$ with k being a constant determined by the parameters of the beam and the FBG and h being the beam deflection. Since the time delay of the fiber grating is given by $\tau = C\Delta\lambda_G$, it follows that $\Delta t = \tau\Delta\lambda/(kh)$ and thus that the incremental delay can be changed by means of changing the value of h. In [44], a five-element TTD system was demonstrated capable of operating in wideband continuous up to radio-frequencies of 10 GHz. The bottom part of Fig. 8.16 plots the delays furnished by the TLCFBG as a function of $1/h$ for the five wavelengths delivered by the laser and an RF frequency of 4 GHz.

An efficient way to combat the limitation of beamforming networks based on FBG prisms and achieve high-resolution or even continuous angle-tuning capability or a higher operation frequency is to incorporate one or more LCFBGs in the structure. In [45], for instance, one LCFBG is incorporated to provide very small time delays that discrete FBGs cannot produce (which according to (8.8.32) results in a potentially higher operation RF frequency) while the rest of the branches in the prism are implemented using discrete FBGs to provide large time delays. The layout shown in the top part of Fig. 8.17 was employed to implement a 5 × 13 fiber grating prism for the tunable beamforming of a 6 GHz signal. Yet this configuration still produces a discrete angle tuning. The most versatile configuration however is that where all the branches in the prism are implemented using LCFBGs. In [46], Liu and co-workers proposed and experimentally demonstrated such a configuration, the layout of which is shown in the bottom part of Fig. 8.17.

The basic principle is the same as with other prisms but in this case, a continuous angle tuning is achieved by wavelength tuning of the input tunable laser. Tuning resolution is limited in this case

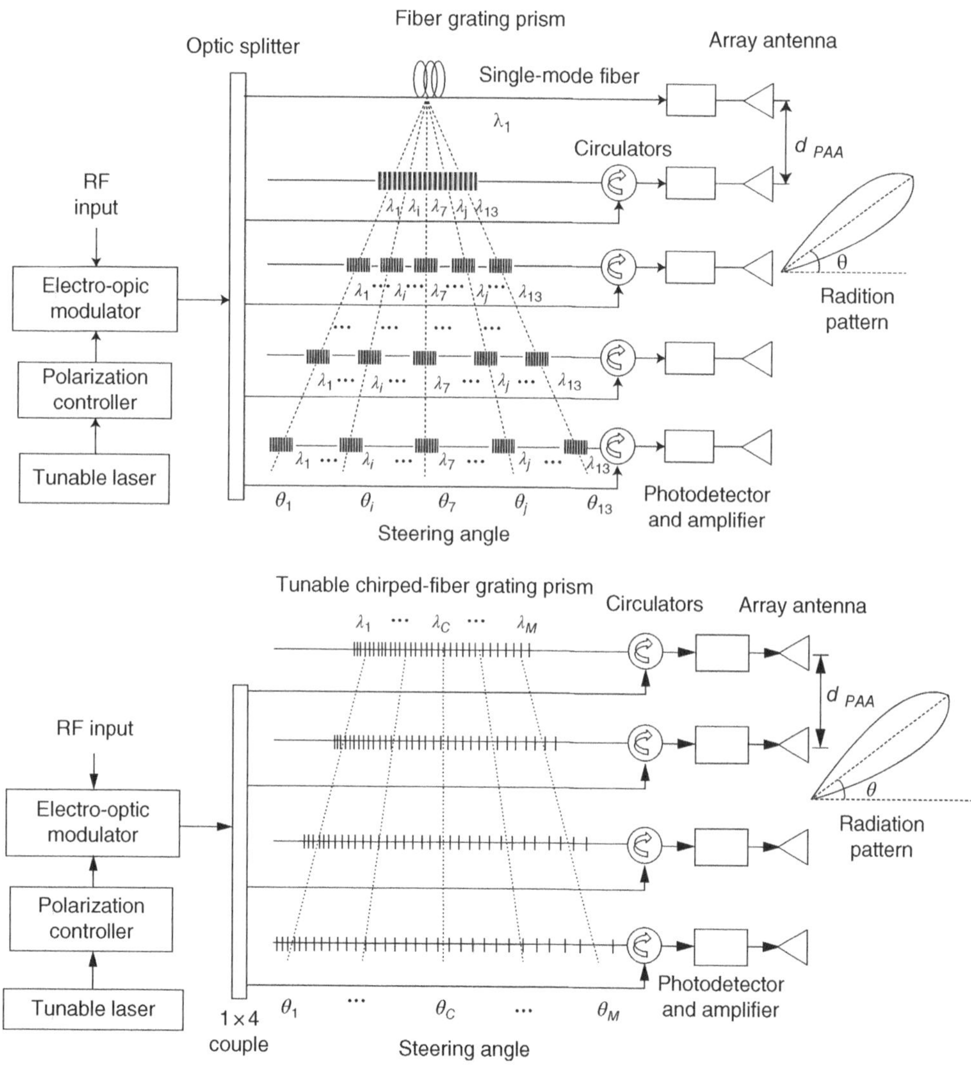

Figure 8.17 (Top) PAA providing high-resolution/continuous angle tuning capability by incorporating an LCFBG in one branch of an FBG and discrete FBG arrays in the rest. *Source:* Liu et al. [45]/with permission of Elsevier. (Bottom) An enhanced version where all the FBGs in the prism branches are linearly chirped. *Source:* Liu et al. [46]/Optical Society of America.

by the wavelength step of the source and not by the minimum separation between gratings. Liu et al. demonstrated a four-element beamformer capable of steering signals of up to 20 GHz RF frequency.

A number of contributions have focused, during the last years, on incorporating innovative devices and/or fiber-based technologies [47–56]. For instance, [47, 48] proposed and evaluated the incorporation of AWG devices in combination with fiber-based dispersive delay lines to the beamforming network part of a PAA. A first fully packaged TTD module for K-band PAA was reported and experimentally demonstrated in [49] based on a mixed fiber and holographic coupling technology. An improved version of the single wavelength perturbed variable delay line proposed in [38] was reported in [50] where a photonic true-time delay based on a tapered FBG with resistive coated was reported and demonstrated. Its main advantage relied on the fact that it was voltage-controlled and provided continuous operation requiring no moving parts and no mechanical perturbation. Raz and co-workers reported the first experimental demonstration of an optical TTD based on high-order mode (HOM) dispersion-compensating fibers (DCFs) [51]. These devices bring up advantages in terms of their very high dispersion (>450 ps/km nm), low insertion losses, and very low phase ripple over a considerable RF bandwidth. Multichannel LCFBGs have been considered as well for the implementation of continuously variable TTDs [52] and switched delay lines have been proposed and experimentally tested in adaptive antennas for fixed and mobile access networks [53] and for 2D PAAs in the X-band [54]. More recently, TTDs have been proposed and demonstrated based on photonic crystal fibers (PCF) [55] and SBS [56]. The use of coherent detection for fiber-based optical beamforming networks has received less attention from the research community, although this tendency is certainly on the way to changing, especially in the context of new integrated optic subsystems which are covered in detail in the next section. The reader can find a comprehensive treatment of the use of coherent detection in TTD PAA in [57].

Several groups have reported activity on the implementation of optical beamforming networks based on integrated photonic components. A variety of approaches have been proposed which include the use of MEM devices [58, 59], silica planar waveguides [60], polymer switches [61], and III–V semiconductor integrated optic components [53, 62, 63]. All the above report hybrid configurations, whereby only a part of the optical beamforming network is implemented using integrated components. The recent development of silicon photonics however opens the way to a complete integrated PAA architecture in this platform as proposed in [64]. Compared to other competing technologies and especially with lithium niobate and III–V semiconductor technologies silicon photonics brings advantages in terms of cost-effectiveness, CMOS compatibility, compactness, and integration. A first step in this evolution path is that reported in [65, 66]. There a PAA architecture is proposed and experimentally demonstrated in which the core is an optical beam-forming network that employs ring resonator-based broadband delays, and coherent optical combining. The general layout of the PAA is shown in the top left part of Fig. 8.18 where the electro-optical conversion is performed by means of single-sideband suppressed carrier modulation, employing a common laser, Mach–Zehnder modulators, and a common optical sideband filter after the OBFN. The unmodulated laser signal is then reinjected in order to perform balanced coherent optical detection, for the optoelectronic conversion.

The optical beamforming network is based on a binary tree topology. In the reported case which is shown in the top right part of Fig. 8.18, it consists of a three-stage architecture with eight inputs, seven tunable combiners and eight ring cavities the resonance positions of which can be tuned by means of eight phase shifters. Delays are provided for the selected RF sideband by the ring cavities which act as Gires-Tournois phase-only filters with each input-output configuration offering a different delay, multiple of a basic value. The integrated waveguides were implemented employing a CMOS-compatible TripleX® technology. The bottom left part of Fig. 8.18 shows a cross section and

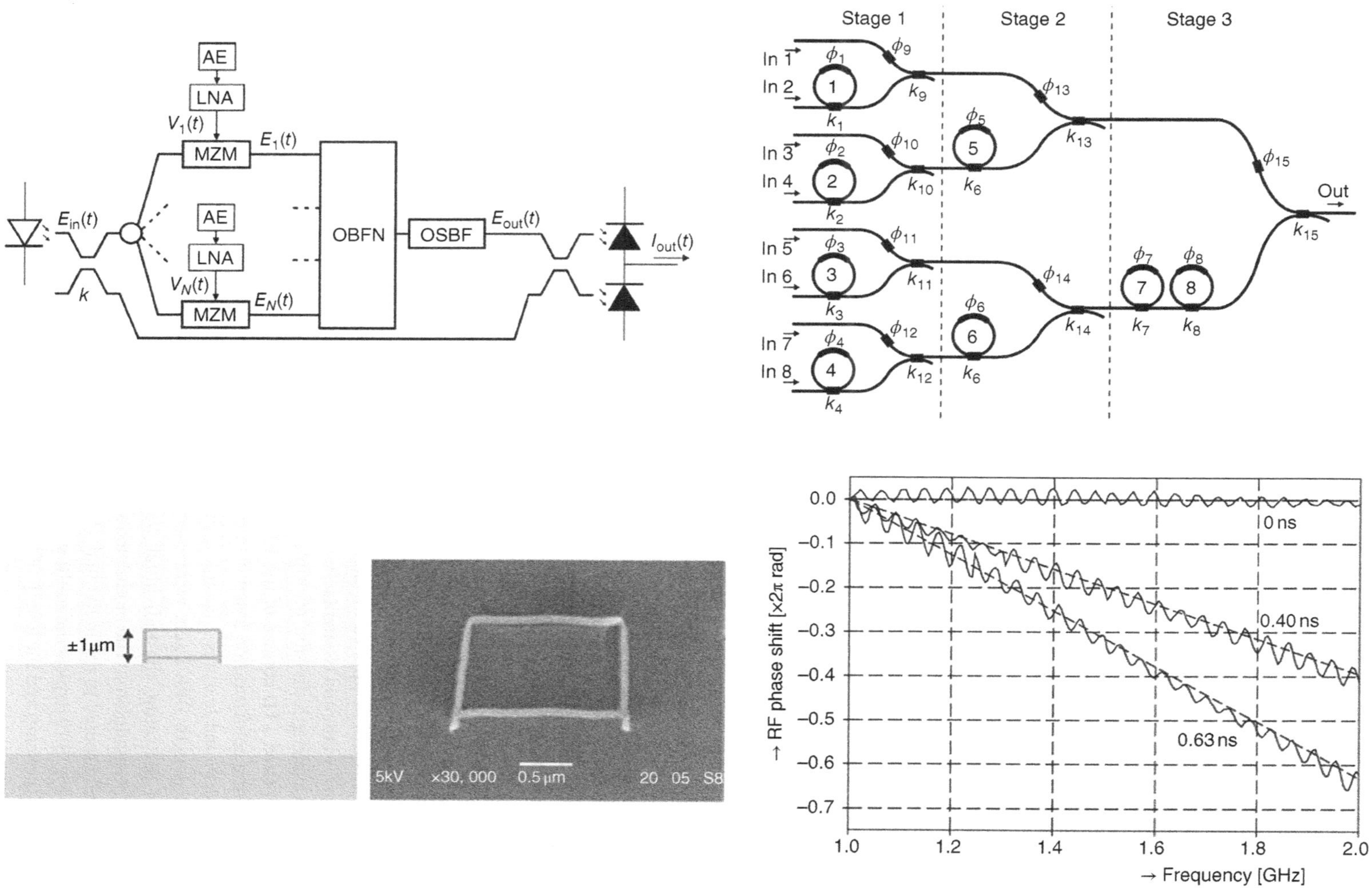

Figure 8.18 (Top). Integrated PAA layout (left) based on an optical beamformer implemented by means of a series of ring resonators disposed in a branch layout (right). (Bottom) Cross-section and image of the Triplex® waveguide profile (left). RF phase shift versus frequency characteristic of the integrated beamformer (right). *Source:* Zhuang et al. [66]/from IEEE.

a microscope image of this waveguide technology. The integrated part of the PAA not only included the design, fabrication, and characterization of the 8 × 1 optical beamforming network but also the integration of the optical sideband filter for single-sideband suppressed carrier modulation, and the carrier reinsertion coupler for balanced optical detection. The chip was designed for a particular application involving satellite television reception using a broadband PAA in a frequency range of 1–2 GHz. Delay tuning was performed acting thermo-optically over the ring cavities, with a switching time of 1 ms. Group delay response, as shown in the bottom right part of Fig. 8.18 and power response measurements confirmed the correct design and operation of the circuit.

8.4 Phase-Shift Optical Beamforming Networks

Beamforming networks based on phase shifters are in principle less attractive since they are narrowband limited due to the beamsquint phenomenon discussed in Section 8.2. Nevertheless, there is an interest in these configurations because the combination of continuous tunable RF phase-shift and digitally programmable true-time delays can provide maximum flexibility in the array factor synthesis [11]. For instance, tunable phase-shift can be used to introduce nulls in the beam at certain angles in order to suppress undesired interference. Moreover, in configurations in which the delays provided by the optical beamforming networks are quantized, these can be supplemented with continuous phase-shift tunability to provide precise beamsteering.

To date, three main approaches have been reported, each one relying on a particular technology to optically implement tunable RF phase shifters: SBS, Slow and Fast light techniques and resonance phase shifting in waveguide resonators. In the first case [67], the technique is based on a method previously developed for wideband RF phase shifting [68] and complex-valued coefficient microwave-photonics filters [69, 70]. The proposed system is schematically shown in the top left part of Fig. 8.19. An optical comb generator is used to generate a multi-line optical signal which is then fed to an optical single sideband (OSSB) modulator that is driven by the RF signal to be transmitted. Each of the spectral lines of the comb provides an optical carrier that is modulated by the RF signal to obtain a multi-wavelength OSSB signal. The number of optical carriers equals the number of antenna elements of the array.

The multi-wavelength OSSB signal is then processed in parallel using SBS. This processing is based on the introduction of an optical phase-shift to the optical carrier [68] so that upon detection, it translates to an RF phase-shift that is frequency-independent [68]. With this configuration, independent phase-shits can be applied to each optical carrier and thus independent RF phase-shift is obtained at the array antenna elements. In the final stage, the different modulated wavelengths are separated in a wavelength division multiplexer (WDM) and routed via programmable ODLs that provide the quantized TTD control of the RF signals for each antenna element.

The experimental configuration assembled to provide a proof of concept is shown in the central top part of Fig. 8.19 and features a beamforming network for a five-element array antenna. Individual fixed-wavelength DFB lasers were used to generate the necessary optical carriers. In the lower branch of the setup, the individual wavelengths are combined using a 200-GHz WDM and fed to an OSSB modulator. The modulated signals were then propagated through a 2-km length of DCF, and finally routed to the individual antenna elements using another 200-GHz WDM. The top right part of Fig. 8.19 depicts the multi-wavelength OSSB signals at the input of the WDM. The upper branch of the setup and the length of DCF were used for parallel SBS processing of the optical carriers. In this case, the Stokes and pump waves from the optical carrier were simultaneously generated by using carrier-suppressed double-sideband modulation (ODSB-SC) in a conventional

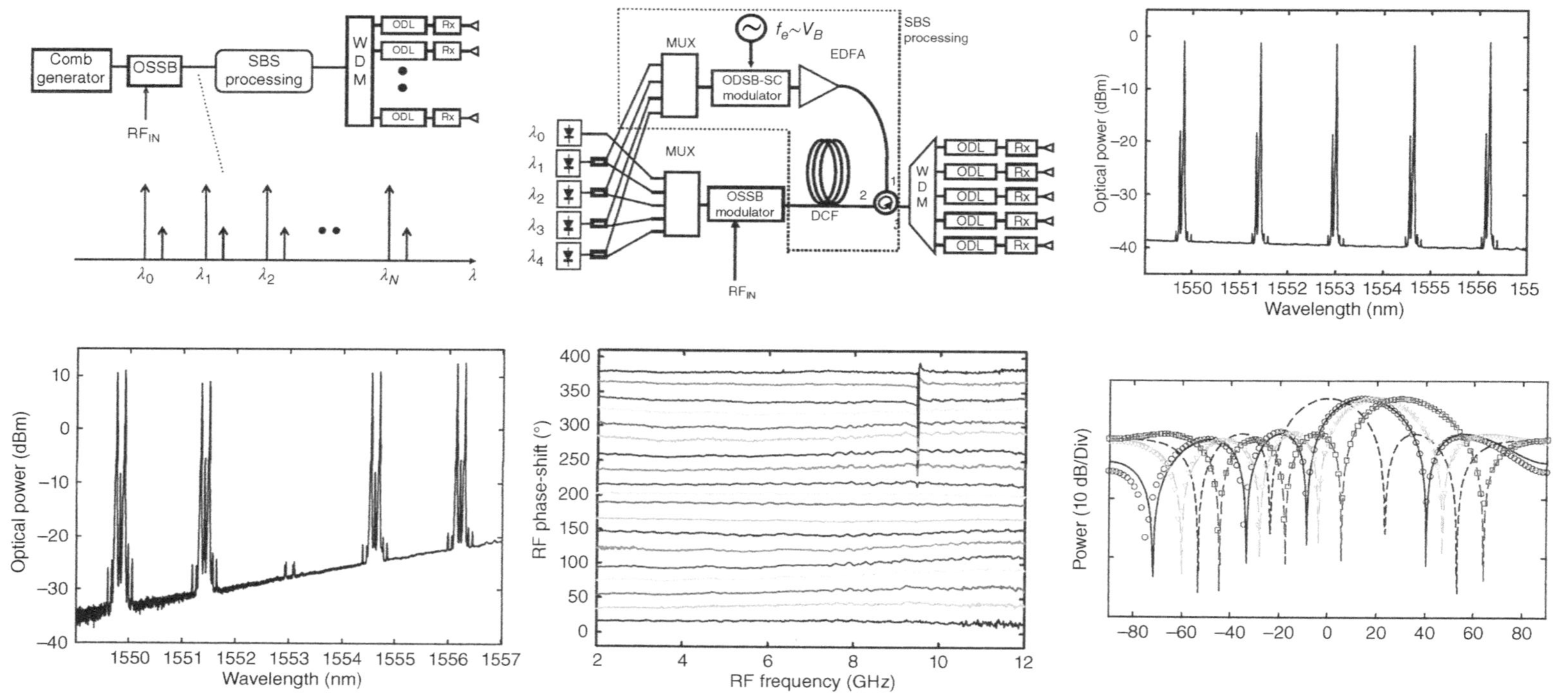

Figure 8.19 (Top) Layout of the phase-shift optical beamformer based on SBS (*Source:* Loayssa and Sagues [67]/IEEE) (left). Experimental configuration of a five-element array (center). Optical spectra of the OSSB (top right) and (bottom left) carrier suppressed optical signals. Phase shift versus frequency characteristic (bottom central) and Broadside radiation diagram for a 7 GHz signal tuned by phase-shifting (bottom right).

Mach–Zehnder modulator biased at minimum transmission. The modulator was driven with an RF tone of a frequency close to the Brillouin frequency shift in the DCF (9.4 GHz).

The bottom left part of Fig. 8.19 clearly shows the generated pump and Stokes waves at the five wavelengths used in the setup for a particular combination of phase-shifts in the array. An optical circulator was used to counter-propagate the pumps and Stokes waves with the optical carriers in the length of fiber. The simultaneous interaction among Stokes wave, optical carrier, and pump wave led to an optical phase-shift being introduced on the optical carrier [68]. This phase shift can be simply tuned by the power of the Stokes and pump at each wavelength [70]. The central bottom part of Fig. 8.19 shows the measured values for the RF phase-shift in one of the antenna's elements as the power of its corresponding Stokes and pump waves power is tuned. A frequency-independent RF phase shift can be noticed which can be continuously tuned in a range larger than 360°. Furthermore, the bottom right part of Fig. 8.19 displays how the measured array pattern at broadside for a 7 GHz signal can be finely tuned by means of the tunable phase shifter.

The recent development of a variety of novel techniques [71–75] to achieve the slowing down of the speed of light has opened a new possibility toward the photonic control of microwave signals conveyed by optical carriers. Indeed, it has been shown that the phase delay of a microwave signal modulating an optical carrier can be controlled by propagation through an alternating electroabsorbing (EA) and amplifying (SOA) semiconductor waveguide [76, 77]. This effect is based on a change in the group velocity of the light and the physical phenomenon which can be described as a wave mixing between the optical carrier and the modulation sidebands is sometimes referred to as coherent population oscillations [73, 76]. In [76], it was specifically shown that EA waveguides can change the phase of the microwave signals, while in [77], the SOA-EA structure has been proposed to improve the range of the change of the phase of the microwave delays. The SOA is used to increase the input power into the EA and, hence, the saturation and the phase shift of the microwave signals. The SOA is also used to compensate losses of the EA and even to obtain net gain. The use of the SOA-EA structure allows a more flexible reconfiguration of the coefficients of the phased-array antenna. By changing the bias of both the SOA and EA, it is possible to adjust both phase and amplitude independently. The application of SOA-EA structures to the implementation of tunable phase shifters for PAA was first proposed in [78]. The top left part of Fig. 8.20 illustrates the scheme of an optically fed phased-array antenna where the phase provided to each antenna is controlled by the use of the alternating SOA and EA sections.

In this particular implementation, the RF signal is modulated onto an optical carrier and transmitted to photodetectors that are connected to individual antennas. The central top part of Fig. 8.20 shows the measured microwave phase shift versus frequency in an EA-SOA-EA structure when the bias voltage of both EAs changes from 0 V to the indicated bias voltages in the range from 0.25 to 2.5 V and the input optical power is fixed at 3.25 dBm. The current of the SOA is 200 mA, the length is around 750 m, the output saturation power is 12 dBm, and the optical wavelength of the laser is 1545 nm. A maximum change of the phase shift of the microwave tone of almost 60° at 10 GHz was achieved which for a three-element PAA results in a 15° shift in the radiation pattern as shown in the top right part of Fig. 8.20. Further investigation [79] has shown that higher phase shifts (over 360°) can be obtained by exploiting the phenomenon of coherent population oscillations in SOA waveguides combined with optical filters to suppress the red-shifted RF sideband. Almost flat broadband operation up to 40 GHz has been demonstrated with a structure comprising five SOAs shown in the bottom left part of Fig. 8.20, and three optical filters. Three of the five SOAs provide the tunable phase shift by adequate control of their injection currents as shown in the bottom right part of Fig. 8.20. This structure has been recently simplified to a single SOA + optical filter [80].

The last approach that we consider here is based on the tunable RF phase-shifting functionality provided by placing the optical carrier of an OSSB signal within the resonance of a ring waveguide cavity shown in the top left part of Fig. 8.21 while leaving the sideband outside of it [82, 83].

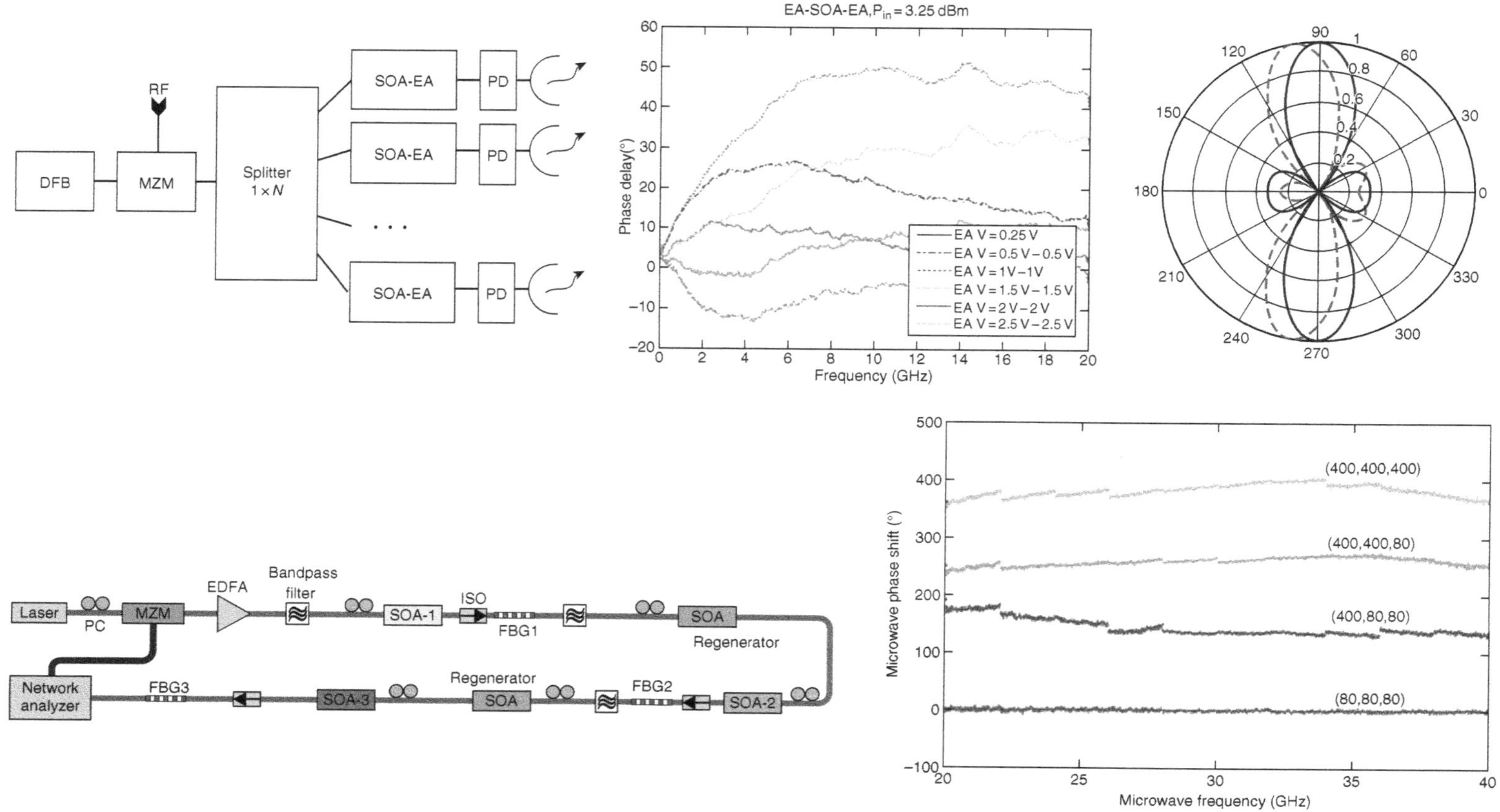

Figure 8.20 (Top) Layout of a phase-shift optical beamformer based on alternating SOA-EA semiconductor waveguides. *Source:* Sales et al. [78]/IEEE. Tunable phase delay versus frequency for an SOA-EA waveguide (Center) and radiation diagram change (right). (Bottom) Broadband phase shifter based on SOA waveguides and optical filters. (left) and tunable microwave phase shift versus frequency as a function of the biasing current of the SOAs. *Source:* Xue et al. [79]/Optical Society of America.

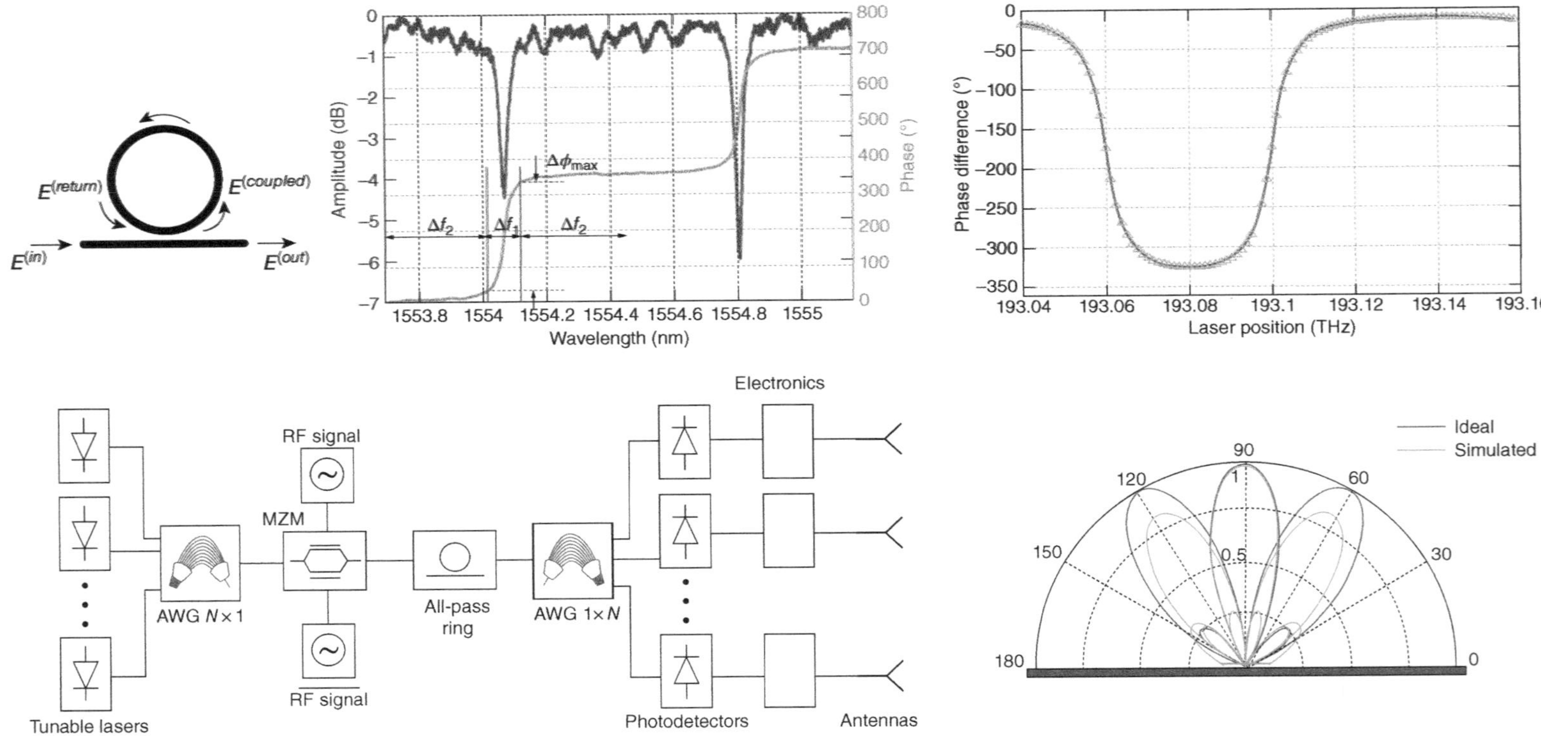

Figure 8.21 (Top) Waveguide ring cavity phase-shifter (left). Periodic spectral amplitude and phase response of a ring cavity phase-shifter (Center). Phase shift versus the position of the optical carrier inside the resonance of a ring cavity (right). (Bottom) Layout of a WDM phase-shift beamformer based on the integrated ring cavity phase shifter. *Source:* Adapted from Domenech et al. [81].

The top central part of Fig. 8.21 shows a typical periodic amplitude and phase spectrum of a ring cavity. Note that in the resonances, a spectral region $\Delta\lambda_1$ suffers a dramatic phase shift while the rest of the spectrum suffers no phase shift. Thus, by placing the optical carrier inside it while leaving the RF sideband outside a phase difference $\Delta\phi$ is impressed between the optical carrier and the Sideband which upon detection is impressed over the RF carrier. This phase shift is tunable up to a maximum value $\Delta\phi_{max}$ just by moving the optical carrier within the resonance as shown in the central right part of Fig. 8.21. Furthermore, since due to the periodic nature of the spectrum, independent phase shifts can be impressed over the same RF signal modulating different optical carriers placed within the filter resonances, the concept can be extended [81] to the implementation of multiple RF phase shifts as required by an optical beamforming network. The bottom part of Fig. 8.21 depicts the proposed configuration where N-tunable lasers are multiplexed to provide the independent phase shifts to the antenna element.

8.5 Summary and Conclusions

This chapter has revised the fundamental concepts and technologies behind the optical implementation of beamforming networks for phase array antennas. After an introduction to the basic system theory and parameters of PAA, we have provided a detailed treatment of optical fiber-based architectures both for TTD and for phase-shift PAA which have been proposed in the literature over the last years showing the principal advantages and limitations of each approach. New developments toward the integration of these architectures in photonics chips have also been described.

Problems

8.1 Show that at a distance very far away from the array, Eq. (8.2) can be approximated by:

$$R_j \approx R - \hat{\mathbf{r}} \cdot \mathbf{r}_j$$

where R is the distance measured from the central element of the array in Fig. 8.2 and $P \equiv (x, y, z)$, $\mathbf{r}_j$ is the position vector of an element j and $\hat{\mathbf{r}}$ is a unit vector in the direction connecting the origin of coordinates and $P \equiv (x, y, z)$.

8.2 Using the result given by Eq. (8.3), show that Eq. (8.2) can be expressed in the form given by Eq. (8.4). Justify your approximations.

8.3 Broadband angle scanning in a PAA is achieved by using a TTD where the weights in the array factor are given by:

$$\begin{aligned} a_s &= |a_s| e^{-jk\mathbf{r}_s \cdot \hat{\mathbf{r}}_o} \\ \hat{\mathbf{r}}_o &= u_o\hat{\mathbf{x}} + v_o\hat{\mathbf{y}} + \cos\theta_o\hat{\mathbf{z}} \\ u_0 &= \sin\theta_o \cos\varphi_o \\ v_o &= \sin\theta_o \sin\varphi_o \end{aligned}$$

Show that in this case, the beam-pointing direction is characterized by angles (θ_o, φ_o) and that this beam-pointing direction is independent of frequency (and thus inherently broadband). Specialize the relevant expressions for 1D operation, where the PAA has N-radiating elements placed at $\mathbf{r}_s = sd\hat{\mathbf{x}}, s = \{0, 1, 2, \rightleftharpoons, N-1\}$.

8.4 Consider a nine-element 1D PAA, where the elements are placed on the x axis according to:

$$\mathbf{r}_s = sd\hat{\mathbf{x}}, \quad s = \{1, 2, 3, \ldots, 9\}$$

Calculate the array factor if all the elements are fed with uniform amplitude $|a_s| = 1$. What is the steering angle?

8.5 Consider the 1D PAA of the former exercise. Design the weighting factors for the radiating element so the new steering angle is 30° while keeping their amplitude uniform and a TTD approach. Calculate the new expression for the array factor.

8.6 A narrowband PAA is implemented using tunable phase shifters instead of TTDs. In this case, the elements are designed to operate at a given wavelength λ_o and the weights are given by:

$$a_s = |a_s| e^{-jk_o \mathbf{r_s} \cdot \hat{\mathbf{r}}_\mathbf{o}} \quad k_o = 2\pi/\lambda_o = \omega_0/c$$
$$\hat{\mathbf{r}}_\mathbf{o} = u_o\hat{\mathbf{x}} + v_o\hat{\mathbf{y}} + \cos\theta_o\hat{\mathbf{z}}$$
$$u_0 = \sin\theta_o \cos\varphi_o$$
$$v_o = \sin\theta_o \sin\varphi_o$$

Derive an expression for the array factor and the pointing angle direction in this case and show that the latter is frequency dependent, leading to the beam squint phenomenon.

8.7 Consider a 1-D uniformly illuminated PAA with 16 elements spaced half wavelength apart and designed to operate at the lowest frequency of the Ku band ($f_o = 12$ GHz). Compute the 3dB beam-width for broadside operation, 10°, 20°, 30°, 45°, and 60° pointing directions. Comment on the results.

8.8 The bandwidth limitation due to beamsquint is given by (8.22). Use the results of exercises 6 and 7 to prove this equation.

8.9 Consider a uniform 1D PAA composed of 6 elements operating at a central frequency placed at the K-band ($f_o = 15$ GHz) and separated by a distance of 1 cm ($d = \lambda/2$). The PAA uses phase shifters so it is subject to beamsquint. Calculate the frequency bandwidth for the PAA for $\theta_o = 30°$, 45°, and 60°. Consider now that the PAA is changed to hold 16 radiating elements. Compute the same bandwidth values for this case.

8.10 An optical dispersive delay line is based on the configuration shown in the top left part of Fig. 8.8 using an SMF-28 singlemode fiber with the following parameters: $L = 100$ m, v_{go} (1550 nm) $= 2 \times 10^8$ m/s, and $D = 17$ ps/(km.nm). If the delay line is operated using a tunable laser with tuning range of 40 nm (from 1550 to 1590 nm), compute the maximum differential delay value that it can provide.

8.11 We wish to design a PAA based on a fiber-optic prism delay line configuration (as shown in Fig. 8.8) where the nominal link length is 2 km and two kinds of fiber ($D = 0$ and $\overline{D} = -80$ ps/km · nm) are employed. The central operating wavelength is 1500 nm and the RF operation frequency is 10 GHz. Complete the design steps if a continuous steering direction from −45° to 45° is desired (choose a radiating element separation equal to half the RF wavelength).

8.12 A linearly chirped FBG (LCFBG) is defined by a local period that changes along its length (in the z coordinate) according to $\Lambda(z) = \Lambda_o + Kz$ where K is the chirp parameter. This results in a local Bragg reflection wavelength given by $\lambda_B(z) = \lambda_B(0) + 2n_{eff}Kz$ with n_{eff} representing the average effective index of the fundamental mode. Derive an expression for the beam-steering direction in terms of the relevant parameters of the LCFBG under the assumption of TTD line operation.

References

1 Brookner, E. (1991). *Practical Phased Array Antenna Systems.* Norwood, MA: Artech House Publishing Co.

2 Mailloux, R.J. (2000). *Phased Array Antenna Handbook*, 2e. Artech House Publishing Co.

3 Zmuda, H. and Toughlian, E. (1994). *Photonic Aspects of Modern Radar.* Norwood, MA: Artech House Publishing Co.

4 Tamburrini, M., Parent, M., Goldberg, L., and Stillwell, D. (1987). Optical feed for a microwave phased array antenna. *Electron. Lett.* 23: 680–681.

5 Toughlian, E.N. and Zmuda, H. (1990). A photonic variable RF delay line for phased array antennas. *IEEE J. Lightw. Technol.* 8: 1824–1828.

6 Riza, N.A. and Psaltis, D. (1991). Acousto-optic signal processors for transmission and reception of phased array antennas. *Appl. Opt.* 30: 3294–3303.

7 Dolfi, D., Michel-Gabriel, F., Bahn, S., and Huignard, J.P. (1991). Two-dimensional optical architecture for time-delay beam forming in phased-array antenna. *Opt. Lett.* 16: 255–257.

8 Riza, N.A. (1992). Liquid crystal-based optical control of phased array antennas. *IEEE J. Lightw. Technol.* 10: 1974–1984.

9 Riza, N.A. (1994). Acousto-optic liquid crystal analog beam former for phased-array antennas. *Appl. Opt.* 33: 3712–3724.

10 Monsay, E.H., Baldwin, K.C., and Cacciutto, M.J. (1994). Photonic true time delay for high-frequency phased-array systems. *IEEE Photonics Tech. Lett.* 6: 118–120.

11 Dolfi, D., Joffre, P., Antoine, J. et al. (1996). Experimental demonstration of a phased-array antenna optically controlled with phase and time delays. *Appl. Opt.* 35: 5293–5300.

12 Frigyes, I. and Seeds, A.J. (1995). Optically generated true-time delay in phased-array antennas. *IEEE Trans. Microw. Theory Tech.* 43: 2378–2386.

13 Ng, W., Walston, A.A., Tangonan, G.L. et al. (1991). The first demonstration of an optically steered microwave phased-array antenna using true time delay. *IEEE J. Lightw. Technol.* 9: 1124–1131.

14 Goutzolis, A.P., Davies, D.K., and Zomp, J.M. (1992). Hybrid electronic fiber-optic wavelength multiplexed systems for true time-delay steering of phased array antennas. *Opt. Eng.* 31: 2312–2322.

15 E. Ackerman, S. Wanuga, D. Kasemset, W. Minford, N. Thorsten and J. Watson (1992). Integrated 6-bit photonic true time delay unit for light 3–6 GHz radar beamformer. *1992 IEEE MTT Microwave Symp. Dig.*, Albuquerque, NM (June 1992), vol. 2, pp. 681–684.

16 Esman, R.D., Monsma, M.J., Dexter, J.-L., and Cooper, D.G. (1992). Microwave true time delay modulator using fibre-optic dispersion. *Electron. Lett.* 28: 1905–1907.

17 Soref, R. (1992). Optical dispersion technique for time-delay beamsteering. *Appl. Opt.* 31: 7395–7397.

18 Esman, R.D., Frankel, M.Y., Dexter, J.L. et al. (1993). Fiber-optic prism true time-delay antenna feed. *IEEE Photonics Technol. Lett.* 5: 1347–1349.

19 Ball, G.A., Glenn, W.H., and Morey, W.W. (1994). Programmable fiber-optic delay line. *IEEE Photonics Technol. Lett.* 6: 741–743.

20 Molony, A., Edge, C., and Bennion, I. (1995). Fibre grating time delay element for phased array antennas. *Electron. Lett.* 31: 1485–1486.

21 Tong, D.K. and Wu, M.C. (1996). A novel multiwavelength optically controlled phased array antenna with a programmable dispersion matrix. *IEEE Photonics Technol. Lett.* 8: 812–814.

22 Tong, D.K. and Wu, M.C. (1996). Programmable dispersion matrix using Bragg fibre grating for optically controlled phased array antennas. *Electron. Lett.* 32: 1532–1533.

23 Molony, A., Zhang, L., Williams, J.A.R. et al. (1997). Fiber Bragg-Grating true time-delay systems: discrete-grating array 3-b delay lines and chirped-grating 6-b delay lines. *IEEE Trans. Microw. Theory Tech.* 45: 1527–1530.

24 Cruz, J.L., Ortega, B., Andres, M.V. et al. (1997). Chirped fibre Bragg gratings for phased-array antennas. *Electron. Lett.* 33: 545–546.

25 Román, J.E., Frankel, M.Y., Matthews, P.J., and Esman, R.D. (1997). Time-steered array with a chirped grating beamformer. *Electron. Lett.* 33: 652–653.

26 Corral, J.L., Marti, J., Fuster, J.M., and Laming, R.I. (1997). True time-delay scheme for feeding optically controlled phased-array antennas using chirped-fiber gratings. *IEEE Photonics Tech. Lett.* 9: 1529–1531. See also Corral, J.L., Martí, J., Regidor, S., et al. (1997). Continuously variable true time-delay optical feeder for phased-array antenna employing chirped fiber gratings. *IEEE Trans. Microwave Theory Tech.* 45: 1531–1536.

27 Minasian, R.A. and Alameh, K.E. (1997). Optical-fiber grating-based beamforming network for microwave phased arrays. *IEEE Trans. Microw. Theory Tech.* 45: 1513–1518.

28 Soref, R.A. (1996). Fiber grating prism for true time delay beamsteering. *Fiber Integr. Opt.* 15: 325–333.

29 Zmuda, H., Soref, R.A., Payson, P. et al. (1997). Photonic beamformer for phased array antennas using fiber grating prism. *IEEE Photonics Tech. Lett.* 9: 241–243.

30 Frankel, M.Y. and Esman, R.D. (1995). True time-delay fiber-optic control of an ultrawideband array transmitter/receiver with multibeam capability. *IEEE Trans. Microw. Theory Tech.* 43: 2387–2394.

31 Matthews, P.J., Frankel, M.Y., and Esman, R.D. (1998). A wide-band fiber-optic true-time-steered array receiver capable of multiple independent simultaneous beams. *IEEE Photonics Tech. Lett* 10: 722–724.

32 Frankel, M.Y., Matthews, P.J., and Esman, R.D. (1996). Two-dimensional fiber-optic control of a true time-steered array transmitter. *IEEE Trans. Microw. Theory Tech.* 44: 2696–2702.

33 Tong, D.T.K. and Wu, M.C. (1998). Multiwavelength optically controlled phased-array antennas. *IEEE Trans. Microw. Theory Tech.* 46: 108–115.

34 Corral, J.L., Marti, J., Fuster, J.M., and Laming, R.I. (1998). Dispersion-induced bandwidth limitation of variable true time delay lines based on linearly chirped fibre gratings. *Electron. Lett.* 34: 209–211.

35 J.L Cruz, B. Ortega, M.V. Andrés, B. Gimeno, J. Capmany and L. Dong, 'Array factor of a phased array antenna steered by a chirped fiber grating beamformer', *IEEE Photonics Tech. Lett.* Vol. 10, pp. 1153–1155, 1998.

36 Corral, J.L. and Martí, J. (1999). Single sideband optical modulation on chirped fibre grating based delay lines for optically controlled phased array antennas. *Electron. Lett.* 35: 761–762.

37 Ortega, B., Cruz, J.L., Capmany, J. et al. (2000). Variable delay line for phased-array antenna based on a chirped fiber grating. *IEEE Trans. Microw. Theory Tech.* 48: 1352–1360.

38 Marti, J., Corral, J.L., Ramos, F. et al. (1999). Millimetre-wave optical beamforming network for phased-array antennas employing optical upconversion and wideband chirped fibre gratings. *Electron. Lett.* 35: 517–518.

39 Corral, J.L., Martí, J., and Fuster, J.M. (1999). Optical up-conversion on continuously variable true-time-delay lines based on chirped fiber gratings for millimetre-wave optical beamforming networks. *IEEE Trans. Microw. Theory Tech.* 47: 1315–1320.

40 Corral, J.L., Martí, J., Matthews, P., and Biernacki, P. (1999). Electro-optical upconversion in chirped fibre grating based true time delay lines. *Electron. Lett.* 35: 651–652.

41 Ortega, B., Cruz, J.L., Capmany, J. et al. (2000). Analysis of a microwave time delay line based on a perturbed uniform fiber Bragg grating operating at constant wavelength. *IEEE J. Lightw. Technol.* 18: 430–436.

42 Wang, Y., Tjin, S.C., Yao, J. et al. (2002). Wavelength-switching fiber laser for optically controlled phased-array antenna. *Opt. Commun.* 211: 147–151.

43 Yao, J., Yang, J., and Liu, Y. (2002). Continuous true-time-delay beamforming employing a multiwavelength tunable fiber laser source. *IEEE Photonics Tech. Lett.* 14: 687–689.

44 Y. Liu, J. Yang and J.P Yao, 'Continuous true-time-delay beamforming for phased array antenna using a tunable chirped fiber grating delay line', *IEEE Photonics Tech. Lett.*, vol 14, pp. 1172–1174, 2002.

45 Liu, Y., Yao, J., and Yang, J. (2002). Wideband true-time-delay unit for phased array beamforming using discrete-chirped fiber grating prism. *Opt. Commun.* 207: 177–187.

46 Liu, Y., Yao, J., and Yang, J. (2002). Wideband true-time-delay beam former that employs a tunable chirped fiber grating prism. *Appl. Opt.* 42: 2273–2277.

47 Vidal, B., Corral, J.L., Piqueras, M.A., and Martí, J. (2002). Optical delay line based on arrayed waveguide grating's spectral periodicity and dispersive media for antenna beamforming applications. *IEEE J. Sel. Top. Quantum. Electron.* 8: 1202–1210.

48 Vidal, B., Madrid, D., Corral, J.L., and Martí, J. (2002). Novel photonic true-time-delay beamformer based on the free-spectra-range periodicity of arrayed waveguide gratings and fiber dispersion. *IEEE Photonics Tech. Lett.* 14: 1614–1616.

49 Chen, Y. and Chen, R.T. (2002). A fully packaged true time delay module for a K-band phased array antenna system demonstration. *IEEE Photonics Tech. Lett.* 14: 1175–1177.

50 Lee, S.S., Oh, Y.-H., and Shin, S.-Y. (2004). Photonic microwave true-time delay based on a tapered fiber Bragg grating with resistive coating'. *IEEE Photonics Tech. Lett.* 16: 2335–2337.

51 Raz, O., Danziger, Y., and Tur, M. (2004). Implementation of Photonic true time delay using high-order-mode dispersion compensating fibers. *IEEE Photonics Tech. Lett.* 16: 1367–1369.

52 Hunter, D., Parker, M.E., and Dexter, J.L. (2006). Demonstration of a continuously variable true-time delay beamformer using a multichannel chirped fiber grating. *IEEE Trans. Microw. Theory Tech.* 54: 861–867.

53 Piqueras, M.A. et al. (2006). Optical beamforming beam-switched adaptive antennas for fixed and mobile broad-band wireless access networks. *IEEE Trans. Microw. Theory Tech.* 54: 887–899.

54 Jung, B.-M., Shin, J.-D., and Kim, B.-G. (2007). Optical true time-delay for two-dimensional X-band phased array antennas. *IEEE Photonics Tech. Lett.* 19: 877–879.

55 Subbaraman, H., Chen, M.Y., and Chen, R.T. (2008). Photonic crystal fiber-based true-time-delay beamformer for multiple RF beam transmission and reception of an X-Band phased-array antenna. *IEEE J. Lightw. Tech.* 26: 2803–2809.

56 Chin, S., Thevenaz, L., Sancho, J. et al. (2010). Broadband true time delay for microwave signal processing, using slow light based on stimulated Brillouin scattering in optical fibers. *Opt. Express* 18: 22599–22613.

57 Xu, L., Taylor, R., and Forrest, S.R. (1995). The use of optically coherent detection techniques for true-time delay phased array and systems. *IEEE J. Lightw. Technol.* 8: 1663–1678.

58 Kaman, V., Zheng, X., Helkey, R.J. et al. (2003). A 32-element 8-bit photonic true-time-delay based on a 288 × 288 3-D MEMS optical switch. *IEEE Photonics Tech. Lett.* 15: 849–851.

59 Shin, J.D., Lee, B.S., and Kim, B.G. (2004). Optical true time-delay feeder for X-band phased array antennas composed of 2 × 2 optical MEMS switches and fiber delay lines. *IEEE Photonics Tech. Lett.* 16: 1364–1366.

60 Horikawa, K., Ogawa, I., Kitoh, T., and Ogawa, H. (1996). Silica-based integrated planar lightwave true-time-delay network for microwave antenna applications. *Optical Fiber Communication Conference*, vol. 2, OSA Technical Digest Series (Optica Publishing Group), pp. 100–101, paper WB4.

61 Howley, B., Chen, Y., Wang, X. et al. (2005). 2-bit reconfigurable true time delay lines using 2x2 polymer waveguide switches. *IEEE Photonics Tech. Lett.* 17: 1944–1946.

62 Charles T. Sullivan, Sayan D. Mukherjee, Mary K. Hibbs-Brenner, Anand Gopinath, Edith Kalweit, Terry Marta, W. Tim Goldberg, and B. Walterson (1992). Switched time delay elements based on AlGaAs/GaAs optical waveguide technology at 1.32 μm for optically controlled phased array antennas. In: Yao, S.K. and Hendrickson, B.M. (eds.) *Proc. SPIE 1703, Optical Technology for Microwave Applications IV and Optoelectronic Signal Processing for Phased-Array Antennas III* (2 December 1992). https://doi.org/10.1117/12.138393.

63 Stulemeijer, J., van Vliet, F.E., Benoist, K.W. et al. (1999). Compact photonic integrated phase and amplitude controller for phased-array antennas. *IEEE Photonics Tech. Lett.* 11: 192–124.

64 Fathpour, S. and Riza, N. (2010). Silicon-photonics-based wideband radar beamforming: basic design. *Opt. Eng.* 49: 018201.

65 Meijerink, A. et al. (2010). Novel ring resonator-based integrated photonic beamformer for broadband phase array receive antennas-Part I: design and performance analysis. *IEEE J. Lightw. Technol.* 28: 3–18.

66 Zhuang, L., Roeloffzen, C.G.H., Meijerink, A. et al. (2010). Novel ring resonator-based integrated photonic beamformer for broadband phase array receive antennas-Part II: experimental prototype. *IEEE J. Lightw. Technol.* 28: 19–31.

67 Sagues, M. and Loayssa, A. (2009). Optical beamforming for phased array antennas using stimulated Brillouin Scattering. *2009 International Topical Meeting on Microwave Photonics*, Valencia, Spain, pp. 1–4.

68 Loayssa, A. and Lahoz, F.J. (2006). Broad-band RF photonic phase shifterbased on stimulated Brillouin scattering and single-sidebandmodulation. *IEEE Photonics Tech. Lett.* 18: 208–210.

69 Loayssa, A., Capmany, J., Sagues, M., and Mora, J. (2006). Demonstration of incoherent microwave photonic filters with all-optical complex coefficients. *IEEE Photonics Tech. Lett.* 18: 1744–1746.

70 Sagues, M., Loayssa, A., and Capmany, J. (2007). Multitap complex-coefficient incoherent microwave photonic filters based on stimulated Brillouin scattering. *IEEE Photonics Technol. Lett.* 19: 1194–1196.

71 Hau, L.V., Harris, S.F., Dutton, Z., and Behroozi, C.H. (1999). Light speed reduction to 17 metres per second in an ultracold atomic GaS. *Nature* 397: 594–598.

72 K. Y. Song, M. G. Herraez, and L. Thevenaz, "Observation of pulse delaying and advancement in optical fibers using stimulated Brillouin scattering," *Opt. Express*, vol. 13, no. 1, pp. 82–88, Jan. 2005.

73 Bigelow, M.S., Lepeshkin, N.N., and Boyd, R.W. (2003). Observation of ultraslow light propagation in a ruby crystal at room temperature. *Phys. Rev. Lett.* 90: 1139031–1139034.

74 Heebner, J.E., Wong, V., Schweinsberg, A. et al. (2004). Optical transmission characteristics of fiber ring resonators. *IEEE J. Quantum Electron.* 40 (6): 726–730.

75 Mok, J.T. and Eggleton, B.J. (2005). Photonics: expect more delays. *Nature* 433 (7028): 811–812.

76 Mørk, J., Kjær, R., Poel, M.v.d., and Yvind, K. (2005). Slow light in a semi- conductor waveguide at gigahertz frequencies. *OSA Opt. Express* 13: 8136–8145.

77 Öhman, F., Yvind, K., and Mørk, J. (2006). Voltage-controlled slow light in an integrated semiconductor structure with net gain. *Opt. Express* 14 (21): 9955–9962.

78 Sales, S., Öhman, F., Capmany, J., and Mork, J. (2007). Controlling microwave signals by means of slow and fast light effects in SOA-EA structures. *IEEE Photonics Tech. Lett.* 19: 1589–1591.

79 Xue, W., Sales, S., Capmany, J., and Mørk, J. (2010). Wideband 360° microwave photonic phase shifter based on slow light in semiconductor optical amplifiers. *Opt. Express* 18: 6156–6163.

80 Sancho, J., Lloret, J., Gasulla, I. et al. (2011). Fully tunable 360° microwave photonic phase shifter based on a single semiconductor optical amplifier. *Opt. Express* 19: 17421–17426.

81 Doménech, J.D., Muñoz, P., and Capmany, J. (2010). Discretely tunable microwave photonics beamformer based on ring resonators and arrayed waveguide gratings. *Proc. SPIE 7719, Silicon Photonics and Photonic Integrated Circuits II*, 77191J (17 May 2010). https://doi.org/10.1117/12.853778.

82 Adams, D. and Madsen, C. (2008). A novel broadband photonic RF phase shifter. *IEEE J. Lightw. Technol.* 26: 2712–2717.

83 Chang, Q., Li, Q., Zhang, Z. et al. (2009). A tunable broadband photonic RF phase shifter based on a silicon microring resonator. *IEEE Photonics Tech. Lett.* 21: 60–62.

9

Photonic-Assisted Instantaneous Frequency Measurements

9.1 Introduction

In the field of electronic warfare (EW), it is of critical importance to analyze an intercepted radio frequency (RF) signal from a hostile radar or communication system. The EW environment is totally different from that of commercial wireless communications, where the carrier frequency, modulation format, and bandwidth of the signal are known. But in an EW environment, not only the information of the signal is not known, but the signal is usually specifically designed to avoid being detected by an RF signal receiver. Typically, the frequency of a radar or communication signal can be varied in a range from several hundreds of MHz to several hundreds of GHz, but an RF signal receiver has to be operated in a very narrow frequency band to minimize the noise and interference. An instantaneous frequency measurement (IFM) module is thus required to instantaneously identify the carrier frequency of an unknown RF signal before passing it to a specialized radio receiver for processing [1–3]. IFM is also of great importance in an EW system to jam an enemy receiver. Typically, a signal emitted by an enemy transmitter would vary in frequency in a predetermined manner. In order to effectively jam such a signal, the frequency of the signal from the enemy transmitter must be accurately and quickly determined. Knowing the frequency, a jammer can be easily configured to concentrate its energy in the desired frequency range [2].

The development of the IFM can date back as early as 1948 [4]. In [4], Earp proposed the first electrical IFM receiver, in which the frequency information was displayed in polar coordinates on a scope. In his IFM receiver, an RF signal with the frequency to be measured was split into two paths; one path was delayed by a constant time with respect to the other one. A phase difference resulted due to the constant time delay. The relative phase angle is frequency-dependent. By measuring this phase angle, the frequency information was obtained. The technique is simple but effective. Most modern electrical IFM receivers are still based on the same principle [5–20]. Figure 9.1(a) shows the block diagram of a typical electrical IFM system [21]. The received RF signal is first passed through a limiting amplifier to remove the amplitude variation. The amplified signal is then divided into two equal portions by an electrical splitter. One portion is delayed relative to the other by a time of τ. The two signals are then mixed in an electrical mixer. After passing through a low-pass filter, a DC signal with a level that is a function of the microwave frequency is obtained. The relationship between the output DC power and the microwave frequency is shown in Fig. 9.1(b). By monitoring the DC power, the microwave frequency can be measured. To avoid

Microwave Photonics, First Edition. Jianping Yao, José Capmany, and Ming Li.

Companion website: www.wiley.com/go/yao/microwavephotonics

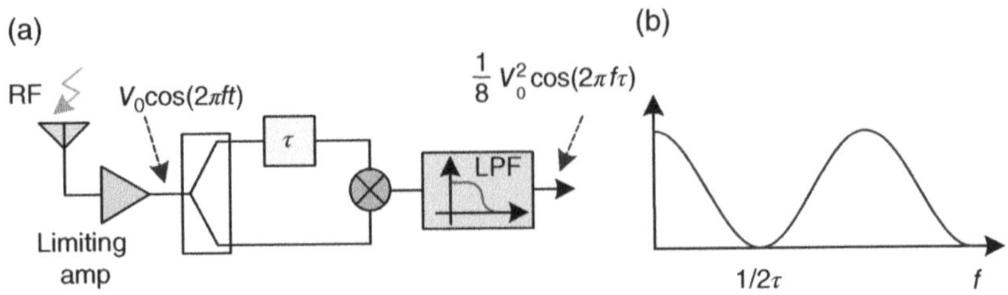

Figure 9.1 A typical electrical approach to IFM. (a) Configuration and (b) received DC power versus microwave frequency.

measurement ambiguity, a bandpass filter may be incorporated at the input of the receiver, and the measurement range is limited to 0–$1/2\tau$.

Thanks to developments in the last few decades, current electrical IFM systems can measure not only the microwave frequency but also the microwave amplitude, pulse width (PW), and time of arrival (TOA). The typical frequency measurement range is 0.5–18 GHz with a measurement resolution of about 3 MHz. The detectable signal level can be from −50 to +10 dBm, and the operating temperature is from −50 to +85 °C [22].

Electrical solutions can achieve high measurement resolution and a large dynamic range, but the frequency measurement range is limited due to the electronic bottleneck. Electrical solutions also suffer from other limitations, such as high power consumption, vulnerability to electromagnetic interference (EMI), and bulky size. To overcome these limitations, photonics-assisted microwave frequency measurements have been proposed [23]. Thanks to the broad bandwidth provided by optics, photonic-assisted IFM can potentially achieve a measurement range up to hundreds of GHz. In addition, photonic-assisted IFM also offers advantages such as low loss, light weight, and immunity to EMI.

Generally, photonics-assisted IFM can be classified into two main categories. In the first category, an optical channelizer [24–27] is employed to divide spectrally an optical microwave signal (a microwave signal modulated on an optical carrier) into many contiguous parallel channels. The channelized optical signals are then detected at an array of low-speed photodetectors (PDs). The optical spectrum of the optical microwave signal is thus obtained, which includes the information of the microwave carrier frequency. In the second category, the frequency of a microwave signal is obtained by power monitoring. The "power" can be either the microwave power or the optical power depending on the configuration of the system. The fundamental concept of the techniques in this category is that a fixed relationship between the microwave frequency and the microwave or optical power is established. By monitoring the microwave [28–38] or optical power [39–41], the microwave frequency is obtained.

In this chapter, a comprehensive review of the techniques for photonics-assisted IFM will be performed, with an emphasis on frequency measurement based on an optical channelizer and frequency measurement based on optical or microwave power monitoring. Other techniques for photonics-assisted IFM, including frequency measurement based on an optical scanning receiver [42], a photonic Hilbert transformer [43], a monolithically integrated echelle diffractive grating (EDG) [44], and a frequency-to-time mapper [45], are also briefly reviewed. The advantages and limitations of these techniques are discussed. The challenges in implementing these techniques for practical IFM are also discussed.

9.2 Frequency Measurement Using an Optical Channelizer

An optical channelizer is a device that was originally developed to instantaneously analyze the spectrum of an arbitrary optical signal. Since the carrier frequency of an RF signal can be retrieved from its spectrum, optical channelizers can also be used for microwave frequency measurement. Figure 9.2 shows a conceptual diagram for microwave frequency measurement using an optical channelizer. A microwave signal with its frequency to be measured is first converted into an optical signal at an intensity modulator, to which an optical carrier emitted from a laser diode (LD) is applied. The optical microwave signal is then split into many contiguous parallel channels using the optical channelizer. Each channel corresponds to a particular microwave frequency band. The split optical signals are then detected by an array of low-speed PDs. The converted electrical signals are sent to a low-speed electrical processing unit, where the spectral information of the microwave signal is obtained.

The key device in this system is the optical channelizer, which can be seen as a group of parallel optical narrow-band filters with a fixed spacing between adjacent center frequencies. A wavelength-dependent splitter or a wavelength division demultiplexer is actually an optical channelizer. In general, a wavelength-division multiplexer (WDM) can be used as a wavelength-division demultiplexer by propagating a light wave in a reversed direction. Since a WDM is a key component in a wavelength division multiplexed optical communication system, it has been well developed in the past few years [46]. The channel spacing determines the measurement resolution. For microwave frequency measurement, a small channel spacing (1 GHz or less) is usually required. Typically, the channel spacing of a WDM for communication applications is tens of GHz. Thus, special designs are needed to reduce the channel spacing. An optical channelizer could be realized using an integrated optical phased array consisting of multiple electro-optic waveguide delay lines [24], a high-resolution free-space diffraction grating [25], an array of phase-shifted gratings [26], or an integrated Bragg grating with a Fresnel lens [27].

The key advantage of an IFM receiver implemented based on an optical channelizer is its capability to simultaneously measure the frequencies of multiple signals. The major difficulty that limits the use of an optical channelizer for practical applications is the limited channel number and small channel spacing. To increase the channel number and to reduce the channel spacing, the fabrication cost would be significantly increased. At the current stage, an optical channelizer with a channel number of 16 would have a channel spacing larger than 1 GHz. In addition, the currently

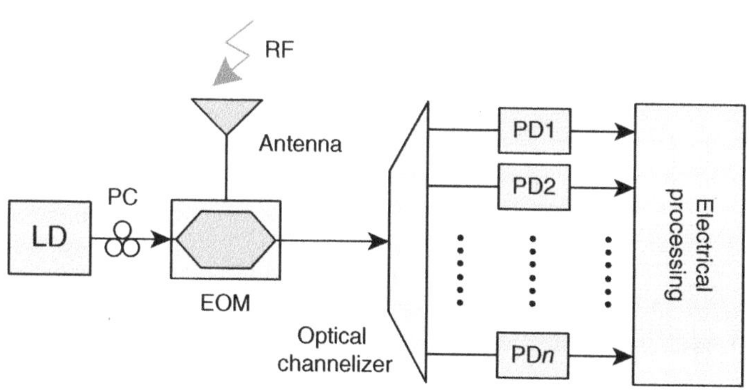

Figure 9.2 Microwave frequency measurement based on an optical channelizer. EOM, electro-optical modulator; LD, laser diode; PC: polarization controller; PD, photodetector; RF, radio frequency.

available optical channelizers are complicated, bulky, unstable, and expensive. In the following, a review of optical channelizers will be presented.

9.2.1 Optical Phased-Array WDM

An integrated optical passive phased-array WDM is also known as an arrayed waveguide grating (AWG). In an AWG, an incoming optical signal is coupled into an array of planar waveguides. The optical signal in each waveguide will experience a different phase shift due to the different path lengths of the waveguides. The phase shifts are also wavelength-dependent because of the frequency dependence of the mode-propagation constant. As a result, the incoming light components at different wavelengths will be deflected at different angles and then coupled again into different output waveguides. To make the channel spacing as small as 1 GHz, the maximum time delay difference between two adjacent waveguides should be as large as 1 ns, which corresponds to a maximum delay length difference of 7.7 cm. Such a long length is difficult to implement when laying the waveguides out on a single photonic integrated chip. A possible solution to this problem is to add electro-optic waveguide delay lines to an optical phased-array WDM. Since the phase of each waveguide delay line is electrically controlled, the phase shift between different waveguides can be increased by introducing different bias voltages to the waveguide delay lines. Such a device with a channel number of 16 was fabricated by Heaton et al. by using 16 high-confinement GaAs/AlGaAs optical waveguides in combination with a multimode interference (MMI) splitter, 128 compact low-loss corners, and 16 electro-optic phase-shifters [24]. The layout of the device is shown in Fig. 9.3. As can be seen, an input light wave is resolved into 16 channels. The channel spacing is 1 GHz. The total chip size is as small as 26 mm × 0.7 mm, which comprises (i) a straight input guide; (ii) a 1-to-16 MMI splitter (96 μm wide and 1884 μm long); (iii) 16 electro-optic phase-shifters with 5 mm-long electrodes; (iv) 16 delay lines; and (v) an output array with parallel guides on a 6 μm pitch. The input, electro-optic and delay line guides are all 3.0 μm wide, and the electrodes are 1.2 μm wide. The delay lines are laid out to achieve a large spread in delay length and to have

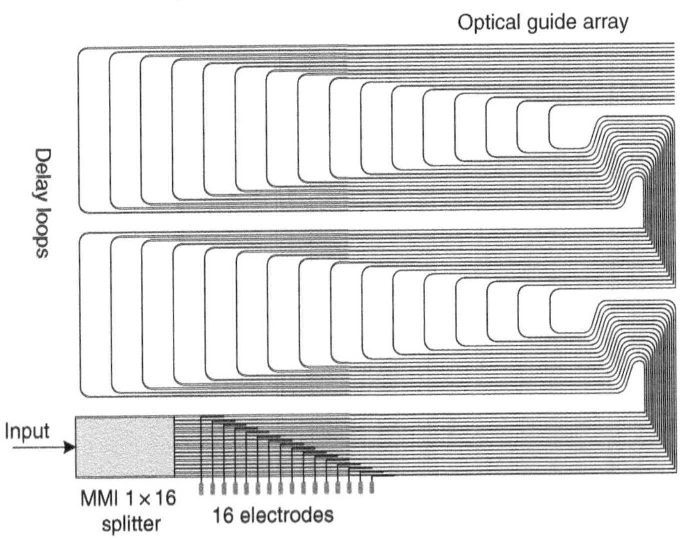

Figure 9.3 The layout of an optical phased-array WDM consisting of multiple electro-optic waveguide delay lines. *Source:* Adapted from Heaton et al. [24].

the same number of corners and S-bends. All of the corners and bends have the same size and shape to ensure that each delay line guide has the same insertion loss.

The key limitation related to the microwave frequency measurement based on this type of optical channelizer is that the measurement resolution and measurement range are limited to 1 GHz and 1∼16 GHz, respectively. Doubling the frequency range to 32 GHz or fining the resolution to 500 MHz might significantly increase the device width and the excess insertion losses.

9.2.2 Free-Space Diffraction Grating

To increase the operating frequency range, optical channelizers based on a high-resolution free-space diffraction grating would be a potential solution. For instance, an instantaneous measurement bandwidth of over 100 GHz with a resolution of 1 GHz is possible by use of a free-space diffraction grating with coherent optical heterodyne detection [25]. Figure 9.4 illustrates the concept of a coherent optical channelizer based on a free-space diffraction grating. The input light wave carrying a microwave signal with an unknown frequency is collimated and incident to the diffraction grating. Different frequency components in the incident signal are dispersed with different angles and subsequently imaged onto a detector array. The position of a detector corresponds to a particular microwave frequency, and the spatial separation of the detectors determines the measurement resolution. When the detected electrical signals are sent to an electrical processing unit, the spectral information of the microwave signal is derived. Since the most advanced technology currently available for post-processing is based on digital electronics, its bandwidth is restricted by the limited speed of the state-of-the-art analog-to-digital (A/D) converters, which is about 1 GHz. Therefore, it is essential to down-convert the central frequencies of all the channel outputs to a common intermediate frequency (IF) that can be handled by a low-speed digital signal

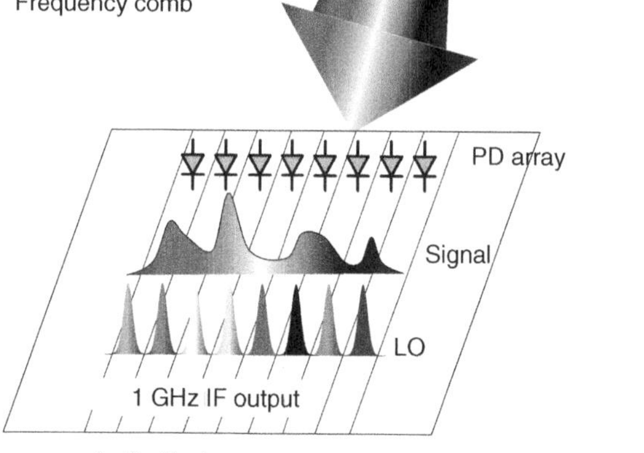

Figure 9.4 A coherent optical channelizing receiver uses a dispersion grating to map the optically carried RF signal and a local oscillator light beam onto a photodetector array. *Source:* Adapted from Wang et al. [25].

processing circuit. Coherent optical heterodyne detection technology is usually applied to perform the down-conversion. A local oscillator (LO) beam, with its spectrum comprising an optical frequency comb with a frequency spacing equal to the channel spacing, is incident to the grating at the same offset angle with respect to the optical microwave signal beam. Such spatial offset will result in a constant frequency difference between the spatially overlapped optical microwave signal beam and the LO comb lines. When the two beams are combined at the PDs, a heterodyne beat tone will be generated at the output of each PD. In this way, every portion of the signal spectrum is translated to the same IF band so that each channel may use the same post-processing electronics. A high degree of coherence between the optical microwave signal and the LO comb is required to minimize the phase noise in the heterodyne RF signal, which can be obtained by using a continuous-wave (CW) light wave as the optical carrier and also using a mode-locked laser that is injection-locked by the same CW light wave as the LO comb.

9.2.3 Phase-Shifted Chirped Fiber Bragg Grating Arrays

The major limitation of phased-array WDM and free-space diffraction grating-based optical channelizers is the large insertion loss. For example, the total insertion loss for a signal path in a coherent optical RF channelizer based on a diffraction grating is about 28 dB [25]. Therefore, to ensure an acceptable signal-to-noise ratio (SNR) at the output of the channelizer, the input signal should have a high power or the frequency measurement error would be large. The high insertion loss can be significantly reduced if an all-fiber optical channelizer is employed. Hunter et al. reported an optical channelizer based on an array of phase-shifted fiber gratings [26]. The schematic diagram is shown in Fig. 9.5. The optical microwave signal is split and applied to a grating array via a coupler tree. The grating array consists of a number of phase-shifted chirped fiber Bragg gratings (FBGs), which have a transmission notch within the reflection spectrum. The center wavelength and the width of the transmission notch can be controlled by adjusting the spatial position of the phase shift along the length of the grating. If the spatial position of the phase shift in each grating is carefully controlled, the transmission notch will be located at a position that corresponds to one of the modulation sidebands. Therefore, the optical energy transmission through any of the fixed notches is associated with a particular RF frequency band, which would channelize the intercepted microwave signal. In an experimental demonstration, a 1 × 16 optical channelizer with an average insertion loss for

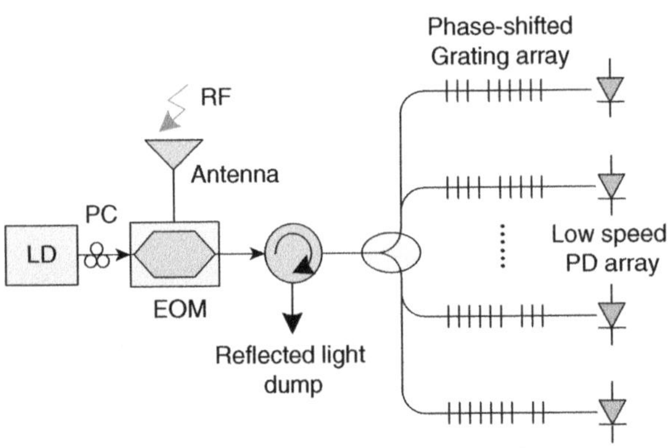

Figure 9.5 Schematic diagram of an optical channelizer based on an array of phase-shifted fiber gratings.

each channel of ~15 dB and a resolution of 2 GHz was realized. The use of the channelizer for the channelization of a 2–18 GHz microwave signal was successfully realized.

9.2.4 Integrated Optical Bragg Grating Fabry–Perot Etalon

An optical channelizer can also be realized based on an integrated optical Bragg grating Fabry–Perot etalon and integrated hybrid Fresnel lens system. The architecture of such a channelizer is depicted in Fig. 9.6 [27]. In the device, the optical phase front of the optical microwave signal is expanded, and the integrated optical lens system would cause this signal to diverge laterally. A Fabry–Perot etalon formed with two Bragg gratings acts upon this wavefront to provide microwave channelization since a resonance with periodic peaks that is a function of the angle and wavelength of the incident light results from the two Bragg gratings. The device is compact, but the insertion loss is very large and dependent on the wavelength of the incident light. The accuracy of the system is severely limited by the finesse of the Fabry–Perot etalon. A demonstration performed by Winnall et al. showed that the channelizer can provide a measurement range of 1–23 GHz with a measurement resolution of 2 GHz [27].

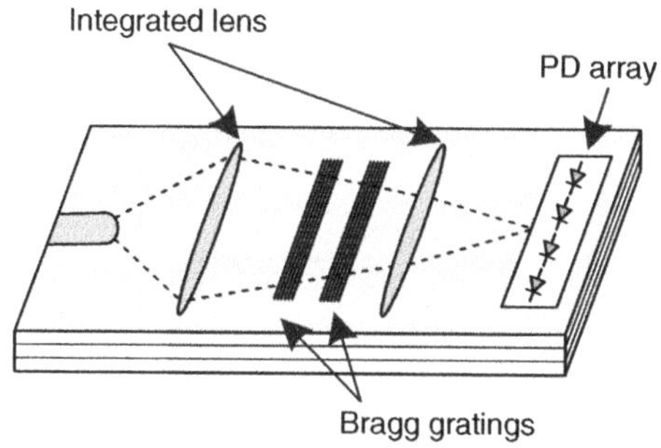

Figure 9.6 Schematic diagram of an optical channelizer based on an integrated optical Bragg grating Fabry–Perot etalon.

9.3 Frequency Measurement Based on Power Monitoring

Since frequency measurement based on an optical channelizer has a poor resolution and a relatively small measurement range, it may not meet the requirements for applications where a wide operating frequency range and high resolution are required. For IFM of a single channel signal, we need not know the detailed spectrum of the microwave signal. As a result, the measurement can be significantly simplified, while the measurement resolution and measurement range are evidently improved. One of such techniques is frequency measurement based on power monitoring. For this type of IFM, if multiple signals are received with one of the signals that have a much higher power, the frequency measurement would give an estimated frequency corresponding to the highest power signal. There are numerous applications that require the frequency information of an RF signal with the highest power. For example, in an adaptive anti-jamming receiver, the jamming signal always has the highest power. If the IFM can obtain the frequency information of the jamming signal, it can be used to control a tunable notch filter to remove the undesirable jamming frequency. Given that the techniques for IFM of multiple signals are not mature, one can use cascaded IFMs based on power monitoring and add/drop multiplexers to receive multiple signals. In each stage of the receiver, the frequency of the RF signal with the highest power is measured. The frequency information is then used to drive an add/drop multiplexer to pass the signal to a specialized receiver.

A typical scheme for microwave frequency measurement based on power monitoring is shown in Fig. 9.7. As usual, the microwave signal is first converted to an optical microwave signal at an electro-optic modulator. Then, the optical microwave signal is split into two channels to undergo different frequency-dependent power penalties. In the electrical processing unit, the

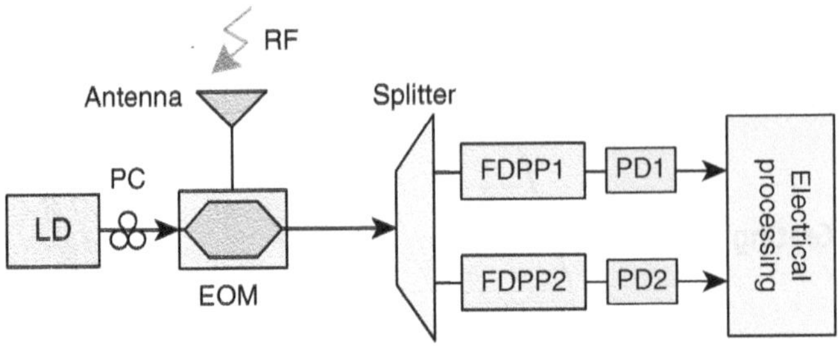

Figure 9.7 A diagram showing microwave frequency measurement based on power monitoring. FDPP, frequency-dependent power penalty.

frequency-dependent powers are monitored and used to retrieve the frequency information of the received microwave signal. The frequency-dependent power functions can be realized using a dispersive element that would introduce chromatic-dispersion (CD)-induced microwave power fading, a photonic microwave filter or an optical filter. In addition, the microwave power penalty functions are dependent on the modulation schemes.

9.3.1 Chromatic Dispersion-Induced Microwave Power Penalty

For conventional double-sideband plus carrier (DSB + C) modulation, a Mach–Zehnder modulator (MZM) is usually employed, which is biased at the quadrature point, thus providing one optical carrier and two optical sidebands. When the optical DSB + C signal is propagating in a dispersive fiber, the optical carrier and sidebands experience different phase shifts due to the CD. The beating of the optical carrier with the upper sideband and the lower sideband at a PD would generate a microwave signal with its power being a function of the CD and the microwave frequency. Mathematically, the microwave power at the output of a PD is given [47].

$$P = R\cos^2\left(\frac{\pi DL\lambda^2 f^2}{c} + \arctan\alpha\right) \tag{9.1}$$

where D is the dispersion parameter, L is the length of the dispersive fiber, λ is the carrier wavelength, f is the microwave frequency, c is the light velocity in vacuum, α is the residual chirp parameter of the modulator, and R is a constant that is related to the optical power, the phase modulation index at the modulator, the insertion losses of the optical devices, and the responsivity of the PD.

As can be seen from (9.1), the microwave power has a direct relationship with the microwave frequency. For a given measurement system, D, L, and λ are fixed and can be considered insensitive to environmental variations. However, R is dependent on the microwave power of the unknown microwave signal, which is usually not known. In addition, R is also sensitive to the losses of the optical path, which may be affected by environmental variations. Based on this consideration, two different carriers (denoted as λ_1 and λ_2) can be selected to obtain two different power fading functions. The ratio of the two power fading functions, referred to as an amplitude comparison function (ACF), is independent of the microwave power. If the ACF is denoted as γ, we have

$$\gamma = \frac{P_1}{P_2} = \frac{R_1\cos^2\left(\left(\left(\pi D_1 L_1 \lambda_1^2 f^2\right)/(c)\right) + \arctan\alpha\right)}{R_2\cos^2\left(\left(\left(\pi D_2 L_2 \lambda_2^2 f^2\right)/(c)\right) + \arctan\alpha\right)} \tag{9.2}$$

R_1 and R_2 can be made identical by adjusting the optical powers of the two optical carriers. Thus, by measuring the ACF, the microwave frequency can be estimated.

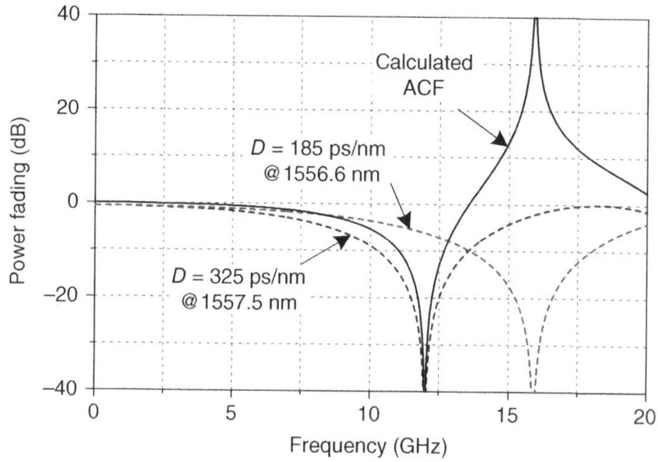

Figure 9.8 Typical power fading functions and the corresponding ACF function. $\lambda_1 = 1556.6$ nm; $\lambda_2 = 1557.5$ nm; $D_1L = 185$ ps/nm; $D_2L = 325$ ps/nm; and $\alpha = 0.4$.

Figure 9.8 shows two power fading functions and the corresponding ACF for a frequency measurement system with two dispersion values of 185 and 325 ps/nm. The residual chirp of the MZM is 0.4. The ACF monotonically decreases for $0 < f < 12$ GHz and monotonically increases for $12 < f < 16$ GHz. Due to the nonmonotone nature of the ACF, frequency measurement ambiguities exist. The problem can be solved by applying a bandpass filter with its passband set to cover the frequency range of interest. Usually, the system is designed to make the first notch the maximum frequency to be measured. So, the bandwidth of the frequency measurement system is from DC to the first notch of the ACF, which is given by [28].

$$BW = \sqrt{\frac{c}{2\left(DL\lambda^2\right)_{\max}}\left(1 - \frac{2}{\pi}\arctan\alpha\right)} \tag{9.3}$$

where $(DL\lambda^2)_{\max}$ is equal to either $D_1L_1\lambda_1^2$ or $D_2L_2\lambda_2^2$, whichever is greater. For practical applications, the lowest measurable frequency cannot reach zero even if the MZM is chirp-free since the slope of the ACF is very small at low frequencies. The system noise would cause a large measurement error in the low-frequency band.

Frequency measurement based on a dispersion-induced microwave power penalty can be implemented based on the system shown in Fig. 9.9 [28]. Two optical carriers from two LDs (1556.71 and 1559.13 nm) are multiplexed and modulated by a microwave signal with its frequency to be measured at an electro-optic modulator. A multichannel chirped fiber grating (MCFG) is employed as the dispersive element. The MCFG is designed to have a dispersion value of 186 ps/nm at 1556.71 nm and a dispersion value of 714 ps/nm at 1559.13 nm. The optical signals from the MCFG are demultiplexed and then converted to microwave signals at the PDs. Two detector logarithmic video amplifiers (DLVAs) are used to detect the microwave powers. The outputs from the DLVAs are digitized by an A/D conversion card. The receiver is controlled through a LabView interface with the frequency information extracted through a simple subtraction operation of the resultant logarithms of the power fading functions. A measurement range of 4–12 GHz with 100-MHz accuracy is achieved [28]. The performance of this IFM receiver is dependent on a number of factors. The frequency resolution is dependent on the resolution of the A/D converter. The accuracy is

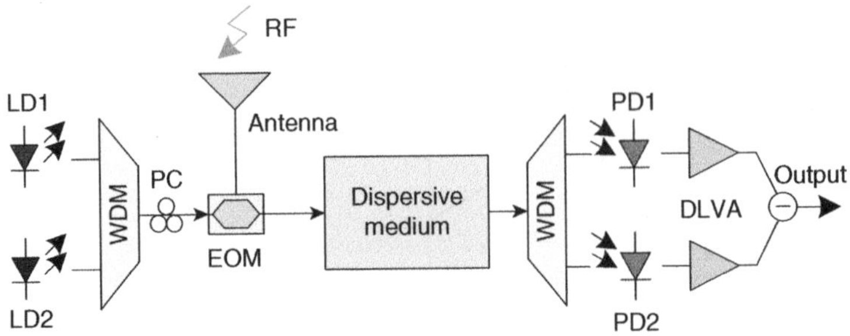

Figure 9.9 A microwave photonic IFM receiver based on a dispersion-induced microwave power penalty. DLVA, detector logarithmic video amplifier. *Source:* Adapted from Nguyen and Hunter [28].

dominated by the ripples present in the ACF, which is attributed to two parts: the RF impedance mismatches between the PDs and the DLVAs and the group delay ripples present in the MCFG.

The work in [28] can be used for frequency measurement for a fixed measurement range and resolution. To achieve frequency measurement with an adjustable measurement range, an optical approach was proposed and demonstrated [29]. The key difference between [28, 29] is that a dispersive fiber was employed to replace the MCFG. Compared with the MCFG, the spacing of the two wavelengths used to achieve the different power fading functions could be tuned to get different dispersion values, resulting in an adjustable frequency measurement range and resolution. In addition, the measurement accuracy is improved since a dispersive fiber has fewer dispersion ripples than an MCFG.

In [29], a 20-km single-mode fiber (SMF) was employed as the dispersive fiber. Two tunable laser sources (TLSs) are used to provide the two wavelengths with a tunable wavelength spacing. Six different wavelengths, 1470, 1500, 1520, 1550, 1580, and 1600 nm, are chosen to get different power penalty functions. The corresponding dispersion coefficients are 234.5, 273, 298, 335, 370, and 393 ps/nm. The power fading functions at these wavelengths are shown in Fig. 9.10(a). It is seen that the location of the first notch of the power fading function is determined by the total dispersion and the wavelength. The measurement range and resolution can thus be tuned by setting the spacing of the two wavelengths. If the two wavelengths are first set at 1470 and 1600 nm, the measurement range is from 5 to 12.1 GHz for an error within ±0.1 GHz, as shown in Fig. 9.10(b).

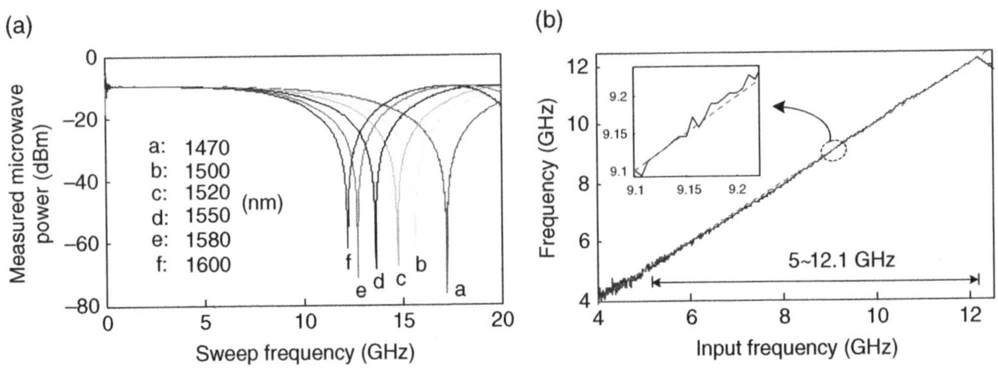

Figure 9.10 (a) Measured power fading functions for six different wavelengths. (b) Frequency measurement range for the system operating at 1470 and 1600 nm. *Source:* Zou and Yao [29]/IEEE.

If the longer wavelength of 1600 nm is tuned to 1520 nm while keeping the shorter wavelength of 1470 nm unchanged, a new measurement range of 9–14.7 GHz is achieved. Meanwhile, as reported, the resolution can be improved by using a smaller wavelength spacing. However, a trade-off between the measurement range and resolution, i.e. a higher resolution is achieved at the cost of a relatively smaller measurement range, has to be considered. To overcome this limitation, one can add a third wavelength. The use of three wavelengths would divide the measurement range into two parts, each having a higher resolution. For example, if three wavelengths at 1470, 1550, and 1600 nm are used, the use of 1470 and 1600 nm would provide a range of 5–11.5 GHz, while the use of 1470 and 1550 nm would offer another range of 11.5–15.6 GHz. Thus, a total measurement range of 5–15.6 GHz is achieved with an error of less than 0.1 GHz.

In [28, 29], the decreasing region of the ACF is used for frequency measurement. In that region, the slope of the ACF is small. The system noise would significantly impact the measurement accuracy. To improve the measurement accuracy without complicating the system, in [30], the increasing region of the ACF is utilized. A new frequency measurement range which exhibits fast variation in the ACF is obtained, leading to an improved measurement range and resolution. With the same experimental setup as used in [29], two wavelengths of 1470 and 1600 nm are used to obtain the ACF shown in Fig. 9.11. The two regions of the ACF to estimate the microwave frequency are evaluated. The variation of the ACF is very low in region *A*, which results in poor measurement resolution, especially when the system has high noise [28, 29]. However, region *B* has a much faster variation since the absolute values of the first-order derivatives in region *B* are larger than those in region *A*. Hence, the use of region *B* would offer an improved measurement resolution. Furthermore, for a given variation rate, region *B* would also provide a larger measurement range than region *A*. For example, the measurement range in region *B* is 12.3–17.1 GHz with a minimum absolute first-order derivative of 6.14 dB/GHz, while the measurement in region *A* is 9.9–12.2 GHz with a minimum absolute first-order derivative of 2.5 dB/GHz.

The frequency-dependent microwave power penalty in the above studies is achieved in an intensity-modulation-based system. The major limitation associated with intensity modulation for the frequency measurement is the bias drifting problem, which may make the system unstable. A sophisticated circuit is thus needed to control the bias voltage, which would increase the system

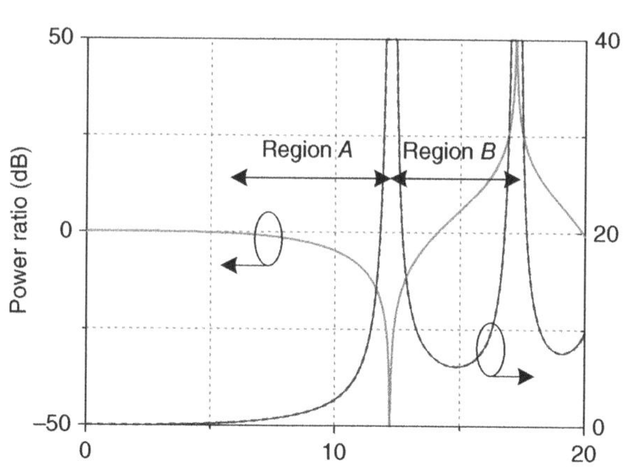

Figure 9.11 Simulated ACF with wavelengths of 1470 and 1600 nm and the absolute value of the first-order derivative of the ACF.

complexity and cost. By using a phase modulator to replace the intensity modulator, the bias drifting problem would be eliminated since a phase modulator needs no bias, which leads to improved system stability. This feature is highly desirable in defense systems, where system stability is one of the major concerns. In addition, the use of an optical phase modulator would make the system simpler with lower insertion loss.

When a phase-modulated microwave signal is transmitted in a dispersive medium and then detected by a PD, the output electrical power is given by [48].

$$P = R\sin^2\left(\frac{\pi DL\lambda^2 f^2}{c}\right) \tag{9.4}$$

Similar to the technique based on intensity modulation, two different microwave power penalty functions can be obtained by using two different wavelengths. The ratio between the two penalty functions can be used to calculate the microwave frequency. Figure 9.12 shows the simulated results of the RF power distribution as a function of the microwave frequency and the power ratio distribution against the microwave frequency, i.e. ACF. To obtain Fig. 9.12, the two optical wavelengths are set at 1520 and 1630 nm; the corresponding values of dispersion are 362 and 512 ps/nm, which correspond to the values of dispersion of a 25-km standard SMF at 1520 and 1630 nm, respectively. To estimate the microwave frequency based on the ACF without ambiguity, one monotone interval should be chosen. The upper frequency bound is dependent on the larger dispersion and its corresponding wavelength, which is given by

$$BW = \sqrt{\frac{c}{(DL\lambda^2)_{\max}}} \tag{9.5}$$

The lower bandwidth bound is limited by the predetermined measurement error tolerance since large errors would result in the low-frequency range due to a relatively slow varying ACF, making the measurement more sensitive to the system noise.

In a proof-of-concept experiment [31], two wavelengths at 1520 and 1620 nm are employed, which provide a microwave frequency measurement range of 7.3–15.05 GHz with a measurement accuracy of ±0.5 GHz. If the two wavelengths are set at 1520 and 1540 nm, the measurement range is 11.75–17.95 GHz for an error within ±0.5 GHz. Again, there is a tradeoff between the

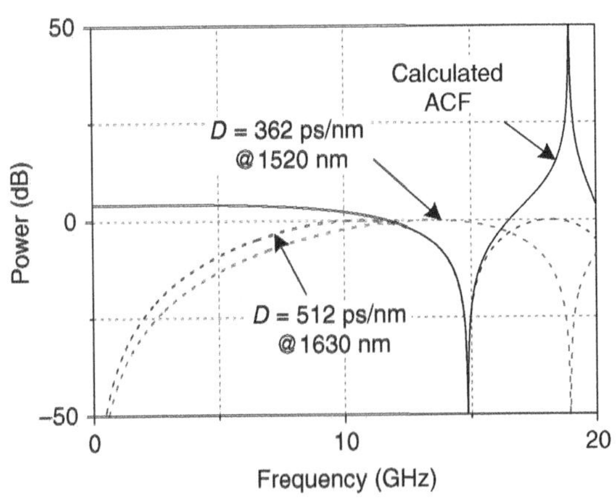

Figure 9.12 Typical power fading functions and the corresponding ACF function in the technique based on phase modulation. λ_1 = 1520 nm; λ_2 = 1630 nm; D_1L = 362 pm/nm; D_2L = 512 ps/nm.

measurement range and the measurement accuracy. A higher measurement resolution results in a smaller measurement range, while a lower measurement resolution leads to a relatively wider measurement range.

9.3.2 Break the Lower Frequency Bound

The ACF calculated based on dispersion-induced microwave power penalty functions at two different wavelengths could only provide IFM with a limited measurement range, typically less than 10 GHz. This measurement range may not be broad enough for practical applications such as EW. There are two possible solutions to extend the frequency measurement range. The first solution is to increase the frequency of the first notch in the ACF (i.e. the upper-frequency bound), but this will compromise the measurement accuracy. The second solution is to break the lower frequency bound. The ACF at low frequencies has a slow-increasing slope, which makes the lower frequency bound at a relatively high frequency. If the lower frequency bound can be smaller, the entire measurement range can then be increased.

A method to increase the slope of the ACF in the low-frequency band is to introduce dispersion-induced power penalties to two frequencies, i.e. f and $f_{LO} - f$, where f_{LO} is a fixed LO frequency, to generate two complementary power fading functions. Figure 9.13(a) shows the diagram of such an IFM scheme. The unknown signal is divided into two paths: one is mixed at a broadband mixer with a fixed LO frequency. The LO frequency is selected to be just beyond the frequency range of operation required by the receiver. The mixing will produce two spectral components at $f_{LO} - f$ and $f_{LO} + f$. Since the $f_{LO} + f$ component lies well outside the bandwidth of interest, only the $f_{LO} - f$ component will be considered. The RF signals from both paths will modulate two independent optical carriers at two MZMs, both biased at the quadrature point, thus providing two optical DSB + C signals. These signals are then combined and passed through a dispersive element such as an optical fiber or a chirped FBG. The dispersed signals are demultiplexed, detected by two PDs and amplified using DLVAs to measure the power of the two signals. The dispersion will introduce a

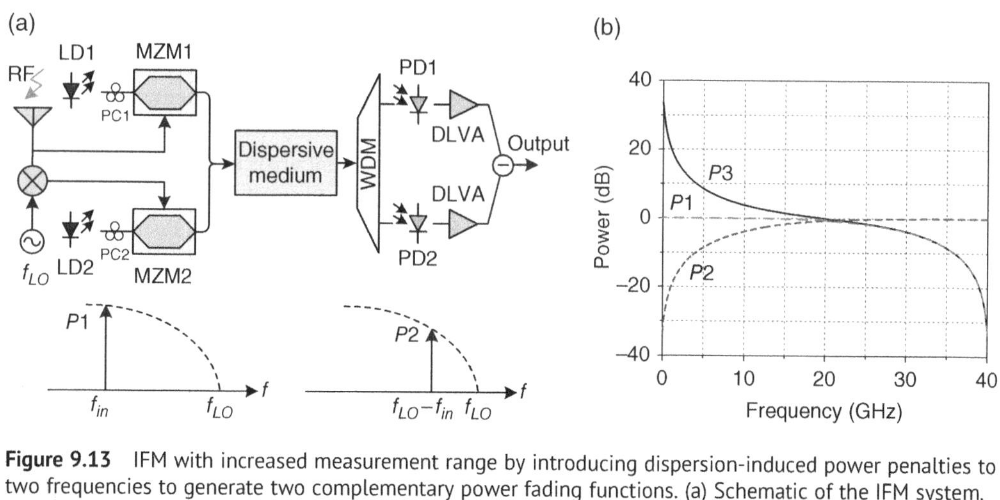

Figure 9.13 IFM with increased measurement range by introducing dispersion-induced power penalties to two frequencies to generate two complementary power fading functions. (a) Schematic of the IFM system. Insets: detected RF frequencies from the two optical paths, the dotted lines show the dispersion-induced microwave power penalty functions. (b) Simulated ACF with a 40-GHz operational bandwidth, DL = 25.5 ps/nm. P1 shows the detected RF power for the upper optical signal path, P2 shows the RF power for the lower optical signal path, and P3 is the ACF. *Source:* Adapted from Attygalle and Hunter [32].

microwave power penalty on each of the signals due to the DSB + C nature of the signals. The ACF can be given by

$$\gamma = \frac{P_1}{P_2} = \frac{R_1 \cos^2\left(\left(\left(\pi D_1 L_1 \lambda_1^2 f^2\right)/c\right) + \arctan\alpha\right)}{R_2 \cos^2\left(\left(\left(\pi D_2 L_2 \lambda_2^2 (f_{LO} - f)^2\right)/c\right) + \arctan\alpha\right)} \tag{9.6}$$

By adjusting the powers of the two optical carriers or the microwave powers to the two MZMs, R_1 and R_2 can be made identical. Figure 9.13(b) shows the typical power fading functions and the corresponding ACF when $f_{LO} = 40$ GHz, $D_1L_1 \approx D_2L_2 = 38.5$ ps/nm, $\lambda_1 = 1549.2$ nm, $\lambda_2 = 1550$ nm, and $\alpha = 0$. The curve named P_3 in Fig. 9.13(b), i.e. ACF, shows a monotonically decreasing function in which each frequency is associated with a unique amplitude in a 40-GHz operational bandwidth. This response, when compared with a calibrated response, will allow the determination of the frequency of the unknown input signal.

An experiment based on the setup in Fig. 9.13 was performed. A mixer operating from 4 to 40 GHz was employed. The LO frequency was 22 GHz. The total dispersion of the dispersive element was ~184 ps/nm. Two DLVAs with a bandwidth of 19 GHz were connected at the outputs of the two PDs [32]. An extended frequency measurement range from 4 to 19 GHz was achieved.

Since a mixer always has a lower operation frequency limit, the IFM system still has a lower frequency bound. In addition, the input RF signal is split and processed in two different paths, and the nonlinearity and uneven frequency responses of the electrical devices and the MZMs inevitably introduce additional frequency measurement errors.

The measurement range can be further extended by reducing the lower bound to DC. For a DSB + C optical signal transmitting in a dispersive medium, the microwave frequency response of the system is low-pass, while for a phase-modulated optical signal transmitting in a dispersive medium, the frequency response of the system is bandpass. Due to the complementary nature of the two frequency responses, the ratio of the two frequency responses, i.e. ACF, monotonically increases or decreases from DC to the frequency of the first notch of the power fading function of the intensity-modulated signal. The use of intensity modulation and phase modulation for microwave frequency measurement was demonstrated in [33]. The approach offers a wide measurement range without the need for tuning the optical carrier wavelength. Figure 9.14(a) shows the experimental setup. Only a laser source is employed. Compared with the technique in [32], the complexity and cost are considerably reduced. Figure 9.14(b) shows the experimentally measured and theoretically calculated ACFs. A frequency measurement range of 0–13.8 GHz is obtained with a measurement accuracy of better than ±0.3 GHz. The measurement error mainly originates from the length difference of the SMFs in the upper and lower arms. Since the SMF lengths can be controlled precisely using a commercially available optical time-domain reflectometer, the measurement error should be greatly reduced. The measurement range can be further extended by reducing the dispersion of the dispersive medium. Since two modulators are used, it is difficult to make the ratio of the modulation indices of the two modulators identical in the whole frequency measurement range, which may introduce additional measurement errors.

The above method using an intensity modulator and a phase modulator can be implemented using a much simpler scheme based on a polarization modulator (PolM) [34]. The schematic diagram of system using a PolM is illustrated in Fig. 9.15. In the system, the microwave signal with its frequency to be measured is applied to the PolM via the RF port to modulate two linearly polarized optical light waves at different wavelengths from two laser sources. The PolM is a special phase modulator that supports both transverse electric (TE) and transverse magnetic (TM) modes but with opposite phase modulation indices [49]. The polarization direction of one light wave is aligned

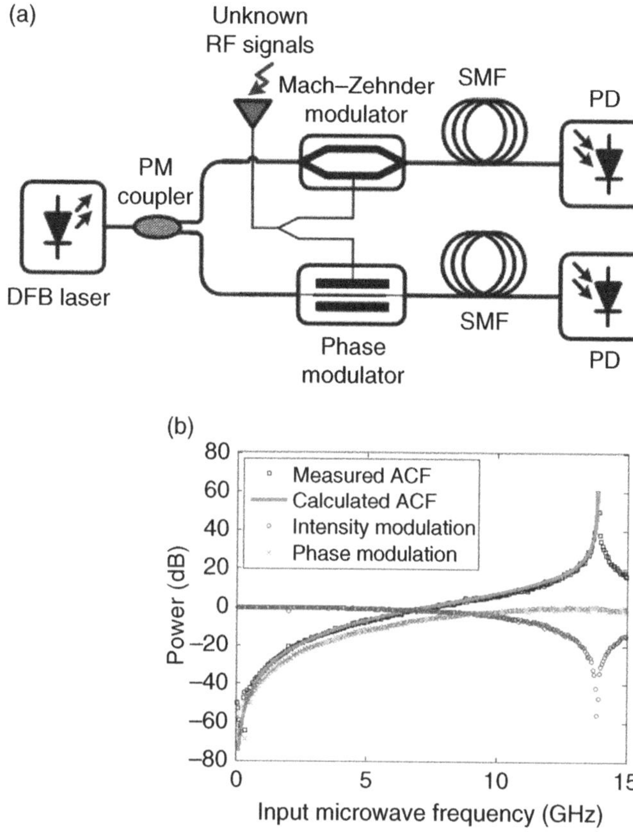

Figure 9.14 A microwave frequency measurement system based on intensity and phase modulation. (a) Schematic diagram of the system, (b) measured and simulated ACF, where $DL = 340$ ps/nm. *Source:* Zhou et al. [33]/IEEE.

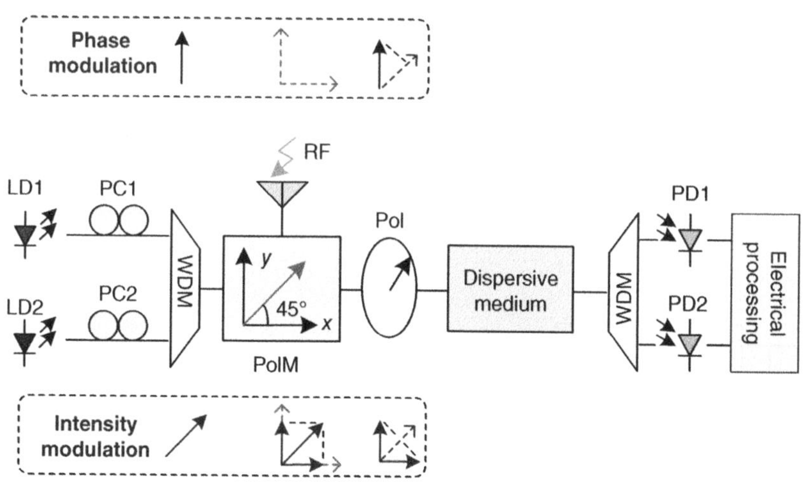

Figure 9.15 A microwave frequency measurement system using a polarization modulator. PolM, polarization modulator. *Source:* Adapted from Zou et al. [34].

with one principal axis of the PolM. Phase modulation is thus imposed on this light wave. The polarization direction of the other light wave is oriented with an angle of 45° with respect to the principal axis of the PolM. Since an optical polarizer is connected with its transmission axis oriented with an angle of 45° to the same principal axis of the PolM, the phase modulated signals along the two principal axes are projected to the transmission axis of the optical polarizer, leading to the generation of an intensity-modulated signal. Both the intensity-modulated and phase-modulated optical signals are then sent to a dispersive element. Similar to the technique in [33], the frequency responses corresponding to the two light waves along the two directions would be low-pass and bandpass. Thus, an ACF similar to that shown in Fig. 9.14 is obtained. It was demonstrated that the system could perform frequency measurements with a measurement range from 3 to 18 GHz with a measurement accuracy of ± 0.2 GHz. The major advantage of this approach over the method in [33] is that only one modulator and one dispersive element are required, which would reduce the complexity and cost of the system.

9.3.3 IFM Based on Photonic Microwave Filters with Complementary Frequency Responses

It is seen from Figs. 9.13 and 9.14 that the broadband frequency measurement range can be achieved if the ACF is calculated using two complementary frequency responses, such as a low-pass frequency response and a bandpass frequency response. Since a photonic microwave filter can be designed to have a low-pass or bandpass frequency response, it is thus possible to perform microwave frequency measurements with an extended frequency measurement range using two photonic microwave filters with complementary frequency responses.

Photonic microwave filters are usually implemented in the optical domain based on a delay-line structure, in which a microwave input signal is modulated on one or multiple optical carriers at an optical modulator; the modulated light waves are then sent to a time-delay device to introduce different time delays; the time-delayed microwave signals are then detected at a PD [50–54]. The schematic diagram of an N-tap photonic microwave delay-line filter is shown in Fig. 9.16. There are two key parameters that determine the frequency response of the filter: the time delay difference τ, which is inversely proportional to the free spectral range (FSR), and the coefficients a_n, which determine the shape of the filter response. Mathematically, the frequency response of an N-tap microwave delay-line filter is given [53].

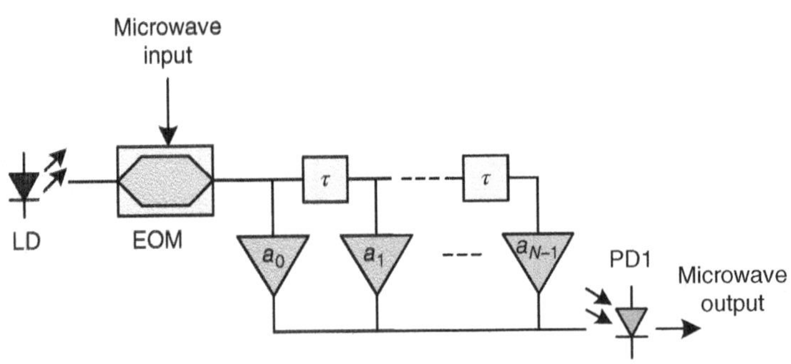

Figure 9.16 Schematic diagram of an N-tap photonic microwave delay-line filter.

$$H(f) = \sum_{n=0}^{N-1} a_n e^{-j2n\pi f\tau} \tag{9.7}$$

For a two-tap filter with coefficients of (1, 1) or (1, −1), the frequency response is given by

$$\left|H_{(1,1)}(f)\right|^2 = 2\cos^2(\pi f\tau) \tag{9.8}$$

$$\left|H_{(1,-1)}(f)\right|^2 = 2\sin^2(\pi f\tau) \tag{9.9}$$

As seen from (9.8) and (9.9), the two frequency responses are complementary, corresponding to the frequency responses of a low-pass and a bandpass filter, respectively. The ratio between the two power spectral responses, again referred to as the ACF, is given by

$$ACF = \frac{\left|H_{(1,1)}(f)\right|^2}{\left|H_{(1,-1)}(f)\right|^2} = \frac{1}{\tan^2(\pi f\tau)} \tag{9.10}$$

From (9.10), we can see that the ACF is monotonically decreasing over a frequency band from DC to $1/2\tau$. Therefore, based on the value of the ACF, we can calculate the frequency of the input microwave signal by the following equation:

$$f = \frac{1}{\pi\tau\tan^{-1}\sqrt{ACF}} \tag{9.11}$$

The maximum measurement range is determined by τ. To have a larger measurement range, we may design the filter pair with a greater FSR. Since the photonic microwave filter pair is generally implemented in the incoherent regime using an incoherent light source [54], the IFM scheme would have a low requirement for the quality of the laser source. The system would also have improved system stability because the wavelength drift of the laser source would not affect the performance of the filter pair due to the incoherent operation of the filter pair. This feature is highly desirable in defense systems.

Figure 9.17 shows the schematic diagram of the microwave frequency measurement system using photonic microwave filters with a complementary frequency response. It consists of an LD, a

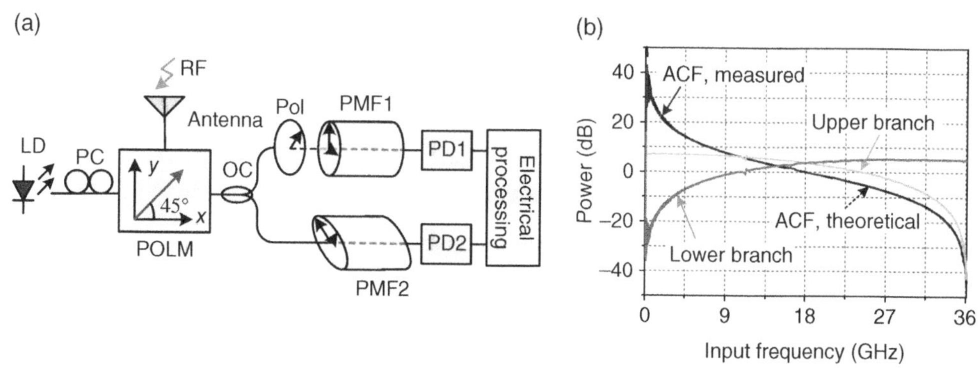

Figure 9.17 (a) Schematic diagram of the microwave frequency measurement system using photonic microwave filters with complementary frequency response. (b) Measured and calculated frequency responses of the two photonic microwave filters and the ACF based on the measured frequency responses. *Source:* Adapted from Pan and Yao [35].

Based on (9.14) and (9.15), we have the ACF, given by

$$ACF(f) = \frac{|I_1|}{|I_2|} = \frac{R_1}{R_2} K^2(\omega_c - \omega_1)\Omega \tag{9.16}$$

R_1 and R_2 can be made identical by adjusting the losses of the optical paths via optical attenuators. As seen from (9.16), the ACF is directly proportional to the frequency of the input microwave signal and is not dependent on the input optical power and the microwave modulation index. Therefore, based on the value of the ACF, the frequency of the input signal can be obtained by the following expression:

$$f = \kappa \cdot ACF \tag{9.17}$$

where $\kappa = 1/(2\pi K^2(\omega_c - \omega_1))$.

It should be noted that κ contains information on the laser wavelength. Therefore, the wavelength drift of the laser source would contribute to the measurement error. For instance, the wavelength of the laser used in our experiment can shift within 5 pm (~600 MHz) in a one-hour observation and $\omega_c - \omega_1 \approx 40$ GHz, which gives a measurement error of $\Delta f \approx 0.015f$. When $f = 20$ GHz, the measurement error can be 300 MHz. One possible way to reduce the influence of the wavelength drift is to have a large $\omega_c - \omega_1$, making the relative wavelength drift smaller, or to perform a calibration before the measurement since the wavelength drift is a slowly varying process.

The concept was demonstrated by a proof-of-concept experiment [38]. The schematic of the microwave frequency measurement system is shown in Fig. 9.19(a). The polarization modulated signal from the PolM is split into two branches. In the upper branch, PM-IM conversion is performed at a polarization beam splitter (PBS), which serves as a polarizer. In the lower branch, one of the phase-modulated signals is selected by another PBS and sent to the OBPF to implement first-order photonic microwave differentiation. The transmission spectrum of the OBPF is shown in Fig. 9.19(b). The left slope from 1549.52 to 1549.91 nm is almost linear. The time delay variation in this range is less than 3 ps, which is small and can be neglected for a measurement range of less than 25 GHz.

Figure 9.20(a) shows the detected microwave powers in the upper and lower branches. The input microwave power is set at 0 dBm. The ACF is almost directly proportional to the frequency of the

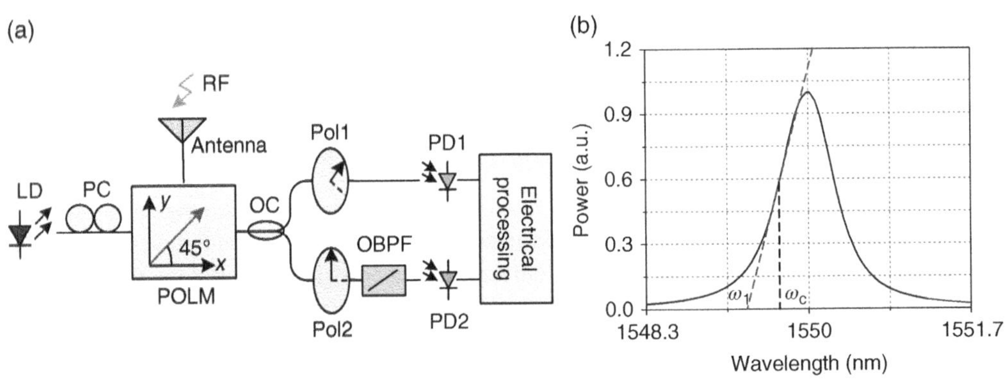

Figure 9.19 (a) Schematic of the microwave frequency measurement system based on a photonic microwave differentiator. (b) Transmission spectrum of the OBPF used in the experiment.

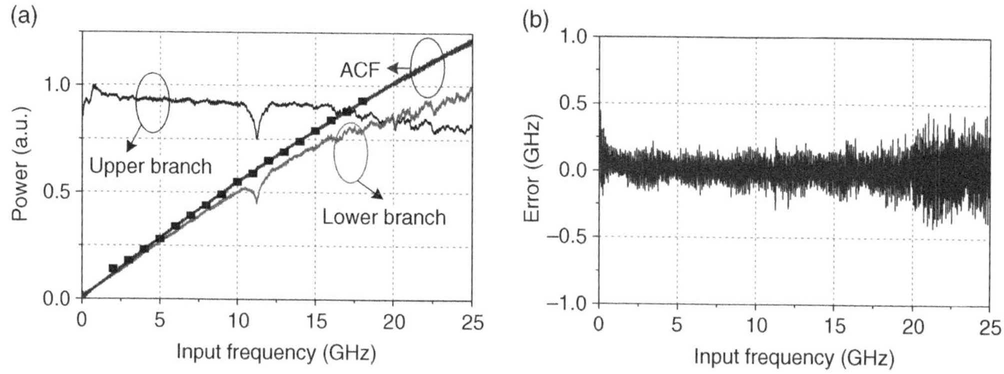

Figure 9.20 (a) Measured microwave powers for the signals in the upper and lower branches and the calculated amplitude comparison function (solid squares are the measured results for the pulsed microwave signal). (b) Measurement error versus input microwave frequency.

input microwave signal, which allows the determination of an unknown input frequency from 0 to 25 GHz. Figure 9.20(b) shows the measurement error versus the frequency of the input microwave signal. In the frequency range from 0.5 to 18 GHz, the measurement error is within ±0.2 GHz. The measurement error is increased to ±0.4 GHz in the frequency range from 20 to 25 GHz, which is due to the imperfections of the electrical devices used in the experiment. In this frequency range, the electrical devices have large insertion losses. The system can be employed to evaluate the carrier frequency of a DSB + C-pulsed microwave signal. The measurement results are shown as the solid squares in Fig. 9.20(a). When the carrier frequency of the pulsed microwave signal (500 Mb/s NRZ signal) is tuned from 3 to 18 GHz, the measurement error is within ±0.4 GHz.

9.3.5 Optical Power Fading Using Optical Filters

The major problem associated with the IFM by monitoring microwave power is that it requires high-frequency PD, microwave transmission lines, and microwave detectors, to detect, transmit, and process the microwave signal. If the optical power instead of the microwave power is monitored for microwave frequency measurement, the system will be easier to implement by using low-frequency devices at a lower cost since only DC optical powers are measured.

Figure 9.21(a) illustrates the schematic diagram of a typical microwave frequency measurement system based on optical power monitoring. Two optical carriers from two LDs are combined at a multiplexer and sent to an MZM through a polarization controller. The MZM is biased such that the optical carriers at the output of the MZM are completely suppressed. An optical filter with a sinusoidal spectral response is placed at the output of the MZM. The wavelengths of the two optical carriers are set at one peak and one valley of the filter spectral response, as shown in Fig. 9.21(b). The carrier-suppressed optical signals are then demultiplexed, with the optical powers measured by two low-frequency PDs. The DC currents from the PDs are digitized and compared in a data processing unit. The ratio between the two detected powers, again referred to as ACF, can be expressed as [39]

$$\gamma = \frac{1-\cos(2\pi f/FSR)}{1+\cos(2\pi f/FSR)} = \tan^2\left(\frac{\pi f}{FSR}\right) \tag{9.18}$$

where FSR denotes the FSR of the optical filter. The ACF is not dependent on the input optical power and the phase modulation index of the modulator. A calculated ACF is given in

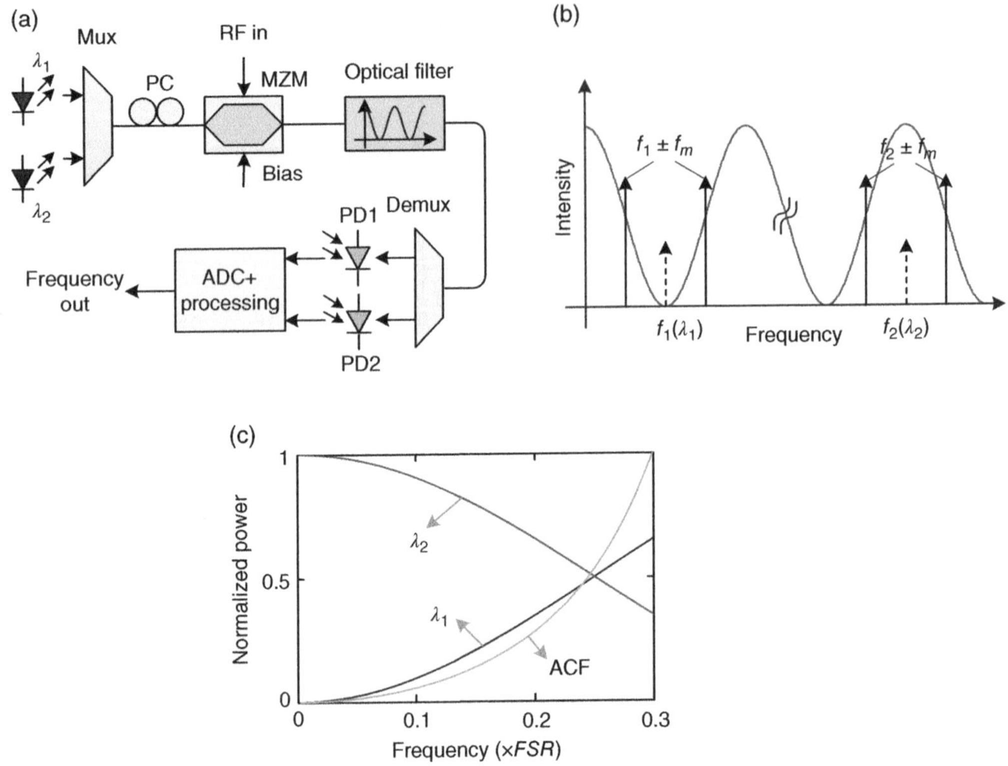

Figure 9.21 Microwave frequency measurement system based on optical power monitoring. (a) Schematic diagram. (b) Operation principle. The wavelengths of the two optical carriers are selected such that one carrier is located at one peak and the other is located at the valley of the spectral response. (c) Calculated ACF. *Source:* Adapted from Chi et al. [39].

Fig. 9.21(c), which monotonically increases over the whole measurement range. This ACF can thus be used to determine the frequency of an unknown input microwave signal. In the experiment reported in [39], a two-tap Sagnac-loop filter (SLF) was used to serve as the sinusoidal filter with an FSR of 50 GHz. A microwave frequency measurement ranges from 1 to 20 GHz with a measurement accuracy of ±0.2 GHz was demonstrated.

The key limitation of this approach is its instability. The measurement error of the system could be significant because of (i) the bias drift of the MZM, (ii) the wavelength fluctuations of the two laser sources existing in the two-wavelength system, and (iii) the poor stability of the SLF. A sophisticated control circuit is thus needed to stabilize the operation.

The technique in [39] can be improved by replacing the SLF with an optical complementary filter pair [40, 41], which simplifies the system since only a single laser source is required. The measurement error due to power fluctuations of the two laser sources is completely eliminated. Figure 9.22(a) shows the schematic diagram. The key component in the system is the complementary filter pair, which can be implemented by using a length of polarization-maintaining fiber in conjunction with two polarization controllers and a PBS, as shown in Fig. 9.22(b). The polarization direction of a linearly polarized light wave is aligned by the first polarization controller with an angle of θ with respect to the fast axis of the PMF, which is projected to the principal axes with one traveling along the fast axis and the other along the slow axis.

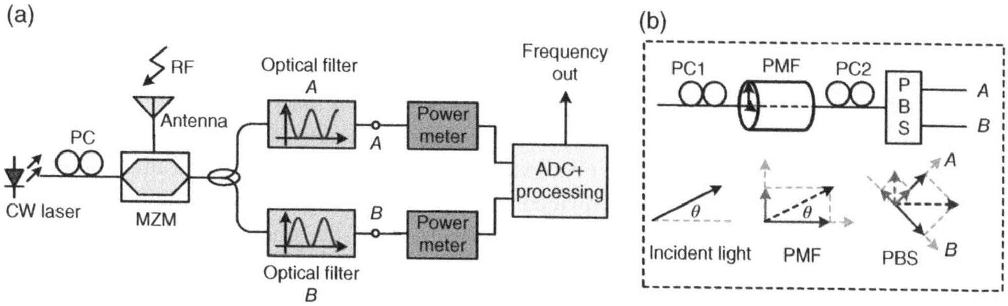

Figure 9.22 A simplified microwave frequency measurement system based on optical power monitoring. (a) Schematic diagram of the measurement system. (b) Diagram of the complementary filter pair. *Source:* Adapted from Zou et al. [40].

Due to the birefringence of the PMF, a time-delay difference between the two orthogonally polarized light waves is generated. A second polarization controller is connected after the PMF to adjust the two light waves to have an angle of 45° with respect to the axes of the PBS. Two complementary transmission responses at ports *A* and *B* are obtained. The ratio of the two transmission responses can be expressed as

$$\gamma = \frac{1 - m\cos(2\pi f/FSR)}{1 + m\cos(2\pi f/FSR)} \tag{9.19}$$

where m is the relative peak-to-notch contrast ratio of the complementary filters, which is related to the incident angle θ by $m = 2 \cdot \tan\theta/(1 + \tan^2\theta)$. In an experiment, the FSR of the complementary filters was 66 GHz. The measurement range was related to the relative peak-to-notch contrast ratio. When $m = 0.6$, 0.825, and 0.99, the estimated measurement errors were less than 0.2 GHz within a measurement range of 1–26, 2–24, and 2–20 GHz. So, a smaller m could provide a larger measurement range.

9.4 Other Methods for Frequency Measurement

In addition to the techniques discussed above, numerous other solutions have also been proposed. In the following, we will discuss four techniques, including microwave frequency measurement: (i) using a Fabry–Perot scanning receiver, (ii) based on photonic Hilbert transform, (iii) using a monolithically integrated EDG, and (iv) based on frequency-to-time mapping.

9.4.1 Fabry–Perot Scanning Receiver

A Fabry–Perot scanning receiver is a high-resolution optical spectrum analyzer that can be implemented using a piezoelectrically scanned fiber Fabry–Perot filter (FFPF) driven by a ramp voltage [42]. With the FFPF scanning over the FSR, the optical spectrum with a high resolution (~90 MHz) can be measured. The microwave frequency is thus derived from the optical spectrum. The measurement requires only low bandwidth post-processing electronics to perform microwave signal processing, but the piezoelectric scanning of the FFPF plate separation is a slow and nonlinear process, which may not meet the near real-time measurement requirement.

9.4.2 Photonic Hilbert Transform

The concept of microwave frequency measurement based on the Hilbert transform was originally demonstrated using pure electronics. A schematic showing the electronic approach is given in Fig. 9.23. Similar to the IFM based on power monitoring, the system is constructed to obtain an ACF in which the frequency of the input microwave signal is only a function of the amplitude data. The input unknown microwave signal is divided into two equal parts and then introduced into two interferometers, i.e. Module 1 and Module 2. In each module, the microwave signal is further divided equally into two arms that have a time delay difference of τ. The output signals from the two arms are then multiplied and DC-filtered. A 90° phase shift is introduced in one arm of Module 2 so that the output of Module 2 has a sine response while that of Module 1 has a cosine response. The 90° phase shift is realized using a Hilbert transformer. The ratio between the two frequency responses can be written as

$$\gamma = \frac{(1/8)V_0^2 \cos(2\pi f \tau)}{(1/8)V_0^2 \sin(2\pi f \tau)} = \cot(2\pi f \tau) \tag{9.20}$$

As seen from (9.20), the ACF is only a function of microwave frequency. Since the signals at the outputs of the two channels are at low-frequency band, this system can be implemented at low cost while still enabling broadband frequency measurement; however, practical electrical implementation of such a system requires broadband microwave delay lines and microwave mixers.

The problem will be solved if the system is implemented using the photonic method. Figure 9.24 shows the photonic implementation of a microwave frequency measurement system based on the Hilbert transform. The microwave signal with its frequency to be measured is converted to the optical domain at an MZM (MZM1). Three optical wavelengths at λ_0, λ_1 and λ_2 are employed to implement a two-tap (λ_1 and λ_2) photonic microwave filter with tap coefficients of (−1, 1) and a reference tap (λ_0). The modulated optical signals at λ_0, λ_1, and λ_2 are reflected by an optical delay line with three FBGs with center wavelengths corresponding to λ_0, λ_1, and λ_2. The time delays are determined by the physical locations of the FBGs. The time-delayed signals are then mixed at a second MZM (MZM2). To achieve a broadband 90° phase shift, a Hilbert transformer is employed, which is implemented by the two-tap photonic microwave filter [56]. The approach was evaluated experimentally. A frequency measurement range from 1 to 10 GHz with a measurement error less than 0.2 GHz was demonstrated. The advantages of the technique are that the time delays were achieved using optical delay lines and mixing was achieved using an MZM, which avoids the use of

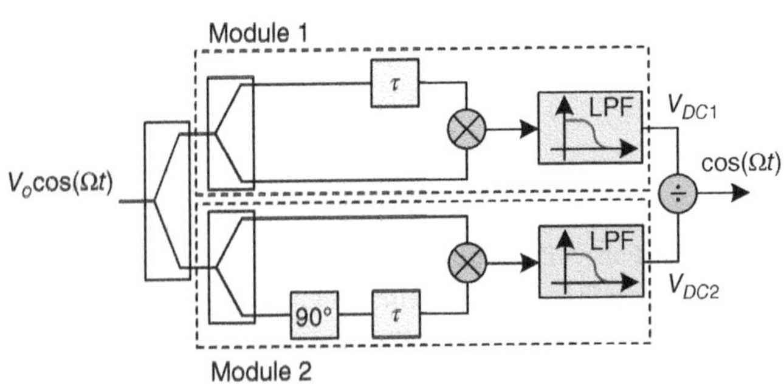

Figure 9.23 Block diagram of the IFM system based on the Hilbert transform. *Source:* Adapted from Emami et al. [43].

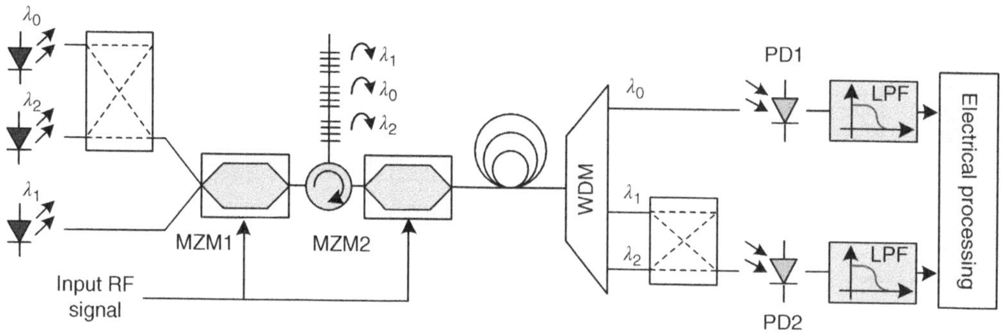

Figure 9.24 Photonic implementation of the microwave frequency measurement using the Hilbert transform. *Source:* Adapted from Emami et al. [43].

broadband microwave delay lines and microwave mixers. However, since numerous electrical and optical devices were employed, the system was still complicated.

9.4.3 Monolithically Integrated EDG

The basic idea for implementing IFM based on a monolithically integrated EDG [57] is to monitor the wavelength difference between one optical sideband and the optical carrier of an optical microwave signal. Usually, one can tune the temperature of an EDG chip to change the center wavelengths of the passband. When the detected light intensity reaches the maximum value by adjusting the EDG chip temperature, the center wavelength of the EDG is the same as that of the injected optical signal. By recording this temperature, the exact wavelength of the input signal can be obtained. Knowing the wavelengths of one sideband and the optical carrier, the frequency of the optical microwave signal can be calculated. An EDG-based frequency measurement system based on the above principle can obtain a measurement resolution of better than 1 pm, corresponding to ~120 MHz in the 1550-nm band. Since the optical carrier and the other sideband may contribute to the received power in the EDG-based measurement system, a carrier-suppressed modulation is required, and a notch filter is employed to filter out the other sideband, which would place a limitation on the measurement range. The measurement would also be sensitive to the bias drift of the MZM. A proof-of-concept experiment based on the system shown in Fig. 9.25 reports a measurement range from 1 to 15 GHz with a measurement error of less than 200 MHz [44]. Since the EDG can be a multichannel grating, the system can be used to measure the frequencies of multiple microwave signals.

9.4.4 Incoherent Frequency-to-Time Mapping

IFM can also be realized based on incoherent frequency-to-time mapping, which provides advantages such as fast frequency measurement since the measurement is performed in the time domain. In addition, the entire spectrum in the frequency band of interest is shown; therefore, the approach is suitable for frequency measurement of multiple simultaneous signals. Figure 9.26 shows an IFM system architecture in which the input signals with different carrier frequencies are mapped to the time domain with different time delays. The mapping operation is achieved by utilizing a chromatic dispersive element [58, 59]. From Fig. 9.26, we can see that the input signals are modulated on an optical carrier at an MZM that is biased at the minimum transmission point to form the carrier-suppressed DSB optical signal. A step-recovery stage in the form of a high-speed optical

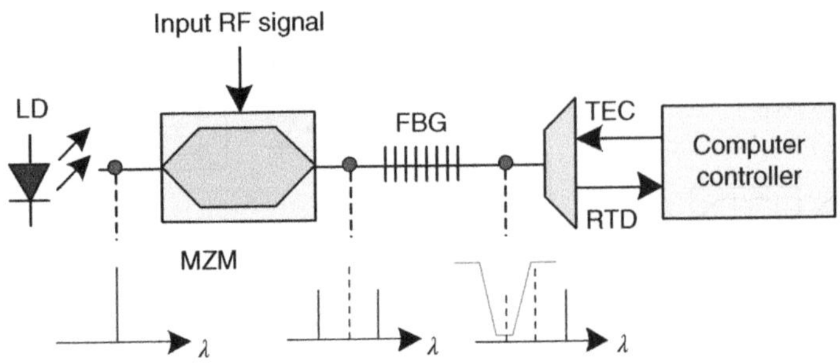

Figure 9.25 Photonic microwave frequency measurement using a monolithically integrated EDG. *Source:* Adapted from Guo et al. [44].

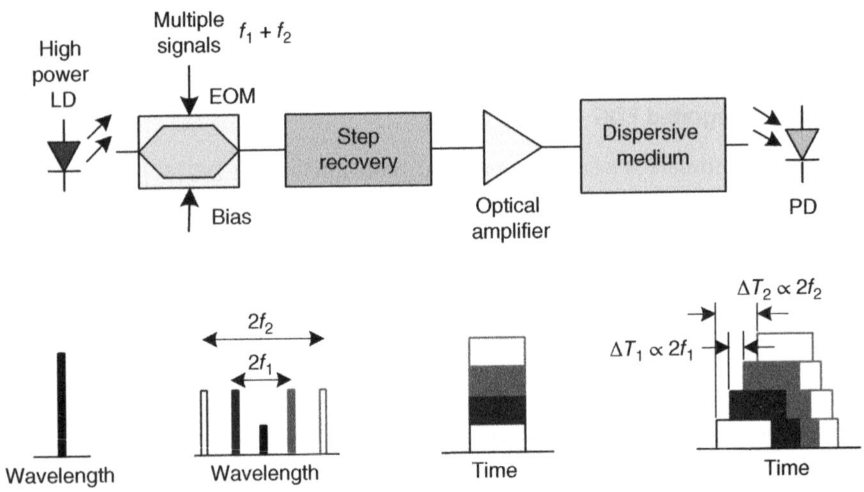

Figure 9.26 Frequency measurement of multiple microwave signals based on frequency-to-time mapping in a dispersive element. The measurements of two signals with frequencies at f_1 and f_2 are shown.

on-off switch is used to gate the optical signal to generate a short pulse train. The pulsed carrier-suppressed DSB optical signal is then propagated through a dispersive element and sent to a PD. When propagating through the dispersive element, each optical signal will propagate at a different relative group velocity. This process maps each coupled pair of sidebands to a distinct time delay centered around the imaginary axis of symmetry provided by the wavelength of the optical carrier, as illustrated in Fig. 9.26. If the time delays are measured, the frequencies of the multiple microwave signals are obtained.

To accurately estimate the time delays, several issues must be considered. First, the gated signal, i.e. the pulsed signal after the step-recovery stage, should have a fast rise time to act as a time measurement reference point. Second, the bandwidth of the step-recovery stage and the PD should have enough bandwidth to replicate the fast gating rise time. Third, the signal processing module should have fast speed to accurately measure the time delays. In a proof-of-concept experiment, two frequencies at 20 and 40 GHz were retrieved using a dispersive element with a dispersion value of 1000 ps/nm and a measurement error of ~1.6 GHz [45].

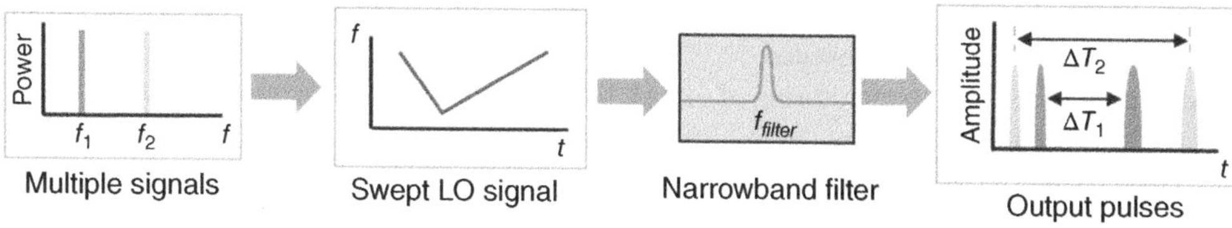

Figure 9.27 Principle of frequency-to-time mapping with a swept reference signal. *Source:* Adapted from Nguyen et al. [62].

In addition to the dispersive medium, using swept frequency signals and narrowband bandpass filters is another important method for frequency-to-time mapping [60–62]. For example, Fig. 9.27 shows the principle of frequency-to-time mapping with a swept reference signal. A bidirectional swept signal is utilized as the LO, the input signals combine with the LO and generate the beating signal. The beating signal is also a swept signal and a narrowband filter is applied to filter out the electrical with certain frequency and generate electrical pulses. The input signal frequency can be obtained by measuring the time interval between pulses. In [60], a Fourier domain mode-locked optoelectronic oscillator (FDML OEO) is established as the bidirectional swept LO, the narrowband electrical bandpass filter has a center frequency of 5 GHz and a 3-dB bandwidth of 60 MHz, the measurement range of 0.5–15 GHz and the measurement error of less than 60 MHz is experimentally demonstrated. With a fixed LO and sweeping the input signal frequency can also realize the frequency-to-time mapping. In [61, 62], frequency shifting recirculating delay line (FS-RDL) is utilized for input signal frequency sweeping, the narrow bandwidth of low-speed photodetector [61] and narrow gain spectrum of stimulated Brillouin scattering (SBS) [62] are utilized as the filter, respectively. The measurement range of 0.1–20 GHz and measurement resolution of 250 MHz is realized.

9.5 Conclusion

Techniques to achieve IFM using photonic techniques have been reviewed. The IFM system based on optical channelizers can estimate the frequencies of multiple simultaneous signals. However, it requires specially designed devices and PD arrays, which are complex and costly. The measurement range (typically <20 GHz) and resolution (typically >1 GHz) are also limited by the fabrication techniques for the optical channelizers, which may not meet the requirements for applications where a wide frequency measurement range and a high measurement resolution are required.

The techniques based on power monitoring are more promising for frequency measurement of a single-channel signal since they can cover a large frequency measurement range (>20 GHz), and can provide better accuracy (<200 MHz). In addition, the system is much simpler with a much lower cost. Although monitoring microwave power is an efficient way to estimate the microwave frequency, there are several issues that still need to be addressed. For instance, the PDs should have a wide bandwidth and a flat response. A wideband DC block is also needed to block the DC component. In addition, a wideband microwave power sensor with a flat response in the whole measurement range, e.g. DLVA, is required to make the energy of the high-frequency microwave signal detectable to low-frequency receivers. The use of these electrical devices will limit the upper bound of the IFM to several tens of GHz. The ripples in the frequency responses of the devices would also deteriorate the measurement accuracy. Moreover, there is a trade-off between the measurement range and the accuracy, which is undesirable for applications where both a large measurement range and a high accuracy are needed.

Techniques based on the monitoring of optical powers only need low-frequency devices. However, the MZMs must be biased at the minimum transmission point to suppress the optical carrier, which would suffer severely from bias drift. The bias drift of the MZMs will make the system unstable, or a sophisticated control circuit is needed to stabilize the operation. Thus, to ensure acceptable measurement accuracy, an LD with high wavelength stability is required since the measurement is highly dependent on the center wavelength of the optical carrier. In addition, the optical filter implemented using a Sagnac loop or an asymmetric Mach–Zehnder interferometer is easily disturbed by environmental variations.

From a practical point of view, the EW environment is always filled with spectral clutter. Therefore, it is required that an IFM system can estimate the frequencies of multiple signals. Although IFM based on optical channelizers may have this capability, its measurement range and accuracy are very limited. IFM based on incoherent frequency-to-time mapping also has the ability to perform IFM for multiple signals, but the accuracy is very low or a high-speed sampling oscilloscope is required. IFM based on power monitoring can provide broadband and high-accuracy frequency measurements, but it is applicable for only single-frequency signals. When multiple signals are received with one signal that has a much higher power, the approach may also provide a good frequency measurement for the strongest signal. However, the system will report an erroneous frequency if the multiple signals have similar powers.

Although significant progress has been made in the last few years for IFM based on photonics, the techniques have limitations in terms of simplicity, cost, measurement range, accuracy and dynamic range. Therefore, photonics-assisted IFM is still a topic of interest, especially its practical applications in warfare systems. The ultimate solution for photonic-assisted IFM with high performance is to use photonic integrated circuits.

We would like to point out that in addition to frequency information of a microwave signal, it is also essential to have knowledge of other parameters of a microwave signal. For example, in a radar system, the emitted signals are usually in the form of a pulse train. Four other important parameters, including pulse amplitude (PA), PW, TOA, and angle of arrival (AOA) [2], are also important for warfare applications. Accurate PA information can be used for inverse gain jamming, where the jammer power is inversely proportional to the PA measured. The PW can be used to coarsely identify the type of radar (for example, weapon radars always have a short PW). The TOA provides information on the pulse repetition rate, which can be used to generate matched jamming pulses with the same pulse repetition rate. It can also be used to predict the arrival time of the next pulse for a jammer. The AOA gives the direction of a hostile radar. If some of the four parameters can be measured together with the frequency information, then the measurement system would provide much detailed information, which is important for practical applications.

References

1 Schleher, D.C. (1999). *Electronic Warfare in the Information Age*. Norwood, MA: Artech House.
2 Tsui, J.B.Y. (1986). *Microwave Receivers with Electronic Warfare Applications*. New York: Wiley.
3 Tsui, J.B.Y. (2001). *Digital Techniques for Wideband Receiver*, 2e. Norwood, MA: Artech House.
4 Earp, C.W. (1948). Frequency indicating cathode ray oscilloscope, US Patent 2,434,914.
5 James, W.G. (1976). Instantaneous frequency measurement system. US Patent 3,939,411.

6 Tsui, J.B.Y. and Schrick, G.H. (1980). Instantaneous frequency measurement (IFM) receiver with capability to separate cw and pulsed signals. US Patent 4,194,206.
7 Cuckson, H. and Curtis, P.D. (1983). Microwave instantaneous frequency measurement apparatus. US Patent 4,414,505.
8 Tsui, J.B.Y. and Shaw, R.L. (1985). Simultaneous signal detection for IFM receivers by transient detection. US Patent 4,547,727.
9 Asendorf, R.H., Fitzpatrick, J.P., and Graves, B.W. (1986). Automatic frequency identifier for radar signals. US Patent 4,612,545.
10 Tsui, J.B.Y. (1986). Instantaneous frequency measurement receiver with digital processing. US Patent 4,633,516.
11 Gagnon, A., McMillan, M., and Gale, P.M. (1988). Method and apparatus for simultaneous instantaneous signal frequency measurement. US Patent 4,791,360.
12 Gale, P.M., McMillan, M., and Gagnon, A. (1989). Apparatus for measuring the frequency of microwave signals. US Patent 4,859,934.
13 Tsui, J.B.Y. and McCormick, W.S. (1990). Instantaneous frequency measurement (IFM) receiver with only two delay lines. US Patent 4,963,816.
14 Sanderson, R.B. and Tsui, J.B.Y. (1992). Instantaneous frequency measurement receiver with bandwidth improvement through phase shifted sampling of real signals. US Patent 5,109,188.
15 Tsui, J.B.Y. and Sharpin, D.L. (1993). Frequency measurement receiver with bandwidth improvement through synchronized phase shifted sampling. US Patent 5,198,748.
16 Choate, D.B. and Tsui, J.B.Y. (1994). Frequency measurement receiver with means to resolve two folding frequencies of complex signals. US Patent 5,323,103.
17 Ward, C.R. (1995). Hybrid instantaneous frequency measurement compressive receiver apparatus and method. US Patent 5,424,631.
18 Schmidt, R.O. (1995). Simultaneous signals IFM receiver using plural delay line correlators. US Patent 5,440,228.
19 Regev, Z. (2002). Method for instantaneous frequency measurement. US Patent 6,433,531.
20 Tsui, J.B.Y., Lopata, S.M., and Ward, C.R. (2007). Wideband digital IFM receiver. US Patent 7,206,707.
21 Tsui, J.B.Y. (1985). *Microwave Receivers and Related Components*. Los Alton, CA: Avionics Laboratory, Air Force Wright Aeronautical Laboratories, Peninsula Publishing.
22 Sullivan, W.B. (2002). Instantaneous frequency measurement receivers for maritime patrol. *J. Electron. Defense* 25 (10): 55–62.
23 Lee, C.H. (2006). *Microwave Photonics*. Boca Raton, FL: CRC Press.
24 Heaton, J.M., Watson, C.D., Jones, S.B. et al. (1998). 16-channel (1- to 16-GHz) microwave spectrum analyzer device based on a phased array of GaAs/AlGaAs electro-optic waveguide delay lines. *Proc. SPIE* 3278 (245): 245–251.
25 Wang, W.S., Davis, R.L., Jung, T.J. et al. (2001). Characterization of a coherent optical RF channelizer based on a diffraction grating. *IEEE Trans. Microw. Theory Tech.* 49 (10): 1996–2001.
26 Hunter, D.B., Edvell, L.G., and Englund, M.A. (2005). Wideband microwave photonic channelised receiver. *Proc. MWP* 2005: 249–252.
27 Winnall, S.T., Lindsay, A.C., Austin, M.W. et al. (2006). A microwave channelizer and spectroscope based on an integrated optical Bragg-grating Fabry–Perot and integrated hybrid fresnel lens system. *IEEE Trans. Microw. Theory Tech.* 54 (2): 868–872.
28 Nguyen, L.V.T. and Hunter, D.B. (2006). A photonic technique for microwave frequency measurement. *IEEE Photon. Technol. Lett.* 18 (9-12): 1188–1190.
29 Zou, X.H. and Yao, J.P. (2008). An optical approach to microwave frequency measurement with adjustable measurement range and resolution. *IEEE Photon. Technol. Lett.* 20 (21-24): 1989–1991.

30 Zou, X.H. and Yao, J. (2009). Microwave frequency measurement with improved measurement range and resolution. *Electron. Lett.* 45 (10): 497–498.

31 Zhang, X.M., Chi, H., Zhang, X.M. et al. (2009). Instantaneous microwave frequency measurement using an optical phase modulator. *IEEE Microw. Wirel. Compon. Lett.* 19 (6): 422–424.

32 Attygalle, M. and Hunter, D.B. (2009). Improved photonic technique for broadband radio-frequency measurement. *IEEE Photon. Technol. Lett.* 21 (1–4): 206–208.

33 Zhou, J.Q., Fu, S.N., Aditya, S. et al. (2009). Instantaneous microwave frequency measurement using photonic technique. *IEEE Photon. Technol. Lett.* 21 (15): 1069–1071.

34 Zou, X.H., Pan, S., and Yao, J.P. (2009). Instantaneous microwave frequency measurement with improved measurement range and resolution based on simultaneous phase modulation and intensity modulation. *J. Lightw. Technol.* 27 (23): 5314–5320.

35 Pan, S.L. and Yao, J.P. Instantaneous microwave frequency measurement using a photonic microwave filter pair. *IEEE Photon. Technol. Lett.* 22: 1437–1439. to be published.

36 Zhou, J.Q., Fu, S.N., Shum, P.P. et al. (2009). Photonic measurement of microwave frequency based on phase modulation. *Opt. Express* 17 (9): 7217–7221.

37 Li, J.Q., Fu, S.N., Xu, K. et al. (2009). Photonic-assisted microwave frequency measurement with higher resolution and tunable range. *Opt. Lett.* 34 (6): 743–745.

38 Pan, S.L. and Yao, J.P. (2009). Instantaneous photonic microwave frequency measurement with a maximized measurement range. *Proc. MWP 2009*, Valencia, Spain (14–16 October 2009), paper Fr 4.3.

39 Chi, H., Zou, X.H., and Yao, J.P. (2008). An approach to the measurement of microwave frequency based on optical power monitoring. *IEEE Photon. Technol. Lett.* 20 (13-16): 1249–1251.

40 Zou, X.H., Chi, H., and Yao, J.P. (2009). Microwave frequency measurement based on optical power monitoring using a complementary optical filter pair. *IEEE Trans. Microw. Theory Tech.* 57 (2): 505–511.

41 Drummond, M.V., Monteiro, P., and Nogueira, R.N. (2009). Photonic RF instantaneous frequency measurement system by means of a polarization-domain interferometer. *Opt. Express* 17 (7): 5433–5438.

42 Winnall, S.T. and Lindsay, A.C. (1999). A Fabry–Perot scanning receiver for microwave signal processing. *IEEE Trans. Microw. Theory Tech.* 47 (7): 1385–1390.

43 Emami, H., Sarkhosh, N., Bui, L.A., and Mitchell, A. (2008). Amplitude independent RF instantaneous frequency measurement system using photonic Hilbert transform. *Opt. Express* 16 (18): 13707–13712.

44 Guo, H.L., Xiao, G.Z., Mrad, N., and Yao, J.P. (2009). Measurement of microwave frequency using a monolithically integrated scannable Echelle diffractive grating. *IEEE Photon. Technol. Lett.* 21 (1-4): 45–47.

45 Nguyen, L.V.T. (2009). Microwave photonic technique for frequency measurement of simultaneous signals. *IEEE Photon. Technol. Lett.* 21 (9-12): 642–644.

46 Agrawal, G.P. (2002). *Fiber-Optic Communication Systems*, 3e. New York: Wiley.

47 Smith, G.H., Novak, D., and Ahmed, Z. (1997). Overcoming chromatic-dispersion effects in fiber-wireless systems incorporating external modulators. *IEEE Trans. Microw. Theory Tech.* 45 (8): 1410–1415.

48 Zeng, F. and Yao, J.P. (2005). Investigation of phase-modulator-based all-optical bandpass microwave filter. *J. Lightw. Technol.* 23 (4): 1721–1728.

49 Bull, J.D., Jaeger, N.A., Kato, H. et al. (2004). 40-GHz electro-optic polarization modulator for fiber optic communications systems. *Proc. SPIE* 5577: 133–143.

50 Capmany, J., Ortega, B., and Pastor, D. (2006). A tutorial on microwave photonic filters. *J. Lightw. Technol.* 24 (1): 201–229.

51 Wang, Q. and Yao, J.P. (2008). Multitap photonic microwave filters with arbitrary positive and negative coefficients using a polarization modulator and an optical polarizer. *IEEE Photon. Technol. Lett.* 20 (1-4): 78–80.

52 Yao, J.P. and Wang, Q. (2007). Photonic microwave bandpass filter with negative coefficients using a polarization modulator. *IEEE Photon. Technol. Lett.* 19 (9-12): 644–646.

53 Wang, Q. and Yao, J.P. (2007). Switchable optical UWB monocycle and doublet generation using a reconfigurable photonic microwave delay-line filter. *Opt. Express* 15 (22): 14667–14672.

54 Yao, J.P. (2009). Microwave photonics. *J. Lightw. Technol.* 27 (3): 314–335.

55 Yao, J.P., Zeng, F., and Wang, Q. (2007). Photonic generation of ultrawideband signals. *J. Lightw. Technol.* 25 (11): 3219–3235.

56 Emami, H., Sarkhosh, N., Bui, L.A., and Mitchell, A. (2008). Wideband RF photonic in-phase and quadrature-phase generation. *Opt. Lett.* 33 (2): 98–100.

57 Chenben, P. (2007, ch. 5,). Wavelength dispersive planar waveguide devices: echelle gratings and arrayed waveguide gratings. In: *Optical Waveguides: From Theory to Applied Technologies* (ed. M.L. Calvo and V. Laksminarayanan), 173–230. Boca Raton, FL: CRC Press.

58 Saperstien, R.E., Alic, N., Pasasenko, D. et al. (2005). Time-domain waveform processing by chromatic dispersion for temporal shaping of optical pulses. *J. Opt. Soc. Am. B* 22 (11): 2427–2436.

59 Azana, J. and Muriel, M.A. (2000). Real-time optical spectrum analysis based on the time-space duality in chirped fiber grating. *IEEE J. Quantum Electron.* 36 (5): 517–526.

60 Hao, T., Tang, J., Li, W. et al. (2018). Microwave photonics frequency-to-time mapping based on a fourier domain mode locked optoelectronic oscillator. *Opt. Express* 26 (26): 33582–33591.

61 Nguyen, T.A., Chan, E.H.W., and Minasian, R.A. (2014). Photonic multiple frequency measurement using a frequency shifting recirculating delay line structure. *J. Lightw. Technol.* 32 (20): 3831–3838.

62 Nguyen, T.A., Chan, E.H.W., and Minasian, R.A. (2014). Instantaneous high-resolution multiple-frequency measurement system based on frequency-to-time mapping technique. *Opt. Lett.* 39 (8): 2419–2422.

10

Microwave Photonic Sensors

10.1 Introduction

Fiber optic sensors have been heavily investigated for the last few decades and numerous techniques have been proposed and demonstrated [1]. It is one of the few fields that has been successfully employed for practical applications such as civil structure health monitoring, biomedical sensing, and quality and process control. In general, an optical sensor is implemented to monitor the wavelength shift due to changes in environmental conditions, such as temperature, strain, or vibration. Fiber Bragg gratings (FBGs) have been widely used as optical sensors thanks to the simplicity in implementation due to the fiber nature of the devices which can be inherently incorporated in a fiber optic link for remote and distributed sensing [2–4]. To monitor the wavelength shift, interrogation techniques have been proposed, including directly monitoring the optical spectrum from a sensor using an optical spectrum analyzer (OSA) [2]. It is simple, but it is costly, and the speed is limited, especially for high-resolution sensing. The use of an edge filter such as a fiber optic interferometer [5,6], to convert the wavelength or phase change to optical intensity change is another approach that can provide fast interrogation, but the power drifts of the light source or the poor stability of the edge filter will affect the accuracy of the measurements [7,8]. The use of a fast-scanning optical filter can also realize high-speed and high-resolution interrogation, but the wavelength scanning range is usually limited, or the resolution is compromised if a wider wavelength scanning range is needed [9]. To achieve fast-speed and high-resolution interrogation, one solution is to translate the optical wavelength shift from the optical domain to a microwave frequency shift in the microwave domain, a technique that is *called microwave photonic interrogation* [10,11]. In this chapter, optical sensing and interrogation techniques developed in the last few years based on microwave photonics will be reviewed, including heterodyne beating of two wavelengths to generate a microwave signal with its frequency being a function of the sensing information, the use of an optoelectronic oscillator (OEO) to generate a microwave signal, to translate the optical wavelength shift to the microwave frequency change, and the use of wavelength-to-time (WTT) mapping to translate the sensing information from the optical wavelength domain to the microwave frequency domain, to increase the interrogation speed and resolution. The use of photonic integrated circuits (PICs) to achieve high-speed and high-resolution microwave photonic sensing is also discussed.

Microwave Photonics, First Edition. Jianping Yao, José Capmany, and Ming Li.

Companion website: www.wiley.com/go/yao/microwavephotonics

10.2 Optical Sensors Based on a Dual-Wavelength Laser Source

A dual-wavelength laser source can be used to implement an optical sensor that has high speed and high resolution. The fundamental concept is to use one wavelength to carry the sensing information while the other wavelength as a reference that is kept unchanged. By beating the two wavelengths at a photodetector (PD), a microwave signal with its frequency change reflecting the wavelength change of the sensing wavelength is generated [11]. By measuring the frequency of the beat signal, the sensing information is obtained at a high speed with high resolution.

Assume that the two wavelengths generated by a dual-wavelength laser are given by

$$E_1(t) = E_{01}\cos(\omega_1 t + \phi_1) \tag{10.1}$$

and

$$E_2(t) = E_{02}\cos(\omega_2 t + \phi_2), \tag{10.2}$$

where E_{01}, E_{02} are the amplitude terms and ϕ_1, ϕ_2 are the phase terms of the two wavelengths. When the two wavelengths are applied to a PD, considering the limited bandwidth of the PD, a beat signal with the frequency corresponding to the wavelength difference between the two wavelengths is generated,

$$i_{RF} = A\cos[(\omega_1 - \omega_2)t + (\phi_1 - \phi_2)], \tag{10.3}$$

where A is a constant which is determined by E_{01}, E_{02} and the responsivity of the PD. As can be seen, a shift in one wavelength would lead to a change in the beat frequency. For example, at the 1550 nm window, if the wavelength shift is 0.1 pm, the beat frequency will have a change of 12.5 MHz, which can be easily and precisely measured by a digital signal processor (DSP).

A dual-wavelength laser can be implemented using fiber optics. Figure 10.1(a) shows a dual-wavelength DBR fiber laser with two matched FBGs inscribed in a rare-earth-doped

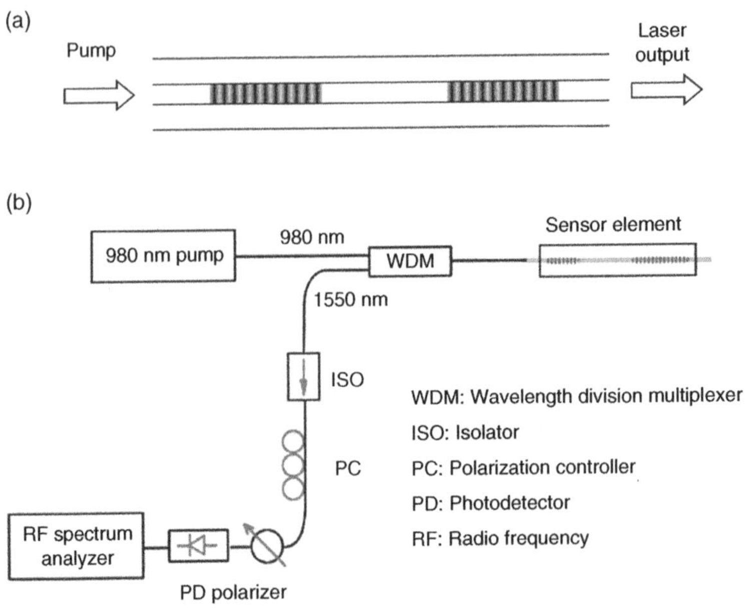

Figure 10.1 (a) A dual-wavelength DBR laser for high-speed and high-resolution sensing. (b) Beat signal generation and its measurement. *Source:* Guan et al. [12]/IEEE.

polarization-maintaining (PM) fiber [12]. By controlling the physical spacing between the two FBGs to make the free spectral range (FSR) greater than the bandwidth of the FBGs, a single longitudinal mode will be supported in the laser cavity, thus mode hoping that is detrimental to the stability of the lasing operation is eliminated. Thanks to the use of a PM fiber, two orthogonal polarization modes are supported with the wavelength spacing determined by the birefringence of the PM fiber. The use of the birefringence of an optical fiber in a laser cavity to generate two orthogonally polarized wavelengths with a wavelength spacing depending on the birefringence has been employed for optical sensing and the approach is known as polarimetric sensing [13, 14]. Compared with the mode hoping between longitudinal modes, the competition between the two polarization modes is much weaker due to the polarization-burning-hole effect [15]. As a result, a beat signal with its frequency corresponding to the two polarization modes is generated, which is used to measure the external perturbations.

Assume that the DBR laser operates in single-longitudinal mode with two orthogonally polarized modes, the two lasing wavelengths can be expressed by

$$\lambda_{x,y} = \frac{2n_{x,y}L_{eff}}{M} \tag{10.4}$$

where $n_{x,y}$ are the effective refractive indices of the two polarization modes corresponding to the two orthogonal polarization directions, L_{eff} is the effective cavity length, and M is the order of the longitudinal mode.

By using a polarizer to project the two orthogonally polarized modes to the same polarization direction and applying the two assigned modes to a PD, as shown in Fig. 10.1(b), a beat signal is generated. The beat frequency is given by

$$f_b = f_x - f_y = \frac{c}{n_0\lambda_0}B \tag{10.5}$$

where n_0 is the average refractive index of the fiber, λ_0 is the Bragg wavelength of the FBGs, and $B = |n_x - n_y|$ is the fiber birefringence.

The use of the DBR laser for lateral force sensing was performed. The lateral force and the beat frequency change have a linear relationship. Figure 10.2 shows the beat-frequency change as a

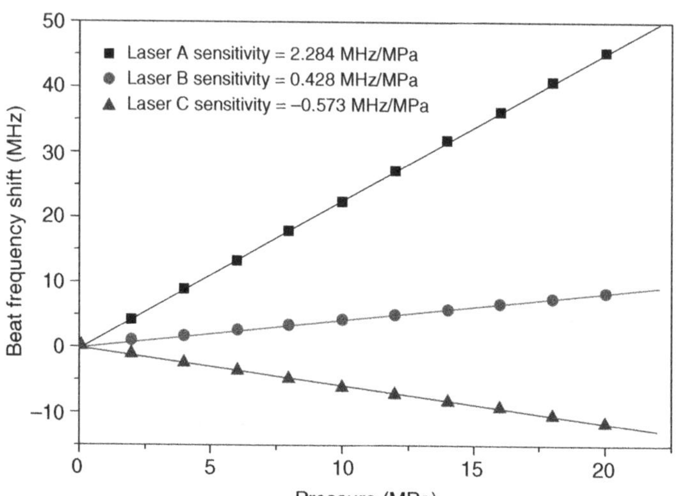

Figure 10.2 The beat-frequency change as a function of the applied lateral force for three different DBR lasers. *Source:* Guan et al. [12]/IEEE.

function of the applied lateral force for three different DBR lasers. As can be seen, the highest sensitivity is 2.28 MHz/MPa. The resolution is determined by the spectral width of the generated microwave signal. In the experiment [12], the 3-dB bandwidth of the generated microwave signal is 2 kHz, corresponding to a very high sensing resolution of 9 kPa.

A similar approach using a dual-wavelength DBR laser for temperature and strain sensing was also reported [16]. A strain sensitivity of 0.15 pm/με over a range of 0–2400 με and a temperature sensitivity of 14.3 pm/°C over a range of 2–230 °C were demonstrated.

An integrated intra-laser-cavity microparticle sensor based on a dual-wavelength distributed-feedback channel waveguide laser in a ytterbium-doped amorphous aluminum oxide on a silicon substrate was also demonstrated [17]. Figure 10.3 shows the experimental setup for microparticle sensing. The laser has a Bragg grating in the cavity with two localized quarter wavelength phase shifts, to induce two resonances that form two longitudinal lasing modes. The wavelength spacing between the two lasing modes is changed due to the particle-induced scattering of laser power, making the population inversion increase, which makes the temperature and refractive index (RI) of the waveguide decrease, thereby altering the thermal chirp of the Bragg grating in the laser cavity, leading to the change in the frequency spacing between the two lasing modes. The results showed that real-time detection and accurate size measurement of single micro-particles with diameters ranging between 1 and 20 μm were achieved.

Instead of using two single-frequency optical wavelengths to beat to generate a microwave signal, high-speed and high-resolution sensing can also be realized by beating two time-delayed linearly frequency-modulated (or chirped) optical waveforms (LFMOW) [18]. As shown in Fig. 10.4(a), a linearly chirped optical waveform generated by a frequency-modulated laser diode (LD) is launched into a linearly chirped FBG (LCFBG) pair consisting of two identical LCFBGs with one serving as a sensing LCFBG and the other as a reference LCFBG. The reflection from the two LCFBGs would lead to two time-delayed linearly chirped optical waveforms. By beating the time-delayed chirped waveforms at a PD, a microwave signal with its beat frequency that is proportional to the time delay difference between the two time-delayed chirped waveforms is generated. By measuring the frequency change of the beat signal, the sensing information is measured [18]. This technique, also called dechirping, has been employed in radar systems for target detection [19]. Figure 10.4(b) shows the relationship between two time-delayed linearly chirped optical waveforms. As can be seen, the frequency difference between the two waveforms is a function of the time delay. By measuring the beat frequency, the time delay is measured.

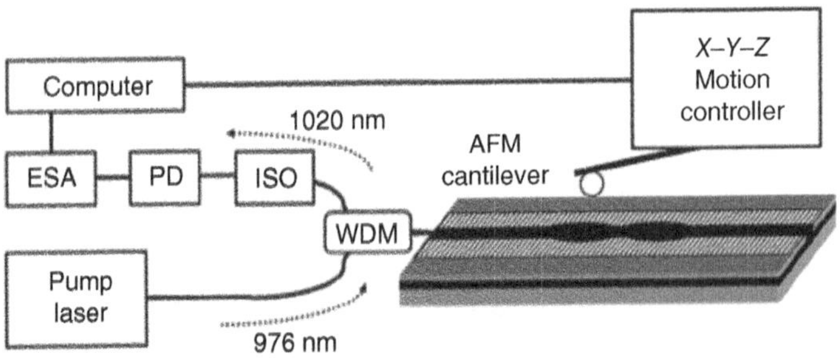

Figure 10.3 Intra-laser-cavity micro-particle dual-wavelength laser for microparticle sensing. *Source:* Bernhardi et al. [17]/John Wiley & Sons.

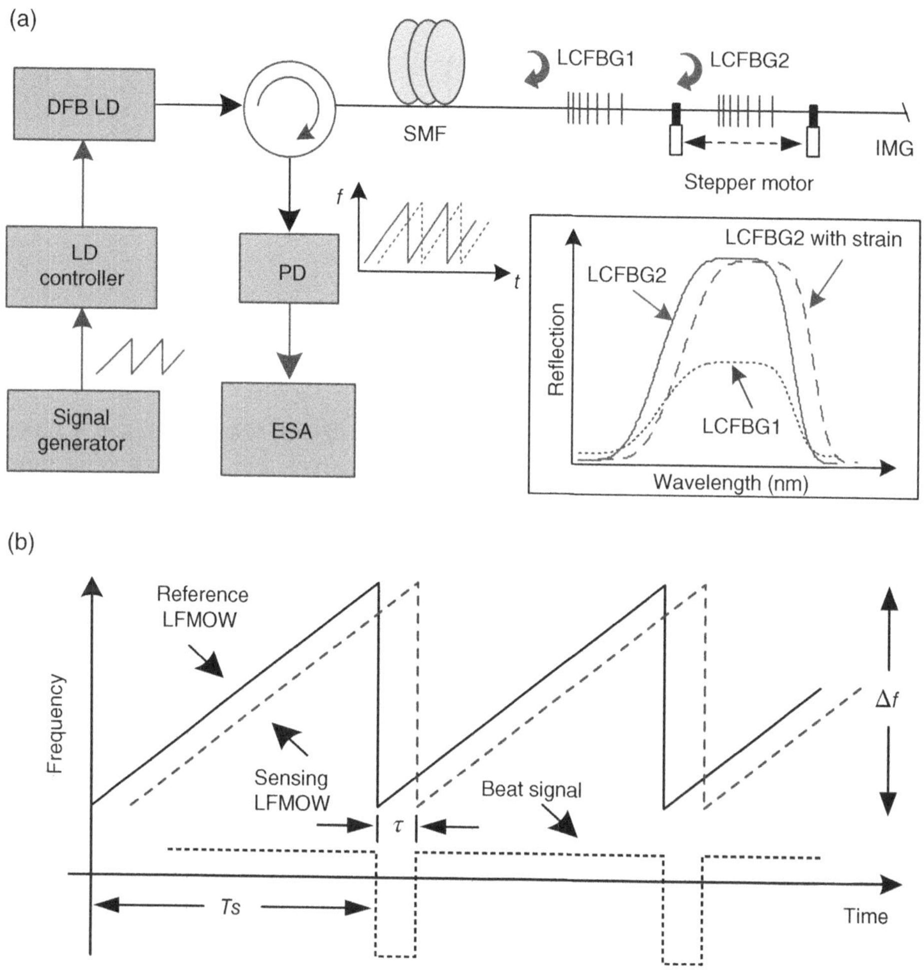

Figure 10.4 (a) A linearly chirped optical waveform generated by a DFB laser is sent two LCFBGs, to generate two time-delayed linearly chirped optical waveforms. A beat signal between the two time-delayed linearly chirped optical waveforms is generated at the output of the PD. (b) Frequency relationship between the two time-delayed linearly chirped optical waveforms and the corresponding beat frequency. *Source:* Wang et al. [18]/Optical Society of America.

For a reference waveform, the instantaneous optical frequency $\omega_1(t)$ is given by

$$\omega_1(t) = \omega_0 + \alpha t \tag{10.6}$$

where ω_0 is the angular frequency at the start of a period of the waveform, and α is the chirp rate.

The optical field of the reference waveform $E_1(t)$ can be written as

$$E_1(t) = A_1 \exp\left[j\left(\omega_0 t + \frac{1}{2}\alpha t^2 + \varphi_0\right)\right] \tag{10.7}$$

where A_1 is the amplitude and φ_0 is the initial phase of the reference waveform.

Similarly, the optical frequency and the optical field of the time-delayed sensing waveform can be written as

$$\omega_2(t,\tau) = \omega_0 + \alpha(t-\tau) \tag{10.8}$$

$$E_2(t) = A_2 \exp\left\{j\left[\omega_0(t-\tau) + \frac{1}{2}\alpha(t-\tau)^2 + \varphi_0\right]\right\} \tag{10.9}$$

where A_2 is the amplitude of the sensing waveform, and τ is the time delay difference between the two waveforms, given by, $\tau = 2n_{eff}L/c$, where L is the distance between the two LCFBGs, n_{eff} is the RI of the fiber, c is the velocity of light in vacuum. When these two waveforms are applied to a high-speed PD, a beat signal is generated, given by

$$\begin{aligned} i(t,\tau) &= R|E_1(t) + E_2(t,\tau)|^2 \\ &= R\left[I_1 + I_2 + 2\sqrt{I_1 I_2}\cos\left(\alpha\tau t + \omega_0\tau - \frac{1}{2}\alpha\tau^2\right)\right] \end{aligned} \tag{10.10}$$

where I_1 and I_2 are the intensities of the reference and sensing signals, respectively, and R is the responsivity of the PD. As can be seen, the beat frequency is $\alpha\tau$, which is proportional to the time delay. By measuring the beat frequency, the time delay is measured.

This technique is classified in the category of heterodyne beating since the sensing information is obtained by measuring the beat frequency between two time-delayed chirped waveforms. The advantage of this approach, compared with the use of a dual-wavelength laser source, is its high stability. The stability of a dual-wavelength laser is usually poor due to mode competition. This problem is fully solved by beating two time-delayed linearly chirped optical waveforms.

10.3 Optical Sensors Based on an Optoelectronic Oscillator

The wavelength change of an optical sensor can also be translated to microwave frequency change using an OEO [20]. An OEO is a hybrid optoelectronic system that consists of a hybrid amplified feedback loop with an optical path and an electrical path. The optical path can have a long optical fiber delay line. Thanks to the use of a long fiber delay line in the loop, the Q factor can be large, which would lead to the generation of a microwave signal with ultra-low phase noise. An OEO with an ultra-low phase noise can find important applications in communications, radar and instrumentation. This was the original motivation for proposing and implementing of an oscillator using a high Q factor hybrid optoelectronic loop for low-phase noise microwave generation. For sensing applications, however, the high Q factor is not a major concern, instead, the stability of microwave generation and the sensitivity of the oscillation frequency to environmental change are the primary concerns.

In general, the oscillation frequency is determined by the center frequency of an optical filter in the hybrid loop. If the wavelength of the optical filter is changing due to the change in the environmental conditions, then the microwave frequency is also changing. By monitoring the microwave frequency, the sensing information is measured. Thus, an optical filter that is sensitive to environmental changes is a critically important device in an sensing OEO. For example, a phase-shifted FBG (PS-FBG) can be used as an optical filter, which can be used for distributed sensing. An integrated ring resonator can also be used as an optical filter, which is particularly useful for microparticle sensing due to the ultra-small size in nature. In the last few years, the use of an OEO for high-speed and high-resolution sensing has been well studied and numerous techniques have been reported. In the following, two OEO-based sensors that were implemented for strain sensing and for temperature-insensitive lateral force sensing will be discussed.

A high-speed and high-resolution strain sensor based on a tunable OEO employing a phase-shifted FBG (PS-FBG) is shown in Fig. 10.5(a) [21]. The system consists of an LD, a polarization

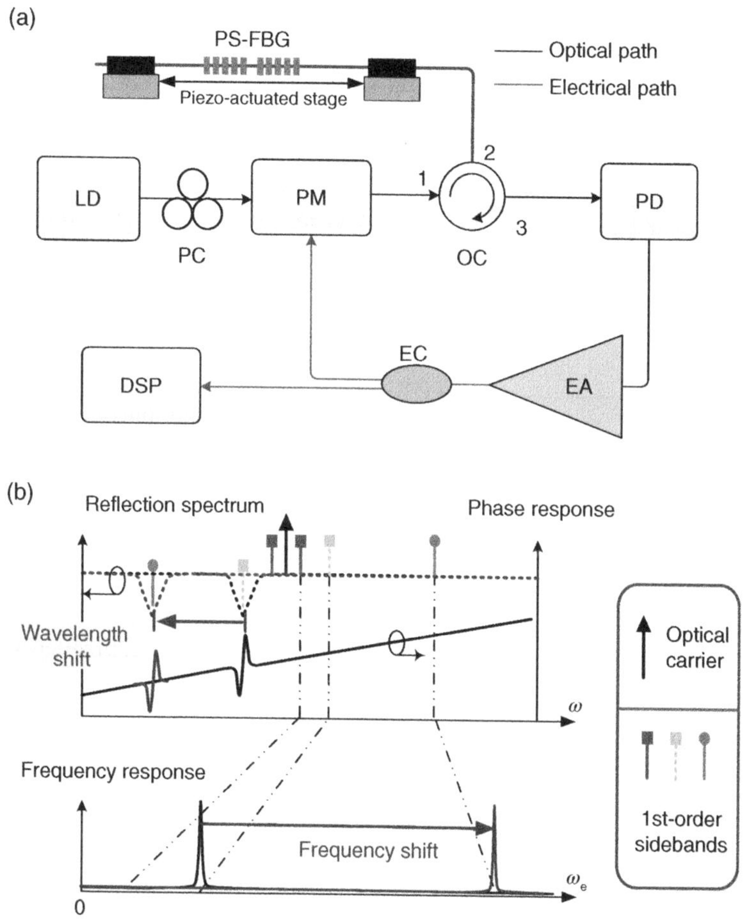

Figure 10.5 (a) A strain sensor based on an OEO employing a PS-FBG. (b) An equivalent microwave bandpass filter in the OEO. *Source:* Li et al. [21]/Optical Society of America.

controller (PC), a phase modulator (PM), a PS-FBG, and a PD. A light wave generated by the LD is sent to the PM via the PC. The PC is used to minimize the polarization-dependent loss at the PM. The phase-modulated light wave is sent to the PS-FBG through an optical circulator (OC), to filter out one sideband, to make the phase-modulated double-sideband plus carrier signal be a single-sideband plus carrier intensity-modulated signal, and the reflected optical signal is sent to the PD where a microwave signal is generated. After electronic amplification by an electronic amplifier (EA), the microwave signal at the output of the PD is split into two paths, with one sent to the PM to close the OEO loop and the other sent to a DSP for frequency measurement. Once the loop net gain is greater than the net loss, microwave oscillation will start and a microwave signal with the frequency determined by the center frequency of an equivalent microwave bandpass filter (MBPF) is generated.

The MBPF is implemented by the joint operation of the LD, the PM, the PS-FBG, and the PD. As can be seen from Fig. 10.5(b), one first-order sideband of the phase-modulated optical signal is filtered out by the notch of the PS-FBG. Thus, the phase-modulated optical signal is converted to an intensity-modulated optical signal, and a MBPF is implemented based on phase-modulation to

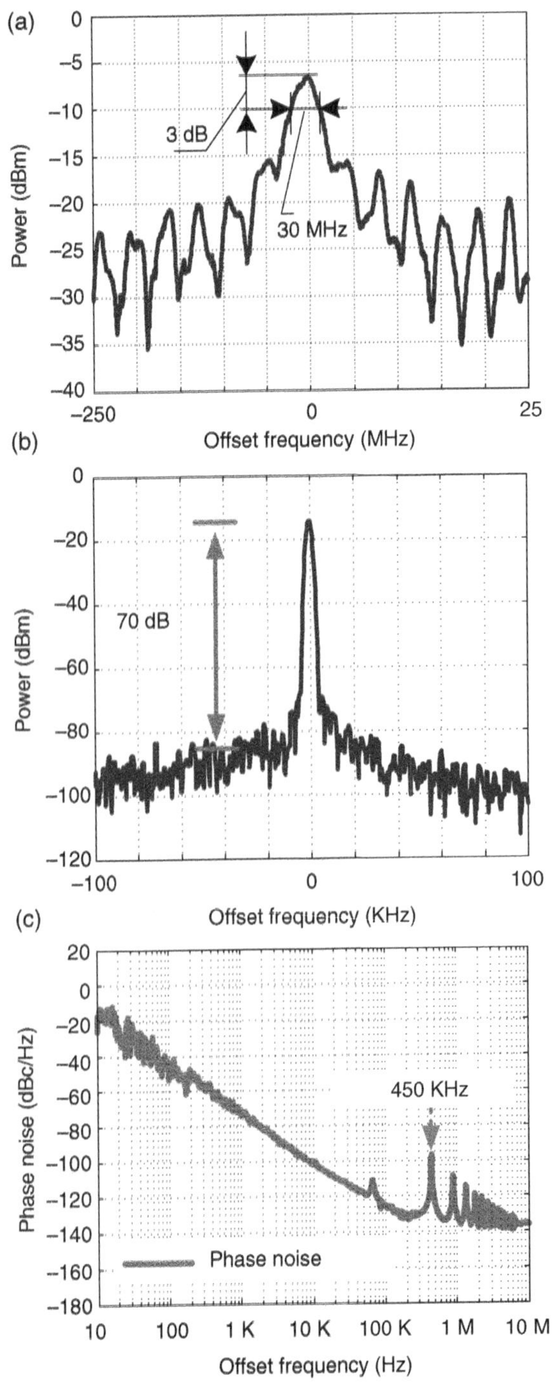

Figure 10.6 (a) The frequency response of the MBPF. The 3-dB bandwidth is 30 MHz. (b) Spectrum of the 10-GHz microwave signal generated from the OEO. (c) Phase noise measurement of the microwave signal. *Source:* Li et al. [21]/Optical Society of America.

intensity-modulation (PM-IM) conversion. The bandwidth of the MBPF is determined by the notch width of the PS-FBG, which is usually a few tens of MHz. As shown in Fig. 10.5(b), when the PS-FBG is stretched, the notch wavelength is shifted due to the variation of the grating pitch. As a result, the central frequency of the MBPF is correspondingly shifted. The OEO system would translate the optical wavelength change to a microwave frequency change, and the optical-domain measurement is translated to an electrical spectrum measurement, ensuring a significantly increased measurement speed and resolution. In addition, the use of an oscillatory scheme has also a positive impact on the signal-to-noise ratio (SNR) performance, making the measurement more accurate.

Figure 10.6(a) shows the frequency response of the MBPF, which was implemented based on phase modulation at the PM and PM-IM conversion in a PS-FBG. The PS-FBG has a bandwidth (full-width at half-maximum or FWHM) of about 30 MHz. Thus, the bandwidth of the MBPF is about 30 MHz. The PS-FBG was fabricated by introducing a phase shift to a uniform FBG during the fabrication process. The PS-FBG was incorporated in the OEO and it was mounted on a piezo-actuated stage for strain measurement. When the OEO loop is closed and the gain is greater than the loss, a microwave signal with a frequency of 10 GHz and a 70-dB sidemode suppression ratio is generated, as shown in Fig. 10.6(b). The 70-dB sidemode suppression ratio is related to the SNR of the wavelength interrogation scheme, well beyond the achievable values by any other techniques. The phase noise performance of the generated 10-GHz microwave signals is shown in Fig. 10.6 (c). At 10 kHz frequency offset, the phase noise is −105 dBc/Hz, confirming the high-phase noise performance of the OEO. Note that multiple peaks are observed at offset frequencies equal or greater than 450 kHz. This value corresponds to the longitudinal mode spacing of the OEO. A shorter OEO loop would make the mode spacing greater but with poorer phase noise performance. When the notch wavelength of the PS-FBG is shifted, the frequency of the generated microwave signal is then shifted accordingly and can be easily measured using a DSP. Figure 10.7(a) shows the measured frequency response of the MBPF for different applied axial strains. The spectra of the generated microwave signals at the OEO output corresponding to the different applied strains are shown in Fig. 10.7(b). The frequency of the generated microwave signal is shifted from about 7–17 GHz. The frequency range can be wider if wider bandwidth components are used. The measurements confirm the expected linear relationship between the applied strain and the frequency of the generated microwave signal, as shown in Fig. 10.7(c).

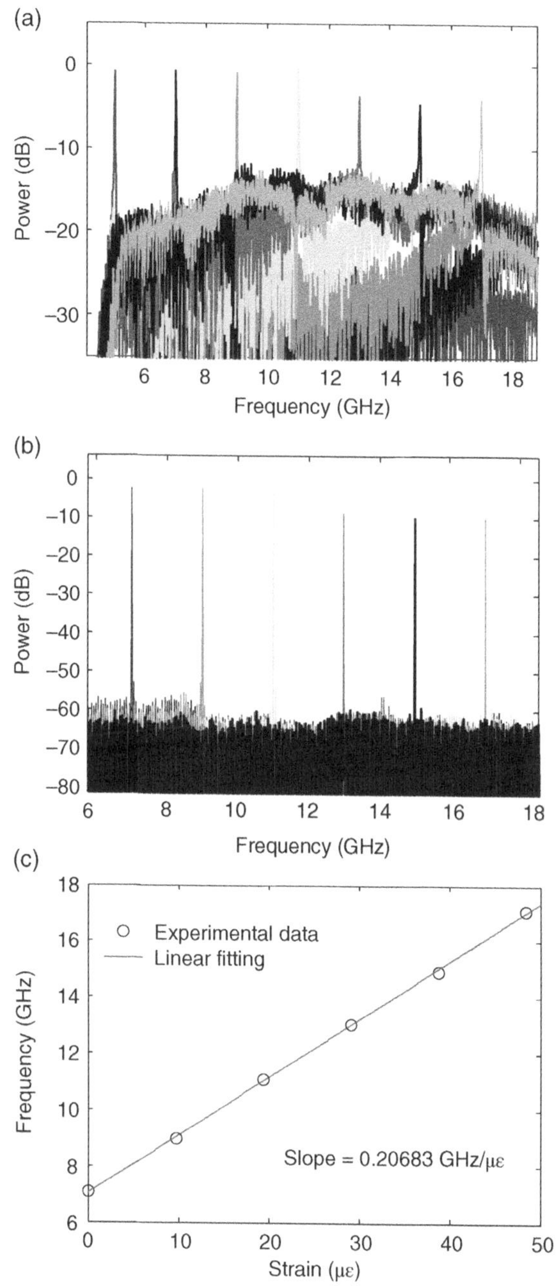

Figure 10.7 (a) Measured frequency response of the PS-FBG-based MBPF for different applied strains. (b) Spectra of the generated microwave signal at the OEO output for different strains. (c) Relationship between the applied strain and the frequency of the generated microwave signal. *Source:* Li et al. [21]/Optical Society of America.

To make the sensor for lateral force sensing that is insensitive to temperature changes, a transverse load is applied to the PS-FBG to introduce a birefringence, to create two orthogonally polarized notches, which would lead to the generation of two microwave signals at two different frequencies. By beating the two microwave frequencies, a third frequency is generated. Since the two microwave frequencies are experiencing the same frequency shift due to temperature change, the third frequency is insensitive to temperature change.

The configuration of an OEO sensor for lateral force sensing that is insensitive to temperature changes is shown in Fig. 10.8(a) [22]. As can be seen, a light wave generated by a laser source is sent to a polarization modulator (PolM) via a polarization controller (PC1). The PolM is a special PM that supports phase modulation along the two orthogonal principal axes with complementary phase modulation indices. For simplicity, here

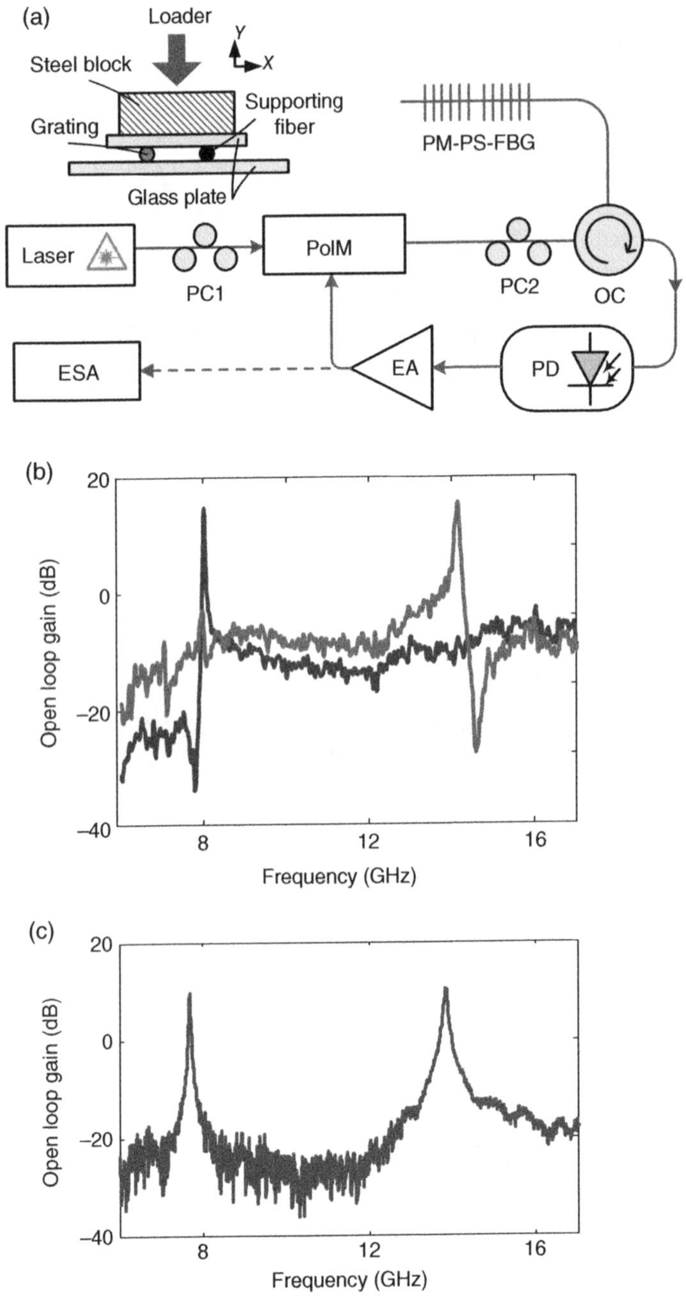

Figure 10.8 (a) Schematic of the temperature-insensitive transverse load sensor based on a dual-frequency OEO employing a polarization-maintaining-PS-FBG. EA, electrical amplifier; ESA, electrical spectrum analyzer; PC, polarization controller; PD, photodetector; PolM, polarization modulator. (b) Single passband photonic microwave filter when the incident light is alighted with an angle of 0° or 90° relative to one principal axis (horizontal or vertical) of the PolM. (c) Dual passband photonic microwave filter when the incident light is alighted with an angle of 45° relative to one principal axis of the PolM. *Source:* Kong et al. [22]/ Optical Society of America.

we assume that the incident light is aligned with one of the principal axes and thus the PolM is operating as a regular PM. The phase-modulated signal is then sent to a PS-FBG through an OC. One sideband of the phase-modulated signal is removed by the notch of the polarization-maintaining PS-FBG, and the phase-modulated signal is converted to a single-sideband intensity-modulated signal and is detected at a PD. The detected electrical signal is sent back to the PM after amplification by an EA, to close the OEO loop.

When the loop gain is greater than the loop loss, the OEO will start to oscillate, and two microwave signals at different frequencies are generated. The beating between the two microwave signals would generate a third microwave signal (a beat signal) with its frequency being a function of the birefringence. Thus, by measuring the beat frequency, the transverse load applied to the PS-FBG can be measured. Since the two frequencies are experiencing the same temperature change, the third frequency is temperature insensitive.

As shown in Fig. 10.8(a), when a linearly polarized incident light with an angle of 45° relative to one principal axis of the PolM is sent to the PolM, the light is equally projected to the two orthogonal polarization axes, a microwave photonic filter with dual passbands with a band separation of 6 GHz is generated. Figure 10.8(b) shows the passband of the filter along the horizontal or vertical polarization direction, measured by aligning the incident light having an angle of 0° or 90° relative to one principal axis of the PolM. Figure 10.8(c) shows the filter frequency response when the incident light having an angle of 45° relative to one principal axis of the PolM. A dual passband filter is realized. When the OEO loop is closed, two microwave signals at two frequencies determined by the two passbands are generated.

Note that due to the nonlinearity of the PolM, a third signal that is the beat note between the two microwave signals is generated, as shown in Fig. 10.9(a). The frequency of the beat note is directly associated with the birefringence introduced by the transverse load to the PS-FBG. Thus, by measuring the beat frequency, the transverse load is measured. By using the typical values of a silica fiber, we have the relationship between the transverse load the beat frequency, which is given by $d\nu/dF \approx -9.9$ GHz/(N/mm).

Then, a transverse load is applied to the PS-FBG. To ensure the system reaches its highest sensitivity and to have a good linearity between the transverse load and the beat frequency, in the experiment, the transverse load is applied to the PS-FBG along the fast axis. A supporting fiber with an identical radius is placed in parallel with the PS-FBG to make the load applied to the PS-FBG transversely, while sharing half of the applied load. By increasing the load applied to the PS-FBG, the beating frequency is shifted linearly towards a smaller frequency, as shown in Fig. 10.9(b). The spectrum of the beat signal is measured by an electrical spectrum analyzer (ESA) (Agilent E4448A), with the spectrum shown in the inset of Fig. 10.9(b). The slope through linear fitting is −9.73 GHz/(N/mm) which agrees well with the theoretical value of 9.9 GHz/(N/mm).

By using two independent PS-FBGs, with one as a sensor and the other as a reference, temperature-insensitive sensing can also be implemented [23].

The concept was then extended with the light source implemented by a dual-wavelength fiber ring laser with improved lasing stability through the use of an injection-coupled OEO [24]. The application of transverse load sensing was evaluated. The OEO-coupled dual-wavelength laser has two mutually coupled loops, the fiber ring loop and the OEO loop. In the fiber ring loop, a polarization-maintaining PS-FBG (PM-PSFBG) was incorporated to generate two optical wavelengths with the wavelength spacing determined by the birefringence of the PM fiber. In the OEO loop, a microwave signal with its frequency also determined by the birefringence of the PM fiber is generated, which is fed into the fiber ring loop to injection lock the dual wavelengths. Due to the injection locking, a very stable dual-wavelength operation is established. The use of the

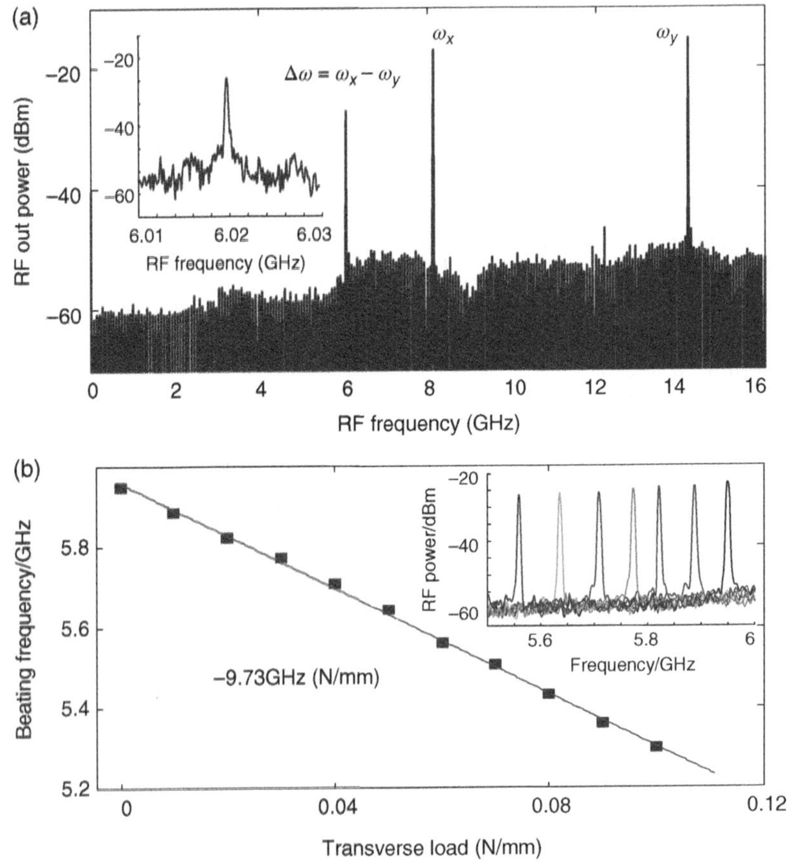

Figure 10.9 (a) The electrical spectrum of the signal generated by the dual-frequency OEO, with two microwave signals at 8.22 and 14.24 GHz and a beat signal at 6.02 GHz. (b) Measured beat frequency as a function of the transverse load and the electrical spectrum with different loads. *Source:* Kong et al. [22]/Optical Society of America.

dual wavelengths for high-resolution and high-speed transverse load sensing was implemented. The sensitivity of the transverse load sensor is measured as high as −9.735 GHz/(N/mm) and +9.754 GHz/(N/mm), along the fast and slow axes, respectively. The high-frequency purity and stability of the generated microwave signal permits very reliable and high-accuracy measurement and the microwave frequency interrogation allows the system to operate at an ultra-high speed.

Instead of using an ultra-narrowband optical filter to implement an MBPF, an OEO-based sensor can also be implemented using a sliced broadband light source [25]. The joint operation of a sliced broadband light source, a Mach–Zehnder interferometer (MZI), a dispersion compensating fiber (DCF), and a PD corresponds to a single passband MBPF [26], as shown in Fig.10.10(a) with the spectrum of the sliced light source in Fig. 10.10 (b) and the spectral response of the single passband MBPF in Fig. 10.10 (c). If one arm of the MZI is used as a sensing arm which is exposed to the temperature variations and the other arm is used as a reference arm, when the temperature is changed, the length difference between the two arms is changed, which leads to the change in the FSR of the MZI. Since the central frequency of the MBPF is a function of the FSR, the oscillation frequency of the OEO is affected by temperature variations. By measuring the frequency change, the

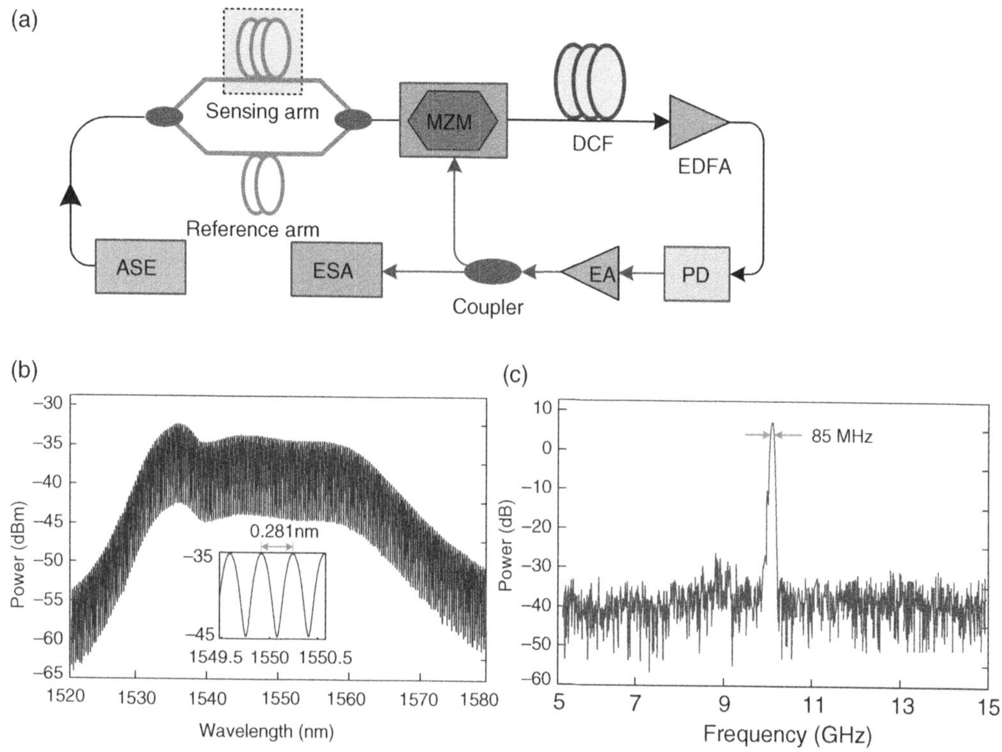

Figure 10.10 (a) An OEO-based sensor based on an MBPF implemented based on a sliced broadband light source. (b) The spectrum of the sliced light source, and (c) The spectral response of the MBPF. ASE: amplified spontaneous emission. DCF: dispersion compensating fiber. EA, electrical amplifier; EDFA, erbium-doped fiber amplifier; ESA, electrical spectrum analyzer; MZM, Mach–Zehnder modulator; PD, photodetector. *Source:* Wang et al. [25]/IEEE.

temperature change to the sensing arm is estimated. High-sensitivity temperature sensing with a sensitivity of 3.7 MHz/°C was experimentally demonstrated [25].

10.4 Optical Sensors Based on Spectrum Shaping and Wavelength-to-Time Mapping

An optical sensor with increased speed and resolution can also be implemented based on optical spectral-shaping and frequency-to-time (SS-WTT) mapping [27], to translate the optical spectrum measurement to the microwave frequency measurement [28, 29]. In addition, through pulse compression, the SNR can also be significantly improved [30], which would increase the measurement accuracy, especially if the sensing information is embedded in strong noise.

Optical WTT mapping can be performed by using a dispersive element, as shown in Fig. 10.11. If a dispersive element is a length of fiber with a value of dispersion of $\ddot{\Phi}$, for an input pulse $g(t)$ with a temporal width of Δt_0, the signal at the output of the dispersive element without considering the fiber loss is given [26]

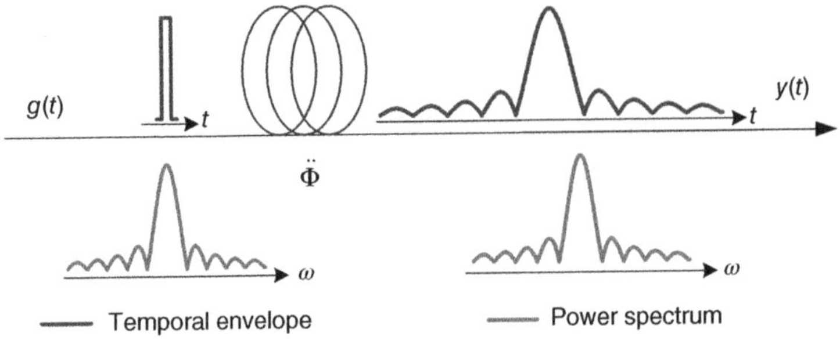

Figure 10.11 Wavelength-to-time mapping in a dispersive element.

$$
\begin{aligned}
y(t) = g(t) \times \exp\left(j\frac{t^2}{2\ddot{\Phi}}\right) &= \int_{-\infty}^{\infty} g(\tau) \times \exp\left[j\frac{(t-\tau)^2}{2\ddot{\Phi}}\right] d\tau \\
&= \exp\left(j\frac{t^2}{2\ddot{\Phi}}\right) \times \int_{-\infty}^{\infty} g(\tau) \times \exp\left(j\frac{\tau^2}{2\ddot{\Phi}}\right) \times \exp\left[-j\left(\frac{t}{\ddot{\Phi}}\right)\tau\right] d\tau \\
&\approx \exp\left(j\frac{t^2}{2\ddot{\Phi}}\right) \times \int_{-\infty}^{\infty} g(\tau) \times \exp\left[-j\left(\frac{t}{\ddot{\Phi}}\right)\tau\right] d\tau \\
&= \exp\left(j\frac{t^2}{2\ddot{\Phi}}\right) \times G(\omega)|_{\omega=\frac{t}{\ddot{\Phi}}}
\end{aligned}
\tag{10.11}
$$

where $G(\omega)$ is the Fourier transform of $g(t)$.

As can be seen, the envelope of the output signal is proportional to the Fourier transform of the input signal envelope. Note that (10.11) is obtained if the duration of the input ultrashort pulse, Δt_0, and the dispersion $\ddot{\Phi}$ of the dispersive element satisfy the condition given by

$$\left|\frac{\Delta t_0{}^2}{2\ddot{\Phi}}\right| \ll 1 \tag{10.12}$$

which means that the phase term $\frac{\tau^2}{2\ddot{\Phi}}$ in (10.11) satisfies $((\tau^2)/(2\ddot{\Phi})) \leq ((\Delta t_0^2)/(2\ddot{\Phi})) \ll 1$. Thus, we have $\exp\left(j\frac{\tau^2}{2\ddot{\Phi}}\right) \approx 1$, and

$$y(t) \approx G\left(\frac{t}{\ddot{\Phi}}\right)$$

Since the pulse compression ratio is determined by the time-bandwidth product (TBWP) of a microwave waveform, the microwave waveform should have a large TBWP. A linearly chirp microwave waveform [31] or a random-coded microwave waveform [32] can have a large TBWP. A linearly chirped microwave waveform (LCMW) can be generated optically based SS-WTT mapping in which a spectral shaper with a wavelength-dependent FSR that is increasing or decreasing with wavelength is needed.

Figure 10.12 shows an optical sensor implemented based on SS-WTT mapping [28]. An LCFBG is incorporated in one arm of an MZI. Due to the wavelength-dependent nature of the length of the arm with the incorporated LCFBG, the MZI would have a spectral response with an increasing or decreasing FSR. An ultra-short optical pulse with a wide spectrum from a mode-locked laser source

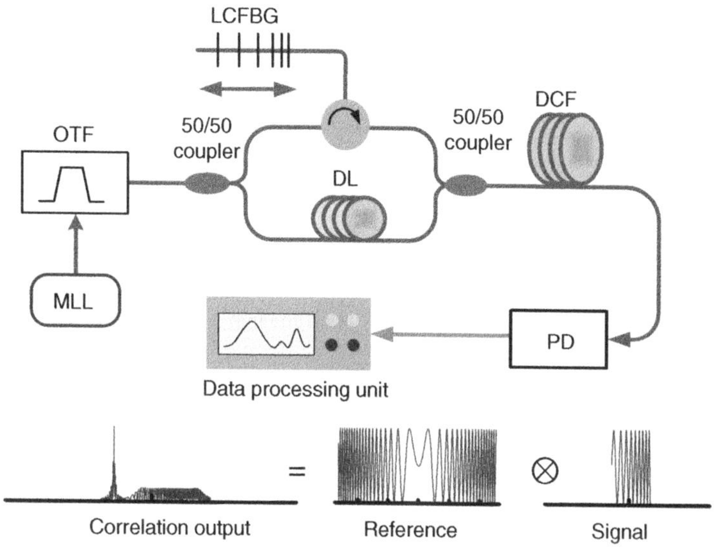

Figure 10.12 An optical sensor system based on SS-WTT mapping. DCF, dispersion compensating fiber; DL, delay line; LCFBG, linearly chirped fiber Bragg grating; MLL, mode-locked laser; OTF, optical tunable filter; PD, photodetector. *Source:* Adapted from Liu et al. [28].

is spectrally shaped by the MZI to generate a spectrum with the shape identical to the spectral response of the MZI. The spectrally shaped spectrum is then mapped to the temporal domain by a dispersive element, such as a DCF or an LCFBG. Due to the linear frequency-to-time mapping, an LCMW with a temporal shape identical to the shape of the spectrally shaped spectrum is generated. The chirped waveform is detected by a PD and then sent to a DSP to perform pulse compression.

It is known that a chirped waveform can be compressed if it is sent to a correlator or matched filter in which a reference waveform identical to the chirped waveform is used. The key significance of performing pulse compression is that the wavelength shift is estimated by measuring the location of the correlation peak, with both improved resolution and SNR. The correlation is implemented in the DSP by building a reference microwave waveform, which is linearly chirped with a chirp rate identical to that of the generated chirped microwave waveform, but with an instantaneous frequency extending from the smallest to the largest possible values corresponding to the generated chirped microwave waveform when the LCFBG is experiencing the largest and the smallest wavelength shift. Therefore, the location of the correlation peak would precisely indicate the wavelength shift.

The concept is evaluated by an experiment with the results shown in Fig. 10.13. A microwave waveform with a chirp rate of –0.068 GHz/ps for $t < 0$ and 0.068 GHz/ps for $t > 0$ is built as a reference microwave waveform, as shown in Fig. 10.13(a). Figure 10.13(b), (c), and (d) shows three LCMWs corresponding to three strains of 71.5, 406.9, and 484.2 $\mu\varepsilon$ applied to the LCFBG, respectively. The correlation of the three LCMWs with the reference microwave waveform is shown in Fig. 10.13(e). The waveforms are highly compressed. The locations of the three peaks indicate the wavelength shifts of the LCFBG are 0.087, 0.495, and 0.589, corresponding to three different strains of 71.5, 406.9, and 484.2 $\mu\varepsilon$, respectively.

The system could also be modified to have the ability to simultaneously measure the temperature and strain by replacing the single-mode LCFBG with a high-birefringence LCFBG (Hi-Bi LCFBG),

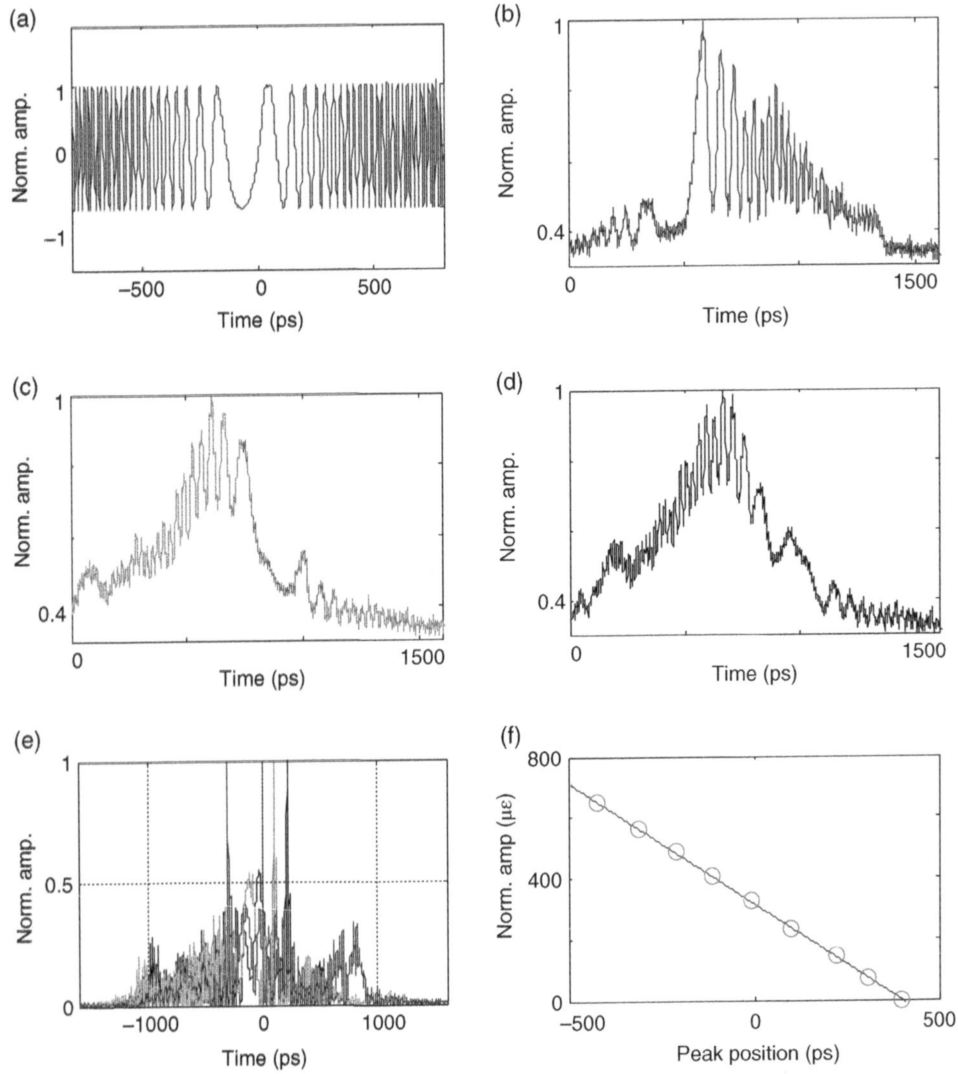

Figure 10.13 (a) Reference microwave waveform. (b) Measured chirped microwave waveform when a strain of 71.5 $\mu\varepsilon$ is applied to the LCFBG. (c) Measured chirped microwave waveform when a strain of 406.9 $\mu\varepsilon$ is applied to the LCFBG. (d) Measured chirped microwave waveform when a strain of 484.2 $\mu\varepsilon$ is applied to the LCFBG. (e) Correlation results for the detected microwave waveforms as shown in (b), (c), and (d). (f) The measured strain vs. the peak position. The circles are the experimental data, and the solid curve is linear fitting of the experimental data. *Source:* Adapted from Liu et al. [28].

as shown in Fig. 10.14 [29]. The strain and temperature information are encoded in the Hi-Bi LCFBG as Bragg wavelength shifts along the two orthogonal axes. Due to the birefringence in the Hi-Bi LCFBG, the MZI has two spectral responses along the fast and slow axes with each having an increasing FSR. The spectrum of an ultra-short optical pulse is spectrally shaped by the MZI. Two spectrum-shaped signals are obtained at the output of the MZI which are mapped to two chirped microwave waveforms by a DCF. Again, by pulse compression, two correlation peaks with the locations containing the strain and temperature information are obtained.

The concept was evaluated in which simultaneous temperature and strain measurements were made. Figure 10.15(a) shows the reference microwave waveform, and Fig. 10.15(b), (c) shows two

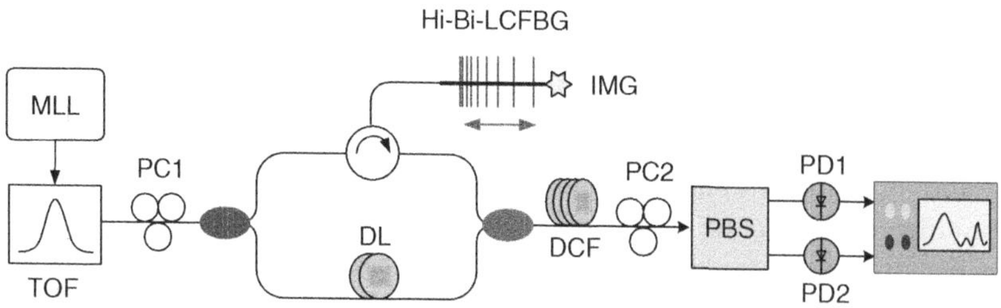

Figure 10.14 Schematic of a Hi-Fi LCFBG sensor interrogation system for simultaneous interrogation of temperature and strain. DCF, dispersion compensating fiber; DL, delay line; IMG, index matching gel; OTF, optical tunable filter; PC, polarization controller; PBS, polarization beam splitter. *Source:* Liu et al. [29]/IEEE.

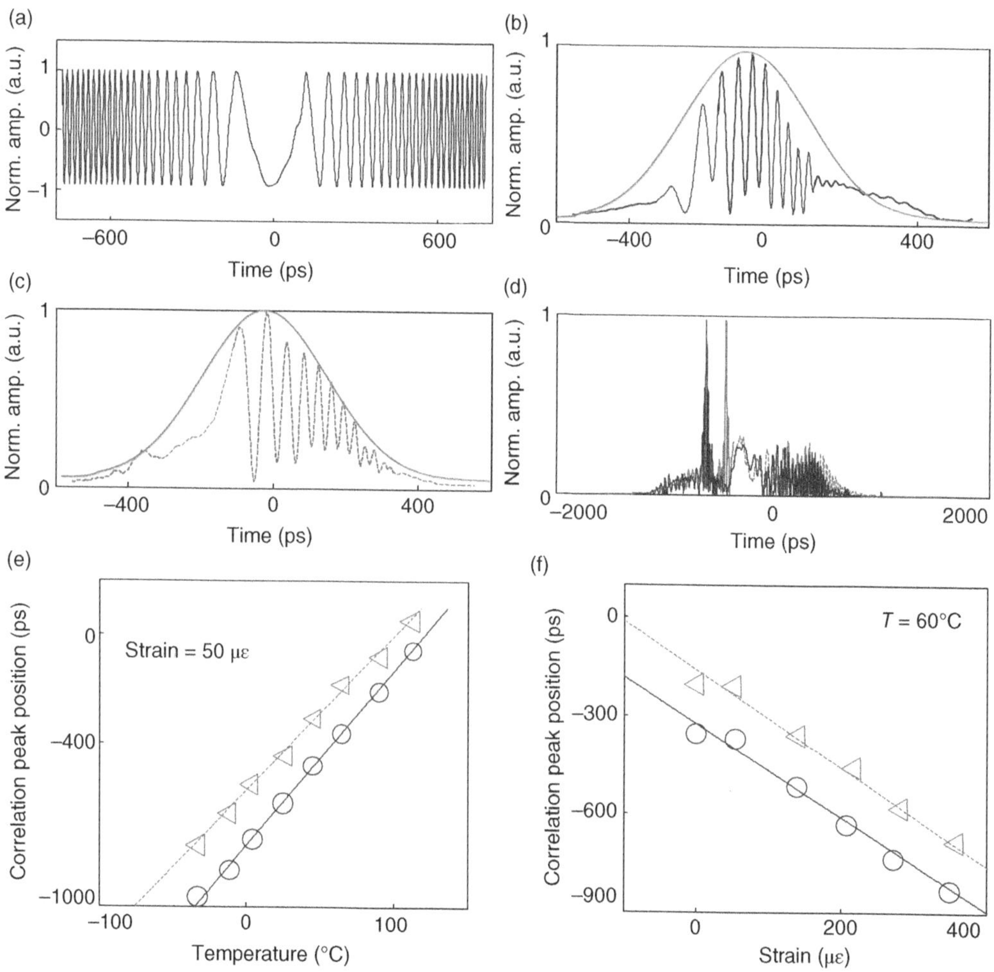

Figure 10.15 Experimental results. (a) The reference microwave waveform. A linearly chirped microwave waveform corresponding to the polarization directions of the ultrashort pulse aligned with (b) the fast axis and (c) the slow axis, when a strain of 50 με is applied to the LCFBG at 25 °C. (d) Correlation of the waveforms in (b) and (c) with the reference waveform. (e) Correlation peak position vs temperature for a given strain of 50 με. (f) Correlation peak position vs. applied strain for a temperature of 60 °C. The triangles and circles indicate the experimental data corresponding to the polarization directions of the ultrashort pulse aligned with the fast axis and slow axis, respectively. *Source:* Liu et al. [29]/IEEE.

LCMWs corresponding to the two polarization directions of the ultrashort pulses aligned with the fast and slow axes of the Hi-Bi LCFBG, respectively, when the strain is 50 $\mu\varepsilon$ and the temperature is 25 °C. The correlation of the two LCMWs with the reference microwave waveform given in Fig. 10.15(a) is shown in Fig. 10.15(d). It can be seen that microwave waveforms are significantly compressed. The locations of the two correlation peaks indicate the wavelength shifts of the Hi-Bi LCFBG and the phase difference due to the birefringence of the Hi-Bi fiber. In the experiment, the relationship between the strain, temperature, and the correlation peak positions was evaluated, which is shown in Fig. 10.15(e), (f).

Note that if a dispersive element has higher-order dispersion, the use of the linear WTT mapping relationship in (10.11) would lead to errors and have to be modified. In [33], WTT mapping incorporating higher order dispersion for high-speed and high-resolution sensing was proposed and studied.

The use of the concept of optical WTT mapping to convert the spectrum of an FBG filtered by two Gaussian filters to perform high-speed and high-resolution sensing was also reported and demonstrated [34]. The spectrums of the FBG after being filtered by the two Gaussian filters were converted to the time domain via WTT mapping. By comparing the temporal waveforms corresponding to the outputs from the two Gaussian filters, the wavelength-shifting information of the FBG was retrieved.

Instead of using an LCMW to achieve pulse compression, the use of a random microwave waveform can also achieve effective pulse compression. Recently, a technique to use a random grating serving as a spectral shaper to generate a random microwave waveform based on SS-WTT mapping for high-speed and high-resolution sensing was reported [35]. The experimental results showed that the sensing resolutions for temperature and strain were 0.23 °C and 2.5 $\mu\varepsilon$, respectively, and the accuracies for temperature and strain were 0.11 °C and 1.2 $\mu\varepsilon$, respectively. One added advantage of this approach is that multiple random gratings can be cascaded for remote and distributed sensing. Since each random grating has its unique random spectrum, through pulse compression, a correlation peak representing a specific random grating at a specific location is generated [36].

10.5 Photonic Integrated Microwave Photonic Sensors

Silicon photonic devices implemented on a silicon-on-insulator (SOI) platform have a high potential for optical sensing owing to their high sensitivity to RI changes, which is very useful for chemical diagnostics and label-free biosensing since the presence of micro- or nanoparticles on the devices can be detected through measuring the resonant wavelength changes [37]. Numerous SOI-based sensors, such as nanocavity sensors [38], microring resonator sensors [39], microdisk resonator sensors [40], and waveguide Bragg grating sensors [41], have been reported. However, the sensing information encoded in these sensors is obtained through direct wavelength shift measurement by an OSA. Due to low interrogation speed and the poor resolution of an OSA, the performance including sensing speed and resolution is limited. A solution is to use microwave photonics techniques to convert the optical spectrum information to the electrical domain, to measure the microwave spectrum using a DSP.

In [42], a high-speed and high-resolution silicon photonic microdisk resonator (MDR) sensor based on microwave photonic filtering and advanced signal processing was proposed. Figure 10.16 shows an MDR which is placed on an Aluminum plate with a temperature control module for thermal tuning. Figure 10.17(a) shows the system configuration. The system consists

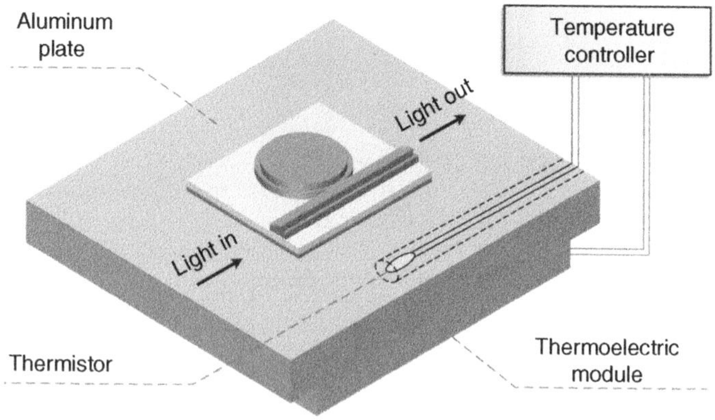

Figure 10.16 An MDR placed on an aluminum plate with a temperature controller for thermal tuning.
Source: Deng et al. [42]/IEEE.

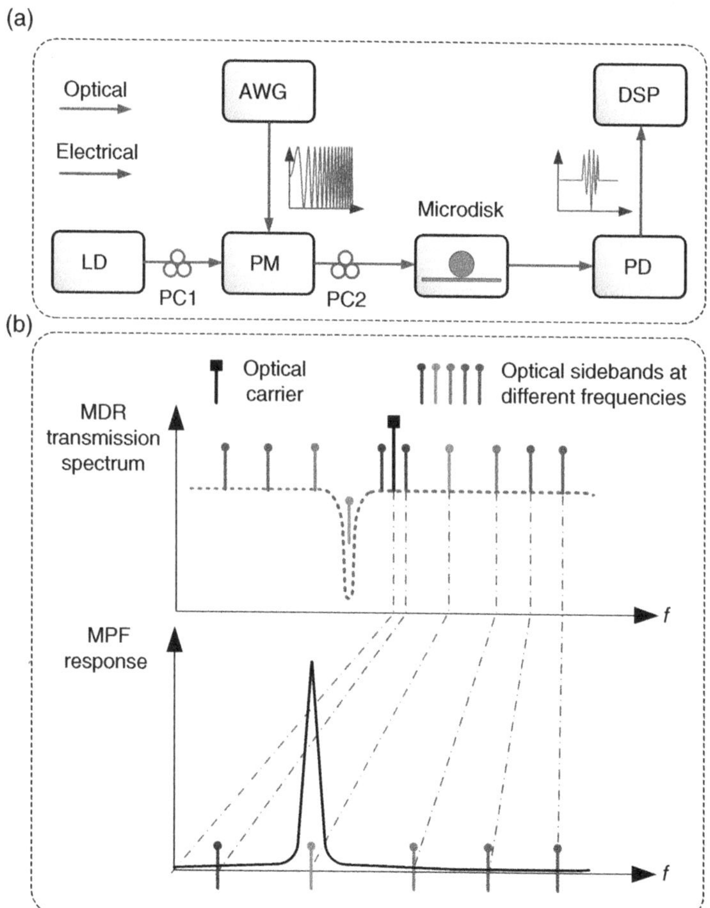

Figure 10.17 (a) The configuration of the interrogation system; (b) An MPF based on phase modulation to intensity modulation conversion. AWG, arbitrary waveform generator; DSP, digital signal processor; LD, laser diode; PC, polarization controller; PD, photodetector; PM, phase modulator.
Source: Deng et al. [42]/IEEE.

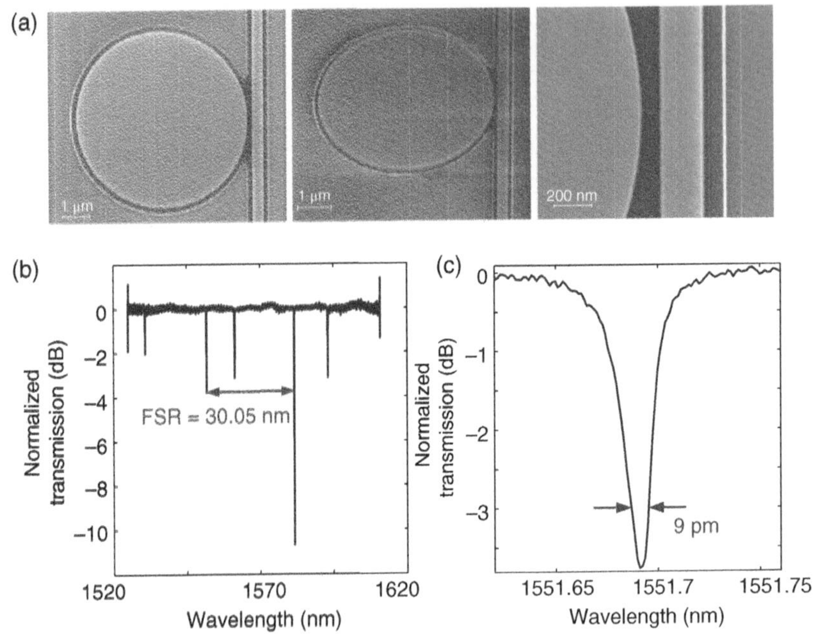

Figure 10.18 (a) SEM micrographs of an MDR (with the left picture taken vertically, the middle taken with a tilted angle of 45°, and the right showing the zoom-in view of the gap); (b) normalized transmission spectrum of the fabricated MDR with air cladding; (c) zoom-in view of the TE_0 mode resonance of the MDR with air cladding. *Source:* Deng et al. [42]/IEEE.

of an LD, a PM, an MDR, and a PD. The integrated MDR with a high Q factor is used as an optical notch filter to filter one of the sidebands, as shown in Fig. 10.17(b). The SEM micrographs of the MDR and its transmission spectrum are shown in Fig. 10.18(a) and (b), respectively. Thanks to the PM-IM conversion in the MDR, a phase-modulated optical signal is converted to an intensity-modulated optical signal, and the entire system is equivalent to a microwave photonic filter with the center frequency of its passband determined by the wavelength of the LD and the notch wavelength of the MDR. Since the center frequency of the microwave photonic filter is a function of the resonant wavelength of the MDR, thus when the temperature or the RI of the microparticles on the MDR is changed, the center wavelength is shifted, leading to the change of the center frequency of the microwave photonic filter. To effectively measure the spectrum change of the microwave photonic filter, a broadband LCMW is applied to the input to generate a filtered microwave waveform. By measuring the temporal location of the filtered microwave waveform, the sensing information is revealed.

The resonant wavelength of an MDR can be expressed as [43]

$$\lambda_{res} = \frac{2\pi R}{m} n_{eff} \tag{10.13}$$

where R is the disk radius, m is the mode order, and n_{eff} is the effective index of the guided mode in the waveguide. When the environmental temperature changes, the effective RI will change due to the thermo-optic effect, and the radius will also change due to the thermal expansion effect. Therefore, the overall wavelength shift is given by [44]

$$\Delta\lambda_{Temp} = \frac{\lambda_{res}}{n_g}\Delta T\left(\alpha_{si}n_{eff} + \frac{\partial n_{eff}}{\partial T}\right) \tag{10.14}$$

where n_g is the group index of the guided mode, ΔT is the temperature change, α_{si} is the thermal expansion coefficient, and $\partial n_{eff}/\partial T$ is the thermo-optic coefficient.

When the RI of the cladding (n_{clad}) changes, the effective index of the guided mode will also change, which would lead to a resonant wavelength shift, which is given by [45]

$$\Delta\lambda_{RI} = \left(\frac{\lambda_{res}}{n_g}\right) \times \Delta n_{clad} \times \left(\frac{\partial n_{eff}}{\partial n_{clad}}\right) \tag{10.15}$$

where Δn_{clad} is the cladding RI change.

Assuming the wavelength of the optical carrier is λ_c, the central frequency of the microwave photonic filter is given by

$$f_{MPF} \approx \left(\frac{c}{n_{avg}}\right) \times \left|\frac{\lambda_c - \lambda_{res}}{\lambda_c^2}\right| \tag{10.16}$$

where c is the velocity of light in vacuum, n_{avg} is the average refractive index for the optical path consisting of fibers and the silicon waveguides. When the resonant wavelength of the MDR shifts, the central frequency of the MPF will change. The frequency change can be expressed as

$$\Delta f_{MPF} = K\Delta\lambda_{res} \tag{10.17}$$

where $K = c/\left(n\lambda_c^2\right)$ and $\Delta\lambda_{res}$ is the wavelength shift due to $\Delta\lambda_{Temp}$ and $\Delta\lambda_{RI}$.

Mathematically, for an LCMW, the instantaneous frequency of the waveform is given by

$$f(t) = f_0 + Ct \tag{10.18}$$

where C is the chirp rate, f_0 is the initial frequency, and t is the time. When the LCMW is fed to a microwave photonic filter, the temporal location of the output waveform is calculated by $t = (f_{MPF} - f_0)/C$. The relationship between the wavelength shift of the MDR and the temporal location is given by

$$\Delta t = \left(\frac{K}{C}\right)\Delta\lambda_{res} \tag{10.19}$$

When the temporal location is measured, using (10.19) with (10.14) or (10.15), the temperature or the RI change can be obtained.

The concept in [42] was extended to enable simultaneous temperature and RI sensing based on a dual-passband microwave photonic filter implemented using two whispering gallery modes of an MDR [46]. Since the two whispering gallery modes are experiencing different wavelength shifts, by measuring the spectral response of the dual-passband microwave photonic filter, to locate the central frequencies of the two passbands, simultaneous temperature and RI sensing was implemented.

10.6 Conclusion

Conventionally, an optical sensor is interrogated in the optical domain to monitor the wavelength shift by either using an OSA to measure the optical spectrum or using an optical edge filter to measure the wavelength-dependent intensity change. The speed is either slow and the resolution is low or the accuracy is poor. The use of microwave photonics techniques to translate the optical wavelength change in the optical domain to the microwave frequency change in the electrical domain is a solution to improve both the sensing speed and the resolution. Since a DSP can resolve a microwave frequency with subHz resolution, the sensing resolution is extremely high. On the other hand, a high resolution is achieved at the cost of limited measurement range. For example, if the wavelength of an optical sensor is shifted by 10 nm, the corresponding microwave frequency shift is 1.25 THz, which is too wide to be measured by a low-cost DSP. A solution is to use an optical comb [47] to divide the large frequency range into multiple spectral segments. Via parallel processing, low-cost DSPs can handle high-speed and high-resolution measurements over a wide measurement range.

References

1 Udd, E. (1995). An overview of fiber-optic sensors. *Rev. Sci. Instrum.* 66 (8): 4015–4030.

2 Kersey, A.D., Davis, M.A., Patrick, H.J. et al. (1997). Fiber grating sensors. *J. Lightw. Technol.* 15 (8): 1442–1463.

3 Othonos, A. and Kalli, K. (1999). *Fiber Bragg Gratings: Fundamentals and Applications in Telecommunications and Sensing*. Norwood: Artech House.

4 Albert, J., Shao, L.Y., and Caucheteur, C. (2013). Tilted fiber Bragg grating sensors. *Laser Photonics Rev.* 7 (1): 83–108.

5 Koo, K.P. and Kersey, A.D. (1995). Bragg grating-based laser sensors systems with interferometric interrogation and wavelength division multiplexing. *J. Lightw. Technol.* 13 (7): 1243–1249.

6 Wang, C. and Yao, J. (2011). Ultrafast and Ultrahigh-Resolution Interrogation of a Fiber Bragg Grating Sensor Based on Interferometric Temporal Spectroscopy. *J. Lightw. Technol.* 29 (19): 2927–2933.

7 Xia, H., Zhang, C., Mu, H., and Sun, D. (2009). Edge technique for direct detection of strain and temperature based on optical time domain reflectometry. *Appl. Opt.* 48 (2): 189–197.

8 Melle, S.M., Liu, K., and Measures, R.M. (1992). A passive wavelength demodulation system for guided-wave Bragg grating sensors. *IEEE Photon. Technol. Lett.* 4 (5): 516–518.

9 Rao, Y.J., Jackson, D.A., Zhang, L., and Bennion, I. (1996). Dual-cavity interferometric wavelength-shift detection for in-fiber Bragg grating sensors. *Opt. Lett.* 21 (19): 1556–1558.

10 Yao, J. (2015). Microwave photonics for high resolution and high-speed interrogation of fiber Bragg grating sensors. *Fiber Integr. Opt.* 34 (4): 230–242.

11 Yao, J. (2009). Microwave photonics. *J. Lightw. Technol.* 27 (3): 314–335.

12 Guan, B., Jin, L., Zhang, Y., and Tam, H. (2012). Polarimetric heterodyning fiber grating laser sensors. *J. Lightw. Technol.* 30 (8): 1097–1112.

13 Ball, G.A., Meltz, G., and Morey, W.W. (1993). Polarimetric heterodyning Bragg-grating fiber-laser sensor. *Opt. Lett.* 18 (22): 1976–1978.

14 Donlagic, D. and Lesic, M. (2006). All-fiber quasi-distributed polarimetric temperature sensor. *Opt. Express* 14 (22): 10245–10254.

15 Lin, X., Liang, Y., Jin, L., and Wang, L. (2019). Dual-polarized fiber laser Sensor for photoacoustic microscopy. *Sensors* 19 (21): 4632.

16 Liu, D., Ngo, N.Q., Tjin, S.C., and Dong, X. (2007). A dual-wavelength fiber laser sensor system for measurement of temperature and strain. *IEEE Photon. Technol. Lett.* 19 (15): 1148–1150.

17 Bernhardi, E.H., van der Werf, K.O., Hollink, A.J.F. et al. (2013). Intra-laser-cavity microparticle sensing with a dual-wavelength distributed-feedback laser. *Laser Photonics Rev.* 7 (4): 589–598.

18 Y. Wang, J. Zhang, O. L. Coutinho, and J. Yao, "Interrogation of a linearly chirped fiber Bragg grating sensor with a high resolution using a linearly chirped optical waveform," *Opt. Lett.*, vo. 40. no. 21, pp. 4923–4926, 2015.

19 Meng, Z., Li, J., Yin, C. et al. (2017). Dual-band dechirping LFMCW radar receiver with high image rejection using microwave photonic I/Q mixer. *Opt. Express* 25 (18): 22055–22065.

20 Yao, J. (2017). Optoelectronic oscillators for high speed and high resolution optical sensing. *J. Lightw. Technol.* 35 (16): 3489–3497.

21 Li, M., Li, W., Yao, J., and Azana, J. (2012). Femtometer-resolution wavelength interrogation of a phase-shifted fiber Bragg grating sensor using an optoelectronic oscillator. *OSA Technical Digest* (online) (23–27 September 2012), Optical Society of America, Burlingame, CA, USA, BTu2E.3.

22 Kong, F., Li, W., and Yao, J. (2013). Transverse load sensing based on a dual-frequency optoelectronic oscillator. *Opt. Lett.* 38 (14): 2611–2613.

23 Xu, O., Zhang, J., Deng, H., and Yao, J. (2017). Dual-frequency optoelectronic oscillator for temperature-insensitive interrogation of a FBG sensor. *IEEE Photon. Technol. Lett.* 29 (4): 357–360.

24 Kong, F., Romeira, B., Zhang, J. et al. (2014). A dual-wavelength fiber ring laser incorporating an injection-coupled optoelectronic oscillator and its application to transverse load sensing. *J. Lightw. Technol.* 32 (9): 1784–1793.

25 Wang, Y., Zhang, J., and Yao, J. (2016). An optoelectronic oscillator for high sensitivity temperature sensing. *IEEE Photon. Technol. Lett.* 28 (13): 1458–1460.

26 Zhang, J., Gao, L., and Yao, J.P. (2014). Tunable optoelectronic oscillator incorporating a single passband microwave photonic filter. *IEEE Photon. Technol. Lett.* 26 (4): 326–329.

27 Yao, J. (2011). Photonic generation of microwave arbitrary waveforms. *Opt. Commun.* 284 (15): 3723–3736.

28 Liu, W., Li, M., Wang, C., and Yao, J. (2011). Real-time interrogation of a linearly chirped fiber Bragg grating sensor based on chirped pulse compression with improved resolution and signal-to-noise ratio. *J. Lightw. Technol.* 29 (9): 1239–1247.

29 Liu, W., Li, W., and Yao, J. (2011). Real-time interrogation of a linearly chirped fiber Bragg grating sensor for simultaneous measurement of strain and temperature. *IEEE Photon. Technol. Lett.* 23 (18): 1340–1342.

30 Richards, M.A. (2005). *Fundamentals of Radar Signal Processing*. McGraw-Hill Education.

31 Li, W. and Yao, J. (2014). Generation of linearly chirped microwave waveform with an increased time-bandwidth product based on a tunable optoelectronic oscillator and a recirculating phase modulation loop. *J. Lightw. Technol.* 32 (20): 3573–3579.

32 Deng, H. and Yao, J. (2018). Photonic generation of pseudo random microwave waveform based on a random fiber grating. *Optical Fiber Communication Conference, OSA Technical Digest* (online) (11–15 March 2018), Optical Society of America, San Diego, CA, USA, paper W2A.37.

33 Xia, H., Wang, C., Blais, S., and Yao, J. (2010). Ultrafast and precise interrogation of fiber Bragg grating sensor based on wavelength-to-time mapping incorporating higher order dispersion. *J. Lightw. Technol.* 28 (3): 254–261.

34 Fernández, M.P., Bulus Rossini, L.A., Cruz, J.L. et al. (2019). High-speed and high-resolution interrogation of FBG sensors using wavelength-to-time mapping and Gaussian filters. *Opt. Express* 27 (25): 36815–36823.

35 Deng, H., Lu, P., Mihailov, S., and Yao, J. (2018). High-speed and high-resolution interrogation of a strain and temperature random grating sensor. *J. Lightw. Technol.* 36 (23): 5587–5592.

36 Liu, J., Lu, P., Mihailov, S.J. et al. (2019). Real-time random grating sensor array for quasi-distributed sensing based on wavelength-to-time mapping and time-division multiplexing. *Opt. Lett.* 44 (2): 379–382.

37 Prabhathan, P., Murukeshan, V.M., Jing, Z., and Ramana, P.V. (2009). Compact SOI nanowire refractive index sensor using phase shifted Bragg grating. *Opt. Express* 17 (17): 15330–15341.

38 Jágerská, J., Zhang, H., Diao, Z. et al. (2010). Refractive index sensing with an air-slot photonic crystal nanocavity. *Opt. Lett.* 35 (15): 2523–2525.

39 De Vos, K., Bartolozzi, I., Schacht, E. et al. (2007). Silicon-on-insulator microring resonator for sensitive and label-free biosensing. *Opt. Express* 15 (12): 7610–7615.

40 Rodriguez, G.A., Hu, S., and Weiss, S.M. (2015). Porous silicon ring resonator for compact, high sensitivity biosensing applications. *Opt. Express* 23 (6): 7111–7119.

41 Passaro, V.M.N., Loiacono, R., D'Amico, G., and De Leonardis, F. (2008). Design of Bragg grating sensors based on sub micrometer optical rib waveguides in SOI. *IEEE Sensors J.* 8 (9): 1603–1611.

42 Deng, H., Zhang, W., and Yao, J. (2018). High-speed and high-resolution interrogation of a silicon photonic microdisk sensor based on microwave photonic filtering. *J. Lightwave Technol.* 36 (19): 4243–4249.

43 Bogaerts, W., De Heyn, P., Van Vaerenbergh, T. et al. (2011). Silicon microring resonators. *Laser Photonics Rev.* 6 (1): 47–73.

44 Kim, G.-D., Lee, H.-S., Park, C.-H. et al. (2010). Silicon photonic temperature sensor employing a ring resonator manufactured using a standard CMOS process. *Opt. Express* 18 (21): 22215–22221.

45 Ackert, J.J., Doylend, J.K., Logan, D.F. et al. (2011). Defect-mediated resonance shift of silicon-on-insulator racetrack resonators. *Opt. Express* 19 (13): 11969–11976.

46 Liu, J., Deng, H., Zhang, W., and Yao, J. (2018). On-chip sensor for simultaneous temperature and refractive index measurements based on a dual-passband microwave photonic filter. *J. Lightw. Technol.* 36 (18): 4099–4105.

47 Hall, J.L. and Ye, J. (2003). Optical frequency standards and measurement. *IEEE Trans. Instrum. Meas.* 52 (2): 227–231.

11

Photonic Analog-to-Digital Conversion

11.1 Introduction

Analog-to-digital converters (ADCs) are circuits that convert continuous-time signals to discrete-time binary-coded forms [1–3]. This operation involves usually two steps. In the first, the input signal is sampled in time, usually at regularly spaced intervals T. In the second, the samples, after being held in time, are quantized or digitized in amplitude. The two main objectives behind AD conversions are: (i) to enable the digital signal processing (DSP) of the input signal and (ii) to enable the digital transmission of the input signal.

Photonic ADC should apply, in principle, to a device where an analog optical signal is input and a digital optical signal is output. However, since there has been very little activity within this direction, it is customary [4] to understand Photonic ADCs as subsystems where the input is an analog electronic radio frequency signal, the intermediate part is a photonic system and the output is an electronic approximation of that signal. Thus, in the context of this chapter, a photonic ADC is understood as a microwave photonic subsystem as shown in Fig. 11.1.

What is the rationale behind considering the incorporation of photonics into ADC configurations for RF signals? First of all, several applications such as radar, surveillance, sensing, and radio-over-fiber telecommunications can benefit from high-resolution ADCs having multi-gigahertz of instantaneous bandwidth [5, 6]. In these systems, the flexibility of receivers can be improved by placing the ADC closer to the antenna that performs more functions in the digital domain by means of DSPs. The possibility of implementing these digital receiver architectures depends on the performance of the ADC, which is limited by several factors [1, 3, 7], including the limited speed and jitter in the sampling stage, the settling time in the sample and hold circuit, the speed of the comparator and mismatches in the transistor thresholds and passive component values. All these factors become more severe at higher frequencies. Typical solutions employed for repetitive signals in time, like sampling oscilloscopes, are not applicable here and the only feasible approach to achieve a real-time capture of these ultrafast electrical signals is to resort to wideband ADCs. In this context, the use of microwave photonic-based solutions provides a very interesting alternative [4–7] as they capitalize on the inherent advantages of photonic components and devices in terms of low losses, broadband operation, and immunity of RF interference. Optical configurations enable, in principle, a very high sampling rate with very low jitter noise and the coupling of the input RF signal and sampling pulses is negligible.

The purpose of this chapter is to present the basic principles and review some of the salient ways in which photonics has been used or proposed for use in ADCs that digitize radio-frequency signals.

Microwave Photonics, First Edition. Jianping Yao, José Capmany, and Ming Li.

Companion website: www.wiley.com/go/yao/microwavephotonics

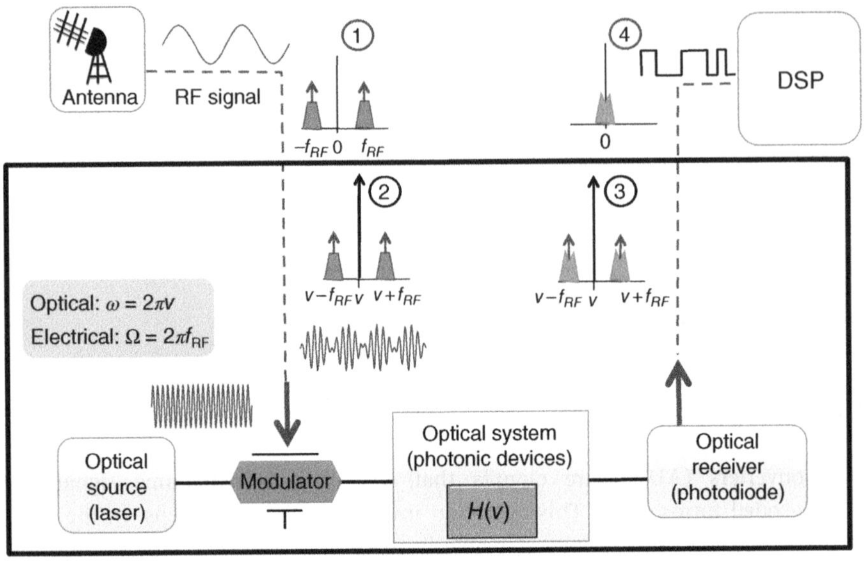

Figure 11.1 ADC block diagram description as a microwave photonics system.

In Section 11.2, we present some basic concepts related to the operation and relevant parameters characterizing ADCs, including a brief discussion on the current state of the art of electronic ADCs. Also, a classification of photonic ADCs is presented. Section 11.3 deals with photonic assisted ADCs in which sampling and quantization are performed electronically and photonic technology plays an auxiliary role to improve their performance. In Section 11.4, we cover photonic sampled and electronic quantized ADCs and in Section 11.5, the reverse option, which is electronically sampled photonic quantized ADCs. Finally, Section 11.6 covers all optical ADCs where both sampling and quantization are performed in the photonic domain. The chapter is summarized and concluded in Section 11.7.

11.2 Basic Concepts on Analog-to-Digital Converters

11.2.1 Types of Converters

The most common type of ADC is the so-called *Nyquist converter* [1, 2] which performs two functions over a time-varying signal: (i) sample and hold it for a specified time and (ii) quantize the held value into a set of discrete levels.

The signal is sampled at a given period T and its inverse is an important parameter known as the *sampling rate* $S_r = 1/T$. The time during which the signal is sampled has to be very short, usually much less than T, while the holding time is usually comparable to T. Another key parameter of the ADC is the number of discrete levels used for quantization. This is usually a power of 2, that is 2^N, where N is an integer and defines its *resolution* expressed in number of bits. For example, a 4096-level ADC is a 12-bit ADC. The top part of Fig. 11.2 shows a diagram of a Nyquist ADC where the sample and hold stage over an input voltage Rf signal is implemented by means of a microwave photonic system and the quantization stage is performed in the electronic domain.

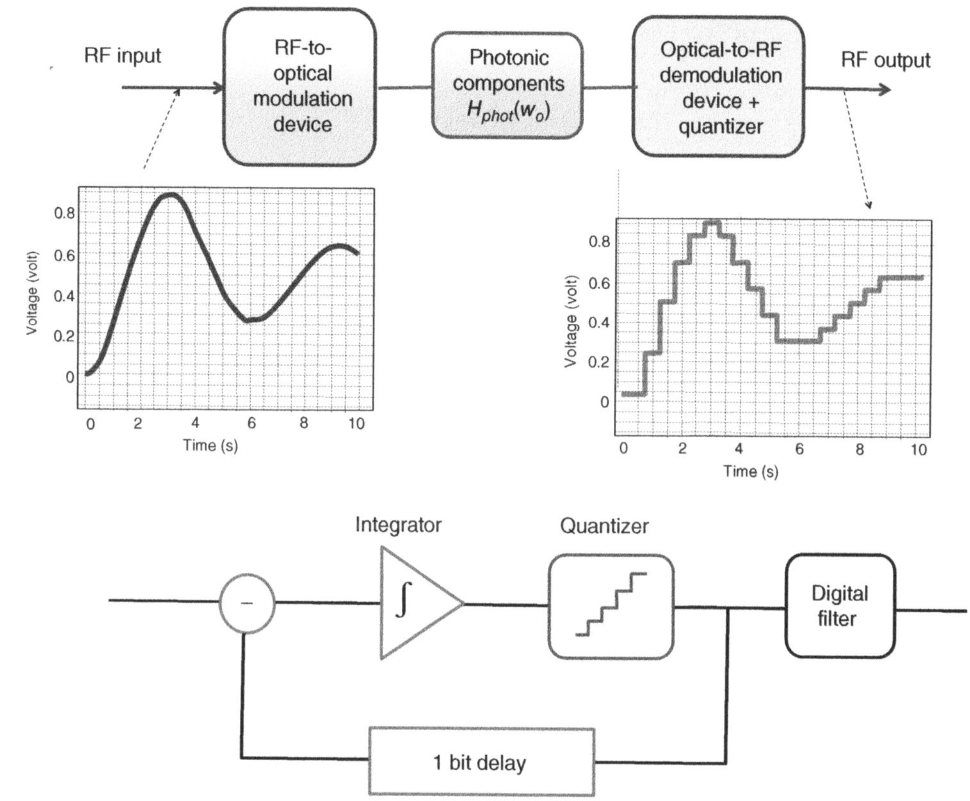

Figure 11.2 (Top) Block diagram of a Nyquist photonic ADC. (Bottom) Block diagram of a sigma-delta ADC.

A second group of ADCs based on a completely different approach are *Sigma-Delta* converters [2, 4] of which the bottom part of Fig. 11.2 shows a schematic diagram. Here, the input signal is summed with the feedback of the previous bit, integrated, quantized, and processed by a digital filter. An external clock triggers the quantizer and the sampling rate is typically much greater than the bandwidth of the input signal, a technique known as oversampling. Sigma-delta converters trade a larger number of samples for smaller numbers of quantization levels. However, none of the key components is, so far, well developed in the optical domain and therefore we will not deal with them further in the chapter.

11.2.2 Operation Principles of the Nyquist ADC

Apart from the sampling rate and the resolution expressed in a number of bits, there are other important ADC parameters [1–3] that we now define. The full-scale voltage V_{fs} gives the difference between the maximum and minimum voltages of the input signal. The quantization or discretization step is defined as the quotient between the full-scale voltage and the number of discrete values, that is $Q = V_{fs}/(2^N - 1)$. Q is therefore the size of the elementary quantization step, which is the least significant bit (LSB) of a binary representation of that value. Finally, the root mean square quantization error is defined as the RMS value of the error between the true value of the input signal when sampled and the assigned discrete value. Assuming a uniform

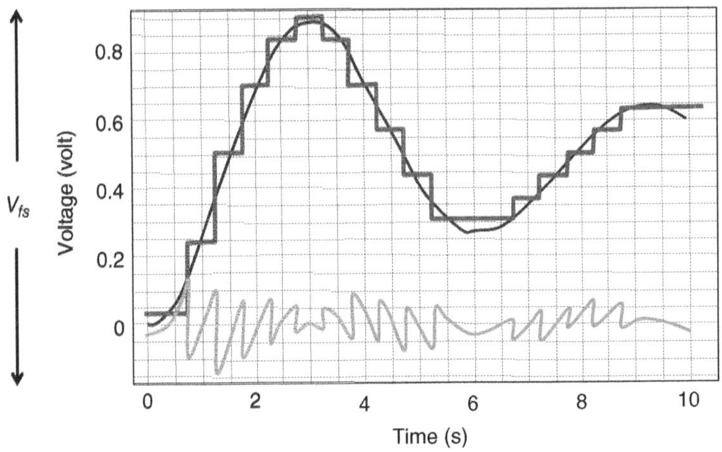

Figure 11.3 Example of input signal to an ADC (solid line), its sampled and quantized version (black wavy line), and the difference between them, also known as quantization error (gray wavy line).

probability density function for the amplitudes in a given quantization region (i.e. taking $[0, Q/2]$ for simplicity), one has [1]:

$$\Delta Q^2 = \int_0^{Q/2} v^2 p(v)dv = \int_0^{Q/2} \frac{2v^2}{Q} dv = \frac{2}{Q}\left[\frac{Q^3}{24}\right] = \frac{Q^2}{12} \Rightarrow \Delta Q = \frac{Q}{\sqrt{12}} \tag{11.1}$$

Figure 11.3 shows as an example the input signal (solid line), its sampled and quantized version (black wavy line), and the difference between them, also known as *quantization error* (gray wavy line). The ADC parameters for this example, taken from [4] are $T = 0.5$ s, $S_r = 2$ samples/s, $V_{fs} = 1$ V, $N = 4$, $Q = 0.067$ V, and $\Delta Q = 0.019$ V.

The performance of an ADC is compromised by several sources or errors, including [7]:

1) Timing errors (caused by random jitter and broadening of the sampling time)
2) Settling time of the sample and hold (S&H) circuit
3) Amplitude errors (due to thermal and shot noise and nonlinearities)
4) The speed of the comparator, also known as comparator *ambiguity*

These errors contribute to a decrease in the effective resolution from the ideal value N to a bottom figure called *ENOB* or *effective number of bits.*

For example, when the only source of impairment is the amplitude quantization error the signal-to-noise ratio of the converter can be related to N as follows [1–3]:

$$\begin{aligned} SNR_Q(\text{dB}) &= 20\log\left[\frac{V_{fs,RMS}}{\Delta Q}\right] = 20\log\left[\frac{V_{fs,RMS}}{(Q/\sqrt{12})}\right] \\ &\approx 20\log\left[\frac{V_{fs}/(2\sqrt{2})}{V_{fs}/(2^N\sqrt{12})}\right] = 20\log\left[2^N\sqrt{6}\right] = 6.02N + 1.76 \end{aligned} \tag{11.2}$$

Or, more conveniently:

$$N = \frac{[SNR_Q(\text{dB}) - 1.76]}{6.02} \tag{11.3}$$

In a more practical scenario, when one has arbitrary sources of noise and nonlinearities, these are characterized by an equivalent signal-to-noise and distortion ratio or *SINAD* and (11.3) can be generalized in order to define the effective number of bits or *ENOB* of the converter [4]:

$$ENOB = \frac{[SINAD(\text{dB}) - 1.76]}{6.02} \tag{11.4}$$

In particular, when timing is the only source of degradation, it is possible to establish a simple relation between the ENOB, the sampling rate S_r, and the root mean square timing jitter σ_j [1, 3]:

$$ENOB_{jitter} = \log_2\left[\frac{1}{\sqrt{3}\pi S_r \sigma_j}\right] \tag{11.5}$$

If limitation is due to thermal noise, then [1–3]:

$$ENOB_{th} = \log_2\left[\sqrt{\frac{V_{fs}^2}{6k_B T R_{eff} S_L}}\right] - 1 \tag{11.6}$$

Where R_{eff} is an effective thermal resistance which lumps together the effects of all noises, k_B is the Boltzmann constant and T is the temperature.

If the ADC performance is limited by the comparator ambiguity, then [1]:

$$ENOB_{Amb} = \frac{2\pi f_T}{13.9 S_r} - 1.1 \tag{11.7}$$

where f_T is the transition frequency associated with the transistors used in the ADC integrated circuit process technology.

In a microwave photonic link of bandwidth Δf, a slightly modified version of Eq. (11.4) has to be employed [4]. Here, the performance is usually described in terms of the carrier-to-noise ratio or CNR:

$$CNR = \frac{(m\Re \text{P})^2/2}{\sigma_{shot}^2 + \sigma_{th}^2 + \sigma_{RIN}^2 + \sigma_{imd}^2} \tag{11.8}$$

where m is the modulation index, $\Re$ the detector responsivity, P the average optical power on the photodiode, and σ_{shot}^2, σ_{th}^2, σ_{RIN}^2, and σ_{imd}^2, the standard deviations associated with shot, thermal, relative intensity noises, and the intermodulation distortion. For the pure noise sources:

$$\begin{aligned}
\sigma_{shot}^2 &= 2e\Re \text{P}\Delta f \\
\sigma_{th}^2 &= \frac{4k_B T F_n \Delta f}{R_L} \\
\sigma_{RIN}^2 &= RIN(\Re \text{P})^2 \Delta f
\end{aligned} \tag{11.9}$$

where e is the electron charge, *RIN* the relative intensity noise of the laser, and F_n and R_L represent the noise figure and the load resistor of the receiver, respectively.

In the context of analog photonic links, Eq. (11.4) is then [4]:

$$ENOB = \frac{[CNR(\text{dB}_c) - 1.76]}{6.02} \tag{11.10}$$

When nonlinearities due to the reduction of the dynamic range are the dominant source, $\sigma_{imd}^2 = (m\Re P)^2/(2SFDR)$ where SFDR represents the spurious free dynamic range. In this case, Eq. (11.6) can be approximated by:

$$CNR \approx \frac{(m\Re P)^2/2}{\sigma_{imd}^2} = SFDR \tag{11.11}$$

So [3]:

$$ENOB \approx \frac{SFDR(\text{dB})}{6.02} \tag{11.12}$$

A universal measure of ADC performance is P, the product of the effective number of quantization level times the sample rate:

$$P = 2^{ENOB} S_r \tag{11.13}$$

A figure of merit that also takes into account the power dissipation P_{diss} is

$$F = \frac{2^{ENOB} S_r}{P_{diss}} \tag{11.14}$$

Typical values of F for available commercial ADCs are around $F = 10^{11}$.

11.2.3 State of the Art of Electronic ADCs

The state of the art in electronic ADCs has been recently reviewed by Walden [1, 3] and more recently by Khilo et al. [8]. Figure 11.4 depicts as a summary the results achieved in terms of ENOB as a function of the ADC bandwidth [3]. The filled circles in the figure show results for existing

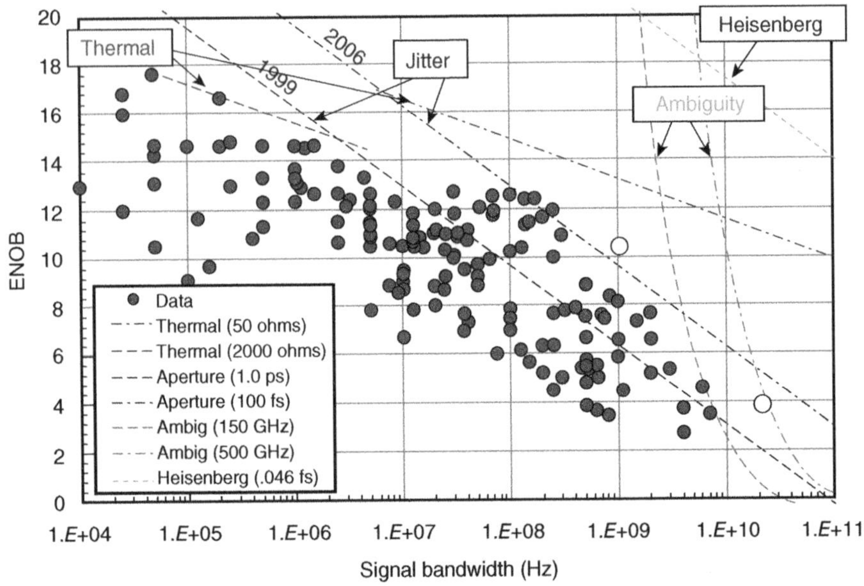

Figure 11.4 Results achieved in terms of ENOB as a function of the ADC bandwidth. *Source:* Adapted from Walden [3].

ADCs and the dashed lines represent the fundamental limits due to jitter for rms apertures of 100 fs and 1 ps (after (11.5)), limits due to comparator ambiguity (Eq. (11.7)) for the cases of transition frequencies given by 100 and 500 GHz, and the limits due to thermal noise (Eq. (11.6)) for load resistances of 50 and 2000 Ω.

Valley [4] indicates that for a photonic ADC to be competitive with an electronic-based solution, it must obtain, at least, an improvement by a factor of 10 in signal bandwidth, or an equivalent reduction in the noise level or break through a fundamental limit. The open circles in Fig. 11.4 show the required ENOB values for 1 and 20 GHz input bandwidth. In the first case, an $ENOB > 11$ is required while in the second case, an $ENOB = 4$ is required to break the ambiguity limit.

For broadband ADCs, it is expected that aperture jitter and ambiguity will become the dominant sources of limitation. Regarding aperture jitter, the current state of the art of electronic ADCs is in the range of 100 fs and it is expected that progress beyond this figure will not be significant within the next decade. On the contrary, photonic-based ADC can potentially provide the required reduced jitter aperture, especially if as discussed in Section 11.3.3, they are based on mode-locked laser sources. Figure 11.5 shows the ENOB vs. the input analog signal frequency for reported ADC architectures from 1999 to 2009 [8]. Filled circles correspond to the ADCs from Walden's survey in 2007 [3] and open circles to other high-performance architectures demonstrated after 2007. Dashed lines correspond to constant jitter aperture values including the year when this value was achieved while dotted lines correspond to constant jitter values not yet achieved but which will be attainable using high-performance wideband photonic ADCs. For instance, reported results for some of these configurations are shown in stars.

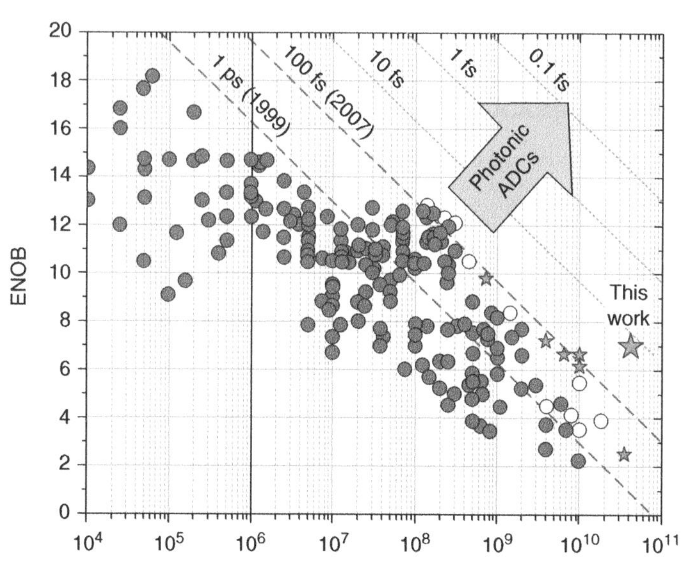

Figure 11.5 ENOB vs. the input analog signal frequency for reported ADC architectures from 1999 to 2009. *Source:* Khilo et al. [8]/Optical Society of America/CC BY 4.0.

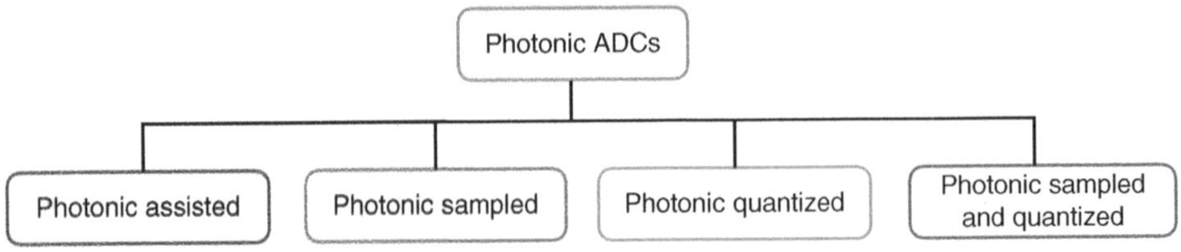

Figure 11.6 Classification groups for reported photonic-based ADCs. *Source:* Adapted from Valley [4].

11.2.4 Classification of Photonic ADCs

A considerable number of photonic approaches for the implementation of ADCs have been reported in the literature. Figure 11.6 shows a diagram that includes the main groups into which the reported configurations can be classified.

Referring to this diagram, photonic ADCs can be grouped into the following classes:

1) *Photonic Assisted ADCs*: These are electronic ADCs using photonics to improve/overcome one or more limiting properties. Both sampling and quantization are performed electronically.
2) *Photonic Sampled ADCs*: In these, sampling is performed in the optical domain and quantization is carried out in the electronic domain.
3) *Photonic Quantized ADCs*: Quantization is performed in the optical domain and sampling in the electronic domain.
4) *Photonic Sampled and Quantized ADCs*: Both sampling and quantization are performed in the optical domain.

In the following sections, we consider each approach independently.

11.3 Photonic-Assisted ADCs

11.3.1 Classification of Photonic-Assisted ADCs

The main approaches reported in this group can be classified into four subgroups as shown by the diagram in Fig. 11.7.

1) *Optically Clocked Track and Hold ADCs*: Where an optical pulse train is employed to gate an electronic ADC.
2) *Optical Replication ADCs*: Where multiple copies of an RF signal are optically generated which allow the use of an electronic ADC at a lower speed.

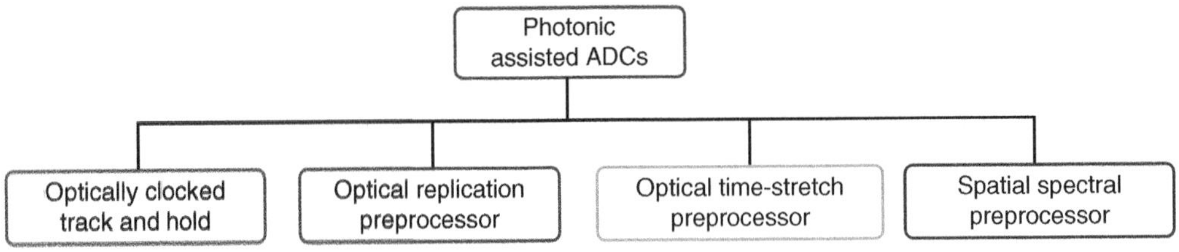

Figure 11.7 Classification of reported photonic assisted ADCs.

3) *Optical Time-Stretch ADCs*: Where a wideband signal is stretched in time so it can be electronically ADC at a lower speed.
4) *Spatial Spectral ADCs*: Where the RF signal is mixed with a reference to generate a spectral hologram in a crystal, which is then compressed (stretched in time) when read by a chirped optical source.

The level of development of each subgroup is different. While considerable work has been reported for optical time-stretched ADCs, spatial-spectral ADCs have received much less attention so far. In the following subsections, we review the major features, advantages, and limitations of each one.

11.3.2 Optically Clocked Track-and-Hold Circuits

The rationale behind this approach is to use periodic optical pulses shorter than 100 ps from an ultra-stable optical mode-locked source to activate fast optoelectronic switches for electronic sampling as these provide faster rise times and lower pulse-to-pulse time jitter as compared to their electronic versions [9–12]. Another advantage is the possibility of isolating the optical clock source from the electronic part of the ADC. Figure 11.8 shows a generic layout representing these ADCs.

Although the main hurdle in implementing these converters resides in the difficulties associated with the integration, at a reasonable cost, of an ultra-stable mode-locked laser source into a commercial product, several groups have reported advances addressing other particular limiting aspects. Figure 11.9 shows the layouts of several interesting configurations. For instance, the top left-hand side of Fig. 11.9 shows an optically clocked diode bridge proposed by Jacobs et al. [13] to overcome the limitations of direct illumination of an optoelectronic switch, especially in track and hold operations as the turn-down times from ON state are not short enough to guarantee high-frequency operation. These limitations arise from the need to charge the capacitor implementing the hold function by means of a weak optical signal and by the slow electronic turn-off response time due to the semiconductor lifetimes. In the configuration, the diode bridge is biased in track mode. Simultaneous illumination of both diodes turns the bridge OFF causing a fast transition from track to hold. This optical bridge circuit is followed by an electronic clocked bridge that extends the hold time of the circuit. The right part of Fig. 11.9 displays a configuration based on GaAs photoconductive switches integrated with CMOS ADCs proposed by Nathawad et al. [14] to implement a time-interleaved ACD. A differential device (bottom right part) is used where one switch is activated by the mode-locked laser. The other switch is actually a dummy. The outputs are subtracted

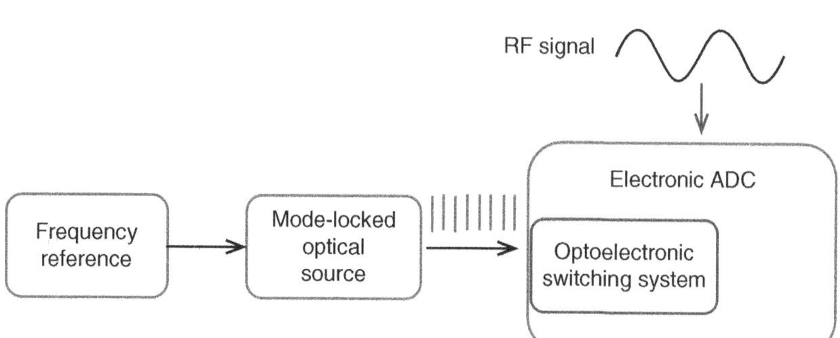

Figure 11.8 A generic layout of an optically clocked track-and-hold-circuit.

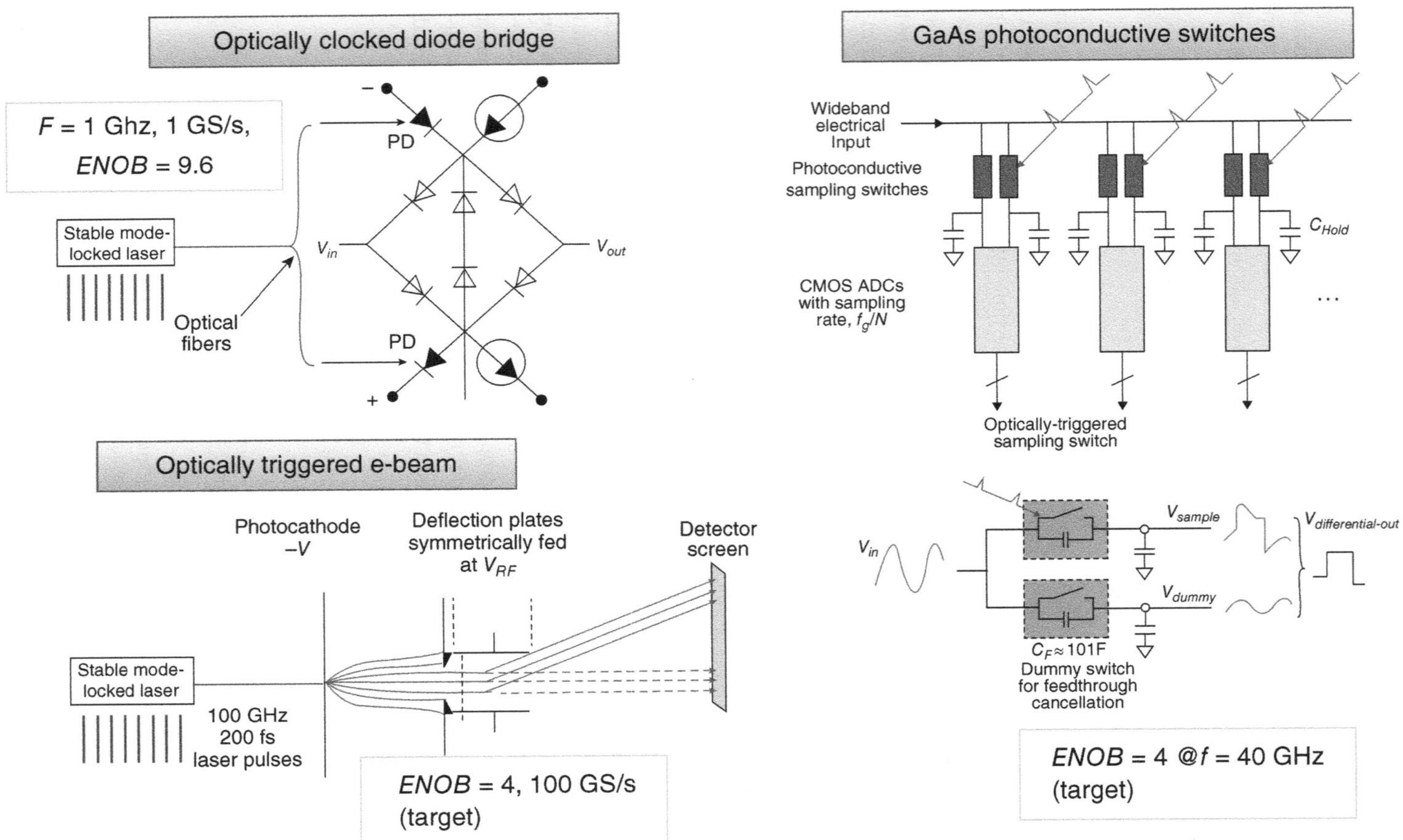

Figure 11.9 Top left: optically clocked diode bridge. *Source:* Adapted from Jacobs et al. [13]. Right: ADC configuration based on GaAs photoconductive switches integrated with CMOS ADCs. *Source:* Adapted from Nathawad et al. [14]. Bottom left: optically triggered electron beam ADC. *Source:* Adapted from Loakeimidi et al. [15].

and the short carrier dynamics of low-temperature GaAs switches are exploited to provide fast operation.

Finally, the bottom left part of Fig. 11.9 shows an optically triggered electron beam ADC [15]. In essence, this configuration makes use of short optical pulses from a high-repetition-rate mode-locked laser which, after being detected by a photocathode, result in a beam of electrons which is collimated and passed through a pair of deflection plates that deflect the beam with an angle which depends on the voltage of the RF signal to be sampled. A detector screen is then used to collect the deflected beam. A special-purpose electronic ADC with spatial diversity must follow to hold and quantize the signal.

11.3.3 Optical Replication Pre-Processors

The concept behind optical replication pre-processors is to generate multiple copies of the same input RF signal in order to allow the use of a lower sampling rate electronic ADC [5, 7, 16, 17]. Replication pre-processors can generate multiple copies in series or in parallel form. The latter option, known as time-interleaving ADC architecture, is the standard approach [7]. Figure 11.10 illustrates its principle of operation for a structure that produces four parallel replicas.

The RF-modulated optical signal, originally requiring a Nyquist sampling rate $S_r = 1/T$ is divided in N parallel replicas ($N = 4$ in the figure) by means of a *1xN* power splitter. The replicas are then captured by a parallel array of N slow digitizers each clocked at a fraction of the Nyquist rate ($S_r/N = 1/T_d$) with a sampling instant delay between neighboring digitizers given by T. Finally, the samples have to be reassembled in order since the Nyquist criterion for the input signal is only satisfied when the signal is reconstructed sample by sample.

A serial approach is also possible as shown in Fig. 11.11 where a fiber loop is used that stores a time-limited microwave signal so that it may be digitized by using a slower, conventional electronic ADC. Under appropriate conditions, the microwave signal degradation is sufficiently small to allow the digitization of a multi-gigahertz signal with a resolution greater than 10 effective bits.

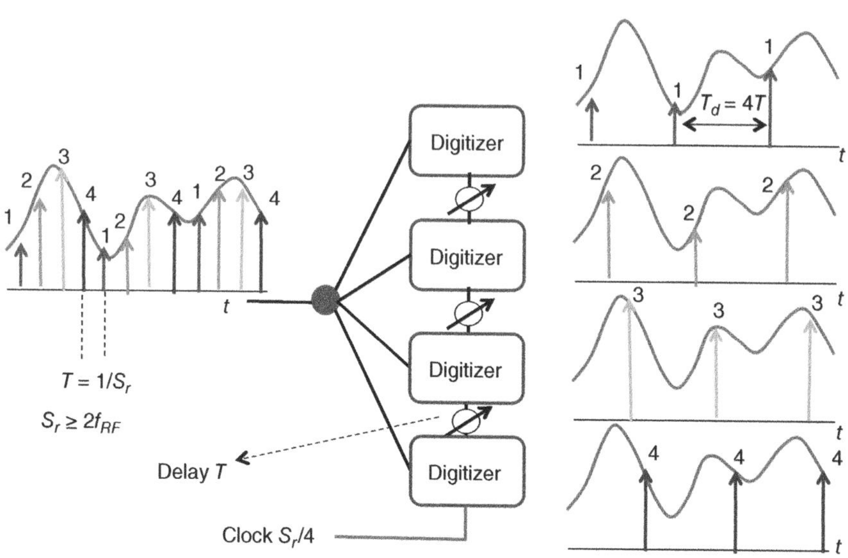

Figure 11.10 Principle of operation of a parallel replication pre-processor.

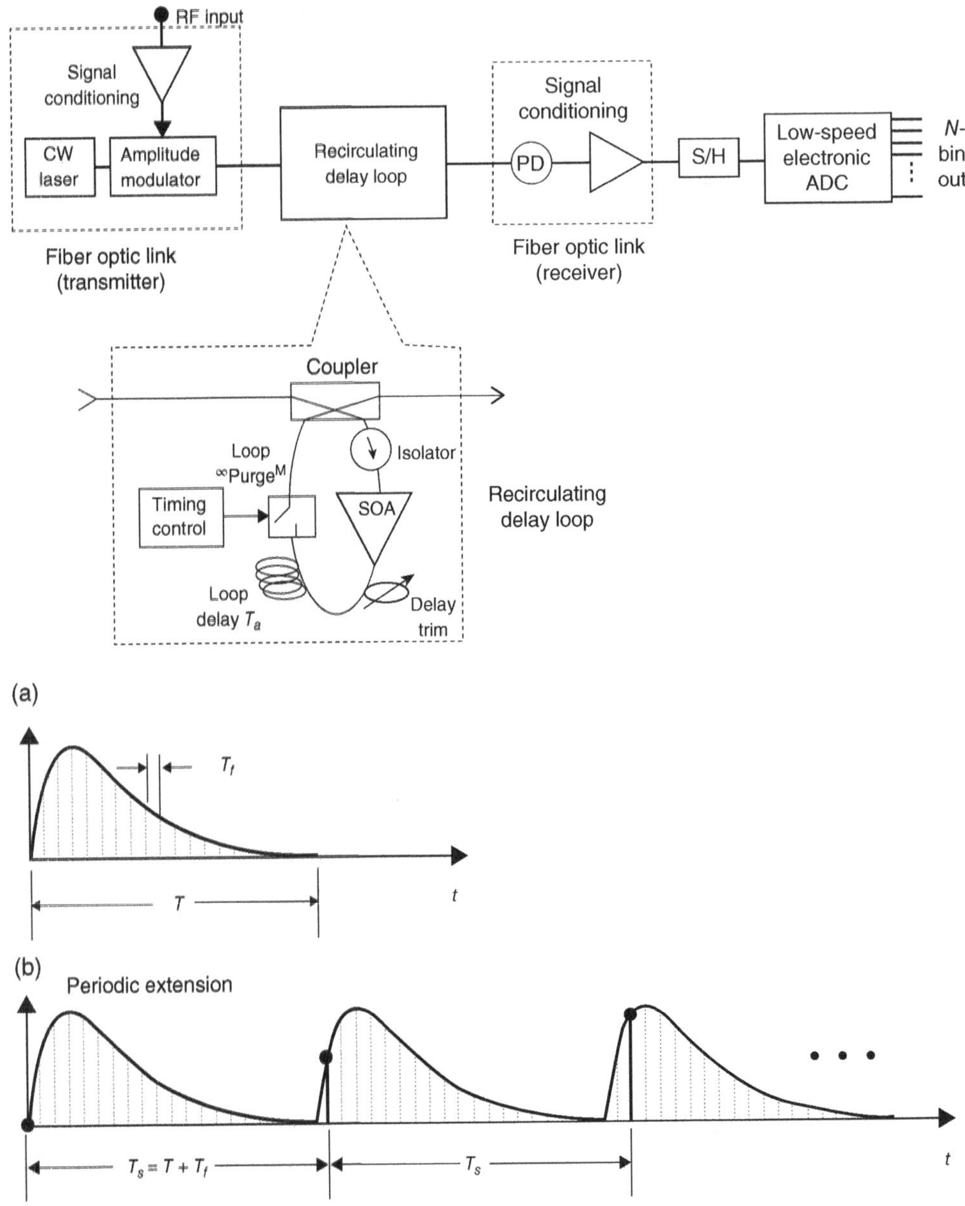

Figure 11.11 Layout and principle of operation of a serial replication pre-processor based on a recirculating delay loop. (a) Sampling of the standalone signal. (b) Period extension by replication of the standalone signal. *Source:* Adapted from Zmuda et al. [18].

In practice, mismatches between digitizers limit the dynamic range and the resolution (ENOB) of sample-interleaved ADCs. The main reason is that the error associated with the jitter in the sampling clock of the digitizer is not reduced as the signal shape is not altered in time. This limitation can be solved if the signal is stretched in time reducing its slew rate. Time-stretched ADCs, which are considered in the next section address this topic.

11.3.4 Optical Time-Stretched Pre-Processors

The objective of a time-stretched pre-processor is to slow down in time by a factor of M the analog signal prior to sampling and quantization by an electronic digitizer [7, 19–27]. Slowing down the signal prior to digitization brings several advantages. For instance, the sampling rate of the electronic digitizer is increased from S_r to MS_r, the effective input bandwidth of the digitizer is also increased by M. Most importantly, the error associated with the jitter in the sampling clock of the digitizer is reduced due to the decrease in the slew rate. This is a distinctive feature as compared to sampled interleaved ADCs.

A practical scheme to implement the time-stretching functionality using photonics technology is shown in Fig. 11.12.

A narrow input pulse from a broadband optical (supercontinuum) source of bandwidth Δf_{opt} is subject to chirp broadening by a first chromatic dispersive element, for example, a singlemode optical fiber link of length L_1 and dispersion parameter D_1. The resulting chirped pulse is modulated by the RF signal in an intermediate electro-optic modulator and subsequently stretched in time in a second dispersive element, for example, a singlemode fiber link of length L_2 and dispersion parameter D_2. Upon detection, the time-stretched RF envelope is obtained.

The most relevant performance parameter of the time-stretched processor is the stretch factor that, in the case of using fibers with identical chromatic dispersion parameter $D_1 = D_2 = D$, is given by [7]:

$$M = \frac{(L_1 + L_2)}{L_1} \tag{11.15}$$

A second parameter of importance is the time aperture available to modulate the chirped pulse. This is given by:

$$T_A = D_1 L_1 \Delta\lambda = 2\pi|\beta_2|L_1 \Delta f_{opt} \tag{11.16}$$

Finally, the RF bandwidth of the pre-processor depends on the modulation scheme employed. For double sideband modulation, one has:

$$\Delta f_{RF} = 1/\sqrt{8\pi|\beta_2|L_2/M} \tag{11.17}$$

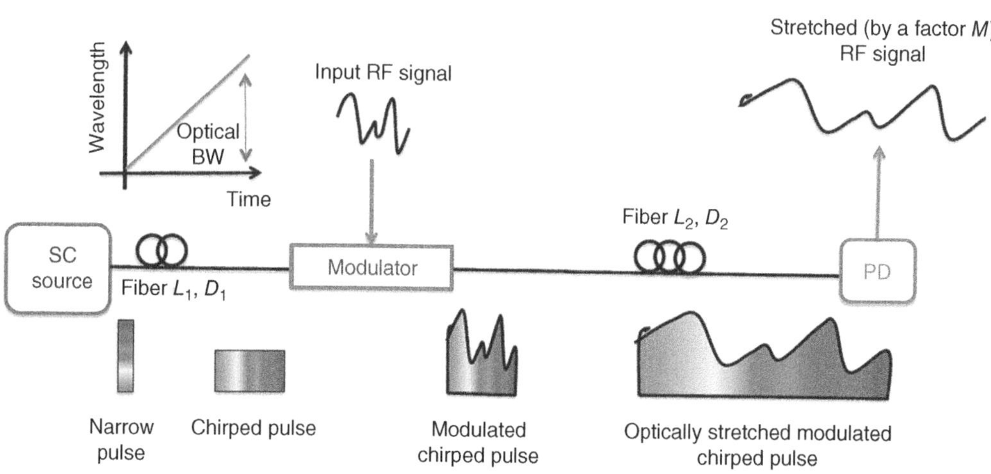

Figure 11.12 Practical for the implementation of the time-stretching pre-processor. *Source:* Adapted from Han and Jalali [7].

It is desirable to maximize the time aperture and, for a fixed optical bandwidth, it is possible to do so and keep a constant stretch factor by increasing the value of L_1 while keeping constant the ratio L_2/L_1. However, this results in a larger dispersion penalty according to (11.17). The consequence of this trade-off is that neither the aperture nor the bandwidth can be employed standalone to assess the performance of the ADC.

The most complete metric to assess such performance is the so-called product of time aperture and RF bandwidth which, for DSB modulation, is given by:

$$TBP_{DSB} = \frac{\Delta f_{opt}}{2\Delta f_{RF}} \tag{11.18}$$

The layout described in Fig. 11.12 corresponds to a single-shot or transient digitizer [7]. However, the time-stretch principle can be applied to implement continuous-time operation as well provided that a multichannel (N channel) system can be incorporated into the ADC architecture, based on using either switching or passive even splitting in the optical domain, as shown in the top part of Fig. 11.13, where a system featuring a stretching factor $M = 4$ and a number of channels $N = 4$ is represented.

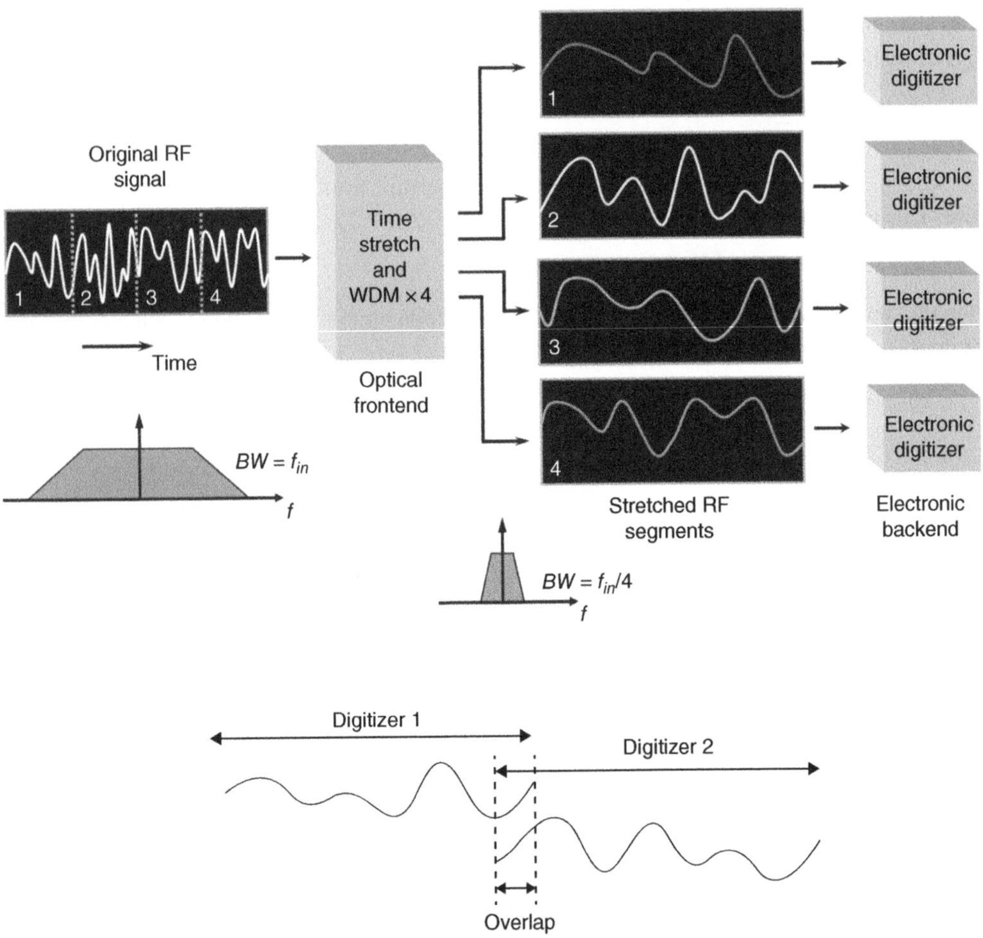

Figure 11.13 Top: Layout of a continuous time operation ADC based on a time-stretched pre-processor. Bottom: Finite overlap between adjacent segments employed for mismatch error correction and calibration purposes. *Source:* Adapted from Han and Jalali [7].

There are two main differences between this scheme and the traditional time-interleaved configuration shown in Fig. 11.10. In the time-interleaved approach, although the sampling rate of each digitizer is reduced from S_r to S_r/N, there is no such reduction on the required input bandwidth (actually the original input signal is simultaneously applied to all digitizers). In other words, the input bandwidth of the system is not improved. In contrast, the signal is slowed down prior to entering the digitizers. The second difference deals with the relative sampling rate of each digitizer. In the time-interleaved system, sampling by each digitizer is done at a fraction of the Nyquist rate and, therefore, the original signal can only be recovered after sample-by-sample reconstruction. In the time stretch approach, as the signal is slowed down before digitization, each digitizer samples the signal at the Nyquist rate. Thus, a time segment of the input signal is available from each digitizer before reconstruction and this feature can be employed as shown in the next paragraph for mismatch error correction.

As mentioned earlier, errors associated with the jitter in the sampling clock are reduced due to the decrease in the signal slew rate. Errors can, however, appear due to inter-channel mismatches as in any time-interleaved architecture, but they can be corrected by exploiting the fact that sampling in each channel meets the Nyquist criteria. This allows the possibility of a finite overlap between adjacent segments as shown in the bottom part of Fig. 11.13. To achieve this overlapping, the number of channels N has to be slightly larger than the multiplication factor M. In such a case, the same RF signal is digitized during an overlap period by adjacent channel ADCs. By comparing the signals from these two adjacent digitizers in the overlap region, mismatch errors can be calculated and corrected in the optical domain. Furthermore, because the error is obtained from the signal itself, there is the possibility of performing real-time error correction.

One key function in continuous-time ADCs is the segmentation of the input analog signal into multiple channels. Two options based on time domain multiplexing are shown in Fig. 11.14.

In the configuration corresponding to the top part of Fig. 11.14, the signal is passively split in the optical domain into N channels and then gated prior to the time-stretch block by sequentially activating different optical switches (one per digitizer). In the configuration shown in the bottom part of Fig. 11.14, the segmentation is carried passively using the time-to-wavelength mapping encoded in a chirped optical carrier (in a similar way as illustrated in the single-shot configuration of Fig. 11.12). A parallel optical filter is then required to perform the wavelength segmentation.

11.4 Photonic Sampled/Electronic Quantized ADCs

Aperture jitter has been a major bottleneck for the implementation of broadband and accurate ACDs required for the conversion of wideband multi-GHz signals [1, 3, 5]. Figure 11.15 shows the allowable timing jitter for a Nyquist ADC as a function of the sampling rate taking the ENOB as a parameter using (11.5). Also shown at the top broken-trace box are the current state-of-the-art limit estimates of aperture jitter in electronic ADCs featuring values between 100 and 250 fs. The part of the graph placed at the right of the broken trace vertical line corresponds to the high-speed sampling rate region (>10 GSa/s) and as it can be appreciated, even the conversion at the lowest speed in this region (10 GSa/s) is not possible at $ENOB > 6$ values for the currently attainable jitter values. The reason behind this limitation stems from the fact that electronic data converters derive their sampling clock from RF oscillators and the timing jitter or phase noise of these devices is governed by the laws of thermodynamics, and more precisely, by the thermal noise of the active elements in the resonators. It is also well known that higher frequency oscillators show lower timing jitter than a lower frequency oscillator with otherwise identical components and noise.

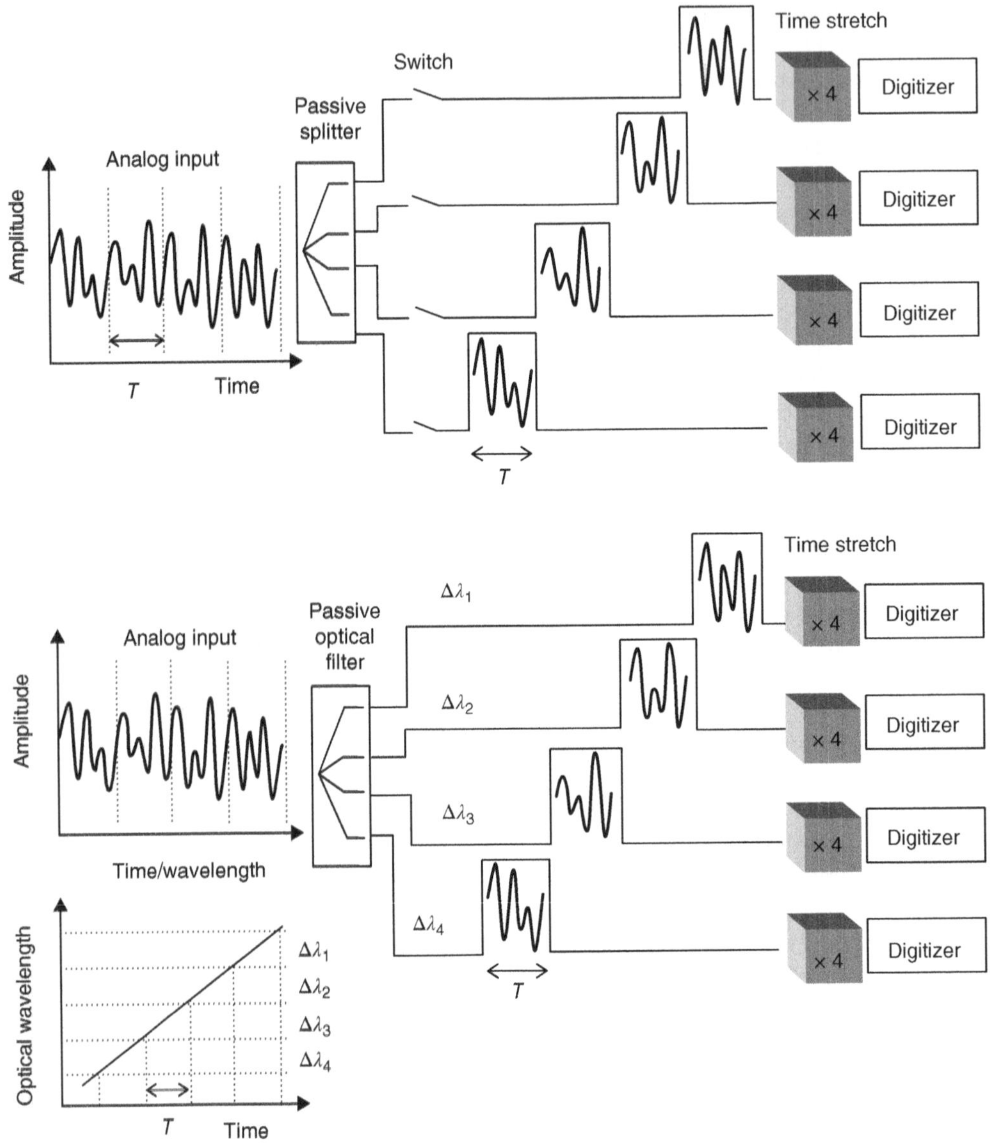

Figure 11.14 Segmentation options for continuous-time operation time-stretched ADCs. Top: Optical time domain multiplexing. Bottom: Wavelength domain multiplexing. *Source:* Han and Jalali [7]/IEEE.

Under these circumstances, it is interesting to consider the use of ultrashort pulses produced by mode-locked lasers as sampling signals since they feature extremely low jitter [28, 29]. Specifically, their microwave phase undergoes a random walk in the laser cavity due to spontaneous emission noise producing a standard deviation for the pulse center of gravity from their nominal position given by [30]:

$$\Delta t \approx \tau \sqrt{\frac{h\nu T_M}{E_p \tau_c}} \tag{11.19}$$

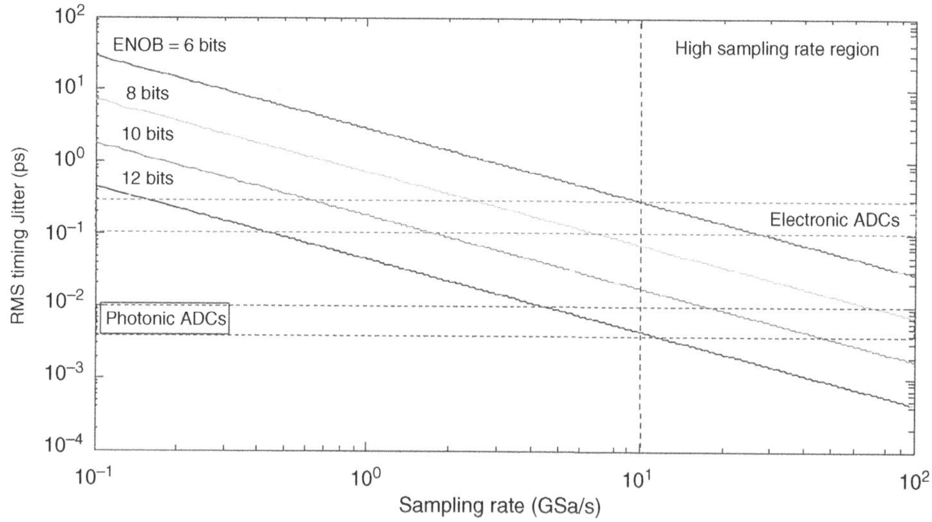

Figure 11.15 Allowable timing jitter for a Nyquist ADC as a function of the sampling rate taking the ENOB as a parameter using [5]. *Source:* Adapted from Juodawlkis et al. [5].

where τ is the pulse duration, E_p the intracavity pulse energy, τ_c the cavity decay time, $h\nu$ the photon energy, and T_M the measurement time. From (11.19), it follows that mode-locked lasers which produce 10–100 fs optical pulses can be employed as sampling clocks featuring three to four lower orders of magnitude in timing jitter as compared to microwave or RF oscillator-driven clocks. The broken-trace box at the bottom in Fig. 11.15 shows the current state-of-the-art range values for mode-locked laser pulses (from around 4 fs for fiber lasers to 10 fs for solid-state lasers). Note that, in this case, conversion at 10 GSa/s should be attainable even for ENOB values of 12.

The use of mode-locked lasers to sample RF signals was proposed as early as in 1975 by Taylor [31–33] and, more recently, it has been the subject of considerable interest by several research groups, even compatible with integrated technology [34–46], especially in connection with the design of the optical sampling stage either standalone or followed by a demultiplexing or time-interleaved configuration to increase the bandwidth of the ADC. Apart from the significant reduction in the aperture jitter optical sampling brings two further advantages. The first is that the sampling process can be made highly linear and with negligible back-coupling between the optical sampling pulses and the electrical signal being sampled. The second advantage is that the potential integration of the mode-locked laser source with other photonic components required for demultiplexing in a silicon photonics platform may clear the path toward the realization of high accuracy ($ENOB > 8$) millimeter and THz band ADCs.

Figure 11.16 shows the basic layout and components of an optically sampled ADC, composed of a short pulse optical source (stable mode-locked laser source), a sampling transducer which is usually implemented by means of an external $LiNbO_3$ Mach–Zehnder electro-optic modulator and a detection and quantization stage.

The sampling occurs when the pulses from the mode-locked laser pass through the electrooptic modulator while the voltage of the signal to be sampled is applied. The output pulse energies represent the RF signal values at the temporal positions of the pulses. Pulses are later converted to electrical signals using photodetectors and quantized using electronic ADCs. The jitter of the optical sampling process is determined by the jitter of the optical pulse train that, as explained above,

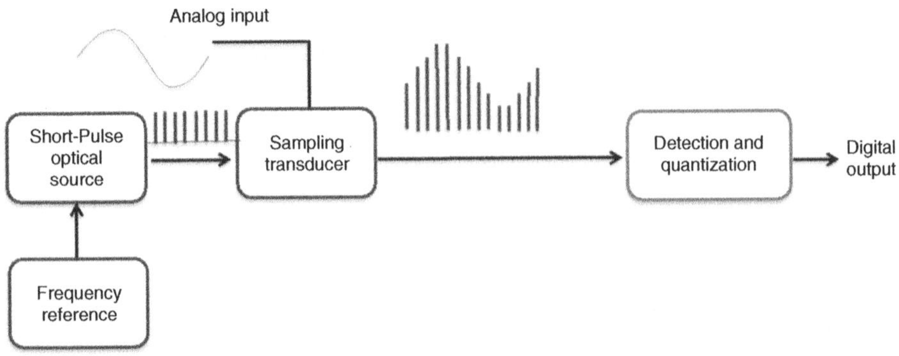

Figure 11.16 Basic layout and components of an optically sampled ADC. *Source:* Adapted from Juodawlkis et al. [5].

can be extremely low. In addition, and as the electro-optic interactions can be very fast, the aperture over which the RF signal is sampled is defined by the pulse durations and can be very short.

Overall, the system in Fig. 11.16 is a microwave photonic link and, therefore, its performance as an ADC can be measured in terms of both end-to-end SNR and SFDR. Apart from low jitter and reduced sampling pulse width, there are other issues of importance like, for instance, the requirement of a linear response of the modulator. Linear operation of the modulator can be achieved using low modulation indexes but this usually puts strict limits on the link power budget. A more interesting solution proposed by Juodawlkis et al. [5] involves the use of 1×2 Mach–Zehnder modulators where the two outputs are digitized and fed to a DSP unit to invert the nonlinear transfer function of the modulator. Another source of potential nonlinearity is due to the dynamic range and response time of the photodiode.

A final crucial issue is related to the fact that the electronic quantization stage must be, in principle, clocked at the same sampling rate of the optical signal or, in other words, it should be capable of handling an enormous flow of data when sampling high-frequency signals. The bandwidth of this technique can be extended by distributing post-sampling pulses to an array of time-interleaved electronic quantizers. Distribution can be performed in the optical domain by means of optical demultiplexing techniques having low channel-to-channel crosstalk as shown in Fig. 11.17.

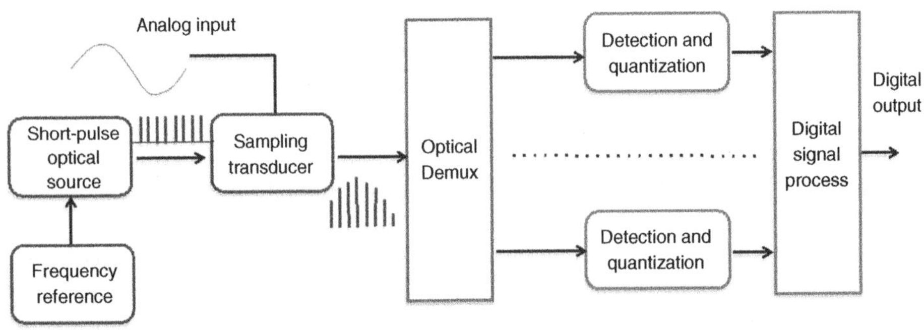

Figure 11.17 Optical domain post-sampling distribution to an array of time-interleaved lower speed electronic quantizers. *Source:* Adapted from Juodawlkis et al. [5].

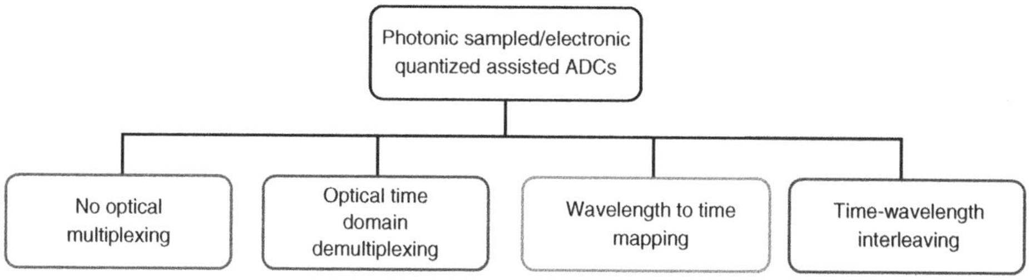

Figure 11.18 Classification of photonic sampled/electronic quantized ADCs.

The different alternatives for photonic sampled and electronic quantized ADCs are shown in the diagram of Fig. 11.18.

The no-optical multiplexing option corresponds to the single-channel configuration shown in Fig. 11.16, where the electronic quantization stage has to work at full-signal speed. The optical time domain multiplexing option is equivalent to time-interleaved approach previously used in a number of all-electronic ADCs [5, 34–38]. The main difficulty in implementing a time-interleaved ADC is, as mentioned earlier in this chapter, to make the interleaved samples from the slow-converter array appear as though they were generated by a single fast converter. Converter-matching errors will produce spurs in the spectra of the time-interleaved ADC output samples. So, in order to achieve a large SFDR, the following conditions must be met:

1) The sampling times of the interleaved ADCs must be uniform.
2) The converter-to-converter gains and offsets must be precisely matched.
3) The crosstalk between ADC converters and sample-to-sample memory effects must be minimal.

For example, to achieve an interleaving SFDR of 80 dB, the converter-to-converter gains must be matched to 0.01%, the offsets must be matched to 0.01% of the signal amplitude, and the converter-to-converter crosstalk must be less than one part in 10^4.

An alternative to the optical time-division demultiplexing approach just described is wavelength-division demultiplexing. The advantage of this scheme is that it allows the replacement of the accurately timed optical switches required in optical time demultiplexing by passive wavelength division demultiplexers. In these architectures, spectrally broad and temporally short optical pulses are either temporally broadened using a dispersive medium [39–41] or separated into N distinct pulses [42–45], each at a different wavelength, with interpulse temporal spacing of T/N, where T is the period of the initial pulse train. Discrete-wavelength pulse separation has been achieved using a wavelength-division multiplexer (WDM) along with fiber delay lines [43, 44] or with an arrayed-waveguide grating (AWG) with integrated feedback waveguides [45–47]. Figure 11.19 shows, for instance, two configurations representing each one of the two approaches.

As can be observed, in these configurations, single-output Mach-Zehnder modulators are employed to perform the sampling function. After sampling the signal, the single temporally broad pulse or discrete pulses are routed to parallel channels of detection electronics using a passive wavelength-division demultiplexer. In the time and wavelength interleaved configuration, the generation of low noise pulses interleaved in wavelength might be as difficult or even more than generating the narrower band identical wavelength pulses needed for the optical time domain multiplexing approach. Furthermore, a considerable number of path-matching, crosstalk and calibration issues, many of which are common to electronic interleaved systems must also be addressed. Some of these requirements can be greatly alleviated if the photonic part can be integrated into an optical chip.

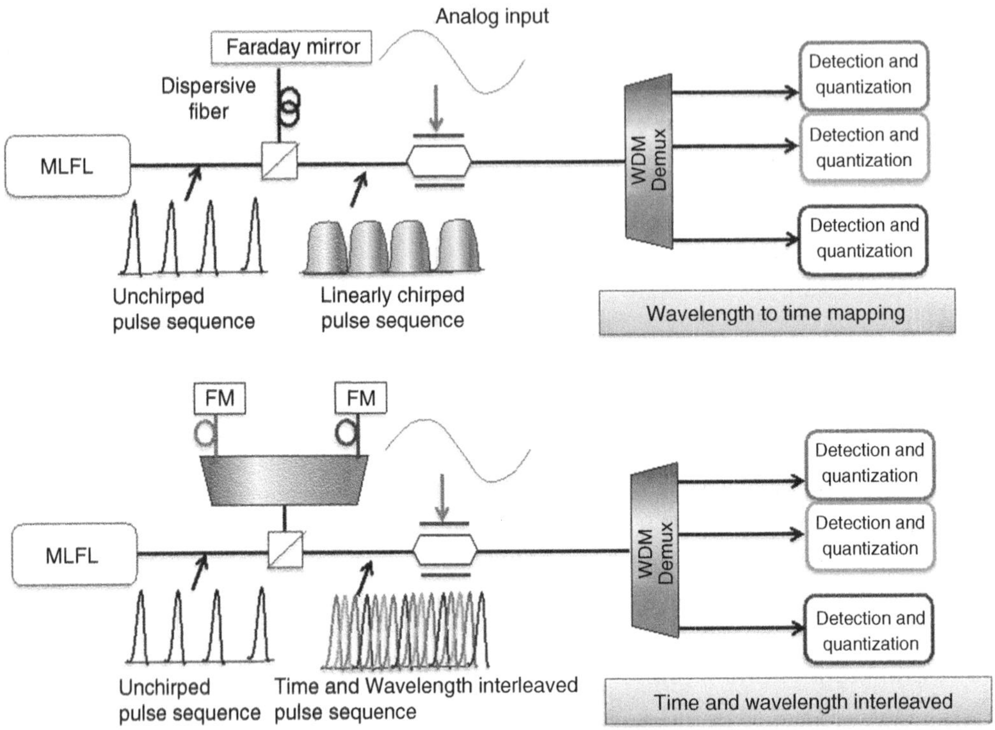

Figure 11.19 Wavelength-to-time mapping (*Source:* Adapted from Frankel et al. [39]) (top) and time and wavelength interleaved (*Source:* Adapted from Kang and Esman [43]) (bottom) ADC configurations based on optical sampling and electronic quantization.

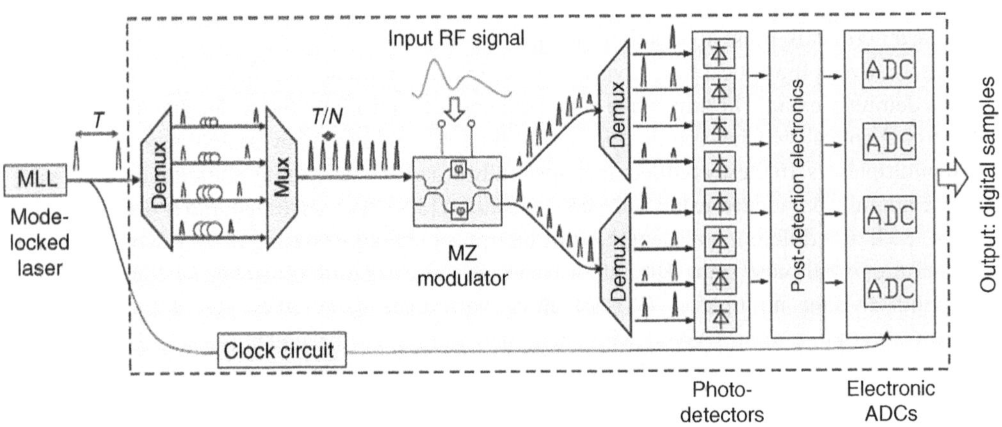

Figure 11.20 Architecture of a silicon photonics chip featuring world-record performance results in broadband photonic optically sampled ADCs. *Source:* Khilo et al. [8]/Optical Society of America/CC BY 4.0.

Figure 11.20 shows the configuration of a reported silicon photonics chip featuring world-record performance results in broadband photonic ADC [8]. In the chip, a modulator, filters, and photodetectors have been integrated and used to sample a 10 GHz signal with $ENOB = 3.5$ (7 in dual output configuration). A sample rate of 2.1 GSa/s was obtained by interleaving two 1.05 GSa/s channels, but the system is readily upgradable to provide 20 channels and therefore the potential to sample a 41 GHz signal with a record value of $ENOB = 7$.

As can be observed, the configuration corresponds to a time and wavelength interleaved architecture where an optical pulse train with repetition period T is split into N trains, each centered at a different wavelength, with a 1-to-N wavelength demultiplexer. These trains passed through optical delay lines, which introduced incremental delays of T/N between them. The trains were then recombined with a multiplexer to produce a pulse train with repetition period T/N, where pulse wavelengths repeated periodically every N pulses. This operation established a discrete time-to-wavelength mapping within the pulse train but a continuous version of time-to-wavelength mapping can also be established using a dispersive fiber.

The next step was the optical sampling carried through the modulation of the optical pulse train with the RF signal to be sampled. In this case, a dual-output Mach–Zehnder modulator is employed to increase linearity [5]. The modulated pulse train was then taken apart into N channels using a wavelength demultiplexer matched to the one used earlier to establish the time-to-wavelength mapping. The pulse trains in all channels were converted to the electrical domain with photodetectors, boosted in amplitude with RF amplifiers, and digitized with electronic ADCs. These ADCs were synchronized with the mode-locked laser and took one sample per pulse, exactly at the pulse peak. Although the electronic ADCs had some aperture jitter, the accuracy of the photonic ADC is insensitive to it because the electronic ADCs sample the relatively flat-top region of the pulses. In the post-processing stage, the samples captured in different channels were compensated for distortions and interleaved to obtain the final digital representation of the RF signal.

The left part of Fig. 11.21 shows a vision of the fully integrated electronic-photonic ADC as proposed in [8]. The chip would include both photonic and electronic components, that is, the dual-output silicon modulator, two matched banks of microring-resonator WDM filters, balanced

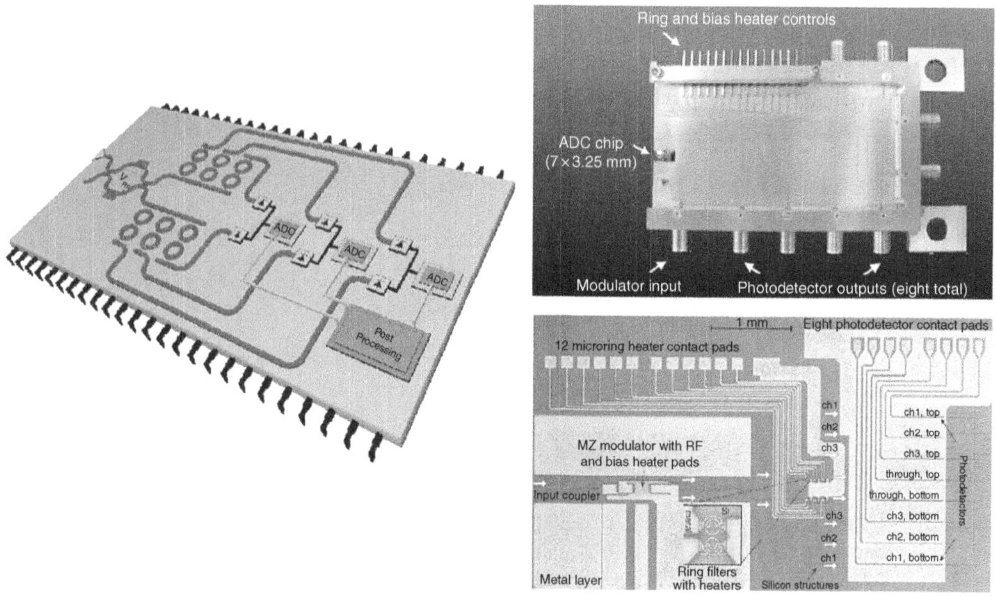

Figure 11.21 Left: Vision of the fully integrated electronic-photonic ADC as proposed in [8]. Right: Photograph of the packaged silicon photonic chip and a top-view photograph of this photonic chip. *Source:* Khilo et al. [8]/reproduced from Optica Publishing Group.

photoreceivers, electronic ADCs, and digital postprocessing circuits. The generation of the wavelength-interleaved pulse train (not shown in the figure) could also be integrated on the same chip. For simplicity, only 3 wavelength channels are shown but channel count can be significantly higher (in fact, up to 20 channel filters are demonstrated in [8]).

The right part of Fig. 11.21 shows a photograph of the packaged silicon photonic chip and a top-view photograph of this photonic chip with metal heaters, wiring, and contact pads fabricated on top of the over-cladding on the silicon layer.

11.5 Electronic Sampled/Photonic Quantized ADCs

These approaches raised some interest in the early years of the present century, but currently, there is practically no activity on them. The reason behind this loss of interest resides in the fact that it does not profit from the advantages that photonic and electronic technologies bring in each process involved in the conversion process, but rather uses a reverse approach, that is, electronics is employed for sampling where photonic schemes have a superior behavior in terms of jitter and photonics is used for quantization where electronics performs much better.

Most of the reported schemes for these converters are based on the work developed by Zmuda and co-workers [48–51] where an electronic sample and hold circuit produces a staircase voltage waveform that is employed to change the wavelength of a tunable semiconductor laser. The output radiation from this source is then directed to a diffractive optical system. So, each wavelength is focused onto a set of different receivers in a photodetector array [52–55]. This wavelength selective routing performs optical quantization since the illumination of a particular set of photodiodes can be coupled to an encoding or quantization process. Figure 11.22 shows a typical example of this configuration [56]. This particular design uses a spectrometer-like setup to separate the wavelength ranges that correspond to different digital values. The analog input signal determines the position of a single spot in the spectrum plane where an array of "fan-out" diffractive optical elements (DOEs) is placed. Each of the DOEs splits the incoming beam and redirects it to the right binary detectors where the digital output is obtained after a set of electronic comparators. For instance, the

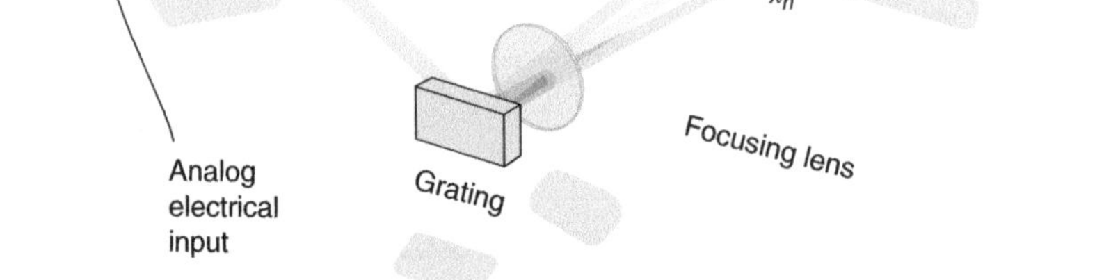

Figure 11.22 Layout of an electronic-sampled photonic-quantized ADC based on a voltage-controlled tunable laser and diffractive optical elements. *Source:* Adapted from Stigwall [56].

wavelength at the center in Fig. 11.22 generates the Gray encoded binary word "110" by splitting the beam into three and illuminating detectors #1, #2, and complementary detector #3. Designing the DOE array cleverly, it is possible to avoid broadening the beam when two neighboring DOEs are hit simultaneously.

Johansson et al. [55] have suggested, though not formally demonstrated, that the number of bits in these schemes using DOEs as encoding devices is limited by the number of resolvable wavelength bands, that is, on the order of 100. This corresponds to an ideal figure of around $N = 6$. However, the expected value of the ENOB should be smaller since degradation sources such as the laser noise and the nonlinear behavior of the voltage-to-wavelength conversion have not been taken into account in this estimation.

11.6 Photonic Sampled/Photonic Quantized ADCs

11.6.1 Classification of Photonic Sampled/Photonic Quantized Converters

Several approaches have been proposed for the implementation of ADCs featuring both sampling and quantization in the photonic domain. Figure 11.23 classifies these into four groups of which we will briefly describe the first three of them since, as mentioned previously, sigma-delta converters are beyond the scope of the present chapter.

11.6.2 Intensity Modulation and Conversion

Among this category are the interferometric ADCs, first proposed by Taylor [31–33], which use an array of Mach–Zehnder modulators with geometrically increasing lengths as shown in Fig. 11.24(a):

Here, the input analog signal modulates an input periodic pulse train from a mode-locked laser and is symmetrically folded by the Mach–Zehnder modulators, with each Mach–Zehnder modulator having an electrode length that is twice that of its nearest more significant bit (NMSB), leading to a doubled folding frequency in its output power vs. input voltage transfer function, as shown in Fig. 11.24(b). The parallel modulation stage is followed by a set of electronic comparators set at the threshold intensity I_t, yielding a Gray code indicated by the grey and white bars below the transfer function curves in Fig. 11.24(b). The folding property in the transfer function imposes a requirement that the half-wave voltage of the Mach–Zehnder modulator at the LSB should be $V_\pi = 2V_{LSB}$, (i.e. 2.5 V for typical $LiNbO_3$ modulators), which is difficult to realize with currently available photonics technology.

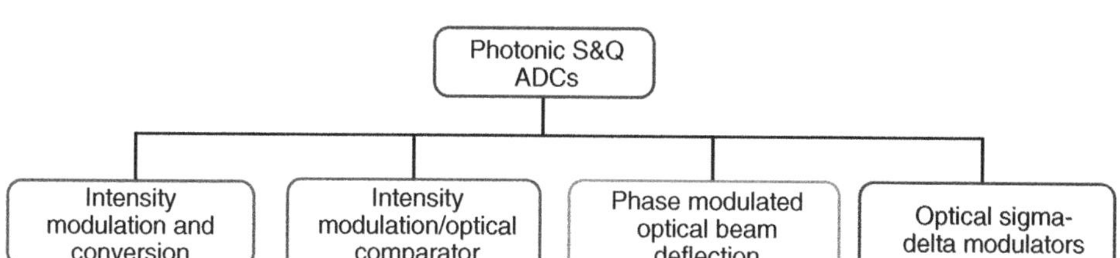

Figure 11.23 Classification of reported photonic-sampled, photonic-quantized ADCs.

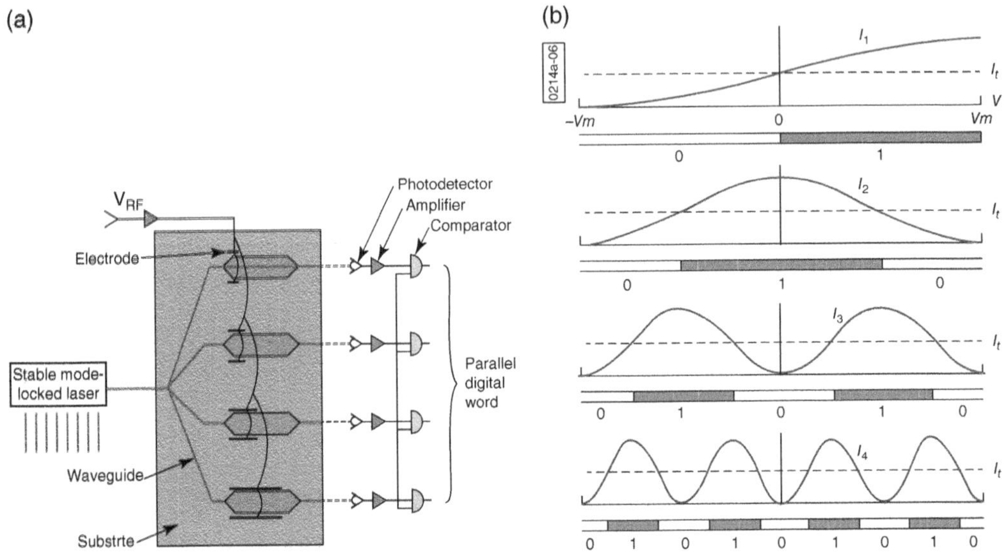

Figure 11.24 (a) Interferometric ADC originally proposed by Taylor [31]. (b) Output power vs. input voltage for each modulator of the interferometric ADC. *Source:* Adapted from Taylor [31].

To avoid using Mach–Zehnder modulators with very low half-wave voltages, a few modified system structures have been proposed including the use of cascaded Mach–Zehnder modulators with identical half-wave voltage [57] and cascaded phase modulators [58]. Recently, a photonic analog-to-digital conversion scheme implemented using an array of Mach–Zehnder modulators with identical half-wave voltages has been demonstrated [59]. The system architecture is identical to the one shown in Fig. 11.24(a), except that the Mach–Zehnder modulators have identical half-wave voltages, as shown in the left part of Fig. 11.25.

The Mach–Zehnder modulators are biased such that the transfer functions of the Mach–Zehnder modulators are laterally shifted, which leads to the generation of a linear binary code to represent the analog input signal. The operation principle is shown in the right part of Fig. 11.25. For an analog-to-digital converter with four channels, the four Mach–Zehnder modulators are biased with their transfer functions shifted laterally with a uniform phase spacing of $\pi/4$, as shown in curve (a). The outputs from the comparators with a threshold level as half of the full scale are shown in (b) while the quantized values of the signal at the output of the 4-channel analog-to-digital converter are shown in (c). The use of the Mach–Zehnder modulators with identical half-wave voltages simplifies greatly the design and implementation but has an important limitation since for an N channel ADC, the code length of this scheme is $2N$, while the code length is 2^N for the schemes based on the proposal by Taylor [31]. Note that this implies that the channel number in this approach does not equal the bit resolution when N is greater than 2. The bit resolution of the approach proposed in [59] is, in fact, given by $\log_2(2N)$. A solution to improve the bit resolution while using small channel numbers is to use multiple comparators for one channel as proposed in [60].

Although several other variations on the original concept proposed by Taylor have been reported [60–67] over the last 30 years, the maximum number of bits obtained with this technique have always been rather modest (*ENOB* < 4). The situation, as pointed out by Valley [4] is not expected to change until ultralow V_π modulators are available in the market.

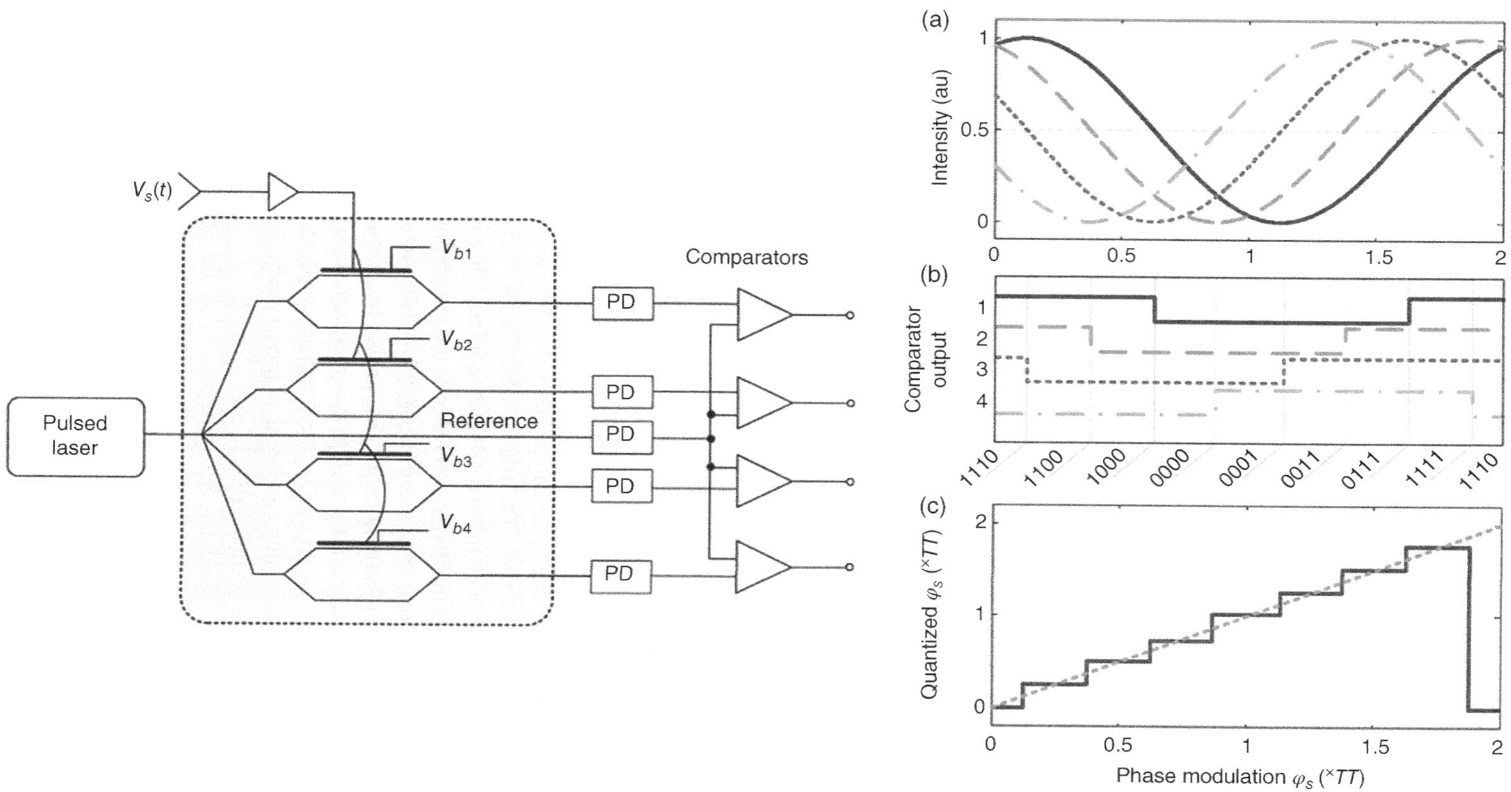

Figure 11.25 Left: An analog-to-digital converter using Mach–Zehnder modulators with identical half-wave voltages. Right: The operation of a four-channel photonic analog-to-digital converter using Mach–Zehnder modulators with identical half-wave voltages. (a) The transfer functions of the four Mach–Zehnder modulators; (b) the linear binary code at the outputs of the comparators; (c) quantized value (solid line) vs. the input phase modulation (dotted line). *Source:* Adapted from Chi and Yao [59].

11.6.3 Intensity Modulation and Optical Comparator

The inability to obtain more than 4 bits using the modulator schemes described in the previous section has pushed research toward more direct-intensity quantization schemes. The advantage of direct quantization is simplicity and potentially lower insertion loss but the disadvantage is that *2N* comparators are needed for *N* bits. Generally speaking, a comparator is a device that compares two input signals and switches its output to on or off (or to 0 or 1 for instance) to indicate which input value is larger. A signal may be digitized to *N* bits with *2N* comparators set at *2N* levels, each one *Q* or the *LSB* larger than the previous level.

Photonics provides a range of processes, devices, and architectures featuring physical responses that emulate that of a comparator. However, in many cases, the fundamental physical processes involved are such that they either do not present sharp transitions, or cannot be tuned to a large number of equally spaced levels, or present operation speeds that are slow compared to their electronics competitors. Referring, for instance, to the chart depicted in Fig. 11.5, it can be seen that to make an ADC that is competitive with electronics, the comparator must be capable of at least 256 levels ($ENOB = 8$ bits) if its response time is longer than about 100 ps, or it must be capable of 16 levels ($ENOB = 4$ bits) with a response time less than 100 ps.

Several groups have attempted to demonstrate optical comparators based on different principles of nonlinear optics. These include, for instance, the use of self electro-optic devices [68], semiconductor saturable absorbers [69–71], nonlinear resonators, optical intensity to phase conversion by means of self- and cross-phase modulation in nonlinear fibers [72–75], four-wave mixing coupled with optical filtering [76], power controlled supercontinuum generation in dispersion flattened [77], and chalcogenide [78] fibers. However, as pointed out in [4], none of them has yet demonstrated the potential to achieve *ENOB* values over 6 or the possibility of becoming commercial.

11.6.4 Phase Modulation and Optical Beam Deflection

In these approaches, photonic quantization is achieved by means of optical phase, frequency, or acousto-optic modulation of the deflection angle of an optical beam or the spatial interference pattern of two optical beams. A first demonstration by Wright et al. [79] was based on an electrically controlled diffraction grating to vary the far-field profile of a beam on an array of photodiodes. By proper setting of the threshold on the electronic comparators that followed the photodiode, they achieved the production of a Gray-scale code. The system can be fed by a short pulse laser source and perform optical sampling first, although in the original demonstration, electronic sampling was employed. A second approach [80] was the first to demonstrate the possibility of voltage-controlled beam deflection to *2N* different positions to quantize the signal. Again, this solution is compatible with an optical sampling stage derived from a mode-locked laser source. Acousto-optic deflection of an optical beam by an angle controlled by an RF voltage was reported as early as in 1991 [81] and improved to feed an array of DOEs. Although an $ENOB = 6$ was achieved, the response time of the acousto-optic modulators limits the conversion bandwidth of these systems to values which are not competitive with those provided by electronic ADCs.

A fast-speed version of the voltage-to-spatial interference pattern quantization scheme, based on phase electro-optic modulators, has been proposed by Stigwal and Galt [82, 83] and is shown in Fig. 11.26. This photonic ADC is essentially a Mach–Zehnder interferometer where optical interference is used to aid the quantization process and optical sampling is used to enable extremely low temporal jitter. Sampling is performed by letting the optical pulses sample the electrical analog input signal as they travel through the electro-optic phase modulator.

After free-space recombination of the modulated arm with the slightly tilted reference arm, the resulting fringe pattern is measured in *N* different positions, each at a slightly different relative phase. It is the output from these measurements that after thresholding constitutes the digital

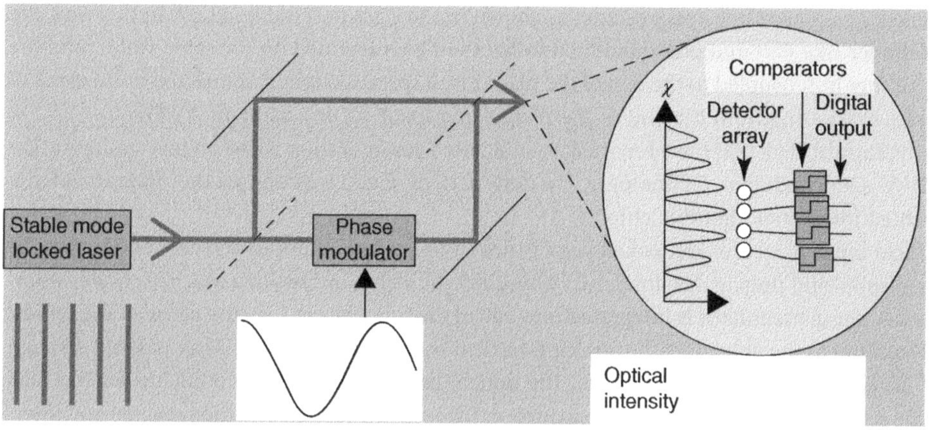

Figure 11.26 Voltage-to-spatial interference pattern quantization scheme, based on a phase electro-optic modulator embedded in a free-space interferometer. *Source:* After Stigwall and Galt [82, 83].

output of the ADC. To illustrate this point and not considering the transverse intensity profile of the beam, the intensity in the fringe pattern varies sinusoidally with the transverse position x as well as with the applied phase modulation φ_m as:

$$I(x, \varphi_m) = a^2 + b^2 + 2ab\cos(k_x x + \varphi_m) \tag{11.20}$$

where a and b are the amplitudes of the two beams, k_x is a constant (phase shift per spatial shift), and $\varphi = \varphi_o(x) + \varphi_m = k_x x + \varphi_m$ is the total phase difference between the two fields in point x. The position of the N detectors should be chosen such that the spatial phase $\Delta\varphi_o = k_x \Delta x$ increases by π/N from one detector to the next. As an example, the output from a three-probe ADC is shown in the left part of Fig. 11.27. Using this architecture, Stigwal and Galt [83] demonstrated an ADC featuring 40 GSa/s with an input frequency of 1.25 GHz and a resolution of 21 digital levels with an *ENOB* of 2.6 and 3.6 for single-sample and averaged measurements, respectively. Furthermore, it was proposed the free-space interference section of the ADC could be replaced by a multimode

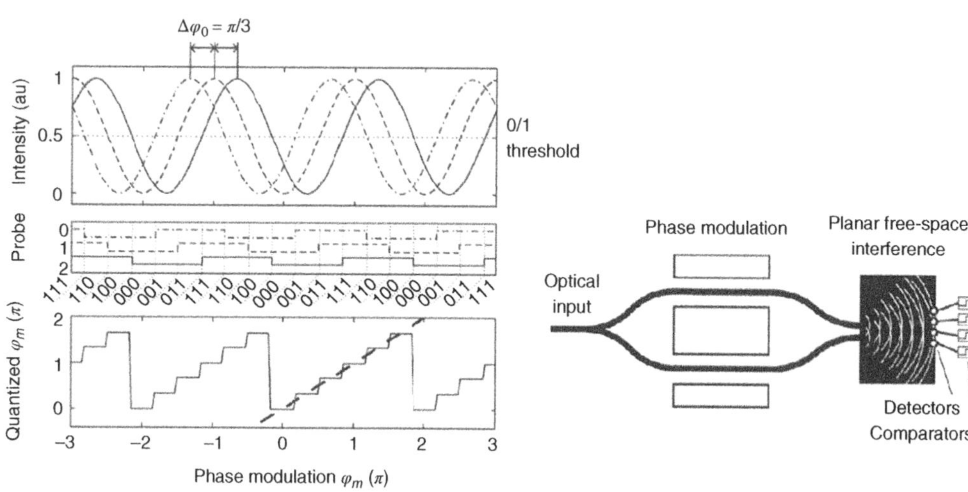

Figure 11.27 Left: Output and encoding from a three-probe ADC based on the configuration of Fig. 11.26. Right: Proposed integrated version, *Source:* Stigwall and Galt [83]/IEEE.

interference, or a planar free-space section as shown in the right part of Fig. 11.27. In this way, the Mach–Zehnder-like part and phase modulator can then be integrated on the same wafer, and the detectors either butt coupled to the end of the planar free-space section or connected by an array of waveguides emanating from the end of the planar free-space section as in an AWG.

Actually, Jarrahi et al. [84] have reported an integrated configuration based on this configuration using III–V semiconductor technology. The left part of Fig. 11.28 shows the diagram of the implemented integrated photonic chip.

An optical input from a mode-locked laser is first coupled into an input waveguide. The optical pulse then splits and propagates down two waveguide branches as part of a Mach–Zehnder interferometer. A phase modulator is integrated into one of the branches to vary the phase of the optical pulses according to the voltage of the analog electrical signal to be digitized. After passing through the half of Mach–Zehnder interferometer, the optical beams from the two branches enter a slab waveguide region, which allows free propagation in the lateral direction and then diverge and interfere. The resulting interference pattern from the two optical pulses forms a spot on an effective image plane, a spot whose position is proportional to the phase difference between the two optical pulses as shown in the right part of Fig. 11.28. In this way, the combination of the phase modulator and the free-propagation region together constitutes an imaging beam-deflection system whose deflection is determined by the phase difference between the two optical pulses entering the free-propagation region, which is set by the modulating electrical signal.

A key point of this configuration is that the integration of photodetector arrays inside the output waveguides at appropriate positions along the image plane enables the measurement of the spatial distribution of optical power. The number of output waveguides $N = 2^b$ determines the total number of resolvable quantization levels. V^j_{bit}, $j = 1, 2 \cdots b$ are resolved by connecting the photodetector arrays in a binary fashion.

The experimental results reported using this scheme have demonstrated an eight-level quantization consuming only 7.2 pJ per quantization with 18-GHz bandwidth, projected to an estimated bandwidth of 30 GHz.

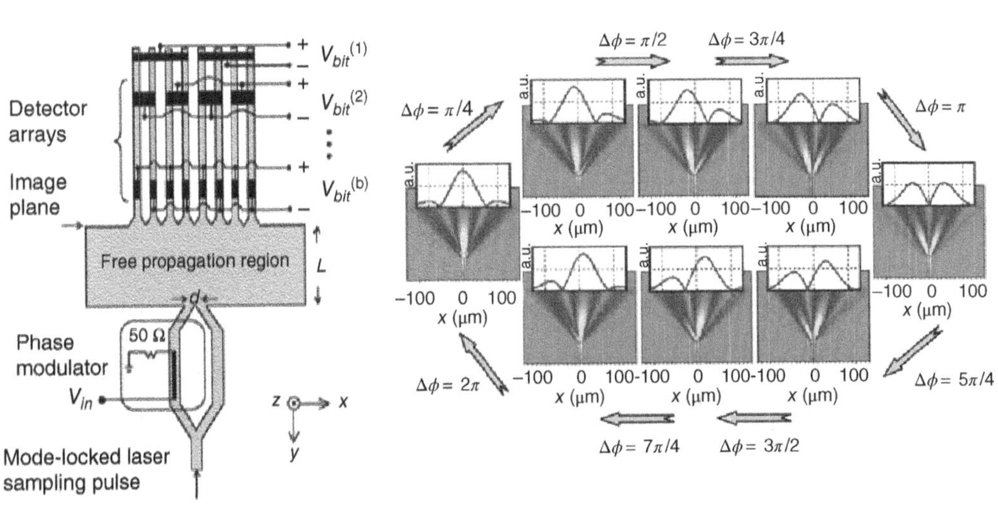

Figure 11.28 Left: Integrated version of the phase modulated spatial interference pattern ADC implemented in III–V semiconductor technology. Right: Spatial interference profile change vs. the interferometer phase shift (phase shift steps $\pi/4$). *Source:* Jarrahi et al. [84]/IEEE.

11.7 Chapter Summary

In this chapter, we have dealt with the photonic implementation of analog-to-digital converters for broadband input RF signals. We have first reviewed the basic generic operation principles and performance parameters of ADCs and the current values of reported state-of-the-art electronic ADCs. Some of the limitations of these devices in terms of bandwidth and resolution bits can be overcome if photonics is employed at different parts of the ADC. We have then reviewed the main approaches reported in the literature over the last 30 years, which have employed photonics either as an auxiliary part, in the sampling or the quantization stages or both at the same time.

So far, as pointed out in the conclusions of the authoritative review by Valley [4], photonics is still not used in any commercial or special-purpose ADC. Furthermore, no complete end-to-end photonic ADC has been reported so far. While it is clear that photonic sampling brings an advantage in terms of jitter noise as compared to electronics, there is no clear technology approach where photonics outperforms electronics in the quantization step. Thus, the evolution of these systems may lead to a hybrid photonic/electronic configuration bringing the best of the two worlds. Photonic sampling can be extended to broadband input signals by time interleaving or by using wavelength-to-time conversion techniques. For this to happen, an extremely careful control of path mismatches, including time delays, filter central wavelengths, and insertion loss calibration will be required. Bulk configurations reported so far can just provide a proof-of-concept but are not adequate for future commercial prototypes. As in many other areas of Microwave Photonics, evolution toward integration on a single chip opens a clear path toward achieving performance metrics currently not available [85]. The recent report of a silicon-integrated optical sampled ADC [8] and deep learning-powered photonics ADC [86] are very promising steps. Furthermore, concentrated low-footprint dispersive delay lines for microwave signals based on photonic crystal waveguide [87] provide a path toward the integration of time-stretched ADCs on a single chip.

Problems

11.1 The spectral noise density seen at an ADC input is composed of various contributions including thermal, shot, $1/f$, and input-referred noises. The thermal and shot components are white while the $1/f$ and input-referred components are frequency dependent. The resulting input-referred noise voltage is obtained by integrating these spectra over the full-Nyquist band $S_r/2$ and is expressed as:

$$\langle v_N^2 \rangle = 4k_B T R_{eff}(S_r/2)$$

where R_{eff} is an effective thermal resistance which lumps together the effects of all noises, k_B the Boltzmann constant, and T the temperature. Using a similar approach to that followed in obtaining the ENOB for an ADC limited by quantization noise derives the value of $ENOB_{th}$ limited by thermal noise given by equation:

$$ENOB_{th} = \log_2\left[\sqrt{\frac{V_{fs}^2}{6k_B T R_{eff} S_L}}\right] - 1$$

11.2 The jitter or aperture uncertainty noise in ADCs happens because the device does not sample the input signal at precisely equal time intervals given by $T = 10/S_r$. Instead, the sampling process is characterized by a mean and a standard deviation with respect to the location in time when sampling occurs. The mean μ_j is the average position of the sampling time and the standard deviation or rms aperture jitter σ_j measures the variation of the variation point. Assuming that we refer the time to the position given by the mean, that is $t' = t - \mu_j$ and that we know the value of σ_j derive an expression for the voltage error rms due to σ_j for the worst case situation, corresponding to a sinusoidal waveform with the highest frequency in the Nyquist band $S_r/2$. To do so, consider that the maximum error will occur when attempting to sample the sinusoid at its zero crossing. Using a similar process to that of problem 1 shows that the $ENOB_{jitter}$ limited by jitter given by equation:

$$ENOB_{jitter} = \log_2\left[\frac{1}{\sqrt{3}\pi S_r \sigma_j}\right]$$

11.3 The ADC performance limitation due to ambiguity is due to the finite speed at which the transistors in the comparators are able to respond to a small voltage difference (0.5 *LSB*). The probability that the *i*th comparator will produce an ambiguous decision as to whether the input signal is above or below the threshold or reference voltage associated with it is given by:

$$P_i = \frac{V_{fs} e^{-t/\tau_{reg}}}{A_o Q}$$

where $t = 0.5$ *clock period* $= 1/(2S_r)$, $\tau_{reg} = 2.5/(\pi f_T)$ is the regeneration time constant, Ao the comparator gain, and Q is the effective LSB voltage due to ambiguity, $Q = V_{fs} 2^{-ENOB_{Amb}}$. Assume that the rms noise power due to quantization and the contribution from all comparators due to ambiguity can be expressed as:

$$\left\langle v_{Q+}^2 \right\rangle = \frac{Q^2}{12}(1 + \alpha) = \frac{Q^2}{12}\left(1 + 2^N P_i\right)$$

where α represents the ambiguity noise expressed as an excess fraction of the quantization noise and 2^N is the number of comparators in the ADC. Express α in terms of the quantity $\Delta N = N - ENOB_{Amb}$ and from it, obtain an expression for $ENOB_{Amb}$. Finally, assuming that $\alpha = 0.25$, $A_o = 2.5$, and $\Delta N = 1.5$, derive Eq. (11.7).

11.4 Compute the P and F parameters of a high-performance 3 dB, Nyquist flash ADC that operates at a sampling rate of 8 Gb/s and consumes 3.5 W.

11.5 The ultimate limit to the value of the ADC resolution sampling rate product P is given by the time-energy Heisenberg uncertainty, which is shown in Fig. 11.4. This uncertainty is written as:

$$\Delta E \Delta t = \frac{h}{2\pi}$$

where ΔE is the energy of the smallest resolvable signal, which is equivalent to half of the LSB (i.e. 0.5 *LSB*), Δt is half of the sampling period (i.e. $T/2$), and h is the Planck constant. Using the Heisenberg uncertainty, derive an upper limit to the value of the ADC resolution sampling rate product P.

11.6 Using the result derived in problem 5 and typical values $R = 50\ \Omega$, $V_{fs} = 1\ \text{V}$, calculate the upper limit to the value of the ADC resolution sampling rate product P and compare to those obtained for the high-performance ADC of problem 4.

11.7 We want to implement a time-stretched ADC pre-processor for 40 Gb/s that will employ 4 Gb/s electronic ADCs. To do so, we can employ standard singlemode fiber featuring a dispersion parameter $D = 17$ps/km.mn and a broadband optical source with bandwidth $\Delta f_o = 15$ nm. Design the system so it meets the requirements of multiplication factor, RF bandwidth and time aperture. Compute the product of time aperture and RF bandwidth for the designed pre-processor.

References

1 Walden, R. (1999). Analog-to-digital converter survey and analysis. *IEEE J. Sel. Areas Commun.* 17: 539–550.

2 Le, B., Rondeau, T.W., Reed, J.H., and Bostian, C.W. (2005). Analog-to digital converters. *IEEE Signal Process. Mag.* 22: 69–77.

3 Walden, R. (2008). Analog-to-digital conversion in the early twenty-first century. In: *Wiley Encyclopedia of Computer Science and Engineering*, 126–138. Wiley.

4 Valley, G. (2007). Photonic analog-to-digital- converters. *Opt. Express* 15: 1955–1982.

5 Juodawlkis, P.W., Twichell, J.C., Betts, G.E. et al. (2001). Optically sampled analog-to-digital converters. *IEEE Trans. Microw. Theory Tech.* 49 (10): 1840–1853.

6 Han, L. and Wu, K. (2010). Radar and radio data fusion platform for future intelligent transportation system. *The 7th European Radar Conference*, Paris, France, pp. 65–68.

7 Han, Y. and Jalali, B. (2003). Photonic time-stretched analog-to-digital-converter: Fundamental concepts and practical considerations. *J. Lightw. Technol.* 21: 3081–3103.

8 Khilo, A., Spector, S.J., Grein, M.E. et al. (2012). Photonic ADC: overcoming the bottleneck of electronic jitter. *Opt. Express* 20: 4454–4469.

9 Low, A.J. and Carroll, J.E. (1978). 10ps optoelectronic sampling system. *Solid-State Electron Dev.* 2: 185–190.

10 Leonberger, F.J. and Moulton, P. (1979). High-speed InP optoelectronic switch. *Appl. Phys. Lett.* 35: 712–714.

11 Cox, C.H. III, Diadiuk, V., Yao, I. et al. (1983). InP optoelectronic switches and their high-speed signal-processing applications. *Proc. SPIE* 439: 164–168.

12 Leonberger, F.J. and Diadiuk, V. (1983). High-speed InP-based photodetectors. *1983 International Electron Devices Meeting*, vol. 29, pp. 460–463.

13 Jacobs, E.W., Sobti, J.B., Vella, V.F. et al. (2004). Optically clocked track-and-hold for high-speed high-resolution analog-to-digital conversion. *2004 IEEE International Topical Meeting on Microwave Photonics (IEEE Cat. No.04EX859)*, Ogunquit, ME, USA, pp. 190–192. http://doi.org/10.1109/MWP.2004.1396871.

14 Nathawad, L.Y., Urata, R., Wooley, B.A., and Miller, D.A.B. (2003). A 40-GHz-bandwidth, 4-bit, time-interleaved A/D converter using photoconductive sampling. *IEEE J. Solid-State Circ.* 38: 2021–2030.

15 Ioakeimidi, K., Leheny, R.F., Gradinaru, S. et al. (2005). Photoelectronic analog-to-digital conversion: sampling and quantizing at 100 Gs/s. *IEEE Trans. Microw. Theory Tech.* 53: 336–342.

16 Johnstone, A., Lewis, M.F., and Hares, J.D. (1998). Optical replication technique for wideband transient waveform digitisation. *Proc. SPIE* 3285: 209–216.

17 Zmuda, H. and Hanna, S. (2004). High-speed, high resolution optically assisted analog-to-digital conversion. *Proc. SPIE* 5435: 153–163.

18 Zmuda, H., Fanto, M., McEwen, T. et al. A photonic recirculating delay line for analog-to-digital conversion and other applications. *Proc. SPIE 6975, Enabling Photonics Technologies for Defense,*

Security, and Aerospace Applications IV, 69750F (24 March 2008). http://doi.org.10.1117/12.783962; http://doi.org.10.1117/12.783962.

19 Coppinger, F., Bhushan, A.S., and Jalali, B. (1998). Time magnification of electrical signals using chirped optical pulses. *Electron. Lett.* 34: 399–400.

20 Jalali, B., Coppinger, F., and Bhushan, A.S. (1998). Photonic time-stretch offers solution to ultrafast analog-to-digital conversion. *Opt. Photon. News* 9 (12): 31–32.

21 Bhushan, A.S., Coppinger, F., Jalali, B. et al. (1998). 150 Gsample/s wavelength division sampler with time-stretched output. *Electron. Lett.* 34: 474–475.

22 Bhushan, A.S., Kelkar, P., and Jalali, B. (2000). 30 Gsample/s time-stretch analogue-to-digital converter. *Electron. Lett.* 36: 1526–1527.

23 Bhushan, A.S., Kelkar, P.V., Jalali, B. et al. (2002). 130-GSa/s photonic analog-to-digital converter with time stretch preprocessor. *IEEE Photon. Technol. Lett.* 14: 684–686.

24 Coppinger, F., Bhushan, A.S., and Jalali, B. (1999). Photonic time stretch and its application to analog-to-digital conversion. *IEEE Trans. Microw. Theory Tech.* 47: 1309–1314.

25 Coppinger, F., Bhushan, A.S., and Jalali, B. (2000). 12 Gsample/s wavelength division sampling analog-to-digital converter. *Electron. Lett.* 36: 316–318.

26 Han, Y. and Jalali, B. (2003). Time-bandwidth product of the photonic time-stretched analog-to-digital converter. *IEEE Trans. Microw. Theory Tech.* 51: 1886–2003.

27 Chou, J., Boyaz, O., and Jalali, B. (2007). Femtosecond real-time single-shot digitzer. *Appl. Phys. Lett.* 91: 161105.

28 Siegman, A.E. and Kuizenga, D.J. (1970). Proposed method for measuring picosecond pulse widths and pulse shapes in CW mode-locked lasers. *IEEE J. Quantum Electron.* QE-6: 212–215.

29 Auston, D.H. (1975). Picosecond optoelectronic switching and gating in silicon. *Appl. Phys. Lett.* 26: 101–103.

30 Haus, H.A. and Mecozzi, A. (1993). Noise of mode-locked lasers. *IEEE J. Quantum Electron.* 29: 983–995.

31 Taylor, H.F. (1975). An electrooptic analog-to-digital converter. *Proc. IEEE* 63: 1524–1525.

32 Taylor, H.F., Taylor, M.J., and Bauer, P.W. (1978). Electro-optic analog-to-digital conversion using channel waveguide modulators. *Appl. Phys. Lett.* 32: 559–561.

33 Taylor, H.F. (1979). An optical analog-to-digital converter—design and analysis. *IEEE J. Quantum Electron.* QE-15: 210–216.

34 Becker, R.A., Woodward, C.E., Leonberger, F.J., and Williamson, R.C. (1984). Wide-band electrooptic guided-wave analog-to-digital converters. *Proc. IEEE* 72: 802–818.

35 Bell, J.A., Hamilton, M.C., and Leep, D.A. (1991). Optical sampling and demultiplexing applied to A/D conversion. *Dev. Opt. Process.* 1562: 276–280.

36 Bell, J.A., Hamilton, M.C., Leep, D.A. et al. (1991). A/D conversion of microwave signals using a hybrid optical/electronic technique. *Opt. Technol. Microw. Appl. V* 1476: 326–329.

37 Twichell, J.C., Wasserman, J.L., Juodawlkis, P.W. et al. (2001). High-linearity 208-MS/s photonic analog-to-digital converter using 1-to-4 optical time-division demultiplexers. *IEEE Photon. Technol. Lett.* 13: 714–716.

38 Williamson, R.C., Juodawlkis, P.W., Wasserman, J.L. et al. (2001). Effects of crosstalk in demultiplexers for photonic analog-to-digital converters. *J. Lightw. Technol.* 19: 230–236.

39 Frankel, M.Y., Kang, J.U., and Esman, R.D. (1997). High-performance photonic analogue–digital converter. *Electron. Lett.* 33: 2096–2097.

40 Kang, J.U., Frankel, M.Y., and Esman, R.D. (1998). Highly parallel pulsed optoelectronic analog–digital converter. *IEEE Photon. Technol. Lett.* 10: 1626–1628.

41 Bhushan, A.S., Coppinger, F., Yegnanarayanan, S., and Jalali, B. (1999). Non-dispersive wavelength-division sampling. *Opt. Lett.* 24: 738–740.

42 Yariv, A. and Koumans, R.G.M.P. (1998). Time interleaved optical sampling for ultra-high speed A/D conversion. *Electron. Lett.* 34: 2012–1023.

43 Kang, J.U. and Esman, R.D. (1999). Demonstration of time-interweaved photonic four-channel WDM sampler for hybrid analogue–digital converter. *Electron. Lett.* 35: 60–61.

44 Clark, T.R., Kang, J.U., and Esman, R.D. (1999). Performance of a time and wavelength-interleaved photonic sampler for analog–digital conversion. *IEEE Photon. Technol. Lett.* 11: 1168–1170.

45 Davenport, M.L., Liu, S., and Bowers, J.E. (2018). Integrated heterogeneous silicon/III–V mode-locked lasers. *Photonics Res.* 6: 468–478.

46 Nishi, H., Tsuchizawa, T., Kou, R. et al. (2012). Monolithic integration of a silica AWG and Ge photodiodes on Si photonic platform for one-chip WDM receiver. *Opt. Express* 20: 9312–9321.

47 Zmuda, H. (2001). Analog-to-digital conversion using high-speed photonic processing. *Proc. SPIE* 4490: 84–95.

48 Zmuda, H., Toughlian, E.N., Li, G., and Likamwa, P. (2001). A photonic wideband analog-to-digital converter. *2001 IEEE Aerospace Conference Proceedings (Cat. No.01TH8542)*, Big Sky, MT, USA, pp. 3/1461–3/1472, vol. 3. http://doi.org/10.1109/AERO.2001.931376.

49 Zmuda, H., Hayduk, M.J., Bussjager, R.J., and Toughlian, E.N. (2002). Wavelength-based analog-to-digital conversion. *Proc. SPIE* 4547: 134–145.

50 Zmuda, H., Toughlian, E.N., Li, G., and LiKamWa, P. (2001). A photonic wideband analog-to-digital converter. *2001 IEEE Aerospace Conference Proceedings (Cat. No.01TH8542)*, Big Sky, MT, USA, vol. 3, pp. 3/1461–3/1472. http://doi.org/10.1109/AERO.2001.931376.

51 Stigwall, J., Galt, S., and Hård, S. (2004). Experimental evaluation of an ultrafast free space optical analog-to-digital conversion scheme using a tunable semiconductor laser. *Proc. SPIE* 5466: 123–130.

52 Johansson, M., Lofving, B., Hard, S. et al. (2000). Study of an ultrafast analog-to-digital conversion scheme based on diffractive optics. *Appl. Opt.* 39 (17): 2881–2887.

53 Galt, S., Magnusson, A., and Hård, S. (2003). Dynamic demonstration of diffractive optic analog-to-digital converter scheme. *Appl. Opt.* 42 (2): 264–270.

54 Stigwall, J. (2004). Photonic analog-to-digital conversion. *Thesis for the Degree of Licentiate of Engineering*. Goteborg, Sweden: Chalmers University of Technology.

55 Jalali, B. and Xie, Y.M. (1995). Optical folding-flash analog-to-digital converter with analog encoding. *Opt. Lett.* 20 (18): 1901–1903.

56 Currie, M. (2005). Optical quantization of microwave signals via distributed phase modulation. *IEEE J. Lightw. Technol.* 23 (2): 827–833.

57 Chi, H. and Yao, J.P. (2008). A photonic analog-to-digital conversion scheme using Mach–Zehnder modulators with identical half-wave voltages. *Opt. Express* 16 (2): 567–572.

58 Pace, P.E. and Styer, D. (1994). High-resolution encoding process for an integrated optical analog-to-digital converter. *Opt. Eng.* 33 (8): 2638–2645.

60 Takizawa, K. and Okada, M. (1979). Analog-to-digital converter: a new type using an electrooptic light modulator. *Appl. Opt.* 18: 3148–3151.

62 Becker, R.A. and Leonberger, F.J. (1982). 2-bit 1 Gsample/s electrooptic guided-wave analog-to-digital converter. *IEEE J. Quantum Electron.* 18: 1411–1413.

63 Leonberger, F.J., Woodward, C.E., and Becker, R.A. (1982). 4-Bit 828-megasample/s electro-optic guided-wave analog-to-digital converter. *Appl. Phys. Lett.* 40: 565–568.

64 Dokhikyan, R.G., Zolotov, E.M., Karinskii, S.S. et al. (1982). Prototype of an integrated-optics four-digit analog-digital converter. *Sov. J. Quantum Electron.* 12: 806–807.

65 King, G.D.H. and Cebulski, R. (1982). Analogue-to-digital conversion using integrated electro-optic interferometers. *Electron. Lett.* 18: 1099–1100.

66 Chang, C.L. and Tsai, C.S. (1983). Electro-optic analog-to-digital converter using channel waveguide Fabry–Perot modulator array. *Appl. Phys. Lett.* 43: 22–24.

67 Becker, R.A., Woodward, C.E., Leonberger, F.J., and Williamson, R.C. (1984). Wideband electrooptic guidedwave analog-to-digital converters. *Proc. IEEE* 72: 802–819.

68 Loh, L.M. and LoCicero, J.L. (1996). Subnanosecond sampling all-optical analog-to-digital converter using symmetric self-electro-optic effect devices. *Opt. Eng.* 35: 457–466.

69 Hayduk, M.J., Bussjager, R.J., and Getbehead, M.A. (2000). Photonic analog to digital conversion techniques using semiconductor saturable absorbers. *Proc. SPIE* 4042: 54–60.

70 Hayduk, M.J., Bussjager, R.J., Getbehead, M.A., and Louthain, J.A. (2000). Recent advancements in photonic converters. *Proc. SPIE* 4112: 28–37.

71 Hayduk, M.J., Bussjager, R.J., Johns, S.T. et al. (2002). Contrast ratio enhancement in a saturable absorber based photonic analog to digital converter. *Proc. SPIE* 4732: 46–52.

72 Sakata, H. (2001). Photonic analog-to-digital conversion by use of nonlinear Fabry–Perot resonators. *Appl. Opt.* 40: 240–248.

73 Jeong, J.-M. and Marhic, M.E. (1992). All-optical analog-to-digital and digital-to-analog conversion implemented by a nonlinear fiber interferometer. *Opt. Commun.* 91: 115–122.

74 Ho, P.P., Wang, Q.Z., Chen, J. et al. (1997). Ultrafast optical pulse digitization with unary spectrally encoded cross-phase modulation. *Appl. Opt.* 36: 3425–3429.

75 Konishi, T., Tanimura, K., Asano, K. et al. (2002). All-optical analog-to-digital converter by use of self-frequency shifting in fiber and a pulse-shaping technique. *J. Opt. Soc. Am. B* 19: 2817–2823.

76 Oda, S., Maruta, A., and Kitayama, K. (2004). All-optical quantization scheme based on fiber nonlinearity. *IEEE Photon. Technol. Lett.* 16: 587–589.

77 Oda, S. and Maruta, A. (2005). A novel quantization scheme by slicing supercontinuum spectrum for all-optical analog-to-digital conversion. *IEEE Photon. Technol. Lett.* 17: 465–467.

78 Panta, R., Xionga, C., Maddenb, S. et al. (2010). Investigation of all-optical analog-to-digital quantization using a chalcogenide waveguide: a step towards on-chip analog-to-digital conversion. *Opt. Commun.* 283: 2258–2262.

79 Wright, S., Mason, I.M., and Wilson, M.G.F. (1974). High-speed electro-optic analogue-digital conversion. *Electron. Lett.* 10: 508–509.

80 Tsunoda, Y. and Goodman, J.W. (1977). Combined optical AD conversion and page composition for holographic memory applications. *Appl. Opt.* 16: 2607–2609.

81 Li, Y. and Zhang, Y. (1991). Optical analog-to-digital conversion using acousto-optic theta modulation and table lookup. *Appl. Opt.* 30: 4368–4371.

82 Stigwall, J. and Galt, S. (2005). Interferometric analog-to-digital conversion scheme. *IEEE Photon. Technol. Lett.* 17: 468–470.

83 Stigwall, J. and Galt, S. (2006). Demonstration and analysis of a 40-Gigasample/s interferometric analog-to-digital converter. *IEEE J. Lightw. Technol.* 24: 1247–1256.

84 Jarrahi, M., Pease, R.F.W., Miller, D.A.B., and Lee, T.H. (2008). Optical spatial quantization for higher performance analog-to-digital conversion. *IEEE Trans. Microw. Theory Tech* 56: 2143–2150.

85 Marpaung, D., Roeloffzen, C., Heideman, R. et al. (2013). Integrated microwave photonics. *Laser Photonics Rev. In press.*

86 Xu, S., Zou, X., Ma, B. et al. (2019). Deep-learning-powered photonic analog-to-digital conversion. *Light Sci. Appl.* 8: 66.

87 Sancho, J., Bourderionnet, J., Lloret, J. et al. (2012). Integrable microwave filter based on a photonic crystal delay line. *Nat. Commun.* 3: 1075 https://doi.org/10.1038/ncomms2092.

12

Novel Optoelectronic Oscillators

12.1 Introduction

An optoelectronic oscillator (OEO) is a paradigmatic microwave photonic oscillator that produces microwave signals with ultra-low phase noise, thanks to the high-quality factor of the OEO cavity that is achieved with the help of optical energy storage elements, such as low-loss optical fiber or a high Q-factor optical resonator [1–8]. Early versions of OEOs can be dated back to 1968 in a study of mode-locked lasers [9], where optoelectronic feedback loops were used to achieve stable mode-locking. Ten years later, the concept of an electro-optical oscillator was proposed by Kersten [10], which was essentially an OEO. The feedback paths were relatively short in the early versions of OEOs. A conceptual breakthrough of the OEO was achieved by Yao and Maleki [1–4] by introducing a long low-loss fiber delay line as the energy storage element, which provided the most attractive high-quality-factor feature of modern OEOs. The authors analyzed in detail the operation of the long fiber-based oscillator and also introduced the acronym "OEO."

One of the key motivations of studies of OEOs is the ultra-low phase noise resulting from the high-quality factor of the OEO cavity, which is achieved by using high-quality-factor optical energy storage elements, such as low-loss optical fiber or a high-quality-factor optical resonator. High-frequency microwave signals can also be easily generated since the bandwidth of an OEO is only limited by the bandwidths of the optoelectronic devices in its cavity, which are as large as 100 GHz. These features are very attractive compared with their electrical counterparts since the features may not be achievable in the electrical domain based on traditional microwave resonator oscillators or systems that are costly and complicated.

In the past few decades, there has been an increasing effort to explore OEOs in various fields. For example, different approaches have been reported to generate single-frequency microwave signals with ultra-low phase noise using a narrowband electrical or microwave photonic filter (MPF), which have been discussed in Chapter 3. Several novel mode control and selection methods have also been proposed and demonstrated in recent years to generate ultra-low phase noise single-frequency microwave signals and other complex microwave waveforms. For instance, parity-time (PT) symmetric OEOs [11–21] have been proposed and demonstrated to produce single-frequency microwave signals based on the principle of PT symmetry, where single-frequency oscillation can be obtained without the use of narrowband filters. The generation of complex microwave signals, such as chirped microwave waveforms using a novel Fourier domain mode-locked (FDML) OEO [22–38], phase-locked stable single-mode and multi-mode signals based on an optoelectronic parametric oscillator (OEPO) [39], and random signals from a broadband random OEO [40] have

Microwave Photonics, First Edition. Jianping Yao, José Capmany, and Ming Li.

Companion website: www.wiley.com/go/yao/microwavephotonics

also been demonstrated by properly controlling the cavity modes of OEOs. In addition, several novel integrated OEOs [41–48] with compact size and low power consumption have been fabricated using Indium phosphide (InP), silicon, and other platforms in recent years, which are key steps toward a new generation of compact microwave sources for demanding applications.

In this chapter, recent advances in novel OEOs will be discussed. We first focus on a model for analyzing the nonlinear dynamic behavior of OEOs, then we introduce single-frequency microwave signal generation using a PT-symmetric OEO, and then we discuss novel OEOs that are capable of producing complex microwave waveforms, including novel FDML OEO, OEPO, and broadband random OEO. In these OEOs, novel mode control methods have been adopted to produce different microwave waveforms, such as chirped signal generation using Fourier domain mode-locking, stable multi-mode oscillation based on a nonlinear parametric process, and random oscillation based on random scattering. Finally, we discuss recent advances in integrated OEOs, which have been fabricated in InP, silicon, and other platforms with compact size and low power consumption.

12.2 Models for Optoelectronic Oscillators

OEOs with long optical fiber delay lines were first reported by Yao and Maleki in 1996 [2, 3], which is a closed feedback loop that is formed with two concatenated optical and electronic branches [8], and has been widely employed to generate ultrapure high-frequency microwaves [3, 49–52].

Figure 12.1(a) shows a fundamental OEO single-loop structure, a narrow-band microwave filter is employed to reject the standard nonlinear effects in the loop [53], which allows OEO to produce

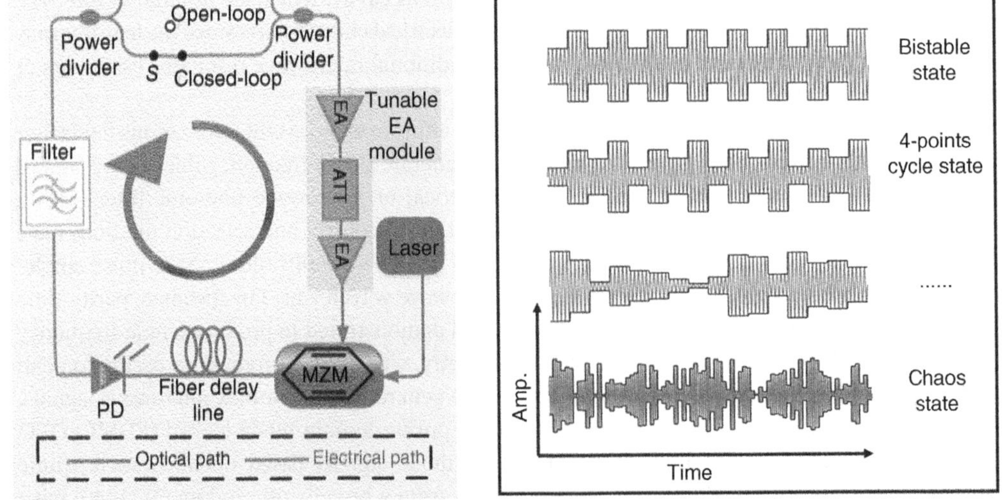

Figure 12.1 (a) Structure of a singlemode OEO and (b) the different output states. *Source:* Xiao et al. [55]/Optical Society of America/CC BY 4.0.

an ultrapure microwave with stable amplitude as shown in the stable state of Fig. 12.1(b). However, even for this simplest OEO structure, the stability is not always satisfied. As shown in Fig. 12.1(b), the output states of the OEO can be bistable, 4-point cycle, etc., and even chaos. What caused the different states in the same OEO loop? This question needs to be resolved not only because the answer is highly important for OEO and other OEO-based applications such as senses and measurement systems [8, 54] and secure chaotic communications [55–57] but also because the nonlinear dynamic behavior in loop itself is interesting and meaningful.

There are two popular theoretical descriptions of the OEO. One is the quasi-linear model which is proposed by Yao and has its roots in linear model (a model for analyzing linear feedback system [58, 59]). This model considers the final output of a feedback system as the summation of all recurrence circulating fields and indicates that the closed-loop gain decreases from small signal gain to unity with the increasing of the oscillating amplitude [2, 3], which ensures stability of the OEO. In this model, the output signal of electrical amplifier can be expressed as

$$V_{out}(t) = V_{ph}\left\{1 - \eta \sin \pi \left[\frac{V_{in}(t)}{V_\pi} + \frac{V_B}{V_\pi}\right]\right\}, \tag{12.1}$$

where $V_{in}(t)$ is the input electrical signal of the EOM. V_π and V_B are the half-wave voltage and bias voltage of the EOM, respectively. η determines the extinction ratio of the modulator $ER = (1+\eta)/(1-\eta)$. $V_{ph} = I_{ph}RG$ is the photovoltage at the output of EA, where $I_{ph} = \rho\alpha P_0/2$ is the photocurrent at the PD, R is the load impedance of the PD, G is the gain of the EA, ρ is the responsibility of the PD, α is the insertion loss of the EOM, and P_0 is the input optical power.

Clearly, Eq. (12.1) describes the transmission of signals in the OEO loop from the input of the EOM to the output of the EA. According to Eq. (12.1), the open-loop gain of the OEO can be expressed as

$$G_S = \left.\frac{dV_{out}}{dV_{in}}\right|_{V_{in}=0} = -\frac{\eta\pi V_{ph}}{V_\pi}\cos\left(\frac{\pi V_B}{V_\pi}\right) \tag{12.2}$$

The magnitude of the open-loop gain must be larger than 1 for a self-sustained oscillation.

Equation (12.1) can be further linearized by assuming the electrical input signal $V_{in}(t)$ of the EOM is a sinusoidal wave and bandwidth of the electrical filter is narrow enough to block all the harmonic components generated by the PD. In this case, the electrical input signal $V_{in}(t)$ can be expressed as $V_{in}(t) = V_0\sin(\omega t + \beta)$, where V_0, ω, and β are the amplitude, angular frequency, and initial phase of the input electrical signal, respectively. The linearized output of the electrical filter after one round-trip of the OEO loop can be written as

$$V_{out}(t) = \mathrm{G}(V_0)V_{in}(t) \tag{12.3}$$

where $G(V_0)$ is the voltage-gain coefficient defined as

$$G(V_0) = G_S\frac{2V_\pi}{\pi V_0}J_1\left(\frac{\pi V_0}{V_\pi}\right) \tag{12.4}$$

where J_1 denotes the first-order Bessel function. As can be seen from Eqs. (12.2) and (12.4), $G(V_0)$ is also a function of frequency ω since V_{ph} and V_π are all frequency dependent. A unitless complex filter function $\tilde{F}(\omega) = F(\omega)\exp[i\phi(\omega)]$ can be introduced to account for the effects of all the frequency-dependent components in the OEO loop in order to extract a frequency-independent factor $G(V_0)$, where $F(\omega)$ is the real normalized transmission function and $\phi(\omega)$ is the frequency-dependent phase. By doing so, Eq. (12.3) can be rewritten as

$$\tilde{V}_n(\omega,t) = \tilde{F}(\omega)G(V_0)\tilde{V}_{n-1}(\omega,t-\tau') \tag{12.5}$$

where $\tilde{V}_n(\omega,t)$ is the complex voltage in the OEO loop after n times of circulation and τ' is the time delay introduced by the physical length of the feedback loop. The initial complex voltage $\tilde{V}_{n=0}(\omega,t) = \tilde{V}_{in}(\omega,t) = \sqrt{2R\rho_N\Delta f}$ is the transient noise, where ρ_N is the power density of the input noise and Δf is the frequency bandwidth.

Once the OEO loop is started from noise, the amplitude of the oscillation is increased until the gain is equal to unity, which would result in a stable oscillation. The total field can be expressed as the summation of all circulating fields. The stable output power can be rewritten as

$$P(\omega) = \frac{\left(G_A^2|\tilde{V}(\omega)|^2\right)/2R}{1+|F(\omega)G(V_0)|^2 - 2F(\omega)|G(V_0)|\cos[\omega\tau' + \phi(\omega) + \phi_0]} \tag{12.6}$$

where $\phi_0 = 0$ when the modulator is positively biased and $\phi_0 = \pi$ when it is negatively biased. As can be seen from Eq. (12.6), the potential oscillation modes of OEO are a serial of frequency-periodic modes whose frequencies are determined by

$$\omega\tau' + \phi(\omega) + \phi_0 = 2k\pi, \quad k = 0, 1, 2... \tag{12.7}$$

The oscillating mode of the OEO is selected by the narrowband electrical filter and only one or several modes are selected. Assuming only one mode k is allowed to oscillate, according to Eq. (12.7), the oscillating frequency and free spectral range (FSR) of OEO can be expressed as

$$f_{osc} = \left(k - \frac{\varphi_0}{2\pi}\right)/\tau \tag{12.8}$$

$$FSR = \frac{1}{\tau} \tag{12.9}$$

where $\tau = \tau' + \left.\frac{d\phi(\omega)}{d\omega}\right|_{\omega=\omega_{osc}}$ is the total cavity delay of the OEO loop.

In addition to the oscillation modes, the phase noise performance is another important figure of merit of an oscillator. The single-sideband phase noise is just the power spectral density of an oscillator in most cases where the amplitude fluctuation is much smaller than the phase fluctuation. The power spectral density of the OEO can be calculated as

$$S_{RF}(f') = \frac{\delta}{(\delta/2\tau)^2 + (2\pi\tau f')^2}, \tag{12.10}$$

where δ is defined as $\delta = \left(\rho_N G_A^2\right)/(P_{osc})$, which is the input noise-to-signal ratio of the OEO. P_{osc}/G_A^2 is the total oscillating power and $f' = (\omega - \omega_{osc})/2\pi$ is the frequency offset from the oscillation peak f_{osc}. As can be seen from Eq. (12.10), the spectral density of the oscillation mode is a Lorentz function, whose full width at half-maximum (FWHM) is $\Delta f_{FWHM} = \delta/2\pi\tau^2$. Thus, the Q-factor of the OEO when the modulator is positively biased can be expressed as

$$Q = \frac{f_{osc}}{\Delta f_{FWHM}} = Q_D\frac{\tau}{\delta} \tag{12.11}$$

where $Q_D = 2\pi f_{osc}\tau$ is the Q-factor of the loop delay line.

Generally, quasi-linear model is good enough for OEOs under stable state. However, it does not apply to analyze unstable states [53, 60]. Another popular OEO model is the delay-differential model proposed by Y. Chembo. The delay-differential model considers the states of the OEO as a function of the open-loop gain and has successfully predicted that the bifurcation phenomena show up when the open-loop small-signal gain is larger than a critical threshold value of 2.31. However, the provided bifurcation diagrams and the threshold value of 2.31 apply to the class of OEOs considering only nonlinear effects of modulator. For a practical OEO structure, the nonlinear properties can be complex due to the saturation characteristics of employed components in addition to modulator such as RF electronic amplifiers (EAs), optical amplifiers (OAs), and photodetectors (PDs). The OEOs with complex nonlinear properties can maintain a stable state even the small-signal gain beyond the threshold value of 2.31. Furthermore, the delay-differential model only considers the stability as a function of small-signal gain. For the class of OEOs with an externally injected microwave [61–64], which is a popular architecture for improving oscillation frequency stability [41, 65–67], the injected amplitude is another variable and deserves to be discussed.

In [55], we proposed a microwave-photonics iterative nonlinear gain (MING) model for analyzing different states of OEOs. After connecting the oscillating process with trajectories of an introduced graphical technique referred to as iterated map, the envelope dynamics can be described. Ming model can be applied to a practical OEO with complex nonlinear effects, and it can also be applied to the class of OEO with an external injected microwave. Figure 12.1 shows the structure under study, which is a single-loop OEO with a switched point S and an injected port. The switched point S allows OEO to be switched between open-loop mode and closed-loop mode, and an external microwave signal can be injected into the OEO loop through the input/injected port. To make the analysis clear, the external microwave signal is named the input signal and the injected signal corresponds to the open-loop and closed-loop, respectively. The employed Mach–Zehnder modulator (MZM) is biased at the positive quadrature and a filter is used to remove the harmonic components.

When point S is disconnected, OEO is in open-loop mode which is a simple two-port optoelectronic link, and the amplitude nonlinear property is clear. Assuming the total time delay corresponding to the physical length of the loop is τ, and the frequency-dependent phase caused by the dispersive component is $\phi(\omega)$. Considering that an input signal with amplitude V is constantly input into the loop, and assuming that input frequency meets the oscillation condition of $\omega_0\tau + \phi(\omega_0) = 2m\pi$, $m = 1, 2, 3, \ldots$ [1], the output amplitude is a function of V and can be given by [3, 68]:

$$\begin{aligned} V_{out} &= G_{loss} 2J_1[G_A f_{EA}(V)(\pi/V_\pi)] \\ &= F_{open}(V), \end{aligned} \tag{12.12}$$

where $G_A f_{EA}(V)$ represents the nonlinear amplified output amplitude of the employed tunable EA module [68]. G_{loss} represents the total voltage loss of the OEO, and J_1 is the first order Bessel function of the first kind. Equation (12.12) shows that the output amplitude of open-loop OEO is a nonlinear function of input amplitude, and this function is named input-to-output amplitude mapping (IOAM) relation in this paper.

Next, we discuss the closed-loop OEO by connecting the switch point S while an injected signal with amplitude of V_{in} and frequency of ω_0 is constantly injected into the loop. Differ from the open-loop, the injected signal is circulated in the closed loop. Thus, the output amplitude envelope is no longer a constant and is separated by the interval of total time delay of OEO τ, which can be given by:

$$\begin{aligned} V_{OSC}(t) &= G_{loss} 2J_1\left\{ G_A f_{EA}[V_{OSC}(t-\tau) + V_{in}]\frac{\pi}{V_\pi}\right\} \\ &= F_{open}[V_{OSC}(t-\tau) + V_{in}]. \end{aligned} \tag{12.13}$$

Equation (12.2) is a typical time-delay nonlinear iterated equation [69, 70], and can be further simplified. Considering amplitude $V_{n-1} = V_{OSC}(t-\tau) + V_{in}$ and $V_n = V_{OSC}(t) + V_{in}$, we have that:

$$V_n - V_{in} = F_{open}(V_{n-1}) \tag{12.14}$$

where n indicates the number of oscillating cycles. Equation (12.3) is a typical nonlinear iterated mapping equation similar to the logistic map and can be analyzed by a graphical method [70–72].

As shown in Fig. 12.2(a), the dotted curve represents the IOAM $F_{open}(V)$, and the black line represents $y = V - V_{in}$. The black point $P^*(V^*, F_{open}(V^*))$ is the intersect between IOAM and $y = V - V_{in}$, and is called fixed point because it meets the relation $V^* - V_{in} = F_{open}(V^*)$. The arrows represent the trajectories for describing the evolution of solution sequence $\{V_n\}$, $n = 0, 1, 2, \ldots$ and the trajectories are produced based on the processes as follows: draw a vertical segment along $V = V_{n-1}$ intersecting with IOAM $F_{open}(V)$ at point P_{n-1}, then draw a horizontal segment from P_{n-1} to the black line. The abscissa of the intersection between the horizontal segment and black line is V_n.

Thus, the envelope dynamics of OEO is transformed into a geometry problem as shown in Fig. 12.2(a). Different OEO loop brings different IOAMs, and different injected amplitude leads to different location of the black line. Generally, the states and the nonlinear degree of the OEO loop can be described by the slope value of IOAM at a fixed point. Here, three linear features are defined to help us analyze the nonlinear degree: (i) Homogeneity: the obtained gain after

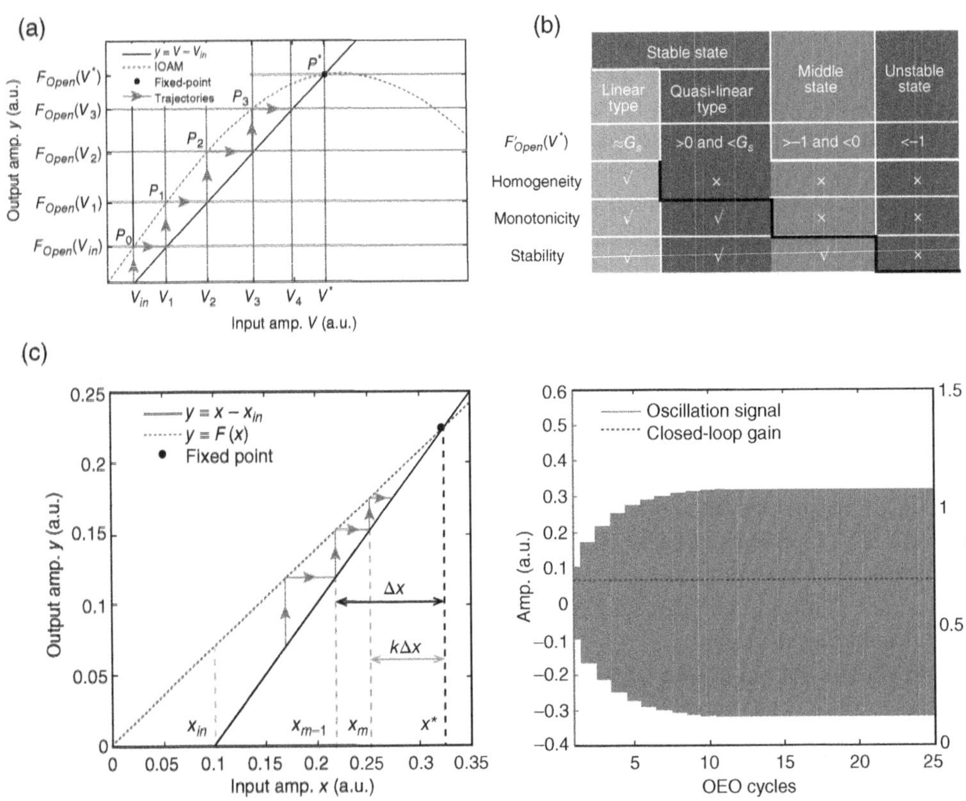

Figure 12.2 (a) A graphical technique for resolving the iterated equation. (b) The general relation between slope value at a fixed point and different OEO states. (c–f) Iterated map and simulated oscillating signal of OEO with (c) linear type, (d) quasi-linear type, (e) middle state, and (f) bistable state. *Source:* Xiao et al. [55]/Optical Society of America/CC BY 4.0.

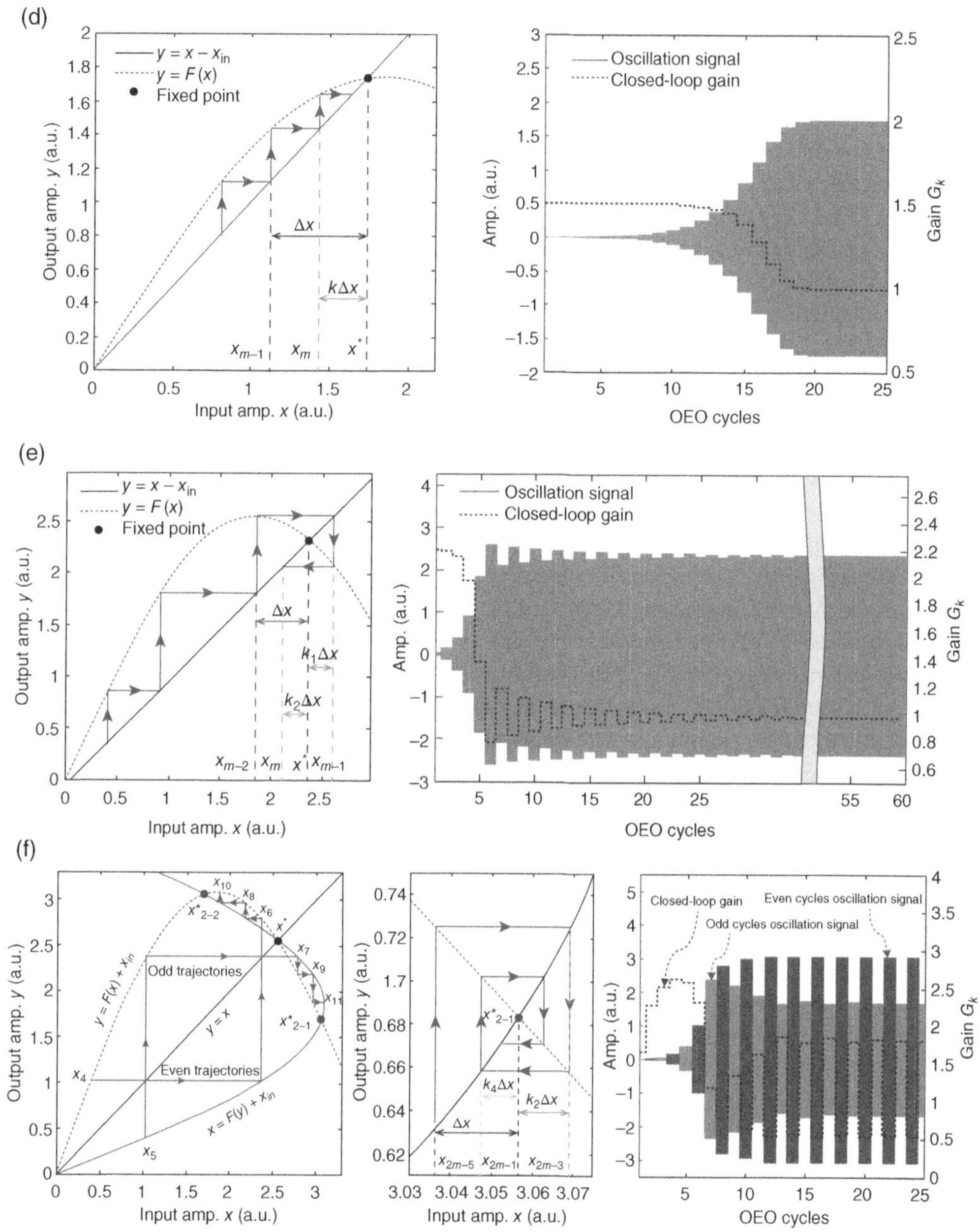

Figure 12.2 (Continued)

oscillating single cycle in the loop for any oscillation microwave signal is identical. (ii) Monotonicity: the oscillation amplitude monotonically increases during the oscillating process. (iii) Stability: the oscillation amplitude is converging.

As shown in Fig. 12.2(b), the nonlinear degree of the OEO loop increases with a decrease of the slope $F'_{open}(V^*)$, and the linear features of homogeneity, monotonicity, and stability are gradually lost. And based on this, OEOs are classified into different types or states. The left picture in Fig. 12.2(c) shows the iterated map related to linear type, due to the slope $F'_{open}(V^*)$, is relatively large, OEO loop shows strong linearity. As shown in the right picture in Fig. 12.2(c), the closed-loop gain (dotted line) is close to a constant which means homogeneity is satisfied. And the amplitude of oscillation signal (filled area) monotonically increases and converges to the fixed point, which means the monotonicity and stability are also satisfied. Then the fixed point moves along the positive direction of horizontal axis with increase of the open-loop gain or injected amplitude, which leads to decrease of the slope. Thus, the nonlinear feature of the loop becomes increasingly significant. Three linear features are gradually lost as shown in Fig. 12.2(d–f).

Based on the above discussion, there are several general rules for OEO: (i) Oscillation amplitude monotonically increases if the slope of IOAM at a fixed point is larger than zero. (ii) OEO is stable when the slope is larger than −1, and the final output amplitude is equal to the fixed point, and more slope is closed −1, more OEO cycles are needed before converging. (iii) OEO is unstable when the slope is less than −1. The above rules can be proved in mathematics (Detailed deviation seen in [55]) and are verified by the following experiments.

Figure 12.3 shows experiments with different oscillating processes. Experimental set up is shown in Fig. 12.1(a). After employing a home-made high-speed optical switch as shown in Fig. 12.3(a) and an oscilloscope, the oscillating processes corresponding to the different injected amplitude are collected. Solid line in Fig. 12.3(b) shows the measured IOAM, and the asterisks are the final amplitude measurements. Asterisk meets the IOAM well and indicates that the oscillation amplitude converges to the fixed point. Furthermore, three cases of the oscillating processes corresponding to points (c–e) shown in Fig. 12.3(b) are provided in upside figures of Fig. 12.3(c–e), and bottom figures show the corresponding STFT results. The slope value at fixed point (c) is positive, which leads to a monotonically increasing amplitude. The slope value at fixed point (d) is negative. Thus, the oscillation amplitude oscillates around the fixed point before converging. The slope value at fixed point (e) is closer to −1 compared with point (c). Thus, the oscillating process shown in Fig. 12.3(e) needs more time before converging. Experiments shown in Fig. 12.3 confirm that the degree of OEO nonlinearity increases with the decrease of the slope value at a fixed point.

Figure 12.4 shows the different states of two OEO loops with different open-loop small signal gain. The experimental setup is shown in Fig. 12.1(a), two EAs with relatively large linear operating range are employed first. Thus, only nonlinear effect needs to be concerned. Figure 12.4(Aa) shows the measured IOAMs with different gain and Fig. 12.4(Ab) shows the corresponding zoom view. Asterisks are the fixed points, and Fig. 12.4(Ac–e) shows the oscillation signals. When the gain is 6.55 dB, the slope is larger than -1, which leads to stable state as shown in Fig. 12.4(Ac). When the gain is 7.55 dB, the slope is less than -1, which leads to bistable state as shown in Fig. 12.4(Ad). Notice that the threshold condition (proposed by Chembo [53]) of voltage gain value of 2.31 (~7.23 dB) is also satisfied here because no additional nonlinear effects are concerned. Then two EAs with relatively small linear operating ranges are employed in the loop. Thus, the saturation characteristics of EAs need to be considered. As shown in Fig. 12.4(Ba), IOAMs are shaped by the saturation characteristics of EAs. The oscillation signal shown in Fig. 12.4(Bb) shows that OEO maintains a stable state with gain of 16.44 dB (which is beyond the threshold gain condition of ~7.23 dB). This can be easily understood by MING model: the slope of IOAM at fixed point (as shown in point(b)) is larger than −1. Thus, the OEO is stable. And slopes at points (c–d) are less than −1, which leads to unstable states as shown in Fig. 12.4(Bc–d).

In addition to the above general rules, Ming model indicates that the bifurcation phenomena in OEO can be caused by injected microwaves. Figure 12.5(A) shows two bifurcation diagrams, and Fig. 12.5(B–C) shows the experiments which meet well with each other. Results show that different injected amplitudes can lead to different OEO states, such as stable state, bistable state, and 4-point cycle state.

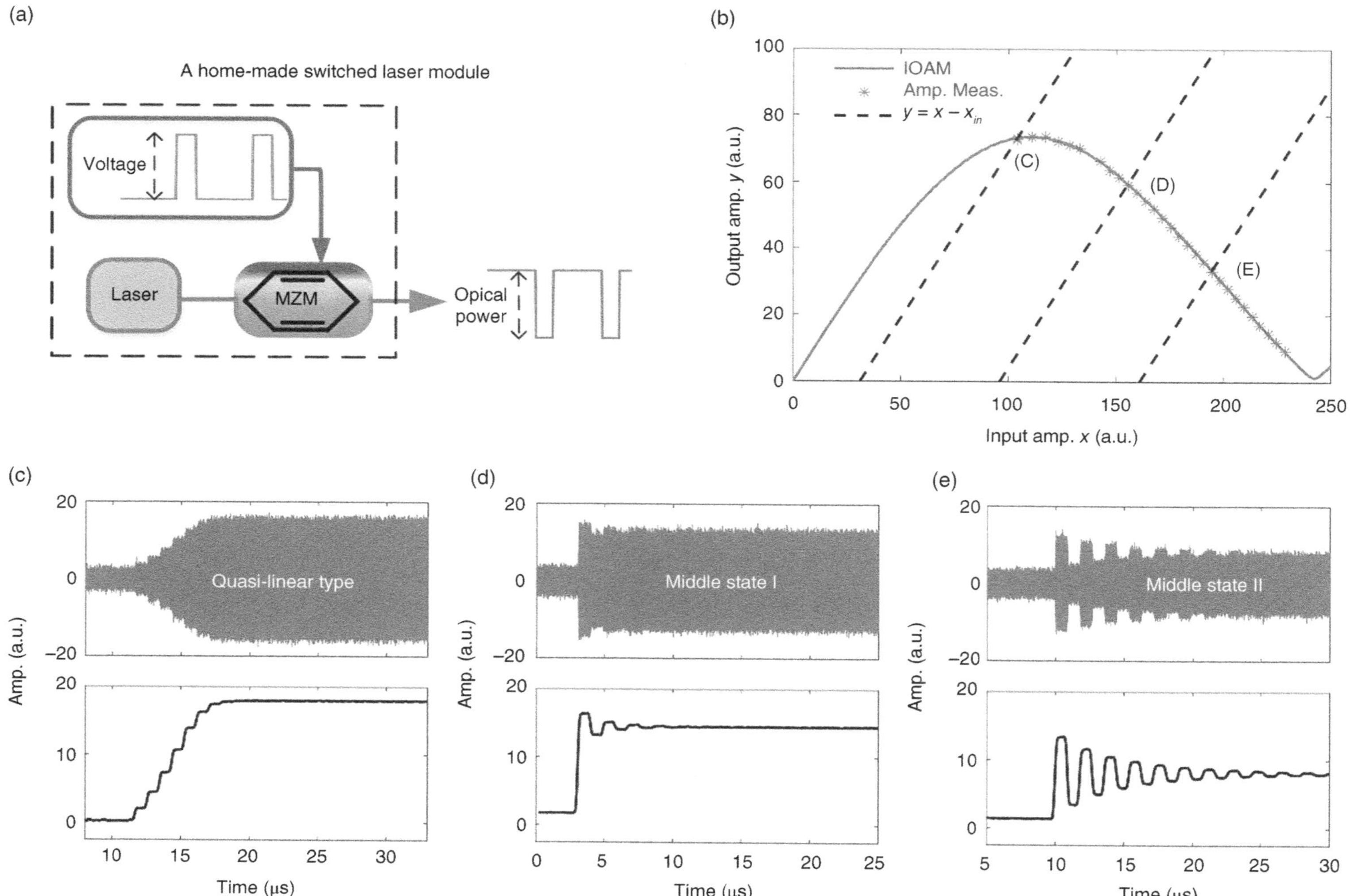

Figure 12.3 Different oscillating types corresponding to different slopes. (a) The structure of the employed switched laser module. (b) The measured IOAM and fixed points. (c–e) Real-time signals (up) and STFT results (bottom) show the oscillating processing corresponding to the different slopes. *Source:* Xiao et al. [55]/Optical Society of America/CC BY 4.0.

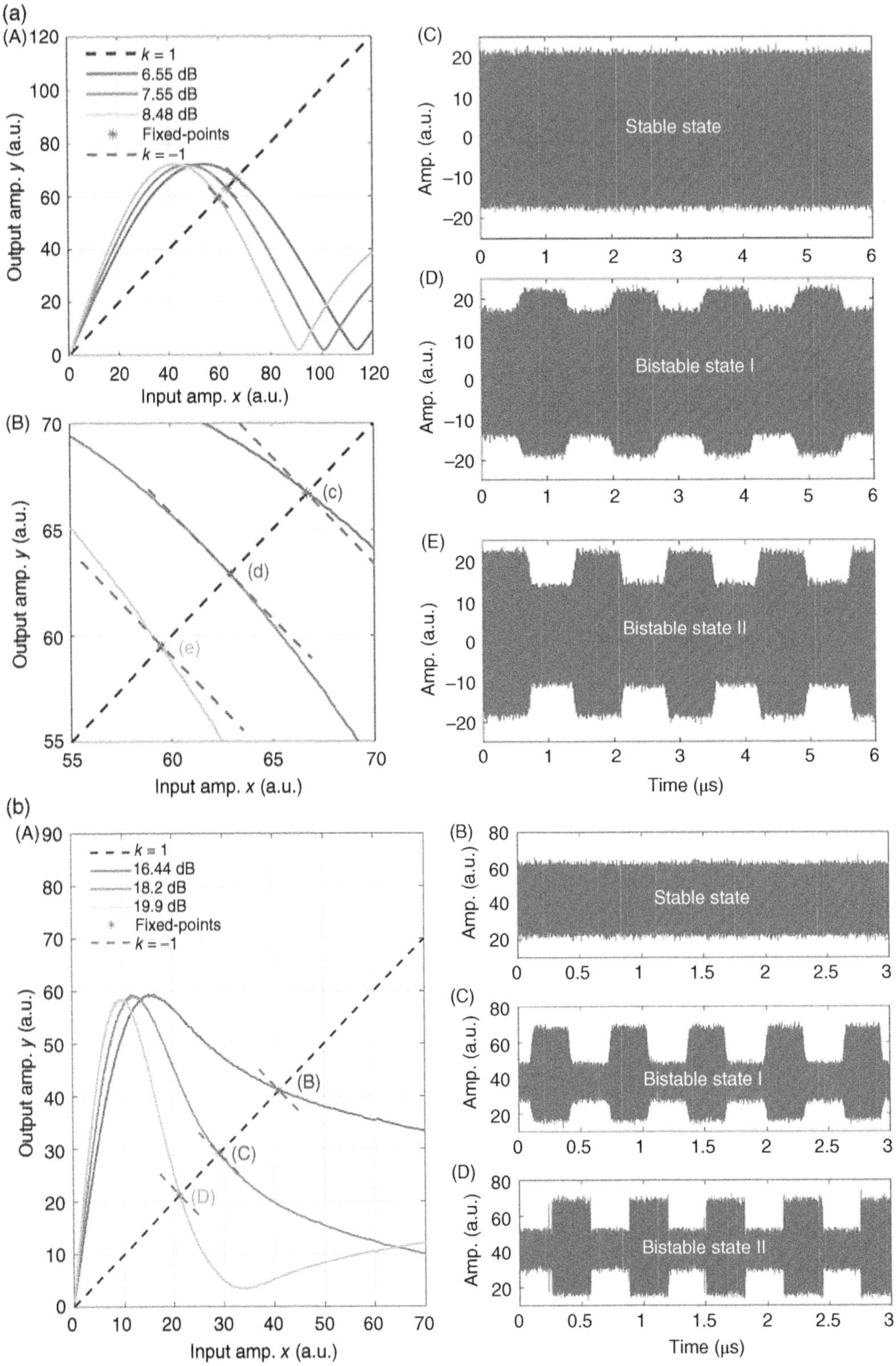

Figure 12.4 Different OEO states corresponding to the loop with simple nonlinear effects (a) and the loop with complex nonlinear effects (b).

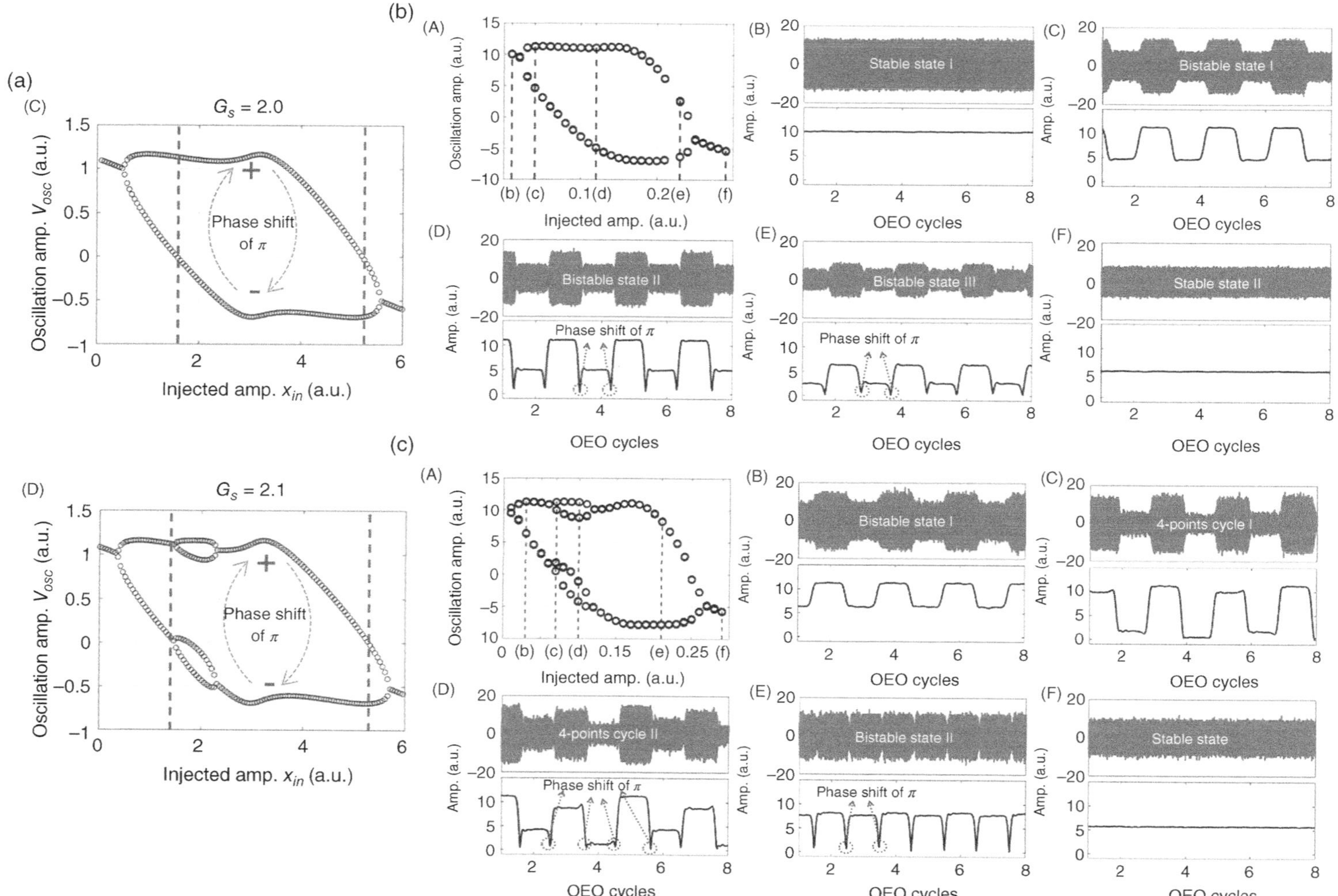

Figure 12.5 Simulated bifurcation diagram (a) caused by injected microwaves and the corresponding experiments (b, c). *Source:* Xiao et al. [55]/Optical Society of America/CC BY 4.0.

Above analysis and experiments show that MING model is a general model for analyzing OEO states. For a practical OEO loop, the working state can be predicted by calculating the slope value after measuring IOAM, which means it is helpful for OEO analysis and design.

12.3 Parity-Time Symmetric OEO

Generally, a long optical fiber delay line is adopted in an OEO to ensure a low phase noise. However, the use of the high Q-factor energy storage element such as a long optical fiber delay line in OEOs would also result in closed-spaced cavity modes. For example, if a 5-km optical fiber delay line is used in the OEO cavity to enhance the Q-factor, the frequency spacing of the two potential oscillation modes is only 40 kHz. Generally, narrowband electrical or optical filters and dual-loop structures are used in traditional OEOs in order to achieve a single-mode operation. However, narrowband electrical and optical filters are hard to achieve especially at high center frequencies, and the loop length must be controlled precisely to maintain the conditions for Vernier effect in the dual-loop structure. Recently, single-mode oscillation has been achieved in PT-symmetric OEOs [11, 12] without using any narrowband optical/electrical filters, which solved the long-existing mode selection problems in traditional OEOs.

PT symmetry fundamentally exploits the fact that non-Hermitian operators can exhibit real eigenvalues associated with nonorthogonal eigenstates. By increasing the level of nonhermiticity, the symmetry can be broken after passing a so-called transition point, leading to nonreal eigenvalues [73–75]. In particular, coupled PT-symmetric structure [76] using two identical feedback loops with one having a gain and the other having a loss of the same magnitude has been widely investigated as a powerful tool for cavity mode selection both in photonic [77–81] and electronic [82] cavities. The OEO is a microwave photonics system with a hybrid optoelectronic cavity, and it is natural to wonder whether coupled PT-symmetric structure can be applied to OEOs for cavity mode selection.

In [11], we have proposed and experimentally demonstrated the use of PT symmetry for mode selection in an OEO with two mutually coupled feedback loops by precisely controlling the interplay between the gain and loss in the two coupled loops. The simplified schematic diagram of the PT-symmetric OEO is shown in the right part of Fig. 12.6. As can be seen, the PT-symmetric OEO has a dual-loop structure with one loop having a gain and the other having a loss of the same magnitude. In theory, the dynamic equations of the nth cavity mode in the two coupled loops are as follows:

$$\frac{da_n^{(1)}}{dt} = \left(i\Delta\omega_n^{(1)} + g\right)a_n^{(1)} - i\mu a_n^{(2)} \tag{12.15}$$

$$\frac{da_n^{(2)}}{dt} = \left(i\Delta\omega_n^{(2)} - \gamma\right)a_n^{(2)} - i\mu a_n^{(1)} \tag{12.16}$$

where $a_n^{(1)}$ and $a_n^{(2)}$ are the amplitudes of the nth cavity mode in the sgain and loss loops, respectively. i is the imaginary unit. $\Delta\omega_n^{(1,2)} = \omega_n - \omega_n^{(1,2)}$ is the detuning frequency of the two loops. ω_n is the angular frequency of the nth cavity mode, and $\omega_n^{(1,2)}$ are the eigenfrequencies of the two loops, respectively. μ is the coupling coefficient between the two loops. g and γ are the gain and loss coefficients of the gain and loss loops, respectively. By solving Eqs. (12.15) and (12.16), we can obtain the eigenfrequencies of the supermodes of the PT-symmetric system

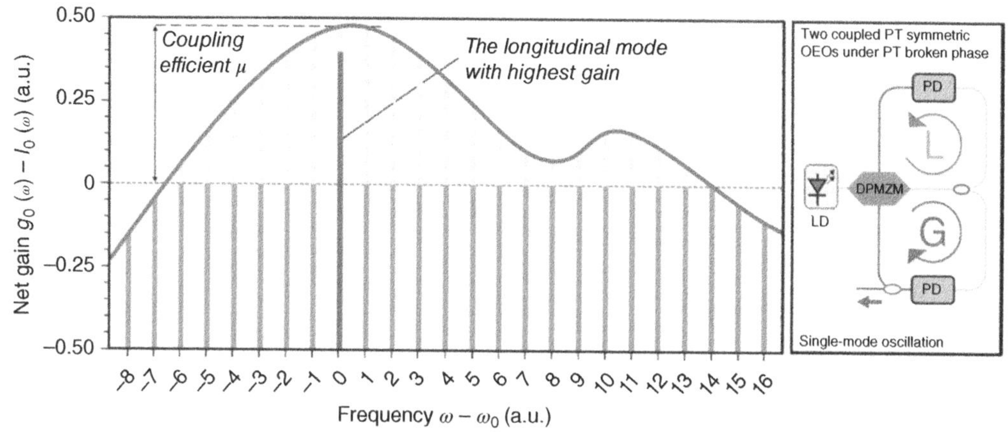

Figure 12.6 PT-symmetric phase of the OEO with two coupled loops. The gain of one loop is balanced by the loss of another loop, and the losses overcompensate the gain for all cavity modes, except for the one with the highest gain. Single-mode oscillation can be achieved in the cavity mode with the highest gain. *Source:* Liu et al. [11]/Springer Nature/CC BY 4.0.

$$\omega_{n\pm} = \frac{\omega_n^{(1)} + \omega_n^{(2)}}{2} + \frac{i(g-\gamma)}{2} \pm \sqrt{\mu^2 - \left(\frac{g+\gamma}{2} - \frac{i\left(\omega_n^{(1)} - \omega_n^{(2)}\right)}{2}\right)^2} \tag{12.17}$$

If the loop lengths of the two loops are the same, we have $\omega_n^{(1)} = \omega_n^{(2)}$. If the PT symmetry condition can be satisfied by precisely manipulating the gain and loss in the two loops, i.e. $g = \gamma$, Eq. (12.17) can be further rewritten as:

$$\omega_{n\pm} = \omega_n \pm \sqrt{\mu^2 - \gamma^2} \tag{12.18}$$

Equation (12.18) indicates a transition point in the PT-symmetric system when the gain/loss coefficient is equal to the coupling coefficient μ. If the gain/loss coefficient is larger than the coupling coefficient, PT symmetry no longer applies, and the frequency difference of the eigenfrequencies of the supermodes becomes imaginary. Thus, a pair of amplifying and decaying modes is generated in each loop with identical frequencies. As a result, single-mode oscillation can be achieved in the gain loop. Otherwise, the two loops oscillate at two slightly different frequencies if the gain/loss coefficient is smaller than the coupling coefficient since the eigenfrequencies of the supermodes are split into two slightly different frequencies.

Figure 12.7(a) shows the measured single-mode oscillation spectrum of the PT-symmetric OEO in the experiment. The gain and loss of the two loops are precisely tuned to the same magnitude of the oscillating signal to satisfy the PT symmetry condition. It should be noted that the OEO would oscillate in multimode without PT symmetry. Figure 12.7(b) shows the measured multimode oscillation spectrum without PT symmetry. Almost at the same time, a PT-symmetric OEO has also been proposed and experimentally demonstrated in [12]. Figure 12.8 shows the schematic diagram and spectra of this PT-symmetric OEO. A coupled dual-loop structure is achieved with the help of a balanced PD (BPD). Singlemode oscillation is also achieved when PT symmetry condition and PT symmetry broken phase condition are satisfied. These two pioneering works achieved

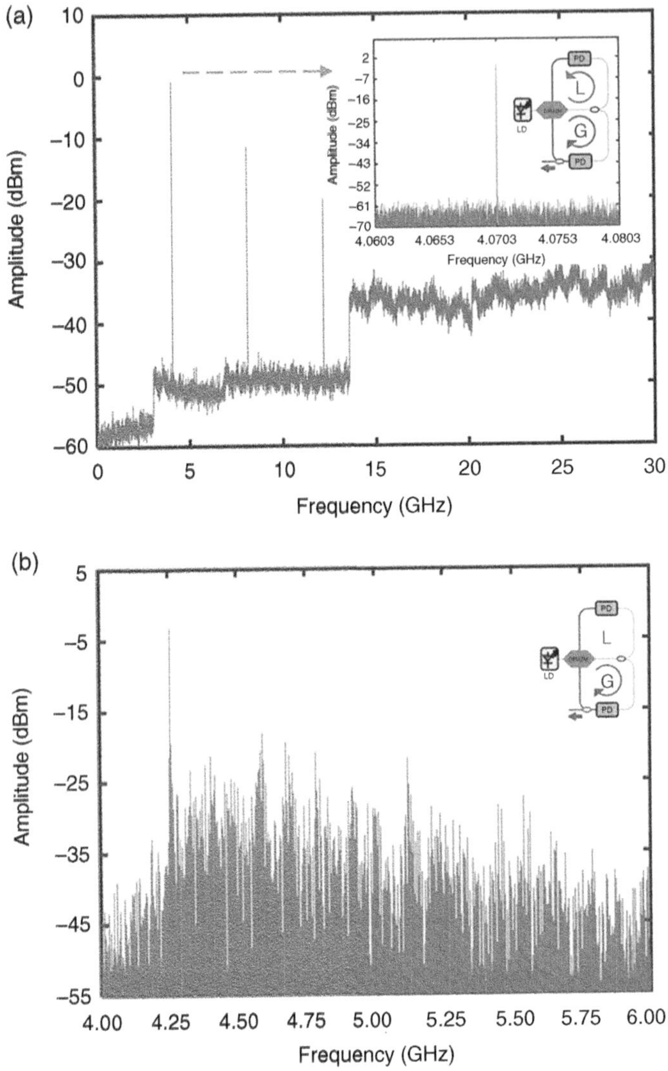

Figure 12.7 Measured spectrum of (a) the singlemode oscillation without PT symmetry and (b) multimode spectrum without PT symmetry. *Source:* Liu et al. [11]/Springer Nature/CC BY 4.0.

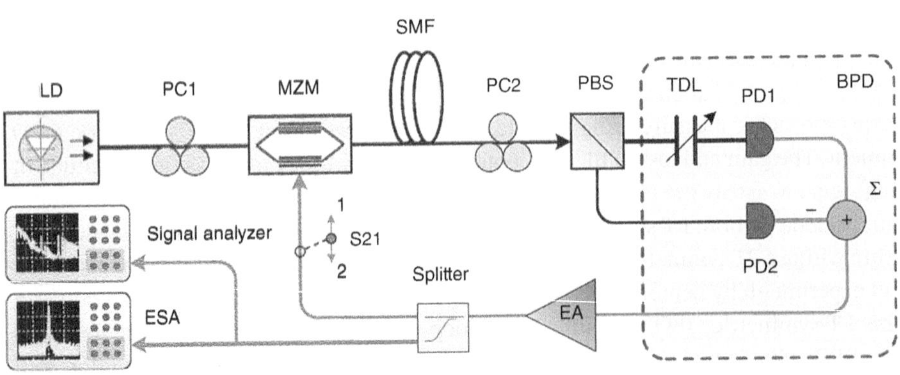

Figure 12.8 Schematic diagram of a PT symmetric OEO using a balanced photodetector (BPD). *Source:* Adapted from Zhang et al. [12].

singlemode oscillations in PT symmetric OEOs without using any narrowband optical/electrical filters, which solved the long-existing mode selection problems in traditional OEOs.

The frequency of the generated microwave signals in the above two works is fixed since no frequency tuning mechanism is incorporated. In many applications, frequency-tunable OEOs are preferred. Several approaches [15–21] have also been proposed to demonstrate frequency-tunable PT symmetric OEOs by using frequency-tunable MPFs for coarse frequency selection. For instance, a frequency-tunable PT symmetric OEO based on a photonic integrated microdisk resonator (MDR) is proposed in [15]. Figure 12.9 shows the schematic diagram and spectra of the frequency-tunable PT symmetric OEO. As can be seen, a coupled dual-loop PT symmetric OEO with one loop having a gain and the other having a loss of the same magnitude is achieved by employing the reciprocity of light propagation in the MDR. A frequency-tunable MPF based on PM-IM conversion is also implemented in the PT symmetric OEO by the joint use of the PM and MDR. Coarse frequency selection is realized by thermally tuning the MDR. As can be seen from Fig. 12.9(c), the frequency of the PT symmetric OEO is tuned from about 2 GHz to 12 GHz.

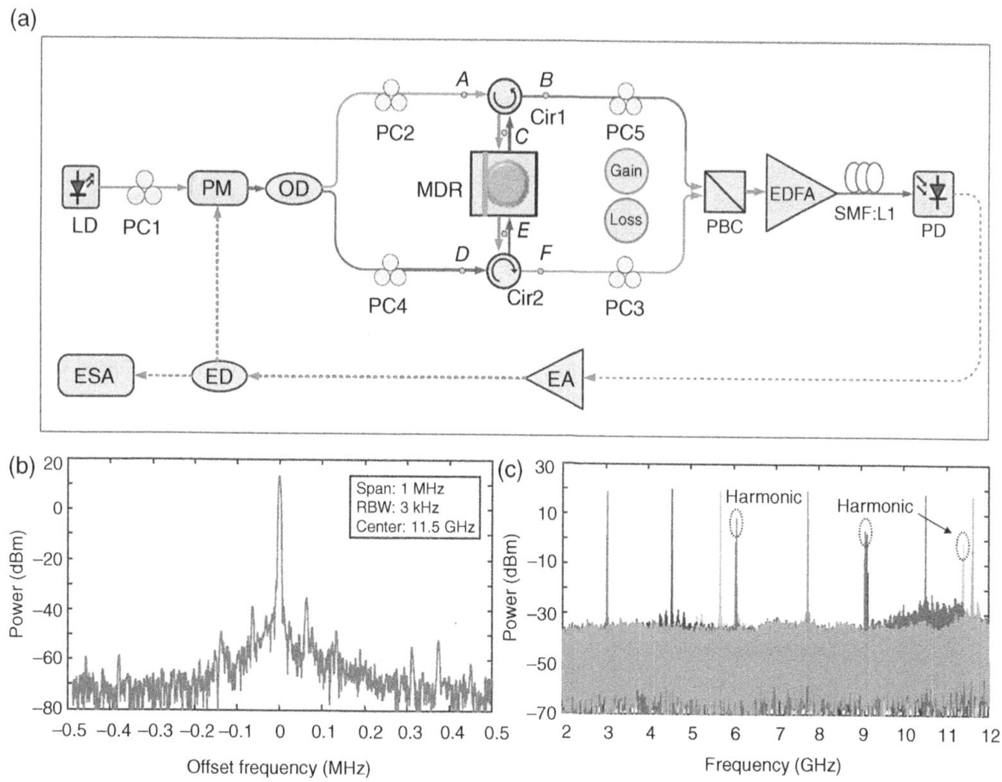

Figure 12.9 Schematic diagram and spectra of a frequency-tunable PT symmetric OEO using a photonic integrated MDR. (a) Schematic diagram of the frequency-tunable PT symmetric OEO. Frequency tuning is achieved by thermally tuning the MDR. (b) Single-mode oscillation measured with a span of 1 MHz and an RBW of 3 kHz. (c) Frequency tunability of the PT-symmetric OEO with a tuning range from about 2–12 GHz. *Source:* Fan et al. [15]/IEEE.

12.4 Fourier Domain Mode-Locked OEO

In addition to the generation of single-frequency microwave signals, another very attractive direction of OEOs is the generation of chirped microwave waveforms with fast frequency scanning, providing a wide range of applications in radar and communication systems. For example, chirped microwave waveforms can be used in modern radar systems to increase the range resolution while maintaining a large detection distance, which overcomes the long-existing tradeoff between range resolution and detection distance in a traditional radar system. Chirped microwave waveforms with tunable parameters have been generated using tunable OEOs in [82–84]. However, the OEO produces single-frequency signals and the chirped microwave waveforms are generated at the outside of the OEO cavity using a baseband electrical chirped microwave source or a recirculating phase modulation loop.

Recently, we proposed and demonstrated an OEO cavity mode control method based on Fourier domain mode-locking [22]. Chirped microwave waveforms were generated directly from the OEO cavity. Figure 12.10(a), (b) shows a comparison of a traditional single-frequency OEO and the FDML OEO. A single passband filter is incorporated in the traditional OEO cavity, and the output signal is a single-frequency sinusoidal wave. In the FDML OEO cavity, a frequency scanning filter is used. The scanning period of the filter is synchronized with the cavity round-trip time of the OEO to enable the FDML operation, i.e.

$$T_{roundtrip} = n \times T_{filter\ drive} \tag{12.19}$$

where $T_{filter\ drive}$ is the cavity round-trip time, $T_{filter\ drive}$ is the scanning period of the filter, and n is an integer. By doing so, the frequency scanning filter is tuned at the same spectral position when the selected mode returns to this position after one cavity round-trip time. For example, the frequency of the selected cavity modes would also be f_1 and f_2 at $t_1 + T_{roundtrip}$ and $t_2 + T_{roundtrip}$, respectively, if the frequency of the selected cavity modes is f_1 and f_2 at t_1 and t_2, respectively. As a result, all the cavity modes can be stored and activated simultaneously in the OEO cavity and are mode-locked with fixed phase relationship if their frequencies are within the frequency scanning range of the filter. The synchronization between the tuning period and the cavity round-trip time in FDML OEO is similar to that of synchronization in traditional active or passive mode-locking, except that the mode-locking is achieved in the frequency domain in the FDML OEO. The output signal of the

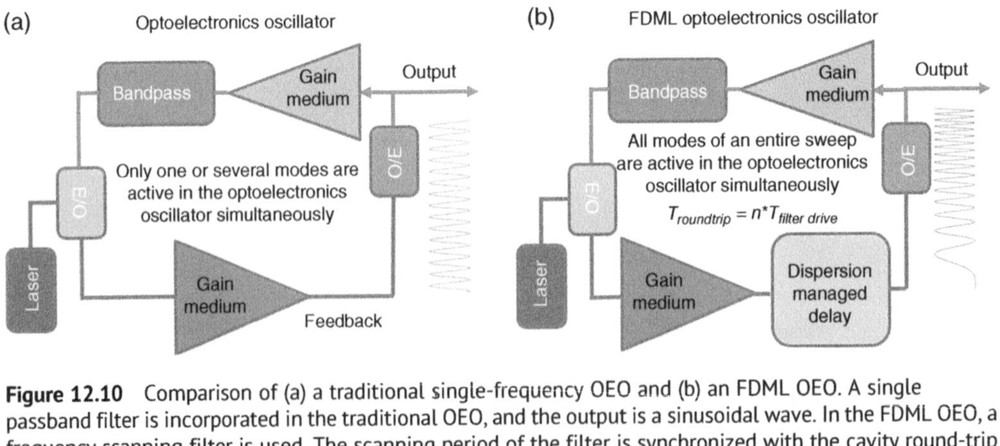

Figure 12.10 Comparison of (a) a traditional single-frequency OEO and (b) an FDML OEO. A single passband filter is incorporated in the traditional OEO, and the output is a sinusoidal wave. In the FDML OEO, a frequency scanning filter is used. The scanning period of the filter is synchronized with the cavity round-trip time to enable the Fourier domain mode-locking operation. The output is a frequency scanning/chirped microwave waveform. *Source:* Hao et al. [22]/Springer Nature/CC BY 4.0.

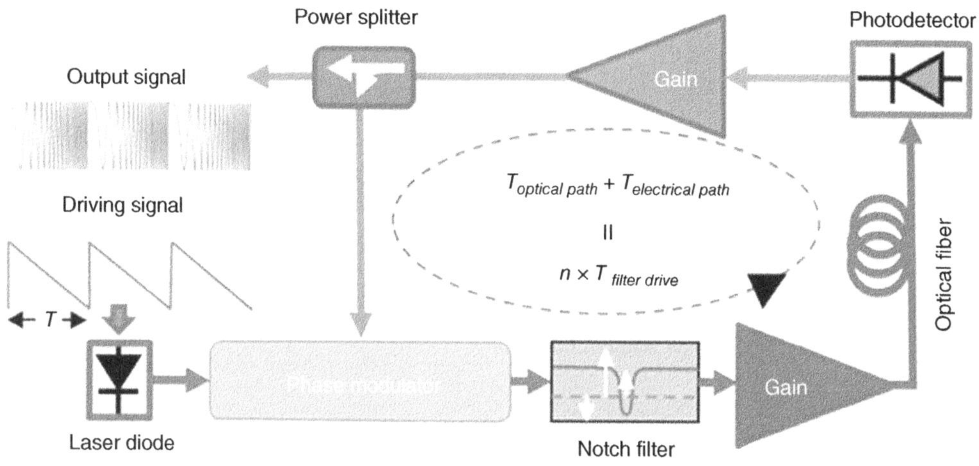

Figure 12.11 Experimental setup of the FDML OEO [22]. The frequency scanning microwave photonic filter consists of a frequency scanning laser diode, a phase modulator, and an optical notch filter.
Source: Hao et al. [22]/Springer Nature/CC BY 4.0.

FDML OEO is a frequency scanning/chirped microwave waveform due to the frequency domain mode-locking.

Figure 12.11 shows the experimental setup of the FDML OEO. A frequency scanning MPF based on PM-IM conversion is used. The MPF consists of a frequency scanning laser diode, a PM, and an optical notch filter. In this scheme, the balance of phase modulation is interrupted by the optical notch filter. The center frequency of the MPF is equal to the frequency difference between the laser diode and the optical notch filter. By sweeping the frequency of the laser diode, a frequency scanning MPF can be achieved. If the system noise is ignored, the frequency scanning oscillation in the OEO loop in the Fourier domain mode-locking operation should satisfy [22]:

$$V_{FDML}^{\Omega}\left(t - T_{round-trip}\right) = S\left(\left|V_{FDML}^{\Omega}\right|\right)\left\{\left[V_{FDML}^{\Omega}(t)e^{i\varphi_{oc}^{T_{round-trip}}(t)}\right] * s_{21}^{open\ loop}(t)\right\}e^{-i\varphi_{oc}^{T_{round-trip}}(t)} \tag{12.20}$$

where $\varphi_{oc}^{T_{round-trip}}$ is the periodic phase variation of the laser diode in the FDML operation, $S\left(\left|V_{FDML}^{\Omega}\right|\right)$ is the saturation factor of the PM-IM conversion process, $s_{21}^{open\ loop}$ is the open-loop response of the OEO when the MPF is not scanning. Equation (12.9) indicates that the stable frequency scanning oscillation repeats itself after each cavity round-trip time. The solution of Eq. (12.9) can be expressed as:

$$V_{FDML}^{\Omega}(t) \propto e^{-i\varphi_{oc}^{T_{round-trip}}} \tag{12.21}$$

The frequency of the generated stable frequency scanning oscillation in the FDML OEO cavity follows the periodic frequency variation of the MPF since the center frequency of the MPF is determined by the frequency of the laser diode. It should be noted that in addition to the PM-IM conversion-based MPF, other kinds of filters can also be used in the FDML OEO for mode selection, as long as the center frequency of the filter can be rapidly tuned to ensure the Fourier domain mode-locking operation. In the FDML operation, chirped microwave signals can be generated directly from the FDML OEO cavity, and the frequency scanning parameters are determined by the loop filter.

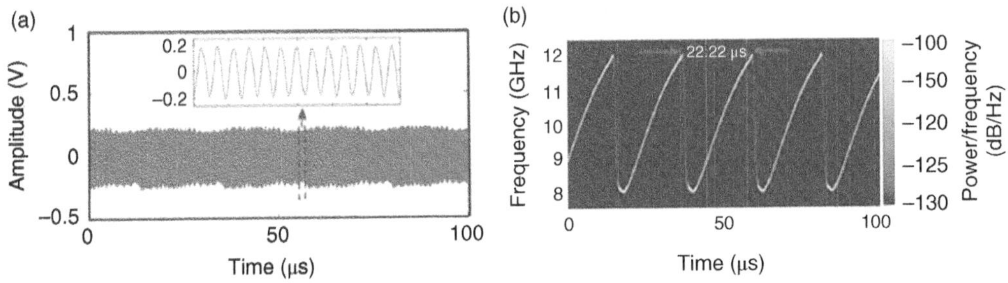

Figure 12.12 (a) Temporal waveform and (b) instantaneous frequency-time of the output signal of the FDML OEO. *Source:* Hao et al. [22]/Springer Nature/CC BY 4.0.

The temporal waveform and instantaneous frequency-time of the generated X-band (8–12 GHz) linearly chirped microwave waveform (LCMW) are shown in Fig. 12.12. The frequency of the generated LCMW ranges from about 8 to 12 GHz, which can be easily tuned by changing the frequency scanning characteristics of the loop filter. The generated LCMW is constant in the time domain, indicating a good mode-locking state. The instantaneous frequency of the generated LCMW increases nearly linearly within one period. The frequency scanning period can also be tuned, for example, by changing the cavity length of the OEO or by harmonic mode-locking. In addition to the LCMW, other kinds of chirped microwave waveforms have also been generated using an FDML OEO, such as dual-chirp microwave waveform, a complementary LCMW pair, phase-coded microwave waveform, phase-coded LCMW, and frequency-doubled LCMW have also been generated using an FDML OEO. Moreover, a microwave photonic radar based on the FDML OEO has also been proposed and experimentally demonstrated, which shows the great potential of FDML OEO in practical applications.

It should also be noted that the FDML OEO can also be used for microwave photonic frequency-to-time mapping. The basic idea is to map the frequency of the unknown microwave signal to the time difference of the output pulses of the FDML OEO, so the microwave spectral information can be measured in the time domain. Figure 12.13 shows the principle of operation of the FDML

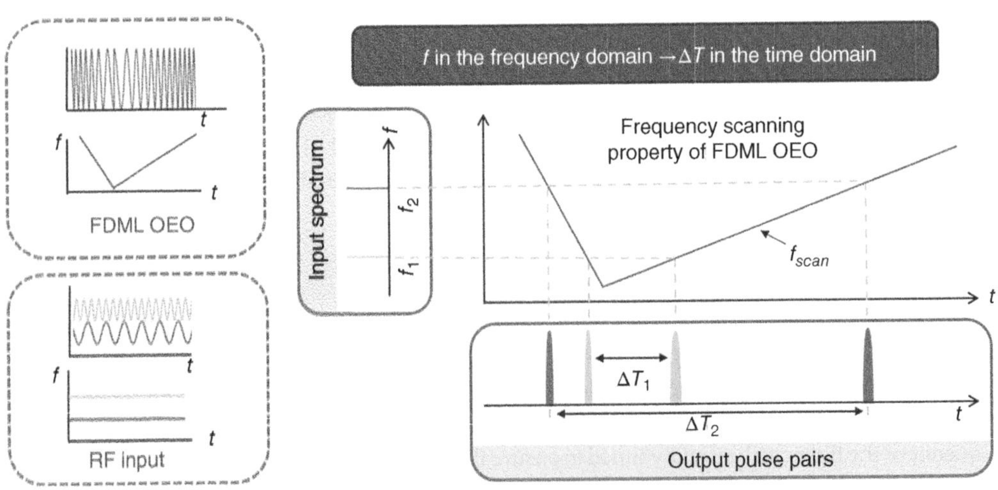

Figure 12.13 Microwave photonics frequency-to-time mapping based on the FDML OEO. *Source:* Hao et al. [38]/Optical Society of America.

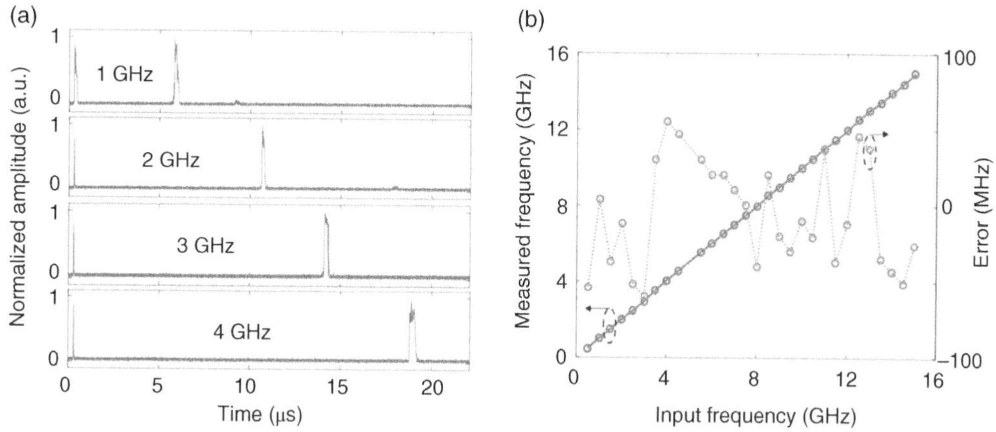

Figure 12.14 (a) Measured pulse envelopes when a single-tone microwave signal with different frequencies is injected into the FDML OEO. (b) Measured results and errors. *Source:* Hao et al. [37]/Optical Society of America/CC BY 4.0.

OEO-based frequency-to-time mapping system [37, 38]. A pair of pulses can be obtained at the output of the system when a single-tone microwave signal is injected into the FDML OEO cavity, thanks to the bidirectional frequency scanning property of the FDML OEO. The time difference ΔT of the output pulses is related to the frequency f of the injected microwave signal, so frequency-to-time mapping is achieved. Multiple pairs of output pulse can be obtained at the output when a multi-tone microwave signal is injected into the FDML OEO, so the proposed frequency-to-time mapping system is capable of measuring multi-tone microwave signals. This feature is highly desired in practical applications since various unknown frequency components may exist in the microwave signals under measurement.

Figure 12.14(a) shows the measured pulse envelopes when a single-tone microwave signal with different frequencies is applied to the FDML OEO-based microwave photonics frequency-to-time mapping system. As can be seen, the time differences of the output pulse pairs are increased with the increase of the applied frequency. The frequency of the single-tone microwave signal is measured based on the frequency-to-time mapping relationship established by the FDML OEO. Figure 12.14(b) shows the measured results and errors. The frequency measurement errors are no more than 60 MHz. The measurement range is only limited by the bandwidth of the PD in our experiment, which can be easily extended to tens of GHz by using large bandwidth optoelectronic devices. The ability to measure the unknown microwave frequency information in the time domain using the proposed frequency-to-time mapping method has the potential to ensure new metrology and signal processing approaches with superior performance in terms of operation speed and real-time bandwidth.

12.5 OEPO

The cavity modes in a traditional OEO are discrete, and the minimum mode spacing is determined by the cavity delay T. This oscillator is called a delay-controlled oscillator, whose stable oscillating signal must repeat itself after one cavity round-trip time if the timing jitter is ignored. As a result, the frequency of the output signal of a traditional OEO is constrained by the cavity delay T.

The frequency must be n/T, where n is an integer. The delay-controlled operation leads to difficulty in frequency tuning. Frequency tuning is discrete unless a phase shifter is used to tune the cavity delay. Moreover, the initial phases of the cavity modes are not determined and the mode competition effective is inevitable in a traditional delay-controlled OEO. Thus, stable multimode oscillation is not possible if no mode-locking operation is applied. Recently, we proposed and experimentally demonstrated a phase-controlled OEPO [39] based on parametric frequency conversion in the optoelectronic cavity, whose stable oscillation is not limited by the cavity delay.

The schematic diagram of the OEPO is shown in Fig. 12.15(a). The major difference between the OEPO and a traditional OEO is that an electrical nonlinear medium (electrical mixer) is introduced

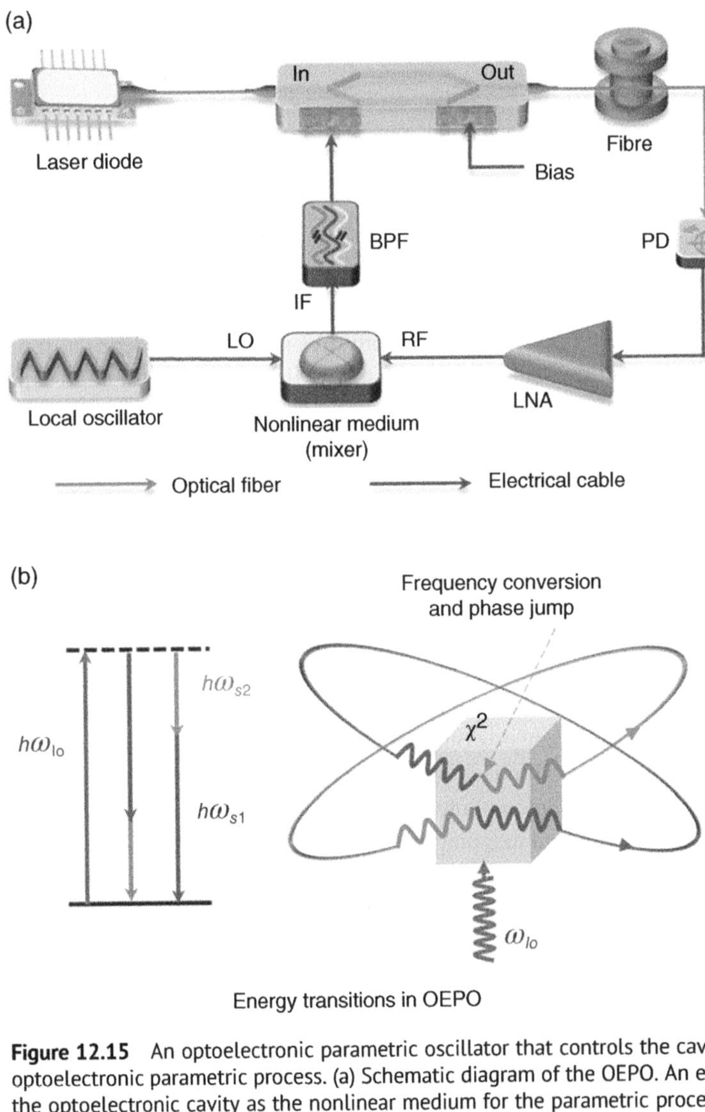

Figure 12.15 An optoelectronic parametric oscillator that controls the cavity modes based on the optoelectronic parametric process. (a) Schematic diagram of the OEPO. An electrical mixer is introduced into the optoelectronic cavity as the nonlinear medium for the parametric process. (b) Energy transitions in the OEPO. A pair of cavity modes are converted into each other by a local oscillator in the electrical nonlinear medium, locking the sum phase of the mode pair thanks to the phase jump in the frequency conversion process. *Source:* Hao et al. [39]/Springer Nature/CC BY 4.0.

into the optoelectronic cavity for the parametric frequency conversion process. The electrical mixer is a nonlinear electrical device that produces new frequencies. The output frequency at the IF port of the mixer is equal to the difference between the two input frequencies at the LO and RF ports. As shown in Fig. 12.10(b), by applying a local oscillator at the LO port, a pair of oscillation modes are converted into each other by the local oscillator; thus, each mode repeats itself after two cavity round-trip times. More importantly, the sum phase of each mode pair is locked, and a phase jump of the oscillating signals occurs in the electrical nonlinear medium, leading to unique mode properties of the OEPO. For example, stable multimode oscillation can be easily achieved in the OEPO since the sum phase of each mode pair is locked. Stable multimode oscillation is difficult to achieve in a traditional OEO due to the unavoidable mode competition effect.

Mathematically, the stable oscillating mode pair in the OEPO should satisfy the following equations:

$$\begin{cases} s(z=0,t) = \alpha_1 p(z=L,t) s^*(z=L,t) + c.c. \\ s(z=L,t) = \alpha_2 s(z=0,t+T) + n_a(t) + c.c. \end{cases} \tag{12.22a, b}$$

where $s(z,t) = e^{-i(\omega_{s1}t - k_{s1}z + \varphi_{s1})} + e^{-i(\omega_{s2}t - k_{s2}z + \varphi_{s2})} + c.c.$ is the oscillating mode pair; ω_{s1}, φ_{s1}, and k_{s1} are the angular frequency, initial phase, and wave vector of one mode, respectively; ω_{s2}, φ_{s2}, and k_{s2} are those of another mode, respectively. z is the spatial position along the OEPO cavity, $z = 0$ denotes the position where the parametric frequency conversion is implemented, and L is the cavity length of the OEPO. α_1 is the frequency conversion loss, α_2 includes all the losses/gains when the oscillating mode pair travels from the output of the mixer to the input via the OEPO cavity and $\alpha_1\alpha_1$ is the total loop gain. $p(z=0,t) = p(z=L,t) = e^{-i\omega_{lo}t} + c.c.$ is the pump signal from the local oscillator, $\omega_{lo} = \omega_{s1} + \omega_{s2}$ is the angular frequency of the local oscillator. Eq. (12.11a) describes the parametric frequency conversion and Eq. (12.22b) shows that the stable oscillating mode pair recovers itself after one cavity round-trip T. In a stable oscillation, the total loop gain $\alpha_1\alpha_1$ is near unity, and the system noise $n_a(t)$ is generally negligible. According to Eq. (12.11), we can derive the following equation describing the angular frequencies and initial phases of the two modes [39]:

$$\begin{cases} \omega_{s1}, \omega_{s2} = \dfrac{\omega_{lo} \pm 2N\pi \times FSR}{2} \\ \varphi_{s1} + \varphi_{s2} = \dfrac{-\omega_{lo}T + 2M\pi}{2} \end{cases} \tag{12.23}$$

where ω_{s1} and φ_{s1} are the angular frequency and initial phase of one mode, respectively, ω_{s2} and φ_{s2} are those of another mode, respectively. $\omega_{lo} = \omega_{s1} + \omega_{s2}$ is the angular frequency of the local oscillator, and M and N are integers. As can be seen from Eq. (12.23), the frequency of the cavity modes of the OEPO is determined by the cavity delay and the local oscillator. The frequencies of each mode pair are symmetrical about half of the frequency of the local oscillator. The minimum mode spacing between the adjacent cavity modes is FSR/2, which is only half of that of a traditional OEO. Moreover, the sum phase of each mode pair is locked to $(-\omega_{lo}T + 2M\pi)/2$, which is a unique phase-controlled operation.

Degenerate oscillation can be further achieved when the frequencies of the two modes in a mode pair are the same, i.e. $\omega_{s1} = \omega_{s2}$. According to Eq. (12.23), the frequency ω_s and initial phase φ_s of the degenerate oscillation can be expressed as:

$$\begin{cases} \omega_s = \dfrac{\omega_{lo}}{2} \\ \varphi_s = -\dfrac{\omega_s T}{2} + K\pi \end{cases} \tag{12.24}$$

where K is an integer. In this case, the oscillating signal is a single-frequency microwave signal whose frequency is free from the cavity delay. The initial phase of the degenerate oscillation is also locked to $-\omega_s T/2 + K\pi$. Due to this phase-locking, i.e. the phase-controlled operation, cavity delay-independent frequency tuning can be achieved [39], which is also not possible in a traditional OEO if the effective cavity delay is not tuned.

The measured multimode and single-mode spectra of the OEPO are shown in Figs. 12.16(a), (b). The frequency of the local oscillator in our experiment is 16 GHz, and the FSR of the OEPO cavity is 1 MHz. We can see that the oscillating modes of the multimode oscillation are symmetrical about half of the frequency of the local oscillator and the minimum mode spacing is FSR/2, which is consistent with our theory. The multimode oscillation is also stable in our experiment once the OEPO is switched on, whereas the multimode oscillation in a traditional OEO is not stable due to the mode competition effect. Degenerate single-mode oscillation is achieved in our experiment by using a loop filter with a narrow-enough bandwidth. A frequency tuning step of 100 Hz is also achieved in our experiment [39], which is much less than the 1-MHz cavity FSR of the OEPO.

In addition to microwave signal generation, the proposed OEPO can also be used for computation and radio-frequency phase-stable transfer. For instance, a microwave photonic Ising machine has been proposed and demonstrated based on the binary-phase oscillation of the OEPO. This method represents a novel approach for the implementation of large-scale, high-coherence Ising machines at room temperature to accelerate the computation of ubiquitous combinatorial optimization problems.

12.6 Broad Random OEO

Typically, a closed optoelectronic feedback loop is adopted in a traditional OEO cavity. Although a high-Q factor or rich dynamic states are ensured by a closed loop, only the frequency components with multiple integers of the FSR can oscillate in the OEO cavity. As we mentioned above, this delay-controlled operation limits the potential oscillation frequency of a traditional OEO once its cavity delay is determined. As a result, the randomness of the generated broadband microwave signals of the broadband OEO and OEPO is limited. Discrete frequency components with frequencies related to the cavity delay can still be observed at the spectrum of the generated signal unless the feedback strength of the optoelectronic loop is ultra-high. One solution to overcome this problem is to use an optoelectronic parametric process in the OEO cavity, which is the method we used in the OEPO. In [40], we also proposed and demonstrated a novel random OEO that overcomes the limitation of the cavity delay in a traditional OEO. Broadband random microwave signals are obtained from the random OEO cavity by using the unique features of random light scattering in an optical fiber as the feedback scheme for the oscillating signal while maintaining an open-loop physical cavity.

The schematic diagram of the proposed random OEO is shown in Fig. 12.17. The signal light generated by a laser diode is sent to a spool of dispersion compensation fiber (DCF) via an optical circulator and a wavelength division multiplexer (WDM). Due to the intrinsic random disorder in the optical fiber, Rayleigh scattering occurs in all directions while the signal light propagates along the fiber. The backscattered part of the signal light is recaptured by the fiber and sent to port 3 of the circulator. A Raman pump signal generated by a pump laser is also sent to the DCF to amplify the incident and backscattered signal light in the DCF and stimulate the Raman amplification process. An EDFA and an EA are also used to further boost the oscillating signal in the optoelectronic cavity.

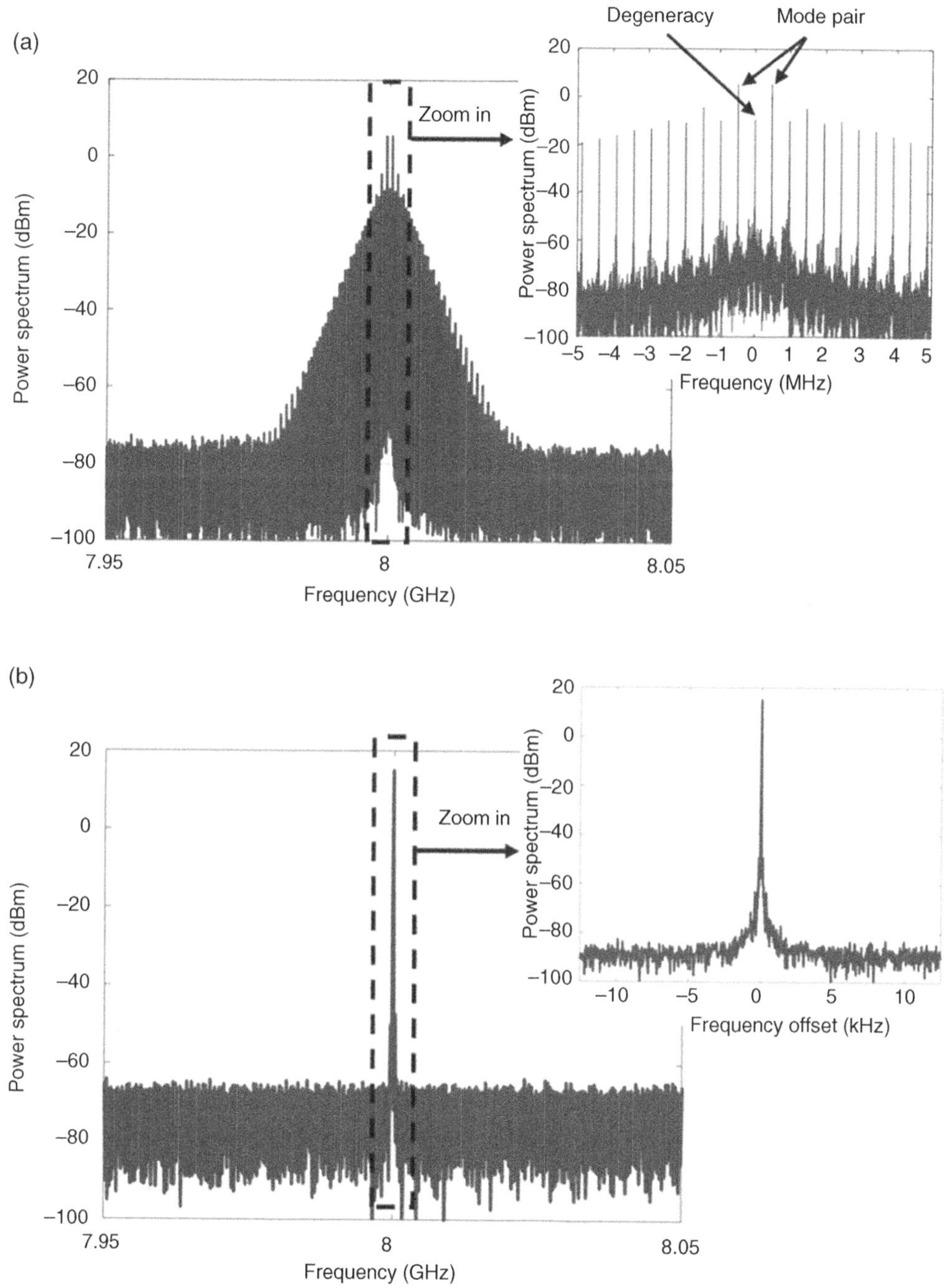

Figure 12.16 (a) Spectrum of the generated stable multi-mode oscillation. (b) Spectrum of the generated single-mode oscillation. *Source:* Hao et al. [39]/Springer Nature/CC BY 4.0.

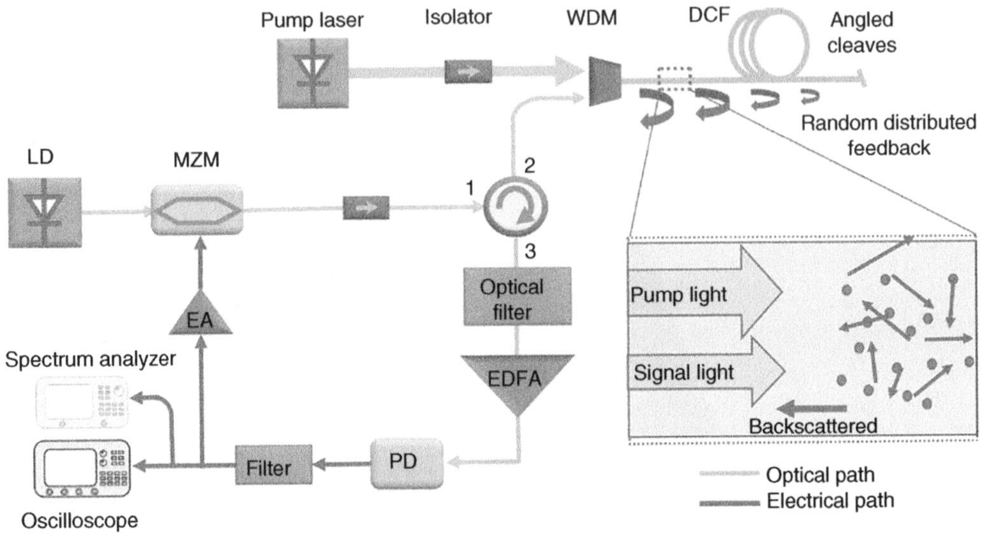

Figure 12.17 A random OEO using Rayleigh light scattering in an optical fiber to provide feedback in an open-loop cavity [40]. The signal light generated by the LD is sent to the DCF via a WDM. Rayleigh scattering occurs as the signal light propagates along the optical fiber due to the intrinsic random disorder in the transmission medium. A Raman pump signal generated by the pump laser is also sent to the DCF to amplify both the incident and backscattered signal light based on the distributed Raman amplification process. The random OEO can be regarded as the sum of singleloop OEOs with randomly distributed cavity lengths since the feedback is proved by the Rayleigh backscattering. As a result, microwave signals with different frequencies can always find the corresponding cavity lengths to oscillate in the random OEO. *Source:* Ge et al. [40]/Springer Nature/CC BY 4.0.

Unlike a traditional OEO that uses a physical closed loop to provide the required feedback for the oscillating signal, Rayleigh backscattering from an open DCF is used to provide the required feedback in the random OEO. At each scatter section z_i in the optical fiber, the instantaneous Rayleigh scattering field $\Delta\varepsilon_b(t, z_i)$ fluctuates randomly in time. The total backscattered field at the input of the DCF is the superimposed field from each scatter section, which can be defined as:

$$\varepsilon_b(t) = \sum_{n=1}^{N} \Delta\varepsilon_b(t, n\Delta l) \tag{12.25}$$

where $N\Delta l = z_L$ is the total length of the DCF. We can see that the total backscattered field also fluctuates randomly in time. A large amount of Rayleigh backscattering can be regarded as different feedback cavity lengths; thus, the random OEO can be regarded as a sum of single-loop OEOs with randomly distributed cavity lengths:

$$L_i = L_0 + 2z_i \tag{12.26}$$

where L_0 is the cavity length of the OEO excluding the DCF, and z_i is the scatter section at the DCF. As a result, microwave signals with different frequencies can always find the corresponding cavity lengths to oscillate in the random OEO and are not confined by the cavity delay as in the traditional OEO. Moreover, the signal power of the microwave signals also fluctuates randomly in time due to the random fluctuation of the backscattered field; thus, random microwave signals can be produced by the proposed random OEO.

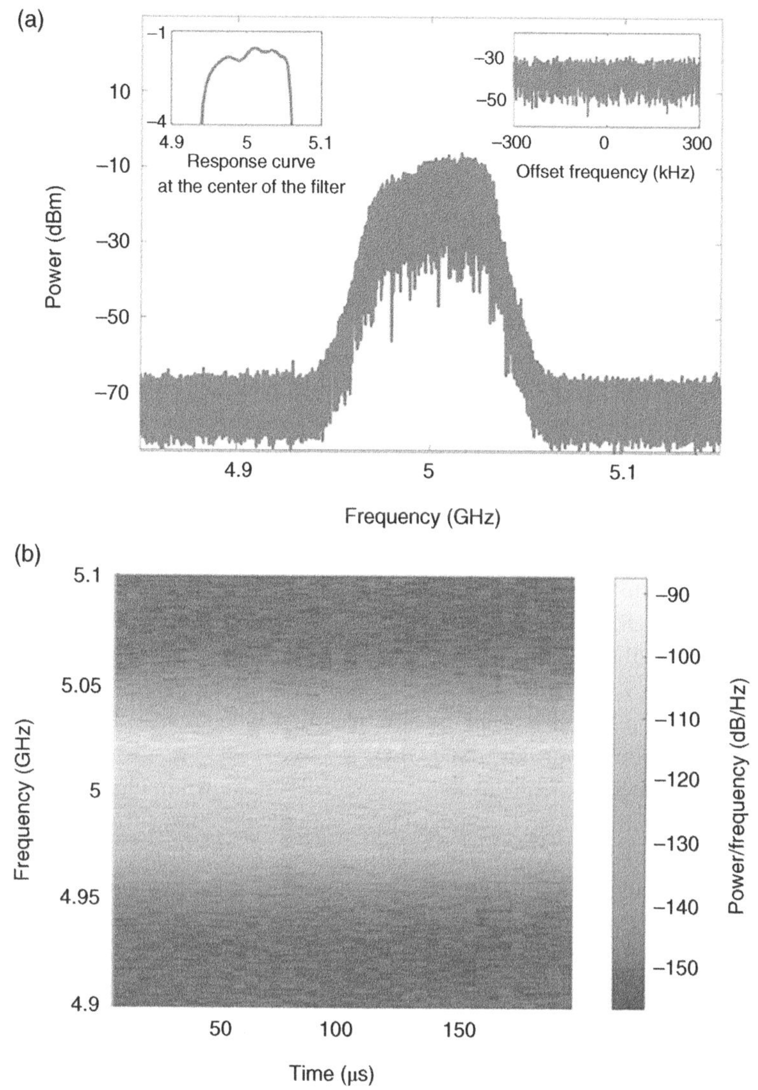

Figure 12.18 (a) Spectrum of the generated random microwave signal when a bandpass filter is used. The inset on the left shows the frequency response of the bandpass filter, and the inset on the right shows the magnification of the spectrum. (b) Instantaneous frequency-time diagram of the generated random microwave signal. *Source:* Ge et al. [40]/Springer Nature/CC BY 4.0.

Figure 12.18(a) shows the spectrum of the generated random microwave signal when a bandpass filter with a center frequency of 5 GHz and a bandwidth of 60 MHz is used. The inset on the left shows the frequency response of the bandpass filter and the inset on the right shows the magnification of the spectrum. All frequencies in the passband of the filter oscillate together without a discrete mode interval. The calculated instantaneous frequency-time diagram using a Hamming window is shown in Fig. 12.18(b). As can be seen, the power of the generated signal varies at different frequencies. The power of a particular frequency changes randomly in each of the two adjacent time windows. Random microwave signals with different center frequencies and bandwidths can also be produced by changing the electrical bandwidth of the random OEO. For example, ultra-wideband random microwave signals with frequencies ranging from DC to about 40 GHz have been successfully produced [40].

12.7 Integrated OEO

Although remarkable developments have been achieved in the past few decades regarding low-phase noise microwave signal generation and cavity mode control of OEOs, the large size and high power consumption still limit the popularization of OEOs in practical applications since most of the aforementioned OEOs are implemented using discrete optical and electrical devices. With the rapid development of photonic integrated circuits (PICs), several partially integrated and even monolithically integrated OEOs [41–48] have been reported in recent years. These devices have a compact size and low power consumption.

Recently, we have proposed and demonstrated an integrated OEO [41, 42] with monolithically integrated photonic parts on an InP substrate and RF parts on a printed circuit board (PCB). Figure 12.19 shows the photograph of the InP integrated OEO. As can be seen, the optical part is monolithically integrated on an InP substrate and the electrical part is fabricated on a PCB. The total size of the InP integrated OEO is 5 cm × 6 cm, which is as small as a coin. Due to the very compact size of the integrated OEO, the loop length is just several centimeters, which corresponds to a large FSR of several GHz. As a result, the requirement of a narrow band filter for mode selection is greatly weakened. The 3-dB bandwidth of the electrical filter is about 7 GHz, which is well enough for singlemode oscillation in our integrated OEO.

The electrical spectra of the generated singlemode microwave signals are shown in Fig. 12.20(a). As can be seen, the frequency is tunable, which is achieved by tuning the injection current of the DML which leads to the tuning of effective cavity length with the help of the dispersion effect. As can be seen, a frequency tuning range of around 20 MHz is achieved in the experiment. We also measured the phase noise performance of the generated microwave signal. As shown in Fig. 12.20(b), the phase noise is −92 dBc/Hz at 1 MHz offset frequency. The phase noise performance is not as good as conventional fiber-based OEOs, which is caused mainly by the small *Q*-factor of the integrated OEO. The phase noise performance can be further enhanced, for example, by the use of a longer low-loss optical delay line. Nevertheless, the InP integrated OEO provides a proof of concept for

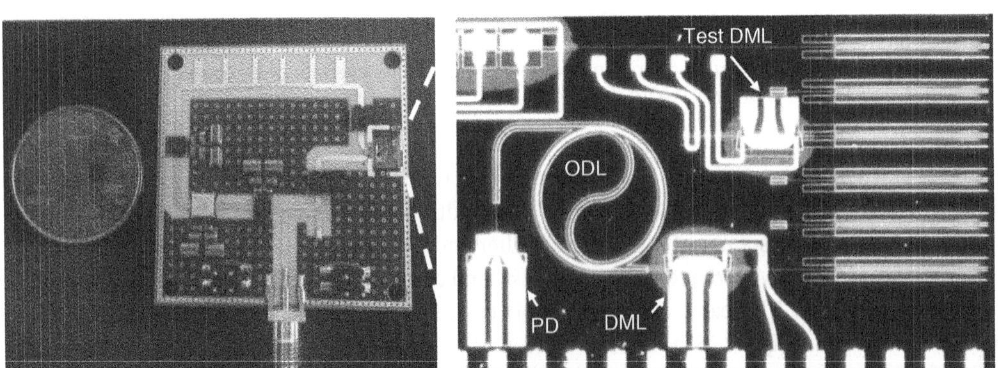

Figure 12.19 An InP integrated OEO with monolithically integrated photonic parts. All the optical devices in the OEO system are integrated into an InP chip. The electrical parts are fabrication on a PCB. *Source:* Hao et al. [45]/from IEEE.

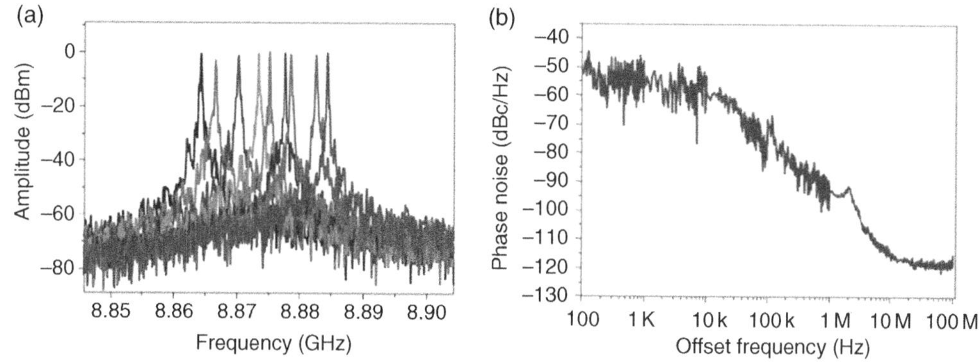

Figure 12.20 (a) Electrical spectra and (b) measured phase noise of the InP integrated OEO. *Source:* Tang et al. [41]/Optical Society of America/CC BY 4.0.

the design of an operational device with a very compact size, which paves the way for future generations of integrated OEOs.

In addition to InP, there coexists a wide diversity of other technology platforms to build PICs, such as silicon, silicon nitrides, lithium niobate, and polymers. Integrated microwave photonic systems including the OEO can be benefited from these technology platforms. For example, a silicon photonic integrated OEO has been proposed and demonstrated in [43, 44]. Figure 12.21 shows the perspective view of the silicon photonic integrated OEO. Three key components of the integrated OEO, including a high-speed PM, a thermally tunable high-selectivity MDR and a high-speed PD are integrated in a silicon photonic chip. The loop filter in the integrated OEO is a silicon photonic integrated MPF based on PM-IM conversion by using the above three key components and an external LD. The closed integrated OEO loop is achieved by amplifying the output signal from the MPF and feeding it back to the input of the MPF. If a sufficiently large gain is provided by an external electrical amplifier, microwave oscillation will start from noise in the integrated OEO. The oscillation frequency is determined by the central frequency of the passband of the MPF, which is equal to the frequency difference between the external LD and the frequency of the notch position of the MDR. By using the thermal-optic effect, the MDR could be thermally tunable. Thus, the central

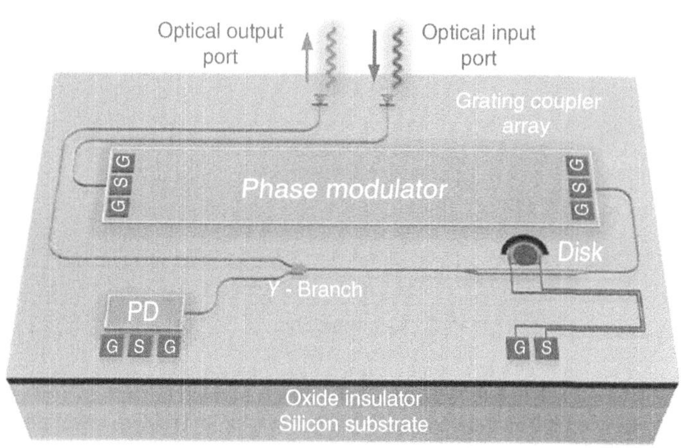

Figure 12.21 A silicon-integrated OEO with several key optical building blocks, including a high-speed PM, a thermally tunable MDR, and a high-speed PD. *Source:* Zhang and Yao [43]/IEEE.

frequency of the passband of the MPF is tuned, which eventually leads to the tuning of the frequency of the generated microwave signal from the integrated OEO. Frequency tuning from about 3–8 GHz is achieved in the experiment, and the measured phase noise performance is around −80 dBc/Hz at 10 kHz offset frequency.

One key advantage of silicon photonics is its compatibility with the mature CMOS technology. It is feasible to realize seamless integration between photonic parts and electronic circuits into the chip. A silicon-integrated OEO formed by hybrid integration of a photonic chip and an electronic chip has also been reported in [46]. Figure 12.22(a) shows the schematic diagram of the

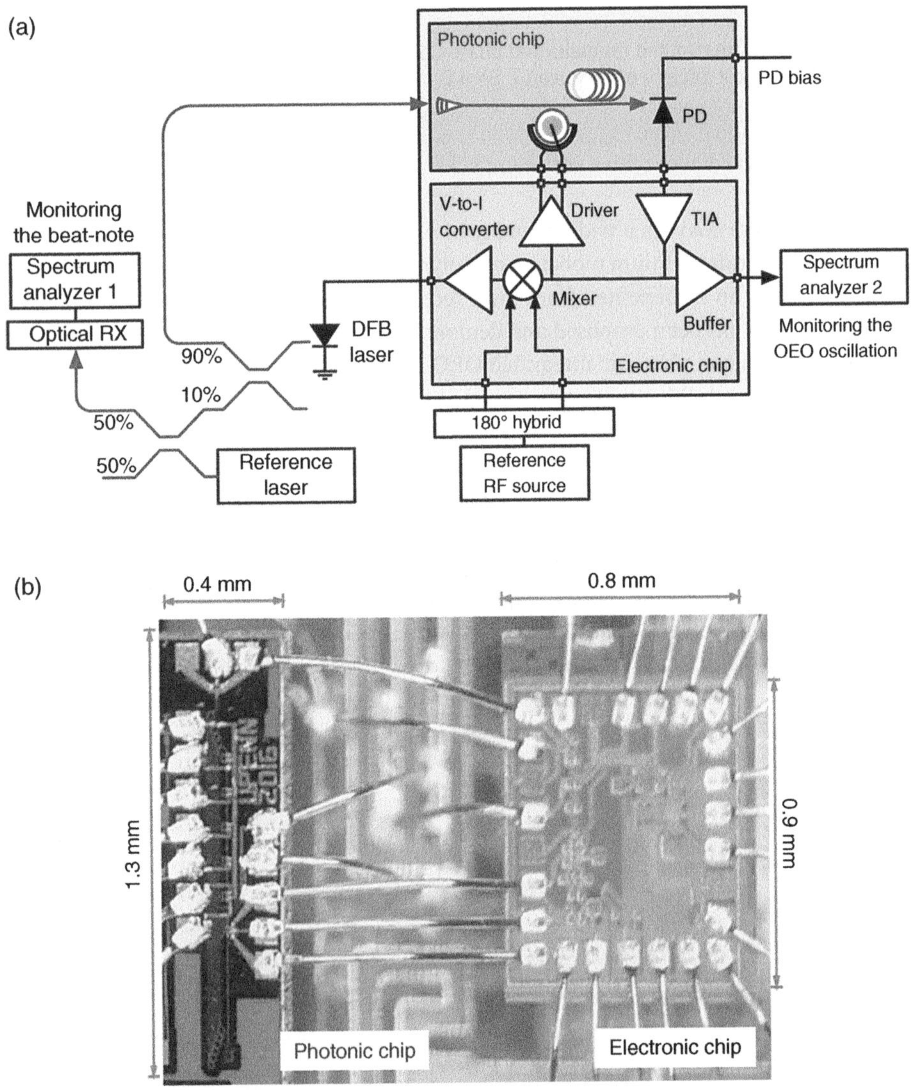

Figure 12.22 A silicon-integrated OEO formed by hybrid integration of a photonic chip and an electronic chip. (a) Schematic diagram. All photonic and electronic building blocks of an OEO are integrated, except for the laser. (b) The micro-photograph of the hybrid-integrated optoelectronic feedback loop. *Source:* Xuan et al. [46]/Optical Society of America/CC BY 4.0.

silicon-integrated OEO. As can be seen, all photonic and electronic building blocks of an OEO are integrated, except for the laser. The photonic chip consists of a grating coupler, a ring modulator, a dispersive optical delay line and a PD. The output of the PD on the photonic chip is wire-bonded to the input of a transimpedance amplifier (TIA) on the electronic chip. The output of the TIA is then split into two parts, one part is fed back to the ring modulator to form a closed OEO loop and is monitored by a spectrum analyzer, and another is used for optoelectronic frequency locking. The micro-photograph of the hybrid-integrated optoelectronic feedback loop is shown in Fig. 12.22(b). The photonic chip is integrated in the IME 180 nm SOI technology and the electronic chip is integrated in the GlobalFoundries 180 nm CMOS SOI technology. The two chips are wire-bonded together. Microwaves oscillation is established in the OEO cavity when an external laser is coupled into the chip and the loop gain exceeds the loss. The measured phase noise is about −103 dBc/Hz at 100 kHz offset frequency. Although the oscillation frequency (about 1 GHz) is low compared with the above-mentioned integrated OEOs, the results demonstrate the great potential of silicon technology for the implementation of fully integrated OEOs.

12.8 Discussion and Conclusion

In this chapter, recent advances in novel OEOs have been discussed. MING model, which is a general model for analyzing OEO states, is proposed. Single-frequency microwave signal generation using a PT-symmetric OEO is first described, where single-frequency oscillation can be obtained without the use of narrowband filters. Novel OEOs that are capable of producing complex microwave waveforms are then discussed, including novel FDML OEO, OEPO, and broadband random OEO. Finally, integrated OEOs with compact size and low power consumption are summarized.

The OEOs are paradigmatic microwave photonics systems for the generation of RF/microwave signals. Various RF/microwave signals, including spectrally pure single-frequency RF signals, wideband chirped microwave waveforms, and chaotic oscillations have been obtained directly from OEO cavity, thanks to the use of various mode control and selection methods. As a delay-line oscillator, the cavity modes of OEOs are a series of frequency-periodic modes whose frequencies are determined by the cavity delay. However, the amplitude and phase relationships of these modes are not bounded by the mode condition. It is still possible to produce other complex or even arbitrary microwave waveforms by selecting specific frequency components from all the potential cavity modes and controlling their amplitude and phase relationships. The above-mentioned PT-symmetric OEO, FDML OEO, OEPO, and broadband random OEO are the latest examples of mode control and selection in OEOs. We can still expect new advances in this area for the generation of different RF/microwave signals in the future.

Integrated OEOs with compact size and low power consumption are highly desired in applications such as satellite communications and 5G networks, where they are expected to generate RF/microwave signals with very low phase noise, for example, bellow −100 dBc/Hz at 10 kHz offset frequency. Although some prototypes have been demonstrated in InP and silicon platforms, the phase noise performances do not meet the high requirements in practical applications. The large phase noise is mainly caused by the reduced Q-factor of the integrated OEOs. This problem could be solved by heterogeneous integration [85–94], where the Q-factor can be enhanced by using ultra-low-loss silicon nitride waveguides [95, 96] or high-Q silicon nitride ring resonators [97]. In addition, to achieve monolithic integration, the electrical parts and a laser must be integrated into the chip for the InP and silicon-integrated OEO, respectively. One potential solution is to combine the strengths of InP platforms and silicon photonics based on the rapidly developed InP membrane on silicon (IMOS) technology [98, 99] to obtain a fully integrated chip consisting of all the active and passive components of an OEO. Moreover, silicon-based lasers may also be possible in the future by modifying the electronic band structure of silicon and silicon-containing alloys [100], which may pave the way for monolithically integrated silicon OEOs.

References

1 Yao, X.S. and Maleki, L. (1994). High frequency optical subcarrier generator. *Electron. Lett.* 30 (18): 1525–1526.

2 Yao, X.S. and Maleki, L. (1996). Optoelectronic oscillator for photonic systems. *IEEE J. Quantum Electron.* 32 (7): 1141–1149.

3 Yao, X.S. and Maleki, L. (1996). Optoelectronic microwave oscillator. *J. Opt. Soc. Am. B* 13 (8): 1725–1735.

4 Yao, X.S. and Maleki, L. (1996). Converting light into spectrally pure microwave oscillation. *Opt. Lett.* 21: 483–485.

5 Hao, T., Liu, Y., Tang, J. et al. (2020). Recent advances in optoelectronic oscillators. *Adv. Photonics* 2 (4): 044001.

6 Li, M., Hao, T., Li, W., and Dai, Y. (2021). Tutorial on optoelectronic oscillator. *APL Photonics* 6: 061101.

7 Maleki, L.2012. The opto-electronic oscillator (OEO): Review and recent progress. *2012 European Frequency and Time Forum* (23–27 April 2012), Gothenburg, Sweden, pp. 497–500.

8 Chembo, Y.K., Brunner, D., Jacquot, M., and Larger, L. (2019). Optoelectronic oscillators with time-delayed feedback. *Rev. Mod. Phys.* 91: 035006.

9 Huggett, G.R. (1968). Modelocking of CW lasers by regenerative RF feedback. *Appl. Phys. Lett.* 13: 186–187.

10 Kersten, R.T. (1978). Ein optisches Nachrichtensystem mit Bauelementen der integrierten Optik für die Übertragung hoher Bitraten. *Arch. Elektrotech.* 60 (6): 353–359.

11 Liu, Y., Hao, T., Li, W. et al. (2018). Observation of parity-time symmetry in microwave photonics. *Light-Sci. Appl.* 7 (1): 38.

12 Zhang, J. and Yao, J.P. (2018). Parity-time symmetric optoelectronic oscillator. *Sci. Adv.* 4 (6): eaar6782.

13 Li, L., Wang, G., Zhang, J., and Yao, J. (2019). A parity-time-symmetric optoelectronic oscillator based on dual-wavelength carriers in a single spatial optoelectronic loop. *2019 International Topical Meeting on Microwave Photonics (MWP)* (7–10 October 2019), Ottawa, ON, Canada, pp. 1–4.

14 Zhang, J., Li, L., Wang, G. et al. (2020). Parity-time symmetry in wavelength space within a single spatial resonator. *Nat. Commun.* 11: 3217.

15 Fan, Z., Zhang, W., Qiu, Q., and Yao, J.P. (2020). Hybrid frequency-tunable parity-time-symmetric optoelectronic oscillator. *J. Lightw. Technol.* 38: 2127–2133.

16 Dai, Z., Fan, Z., Li, P., and Yao, J. (2020). Frequency-tunable parity-time-symmetric optoelectronic oscillator using a polarization-dependent Sagnac loop. *J. Lightw. Technol.* 38 (19): 5327–5332.

17 Teng, C., Zou, X., Li, P. et al. (2020). Fine tunable PT-symmetric optoelectronic oscillator based on laser wavelength tuning. *IEEE Photon. Technol. Lett.* 32 (1): 47–50.

18 Li, P., Dai, Z., Fan, Z. et al. (2020). Parity–time-symmetric frequency-tunable optoelectronic oscillator with a single dual-polarization optical loop. *Opt. Lett.* 45: 3139–3142.

19 Zou, F., Zou, L., Lai, Y. et al. (2021). Parity-time symmetric optoelectronic oscillator based on an integrated mode-locked laser. *IEEE J. Quantum Electron.* 57 (2): 5000209.

20 Ding, Q., Wang, M., Zhang, J. et al. (2020). A precisely frequency-tunable parity-time-symmetric optoelectronic oscillator. *J. Lightw. Technol.* 38: 6569–6577.

21 Teng, C., Zou, X., Li, P. et al. (2020). Wideband frequency-tunable parity-time symmetric optoelectronic oscillator based on hybrid phase and intensity modulations. *J. Lightw. Technol.* 38 (19): 5406–5411.

22 Hao, T., Cen, Q., Dai, Y. et al. (2018). Breaking the limitation of mode building time in an optoelectronic oscillator. *Nat. Commun.* 9: 1839.

23 Hao, T., Tang, J., Li, W. et al. (2018). Tunable Fourier domain mode locked optoelectronic oscillator using stimulated Brillouin scattering. *IEEE Photon. Technol. Lett.* 30 (21): 1842–1845.

24 Hao, T., Tang, J., Li, W. et al. (2018). Fourier domain mode locked optoelectronic oscillator based on the deamplification of stimulated Brillouin scattering. *OSA Contin.* 1 (2): 408–415.

25 Hao, T., Tang, J., Li, W. et al. (2019). Harmonically Fourier domain mode-locked optoelectronic oscillator. *IEEE Photon. Technol. Lett.* 31 (6): 427–430.

26 R. Liu, P. Du, C. Fu, et al.(2019). Generating ultra-wideband LFM waveforms with large time duration based on frequency-sweeping optoelectronic oscillation. *18th International Conference on Optical Communications and Networks (ICOCN)* (5–8 August 2019), Huangshan, China, pp. 1–3.

27 Hao, T., Tang, J., Shi, N. et al. (2019). Dual-chirp Fourier domain mode-locked optoelectronic oscillator. *Opt. Lett.* 44 (8): 1912–1915.

28 Zhang, L., Zeng, Z., Zhang, Y. et al. (2020). Frequency-sweep-range-reconfigurable complementary linearly chirped microwave waveform pair generation by using a Fourier domain mode locking optoelectronic oscillator based on stimulated Brillouin scattering. *IEEE Photonics J.* 12 (3): 5501010.

29 Li, Y., Hao, T., Li, G. et al. (2021). Photonic generation of phase-coded microwave signals based on Fourier domain mode locking. *IEEE Photon. Technol. Lett.* 33: 433–436. under revision).

30 Liu, R., Wang, A., Du, P. et al. (2020). Simultaneous generation of ultra-wideband LFM and phase-coded LFM microwave waveforms based on an improved frequency-sweeping OEO. *Opt. Commun.* 459: 124938.

31 Zhu, S., Fan, X., Xu, B. et al. (2020). Polarization manipulated Fourier domain mode-locked optoelectronic oscillator. *J. Lightw. Technol.* 38 (19): 5270–5277.

32 Zeng, Z., Zhang, L., Zhang, Y. et al. (2020). Frequency-definable linearly chirped microwave waveform generation by a Fourier domain mode locking optoelectronic oscillator based on stimulated Brillouin scattering. *Opt. Express* 28 (9): 13861–13870.

33 Zhang, H., Zhang, F., Pan, S. et al. (2020). Photonic generation of linearly chirped microwave waveforms with tunable parameters. *IEEE Photon. Technol. Lett.* 32 (17): 1037–1040.

34 Hao, P., Lu, H., Han, R. et al. (2020). Fourier domain mode-locked opto-electronic oscillator with a diode-tuned bandpass filter. *Opt. Express* 28 (16): 23454–23466.

35 Wang, Y., Li, X., Zhang, J., and Wo, J. (2021). Spurious level and phase noise improved Fourier domain mode-locked optoelectronic oscillator based on a self-injection-locking technique. *Opt. Express* 29: 7535–7543.

36 Zhang, X., Zeng, H., Yang, J. et al. (2020). Novel RF-source-free reconfigurable microwave photonic radar. *Opt. Express* 28 (9): 13650–13661.

37 Hao, T., Tang, J., Li, W. et al. (2018). Microwave photonics frequency-to-time mapping based on a Fourier domain mode locked optoelectronic oscillator. *Opt. Express* 26 (26): 33582–33591.

38 Hao, T., Tang, J., Shi, N. et al. (2019). Multiple-frequency measurement based on a Fourier domain mode-locked optoelectronic oscillator operating around oscillation threshold. *Opt. Lett.* 44 (12): 3062–3065.

39 Hao, T., Cen, Q., Guan, S. et al. (2020). Optoelectronic parametric oscillator. *Light Sci. Appl.* 9: 102.

40 Ge, Z., Hao, T., Capmany, J. et al. (2020). Broadband random optoelectronic oscillator. *Nat. Commun.* 11: 5724.

41 Tang, J., Hao, T., Li, W. et al. (2018). Integrated optoelectronic oscillator. *Opt. Express* 26 (9): 12257–12265.

42 Tang, J., Hao, T., Li, W.et al. (2017). An integrated optoelectronic oscillator. *2017 International Topical Meeting on Microwave Photonics (MWP)*, Beijing, China, pp. 1–4.

43 Zhang, W. and Yao, J.P. (2018). Silicon photonic integrated optoelectronic oscillator for frequency-tunable microwave generation. *J. Lightw. Technol.* 36 (19): 4655–4663.

44 Zhang, W. and Yao, J.P.(2017).A silicon photonic integrated frequency-tunable optoelectronic oscillator. *2017 International Topical Meeting on Microwave Photonics (MWP)* (23–26 October 2017), Beijing, China, pp. 1–4.

45 Hao, T., Tang, J., Domenech, D. et al. (2018). Toward monolithic integration of OEOs: from systems to chips. *J. Lightw. Technol.* 36 (19): 4565–4582.

46 Xuan, Z., Du, L., and Aflatouni, F. (2019). Frequency locking of semiconductor lasers to RF oscillators using hybrid-integrated opto-electronic oscillators with dispersive delay lines. *Opt. Express* 27 (8): 10729–10737.

47 Merklein, M., Stiller, B., Kabakova, I.V. et al. (2016). Widely tunable, low phase noise microwave source based on a photonic chip. *Opt. Lett.* 41 (20): 4633–4636.

48 Do, P.T., Alonso-Ramos, C., Le Roux, X. et al. (2020). Wideband tunable microwave signal generation in a silicon-micro-ring-based optoelectronic oscillator. *Sci. Rep.* 10: 6982.

49 Maleki, L. (2011). The optoelectronic oscillator. *Nat. Photonics* 5 (12): 728–730.

50 Strekalov, D., Aveline, D., Yu, N. et al. (2003). Stabilizing an optoelectronic microwave oscillator with photonic filters. *J. Lightw. Technol.* 21 (12): 3052.

51 Ji, Y., Yao, X., and Maleki, L. (1999). Compact optoelectronic oscillator with ultra-low phase noise performance. *Electron. Lett.* 35 (18): 1554–1555.

52 Davidson, T., Goldgeier, P., Eisenstein, G., and Orenstein, M. (1999). High spectral purity CW oscillation and pulse generation in optoelectronic microwave oscillator. *Electron. Lett.* 35 (15): 1260–1261.

53 Chembo, Y.K., Larger, L., and Colet, P. (2008). Nonlinear dynamics and spectral stability of optoelectronic microwave oscillators. *IEEE J. Quantum Electron.* 44 (9): 858–866.

54 Bao, Y., Banyas, E., and Illing, L. (2018). Periodic and quasiperiodic dynamics of optoelectronic oscillators with narrow-band time-delayed feedback. *Phys. Rev. E* 98 (6): 062207.

55 Xiao, Y., Dai, Y., Shi, N. et al. (2022). Microwave-photonics iterative nonlinear gain model for optoelectronic oscillators. *Opt. Express* 30 (7): 12131–12149.

56 Callan, K.E., Illing, L., Gao, Z. et al. (2010). Broadband chaos generated by an optoelectronic oscillator. *Phys. Rev. Lett.* 104 (11): 113901.

57 Goedgebuer, J.-P., Levy, P., Larger, L. et al. (2002). Optical communication with synchronized hyperchaos generated electrooptically. *IEEE J. Quantum Electron.* 38 (9): 1178–1183.

58 Black, H.S. (1934). Stabilized feedback amplifiers. *Bell Syst. Tech. J.* 13 (1): 1–18.

59 Oppenheim, A.V., Willsky, A.S., Nawab, S.H., and Hernández, G.M. (1997). *Signals & Systems*. Pearson Educación.

60 Chembo, Y.K., Larger, L., Tavernier, H. et al. (2007). Dynamic instabilities of microwaves generated with optoelectronic oscillators. *Opt. Lett.* 32 (17): 2571–2573.

61 Banerjee, A., de Britto, L.A.D., and Pacheco, G.M. (2019). Analysis of injection locking and pulling in single-loop optoelectronic oscillator. *IEEE Trans. Microw. Theory Tech.* 67 (5): 2087–2094.

62 Bogataj, L., Vidmar, M., and Batagelj, B. (2014). A feedback control loop for frequency stabilization in an opto-electronic oscillator. *J. Lightw. Technol.* 32 (20): 3690–3694.

63 Saruwatari, M. and Kawanishi, S.Ultrawide frequency response measurement system using optical heterodyne detection method. *1988 Conference on Precision Electromagnetic Measurements* (7–10 June 1988), Ibaraki, Japan, pp. 403–405.
64 Zhenghua, Z., Chun, Y., Zhewei, C. et al. (2015). An ultra-low phase noise and highly stable optoelectronic oscillator utilizing IL-PLL. *IEEE Photon. Technol. Lett.* 28 (4): 516–519.
65 Lee, K.-H., Kim, J.-Y., and Choi, W.-Y. (2008). Injection-locked hybrid optoelectronic oscillators for single-mode oscillation. *IEEE Photon. Technol. Lett.* 20 (19): 1645–1647.
66 Fu, R., Jin, X., Zhu, Y. et al. (2017). Frequency stability optimization of an OEO using phase-locked-loop and self-injection-locking. *Opt. Commun.* 386: 27–30.
67 Zhang, X., Zhou, X., and Daryoush, A.S. (1992). A theoretical and experimental study of the noise behavior of subharmonically injection locked local oscillators. *IEEE Trans. Microw. Theory Tech.* 40 (5): 895–902.
68 Urick, V.J., Williams, K.J., and McKinney, J.D. (2015). *Fundamentals of Microwave Photonics*. Wiley.
69 Larger, L. (2013). Complexity in electro-optic delay dynamics: modelling, design and applications. *Philos. Trans. R. Soc. A* 371 (1999): 20120464.
70 Neyer, A. and Voges, E. (1982). Dynamics of electrooptic bistable devices with delayed feedback. *IEEE J. Quantum Electron.* 18 (12): 2009–2015.
71 Feigenbaum, M.J. (1983). Universal behavior in nonlinear systems. *Phys. D Nonlinear Phenom.* 7 (1–3): 16–39.
72 Collet, P. and Eckmann, J.-P. (2009). *Iterated Maps on the Interval as Dynamical Systems*. Springer Science & Business Media.
73 Bender, C.M. and Boettcher, S. (1998). Real spectra in non-Hermitian hamiltonians having PT symmetry. *Phys. Rev. Lett.* 80: 5243.
74 Hatano, N. and Nelson, D.R. (1996). Localization transitions in non-Hermitian quantum mechanics. *Phys. Rev. Lett.* 77: 570.
75 Zheng, C., Hao, L., and Long, G.L. (2013). Observation of a fast evolution in a parity-time-symmetric system. *Philos. Trans. R. Soc. A Math. Phys. Eng. Sci.* 371: 20120053.
76 Feng, L., El-Ganainy, R., and Ge, L. (2017). Non-Hermitian photonics based on parity–time symmetry. *Nature Photon.* 11: 752.
77 Hodaei, H., Miri, M.-A., Heinrich, M. et al. (2014). Parity-time-symmetric microring lasers. *Science* 346 (6212): 975–978.
78 Feng, L., Wong, Z.J., Ma, R.-M. et al. (2014). Single-mode laser by parity-time symmetry breaking. *Science* 346 (6212): 972–975.
79 Liu, W., Li, M., Guzzon, R.S. et al. (2017). An integrated parity-time symmetric wavelength-tunable single-mode microring laser. *Nat. Commun.* 8: 15389.
80 Chang, L., Jiang, X., Hua, S. et al. (2014). Parity–time symmetry and variable optical isolation in active–passive-coupled microresonators. *Nat. Photonics* 8: 524–529.
81 Peng, B., Özdemir, Ş.K., Lei, F. et al. (2014). Parity–time-symmetric whispering-gallery microcavities. *Nat. Phys.* 10: 394–398.
82 Schindler, J., Li, A., Zheng, M.C. et al. (2011). Experimental study of active LRC circuits with PT symmetries. *Phys. Rev. A* 84: 040101.
83 Li, W., Kong, F., and Yao, J. (2013). Arbitrary microwave waveform generation based on a tunable optoelectronic oscillator. *J. Lightw. Technol.* 31 (23): 3780–3786.
84 Li, W. and Yao, J. (2014). Generation of linearly chirped microwave waveform with an increased time-bandwidth product based on a tunable optoelectronic oscillator and a recirculating phase modulation loop. *J. Lightw. Technol.* 32 (20): 3573–3579.
85 Marpaung, D., Yao, J., and Capmany, J. (2019). Integrated microwave photonics. *Nat. Photonics* 13: 80–90.

86 de Valincourt, G., Chang, C.M., Eggleston, M.S. et al. (2018). Photonic integrated circuit based on Hybrid III-V/Silicon Integration. *J. Lightw. Technol.* 36 (2): 265–273.

87 Smit, M., Leijtens, X., Ambrosius, H. et al. (2014). An introduction to InP-based generic integration technology. *Semicond. Sci. Technol.* 29: 083001.

88 Thomson, D., Zilkie, A., Bowers, J.E. et al. (2016). Roadmap on silicon photonics. *J. Opt.* 18: 073003.

89 Zhang, W. and Yao, J. (2016). Silicon-based integrated microwave photonics. *IEEE J. Quantum Electron.* 52: 0600412.

90 Leuthold, J., Koos, C., and Freude, W. (2010). Nonlinear silicon photonics. *Nat. Photonics* 4: 535–544.

91 Moss, D.J., Morandotti, R., Gaeta, A.L., and Lipson, M. (2013). New CMOS compatible platforms based on silicon nitride and Hydex for nonlinear optics. *Nat. Photonics* 7: 597–607.

92 Park, H., Zhang, C., Tran, M.A., and Komljenovic, T. (2020). Heterogeneous silicon nitride photonics. *Optica* 7 (4): 336–337.

93 Q. Yu, J. Gao, N. Ye, et al. (2020). Heterogeneous photodiodes on silicon nitride waveguides with 20 GHz bandwidth. *OFC 2020* (8–12 March 2020), San Diego, paper W4G.1.

94 Roeloffzen, C.H.G., Zhuang, L., Taddei, C. et al. (2013). Silicon nitride microwave photonic circuits. *Opt. Express* 21 (19): 22937–22961.

95 Bauters, J.F., Heck, M.J.R., John, D.D. et al. (2011). Planar waveguides with less than 0.1 dB/m propagation loss fabricated with wafer bonding. *Opt. Express* 19 (24): 24090–24101.

96 Bauters, J.F., Heck, M.J.R., John, D.D. et al. (2011). Ultra-low-loss high-aspect-ratio Si3N4 waveguides. *Opt. Express* 19 (4): 3163–3174.

97 Xiang, C., Davenport, M.L., Khurgin, J.B. et al. (2018). Low-loss continuously tunable optical true time delay based on Si3N4 ring resonators. *IEEE J. Sel. Topics Quantum Electron.* 24 (4): 5900109.

98 van der Tol, J.J.G.M., Jiao, Y., Van Engelen, J.P. et al. (2020). InP membrane on silicon (IMOS) photonics. *IEEE J. Quantum Electron.* 56 (1): 6300107.

99 Kish, F., Lal, V., Evans, P. et al. (2018). System-on-chip photonic integrated circuits. *IEEE J. Sel. Topics Quantum Electron.* 24 (1): 6100120.

100 Fadaly, E.M.T., Dijkstra, A., Suckert, J.R. et al. (2020). Direct-bandgap emission from hexagonal Ge and SiGe alloys. *Nature* 580: 205–209.

13

Integrated Microwave Photonics

13.1 Introduction

Throughout this book, we have addressed microwave photonics (MWP) from a holistic point of view, including the main technologies, devices, systems, and subsystem configurations and also the most relevant analysis and design techniques. It should now be clear to the reader that MWP deals with the generation, processing, and distribution of microwave and millimeter-wave signals by optical means benefiting from the unique advantages inherent to photonics, such as low-loss (independent of frequency), high-bandwidth, and immunity to electromagnetic interference [1–4]. On top of these, MWP brings the fundamental added value of enabling the realization of key functionalities in microwave systems, such as reconfigurable filtering, arbitrary waveform generation (AWG), beamforming, agile frequency up/down conversion and multiplication, phase shifting, true-time delaying, instantaneous measurement, and analog-to-digital conversion, which are either complex or even not directly possible in the radiofrequency (RF) domain. In addition, MWP creates new opportunities for information and communication technology (ICT) systems and networks.

While initially the research activity in MWP was focused toward defence applications, it has now expanded to civil scenarios, spanning conventional areas such as cellular [5], wireless [6] and satellite [7] communications, distributed antenna systems [8] and optical signal processing [9], as well as to emerging fields such as converged fiber-wireless [10] and in-home networks [11], medical imaging systems using terahertz (THz) waves, and instrumentation [4]. However, and despite the tremendous potential of MWP unveiled by the former areas, the widespread use and application of this technology are currently limited by the high-cost, bulky, complex, and power-consuming nature of its systems. Thus, reducing their cost as well as size, weight, and power consumption (SWAP) is a major challenge in MWP research. In addition, emerging paradigms such as 5G communications [12, 13], the Internet of Things (IoT) [14], car-to-car communications [15], wireless body and personal area networks (WB/PANs) [16], and high-resolution sensing [17] are expected to push this pressure even further (see Fig. 13.1). The requirements demanded by most of these scenarios call for novel technology developments in both the physical layer and the network architectures. For instance, 5G wireless communications targets a 1000-fold increase in capacity, connectivity for over 1 billion users, strict latency control, and network software programming.

Integrated photonics has the potential to change the power-scaling laws of high-bandwidth systems through proper architectural choices that combine photonics with electronics to optimize performance, power, footprint, and cost [18]. In particular, MWP has a qualitatively different

Microwave Photonics, First Edition. Jianping Yao, José Capmany, and Ming Li.

Companion website: www.wiley.com/go/yao/microwavephotonics

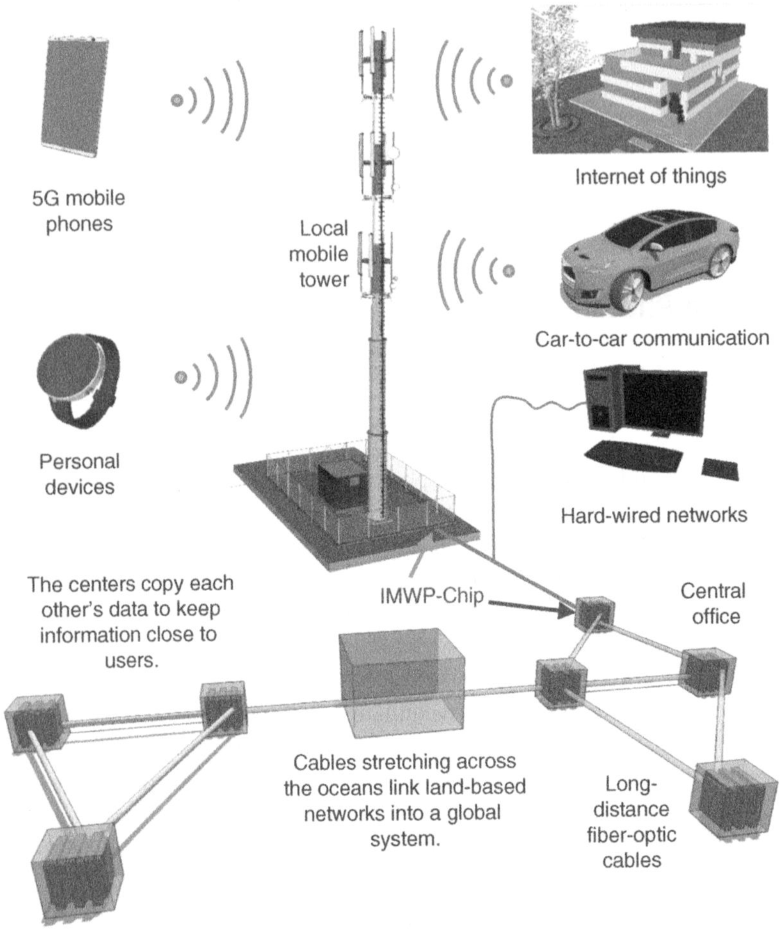

Figure 13.1 Emerging ICT applications requiring smooth transition between the fiber and wireless segments of communication infrastructures. *Source:* Adapted from Hetch [15].

behavior compared to digital electronics since the energy per analog task is dominated by the steady-state bias power and does not increase significantly as the bandwidth increases [18]. Furthermore, most photonic devices are currently highly temperature-dependent and, therefore, temperature regulation is required, which consumes the majority of bias power. As integrated photonics favors alternative means of temperature control, the power consumed by the photonic devices can be drastically reduced. Integrated MWP (IMWP) [19–21] that aims at the incorporation of MWP components/subsystems in monolithic or hybrid photonic circuits is instrumental to achieving the aforementioned evolution objectives.

This chapter outlines the principles of IMPW and reviews some of the more salient results reported in the recent literature. The main available integration technologies are covered in Section 13.2 providing the reader with a general background of the pros and cons of each one of these platforms. To date, the activity in IMWP has been almost exclusively focused toward the so-called application-specific photonic integrated circuits (ASPICs) where a particular circuit and chip configuration is designed to optimally perform a particular MWP functionality. Section 13.3 reviews some of the most important ASPIC results reported in the literature by functionality.

The ASPIC approach however results in almost as many technologies as applications and, due to this considerable fragmentation, the market for many of these application-specific technologies is too small to justify their further development into cost-effective industrial mass-volume manufacturing processes. A radically different approach is to design a universal MWP signal processor architecture that can be integrated on a chip and is capable of performing all the main functionalities by suitable software programming of its control signals. This is inspired by the flexibility of digital signal processors, and field programmable gate arrays (FPGAs), where a common hardware is shared by multiple functionalities through a software-defined approach (or programmability), leading to significant cost reduction in the hardware fabrication. Section 13.5 presents the concept of the universal MWP programmable processor, including its architecture and design principles. Finally, Section 13.6 concludes the chapter and provides future directions for research.

13.2 Integration Technologies and Platforms

Several material platforms have been employed to implement basic building blocks (BBBs), devices, and ASPICs for MWP applications. Active materials, i.e. those able to generate/amplify light, such as GaAs and InP (and combinations) excel in light generation, detection, and modulation. Silicon materials may well be split into three subfamilies: silica (fiber-like) waveguide technology, silicon (oxy)nitride and silicon on insulator (SOI). While nitride-based technology provides superb filtering and packaging solutions, SOI provides light detection, modulation, and natural merge with electronics. Lithium niobate (LN) is mainly employed in modulator technology and, recently, chalcogenide materials have also been employed. Of all these material platforms, only three: (i) indium phosphide (InP), (ii) silicon on insulator (SOI), and (iii) silicon (oxy)nitride (Si_4N_3—SiO_2) have reached the required degree of maturity to be currently considered viable options for the implementation of complex MWP-integrated circuits as they are aligned with generic integration (GIM) and generic foundry (GFM) models [22]. In the GIM, a small set of standardized BBBs is provided to control the basic properties of light and more complex circuits or compound building blocks (CBBs) are then built by connecting several BBBs. In the GFM, the foundry provides shared open access to its GI process through multiproject wafers, where different designs or ASPICs from different users are combined on the same wafer providing cost sharing and additional advantages such as on-wafer testing. Since GIM and GFM allow a broad range of functionalities to be synthesized from a small set of BBBs, they have been identified as the only path leading to mass production and subsequent cost reduction for PIC manufacturing, where figures in the order of 10 €/mm^2 can already be reached at volumes of 1000 chips for InP technology [22].

13.2.1 Indium Phosphide

InP is a semiconductor material and the only technology capable of the monolithic integration of active (i.e. featuring optical amplification) and passive photonic components [23, 24]. A variety of techniques, including butt-joint regrowth, selective area growth, offset, dual quantum well placing, and quantum well intermixing, can be employed to integrate regions with different absorption/gain properties along a single waveguide. Waveguide types include shallow (low loss and large bending radius) and deeply etched (higher losses and lower bending radius) designs (see top row in Fig. 13.2). Shallow waveguides have typical widths of 2 μm and this figure is approximately 1.5 μm for deeply etched designs. Typical losses are around 1.5 and 3 dB/cm for shallow and

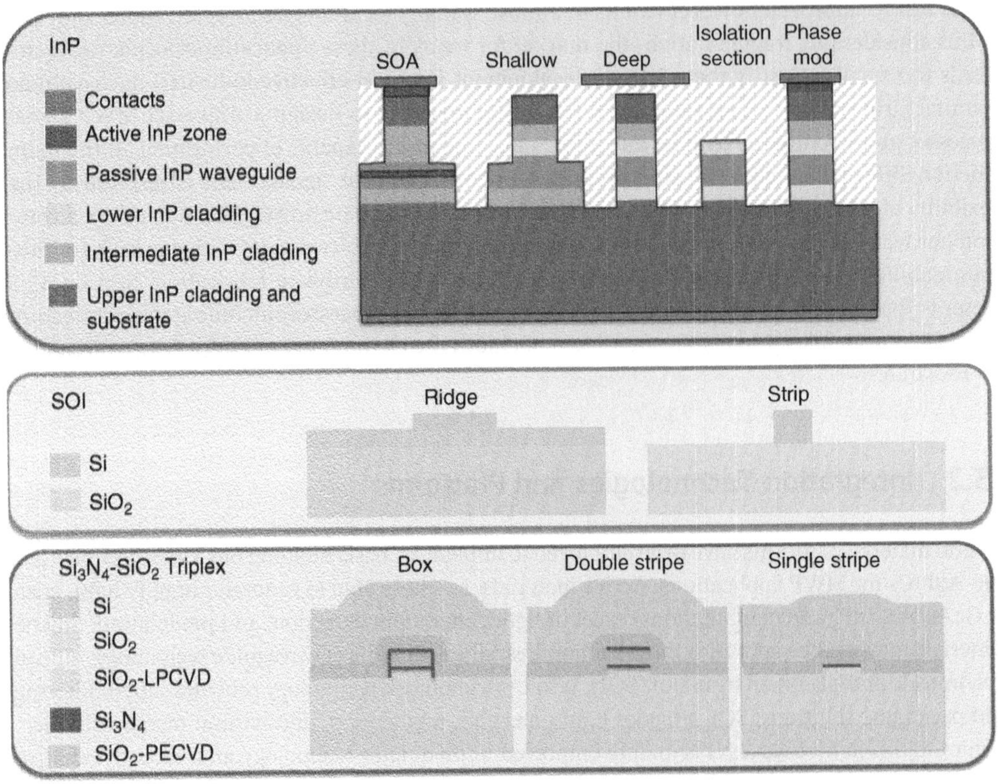

Figure 13.2 Waveguide structures for the InP (top row), SOI (intermediate row), and Si_3N_4–SiO_2 (bottom row) platforms.

deeply etched waveguides, respectively, and this is usually quoted as one of the disadvantages of this technology. These losses can be reduced however to figures below 1 dB/cm by removing the p-type top layer in the nonactive parts of the chip. Minimum bending radii are in between 10 and 500 μm for deeply etched and shallow etched waveguides, respectively. Another quoted limitation of the InP platform is related to the difficulty of its integration with electronics into the same chip. However, highly promising work is underway at the COBRA research center of the Technical University of Eindhoven (TU/e) toward the integration of the full photonic functionality in a single InP Membrane on Silicon (IMO) that could be implemented both in a CMOS or an InP fab, merging InP photonics with silicon electronics.

InP provides the most complete list of available components for integration including [22]: passives, optical filters, arrayed waveguide gratings, semiconductor optical amplifiers (SOAs), saturable absorbers, optical sources (Fabry–Perot, distributed Bragg reflector (DBR), distributed feedback (DFB), and modelocked lasers), detectors (pin single and balanced configurations), amplitude and phase modulators, and polarization-handling devices. Moreover, it has shown record performance in integrated component count on the same chip with figures exceeding 450 and now targeting 1000 [24]. GIM and GFM platforms have been developed with Europe taking the lead as evidenced by several initiatives including the pioneering ePIXnet network of excellence and the EuroPIC and PARADIGM projects [22]. In fact, access to InP-based generic processes is organized by the JePPIX platform that apart from bringing together Europe's key players in the field offers semicommercial access.

13.2.2 Silicon Photonics

Also known as silicon on insulator (SOI) [25–28], silicon (Si) photonics is a semiconductor technology where components are etched/patterned/fabricated in a 180–220 nm Si layer placed on top of a 1–3 μm insulator layer [25–27]. Si passives are formed by initial few mask layers through partial and/or full Si-etching steps after which multiple ion implantations are conducted for "active" devices such as Ge photodetectors and Si modulators. Coupling into and outside of the chip can be performed via edge couplers (with typical losses of 1 dB/facet) or vertically, via Si surface gratings (3.5–6 dB/coupler with 40–70 nm 3-dB bandwidth). The main advantage of SOI technology resides in its compatibility with CMOS fabrication processes and the infrastructure used in microelectronics and thus in the potential for monolithic integration of the electronic and photonic parts of the chip. Refractive index contrast is over 100% ($n = 3.4$ for Si and $n = 1.45$ for SiO_2), leading to small footprint circuits. Two main types of waveguides are available (see intermediate row in Fig. 13.2): ridge or shallow (1–8 μm width), which exhibit relatively low losses down to 0.1–0.5 dB/cm, but are limited in bending radius to around 100 μm, and strip waveguides (500 nm width) that exhibit much higher losses (1–3 dB/cm) but support lower values for minimum bending radius (5–20 μm). Integration density on a chip is currently below 100 components [28], but the component count integration trend is exceeding the rate given by Moore's law indeed. Several basic photonic components are available in monolithic SOI, including passives, such as arrayed waveguide gratings and optical filters, Ge photodetectors, and ring and travelling-wave electrorefractive modulators (up to 50 GHz). GFM and GIM are being developed in Europe where access is granted to IMEC, CEA-LETI, and IHP foundries through the Europractice platform (www.europractice-ic.com) which also provide several BBBs. The main disadvantage of monolithic SOI technology is that it does not support optical sources and other active components such as optical amplifiers. To overcome this limitation, III–V functionalities have to be integrated into the SOI platform by means of either molecular [29] or adhesive [30] wafer bonding. This approach, known as hybrid silicon or hybrid integration technology, has succeeded in incorporating DFB and modelocked lasers, optical amplifiers, electroabsorption (>70 GHz), Mach–Zehnder (>25 GHz), modulators, photodetectors (>35 GHz), and polarization-handling components into SOI chips. However, GFM and GIM have not yet been developed for hybrid integration technology.

13.2.3 Silicon Nitride

This waveguide technology is based on a combination of stoichiometric silicon nitride (Si_3N_4) as waveguide layers, filled by and encapsulated with silica (SiO_2) as cladding layers grown on a silicon wafer [31, 32]. SiO_2 and Si_3N_4 layers are fabricated with CMOS-compatible industrial standard low-pressure chemical vapor deposition (LPCVD) equipment that enables cost-effective volume production. A special technology known as TriPleX™ [31], developed by the Dutch company LioniX (www.lionixbv.nl), allows the fabrication of waveguides with minimized surface roughness allowing high refractive index contrast (20–30%) and low scattering losses. Three different waveguide cross-sectional geometries are standardized [31] (see bottom row in Fig. 13.2). Their salient characteristics @1550 nm are: box-shaped (1 μm × 1 μm) that features reduced birefringence, ≤0.2 dB/cm losses [19, 31] and minimum bending radius around 150 μm; double-stripe (1 μm × 1 μm, suited for sharp bends) obtained by removing the sidewalls of the box configuration, which leads to a significant reduction of the waveguide losses that are now in the 0.095–0.12 dB/cm range and minimum bending radii in between 50 and 70 μm; and single-stripe (suited for ultra-low loss propagation) that is highly birefringent (core thickness of 40 nm and width of 13 μm) and features ultralow losses of 0.00045 dB/cm (current world record). In-/out-coupling is achieved by means of adiabatically tapered spot-size converters with <1 dB coupling loss. Several fundamental building blocks are available including the optical waveguide, tuning elements, and directional and multimode interference couplers. From these, more complex subsystems have been demonstrated. For the double stripe geometry, a library of standard optical components with predictable characteristics is available and is currently being offered as a multiproject wafer service run by JePPIX platform. The main disadvantage of this technology is that no optical sources, detectors, amplifiers, and modulators are available. The integration of these components requires a hybrid approach with separately fabricated InP platform chips [32]. Coupling these two platforms together with low insertion loss is targeted in the roadmap of JePPIX [33].

13.2.4 Other Platforms

Besides the three platforms discussed above, IMWP has been demonstrated with other materials, including GaAs, $LiNbO_3$, polymers, and chalcogenide glasses (ChG). LN has been employed for developing linear, high-performance electro-optic (EO) modulators. In the context of MWP, it has been used in beamforming applications, in optoelectronic converters [34–36] as well as in ultrawideband (UWB) signal generation. ChGs are amorphous semiconductors that contain as a major constituent one or more of the chalcogen elements from group 6A of the periodic table (sulfur, selenium, and tellurium, but excluding oxygen), covalently bonded to other elements such as arsenic, germanium, antimony, gallium, silicon, or phosphorus [37, 38]. They offer unconventional properties such as a high refractive index of 2–3, transparency up to the midinfrared wavelengths, and a high nonlinear refractive index n_2 up to a thousand times that of silica. In addition, ChGs are free from nonlinear losses such as two-photon absorption. A typical ChG optical waveguide consists of an arsenic trisulfide (As_2S_3) core on silica, with silica or polymer overclad. The most common waveguide geometry is a rib waveguide with a cross section of 2–4 nm by 0.85 nm. In addition to high Kerr nonlinearities, the As_2S_3 waveguide has been shown to support gigahertz frequency acoustic modes. This gives rise to stimulated Brillouin Scattering (SBS). Harnessing these nonlinear phenomena in centimeter-scale optical waveguides (instead of several kilometers of optical fibers) leads to the potential miniaturization of key MWP functionalities, as recently reported [39–41]. Silica (fiber-like) waveguide technology provides an excellent platform for the implementation of passive devices. However, owing to its relatively low refractive index, contrast leads to a low density of integration.

13.2.5 Comparative Analysis

Table 13.1 gives a general overview per material technology of the building blocks available, that are required in an MWP system [42]. Table 13.2 gives a technical overview per material technology. The first two rows, refractive index contrast and bend radius, are related to the circuit footprint. The third entry, attenuation coefficient, together with the two previous ones, is related to the quality of optical filters.

Remarkable work resulted in the hybrid integration of InP and silicon material technologies [42]. The last column in Table 13.1 (after [42]) remarks all the required building blocks for MWP system can be built within this material technology, i.e. sources (DFB and mode-locked lasers), amplifiers, electroabsorption modulators, all sorts of passives (filters, couplers, polarization-handling blocks ...) as well as high-speed photodetectors [29]. How this hybrid technology is of application to MWP is reviewed in [42], where the focus is not to beat traditional discrete component aggregation in MWP systems (on some aspects impossible, i.e. Watt operating fiber lasers, see below) but rather on exploiting novel system concepts enabled by this hybrid integration scheme. Apart from the works by [42], other groups are pursuing the integration of gain media on silicon technologies, and they have demonstrated amplifiers and lasers with aluminum oxide erbium-doped silicon waveguides [43–45].

Table 13.1 Building block features available per material technology.

	GaAs	InP	$LiNbO_3$	Si(O)N, TriPleX	SOI	Hybrid InP/SOI
TL/BS	+++	+++				+++
PD	+++	+++			++	+++
MOD	++	++	+++		+	++
Passive	+	+	+	+++	++	+++
WPKG				+++	+++	+++
ELECT					+++	+++

ELECT, Electronic System-on-Chip and System-in-Package integration; MOD, modulator; Passive, passive devices (couplers, filters, ...); PD, photodetector; TL/BS, Tunable laser or broadband source; WPKG, wafer-level packaging. silica on silicon not shown, would excel in passive devices.
Source: Adapted from [42].

Table 13.2 Technology features per material technology.

	GaAs/InP	$LiNbO_3$	Si(O)N	TriPleX	SOI
Δn (%)	~10	~~5	<5	>25	>100
r_b (mm)	0.1	0.2	0.2-0.8	0.05	0.05
α (dB/cm)	2.5	0.2	0.05-0.3	0.01	<2
Transparency	NIR	NIR	VIS-NIR	VIS-IR	NIR
Fiber-chip coupling	–	+	+	+++	–
Function integration	++++	+++	+	+++	++
Fabrication cost	––	––	++	+++	++

Δn = refractive index contrast, r_b = bend radius (related to overall circuit footprint), α = attenuation. IR = infrared, VIS = visible, NIR = near infrared. "+" good, positive, "−" worse, negative. Note Table 13.1 includes a hybrid InP/SOI column not shown in this table, since the hybrid combination may benefit from the positive aspects of both technologies, shown in separate columns.
Source: Adapted from [32].

The active functions for the building blocks in Table 13.1 (laser, modulator, and photodetector) are subject to demanding requirements in an MWP system. The basic devices for signal generation in MWP systems are high-power, low-relative intensity and phase noise optical sources, and broadband linear external modulators. On-chip semiconductor lasers provide lower-power values (300 mW) compared to watt-class Er-doped fiber lasers (3 W) but can feature very competitive performance in terms of RIN. For instance, an InGaAlAs/InP quantum-well, high-power, low-noise packaged semiconductor external cavity laser (ECL) operating at 1550 nm has been recently reported [46] capable of providing 370 mW of fiber-coupled power with a linewidth of 1 kHz, and relative intensity noise −160 dB/Hz from 200 kHz to 10 GHz.

High RF power-handling broadband linear external modulators have been reported in III–V semiconductor technologies. For instance, both electroabsorption (EAM) and Mach–Zehnder modulators coupling with low noise (RIN < −160 dB/Hz) tunable sampled grating DBR InP lasers have been reported. EAM-based transmitters have been reported featuring a dynamic range of 126.3 dB $Hz^{4/5}$, while figures in excess of 115 dB $Hz^{4/5}$ can be achieved for MZ transmitters.

Both values are over the required level with very low $V\pi$ values. For instance, in [47], an advanced semiconductor design with a transfer substrate has been reported featuring a 0.3 V $V\pi$ value, capable of handling an RF power of up to 200 mW and providing a modulation bandwidth in excess of 40 GHz.

Finally, photodetectors delivering high RF power are required in order to enhance the signal-to-noise ratio and dynamic range. Among the possible photodetector design, the unitravelling carrier (UTC) configuration exhibits high-saturation current at high frequencies [48, 49]. High-power UTC detectors delivering up to 31.7 dBm (1.5 W) at a 3 dB bandwidth of 8 GHz have been reported [48].

13.3 Application-Specific Photonic Integrated Circuits for Microwave Photonics

IMWP with a variable degree of complexity arising from the combination of building blocks have been reported for a variety of functionalities during the last years. In particular, ASPICS have been demonstrated. For some of these functions, an excerpt of relevant works is provided in the following subsections and summarized in Table 13.3. For further details, refer to [19, 20, 50]. We now review some salient results:

Table 13.3 Excerpt of recent IMWP devices.

Function	Material	FSR	CF	BW	R
Filter	InP	23.5 GHz or 47 GHz	28 GHz	1.9–5.4 GHz	32 dB
Filter	InP	—	10–40 GHz	3.9–7.1 GHz	40 dB
Filter	TriPleX	—	2–8 GHz	247–840 MHz	60 dB
Function	**Material**	**Pulse rate**	**Pulse width**	**BW**	**FCC**
AWG	SiN	595 GHz	300 fs	—	—
AWG	As2S3 Chalc.	—	—	4–15 GHz	Yes
AWG	InP	—	—	4–15 GHz	Yes
AWG	TriPleX	—	—	4–15 GHz	Yes
Function	**Material**	**PS**	**PS BW**		
TBPS	SOI	0–90°	16–20 GHz	—	—
Function	**Material**	**Delay**	**Delay BW**		
TTTD	SOI	0–25.7 ps	8 GHz	—	—
TTTD	TriPleX	0–396 ps	1 GHz	—	—
TTTD	SOI	21–96 ps	—	—	—
Function	**Material**	**Range**	**Error**	**Power**	
IFM	InP	5–15 GHz	200 MHz	[−3, 7] dBm	—
IFM	TriPleX	0.5–4 GHz	93.6 MHz	—	—
Function	**Material**	**Lock range**	**Loop BW**		
OPLL	InP	[−9, 7.5] GHz	400 MHz	—	—
Function	**Material**	**RF band**	**Inst. BW**	**# AE**	
BFM	TriPleX	2–10 GHz	8 GHz	16	—

AE, antenna element; BW, bandwidth; CF, center frequency; FCC, US Federal Communications Commission compliance; FSR, free spectral range; Inst, instantaneous; PS, phase shift; R, rejection.
Source: After Marpaung et al. [19] and Muñoz et al. [50].

13.3.1 Filters

IMWP filters have been reported both for coherent and incoherent operation. However, the first approach leads to more compact and versatile configurations. So, we will only discuss them here. Relevant results for coherent filters have been reported in [42, 51–66]. Many of the preliminary approaches have been based mainly on single-cavity ring resonators. A few however have also focused on more elaborate designs, which involve more than one cavity and programmable features. These filters can be useful particularly when the RF information has already been modulated onto the lightwave carrier and it might be advisable to perform some prefiltering in the optical domain prior to the receiver or either by direct translation of the optical filter characteristic to the RF domain by beating the (unfiltered) optical carrier with the filtered sideband using an SSB-modulated signal. Representative results from one cavity filter can be found in [42, 51, 52]. For instance, Norberg [52] reports the results for a unit cell, shown in the top part of Fig. 13.3 that could be an element of more complex lattice filters. This unit cell, integrated in InP-InGaAsP, is composed of two forward paths and contains one ring. By selectively biasing one SOA and phase modulators placed in the arms of the unit cell, filters with a single pole, a single zero, or a combination of both can be programmed as shown in the bottom part of Fig. 13.3. In particular, for the design reported in [52], the frequency-tuning range spans around 100 GHz. A hybrid version incorporating silicon waveguides has also been reported [53] that combines III–V quantum well layers bonded with low-loss passive silicon waveguides. Low-loss waveguides allow for long loop delays while III–V quantum devices provide active tuning capability. The same group involved in [54] subsequently reported results of more complex tunable designs involving second- and third-order filters as well as other different unit cell configurations [53].

Silicon photonics second-order and fifth-order ring resonator filters were reported in [58, 59] providing a very high extinction ratio (>50 dB) and narrowband (1–2 GHz) bandpass characteristics, which make them especially useful in microwave/RF signal processing.

The top part of Fig. 13.4 shows the filter layouts while their spectral response can be found in the bottom part of the same figure.

CMOS-compatible reconfigurable filters using cascades of unit cells have also been reported in SOI [63]. Figure 13.5 illustrates the configuration of such filters, which employ a ring-loaded MZI configuration as a unit cell. Phase shifters (PSs), based on free-carrier plasma dispersion effect [14] in silicon photonics, are the basic tuning elements for each unit cell.

For lower RF frequencies, implementations have been reported in TriPleX technology, for the 2–8 GHz band, with bandwidths in 300–800 MHz and 60 dB rejection. Furthermore, the low-propagation loss of this technology allows for longer cavity perimeters and thus more selective filters [32, 65, 66].

To date, most filter implementations are incomplete in the sense that not all the required elements are included in the chip. The exception is a recently reported monolithic MWP filter implemented in InP technology [64].

13.3.2 Microwave Signal Generators

IMPW ASPICs have been reported, which implement a wide variety of microwave generation possibilities ranging from simple continuous wave (CW) high-purity single by means of optoelectronic oscillations (OEOs), comb carriers, and impulse radio ultra-wideband (IR-UWB) signals [67–73] to the most versatile AWG.

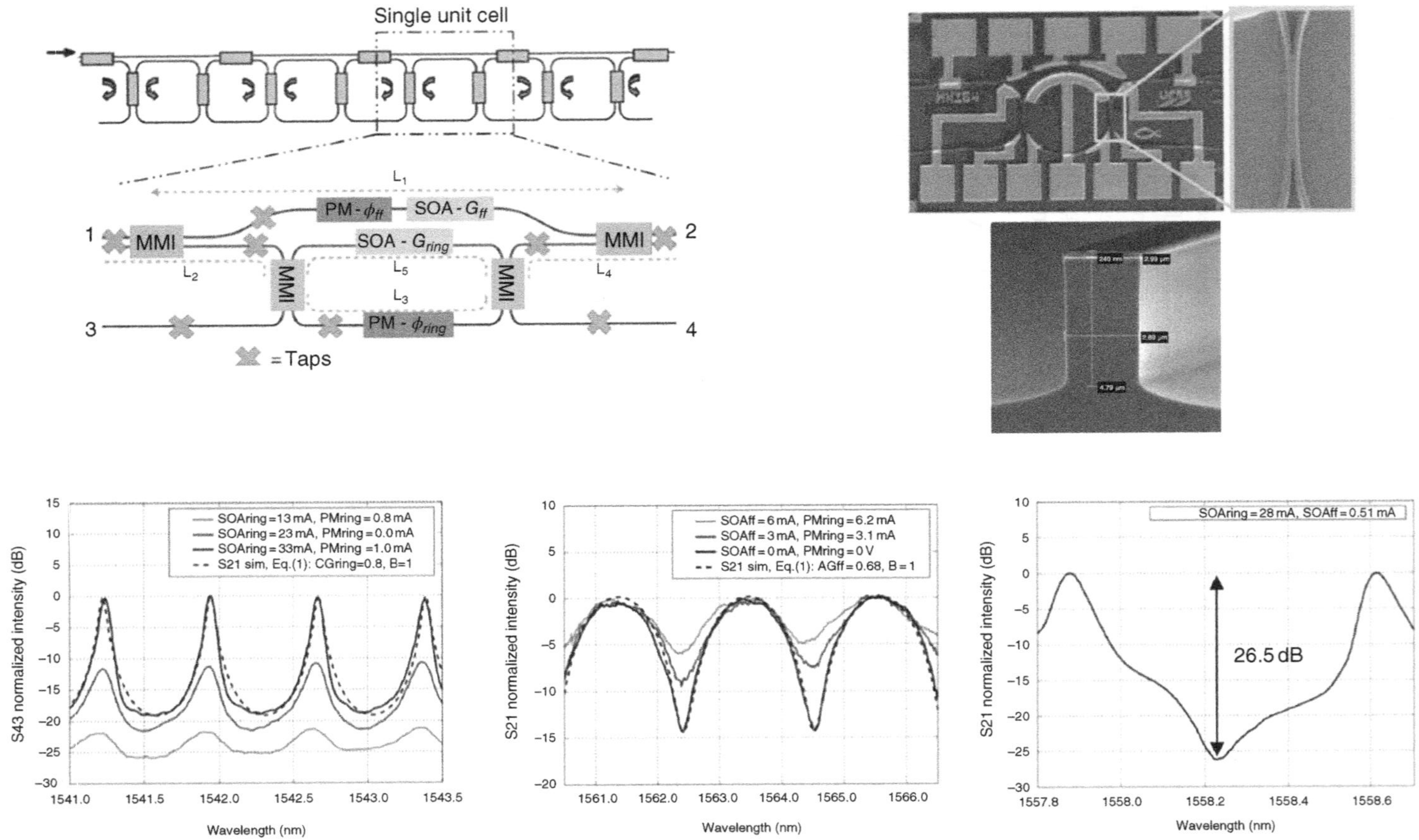

Figure 13.3 Integrated InP-InGaAsP first-order MWP coherent filter providing one pole and one zero reported in [52]. (Top) Schematic of a unit cell configuration and SEM images of the contacts and waveguides. (Bottom) Measured transfer functions for one pole (left), one zero (center), and one pole-one zero (right) configurations. *Source:* Norberg et al. [52]/IEEE.

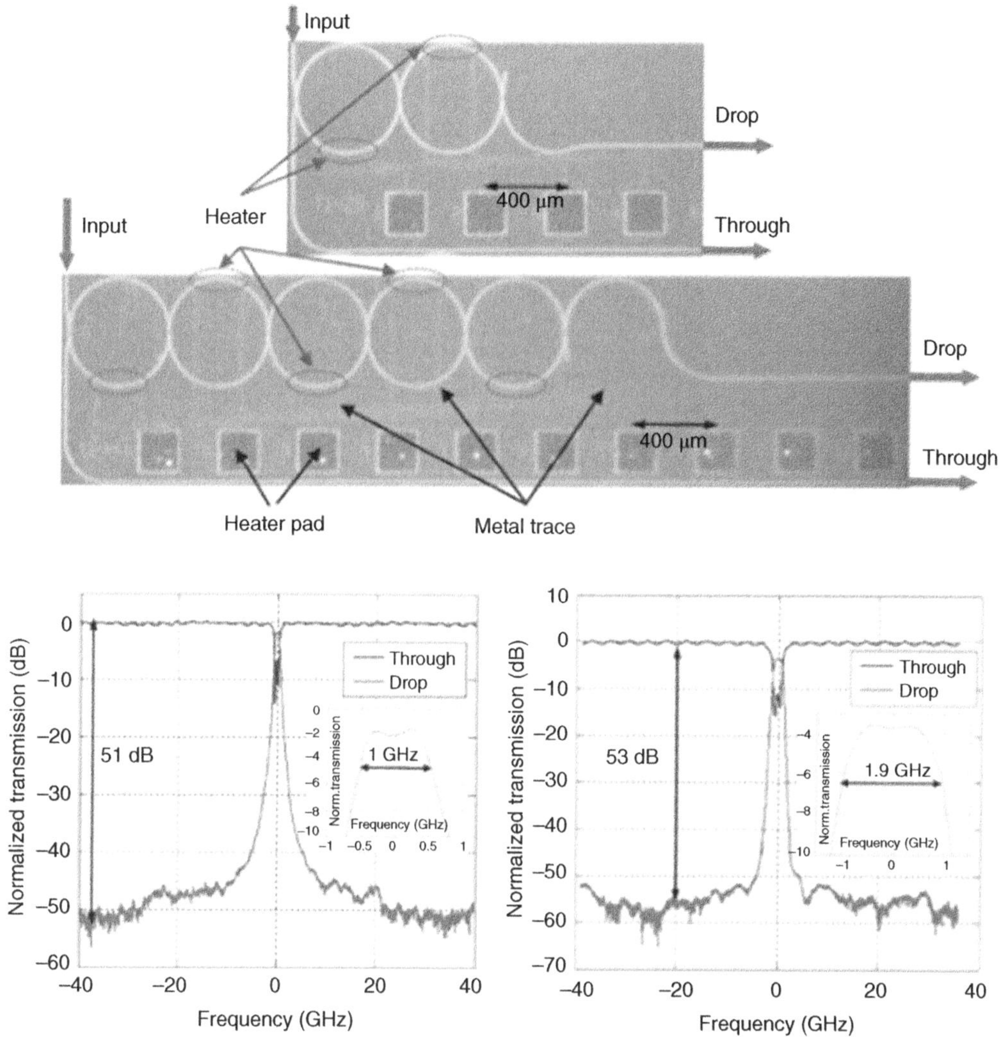

Figure 13.4 Integrated two (top) and five (middle) cavity silicon MWP-coherent filters. Details of measured passbands (bottom) for the two (left) and five (right) cavity structures as reported in [59]. *Source:* Reproduced from Feng et al. [59]/Optica Publishing Group.

13.3.2.1 Optoelectronic Oscillators

The spectral purity of the signal in an OEO is directly related to the Q-factor of the loop and so far, most OEOs utilize a long length of fiber to achieve high spectral purity. A disadvantage of using a fiber loop is the production of super modes that appear in the phase noise spectrum caused by the propagation of waves multiple times around the OEO loop. In addition, a fiber delay line is bulky, so the oscillators cannot be considered an optimal solution for the implementation of a transportable microwave source. In addition, this bulky delay line element has to be temperature-stabilized, a feedback control process which is energy-consuming. A solution to circumvent all these disadvantages has been proposed which also leads to the possibility of full integration of OEOs. It consists in replacing the optical fiber loop with a high Q-factor cavity implemented by means of a whispering

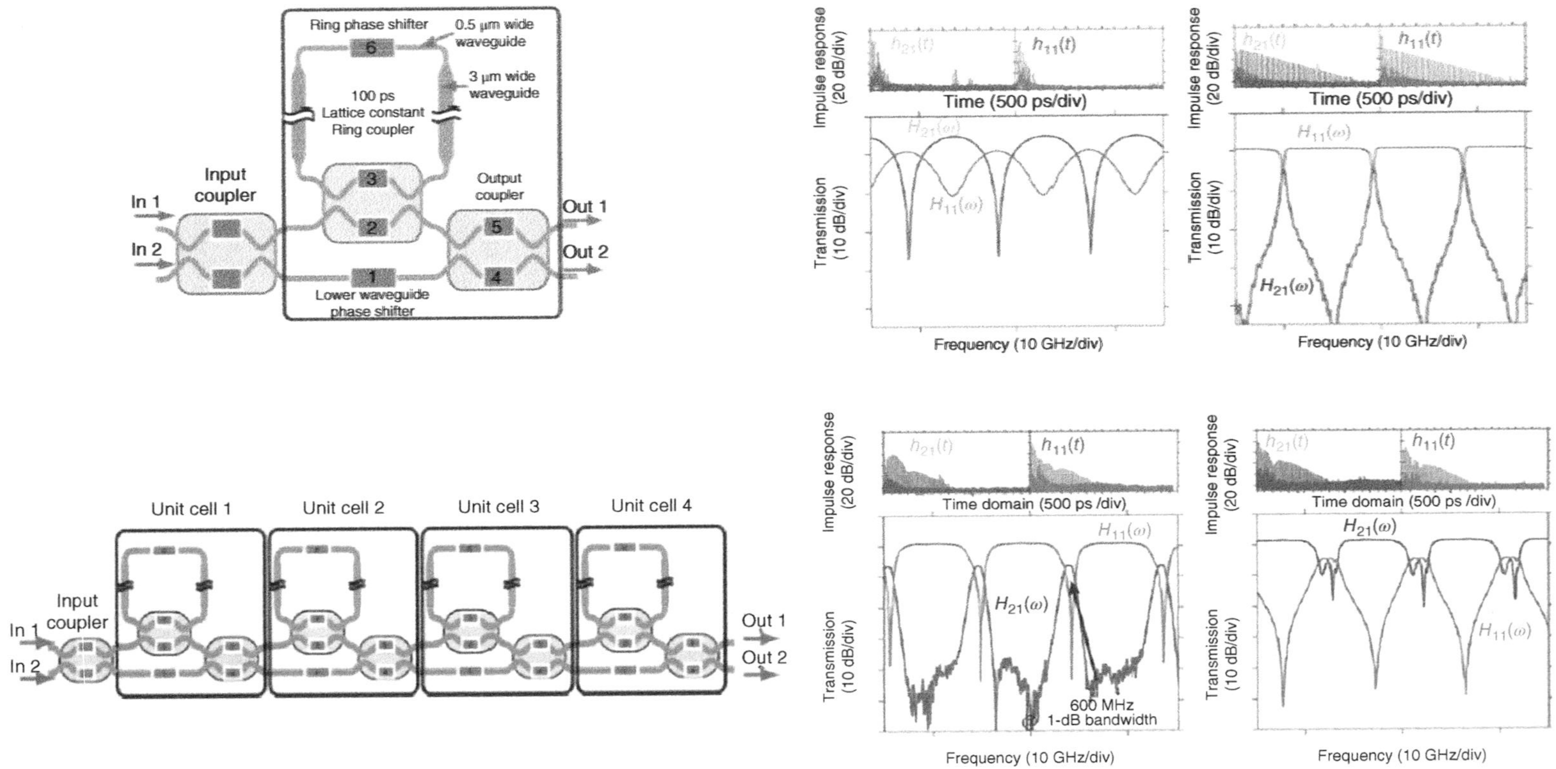

Figure 13.5 Integrated ICMOS-compatible SOI-tunable filter. Unit cell (top left). Zero-only and zero-pole response for the unit cell (top right). Four-cell filter configuration (bottom left) and several examples of tunable responses by changing the pole-zero location by means of the tunable phase shifters.

gallery mode resonator (WGMR) [74–77]. WGMRs ranging in size from a few hundred micrometers to a few millimeters can be fabricated from a wide variety of optically transparent materials reaching Q-factors in the range of 3×10^{11}. OEOs based on high-Q WGMRs made from EO materials [74] can provide high performance in a miniaturized package smaller than a coin, operating in frequency ranges from 10 to 40 GHz and featuring instantaneous linewidths below 200 Hz. Furthermore, in this configuration, the resonator serves both as the high-Q element and as the modulator in the OEO loop.

13.3.2.2 Comb Sources

Optical Comb generation on a chip [78] enables several applications like the precise measurement of optical frequencies through direct referencing to microwave atomic clocks [79] and the production of multiple taps for high sidelobe rejection signal processors [80]. Recently, the generation of optical frequency combs has been reported by means of nonlinear process of CW optical-parametric oscillation using microscale resonators. The devices implemented through this approach have the potential to yield highly compact and frequency-agile comb sources. For instance, Foster et al. [79] report the generation of optical frequency combs from a compact CMOS-compatible integrated microresonator optical parametric oscillator where both the microresonator and the coupling waveguide are fabricated monolithically in a single silicon nitride layer using electron-beam lithography and subsequently clad with silica. This approach brings the advantage of providing a fully monolithic and sealed device with coupling and operation that is insensitive to the surrounding environment.

13.3.2.3 IR-UWB Generators

Liu et al. reported the use of an SOI ORR as a temporal differentiator to shape input Gaussian electrical pulses into their first-order derivatives (i.e. monocycles) [73]. Inspired by a similar technique, a silicon add-drop ORR has been used to generate monocycles from 12.5 Gb/s NRZ signals [67]. A main drawback of Gaussian monocycles is that they cannot fill the FCC spectral mask with high efficiency. For this reason, higher-order derivatives of the Gaussian pulses are often desired. Marpaung et al. reported the use of a cascade of two add-drop ORRs in TriPleX technology to generate the second-order derivatives (doublets and modified doublets) of the input Gaussian pulses used to modulate the phase of the optical carrier [69]. The PM-IM transfer using the cascaded ORRs forms an MWP bandpass filter that shapes the input Gaussian spectrum accordingly. This is the integrated photonics implementation of the MWP delay-line filtering technique. An example of the spectral filtering to generate the Gaussian doublet is shown in Fig. 13.6.

An important aspect of IR-UWB pulse shaping is to fill the spectral mask with a high-power efficiency. Mirshafiei et al. [69] have demonstrated an output pulse obtained from a linear combination of a Gaussian pulse and its copy, filtered using a silicon ORR. By careful adjustment of the amplitude and relative time delay between the pulses, an output pulse with a power efficiency of 52% has been obtained.

InP-integrated chips have also been reported for the generation of (IR-UWB). In this case, the principle of operation does not rely on a filtering structure but rather on exploiting the cross-phase modulation present in an SOA-unified structure. The key components of this system are an integrated MZI with two SOAs in its branches and an optical processing unit. The combination of these two components permits the generation and customization of UWB monocycle pulses. Finally, on-chip nonlinear optics have been used for IR-UWB generation. These include a monocycle generation based on two-photon absorption in a silicon waveguide [68] and monocycle generation exploiting the parametric attenuation effect of sum-frequency generation (SFG) [71] or by using

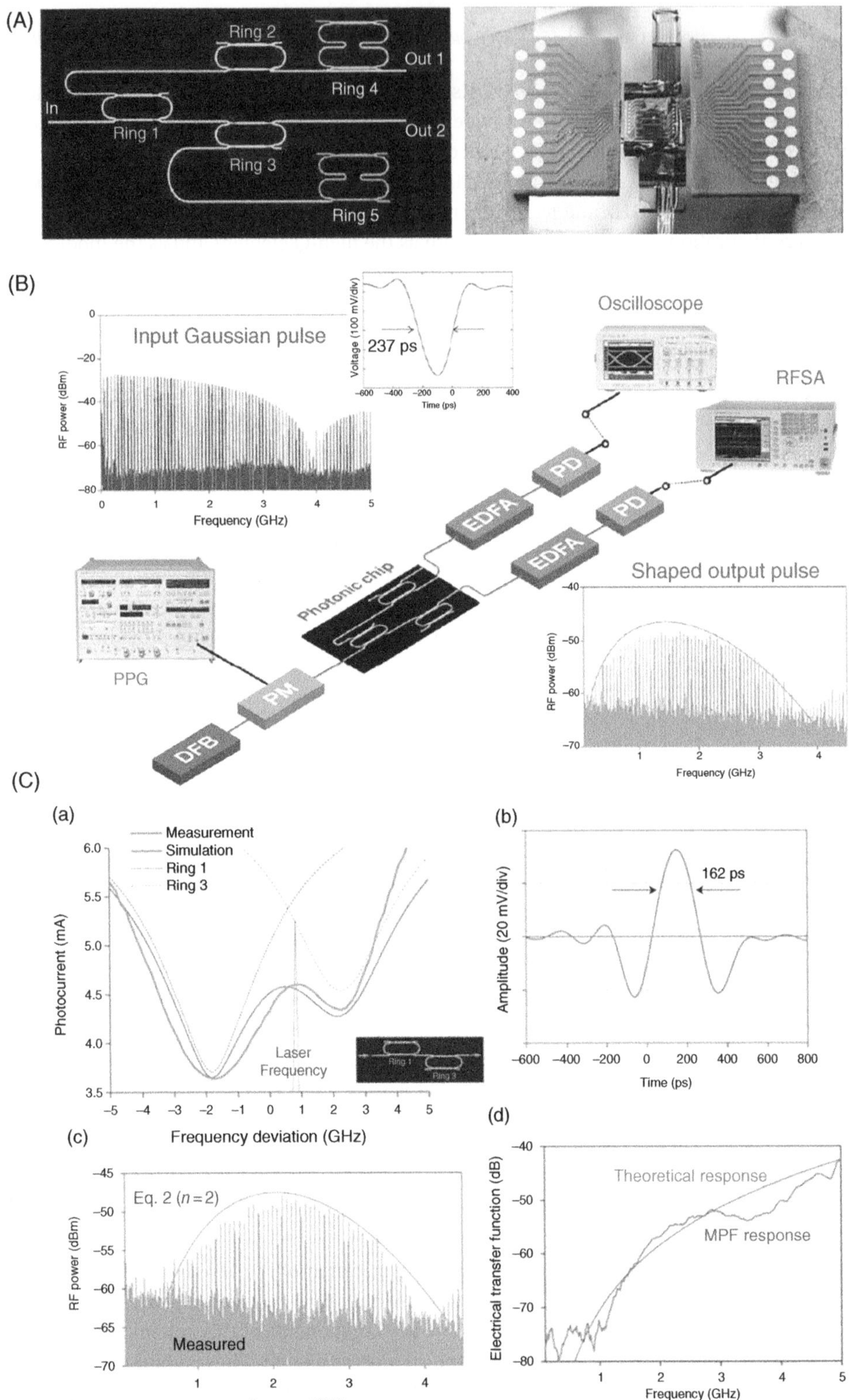

Figure 13.6 Integrated TriPLeX IR-UWB doublet signal generator. (A) Circuit layout and PCB-mounted device. (B) Layout of the measurement setup. (C) Detail of the measured two ORRs through responses depicted together with simulation results, the generated doublet pulse, power spectral density of the generated doublet theoretical, and measured transfer function for the doublet generation. *Source:* From Marpaung et al. [70]/Optical Society of America/CC BY 4.0

the quadratic nonlinear interaction seeded by dark pulses [72] in a periodically poled LN (PPLN) waveguide.

13.3.2.4 Arbitrary Waveform Generators

On-chip integrated pulse shapers are a desirable solution to overcome the limitations commonly associated with conventional bulk optics pulse shapers treated in Chapter 7. Several designs with increasing degrees of complexity and integration have been reported during the last years. For instance, researchers at Purdue University demonstrated an integrated ultra-broadband arbitrary microwave waveform generator that incorporates a fully programmable spectral shaper fabricated on a silicon photonic chip [81, 82]. The spectral shaper is a reconfigurable filter consisting of eight add/drop microring resonators on a silicon photonics platform. Ultra-compact cross-section (500 nm × 250 nm) silicon nanowires were used to fabricate the rings [81]. The typical bending radius was 5 μm and the waveguide propagation loss was around 3.5 dB/cm [83]. The spectral shaper programmability was achieved by thermally tuning both the resonant frequencies and the coupling strengths of the microring resonators. Figure 13.7 shows a cartoon of the spectral shaper and an optical image showing two channels of ring resonators with micro-heaters are shown in Figs. 13.7(a), (b), respectively.

The principle of the photonic arbitrary microwave waveform generation system implemented here is shown in Fig. 13.7(a). It is based on the frequency-to-time mapping technique described in Chapter 7. The spectral shaper is used to modify the spectrum emitted from a mode-locked laser. The shaped spectrum then undergoes wavelength-to-time mapping in a dispersive device, which in this case is a length (5.5 km) of optical fiber, before being converted to the electrical domain, a microwave waveform using a high-speed photodetector. By incorporating the spectral shaper into a variety of configurations, different waveforms are generated, including those with an apodized amplitude profile, multiple π phase shifts (Fig. 13.7(c)), two-tone waveforms, and frequency-chirped waveforms at the central frequency of 60 GHz. This configuration is not a fully integrated device as it still requires the use of a dispersive fiber link.

The same group has recently reported a compact reconfigurable RF-AWG architecture integrated on a silicon chip [84]. Figure 13.8 shows a cartoon describing the operation of the chip. The device works using the frequency-to-time-mapping technique as well and can generate programmable RF bursts or continuous waveforms with only the light source, electrical drives/controls, and detectors being off-chip. In this case, no external dispersive element is required as tunable delay lines are provided for each spectral region, as shown in the top left part of Fig. 13.8. Tunable rings are employed to select the spectral part of the input pulse and adjust its output amplitude after being delayed and before final combination. The device can synthesize both burst and CW signals and tuning speed is provided by silicon modulators featuring faster reconfiguration speeds by up to two to three orders of magnitude as compared to tunable schemes using thermal heaters.

A particular waveform of interest for radar and communication applications is the linearly chirped microwave waveform (LCMW). For LCMW generation, the optical spectral shaper should have a spectral response with an increasing or decreasing free spectral range (FSR). The approach based on spectral shaping and wavelength-to-time (SS-WTT) mapping provides large flexibility since the spectral response of the spectral shaper can be easily reconfigured.

The key component in an SS-WTT mapping system is the spectral shaper. In fact, the configuration proposed in [82] can provide an increasing or decreasing FSE by thermal tuning of the MRR. However, the chirp rate that can be obtained is quite limited. To generate an LCMW with a large chirp rate, other approaches have been recently proposed and demonstrated in silicon photonics and fabricated using a CMOS-compatible technology with 193-nm deep ultraviolet lithography.

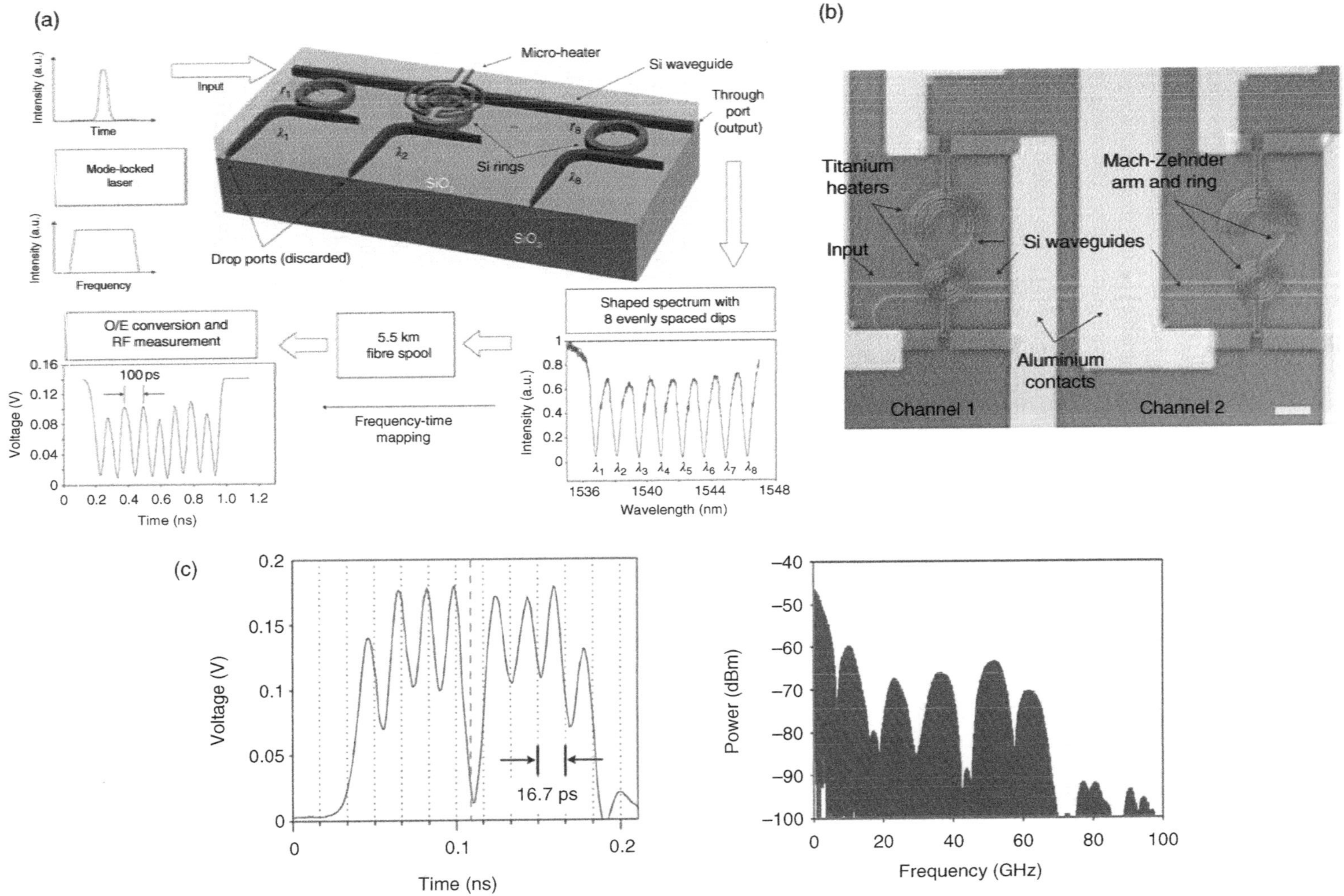

Figure 13.7 Integrated on-chip arbitrary waveform generation using cascades of eight microring resonators in SOI platform. (a) Waveform generator system. (b) Optical image of the pulse shaper. (c) Example of waveform generation at 60 GHz. *Source:* Adapted from Khan et al. [82].

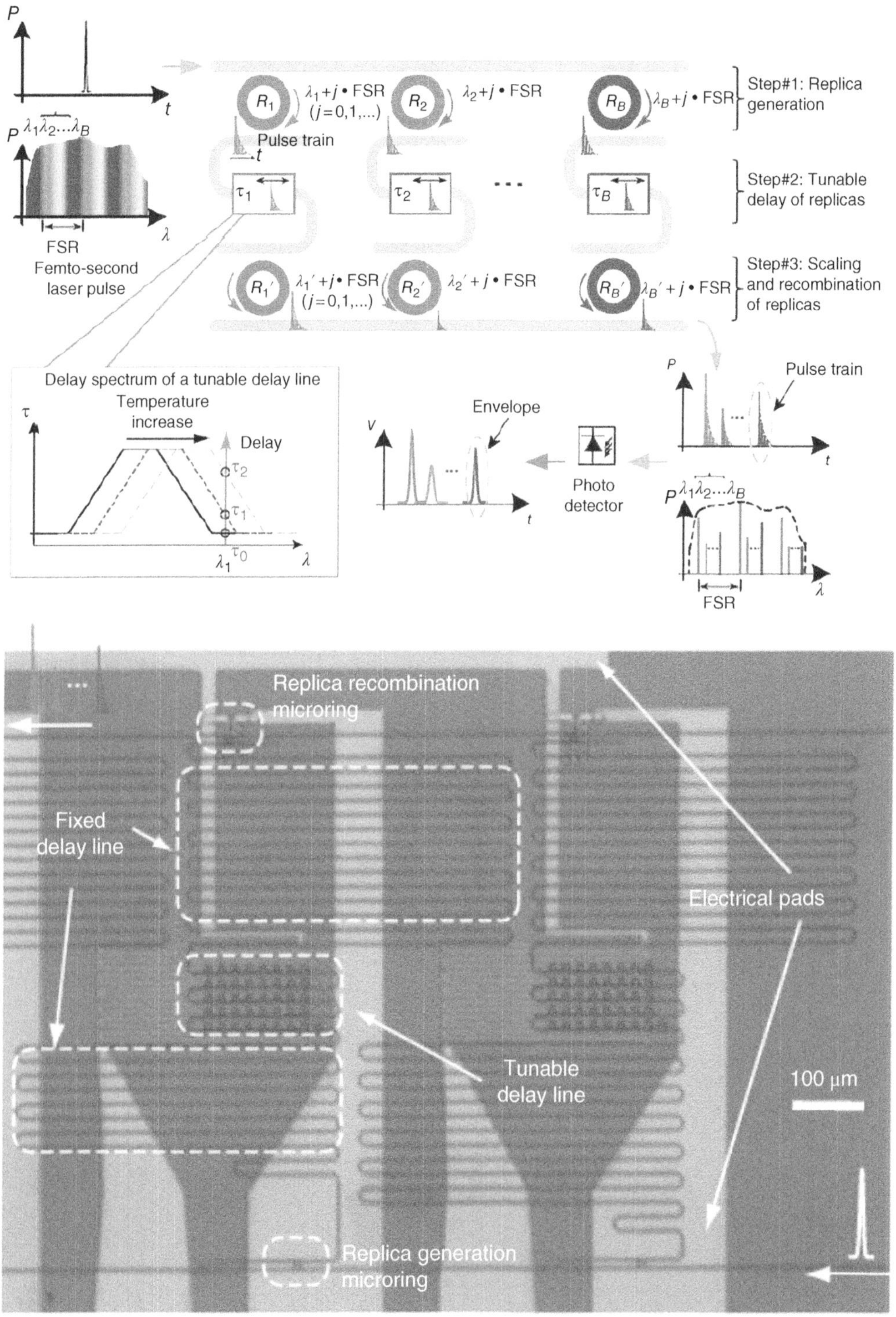

Figure 13.8 (Top left) Compact silicon-integrated on-chip arbitrary waveform generator architecture, (top right) optical image of a fabricated pulse shaper. (Bottom) Experimentally demonstrated RF bursts and their spectra. (a, b) Experimentally demonstrated a 40-GHz RF burst and its spectrum calculated via Fourier transform, where the mainlobe–sidelobe suppression ratio (MSSR) is B8 dB. (c, d) An apodized 40-GHz RF burst and its calculated spectrum, where the MSSR increases to B14 dB. (e, f) A 30-GHz RF burst and its calculated spectrum. (g) *p* phase shift in a 40-GHz RF burst. (h) An upchirping signal with frequencies ranging from 30 to 50 GHz. *Source:* Wang et al. [84]/Springer Nature/CC by 4.0.

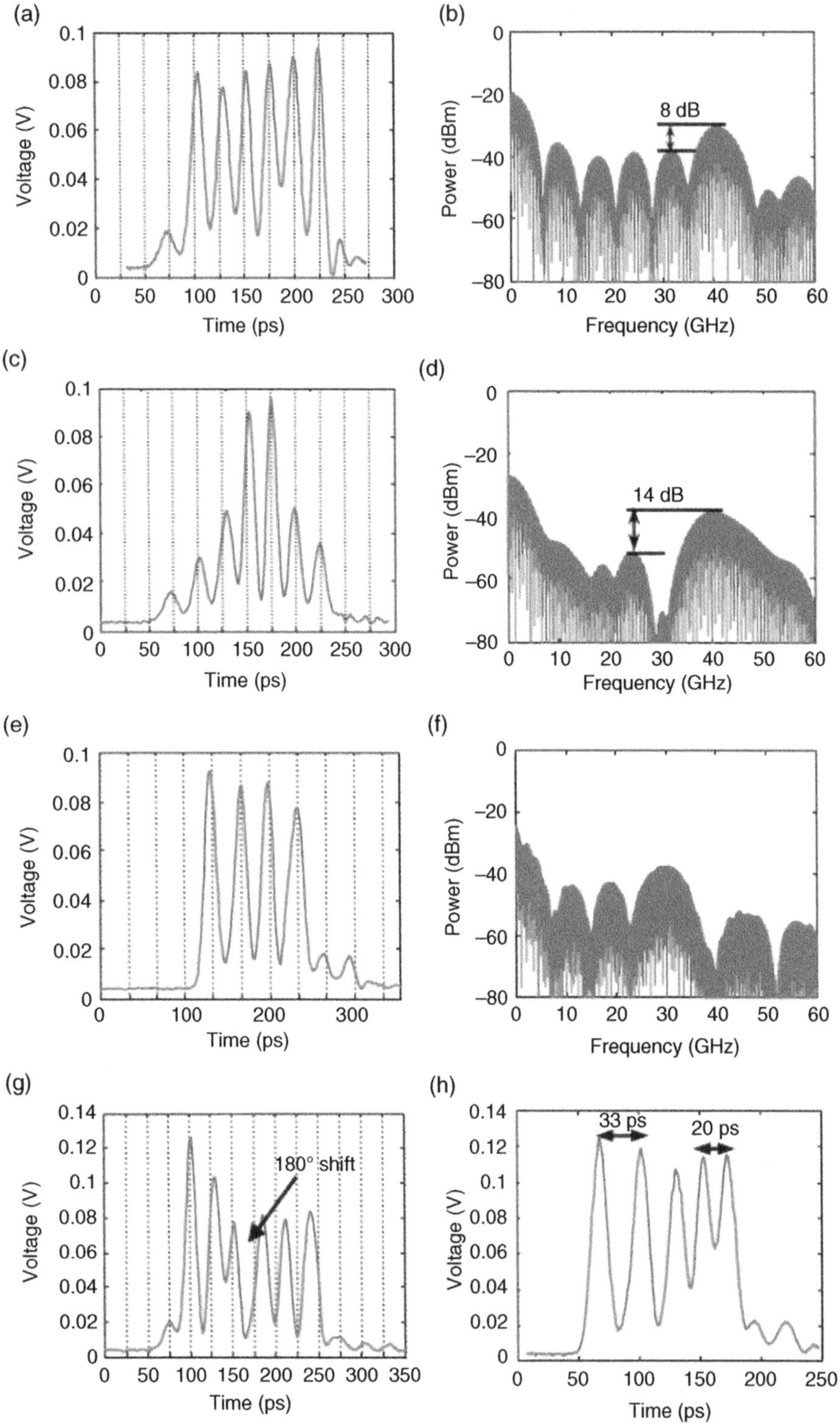

Figure 13.8 (Continued)

The top left part of Fig. 13.9(a) shows the layout of a spectral shaper [85] that consists of multiple cascaded MRRs with largely different radii. It has an MZI structure incorporating multiple cascaded MRR. The incident light beam is split by the first adiabatic 3-dB coupler into two beams that travel through the upper and lower arms of the MZI. Each ring selectively transfers the optical power at its resonance wavelength from the through-port waveguide to its drop-port waveguide (note that different colors of the rings shown in the figure indicate that each ring is designed to have a different radius and therefore a different resonant wavelength). The spacing between two neighboring rings is large enough to avoid mutual interferences. In the upper and lower waveguides, S-shaped waveguide bends, which are designed using Bezier curves to minimize the mode mismatch and thus reduce the waveguide-bending loss, are added to accommodate the change of the rings. The resonant wavelengths from both waveguides are then recombined at the first adiabatic 3-dB coupler and constructively interfere as reflected signals while nonresonant wavelengths are recombined at the second adiabatic 3-dB coupler at the transmission port.

In order to achieve maximum reflection power, each ring is designed to work in the critical coupling condition. By carefully designing the ring radii, at the reflection port, the reflection peaks with a linearly increasing spacing can be achieved. Figure 13.9(b), (c), and (d) shows, respectively, the spectral response, the generated waveform, and its spectrogram at the reflection port of an on-chip spectral shaper consisting of five MRRs.

While compared with the work in [82], the distinct feature of this optical spectral shaper is that the MRRs are designed to have largely different radii for the generation of an LCMW with a much higher chirp rate (17.2 GHz/ns compared to 8 Ghz/ns). The time-bandwidth product (TBWP) of this configuration is quite small (18.7). Hence, for practical applications, where large pulse compression is required, an LCMW with a much greater TBWP is needed.

A spectral shaper designed to provide a much larger TBWP [86] is shown in Fig. 13.9(e). The shaper layout and photograph are shown in Fig. 13.9(f). It has an MZI structure that incorporates two identical linearly chirped waveguide Bragg gratings (LC-WBGs) with opposite chirp rates in its two arms. An input light wave is split by a compact Y-branch into two beams to travel through the LC-WBGs in the upper and lower arms. Spectral components of the light waves of different wavelengths are reflected from different positions in the LC-WBGs. By the second and the third Y-branches, the reflected light waves are collected and recombined at the fourth Y-branch. The combined light wave is guided to an output grating coupler to couple the light out of the chip. When recombined, due to the optical interference, the optical spectral shaper has a spectral response with a wavelength-dependent FSR, which is required to generate an LCMW. The central frequency of the generated LCMW can be controlled by using an offset waveguide added in the lower arm of the spectral shaper to control the length difference between the two arms. By properly designing the LC-WBGs and choosing the length of the offset waveguide, the spectral response of the spectral shaper can be controlled to have a symmetrical, uniformly increasing or decreasing FSR. The two LC-WBGs are the key components in the spectral shaper. These gratings are realized by introducing periodic sidewall corrugations on the slab as shown in Fig. 13.9(g). Keeping the grating period uniform and linearly increasing the width of the rib along the grating, a linear chirp is produced since the effective refractive index is linearly increasing as the rib width increases in a definite range. Using this design, LCMW with TBWP values over 200 have been reported.

13.3.3 Tunable True Time Delay Lines and Phase Shifters

A number of integrated photonic approaches for the implementation of basic blocks providing optical true time delay lines (OTTDLs) and phase shifters (PS) have been reported in the last years. In the case of OTTDLs, the most basic approach is the so-called length-variable principle, where

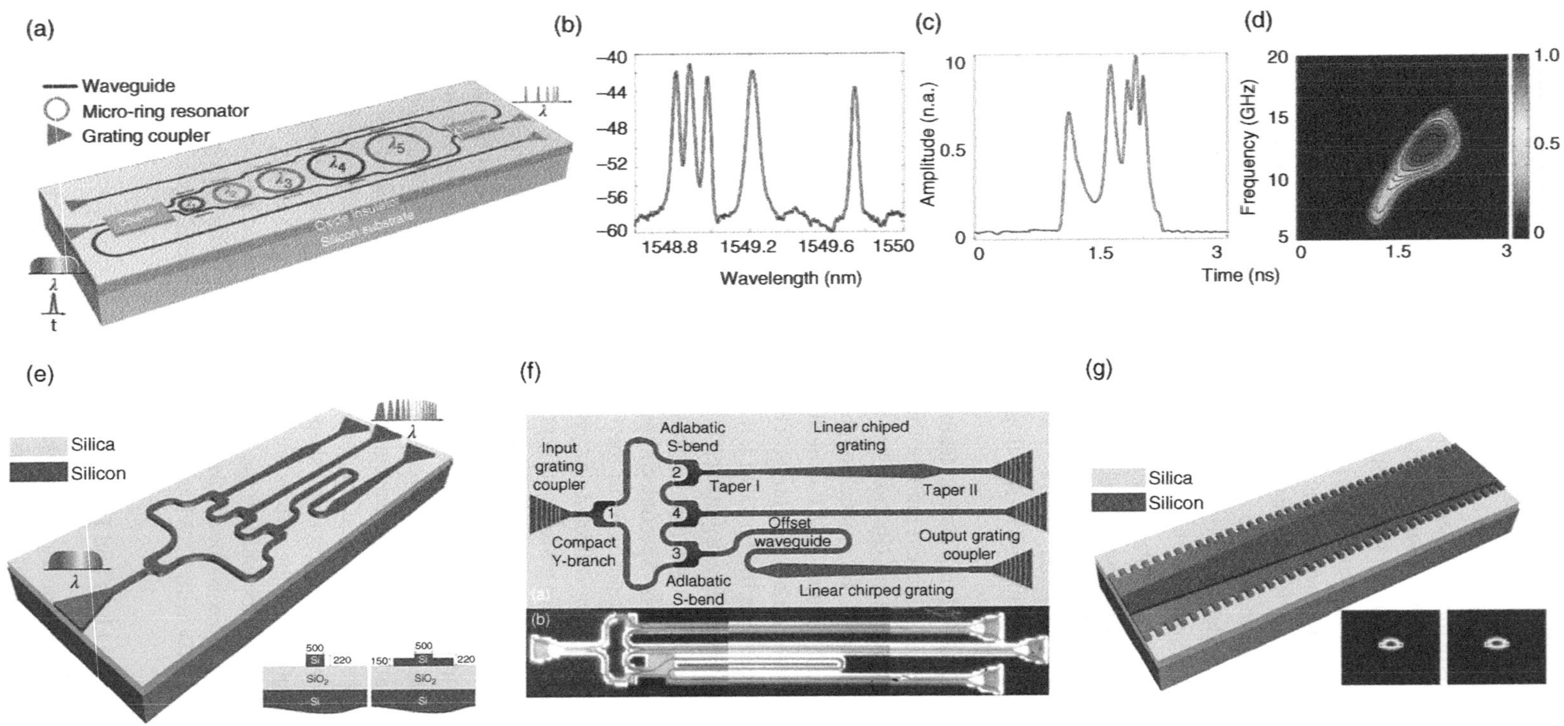

Figure 13.9 Silicon photonics chirped microwave signal generators reported in [21]. (a) Schematic view of a silicon-based on-chip spectral shaper consisting of five MRR. (b) Measured spectral response of the spectral shaper consisting of five cascaded MRRs. (c, d) Generated LCMW and its spectrogram when for the spectral shaper consisting of five cascaded MRRs. (e) Perspective view of the silicon-based on-chip optical spectral shaper incorporating LC-WBGs. (Inset: [left] Wire waveguide and [right] rib waveguide). (f) Schematic layout of the designed on-chip spectral shaper; and image of the fabricated spectral shaper with the length of the offset waveguide equal to the length of the LC-WBG captured by a microscope camera. (g) View of an LC-WBG. (Inset: Simulated fundamental TE mode profile of the rib waveguide with the rib width of 500 nm [left] and 650 nm [right]). *Source:* Zhang and Yao [21]/IEEE.

optical switches are employed to provide discretely tunable delay by means of selecting waveguides with different propagation lengths. This approach has been demonstrated using devices in silica [87] and in polymer [88].

Tunable delay lines have been proposed based on the wavelength-variable or dispersive delay line (DDL) principle. Here one can exploit the group delay change in the resonance of optical filters [89] obtaining a tunable OTTDL by changing the wavelength of the optical source at the filter input. For example, cascaded ORRs have been demonstrated for tunable delays in silica [87], TriPleX [90], silicon oxynitride (SiON) [91], and SOI [92, 93]. The same concept can also be exploited using integrated Bragg gratings in SOI that can be either electrically [94] or thermally tuned [95].

Finally, a novel concept to implement OTTDLs is based on the principle of index-variable delay lines. Here true time delay control can be realized through varying the group index/propagation velocity in the waveguides. One efficient way to achieve this is by the inscription of subwavelength gratings (SWGs) [96] in SOI waveguides by alternating sections of silicon and silica. These waveguides can provide a variable group delay by changing the duty cycle relationship between the silicon section within the grating period. In [96], a 4-tap OTTDL is reported where each tap is based on one SWG waveguide, and all waveguides have the same length of $L = 8$. The SWG waveguides are realized by alternating periodically segments of silicon and silica, with a period of $\Lambda = 250$ nm along the propagation direction. Figure 13.10 shows the schematic and performance operation for an input 1 ps Gaussian pulse featuring a 9 ps incremental delay between waveguides by choosing four different values of the SWG duty cycles (60, 50, 40, and 30%).

Phase shifting is also attractive for a number of signal-processing applications. For example, as was seen in Chapter 8, PSs are a fundamental building block in narrowband beamformers. Integrated PS SOI ring resonators have been reported to provide widely tunable operation [97, 98]. Other approaches have also been reported based on coherent population oscillation in SOA waveguides [99, 100], or in tunable microdisks in hybrid III–V/SOI platform [101].

Both OTTDL and PS functionalities can however be achieved within the same chip enabling complete signal-processing capabilities by means of the so-called separate carrier tuning (SCT) scheme, proposed by Morton and Khurgin [102]. Burla et al. [103] demonstrated the SCT scheme together with optical single sideband filtering monolithically integrated into a single chip. The unit, shown in Fig. 13.11, consists of a reconfigurable optical delay line, an SCT unit, and an optical sideband filter (OSBF). The OSBF, which is a Mach–Zehnder interferometer loaded with an optical ring resonator (ORR) in one of its arms, removes one of the RF sidebands of a double-sideband intensity-modulated optical carrier. The OTTDL and SCT units are individually implemented using a pair of cascaded ORRs.

Changing the group delay of the signal sideband by tuning the resonance frequencies and the coupling factor of the ORRs in OTTDL, while also applying a full $0 - 2\pi$ carrier phase shift in SCT, allowed the demonstration of a two-tap MWP filter whose notch position can be shifted by 360° over a bandwidth of 1 GHz.

13.3.4 Optical Beamforming

In Chapter 8, it was shown that narrowband beamforming can be achieved using optical PSs instead of OTTDs. Using a coherent detection scheme, the phase and amplitude conveyed by an optical signal can be directly transferred to a microwave signal by heterodyning this signal with an optical local oscillator signal. In this way, phase modulation of a microwave signal can be performed using optical PSs. This concept was employed as early as in 1995 by Horikawa and co-workers [104], who proposed a self-heterodyning beamformer based on $LiNbO_3$. An InP PIC beamformer was proposed

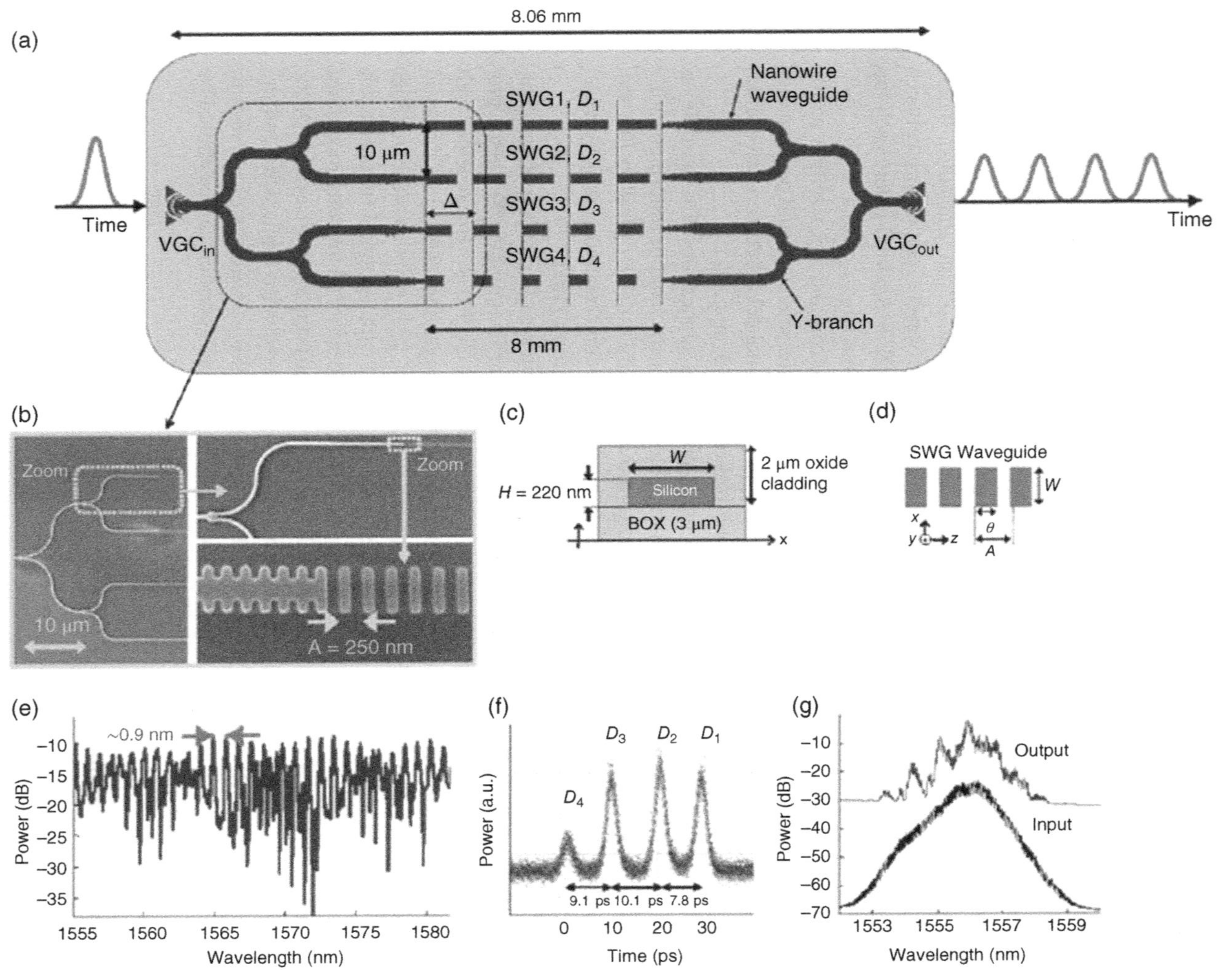

Figure 13.10 (a) Schematic and (b) an SEM image of our fabricated 4-tap OTTDL structure based on SWG waveguides in SOI reported in [96]. (c) The waveguide cross-section and (d) parameters of the SWG waveguides. (e) Measured power spectral response of the fabricated OTTDL. (f) Generated time-domain pulse train at the output of the fabricated OTTDL device in response to a single input optical pulse. (g) Input and amplified output signal spectra. *Source*: Wang et al. [96]/Springer Nature/CC by 4.0.

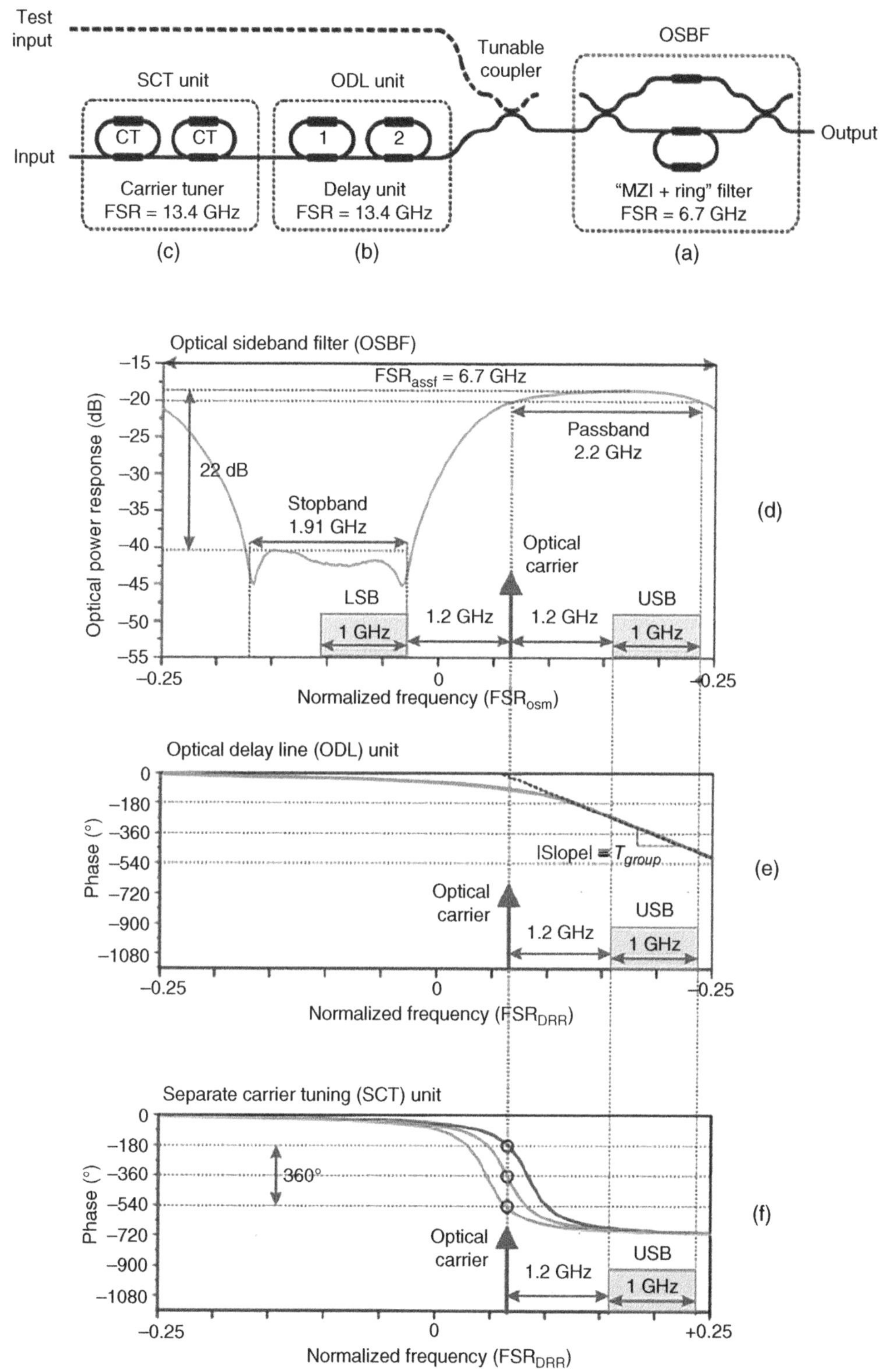

Figure 13.11 Separate Carrier Tuning (SCT) unit providing both OTTDL and PS operation. The unit consists of (a) an optical sideband filter (OSBF), (b) optical delay lines (OTTDL), and (c) carrier phase tuner (SCT). (d–f) Denote the measured responses of the OSBF, OTTDL and SCT, respectively. *Source:* Burla et al. [103]/Optical Society of America/CC BY 4.0.

where phase modulators and variable attenuators are employed to control the amplitude and phase of the RF signal, featuring an ultracompact footprint of $8.5 \times 8\ mm^2$ for a 16×1 device. Narrowband beamforming was also been reported in silica PLC [105], where the thermo-optic effect was used to achieve phase shifting and where amplitude control was performed using PLC-type MZIs with two 3 dB multimode interferometer (MMI) couplers (50 : 50 ratio) and independent thermo-optic PSs on both waveguide arms. The chip size of the realized eight-channel OBFN was about 4 mm × 65 mm. Beamforming at 60 GHz was demonstrated [106].

Wideband beamformers require OTTDLs as shown in Chapter 8. In particular, continuously tunable OTTDLs are desired in many practical applications such as satellite and airborne communications. A wide variety of integrated broadband beamformers based on cascaded tunable ORRs have been proposed and developed by researchers at the University of Twente during the last years [107–113]. For example, a broadband and continuously tunable integrated optical beamforming network (IOBFN) capable of providing continuously tunable true time delay up to 236 ps over the entire DVB-S band (10.7–12.75 GHz) and realized with a CMOS-compatible process was reported in [107]. In [108, 109], the suitability of the beamformer was theoretically and practically investigated for an application to satellite communications in the Ku-band frequency range (10–12.5 GHz). The tunable delays were based on reconfigurable ORRs in conjunction with a single OSBF integrated on the same optical chip. The delays and filter responses were software-programmable. Four tunable delay lines were integrated on a single chip and configured to feed a 16-element linear antenna array as shown in Fig. 13.12(a). The broadband beamsteering capability of the proposed IOBFN was demonstrated and the squint-free antenna was guaranteed by the linear dependence of the phase delays in each arm (see Fig. 13.12(b), (c)). Furthermore, the chip could be operated to provide phase-shift-based and true-time delay-based beamsteering. In the first case, beamsquint was observed, while in the second, broadband operation was demonstrated immune from that deleterious effect (see Fig. 13.12(d), (e)).

Tunable broadband PS and tunable true time delays have been demonstrated mainly in silicon technologies (SOI and TriPleX). TBPSs exhibit tunable phase shifts in the range of 0–90° in a 16–20 GHz bandwidth [23]. TTTDs provide tunable delays of up to 25 ps in 8 GHz [114] or up to 400 ps limited to 1 GHz bandwidth [115]. Instantaneous frequency measurement chips are reported in InP [116] and TriPleX technologies [117], featuring RF frequency determination ranges (RMS error) of 5–15 GHz (200 MHz) and 0.4–5 GHz (93.6 MHz). Optical-phase-locked loops with locking range in the (−9.5, 7) GHz range and locking bandwidth of 400 MHz have been reported in InP [118]. Finally, beamformer chips have been demonstrated in TriPleX technology, capable of addressing 16 antenna elements, in the 2–10 GHz band, and with instantaneous bandwidth of 8 GHz [119].

13.4 Multifunctional Circuits

Despite the fact that most IMWP circuits have been designed and fabricated having in mind a single application or functionality, there are some examples of design, for which several functionalities have been demonstrated. These so-called multifunctional circuits feature optimum operation for the tasks they were conceived but, nevertheless can support others with an acceptable degree of performance. The first example is the Si_3N_4 chip shown in Fig. 13.6. It was initially designed to operate as a frequency discriminator for phase modulation direct-detection (PM-DD) MWP links [120]. Figure 13.13 displays two multifunctional circuits fabricated using InP technology.

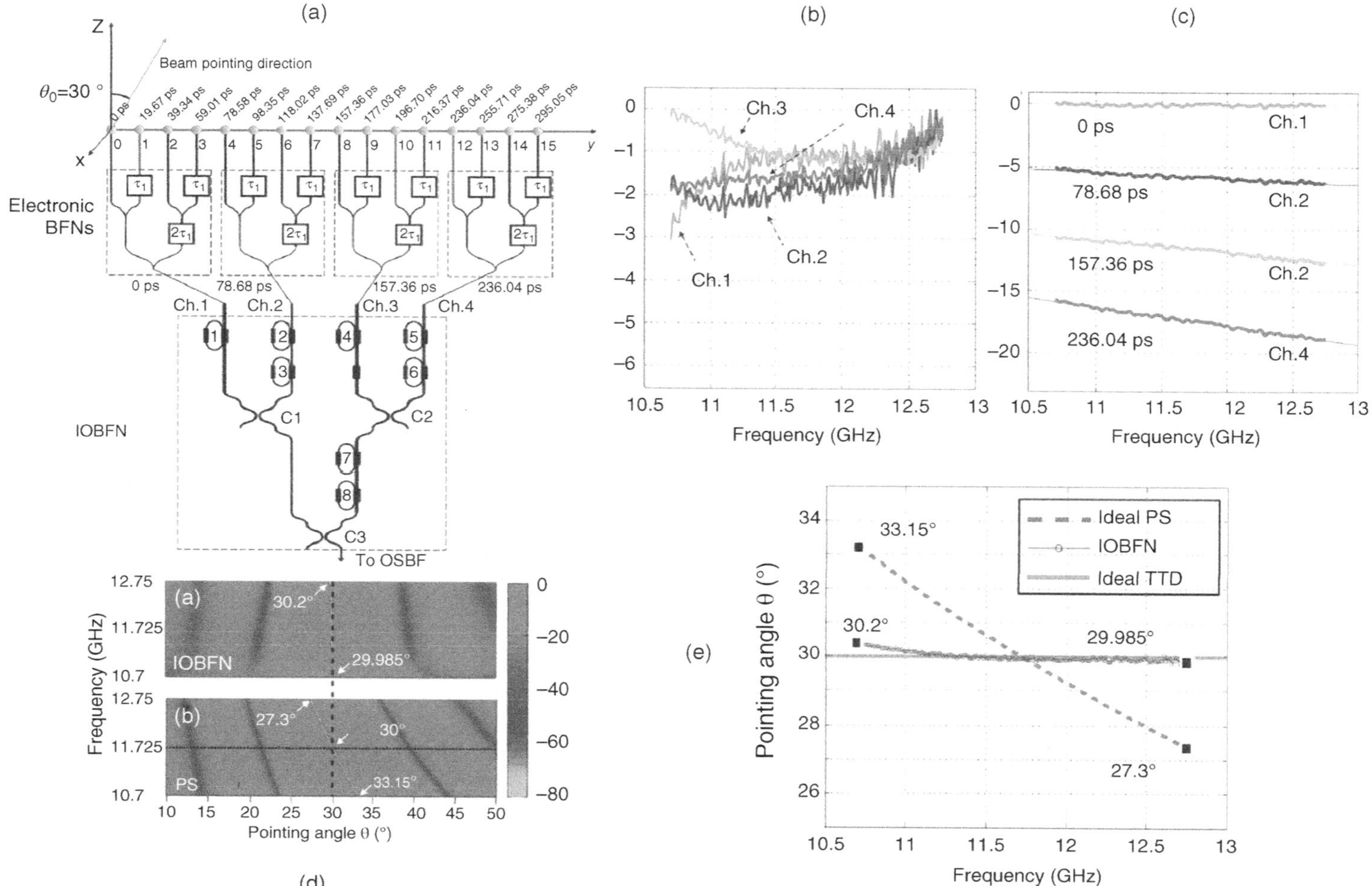

Figure 13.12 TripleX-based optical beamformer. (a) Chip layout. (b) Group delay versus frequency relationship for the four optical channels in the chip. (c) Phase delay versus frequency relationship for the four optical channels in the chip. (d) Pointing angle variation versus frequency for the beamformer action as a true time delay and as a phase shifter. (e) Pointing angle versus frequency around a steering angle of 30° for the beamformer operating as a true time delay line and comparative results for the ideal true time delay line and phase shifter.

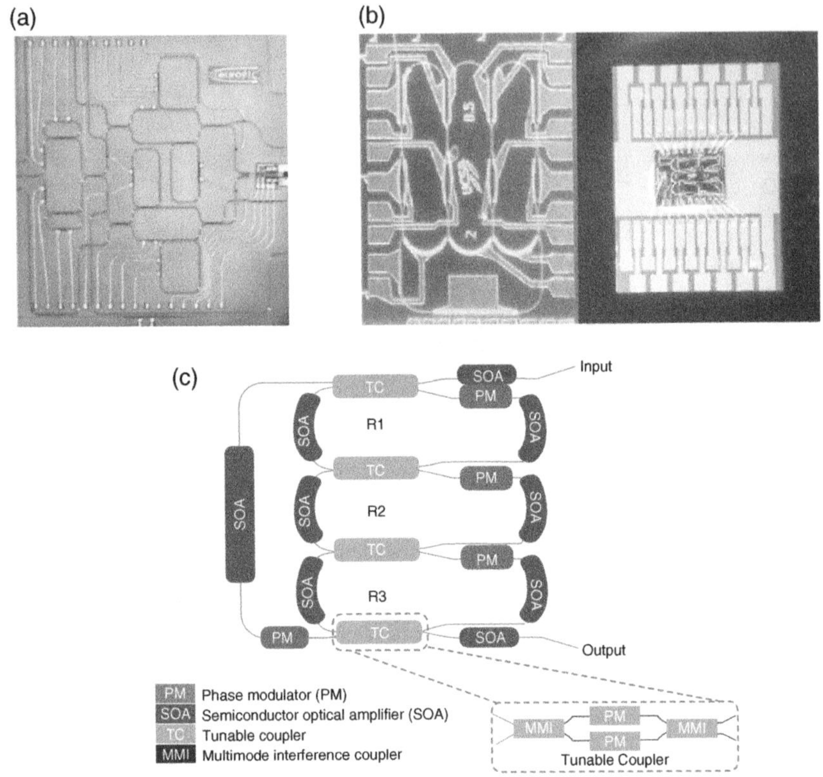

Figure 13.13 Examples of multifunctional circuits fabricated in InP technology: (a) a two-stage ring-loaded MZI filter. *Source:* Fandiño et al. [121]/Optica Publishing Group. (b) A three-cavity tunable filter. *Source:* Liu et al. [122]/from Springer Nature. (c) Represents the layout of (b).

The structure in Fig. 13.13(a) shows a circuit composed of a two-stage ring-loaded MZI units with an integrated balanced detector. It was initially designed to operate as a frequency discriminator for PM-DD and MWP links but its use has been extended to other applications such as instantaneous frequency measurement [123] and SSB modulation [124]. The structure in Fig. 13.13(b) was originally reported as a tunable bandpass filter [52] but its use can be expanded as it has been recently demonstrated to act as a signal integrator, differentiator, and a Hilbert transformer [122].

13.5 Universal Microwave Photonic Processors

The fact that several ASPIC designs have proven to operate successfully as multifunctional circuits raises the question of whether it would be possible to design a simple and replicable photonic hardware architecture that under suitable programming could be able to implement the different functionalities required in MWP. This so-called universal MWP processor would be inspired on similar grounds as the FPGA device in electronics or the digital signal processor (DSP) in advanced communications. In this section, we provide some insight into the recent and currently undergoing efforts in this field.

13.5.1 Early Designs

The starting point for the MWP programmable processor can be traced back to the proposal of the MWP transistor [125] shown in Fig. 13.14, which is composed of subsystems, each of which is a collection of connected fixed-and-variable components.

The transistor configuration is obtained by opening the feedback loop of a generalized optoelectronic oscillator [74]. As it was shown in [125], this architecture is in principle capable of performing few, but important, required functionalities in MWP provided that its optical subsystems can be reconfigured by means of suitable control signals. The main limitation of the MWP transistor is however that it does not support the reconfiguration of interconnections between its internal optical subsystems. This results in an intermediate performance between ASPIC and a universal MWP processor.

As a first alternative to go one step further in terms of flexibility, the former approach can be extended to implement a programmable MWP processor architecture [126], which incorporates optical routing and switching elements (ORSEs) into the MWP transistor layout and replaces the reconfigurable optical filtering system by a more powerful and versatile optical core. The top part of Fig. 13.15 schematizes the concept.

The central element of the programmable optical processor is the optical core, where the main signal-processing tasks are carried out in the photonic domain. Ideally, the optical core should be built upon a versatile architecture capable of implementing different functionalities in response to different electronic control signals. In practice, among the reconfigurable photonic circuits proposed in the literature, the ones that could especially emerge as candidates for this core are either based on the cascade of finite (i.e. Mach–Zehnder interferometers) or infinite (i.e. ring cavity) impulse response cells or a combination of both. These structures feature periodic filters and allow for bandwidth reconfiguration and notch tunability by moving the zeros and poles along the z-plane. Thus, to further increase the degree of flexibility of the processor, the former configuration

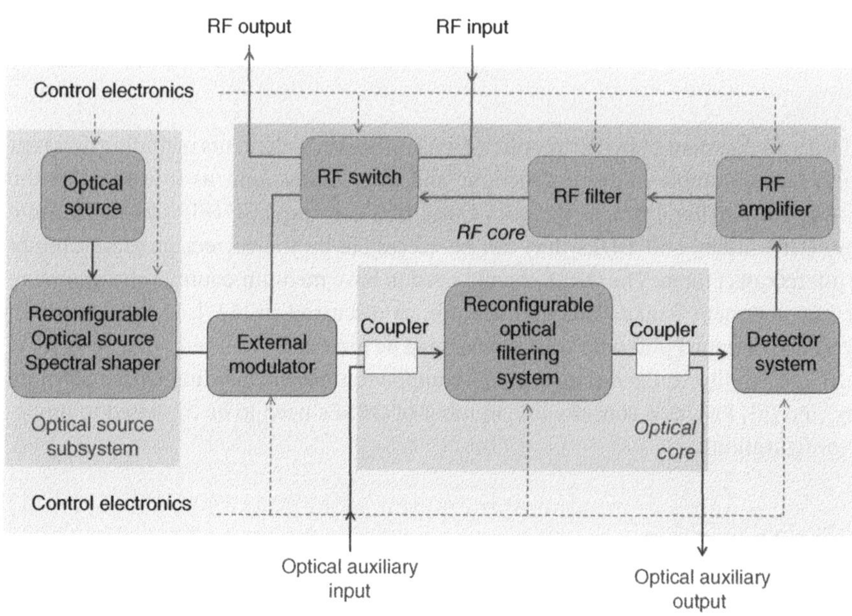

Figure 13.14 Layout of a MWP transistor. *Source:* Adapted from Capmany et al. [125].

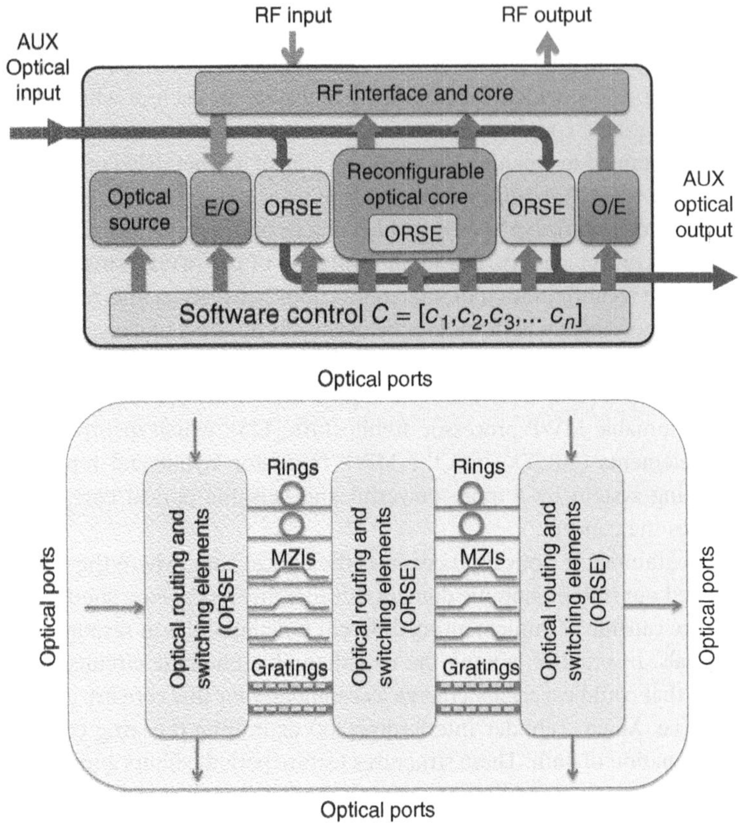

Figure 13.15 (Top) Layout of the universal processor. (Bottom) Particular implementation of the optical core using ad-hoc photonic components such as Mach–Zehnder interferometers (MZIs), optical ring resonators (ORRs) and gratings connected by several optical routing and switching elements (ORSEs). *Source:* Adapted from Iezekiel et al. [20].

can be extended with a subsystem of switched optical signal processing elements including different types of fixed and reconfigurable filtering structures and interconnections, as illustrated in the bottom part of Fig. 13.15. This example includes finite impulse response (FIR) filters, infinite impulse response (IIR) filters, and DDLs that can be accessible by the correct programming of the ORSEs that interconnect them. The ORSEs would need to have medium-count ports depending on the number of component banks. This architecture, as discussed in [127], is still limited in (i) flexibility, as ad-hoc designed photonic subsystems must be included and interconnected within the optical core, (ii) scalability, as fixed-size photonic components are the building blocks for more complex circuits, and (iii) loss, as a considerable number of ORSEs need to be traversed to implement complex configurations.

13.5.2 Waveguide Mesh Core Processors

A more versatile optical core architecture can be obtained by following similar principles as those of the FPGAs in electronics [128]. The central concept is based on breaking down complex circuits into a large network of identical two-dimensional (2D) unit cells implemented by means of an MZI

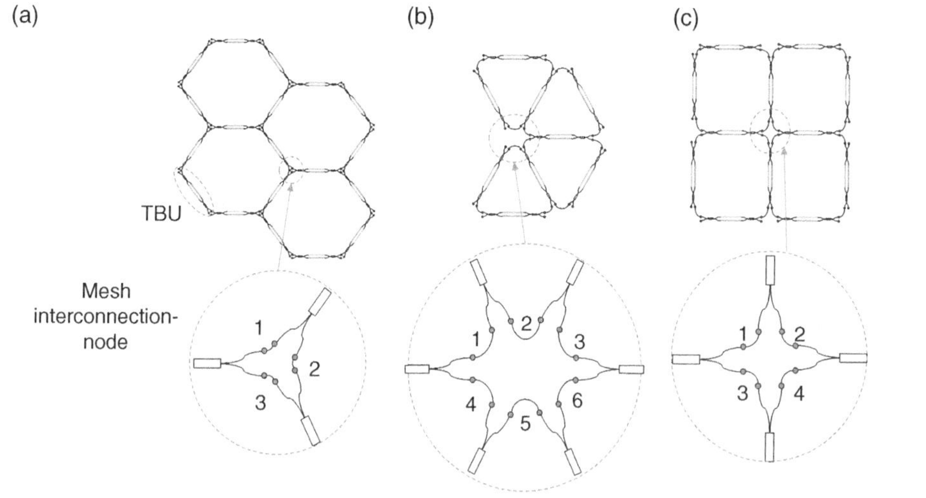

Figure 13.16 Reconfigurable Mesh designs (top): (a) Hexagonal type, (b) triangular type, and (c) square type, and their associated interconnection points (bottom). *Source:* Pérez et al. [130]/Optical Society of America/CC BY 4.0.

waveguide mesh or lattice. Zhuang and co-workers [129] were the first to propose a programmable optical chip architecture connecting MZI devices in a square-shaped mesh network grid. The distinctive feature of this approach is that it enables both feedforward and feedbackward filtering configurations, selecting the adequate path through the mesh and providing independent tuning of circuit parameters to complex-valued coefficients by introducing phase-tuning elements in both arms of the MZIs to enable independent control of amplitude and phase of light at coupler outputs. Further to the square-shaped mesh, other configurations, such as the triangular and hexagonal geometries, have been reported [130]. The operation principles of these and their comparative performance are discussed in the next subsection.

Figure 13.16 shows three different types of waveguide mesh designs. Figure 13.16(a)–(c) illustrates a four-cell arrangement for the hexagonal, triangular, and square-type meshes, respectively.

The BBB of these meshes is a tunable coupler that must provide, independently, a complete splitting ratio tuning and phase response. This switching/tapping/de-phasing mechanism can be obtained either by exploiting the electro-refraction and/or the electroabsorption effect, or by means of the thermo-optic effect in a broadband 3-dB-balanced MZI. By configuring each tunable coupler placed at each side of the cell perimeter as a switch (in either its cross or bar state) or as a tunable coupler, one can synthesize a given photonic integrated circuit topology, as shown in [129].

The top part of Fig. 13.17 shows the tunable basic unit (TBU) composed of the tunable coupler and its access (input/output) waveguides. The geometry of the latter is a function of the bending radius and varies for each mesh topology due to different angles between elements. The basic unit length (BUL) is:

$$BUL = L_{access} + L_{Tunable-Coupler} \tag{13.1}$$

where L_{access} is the overall length of the access waveguide segment and $L_{Tunable\text{-}Coupler}$ is the length of the tunable coupler.

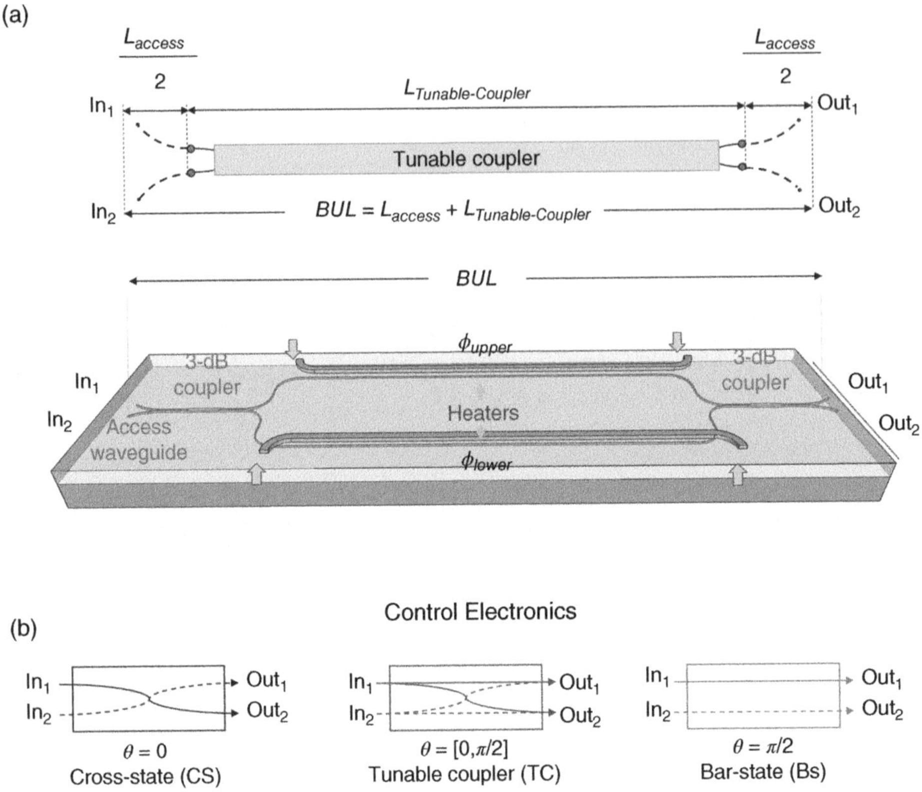

Figure 13.17 (a) (Top) Labeled schematic of a general tunable coupler acting as the basic building block of the mesh. The basic unit length (BUL) is illustrated as the sum of the tunable coupler length and the arc length of the access waveguides. (Bottom) Particular case of an integrated balanced MZI-based tunable coupler. (b) Signal flow for the different tunable basic unit (TBU) configuration states.

Referring to Fig. 13.17(b), the TBU can implement three different states: cross-state switch (light path connects in1 to out2 and in2 to out1), bar state switch (light path connects in1 to out1 and in2 to out2) and tunable splitter. For a balanced MZI loaded with heaters on both arms, the splitting ratio is obtained by increasing the effective index due to the Joule effect in the upper or lower arm, producing a ϕ_{upper} and ϕ_{lower} phase shift, respectively. Once set, a common drive in both heaters will provide a common phase shift, leading to independent control of the amplitude ratio and the phase. The device matrix is defined by:

$$h_{TBU} = je^{j\Delta}\begin{pmatrix} \sin\theta & \cos\theta \\ \cos\theta & -\sin\theta \end{pmatrix}\gamma \tag{13.2}$$

where θ is $(\phi_{upper} - \phi_{lower})/2$ and Δ is $(\phi_{upper} + \phi_{lower})/2$. The coupling factor K is then defined as the $\cos^2(\theta)$ and the general loss term γ includes the effect of propagation losses in the access waveguides, the tunable coupler waveguide, and the insertion losses for both 3-dB couplers.

The different waveguide mesh geometries have been analyzed and compared against a set of different figures of merit in [130], concluding that the hexagonal lattice yields the best performance for almost all the performance metrics. Table 13.4 summarizes the results of that analysis.

Table 13.4 Summary of values for the figures of merit of the different mesh designs (see definitions in [23]).

Figure of merit	Triangular	Square	Hexagonal
ORR cavity spatial tuning resolution step in BUL units (the lower the better)	3	4	2* The first and second step has a resolution of 6 and 4.
MZI arm imbalance spatial tuning resolution step in BUL units (the lower the better)	3	4	2
ORR reconfiguration performance (the higher the better) (for $X = 25$ *BUL*)	8	6	9
MZI reconfiguration performance (for $X = 25$ *BUL*)	8	6	12
Switching elements per unit area compared to *square* % (the lower the better for a fixed value of reconfiguration performance)	+65.00%	0.00%	−36.66%
Replication ratio for ORR structures up to 16 BUL cavity length (the higher the better).	1	2.68	1.31
Replication ratio for MZI structures up to 12 BUL interferometric length (the higher the better).	1	3	3.36
$L_{access}/L_{access\ square}$ % for a fixed radius of curvature *Ra* (the lower the better)	+33.33%	+0.00%	−33.33%
Ra/Ra_{square} % for a fixed BUL value (the higher the better)	−25.00%	+0.00%	+50.00%

Source: Adapted from Coldren et al. [23].

For reference, the spatial tuning resolution step quantifies which is the minimum step in BUL units by which the arm length mismatch or the cavity length can be increased/decreased, the reconfiguration performance of the mesh is given by the number of filters with different spectral period values that can be implemented given a maximum value X (in BUL units) for the path imbalance in unbalanced Mach–Zehnder interferometers (UMZIs) or the cavity length in ORRs. The replication flexibility gives the number of possible alternative geometries for a specific filter implementation, while the switching elements per unit area denote the number of required BULs per surface unit to achieve equivalent functionality. Finally, the two last rows in Table 13.1 quantify the overall losses in the TBUs of triangular and hexagonal waveguide meshes as compared to those in the square mesh configuration. A complete definition of the figures of merit can be found in [130].

With suitable programming of a selected set of TBUs in the 2D waveguide mesh, different circuit layouts of variable complexity can be programmed over the same hardware. For example, the top and bottom parts of Fig. 13.18 show two implementation examples of simple UMZI FIR filters and ORR IIR filters with different path unbalance and cavity lengths, respectively. Filters with an increased degree of complexity can be built upon cascading of these simple units.

Since waveguide mesh processors are a quite recent area of research, there are only a few experimental results reported based on silicon nitride [129] and silicon on insulator (SOI) [131] technologies.

Figure 13.19(a) shows the basic layout and photograph of the programmable optical chip architecture connecting MZI devices in a square-shaped mesh network grid proposed by Zhuang and co-workers [129]. The structure, fabricated in Si_3N_4, comprised two square cells and was employed

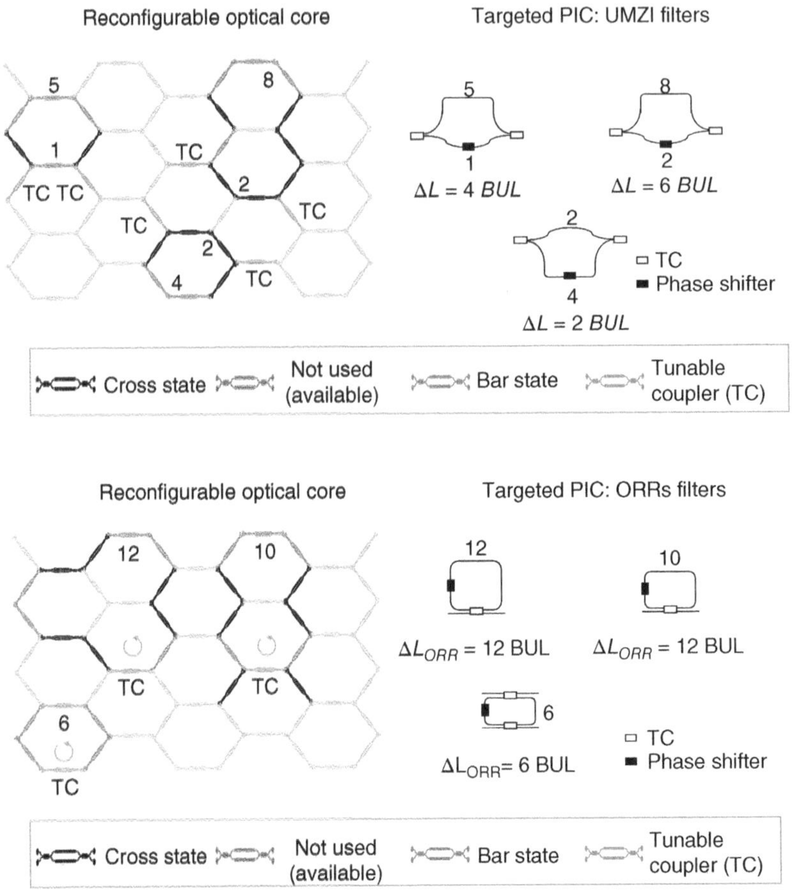

Figure 13.18 (Top) FIR filter implementations using a hexagonal waveguide mesh. (Left) mesh setting for (top/right) three different targeted UMZI filters, (bottom/right) three different targeted ORRs.

to demonstrate simple FIR and IIR impulse response filters with single and/or double input/output ports of synthetized ORRs. The processor features an FSR of 14 GHz and is fully programmable. By appropriate programming of this processor, Zhuang et al. have demonstrated bandpass filters with a tunable center frequency that spans two octaves (1.6–6 GHz) and a reconfigurable band shape (including flat-top resonance with up to passband–stopband 25 dB extinction). They also demonstrated notch filters with up to 55 dB rejection ratio, Hilbert transformers and tunable delay lines as shown in Fig. 13.19(b).

A recent contribution has reported the results of a waveguide mesh composed of seven hexagonal cells fabricated in silicon on an insulator [131]. The chip photograph is shown in Fig. 13.20. The device was fabricated at the Southampton Nanofabrication Centre at the University of Southampton. SOI wafers with a 220 nm-thick silicon overlayer and a 3 μm-thick buried oxide layer were used (for more details on fabrication and testing, see [131]).

The overall structure comprised 30 independent MZI devices and 60 thermo-optic heaters. The waveguide mesh MZI devices were independently tuned in power-splitting ratio and overall phase shift by means of current injection to the heaters deposited on top of the waveguides implementing

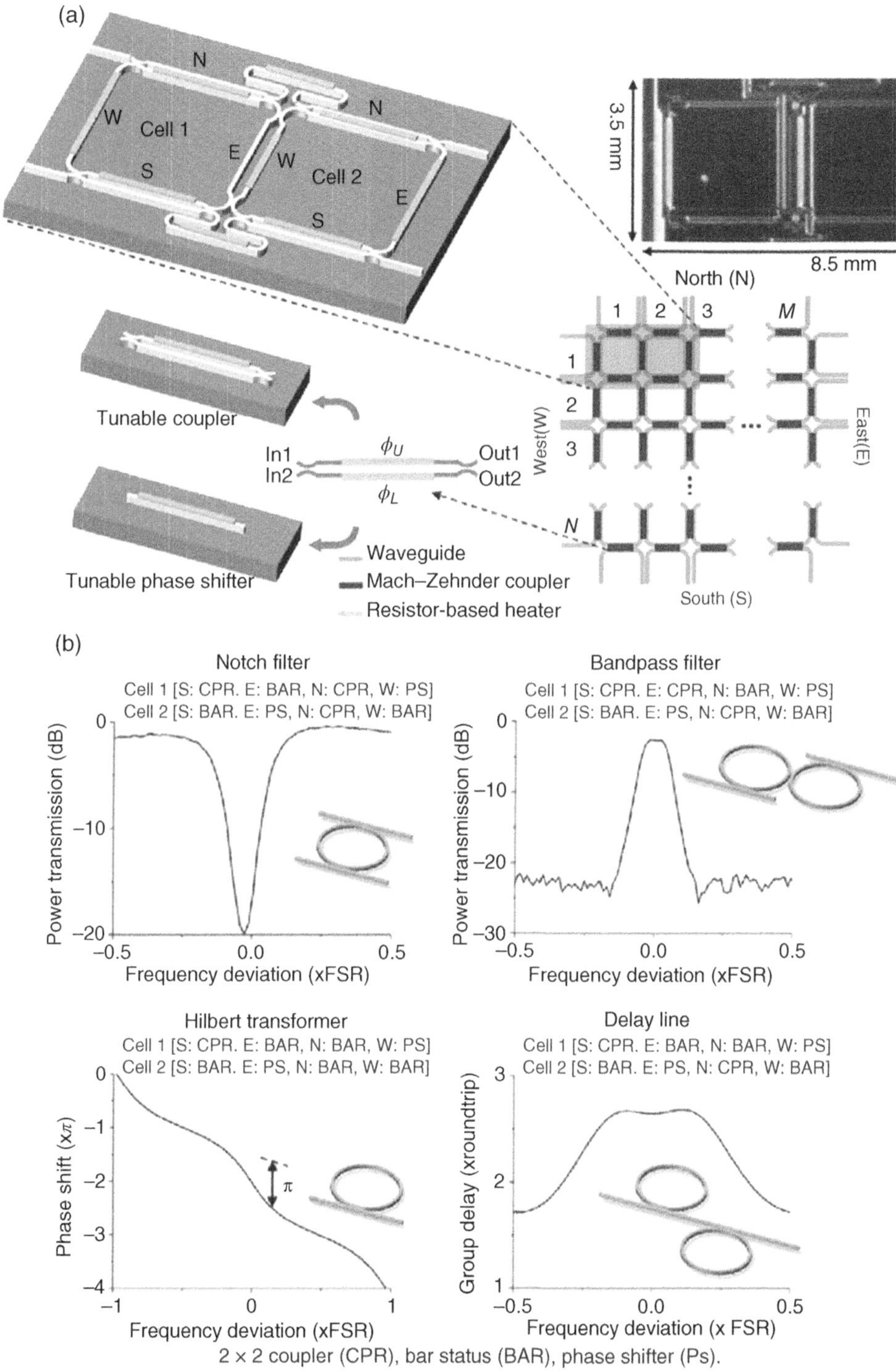

Figure 13.19 (a) Schematic and a photo of the Si3N4 waveguide technology (TriPleX) chip implementing a two square cell waveguide mesh reported in [129]. *Source:* Reproduced from Zhuang et al. [129]/Optica Publishing Group. (b) Different programmed circuit configurations obtained by varying phase-tuning elements in the chip and the measurements of their corresponding frequency responses.

Figure 13.20 Photograph of the seven-hexagonal cell waveguide mesh core fabricated in SOI and reported in [131]. *Source:* Pérez et al. [131]/Springer Nature/CC BY 4.0.

the interferometer arms. Despite the simplicity of the layout depicted in Fig. 13.20, the 7-cell structure is capable of implementing over 100 different circuits. For MWP filtering applications, basic MZI, FIR transversal filters, basic tunable ring cavities, and IIR filters, as well as compound structures such as coupled-resonator optical waveguides (CROWs) and side-coupled integrated spaced sequence of resonators (SCISSORs) are of particular interest. In [131], a different configuration featuring a wide variety of FSR values was reported. Figure 13.21 shows, for instance, five examples of RF photonic FIR and IIR filters synthesized with the same hardware just by changing the operation state point of the different MZI elements in the mesh.

Another application of interest in RF photonics is true time delay lines as these are the BBBs of key functionalities, including AWG and optical beamforming. The hexagonal waveguide mesh core can implement both dispersive as well as discrete-time programmable true time delay lines [132].

13.5.3 Waveguide Mesh MWP Universal Processors

The incorporation of the waveguide mesh structure in the MWP processor core enables a more versatile processor architecture, which is the one we propose for future development. In this architecture, which we show in Fig. 13.22, all the subsystems are directly connected to the reconfigurable optical core. This architecture represents, to the best of our knowledge, the first proposal for a generic-purpose software-defined MWP processor that incorporates the possibility of modifying the control signal flow to reconfigure the processing and interconnecting subsystems as required. It supports four different modes of operation as far as the input/output signals are concerned: electrical/electrical, electrical/optical, optical/electrical, and optical/optical operations.

Electrical/electrical operations are typically employed in MWP functionalities such as RF filtering, instantaneous frequency measurement, frequency mixing, RF, and millimeter-wave arbitrary signal generation to cite a few. It requires the processor to enable an optical source, EO and optoelectronic (OE) converters as well as the reconfigurable optical core. Figure 13.22(a) illustrates the signal flow for these operations. Note that if a second modulator is integrated, it can be enabled to perform frequency mixing operations based on the cascade of two EO modulators.

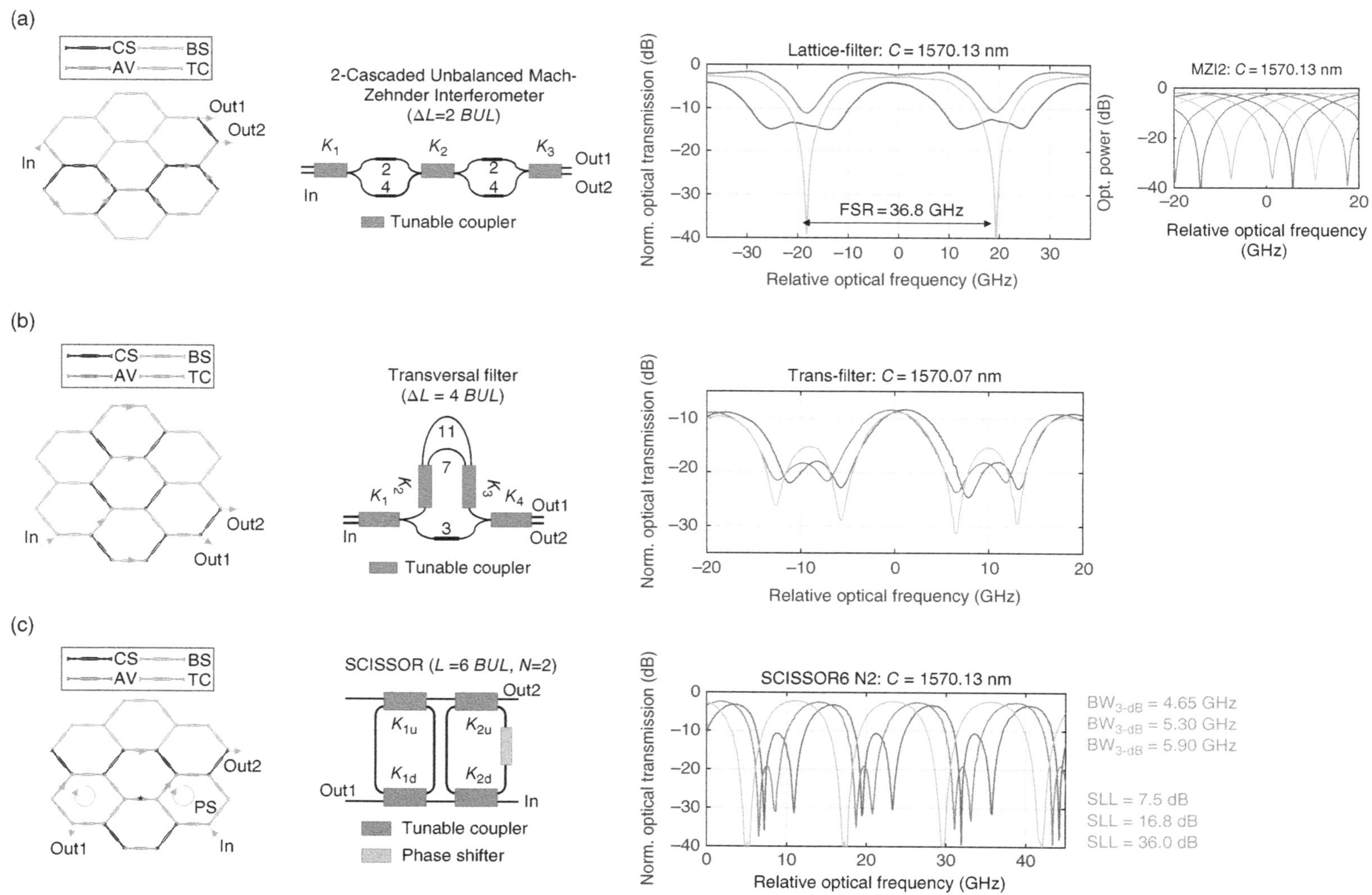

Figure 13.21 Five examples of synthesized RF photonic filters using the same seven-cell hexagonal mesh. In each case, the left part shows the TBU settings in the mesh. The intermediate part shows the circuit layout and the right part shows the measured results for the RF transfer function modulus. (a) Two-cascaded unbalanced 2-MZI FIR filter, (b) three-tap transversal FIR filter, (c) two-cavity SCISSOR FIR filter, (d) two-cavity CROW filter, and (e) three-cavity CROW filter. MZI programming code: CS, cross-state switch, BS, bar-state switch, TC, tunable coupler, and AV, available. *Source:* Pérez et al. [131]/Reproduced from Springer Nature/CC BY 4.0.

(d)

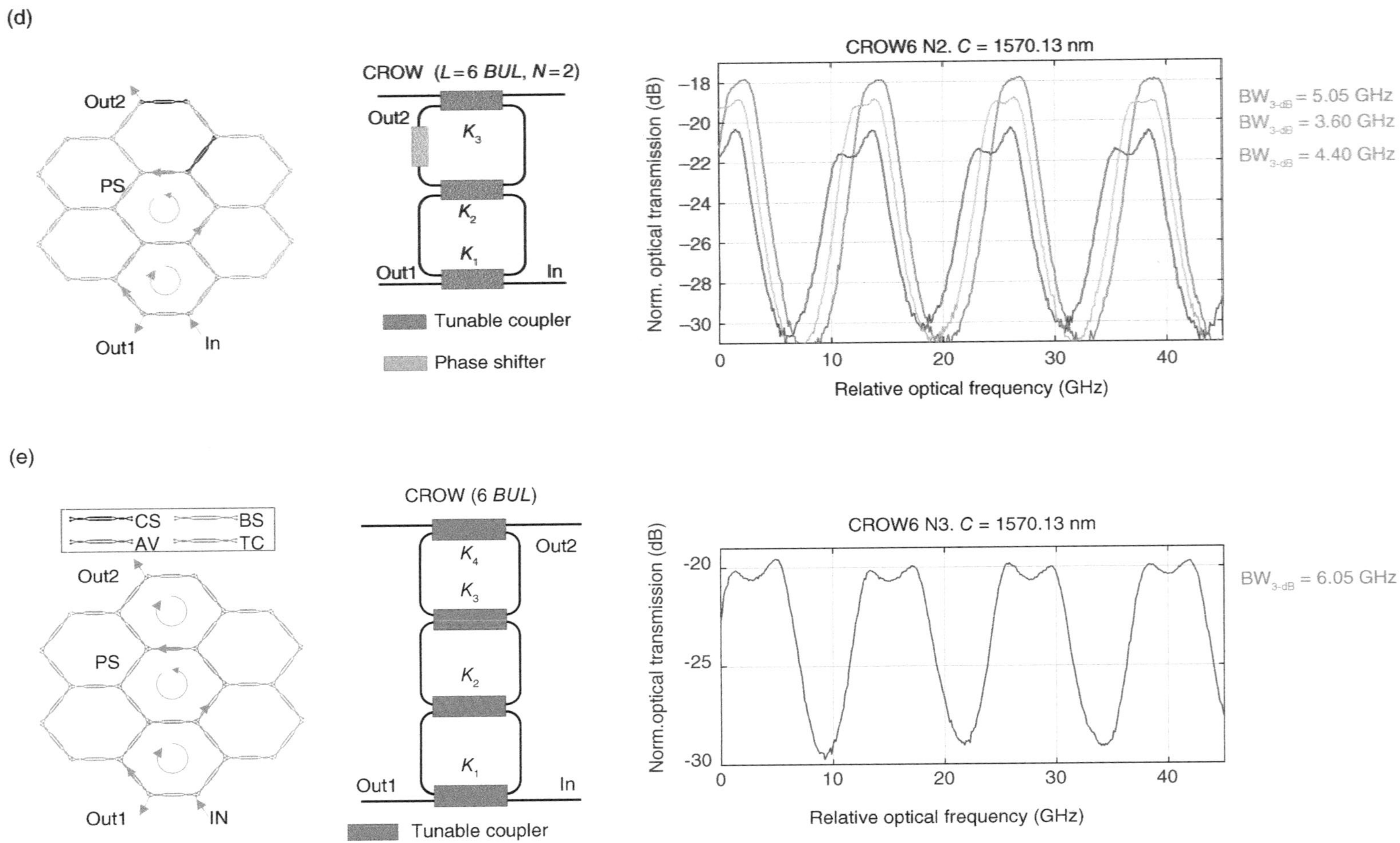

Figure 13.21 (Continued)

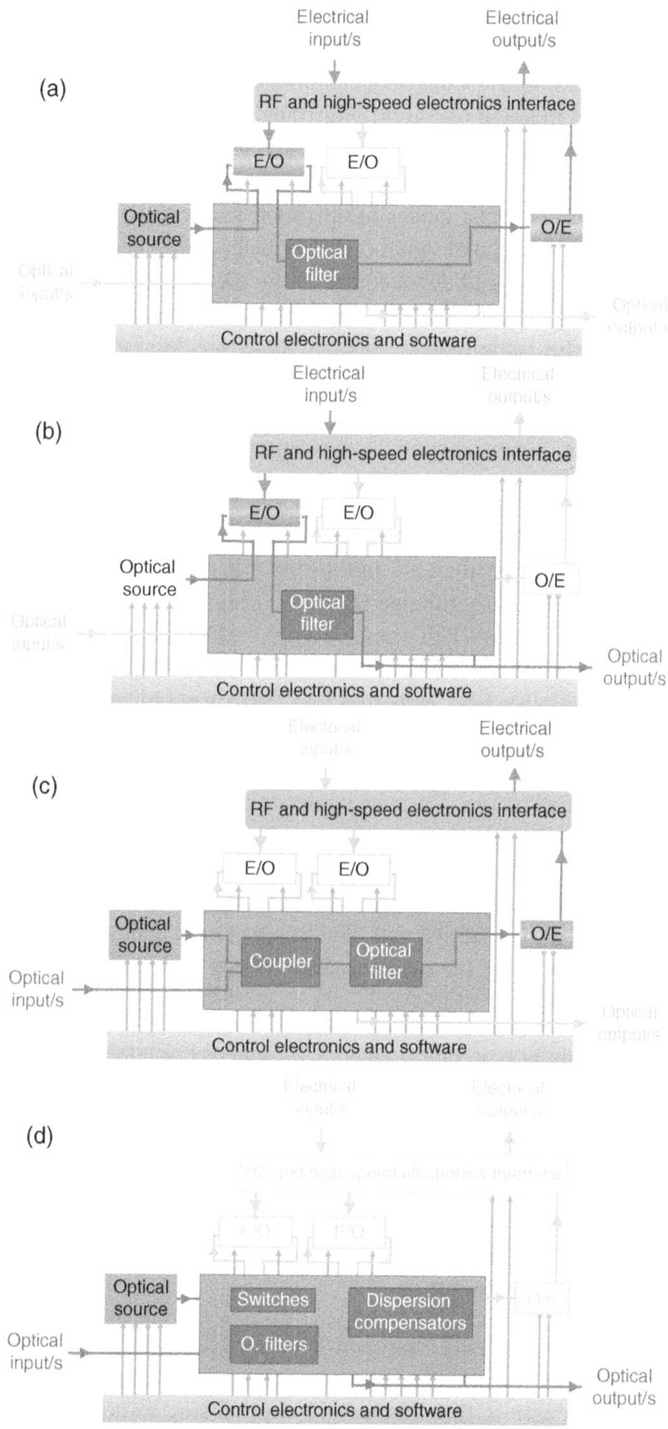

Figure 13.22 Generic purpose MWP processor architecture programmed different modes of operation: (a) electrical/electrical, (b) electrical/optical, (c) optical/electrical and (d) optical/optical. (After [132].)

Sometimes, the processed signal has to be distributed over a particular fiber link length after generation and/or processing. The processor can leverage the inherent properties of optical fibers for distribution purposes. The electrical/optical mode is widely employed in radio over fiber MWP links. At the receiver point of the link, another multipurpose MWP processor can be employed.

In this case, the receiver would be working in optical/electrical mode, processing the signal before the photo-detection. Optionally, the receiver can enable its own optical source to act as a local oscillator as well. Both modes of operation are displayed in Fig. 13.22(b), (c), respectively.

The last mode of operation, illustrated in Fig. 13.22(d), is the optical/optical. In this case, the input signal can be processed directly in the optical domain. Optical channel management can then perform common optical processing operations such as add/drop, switching and broadcasting. Note that all the previous modes of operation may coexist for a certain multi-task functionality. For example, a modulated signal could be divided after being processed and both distributed through the optical ports and down-converted by the photodetectors.

In the previous section, we have shown that the hexagonal waveguide core is the best candidate to perform the reconfigurable optical core task: featuring subsystems interconnection as well as enabling versatile optical processing operations. By suitable programming each element on the processor, including the reconfigurable optical core, the same hardware can be configured to perform all the main functionalities required in MWP, as we illustrate in Figs. 13.23 and 13.24.

True Time Delay Lines (DL): Together with tunable RF PSs are a key functionality required in complex signal processing applications. In particular, they constitute an essential building block in a considerable number of filtering and optical beamforming schemes. Discretely and continuously tunable DLs [133, 134] have been demonstrated, both of which can be supported by the MWP processor. Figure 13.23(a) illustrates, for example, the processor settings for the implementation of a discrete true time delay based on the Single Carrier Tuning (SCT) approach [93, 134]. Note that four ORRs need to be enabled for the carrier tuning and optical delay line sections, as well as a double-loaded MZI for the implementation of the optical single side-band filter.

Filtering Applications: The same hardware can be programmed as well to perform versatile RF photonic tunable filtering. Here, the spectral characteristics of the optical filter implemented by the optical waveguide mesh core are directly translated into the RF spectrum by means of a self-beating homodyne technique, where part of the un-modulated optical carrier is sent separately and combined with the RF-modulated and optically filtered prior to photodetection, as illustrated in Fig. 13.23(b) [135]. In this example, the synthetized RF filter is implemented by means of 6 cascaded ORRs.

Microwave and mm-wave tone generation: This functionality enables the photonic generation and distribution of high-frequency RF signals [74, 136]. Among the different approaches reported, we illustrate here the implementation of two. Figure 13.23(c) depicts the programmable processor employing the external modulator with different modulation biasing points together with an optical filter to suppress the optical carrier and the undesired (odd or even) modes. The filter is implemented here with a fourth-order lattice of MZIs. The second approach consists in programming an optoelectronic oscillator by closing the feedback loop of the RF section, as shown in Fig. 13.23(d) [74]. In this example, the reconfigurable optical core implements a High-Q optical storage section with seven ORRs.

Arbitrary Waveform Generation: This application consists in the generation of ultra-broad bandwidth RF waveforms with arbitrary and reconfigurable phase or amplitude characteristics. Wavelength-to-time mapping is a promising solution and can be integrated on a chip [82]. A broadband optical source is filtered by a spectral shaper. The spectral shape can be translated to the time domain by a dispersive element. Figure 13.24(a) illustrates the processor settings to configure the spectral shaper by means of five cascaded ORRs [82], and a dispersive element implemented by a fourth-order lattice filter [137]. Note that versatility is the most valuable

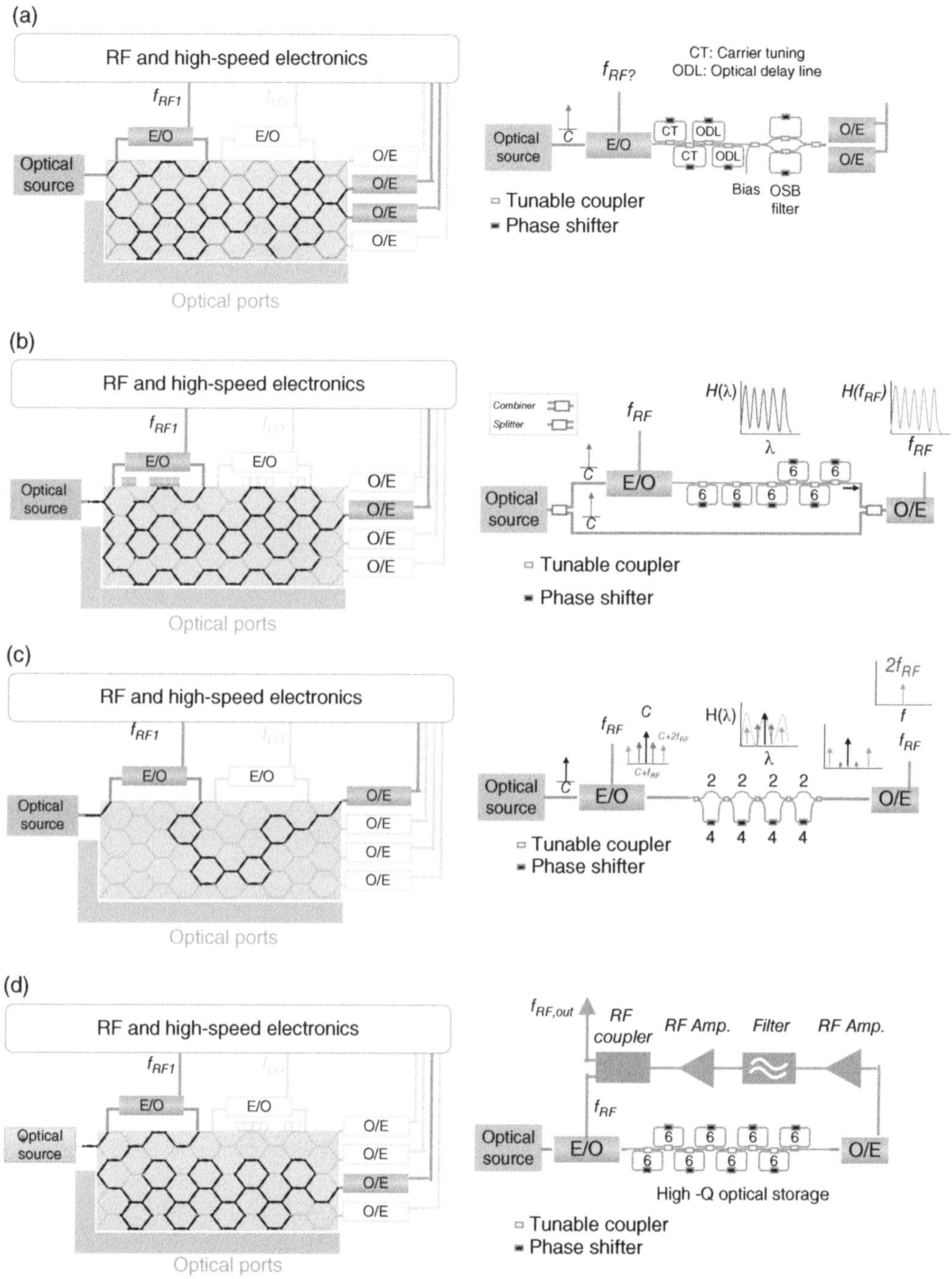

Figure 13.23 (Left) Examples of MWP processor settings for the implementation of different MWP functionalities. For each functionality the circuit schemes are shown in the right-hand side: (a) true time delay line employing the Single Carrier Tuning (SCT) technique, (b) self-homodyne filter FIR, microwave and mm-wave tones generation based on (c) external modulation, and (d) optoelectronic oscillation. (After [132].)

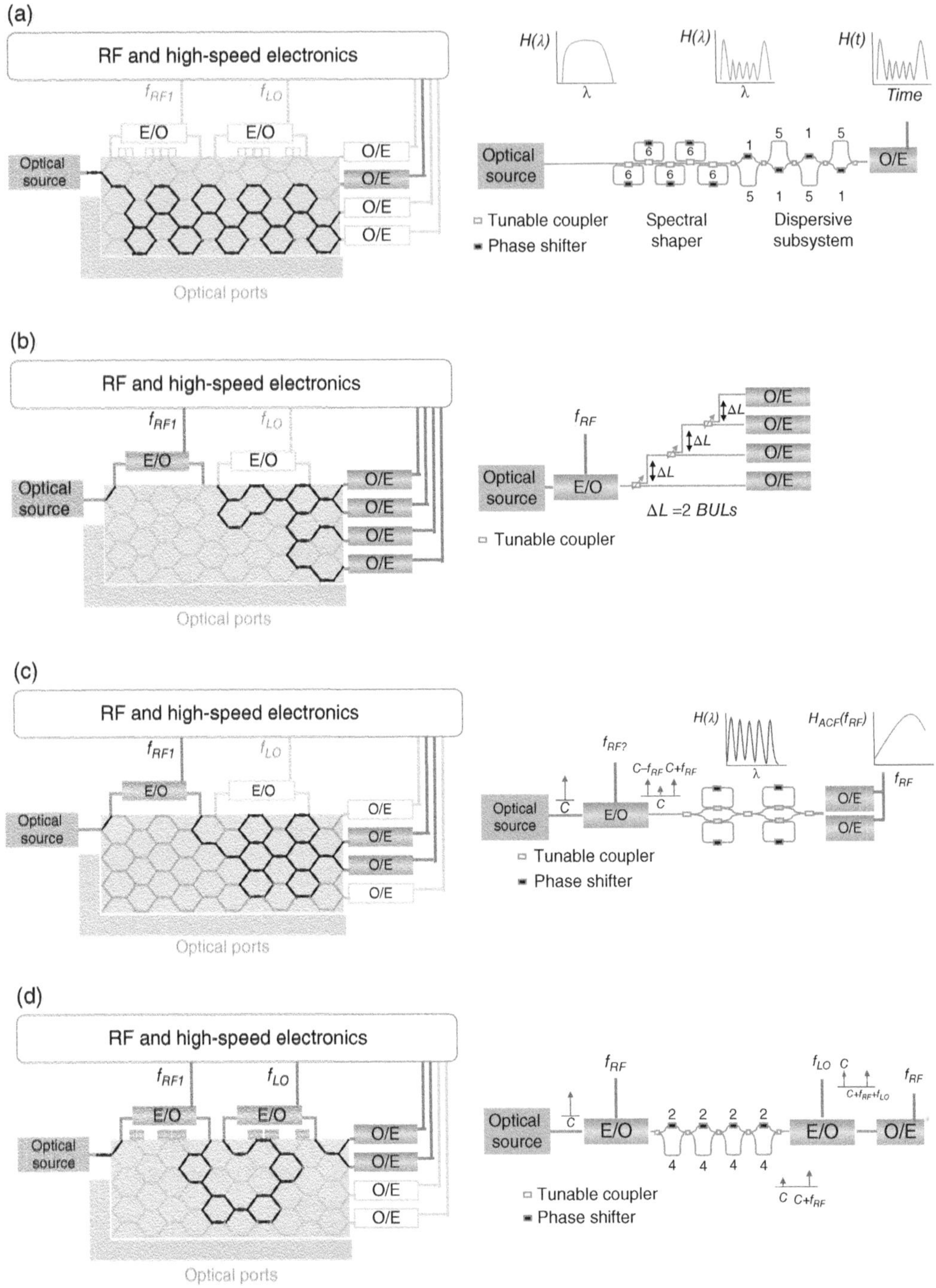

Figure 13.24 (Left) General-purpose photonic processor settings for the implementation of a particular MWP scheme (right): (a) arbitrary waveform generation based on wavelength-to-time mapping, (b) beamforming networks based on true time delay, (c) instantaneous frequency measurement, and (d) RF-mixing based on two cascaded modulators. (After [132].)

property of the spectral shaper, and the hexagonal waveguide processor core suits perfect for this task, as demonstrated in the previous section. Higher dispersive delays might be required and could be obtained by employing the optical ports for the connection of dispersive fiber reels.

Beamforming Networks: These systems are widely used to feed the radiating elements of phased array antennas. The radiation characteristics of an array can be modified by changing their electrical excitations [113]. The use of tunable delay lines allows a stable pointing angle independent of the transmitted frequency, enabling broadband operation [113]. Figure 13.24(b) illustrates the implementation of a beamformer based on the natural discrete delay lines of the hexagonal core. By modifying the length of the paths, the differential delay can be changed. The tilt angle tuning will be limited to discrete values in the range of $[-\pi, \pi]$.

Instantaneous Frequency Measurement: These systems are employed to measure the frequency of microwave signals in real-time applications. Figure 13.24(c) depicts one of the approaches that consist in mapping the unknown microwave frequency to an optical power ratio by establishing a unique relationship, commonly referred as the amplitude comparison function (ACF), that is independent of both the laser and input RF powers. The ACF can be performed by a 2×2 complementary filter [123, 138]. In this case, the filter is implemented by means of two cascaded balanced MZIs loaded with a total of 4 ORRs. Sharper filter functions lead to an increase in the resolution at the expense of reducing the frequency range. A key advantage of the reconfigurable processor is that the ACF can be made reconfigurable, so multiple measurements can be done while reconfiguring an ACP for each frequency range of measurement.

Frequency Mixing: Frequency up-and/or down-conversion of RF signals is typically required in many radio-over-fiber and intermediate-frequency-over-fiber systems [139]. Figure 13.24(d) illustrates the processor configuration to achieve this functionality, where an optional optical filter, this time implemented by a fourth-order lattice filter, is employed to suppress the possible intermodulation contributions prior to the second modulator input.

For some of the previous functionalities, the integration of an optical amplifier subsystem or an optical amplifier array prior to photodetection would be required to overcome the EO and optoelectronic conversion losses.

13.6 Conclusions and Future Prospects

We have reviewed the basic principles and some salient results reported in the area of integrated MWP. Although IMWP is still in its infancy, it provides a unique solution toward the manufacturing of low-cost MWP systems with the reduced SWaP metrics required by most emergent applications such as 5G communications, IoT, and autonomous driving systems, which are called to play a key role in our middle-term future. MWP as a technology enabler for the smooth interfacing between the fiber and wireless segments of networks supporting the aforementioned applications needs to incorporate photonic integration technologies to fulfill these highly demanding requisites.

The chapter has focused mainly on IMWP ASPICs as they represent 95% of the work reported on the field. Yet, we have also covered the new emerging topic of multifunctional circuits and the no less exciting area of universal programmable MWP processors, where a single hardware architecture can be programmed to carry out the different functionalities required in MWP applications.

References

1 Capmany, J. and Novak, D. (2007). Microwave photonics combines two worlds. *Nat. Photonics* 1: 319–330.

2 Seeds, A. (2002). *IEEE Trans. Microw. Theory Tech.* 50: 877.

3 Yao, J. (2009). Microwave photonics. *J. Lightw. Technol.* 27: 314–335.

4 Seeds, A.J. and Williams, K.J. (2011). Technology focus on microwave photonics. *Nat. Photonics* 5: 723.

5 Pizzinat, A., Chanclou, P., Saliou, F., and Diallo, T. (2015). Things you should know about fronthaul. *J. Lightw. Technol.* 33 (5): 1077–1083 https://doi.org/10.1109/JLT.2014.2382872.

6 Mjeku, M. and Gomes, N.J. (2008). Performance analysis of 802.11e transmission bursting in fiber-fed networks. *2008 IEEE Radio and Wireless Symposium*, Orlando, FL, USA, pp. 133–136. http://doi.org/10.1109/RWS.2008.4463446.

7 Sotom, M., Benazet, B., Le Kernec, A., and Maignan, M. (2009). Microwave photonic technologies for flexible satellite telecom payloads. *2009 35th European Conference on Optical Communication*, Vienna, Austria, 2009, pp. 1–4.

8 Crisp, M., Penty, R.V., White, I.H., and Bell, A. (2010). Wideband radio over fiber distributed antenna systems for energy efficient in-building wireless communications. *2010 IEEE 71st Vehicular Technology Conference*, Taipei, Taiwan, pp. 1–5. http://doi.org/10.1109/VETECS.2010.5493764.

9 Capmany, J., Ortega, B., and Pastor, D. (2006). A tutorial on microwave photonic filters. *J. Lightw. Technol.* 24: 201–229.

10 Popov, M. (2010). The convergence of wired and wireless services delivery in access and home networks. *Invited, Conference on Optical Fiber Communication (OFC/NFOEC), San Diego USA* OWQ6.

11 Koonen, A.M.J., Larrodé, M.G., Ngoma, A. et al. (2008). Perspectives of radio over fiber technologies. *Optical Fiber Communication Conference/National Fiber Optic Engineers Conference*, OSA Technical Digest (CD), Optica Publishing Group, paper OThP3.

12 Waterhouse, R. and Novak, D. (2015). Realizing 5G: microwave photonics for 5G mobile wireless systems. *IEEE Microw. Mag.* 16: 84–92.

13 Andrews, J.G., Buzzi, S., Wan, C. et al. (2014). What will 5G be? *IEEE J. Sel. Areas Commun.* 32: 1065–1082.

14 Al-Fuqaha, A., Guizani, M., Mohammadi, M. et al. (2015). Internet of things: a survey on enabling technologies, protocols, and applications. *IEEE Communications Surveys & Tutorials* 17: 2347–2376.

15 Hetch, J. (2016). The bandwidth bottleneck. *Nature* 536: 139–142.

16 Movassaghi, S., Abolhasan, M., Lipman, J. et al. (2014). Wireless body area networks: a survey. *IEEE Commun. Surv. Tutor.* 16: 1658–1686.

17 Hervás, J., Ricchiuti, A.L., Li, W. et al. (2017). Microwave photonics for optical sensors. *J. Sel. Topics Quantum Electron.* 23 (2): 327–339.

18 Blumenthal, D.J., Barton, J., Beheshti, N. et al. (2011). Integrated photonics for low-power packet networking. *IEEE Journal of Selected Topics in Quantum Electronics* 17: 458–471.

19 Marpaung, D.A.I., Roeloffzen, C.G.H., Heideman, R.G. et al. (2013). Integrated microwave photonics. *Laser Photonics Rev.* 7: 1–33.

20 Iezekiel, S., Burla, M., Klamkin, J. et al. (2015). RF engineering meets optoelectronics. *IEEE Microw. Mag.* 16: 18–45.

21 Zhang, W. and Yao, J. (2016). Silicon-based integrated microwave photonics. *IEEE J. Quantum Electron.* 52: 1–12.

22 Smit, M.K., Leijtens, X., Ambrosius, H. et al. (2014). An introduction to InP-based generic integration technology. *Semicond. Sci. Technol.* 29 (8): 083001.

23 Coldren, L.A., Nicholes, S.C., Johansson, L. et al. (2011). High performance InP-based photonic ICs—a tutorial. *IEEE J. Lightw. Technol.* 29: 554–570.

24 Kish, F., Nagarajan, R., Welch, D. et al. (2013). From visible light-emitting diodes to large-scale III–V photonic integrated circuits. *Proc. IEEE* 101: 2255–2270.

25 Hochberg, M. and Baehr-Jones, T. (2010). Towards fabless silicon photonics. *Nat. Photonics* 4: 492–494.

26 Bogaerts, W., Fiers, M., and Dumon, P. (2014). Design challenges in silicon photonics. *IEEE Journal of Selected Topics in Quantum Electronics* 20: 1–8.

27 Soref, R. (2006). The past, present, and future of silicon photonics. *IEEE Journal of Selected Topics in Quantum Electronics* 12: 1678–1687.

28 Chrostowski, L. and Hochberg, M. (2015). *Silicon Photonics Design: From Devices to Systems.* Cambridge University Press.

29 Heck, M.J.R., Bauters, J.F., Davenport, M.L. et al. (2013). Hybrid silicon photonic integrated circuit technology. *IEEE J. Sel. Top. Quant. Electron.* 19: 6100117.

30 Keyvaninia, S., Muneeb, M., Stanković, S. et al. (2013). Ultra-thin DVS-BCB adhesive bonding of III–V wafers, dies and multiple dies to a patterned silicon-on-insulator substrate. *Opt. Mater. Express* 3: 35–46.

31 Heideman, R.G., Hoekman, M., and Schreuder, F. (2012). TriPleX™-based integrated optical ring resonators for lab-on-a-chip- and environmental detection. *IEEE J. Sel. Top. Quantum Electron.* 18: 1583–1596.

32 Roeloffzen, C.G.H., Zhuang, L., Taddei, C. et al. (2013). Silicon nitride microwave photonic circuits. *Opt. Express* 21: 22937–22961.

33 (2015). JePPIX Roadmap 2021–2025. http://www.jeppix.eu.

34 Mitchell, A. (2007). $LiNbO_3$ devices for microwave photonics. In: *Annual Meeting of the IEEE Lasers and Electro-Optics Society (LEOS 2007)* (ed. E. Golovchenko, M.S. Ünlü, and S. Woodward). Lake Buena Vista, USA: (21–25 October 2007).

35 Ilchenko, V.S., Matsko, A.B., Solomatine, I. et al. (2008). Ka–band all-resonant photonic microwave receiver. *IEEE Photon. Technol. Lett.* 20: 1600–1612.

36 Wijayanto, Y.N., Murata, H., and Okamura, Y. (2013). Electrooptic millimeter-wave–lightwave signal converters suspended to gap-embedded patch antennas on low-k dielectric materials. *IEEE J. Sel. Top. Quantum Electron.* 19 (6): 3400709.

37 Pant, R., Byrnes, A., Poulton, C.G. et al. (2012). Photonic-chip-based tunable slow and fast light via stimulated Brillouin scattering. *Opt. Lett.* 37: 969–971.

38 Byrnes, A., Pant, R., Li, E. et al. (2012). Photonic chip based tunable and reconfigurable narrowband microwave photonic filter using stimulated Brillouin scattering. *Opt. Express* 20: 18836–18845.

39 Pagani, M., Morrison, B., Zhang, Y. et al. (2015). Low-error and broadband microwave frequency measurement in a silicon chip. *Optica* 2: 751.

40 Ma, M., Adams, R., and Chen, L. (2017, 2017). Integrated photonic chip enabled simultaneous multichannel wideband radio frequency spectrum analyzer. *J. Lightw. Technol.* 35 (13): 2622–2628. https://doi.org/10.1109/JLT.2017.2694003.

41 Marpaung, D., Morrison, B., Pagani, M. et al. (2015). Low power, chip-based stimulated Brillouin scattering microwave photonic filter with ultrahigh selectivity. *Optica* 2: 76.

42 Chen, H.-W., Fang, A.W., Peters, J.D. et al. (2010). Integrated microwave photonic filter on a hybrid silicon platform. *IEEE Trans. Microw. Theory Tech.* 58: 3213–3219.

43 Bradley, J.D.B., Purnawirman, P., Hosseini, E.S. et al. (2013). C- and L-B and erbium-doped aluminum oxide lasers with silicon nitride distributed Bragg reflector cavities. *CLEO: 2013, OSA Technical Digest* (Online), Optical Society of America, paper CF1I.4.

44 Belt, M., Huffman, T., Davenport, M. et al. (2013). Arrayed narrow linewidth erbium-doped waveguide-distributed feedback lasers on an ultra-low-loss silicon-nitride platform. *Opt. Lett.* 38: 4825–4828.
45 Bradley, J.D. and Pollnau, M. (2011). Erbium-doped integrated waveguide amplifiers and lasers. *Laser Photonics Rev.* 5: 368–340.
46 Loh, W., O'Donnell, F.J., Plant, J.J. et al. (2011). Packaged, high-power, narrow-linewidth slab-coupled optical waveguide external cavity laser (SCOWECL). *IEEE Photon. Tech. Lett.* 23: 974–977.
47 Shin, J.H., Chang, Y.-C., and Dagli, N. (2008). 0.3 V drive voltage GaAs/AlGaAs substrate removed Mach–Zehnder intensity modulators. *Appl. Phys. Lett.* 92: 201103.
48 Qiugui Zhou, A.S., Cross, Y.F., Beling, A. et al. (2013). Balanced InP/InGaAs photodiodes with 1.5-W output power. *IEEE Photon. J.* 5 (3): 6800307.
49 Zhou, Q., Cross, A.S., Beling, A. et al. (2013). High-power V-band InGaAs/InP photodiodes. *Photon. Technol. Lett. IEEE* 25 (10): 907–909.
50 Muñoz, P., Capmany, J., Pérez, D. et al. (2014). Integrated microwave photonics: state of the art and future trends. *Proc. Int. Conf. on Transparent Opt. Networks (ICTON)*, Graz, Austria (6–10 July 2014).
51 Norberg, E., Guzzon, R., Nicholes, S. et al. (2010). Programmable photonic lattice filters in InGaAsP–InP. *IEEE Photon. Technol. Lett.* 22: 109–111.
52 Norberg, E., Guzzon, R., Parker, J. et al. (2011). Programmable photonic microwave filters monolithically integrated in InP–InGaAsP. *J. Lightw. Technol.* 29: 1611–1619.
53 Guzzon, R.S., Norberg, E.J., Parker, J.S. et al. (2011). Integrated InP-InGaAsP tunable coupled ring optical bandpass filters with zero insertion loss. *Opt. Express* 19: 7816–7826.
54 Guzzon, R., Norberg, E., and Coldren, L. (2012). Spurious-free dynamic range in photonic integrated circuit filters with semiconductor optical amplifiers. *IEEE J. Quantum Electron.* 48: 269–278.
55 Rasras, M.S., Tu, K.Y., Gill, D.M. et al. (2009). Demonstration of a tunable microwave photonic notch filter using low-loss silicon ring resonators. *J. Lightw. Technol.* 27: 2105–2110.
56 Ibrahim, S., Fontaine, N.K., Djordjevic, S.S. et al. (2011). Demonstration of a fast-reconfigurable silicon CMOS optical lattice filter. *Opt. Express* 19: 13245–13256.
57 Rasras, M., Gill, D., Patel, S. et al. (2007). Demonstration of a fourth-order pole-zero optical filter integrated using CMOS processes. *J. Lightw. Technol.* 25: 87–92.
58 Dong, P., Feng, N.N., Feng, D. et al. (2010). GHz-bandwidth optical filters based on high-order silicon ring resonators. *Opt. Express* 18: 23784–23789.
59 Feng, N.N., Dong, P., Feng, D. et al. (2010). Thermally-efficient reconfigurable narrowband RF-photonic filter. *Opt. Express* 18: 24648–24653.
60 Djordjevic, S., Luo, L., Ibrahim, S. et al. (2011). Fully reconfigurable silicon photonic lattice filters with four cascaded unit cells. *IEEE Photon. Technol. Lett.* 23 (1): 42–44.
61 Alipour, P., Eftekhar, A.A., Atabaki, A.H. et al. (2011). Fully reconfigurable compact RF photonic filters using high-Q silicon microdisk resonators. *Opt. Express* 19: 15899–15907.
62 Tu, K.Y., Rasras, M.S., Gill, D.M. et al. (2010). Silicon RF-photonic filter and down-converter. *J. Lightw. Technol.* 28: 3019–3028.
63 Guan, B., Djordjevic, S.S., Fontaine, N.K. et al. (2014). CMOS compatible reconfigurable silicon photonic lattice filters using cascaded unit cells for RF-photonic processing. *IEEE Journal of Selected Topics in Quantum Electronics* 20: 359–368.
64 Fandiño, J.S., Domenech, J.D., Muñoz, P., and Capmany, J. (2017). A monolithic integrated photonic microwave filter. *Nat. Photonics* 11: 124–129.
65 Marpaung, D., Morrison, B., Pant, R. et al. (2013). Si_3N_4 ring resonator-based microwave photonic notch filter with an ultrahigh peak rejection. *Opt. Express* 21: 23286–23294.

66 Zhuang, L., Zhu, C., Corcoran, B. et al. (2016). Sub-GHz-resolution C-band Nyquist-filtering interleaver on a high-index-contrast photonic integrated circuit. *Opt. Express* 24: 5715–5727.

67 Ding, Y., Peucheret, C., Xu, J. et al. (2011). Ultra-wide band signal generation using a silicon micro-ring resonator. *IEEE Photonic Society 24th Annual Meeting*, Arlington, VA, USA, pp. 258–259. http://doi.org/10.1109/PHO.2011.6110524.

68 Yue, Y., Huang, H., Zhang, L. et al. (2012). UWB monocycle pulse generation using two-photon absorption in a silicon waveguide. *Opt. Lett.* 37: 551–553.

69 Mirshafiei, M., LaRochelle, S., and Rusch, L.A. (2012). Optical UWB waveform generation using a micro-ring resonator. *IEEE Photon. Technol. Lett.* 24 (15): 1316–1318.

70 Marpaung, D., Chevalier, L., Burla, M., and Roeloffzen, C. (2011). Impulse radio ultrawideband pulse shaper based on a programmable photonic chip frequency discriminator. *Opt. Express* 19: 24838–24848.

71 Wang, J., Sun, Q., Sun, J., and Zhang, W. (2009). All-optical UWB pulse generation using sum-frequency generation in a PPLN waveguide. *Opt. Express* 17 (5): 3521–3530.

72 Wang, J. and Sun, J. (2010). All-optical ultrawideband monocycle generation using quadratic nonlinear interaction seeded by dark pulses. *IEEE Photon. Technol. Lett.* 22 (3): 140–142.

73 Liu, F., Wang, T., Zhang, Z. et al. (2009). On-chip photonic generation of ultra-wideband monocycle pulses. *Electron. Lett.* 45 (24): 1247–1248.

74 Maleki, L. (2011). Sources: the optoelectronic oscillator. *Nat. Photonics* 5: 728–730.

75 Liang, W., Ilchenko, V.S., Savchenkov, A.A. et al. (2010). Whispering-gallery-mode-resonator-based ultranarrow linewidth external-cavity semiconductor laser. *Opt. Lett.* 35: 2822–2824.

76 Devgan, P., Pruessner, M., Urick, V., and Williams, K. (2010). Detecting low-power RF signals using a multimode optoelectronic oscillator and integrated optical filter. *IEEE Photon. Technol. Lett.* 22: 152–154.

77 Volyanskiy, K., Salzenstein, P., Tavernier, H. et al. (2010). Compact optoelectronic microwave oscillators using ultra-high Q whispering gallery mode disk-resonators and phase modulation. *Opt. Express* 18: 22358–22363.

78 Kippenberg, T.J., Holzwarth, R., and Diddams, S.A. (2011). Microresonator-based optical frequency combs. *Science* 332: 555–559.

79 Foster, M.A., Levy, J.S., Kuzucu, O. et al. (2011). Silicon-based monolithic optical frequency comb source. *Opt. Express* 19: 14233–14239.

80 Ferdous, F., Miao, H., Leaird, D.E. et al. (2011). Spectral line-by-line pulse shaping of on-chip microresonator frequency combs. *Nat. Photonics* 5: 770–776.

81 Shen, H., Khan, M.H., Fan, L. et al. (2010). Eight-channel reconfigurable microring filters with tunable frequency, extinction ratio and bandwidth. *Opt. Express* 18: 18067–18076.

82 Khan, M.H., Shen, H., Xuan, Y. et al. (2010). Ultrabroad-bandwidth arbitrary radiofrequency waveform generation with a silicon photonic chip-based spectral shaper. *Nat. Photonics* 4: 117–122.

83 Xiao, S., Khan, M.H., Shen, H., and Qi, M. (2007). Compact silicon microring resonators with ultra-low propagation loss in the C band. *Opt. Express* 15: 14467–14475.

84 Wang, J., Shen, H., Fan, L. et al. (2015). Reconfigurable radio-frequency arbitrary waveforms synthesized in a silicon photonic chip. *Nature Commun.* 6957.

85 Zhang, W., Zhang, J., and Yao, J.P. (2014). Largely chirped microwave waveform generation using a silicon-based on-chip optical spectral shaper. *MWP2014*, Sapporo, Japan (20–23 October 2014).

86 Zhang, W. and Yao, J.P. (2015). Photonic generation of linearly chirped microwave waveform with a large time-bandwidth product using a silicon-based on-chip spectral shaper. *MWP2015*, Pathos, Cyprus (26–29 October 2015).

87 Rasras, M., Madsen, C., Cappuzzo, M. et al. (2005). Integrated resonance-enhanced variable optical delay lines. *IEEE Photon. Technol. Lett.* 17: 834–836.

88 Howley, B., Chen, Y., Wang, X. et al. (2005). 2-Bit reconfigurable true timedelay lines using 2×2 polymer waveguide switches. *IEEE Photon. Technol. Lett.* 17 (9): 1944–1946.

89 Lenz, G., Eggleton, B., Madsen, C., and Slusher, R. (2001). Optical delay lines based on optical filters. *IEEE J. Quantum Electron.* 37 (4): 525–532.

90 Zhuang, L., Roeloffzen, C., Heideman, R. et al. (2007). Single-chip ring resonator-based 1×8 optical beam forming network in CMOS-compatible waveguide technology. *IEEE Photon. Technol. Lett.* 19: 1130–1132.

91 Melloni, A., Morichetti, F., Ferrari, C., and Martinelli, M. (2008). Continuously tunable 1 byte delay in coupled-resonator optical waveguides. *Opt. Lett.* 33: 2389–2391.

92 Cardenas, J., Foster, M.A., Sherwood-Droz, N. et al. (2010). Wide-bandwidth continuously tunable optical delay line using silicon microring resonators. *Opt. Express* 18 (25): 26525–26534.

93 Morton, P.A., Cardenas, J., Khurgin, J.B., and Lipson, M. (2012). Fast thermal switching of wideband optical delay line with no long-term transient. *IEEE Photon. Technol. Lett.* 24 (6): 512–514.

94 Khan, S., Baghban, M.A., and Fathpour, S. (2011). Electronically tunable silicon photonic delay lines. *Opt. Express* 19 (12): 11780–11785.

95 Giuntoni, I., Stolarek, D., Kroushkov, D.I. et al. (2012). Continuously tunable delay line based on SOI tapered Bragg gratings. *Opt. Express* 20 (10): 11241–11246.

96 Wang, J., Ashrafi, R., Adams, R. et al. (2016). Subwavelength grating enabled on-chip ultra-compact optical true time delay line. *Sci. Rep.* 6: 30235.

97 Chang, Q., Li, Q., Zhang, Z. et al. (2009). A tunable broadband photonic RF phase shifter based on a silicon microring resonator. *IEEE Photon. Technol. Lett.* 21: 60–62.

98 Lloret, J., Sancho, J., Pu, M. et al. (2011). Tunable complex-valued multi-tap microwave photonic filter based on single silicon-on-insulator microring resonator. *Opt. Express* 19: 12402–12407.

99 Xue, W., Sales, S., Capmany, J., and Mørk, J. (2009). Microwave phase shifter with controllable power response based on slow- and fast-light effects in semiconductor optical amplifiers. *Opt. Lett.* 34: 929–931.

100 Xue, W., Sales, S., Capmany, J., and Mørk, J. (2010). Wideband 360° microwave photonic phase shifter based on slow light in semiconductor optical amplifiers. *Opt. Express* 18: 6156–6163.

101 Lloret, J., Morthier, G., Ramos, F. et al. (2012). Ultracompact electro-optic phase modulator based on III–V-on-silicon microdisk resonator. *Opt. Express* 20 (10): 10796–10806.

102 Morton, P.A. and Khurgin, J.B. (2009). Microwave photonic delay line with separate tuning of the optical carrier. *IEEE Photon. Technol. Lett.* 21 (22): 1686–1688.

103 Burla, M., Marpaung, D., Zhuang, L. et al. (2011). On-chip CMOS compatible reconfigurable optical delay line with separate carrier tuning for microwave photonic signal processing. *Opt. Express* 19: 21475–21484.

104 Horikawa, K., Nakasuga, Y., and Ogawa, H. (1995). Self-heterodyning optical waveguide beam forming and steering network integrated on Lithium Niobate substrate. *IEEE Trans. Microw. Theory Tech.* 43: 2395–2401.

105 Grosskopf, G. (2003). Silica based optical beam former for 60 GHz array antennas. *Fiber Integr. Opt.* 22: 35–46.

106 Grosskopf, G., Eggemann, R., Ehlers, H. et al. (2003). Maximum directivity beam-former at 60 GHz with optical feeder. *IEEE Trans. Antennas Propag.* 51 (11): 3040–3046.

107 Burla, M., Marpaung, D., Zhuang, L. et al. (2013). Integrated photonic Ku-band beamformer chip with continuous amplitude and delay control. *IEEE Photon. Technol. Lett.* 25: 1145–1148.

108 Meijerink, A., Roeloffzen, C.G.H., Meijerink, R. et al. (2010). Novel ring resonator-based integrated photonic beamformer for broadband phased array receive antennas—Part I: design and performance analysis. *J. Lightw. Technol.* 28: 3–18.

109 Zhuang, L., Roeloffzen, C.G.H., Meijerink, A. et al. (2010). Novel ring resonator-based integrated photonic beamformer for broadband phased array receive antennas—Part II: experimental prototype. *J. Lightw. Technol.* 28: 19–31.

110 Marpaung, D., Zhuang, L., Burla, M. et al. (2011). Towards a broadband and squint-free Ku-band phased array antenna system for airborne satellite communications. *Proceedings of the 5th European Conference on Antennas and Propagation (EuCAP 2011)*, Rome, Italy, pp. 2623–2627.

111 Marpaung, D., Zhuang, L., Burla, M. et al. (2011). Photonic integration and components development for a Ku-band phased array antenna system. *Proceedings of the IEEE Topical Meeting on Microwave Photonics (MWP 2011)*, Singapore, pp. 458–461.

112 Burla, M., Roeloffzen, C.G.H., Zhuang, L. et al. (2012). System integration and radiation pattern measurements of a phased array antenna employing an integrated photonic beamformer for radio astronomy applications. *Appl. Opt.* 51 (7): 789–802.

113 Burla, M., L. Zhuang, D. Marpaung et al. (2013). On-chip, CMOS-compatible, hardware-compressive integrated photonic beamformer based on WDM. *2013 International Topical Meeting on de Microwave Photonics (MWP)*, Alexandria, VA, USA, pp. 41–44. http://doi.org/10.1109/MWP.2013.6724014.

114 Smit, M., van der Tol, J., and Hill, M. (2012). Moore's law in photonics. *Laser Photonics Rev.* 6: 1–13.

115 Kish, F., Welch, D., Nagarajan, R. et al. (2011). Current status of large-scale InP photonic integrated circuits. *IEEE J. Sel. Topics Quantum Electron.* 17: 1470–1489.

116 Orcutt, J.S., Khilo, A., Holzwarth, C.W. et al. (2011). Nanophotonic integration in state-of-the-art CMOS foundries. *Opt. Express* 19: 2335–2346.

117 Orcutt, J.S., Moss, B., Sun, C. et al. (2012). Open foundry platform for high-performance electronicphotonic integration. *Opt. Express* 20: 12222–12232.

118 Koehl, S., Liu, A., and Paniccia, M. (2011). Integrated silicon photonics: harnessing the data explosion. *Opt. Photonics News* 22: 24–29.

119 Dumon, P., Bogaerts, W., Baets, R. et al. (2009). Towards foundry approach for silicon photonics: silicon photonics platform ePIXfab. *Electron. Lett.* 45: 581–582.

120 Marpaung, D., Roeloffzen, C., Leinse, A., and Hoekman, M. (2010). A photonic chip based frequency discriminator for a high performance microwave photonic link. *Opt. Express* 18: 27359–27370.

121 Fandiño, J.S., Doménech, J.D., Muñoz, P., and Capmany, J. (2013). Integrated InP frequency discriminator for phase-modulated microwave photonic links. *Opt. Express* 21: 3726–3736.

122 Liu, W., Li, M., Guzzon, R.S. et al. (2016). A fully reconfigurable photonic integrated signal processor. *Nat. Photonics* 10: 190–195.

123 Fandiño, J.S. and Muñoz, P. (2013). Photonics-based microwave frequency measurement using a double-sideband suppressed-carrier modulation and an InP integrated ring-assisted Mach–Zehnder interferometer filter. *Opt. Lett.* 38 (21): 4316–4319.

124 Fandino, J.S., Rius, M., Mora, J. et al. (2013). Optical single sideband transmitter using phase modulation and a photonic integrated filter. *2013 IEEE International Topical Meeting on Microwave Photonics (MWP)*, Alexandria, VA, USA, pp. 154–157. http://doi.org/10.1109/MWP.2013.6724043.

125 J. Capmany, J. Mora, P. Muñoz, S. Sales, "A microwave photonics transistor," de IEEE Topical meeting on MWP, Alexandria, USA, (2013).

126 Pérez, D., Gasulla, I., and Capmany, J. (2015). Software-defined reconfigurable microwave photonics processor. *Opt. Express* 23: 14640–14654.

127 D. Perez, I. Gasulla and J. Capmany, "Honeycomb lattice meshes for reconfigurable universal microwave photonics processors", SPIE/COS Photonics Asia, (2016).

128 Capmany, J., Gasulla, I., and Pérez, D. (2016). Microwave photonics: the programmable processor. *Nat. Photonics* 10: 6–8.

129 Zhuang, L., Roeloffzen, C.G.H., Hoekman, M. et al. (2015). Programmable photonic signal processor chip for radiofrequency applications. *Optica* 2: 854–859.

130 Pérez, D., Gasulla, I., Capmany, J., and Soref, R.A. (2016). Reconfigurable lattice mesh designs for programmable photonic processors. *Opt. Express* 24: 12093–12106.

131 Pérez, D., Gasulla, I., Crudgington, L. et al. (2017). General-purpose silicon photonics signal processor core. *Nat. Commun.*, 8: 636.

132 Pérez, D., Gasulla, I., and Capmany, J. (2019). Towards programmable microwave photonics processors," IEEE. *J. Lightw. Technol.* 36: 519–532.

133 Wang, X., Zhou, L., Lie, R., and Xie, J. (2017). Continuously tunable ultra-thin silicon waveguide optical delay line. *Optica* 4: 507–515.

134 Burla, M., Khan, M.R.H., Marpaung, D.A.I. et al. (2011). Separate carrier tuning scheme for integrated optical delay lines in photonic beamformers. *2011 International Topical Meeting on & Microwave Photonics Conference on Microwave Photonics*, Singapore, pp. 65–68. http://doi.org/10.1109/MWP.2011.6088670.

135 Pérez, D., Gasulla, I., Capmany, J. et al. (2016). Figures of merit for self-beating filtered microwave photonic systems. *Opt. Express* 24: 10087–10102.

136 Qi, G., Yao, J., Seregelyi, J. et al. (2005). Generation and distribution of a wide-band continuously tunable millimeter-wave signal with an optical external modulation technique. *IEEE Trans. Microwave Theory Tech.* 53 (10): 3090–3097.

137 Jinguji, K. and Kawachi, M. (1995). Synthesis of coherent two-port lattice-form optical delay-line circuit. *J. Lightw. Technol.* 13: 73–82.

138 Shao, H., Yu, H., Jiang, X. et al. (2014). Large bandwidth and high accuracy photonic-assisted instantaneous microwave frequency estimation system based on an integrated silicon micro-resonator. *11th International Conference on Group IV Photonics (GFP)*, Paris, France, pp. 47–48. http://doi.org/10.1109/Group4.2014.6962043.

139 Ho, K.-P., Liaw, S., and Lin, C. (1997). Efficient photonic mixer with frequency doubling. *IEEE Photonics Technol. Lett.* 9: 511–513.

14

Quantum Microwave Photonics

14.1 Introduction

The main building blocks of a typical microwave photonic system [1–4] include a laser source, an electro-optic modulator (EOM), an optical signal processor, and a photodetector (PD). The optical signal from the laser source is first modulated by an input microwave signal in the EOM, and the modulated optical signal is then processed by the optical signal processor. Finally, the processed optical signal is detected by the PD to recover the desired microwave signal. The key advantage of such a microwave photonic system is the wide processing bandwidth obtained due to the wide bandwidth of optical technology [5–13]. However, the bandwidth of the whole microwave photonic system is still limited by the microwave devices after the PD, such as the coaxial cable and the analog-to-digital converter (ADC), since the recovered microwave signal should be sent to these microwave devices for further processing. As a result, the bandwidth of the whole microwave photonic system is generally limited to about 10 GHz, although the processing bandwidth of the optical technology can be as large as hundreds of THz. That is to say, the wide bandwidth feature of optical technology has not been fully exploited in microwave photonic systems yet.

In addition, there is a tradeoff between the bandwidth and quantum efficiency of the PD [14, 15] in the microwave photonic system. To obtain a high quantum efficiency, the barrier width of the PD should be wider to absorb more incoming photons. However, the wider barrier width would also result in decreased bandwidth of the PD since the carrier transit time is increased due to the enlargement of barrier width. With the rapid increase of the processing bandwidth of the microwave photonic system, the quantum efficiency of the PD would be decreased, which limits the sensitivity or minimum detectable optical power of the microwave photonic system. Although optical amplifiers can be adopted to boost the optical power, this process would also introduce extra unwanted noise to the system. Therefore, the PD in current microwave photonic systems cannot be applied to scenarios with ultra-low power and low signal-to-noise ratio.

Since the above two limitations in traditional microwave photonic systems come from the detecting part of the whole system, it is possible to overcome the above limitations by exploring new detecting schemes. In recent years, photonic quantum technology has been rapidly developed and provided promising implementation paths for modern quantum technology [16–20]. In photonic quantum systems, optical signals with ultra-weak power that are down to a single photon level can be detected by using a highly sensitive single-photon detection scheme [21–27]. The single-photon detection scheme also avoids the requirement of large bandwidth for successive electronic devices, such as the coaxial cable and ADC, behind the single-photon detector (SPD). Thus, the bandwidth and sensitivity limitations in traditional microwave photonic systems can be solved

Microwave Photonics, First Edition. Jianping Yao, José Capmany, and Ming Li.

Companion website: www.wiley.com/go/yao/microwavephotonics

by harnessing the wide processing bandwidth and high-sensitivity advantages of quantum single-photon detection scheme [28].

Very recently, we have proposed a quantum microwave photonic signal processing system by replacing the signal detection scheme in the traditional microwave photonic link with the quantum single-photon detection scheme [29]. Microwave photonic signal processing, including RF phase shifting and filtering, has been successfully demonstrated with an ultra-high sensitivity and a bandwidth larger than that of the coaxial cable. Several novel microwave photonic systems have also been proposed and demonstrated based on the quantum single-photon detection scheme in recent years for weak microwave signal detection [30], nonlocal frequency-to-time mapping [31], RF-compressed sensing [32], as well as quantum key distribution (QKD) [33–35], which show clearly that quantum technology brings new opportunities to microwave photonics, or vice versa. It should be noted that the names of microwave photonics and quantum microwave photonics have also been adopted in quantum technology, for example, in superconducting quantum circuits where the quantum properties of microwave fields generated from these circuits are studied [36, 37]. We would like to clarify that this content is out of the scope of this chapter and we only focus on the novel quantum microwave photonic systems that combine microwave photonics and quantum technology with features or functions that are very complex or even not possible using traditional microwave photonic or quantum technology.

In this chapter, we discuss the emerging advances in quantum microwave photonics that use quantum technology to improve the performance of classical microwave photonic systems and verse visa. We first provide the basic operation principle of the quantum single-photon detection scheme, which is the key to current quantum microwave photonic systems. Then we summarize the emerging advances in quantum microwave photonics in recent years, including weak microwave signal detection, microwave photonic phase shifting and filtering, nonlocal frequency-to-time mapping, RF-compressed sensing, and QKD. Finally, we discuss the challenges and future prospects of quantum microwave photonics.

14.2 Principle of the Single-Photon Detection Scheme

Comparison between a classic microwave photonic link and a quantum single-photon microwave photonic link is shown in Fig. 14.1. As can be seen, a conventional high-power light source and high-speed PD are used in a classic microwave photonic link. The processed optical signal is detected by the high-speed PD and the recovered RF signal is sent to the successive microwave devices for post-processing. In a quantum single-photon microwave photonic link, a single-photon source and SPD are used. The RF signal is modulated on the photon signal from the single-photon source. The processed optical signal is first detected by a low-jitter SPD and then sent to a low-jitter recovering system, i.e. the time-correlated single-photon counting (TCSPC) module, to recover the RF signal. Clearly, the SPD and TCSPC modules are keys to the single-photon detection scheme in the quantum microwave photonic link.

In order to understand the principle of the quantum microwave photonic link in Fig. 14.1(b), it is necessary to briefly discuss the principle of the single-photon detection scheme [21–27]. The schematic diagram and principle of the single-photon detection scheme are shown in Fig. 14.2(a), (b), respectively. As can be seen, there is a periodic trigger signal whose phase is synchronized with the signal under test for the "start" port of the TCSPC module, and the "stop" port of the TCSPC module is connected to the output of the SPD. The output single-photon response of the SPD is a train of

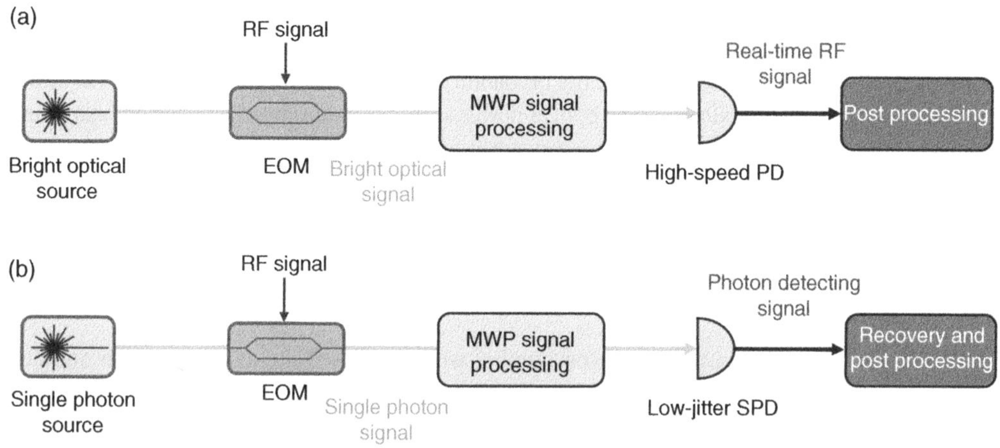

Figure 14.1 Comparison between a typical microwave photonic link (a) and a quantum single-photon microwave photonic link (b). The modulated optical signal is detected using a low-jitter SPD in the quantum single-photon microwave photonic link. The bandwidth and sensitivity limitations in traditional microwave photonic links can be solved by harnessing the advantages of a quantum single-photon detection scheme. EOM, electro-optic modulator; MWP, microwave photonic; RF, radiofrequency; SPD, single-photon detector.

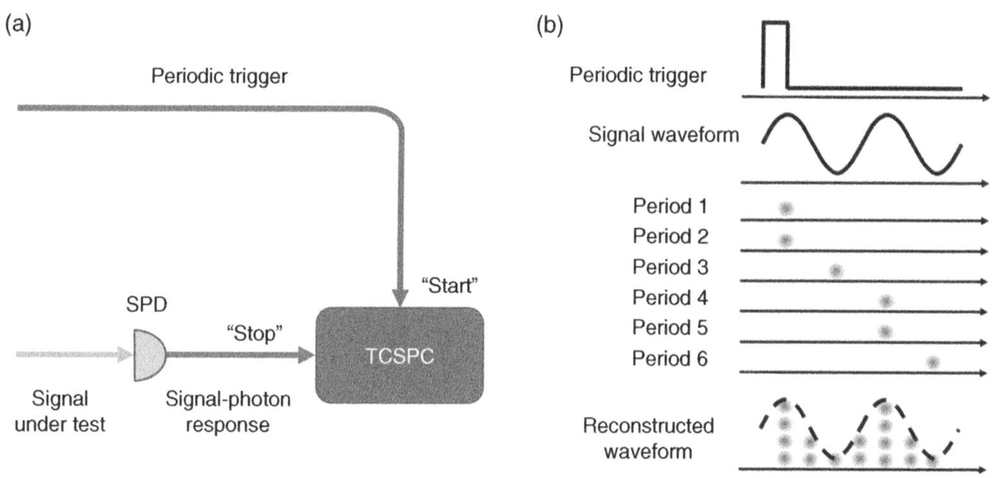

Figure 14.2 Single-photon detection scheme. (a) Schematic diagram. (b) Principle. The time-correlated single-photon counting (TCSPC) module records the detection times in each trigger cycle or period. After a certain number of accumulations, the waveform of the signal under test is reconstructed as the histogram of the detection times. It should be noted that trigger periods without detection events are also possible and these periods are not shown in (b).

pulses with a density that represents appearance of the photons. The TCSPC module measures the probability of the appearance of the photons at certain detection times. Specifically, when the TCSPC module detects the arrival of the trigger, it is ready to capture the single-photon response from the SPD in that trigger cycle. If the single-photon response is captured, the TCSPC module would record the detection time, i.e. the time difference between the arrival time of the single-photon response and the trigger signal. Then, the TCSPC module would repeat the above process, and a certain number of detection times are recorded after a certain number of measurements. Finally, the TCSPC module would count the results of the recorded detection times and provide a histogram of the detection times. Since the probability of the appearance of the photons at a certain detection time is approximately the optical power at this certain time, the resulting histogram of the detection times indeed presents the waveform of the signal under test.

In theory [24–26], the signal under test of the single-photon detection system satisfies Poisson distribution. The probability that there are g photons in test time T can be expressed as

$$p_g = \frac{(\mu_T)^g}{g!}\exp(-\mu_T) \tag{14.1}$$

where $\lambda(t)$ represents the change of photon number with time, μ_T is the average photon number within the test time.

$$\mu_T = \int_0^T \lambda(t)dt \tag{14.2}$$

The photons are detected by the SPD, and the output of SPD also satisfies Poisson distribution except that the detected photons are related to the quantum efficiency of the SPD. The TCSPC module is generally used in conditions where the photon number in one trigger cycle is close or much less than one, thus the cases when two or more photons are detected in one trigger cycle can be neglected. As a result, we can divide the test time (0, T) into q pieces and assume that there is a photon in time interval (t_i, $t_i + \Delta t$), while there is no photon in time interval ($t_i + \Delta t$, t_{i+1}). The probability of one photon in time interval (t_i, $t_i + \Delta t$) and no photon in time interval ($t_i + \Delta t$, t_{i+1}) can be expressed as

$$P_i = P_{(t_i, t_i+\Delta t)}(1) \cdot P_{(t_i+\Delta t, t_{i+1})}(0) = \mu_{i1}e^{-\mu_{i1}} \cdot e^{-\mu_{i2}} \tag{14.3}$$

where $\mu_{i1} = \int_{t_i}^{t_i+\Delta t}\lambda(t)dt$ and $\mu_{i2} = \int_{t_i+\Delta t}^{t_{i+1}}\lambda(t)dt$ are the photon number in time intervals (t_i, $t_i + \Delta t$) and ($t_i + \Delta t$, t_{i+1}), respectively. Thus, the total probability of one photon in time intervals (t_1, $t_1 + \Delta t$), (t_2, $t_2 + \Delta t$), ..., (t_g, $t_g + \Delta t$), respectively, and no photon in other time intervals can be written as

$$\begin{aligned} P &= \prod_{i=1}^{g} P_i = \prod_{i=1}^{g} \exp\left[-\int_{t_i}^{t_{i+1}} \lambda(t)dt\right] \cdot \prod_{i=1}^{g} \int_{t_i}^{t_i+\Delta t} \lambda(t)dt \\ &= \exp\left[-\int_0^T \lambda(t)dt\right] \cdot \prod_{i=1}^{g} \int_{t_i}^{t_i+\Delta t} \lambda(t)dt \\ &= e^{-\mu_T}\prod_{i=1}^{g}\lambda(t_i)\Delta t \end{aligned} \tag{14.4}$$

Here we assume that Δt is small enough so that $\int_{t_i}^{t_i+\Delta t}\lambda(t)dt = \lambda(t_i)\Delta t$ is satisfied. According to this assumption, we can also know that the probability density when the photons are in time interval (t_i, $t_i + \Delta t$) is $P_{t_i}(t_1, t_2, ..., t_g)\Delta t$. On the other hand, the above situation can also be considered as a situation that g events happened in g time intervals and there is only one event in each time interval. Thus, the total probability P can also be expressed as

$$P = g!P_{t_i}(t_1, t_2, ..., t_g)(\Delta t)^g \tag{14.5}$$

According to Eqs. (14.4) and (14.5), we can have

$$P_{t_i}(t_1, t_2, ..., t_g) = \frac{e^{-\mu_T}\prod_{i=1}^{g}\lambda(t_i)}{g!} \tag{14.6}$$

Dividing Eq. (14.6) by Eq. (14.1), we can get

$$P_{t_i}(t_1, t_2, ..., t_g|g) = \frac{\prod_{i=1}^{g}\lambda(t_i)}{(\mu_T)^g} \tag{14.7}$$

Each of the g events is independent since they satisfy Poisson distribution. Thus, we can have

$$P_{t_i}(t_1, t_2, ..., t_g|g) = P_{t_1}(t_1)P_{t_2}(t_2)\cdots P_{t_g}(t_g) \tag{14.8}$$

According to Eqs. (14.7) and (14.8), we can finally have

$$P_{t_i}(t) = \frac{\lambda(t)}{\mu_T} = \frac{I(t)}{\mu_T h\nu} \propto I(t) \tag{14.9}$$

where $I(t)$ is the optical power, h is the Planck constant, and ν is the optical frequency. As can be seen from Eq. (14.9), the probability of the appearance of the photons at time t is directly proportional to the optical power $I(t)$. Thus, the histogram of the detection times of the TCSPC module indeed presents the waveform of the RF signal under test.

14.3 Weak Signal Detection

A direct and promising application of the quantum single-photon microwave photonic link is for weak microwave and RF signal detection since low-power photons that are modulated by the microwave or RF signal can be detected by the SPD and TCSPC module. In [30], a quantum single-photon microwave photonic link for weak microwave signal detection is proposed and experimentally demonstrated. The schematic diagram and reconstructed waveforms of the proposed scheme are shown in Fig. 14.3. As can be seen, the signal under test is first mixed with a subcarrier from the signal generator for preprocessing, and then converted into the optical domain using double-sideband suppressed carrier modulation on an EOM to make optical sidebands at

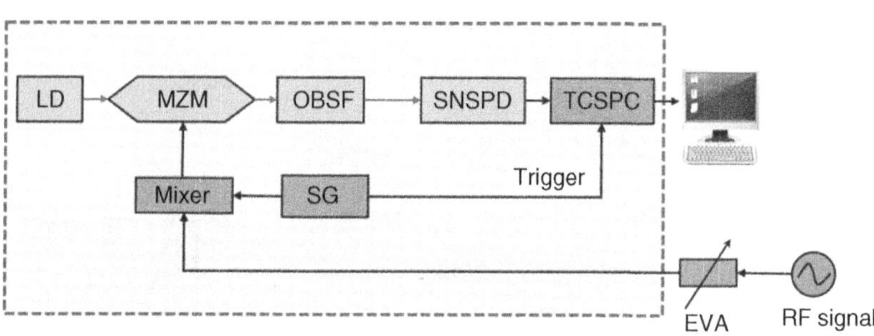

Figure 14.3 Quantum single-photon microwave photonic system for weak RF and microwave signal detection. EVA: electrical variable attenuator; LD, laser diode; MZM, Mach–Zehnder; OBSF, optical band-stop filter; SG, signal generator; SNSPD, superconducting nanowire SPD. *Source:* Chen et al. [30]/IEEE.

few-photon level. The optical signal is then filtered by an optical band-stop filter to further suppress the optical carrier and the filtered signal is finally detected and reconstructed by a low-jitter superconducting nanowire single-photon detector (SNSPD) and TCSPC module. The detection and reconstruction of single-frequency and linearly chirped microwave waveforms are successfully demonstrated. A minimum detectable signal power as low as −112.8 dBm is obtained without the use of electrical amplifiers, showing the ultra-high sensitivity of the quantum microwave photonic system. As a comparison, although weak microwave and RF signal detection can also be achieved using traditional microwave photonic systems such as microwave photonic front-ends [38, 39] and optoelectronic oscillators [40–43], but the high sensitivity requires the cascades of high-gain amplifiers, which increases the noise and cost of the whole system. Moreover, the traditional PD in traditional microwave photonic systems also introduces additional noise in the photodetection process and thus deteriorates the signal quality of the whole system.

14.4 Quantum Microwave Photonic Signal Processing

To overcome the bandwidth and sensitivity limitations mentioned in the introduction section in traditional microwave photonic signal processing systems, we have proposed and demonstrated a quantum microwave photonic signal processing system by introducing the single-photon detection scheme to the microwave photonic link [29]. The schematic diagram of the proposed quantum microwave photonic signal processing system is shown in Fig. 14.4. The RF signal being processed is modulated onto the intensity of the optical signal generated by a programmable light source. The modulated optical signal is first attenuated to the single-photon level (4×10^5 photons/s) by a variable optical attenuator and then processed by a microwave photonic signal processing unit. The processed optical signal is detected by an SNSPD and the processed RF waveform is reconstructed with the help of the TCSPC module. To evaluate the performance of the proposed quantum microwave photonic system, two typical microwave photonic signal processing functions, i.e. RF phase shifting and filtering, are demonstrated by using a dispersion compensation module (DCM) as the

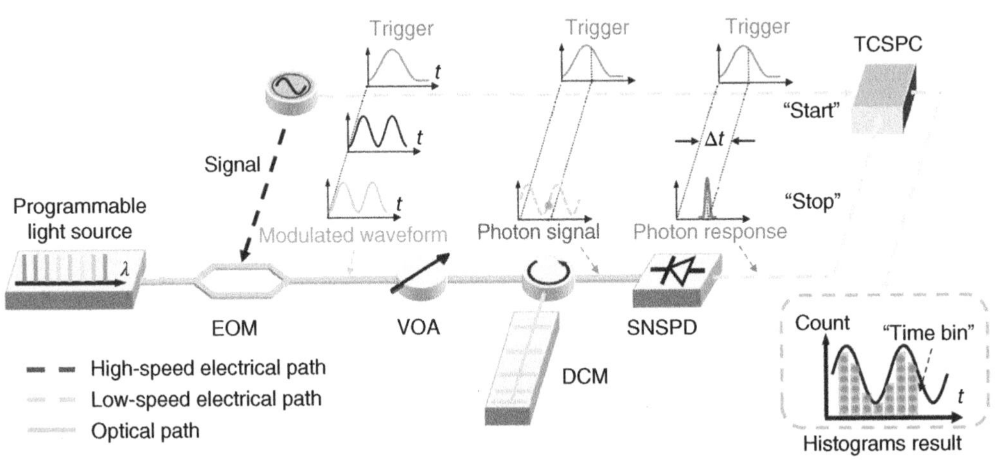

Figure 14.4 Quantum single-photon microwave photonic signal processing system with an optical sensitivity down to the single-photon level and processing bandwidth larger than that of the coaxial cable. DCM, dispersion compensation module; EOM, electro-optic modulator; VOA, variable optical attenuator. *Source:* Yang et al. [29]/with permission of Elsevier.

microwave photonic signal processing unit. By changing the center wavelength of the emitted single-frequency optical signal of the programmable light source, RF phase shifting can be achieved. RF filtering is obtained by programming the light source into a multi-frequency one to construct a multi-source microwave photonic filter.

Similar to the microwave photonic weak signal detection scheme in [30], a direct advantage of the proposed quantum microwave photonic signal processing system is the ultra-high sensitivity due to the use of the quantum single-photon detection scheme. Meanwhile, different from a traditional microwave photonic system where the signal processing bandwidth is limited by the bandwidth of the PD and successive microwave devices, the signal processing bandwidth of the proposed quantum microwave photonic system is only limited by the timing jitter of the SPD with a relationship of BW × jitter = 0.44 [29]. Since SPDs with timing jitter lower than 3 ps are available [27], a processing bandwidth larger than 100 GHz can be expected. In our experiment, the timing jitter of the SPD is about 35 ps, corresponding to a maximum bandwidth of 12.5 GHz. Figure 14.5(a) shows phase shifting results for 5 and 8-GHz RF signals under different dispersions. It should be noted that the bandwidth of the coaxial cable behind the SPD is only about 3 GHz. Thus, the phase-shifting results successfully demonstrate the bandwidth advantage of the proposed quantum microwave photonic system. RF filtering with a bandwidth as large as 8 GHz is also demonstrated by using a 3-GHz electrical cable, and the results are shown in Fig. 14.5(b). Moreover, the proposed system also has a high anti-interference capability [29]. The reason is that the TCSPC module relies on the trigger to reconstruct the desired RF signal, and the interferences that are not phase-locked to the trigger cannot be recovered.

14.5 Nonlocal Frequency-to-Time Mapping

Frequency-to-time mapping technique has been widely investigated in microwave photonics. The basic idea is to map the spectrum of the optical pulse to the intensity of the temporal waveform. So, the spectrum information can be measured at the output of a PD using the temporal information at a fast speed [44–46]. By stretching an ultrafast pulse in time using chromatic group velocity dispersion (GVD) in a single-mode fiber, the spectral intensity distribution of a pulse laser-pumped single-photon source has been measured [47]. Nevertheless, this scheme is not suitable for continuous wave (CW) pumped single-photon sources. By combining the entanglement property of quantum entangled photon pair with frequency-to-time mapping technique in microwave photonics, our group has demonstrated a nonlocal quantum microwave photonic frequency-to-time mapping scheme in [31], where the mapping and measurement of CW pumped single-photon sources can also be obtained. The schematic diagram of the nonlocal frequency-to-time mapping scheme is shown in Fig. 14.6. CW-pumped entangled photon pairs generated from spontaneous parametric down-conversion process [48, 49] are used. Thanks to the property of frequency-correlated entangled photon pairs and the coincidence count process, the spectral filtering pattern experienced by one photon in the entangled photon pair is nonlocally mapped into the time domain when another photon is propagated through a dispersion-compensation fiber (DCF) with large GVD [31, 50–52]. In addition to the capability of measuring CW-pumped single-photon sources, the nonlocal frequency-to-time mapping system also separates the spectra information and the dispersion medium into different physical paths, thus avoiding the possible distortion of the dispersion medium to the measured spectra information, and allows flexible management of the dispersion without influencing the spectra information in another physical path.

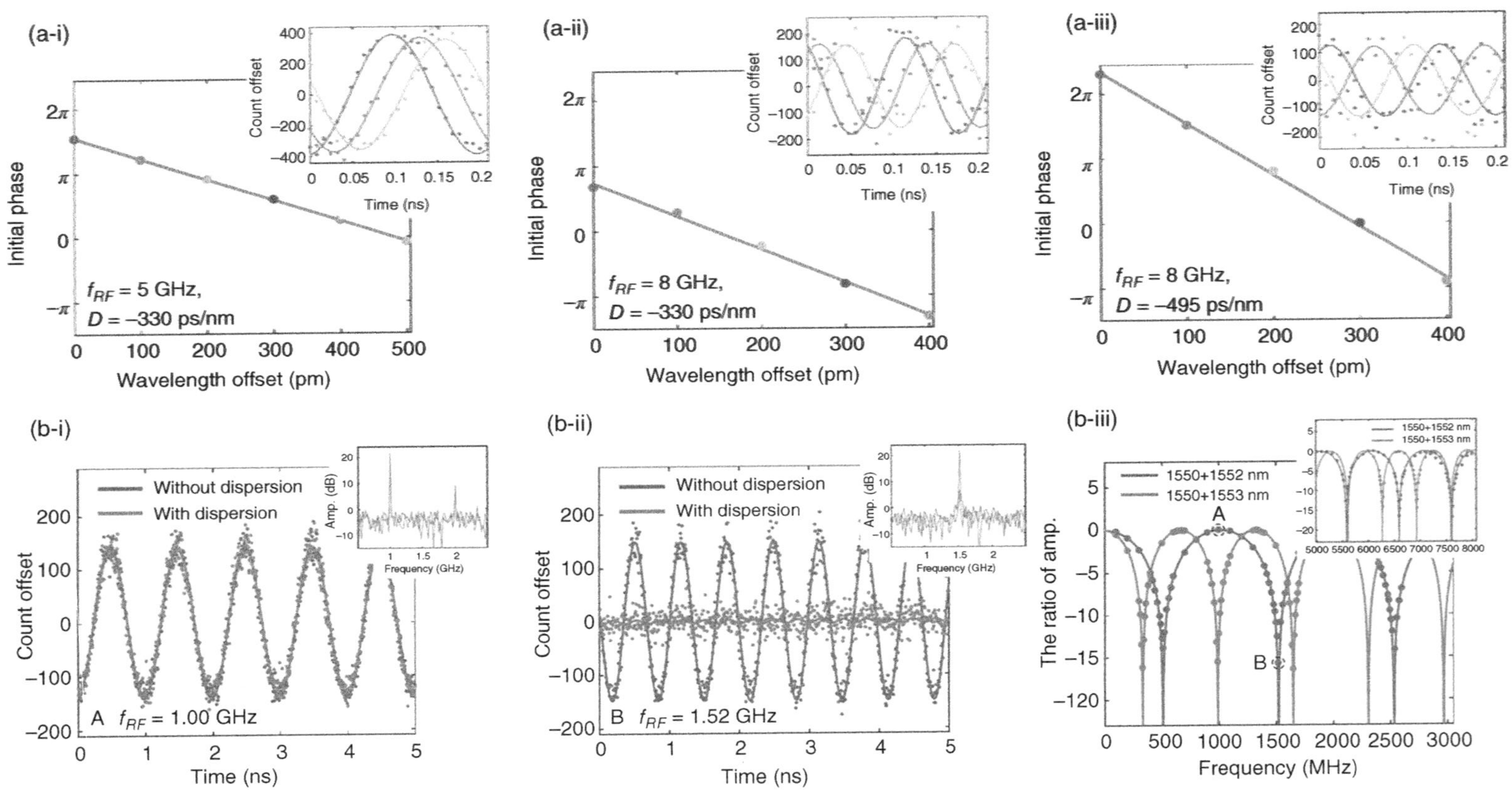

Figure 14.5 (a) Phase shifting results for 5- and 8-GHz RF signals under different dispersions. The bandwidth of the electrical cable is only about 3 GHz. Inset: reconstructed waveforms with different optical wavelengths. (b) Filtering for RF signals with different frequencies. (b-i) and (b-ii) show the reconstructed waveforms for the 1 and 1.52 GHz RF signals, respectively. Inset: spectra of reconstructed signals. (b-iii) Frequency response of the filter. Inset: frequency response from 5 to 8 GHz. *Source:* Yang et al. [29]/with permission of Elsevier.

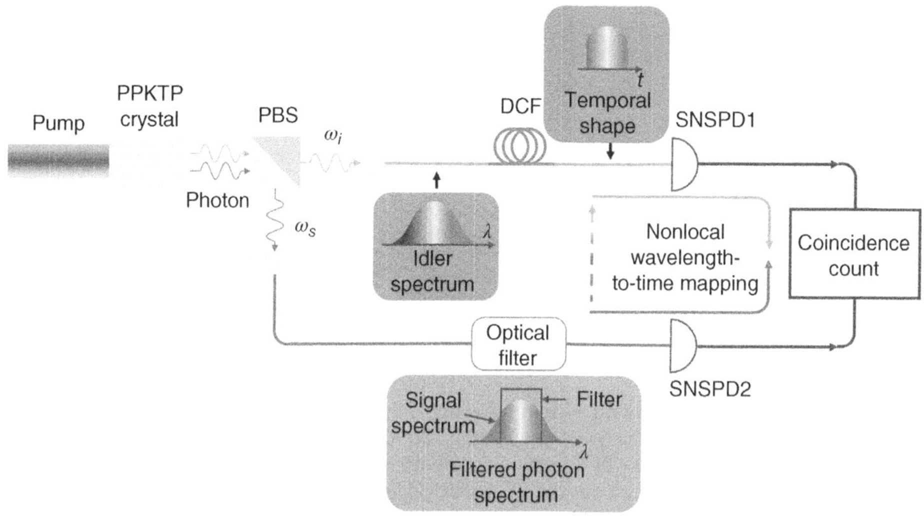

Figure 14.6 Nonlocal frequency-to-time mapping system. Entangled photon pairs are used. The entangled photon pairs are generated from spontaneous parametric down-conversion process. One photon of the entangled photon pair is filtered by an optical filter, while the other photon is propagated through a DCF. The spectral filtering pattern experienced by one photon is nonlocally mapped into the time domain due to the property of frequency-correlated entangled photon pairs. PBS, polarization beam splitter; PPKTP, periodically poled potassium titanyl phosphate. *Source:* Yang et al. [31]/Optical Society of America/CC BY 4.0.

14.6 Compressed Sensing

Compressed sensing for wideband microwave and RF signals has been widely studied in various signal processing systems to recover the sparse signal with a sampling rate below the Nyquist sampling rate [53–55]. Generally, a random matrix is required in compressed sensing to reconstruct the desired signal. In [32], a quantum microwave photonic compressed sensing system is proposed and demonstrated, where a random matrix is obtained based on the consumption of coherence in the quantum photon detection process. The schematic diagram of the quantum microwave photonic compressed sensing system is shown in Fig. 14.7. The linearly polarized coherent photons from the laser are modulated by the RF signal under test in an electro-optical (EO) crystal. The polarization state of the output signal of the EO crystal is adjusted by the RF signal and the RF signal is transferred to the probability of detecting the photons with the help of a polarizer. An SPD and a time-to-digital converter are used to read out the RF signal. Through the above process, the RF signal under test is modulated on the wave function or the probability amplitude of the quantum state. Thus, a random matrix for compressed sensing can be constructed due to the fact that the wave function of the coherent photons is collapsed and provided natural truly random numbers in the detection process [32, 56]. The recovery of 10-GHz RF spectrum and multi-tone spectrum with a 5- and 10-GHz frequency components using the quantum microwave photonic compressed sensing system are demonstrated, which show the wideband measurement capability of the proposed system. In theory, the real-time analysis bandwidth (RTBW) of the proposed system is related to the bandwidth of the EO crystal and the timing jitter of the system. By using a wideband EO crystal and a low-timing jitter SPD, the RTBW of the proposed quantum microwave photonic compressed sensing system can be larger than 100 GHz, which is very promising for practical applications.

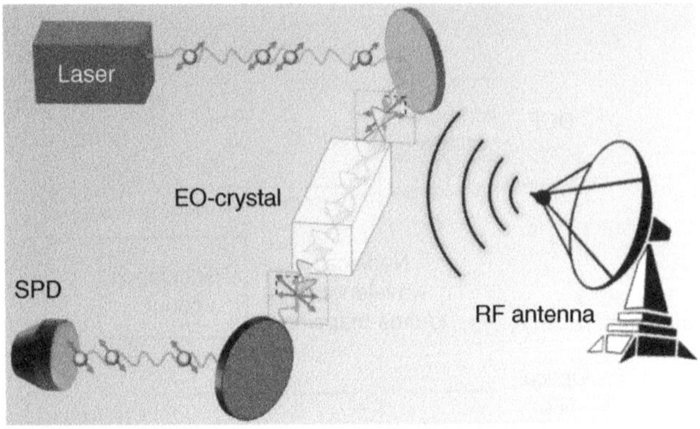

Figure 14.7 Quantum compressed sensing system for wideband microwave and RF signals. An electro-optical (EO) crystal is used as the sensing element. The RF signal under test is applied to the crystal and transferred to the probability of detecting the photons using a polarizer. An SPD and a time-to-digital converter are used to read out the RF signal. *Source:* Hu et al. [32] /arXiv/CC BY 4.0.

14.7 Microwave Photonic Quantum Key Distribution

In addition to the use of photonic quantum technology to overcome the drawbacks of conventional microwave photonic systems, the use of microwave photonic technology to boost the performance of photonic quantum systems has also been proposed and demonstrated, for example, for QKD [57–59]. The aim of QKD is to share a random secret key between two parties, Alice and Bob, where the key is known only to them and any eavesdropping of the third party can be detected due to the introduction of detectable anomalies in quantum systems. Long-distance QKD through optical fiber [60–64] has been proposed and demonstrated for point-to-point links as well as for multiuser operations based on wavelength division multiplexing (WDM). However, the spectral efficiency is low even for the WDM-based scheme since each key consumes a full wavelength channel.

In [33–35], subcarrier multiplexed QKD (SCM-QKD) has been proposed and demonstrated using a technique borrowed from microwave photonics, i.e. subcarrier multiplexing that can distribute more than one key per wavelength channel. The schematic diagram of a SCM-QKD system with two subcarriers is shown in Fig. 14.8. As can be seen, the SCM-QKD system consists of four parts: a quantum transmitter (Alice), a quantum receiver (Bob), a classical reference channel, and an electronic control system. At Alice's transmitter, two RF subcarriers with frequencies of f_{RF1} and f_{RF2} transmit two keys that are generated by two randomly phase-modulated voltage-controlled oscillators. Four possible phase values of 0, $\pi/2$, π, and $3\pi/2$ are used to implement the Bennet-Brassard BB84 protocol. A faint pulse laser source is modulated by the two RF subcarriers, so two independent keys are multiplexed on two RF subcarriers in a single wavelength channel. The Alice and Bob are interconnected by an 11-km optical fiber link. The classical reference channel is used to convey a synchronization signal from Alice to Bob and also stabilize the whole system against fiber length fluctuations. At Bob's receiver, independent random phase shifts with values of 0 or π is synchronously inserted using a phase modulator (PM) to measure the transmitted qubit. Narrowband filters are employed after the PM to discriminate each subcarrier, and the filtered signals are detected by single photon avalanche detectors.

The measured evolution of the sifted key rate and corresponding quantum bit error rate are shown in Fig. 14.9(a), (b), respectively. A sifted key rate of 10 kb/s/channel with quantum bit error

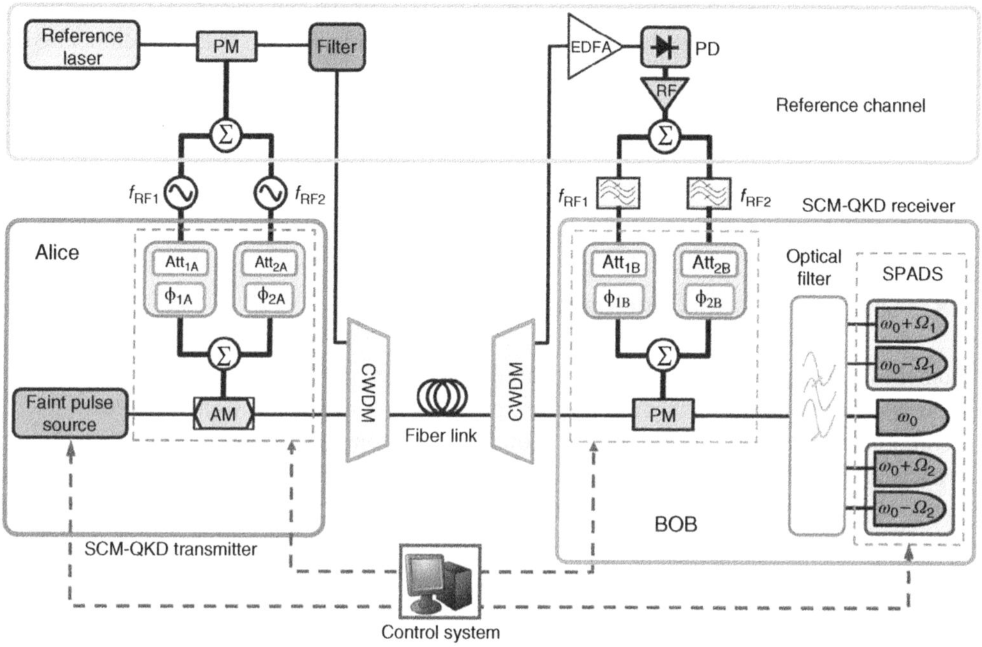

Figure 14.8 Subcarrier multiplexed quantum key distribution (SCM-QKD) system. The SCM-QKD system consists of a quantum transmitter (Alice), a quantum receiver (Bob), a classical reference channel, and an electronic control system. Different independent keys are multiplexed on two RF subcarriers with frequencies of f_{RF1} and f_{RF2}. AM, amplitude modulator; Att: electrical attenuator; CWDM, coarse wavelength division multiplexing; EDFA, Erbium-doped fiber amplifier; PM, phase modulator; SPAD, single photon avalanche detector; Φ, RF phase shifter. *Source:* Ruiz-Alba et al. [34]/IEEE.

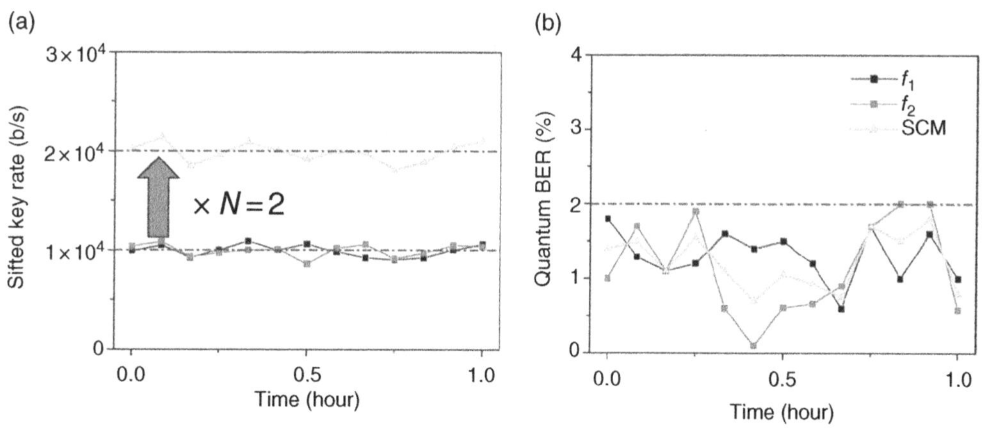

Figure 14.9 (a) Evolution of the sifted key rate for each single channel and the SCM sifted key rate. The sifted key rate is doubled thanks to the use of microwave photonic subcarrier multiplexing technique. (b) Measured corresponding quantum bit error rate for each single channel and for the multiplexed rate. *Source:* Ruiz-Alba et al. [34]/IEEE.

rate <2% is achieved. Thanks to the introduction of the microwave photonic subcarrier multiplexing technique, the SCM-QKD system has a spectral efficiency higher than that of traditional QKD systems. The system complexity and cost are also reduced since only a single optical source that is shared by all the multiplexed channels is required. Moreover, the subcarrier multiplexing approach can also be combined with other multiplexing approaches such as the WDM to further increase the number of parallel keys of the QKD system [34].

14.8 Discussion and Conclusion

By combining microwave photonics and quantum technology, quantum microwave photonics can achieve features or functions that are very complex or even not possible using traditional microwave photonic technology and vice versa. However, the promising features of current quantum microwave photonic systems are obtained at the expense of real-time processing capability. In traditional microwave photonic systems, the microwave signal is detected in real-time by the PD and processed at a fast speed using successive devices after the PD. In current quantum microwave photonic signal detection and processing systems, an accumulation time for the recovering process is required to reconstruct the waveform in the single-photon detection scheme. Thus, the signal detection and processing process are pretty long. Besides, the obtained waveform of the TCSPC module is indeed a reconstructed waveform represented by the histogram of the detection times. The histogram should be converted to a real microwave waveform for some application scenarios. One way to achieve this goal is to use the histogram of the detection times as the input waveform sequence of an electrical arbitrary waveform generator, whose output waveform follows that of the input waveform sequence. It should also be noted that the current quantum microwave photonic approaches may not be effective for abrupt or single-shot signals or events, due to the loss of real-time capability. It is highly desirable to develop new approaches capable of dealing with abrupt or single-shot signals or events since such signals or events widely exist in practical applications. Nevertheless, the obtained ultra-high sensitivity of the quantum microwave photonic signal detection and processing systems is still very promising in certain applications, such as deep-space and underwater optical communication systems [65, 66], as well as testing systems for integrated photonic devices [67]. The bandwidth of the quantum microwave photonic systems is also no longer limited by the electronic backend instruments as in classic microwave photonic systems since the high-speed microwave signal is transformed to a slow-speed SPD response with a low timing jitter.

On the other hand, the emerging advances in quantum microwave photonics in recent years are generally simple prototypes, which demonstrate the feasibility of combining microwave photonics and quantum technology [30–35, 68, 69]. More extensive and in-depth researches should be conducted with a deep understanding of the whole system or explore new applications. For instance, quantum microwave photonic systems are generally constructed by a combination of traditional microwave photonic and quantum devices. The operation of quantum devices under microwave photonic regimes as well as the behavior of microwave photonic devices in quantum conditions should be further studied [70, 71], although the property of these devices is well known under their original application regimes. Novel applications of quantum microwave photonics can also be explored. For example, liquid microsphere measurement can be implemented based on the nonlocal frequency-to-time mapping system [26, 31]. The stability of the microsphere can be enhanced thanks to the reduction of the evaporation effect since liquid microsphere is coupled with a low-power single-photon light source. The interaction between light and the constituent elements of the microsphere can also be measured and analyzed using the highly sensitive SNSPD. Moreover, quantum microwave photonics may also be propelled to new heights with the help of integrated photonic or electronic circuits. The most direct benefit is the reduction of footprint and complexity since the key elements of the current quantum microwave photonic systems are still bulky discrete devices. Moreover, novel technological tools and concepts are also possible for integrated quantum microwave photonics, considering the fact that the interaction of light and microwave with matter will be enhanced since they are confined in a small mode volume.

References

1 Marpaung, D., Yao, J., and Capmany, J. (2019). Integrated microwave photonics. *Nature Photonics* 13: 80–90.

2 Yao, J. (2009). Microwave photonics. *J. Lightw. Technol.* 27 (3): 314–335.

3 Capmany, J. and Novak, D. (2007). Microwave photonics combines two worlds. *Nature Photonics* 1: 319–330.

4 Seeds, A.J. and Williams, K.J. (2006). Microwave photonics. *J. Lightw. Technol.* 24 (12): 4628–4641.

5 Capmany, J., Mora, J., Gasulla, I. et al. (2013). Microwave photonic signal processing. *J. Lightw. Technol.* 31 (4): 571–586.

6 Pan, S. and Zhang, Y. (2020). Microwave photonic radars. *J. Lightw. Technol.* 38 (19): 5450–5484.

7 Hunter, D.B., Edvell, L.G., and Englund, M.A. (2005). Wideband microwave photonic channelised receiver. *Int. Topical Meeting on Microwave Photonics Tech. Dig.* (14 October 2005), Seoul, Korea, pp. 249–252.

8 Hao, T., Tang, J., Domenech, D. et al. (2018). Toward monolithic integration of OEOs: from systems to chips. *J. Lightw. Technol.* 36 (19): 4565–4582.

9 Yao, J.P. (2015). A fresh look at microwave photonics filters. *IEEE Microw. Mag.* 16 (8): 46–60.

10 Zou, X., Lu, B., Pan, W. et al. (2016). Photonics for microwave measurements. *Laser Photonics Rev.* 10 (5): 711–734.

11 Zhang, W. and Yao, J. (2016). Silicon-based integrated microwave photonics. *IEEE J. Quantum Electron.* 52: 0600412.

12 Hao, T., Cen, Q., Dai, Y. et al. (2018). Breaking the limitation of mode building time in an optoelectronic oscillator. *Nat. Commun.* 9: 1839.

13 Hao, T., Cen, Q., Guan, S. et al. (2020). Optoelectronic parametric oscillator. *Light Sci. Appl.* 9: 102.

14 Matavulj, P.S., Gvozdic, D.M., and Radunovic, J.B. (1997). The influence of nonstationary carrier transport on the bandwidth of p-i-n photodiode. *J. Lightw. Technol.* 15 (12): 2270–2277.

15 Rakshit, P. and Das, N.R. (2020). Effect of device parameters on improving the quantum efficiency of a lateral Si p–i–n photodetector. *Opt. Quant. Electron.* 52: 371.

16 Elshaari, A.W., Pernice, W., Srinivasan, K. et al. (2020). Hybrid integrated quantum photonic circuits. *Nature Photonics* 14: 285–298.

17 O'Brien, J., Furusawa, A., and Vučković, J. (2009). Photonic quantum technologies. *Nature Photonics* 3: 687–695.

18 Gisin, N. and Thew, R. (2007). Quantum communication. *Nature Photonics* 1: 165–171.

19 Degen, C.L., Reinhard, F., and Cappellaro, P. (2017). Quantum sensing. *Rev. Mod. Phys.* 89 (3): 035002.

20 Goyal, S.K., Boukama-Dzoussi, P.E., Ghosh, S. et al. (2014). Qudit-teleportation for photons with linear optics. *Sci. Rep.* 4: 4543.

21 Shcheslavskiy, V., Morozov, P., Divochiy, A. et al. (2016). Ultrafast time measurements by time-correlated single photon counting coupled with superconducting single photon detector. *Rev. Sci. Instrum.* 87 (5): 053117.

22 Natarajan, C.M., Tanner, M.G., and Hadfield, R.H. (2012). Superconducting nanowire single-photon detectors: physics and applications. *Supercond. Sci. Technol.* 25: 063001.

23 Holzman, I. and Ivry, Y. (2019). Superconducting nanowires for single-photon detection: progress, challenges, and opportunities. *Adv. Quantum Technol.* 2: 1800058.

24 Becker, W. (2005). *Advanced Time-Correlated Single Photon Counting Techniques*. New York, NY: Springer-Verlag.

25 Wang, H. (2011). Experimental study on laser ranging with time-correlated single photon counting techniques. Dissertation of Harbin Institute of Technology (in Chinese).

26 Yang, Y. (2021). Microwave photon signal processing technology based on single photon detection. Dissertation of University of Chinese Academy of Sciences [in Chinese].

27 Korzh, B., Zhao, Q.Y., Allmaras, J.P. et al. (2020). Demonstration of sub-3 ps temporal resolution with a superconducting nanowire single-photon detector. *Nature Photonics* 14: 250–255.

28 Hao, T., Yang, Y., Jin, Y. et al. (2022). Quantum microwave photonics. *J. Lightw. Technol.* 67 (20): 6616–6625.

29 Yang, Y., Jin, Y., Xiang, X. et al. Single-photon microwave photonics. *Sci. Bull.* 40, 2022 (7): 700–706.

30 Chen, X., Li, S., Xue, X., and Zheng, X. (2019). Weak microwave signal detection based on microwave photonics-enabled single-photon technology. *Asia Communications and Photonics Conference* (2–5 November 2019), Beijing, China, paper number: M4A, 329.

31 Yang, Y., Xiang, X., Hou, F. et al. (2020). Inherent resolution limit on nonlocal wavelength-to-time mapping with entangled photon pairs. *Opt. Express* 28 (5): 7488–7497.

32 Hu, J., Han, Y., Li, W. et al. (2021). Quantum compressed sensing of ultra-wideband radio-frequency signal. arXiv: 2106.13668.

33 Ortigosa-Blanch, A. and Capmany, J. (2006). Subcarrier multiplexing optical quantum key distribution. *Phys. Rev. A* 73 (2): 024305.

34 Ruiz-Alba, A., Mora, J., Amaya, W. et al. (2012). Microwave photonics parallel quantum key distribution. *IEEE Photonics J.* 4 (3): 931–942.

35 Mora, J., Ruiz-Alba, A., Amaya, W. et al. (2012). Experimental demonstration of subcarrier multiplexed quantum key distribution system. *Opt. Lett.* 37 (11): 2031–2033.

36 Gu, X., Kockum, A.F., Miranowicz, A. et al. (2017). Microwave photonics with superconducting quantum circuits. *Phys. Rep.* 718: 1–102.

37 Eder, P., Ramos, T., Goetz, J. et al. (2018). Quantum probe of an on-chip broadband interferometer for quantum microwave photonics. *Supercond. Sci. Technol.* 31: 115002.

38 Scotti, F., Laghezza, F., Serafino, G. et al. (2014). In-field experiments of the first photonics-based software defined coherent radar. *J. Lightw. Technol.* 32 (20): 3365–3372.

39 Laghezza, F., Scotti, F., Serafino, G. et al. (2015). Field evaluation of a photonicsbased radar system in a maritime environment compared to a reference commercial sensor. *IET Radar Sonar Navig.* 9 (8): 1040–1046.

40 Li, M., Hao, T., Li, W., and Dai, Y. (2021). Tutorial on optoelectronic oscillator. *APL Photonics* 6: 061101.

41 Wang, G., Hao, T., Li, W. et al. (2019). Detection of wideband low-power RF signals using a stimulated Brillouin scattering-based optoelectronic oscillator. *Opt. Commun.* 439: 133–136.

42 Hao, T., Liu, Y., Tang, J. et al. (2020). Recent advances in optoelectronic oscillators. *Adv. Photonics* 2 (4): 044001.

43 Shao, Y., Han, X., Li, M., and Zhao, M. (2018). RF signal detection by a tunable optoelectronic oscillator. *Opt. Lett.* 43 (6): 1199–1202.
44 Goda, K. and Jalali, B. (2013). Dispersive Fourier transformation for fast continuous single-shot measurements. *Nature Photonics* 7: 102–112.
45 Goda, K., Solli, D.R., and Jalali, B. (2008). Real-time optical reflectometry enabled by amplified dispersive Fourier transformation. *Appl. Phys. Lett.* 93 (3): 031106.
46 Solli, D., Chou, J., and Jalali, B. (2008). Amplified wavelength–time transformation for real-time spectroscopy. *Nature Photonics* 2: 48–51.
47 Avenhaus, M., Eckstein, A., Mosley, P.J., and Silberhorn, C. (2009). Fiber-assisted single-photon spectrograph. *Opt. Lett.* 34 (18): 2873–2875.
48 Shih, Y. (2003). Entangled biphoton source - property and preparation. *Rep. Prog. Phys.* 66: 1009–1044.
49 Horodecki, R., Horodecki, P., and Horodecki, M. (2009). Quantum entanglement. *Rev. Mod. Phys.* 81: 865–942.
50 Baek, S.Y., Kwon, O., and Kim, Y.H. (2008). Nonlocal dispersion control of a single-photon waveform. *Phys. Rev. A* 78 (1): 013816.
51 Dong, S., Zhang, W., Huang, Y., and Peng, J. (2016). Long-distance temporal quantum ghost imaging over optical fibers. *Sci. Rep.* 6 (1): 26022.
52 Yao, X., Zhang, W., Li, H. et al. (2018). Long-distance thermal temporal ghost imaging over optical fibers. *Opt. Lett.* 43 (4): 759–762.
53 Donoho, D. (2006). Compressed sensing. *IEEE Trans. Inform. Theory* 52 (4): 1289–1306.
54 Eldar, Y. and Kutyniok, G. (2012). *Compressed Sensing: Theory and Applications*. Cambridge Univ. Press.
55 Bosworth, T. and Foster, M.A. (2013). High-speed ultrawideband photonically enabled compressed sensing of sparse radio frequency signals. *Opt. Lett.* 38 (22): 4892–4895.
56 Li, Y.-H., Han, X., Cao, Y. et al. (2019). Quantum random number generation with uncharacterized laser and sunlight. *NPJ Quantum Inf.* 5: 97.
57 Gisin, N., Ribordy, G., Tittel, W., and Zbinden, H. (2002). Quantum cryptography. *Rev. Mod. Phys.* 74 (1): 145–195.
58 Scarani, V., Bechmann-Pasquinucci, H., Cerf, N.J. et al. (2009). The security of practical quantum key distribution. *Rev. Mod. Phys.* 81 (3): 1301–1350.
59 Ekert, A. and Renner, R. (2014). The ultimate physical limits of privacy. *Nature* 507 (7493): 443–447.
60 Gobby, C., Yuan, Z.L., and Shields, A.J. (2004). Quantum key distribution over 122km standard telecom fiber. *Appl. Phys. Lett.* 84 (19): 3762–3764.
61 Rosenberg, D., Harrington, J.W., Rice, P.R. et al. (2007). Long-distance decoy-state quantum key distribution in optical fiber. *Phys. Rev. Lett.* 98 (1): 010503.
62 Yin, H.-L., Chen, T.Y., Yu, Z.W. et al. (2016). Measurement-device-independent quantum key distribution over a 404 km optical fiber. *Phys. Rev. Lett.* 117 (19): 190501.
63 Boaron, A., Boso, G., Rusca, D. et al. (2018). Secure quantum key distribution over 421 km of optical fiber. *Phys. Rev. Lett.* 121 (19): 190502.
64 Ciurana, A., Martínez-Mateo, J., Peev, M. et al. (2014). Quantum metropolitan optical network based on wavelength division multiplexing. *Opt. Express* 22 (2): 1576–1592.
65 Hemmati, H., Biswas, A., and Djordjevic, I.B. (2011). Deep-space optical communications: future perspectives and applications. *Proc. IEEE* 99 (11): 2020–2039.
66 Hanson, F. and Radic, S. (2008). High bandwidth underwater optical communication. *Appl. Opt.* 47 (2): 277–283.

67 J. De Coster, P. De Heyn, M. Pantouvaki, et al. (2016). Test-station for flexible semi-automatic wafer-level silicon photonics testing. *Proc. IEEE Eur. Test Symp.* (23–27 May 2016), Amsterdam, Netherlands, pp. 1-6.

68 Li, M., Hao, T., Yang, Y. et al. (2021)). Recent advances in optoelectronic oscillators and quantum microwave photonics. *2021 International Topical Meeting on Microwave Photonics (MWP)* (15–17 November 2021), Pisa, Italy, pp. 1–3.

69 Y. Jin, Y. Yang, H. Hong, et al. (2022). A proof-of-principle demonstration of quantum microwave photonics. arXiv: 2201.12106.

70 Capmany, J. and Fernndez-Pousa, C.R. (2011). Quantum modelling of electro-optic modulators. *Laser Photonics Rev.* 5 (6): 750–772.

71 Capmany, J., Mora, J., Fernández-Pousa, C.R., and Muñoz, P. (2013). Quantum model of light transmission in array waveguide gratings. *Opt. Express* 21 (12): 14841–14852.

15

Future and Perspectives

15.1 Introduction

As we have discussed in previous chapters, microwave photonics (MWP) has been widely studied using discrete devices for the generation, processing, control, and distribution of microwave signals with the help of photonics technology [1–8]. Unique features such as broad bandwidth, low loss, lightweight, fast reconfigurability, and immunity to electromagnetic interference have been achieved, which are not possible even for state-of-the-art electronics. Nevertheless, the applications of discrete MWP systems are still limited due to factors such as large size and high power consumption. In recent years, integrated MWP (IMWP) has been actively researched, which is a solution to reduce the size and power consumption of an MWP system [9–15]. Various integrated microwave photonic subsystems, such as integrated microwave photonic signal generators, filters, and signal processors, have been demonstrated, which is a key step toward a new class of novel microwave photonic systems for practical applications.

On the other hand, the study of IMWP is still in its early stage, in-depth and extensive research is still required to show the full potential of this area. Moreover, with the rapid development of the theory and technology of MWP, cross-research between MWP and other disciplines has also been the key to its direction of development [16]. This multidisciplinary research approach between MWP and other disciplines can be conducted by combining MWP with technologies such as laser technology, quantum technology, and artificial intelligence (AI), which would provide new opportunities for MWP.

In this chapter, we briefly discuss the future and perspectives of MWP. An outlook about future prospects of IMWP is given, including the integration of single functional devices, hybrid integration of electronics and photonics, as well as reconfigurable multi-function and multi-channel integrated circuits. Prospects of cross-fertilization between MWP and other disciplines, such as laser technology, quantum technology, and AI will also be discussed.

15.2 Future and Perspectives

MWP has great potential for both civil and defense applications. The performance of application systems can be boosted by the attractive features of MWP such as broad bandwidth, low loss, light weight, fast reconfigurability, and immunity to electromagnetic interference. However, the size,

Microwave Photonics, First Edition. Jianping Yao, José Capmany, and Ming Li.

Companion website: www.wiley.com/go/yao/microwavephotonics

power consumption, stability, and cost of traditional microwave photonic systems are still not as competitive as their electrical counterparts. Thus, by integrating MWP systems on a chip, IMWP with small size, low power consumption, high stability, and low cost will provide a key solution for future application scenarios such as ultra-wideband wireless communication, high-performance radar, and electronic warfare systems. Various integrated microwave photonic subsystems have been proposed and demonstrated in recent years with attractive features, and there is still a lot of room for the development of IMWP in at least the following three aspects [15].

The first one is the integration of single functional devices. The core devices of MWP systems include low noise and high power semiconductor lasers, broad bandwidth, high linearity, and low half-wave voltage electro-optical modulators, and high saturation power, broad bandwidth, and high sensitivity photodetectors. Further enhancement of the performance of these core devices in terms of power, noise, bandwidth, linearity, and insertion loss, as well as further improvement of the integration and industrialization levels, are key to future research efforts. For example, there is an urgent need to develop laser chips with high power, low noise, and narrow linewidth [17], photodetectors with broad bandwidth, high saturation power, and high responsivity [18], and electro-optical modulators with wide bandwidth, low half-wave voltage, and high linearity [19]. Besides active optical components, several kinds of passive optical components are also the key to realizing high-performance IMWP such as high Q-factor optical filters, high extinction ratio optical switches, and low loss on-chip delay lines. The development of these devices will help enhance the performance and types of functional units of integrated microwave photonic chips, thus gradually replacing all kinds of core devices in traditional discrete MWP systems, and laying a solid foundation for the practical application and industrialization of IMWP [15].

The second aspect is the hybrid integration of electronics and photonics. The hybrid integration of electronics and photonics can be combined with the mature and powerful integrated microelectronics technology, and highly integrated microwave photonic chips ranging from single-material systems to hybrid integrated multi-material systems can be achieved [9]. The main research focus of this field is heterogeneous integrated microwave photonic chips, which would enable the realization of monolithic integrated high-power low-noise laser source, high-efficiency electro-optical modulator, and high-saturation power photodetector. Hybrid packaging and testing of electronics and photonics, as well as key drive and control technologies of large-scale chips, should also be further explored. Moreover, considerable efforts should be devoted to the study of new materials, structures, and mechanisms of the hybrid integrated electronic and photonic chips, as well as to reduce or even eliminate the optical, electrical, and thermal crosstalks of the integrated chips [15].

The last one is system-level integration of reconfigurable multi-function and multi-channel integrated circuits. IMWP chips with different functions should be developed, including integrated beam forming, analog optical signal processing, optoelectronic oscillator, optical frequency comb, arbitrary waveform generation, frequency mixing and cancellation, optical analog-to-digital conversion, and analog signal transceiver. The degree of integration level of different functional chips should also be enhanced for the realization of multifunctional IMWP chips. At the same time, multi-channel and multi-band chips and corresponding packaging technology should also be developed to meet the requirements of large-scale integrated MWP arrays. In addition, the versatility of the IMWP chips can be further improved by the large-scale integration of a large number of active or passive tunable devices [20]. Thus, reconfigurable IMWP chips can be realized for general-purpose microwave photonic signal processing and computing (e.g. photonic FPGAs, and photonic computers) [15, 21].

Cross-research between MWP and other disciplines, such as laser technology, quantum technology, and AI, is another important and promising research direction of MWP [16]. On one hand,

MWP itself is a product of the cross-research between microwave and photonics technologies. Through further synergies between MWP and other disciplines, new mechanisms for the interaction of microwaves and light waves with other physical fields, as well as new manipulation and control methods for microwaves and light waves might be further developed. On the other hand, the performance of current microwave photonic systems can be further enhanced by effort combination between MWP and other disciplines, which can meet the urgent demand for practical applications such as wireless communications, radar, electronic warfare, measurement, and sensing systems. As shown in Fig. 15.1, scientifically, MWP is expected to develop from the cross-research between microwave and photonics technologies to the multidisciplinary collaboration between a broader range of disciplines, including photonics, electronics, quantum science, acoustics, magnetics, mechanics, and so on, to further expand the boundaries of knowledge. Technically, photonic/optoelectronic breadboards and microwave photonic chiplets that have a weak correlation with IP/process are expected to achieve the best performance under confined circumstances and realize quick development [16].

For instance, new microwave photonic signal manipulation methods can be further borrowed from laser technology that is used for the control of light waves. In fact, MWP has been rapidly developed in the last 30 years, whereas laser technology has a first-move advantage of tens of years over MWP. There is still a lot of room for cross-research with laser technology. New mechanisms and control methods can still be borrowed from the optical domain, and new breakthroughs in key fields of MWP, such as arbitrary waveform generation, broadband complex microwave signal processing, and the interactions between microwave and light waves, can be further expected [16].

The cross-research between MWP and quantum technology is an emerging development direction of MWP, which is expected to overcome some of the existing bottlenecks of current microwave photonic systems and realize novel high-speed, high-sensitivity quantum microwave photonic systems [22]. For conventional MWP systems, the required signal processing bandwidth is gradually increasing, and the contradiction between the system bandwidth and the lowest detectable optical power limits the development of novel MWP systems with both high-speed and low-power detection capabilities. With the rapid development of quantum technology, the process of single-photon signals with very low optical power is possible, and low-jitter single-photon detectors can realize

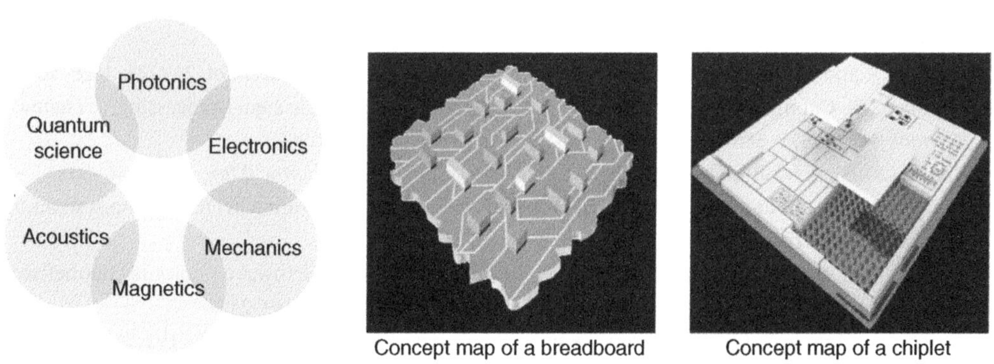

Figure 15.1 Sketch map of cross-research between MWP and multidiscipline. Scientifically, MWP is expected to be fused and learnt from each other with more disciplines, such as photonics, electronics, mechanics, magnetics, acoustics, and quantum science, to further expand the boundary of knowledge. Technically, photonic/optoelectronic breadboard and microwave photonic chiplet that have a weak correlation with IP/process are expected to achieve the best performance under confined circumstances and realize quick development. *Source:* Li et al. [16]/Infrared and Laser Engineering.

high-resolution detection of single-photon signals. Thus, the detection of high-speed microwave signals with ultra-low power can be realized by the cross-research between MWP and quantum technology. The generation and processing of nonclassical microwave and optical signals and their possible application scenarios can also be further explored by the combination of classical microwave signals, classical optical signals, and quantum optical signals [16].

The cross-research between MWP and AI is also an interesting emerging direction of MWP, which can be used to optimize the performance of MWP systems by means of AI [23]. AI is the intelligence exhibited by artificially created machines that is able to learn and adjudicate intelligently. By combining the high center frequency, fast reconfigurability, and broadband signal processing capabilities of MWP with AI schemes or algorithms such as optical neural networks, performance-enhanced, or intelligently optimized MWP devices and systems might be achieved. Currently, cross-research between MWP and AI is still in its infancy. Novel AI-enhanced MWP devices and systems can be expected in the near future by fully combining MWP with AI [16].

15.3 Discussion and Conclusion

We have provided an overview of the possible research directions of MWP. Two key research directions, including IMWP and cross-research between MWP and other disciplines, have been outlined. For IMWP, the integration of high-performance single-functional devices, hybrid integration of electronics and photonics, as well as reconfigurable multi-function and multi-channel integrated circuits can be further explored. Cross-research between MWP and other disciplines, such as laser technology, quantum technology, and AI may also be further conducted for the development of a new generation of MWP devices and systems with novel characteristics.

References

1 Yao, J. (2009). Microwave photonics. *J. Lightw. Technol.* 27 (3): 314–335.
2 Capmany, J. and Novak, D. (2007). Microwave photonics combines two worlds. *Nature Photonics* 1: 319–330.
3 Seeds, A.J. and Williams, K.J. (2006). Microwave photonics. *J. Lightw. Technol.* 24 (12): 4628–4641.
4 Capmany, J., Capmany, J., Mora, J. et al. (2013). Microwave photonic signal processing. *J. Lightw. Technol.* 31 (4): 571–586.
5 Yao, J. (2015). A fresh look at microwave photonics filters. *IEEE Microw. Mag.* 16 (8): 46–60.
6 Hao, T., Tang, J., Domenech, D. et al. (2018). Toward monolithic integration of OEOs: from systems to chips. *J. Lightw. Technol.* 36 (19): 4565–4582.
7 Hunter, D.B., Edvell, L.G., and Englund, M.A. (2005). Wideband microwave photonic channelised receiver. *Int. Topical Meeting on Microwave Photonics Tech. Dig.*, pp. 249–252.
8 Pan, S. and Zhang, Y. (2020). Microwave photonic radars. *J. Lightw. Technol.* 38 (19): 5450–5484, 1.
9 Marpaung, D., Yao, J., and Capmany, J. (2019). Integrated microwave photonics. *Nature Photonics* 13: 80–90.
10 Zhang, W. and Yao, J. (2016). Silicon-based integrated microwave photonics. *IEEE J. Quantum Electron.* 52: 0600412.
11 Marpaung, D., Roeloffzen, C., Heideman, R. et al. (2013). Integrated microwave photonics. *Laser Photonics Rev.* 7: 506–538.

12 Maram, R., Kaushal, S., Azaña, J., and Chen, L. (2019). Recent trends and advances of silicon-based integrated microwave photonics. *Photonics* 6 (1): 13.

13 Marpaung, D., Pagani, M., Morrison, B., and Eggleton, B.J. (2014). Nonlinear integrated microwave photonics. *J. Lightw. Technol.* 32: 3421–3427.

14 Liu, Y., Choudhary, A., Marpaung, D., and Eggleton, B.J. (2020). Integrated microwave photonic filters. *Adv. Opt. Photon.* 12: 485–555.

15 Li, M., Hao, T., Pan, S. et al. (2021). Frontiers and prospects of integrated microwave photonics. *Infrared Laser Eng.* 50: 20211048.

16 Li, M., Hao, T., and Li, W. (2021). Prospects of cross research between microwave photonics and multidiscipline. *Infrared Laser Eng.* 50: 20211042.

17 Xiang, C., Morton, P., and Bowers, J. (2019). Ultra-narrow linewidth laser based on a semiconductor gain chip and extended Si3N4 Bragg grating. *Opt. Lett.* 44: 3825–3828.

18 Lin, Y., Lin, Y., Lee, K.H. et al. (2017). High-efficiency normal-incidence vertical P-I-N photodetectors on a germanium-on-insulator platform. *Photonics Res.* 5: 702–709.

19 Haffner, C., Haffner, C., Chelladurai, D. et al. (2018). Low-loss plasmon-assisted electro-optic modulator. *Nature* 556: 483–486.

20 Liu, S. and Khope, A. (2021). Latest advances in high-performance light sources and optical amplifiers on silicon. *J. Semicond.* 42: 041307.

21 Pérez, D., Gasulla, I., and Capmany, J. (2018). Toward programmable microwave photonics processors. *J. Lightw. Technol.* 36: 519–532.

22 Hao, T., Yang, Y., Jin, Y. et al. (2022). Quantum microwave photonics. *J. Lightw. Technol.* 40: 6616–6625.

23 Xu, S., Zou, X., Ma, B. et al. (2019). Deep-learning-powered photonic analog-to-digital conversion. *Light Sci. Appl.* 8: 66.

Index

Microwave Photonics, First Edition. Jianping Yao, José Capmany, and Ming Li.

Companion website: www.wiley.com/go/yao/microwavephotonics

n

q

r

s

t

u

Printed and bound by CPI Group (UK) Ltd, Croydon, CR0 4YY
17/09/2026
14964313-0005